Grundlehren der mathematischen Wissenschaften 330

A Series of Comprehensive Studies in Mathematics

To Margaret, as always

Preface

Namely, because the shape of the whole universe is most perfect and, in fact, designed by the wisest creator, nothing in all of the world will occur in which no maximum or minimum rule is somehow shining forth.

Leonhard Euler (1744)

We can treat this firm stand by Euler [411] ("... nihil omnino in mundo contingint, in quo non maximi minimive ratio quapiam eluceat") as the most fundamental principle of *Variational Analysis*. This principle justifies a variety of striking implementations of *optimization/variational* approaches to solving numerous problems in mathematics and applied sciences that may not be of a variational nature. Remember that optimization has been a major motivation and driving force for developing differential and integral calculus. Indeed, the *very concept of derivative* introduced by Fermat via the tangent slope to the graph of a function was motivated by solving an optimization problem; it led to what is now called the *Fermat stationary principle*. Besides applications to optimization, the latter principle plays a crucial role in proving the most important calculus results including the mean value theorem, the implicit and inverse function theorems, etc. The same line of development can be seen in the infinite-dimensional setting, where the Brachistochrone was the first problem not only of the calculus of variations but of all functional analysis inspiring, in particular, a variety of concepts and techniques in infinite-dimensional differentiation and related areas.

Modern variational analysis can be viewed as an outgrowth of the calculus of variations and mathematical programming, where the focus is on optimization of functions relative to various constraints and on sensitivity/stability of optimization-related problems with respect to perturbations. Classical notions of variations such as moving away from a given point or curve no longer play a critical role, while concepts of problem *approximations* and/or *perturbations* become crucial.

One of the most characteristic features of modern variational analysis is the intrinsic presence of *nonsmoothness*, i.e., the necessity to deal with nondifferentiable functions, sets with nonsmooth boundaries, and set-valued mappings. Nonsmoothness naturally enters not only through initial data of optimization-related problems (particularly those with inequality and geometric constraints) but largely via *variational principles* and other optimization, approximation, and perturbation techniques applied to problems with even smooth data. In fact, many fundamental objects frequently appearing in the framework of variational analysis (e.g., the distance function, value functions in optimization and control problems, maximum and minimum functions, solution maps to perturbed constraint and variational systems, etc.) are inevitably of nonsmooth and/or set-valued structures requiring the development of new forms of analysis that involve *generalized differentiation*.

It is important to emphasize that even the simplest and historically earliest problems of *optimal control* are *intrinsically nonsmooth*, in contrast to the classical calculus of variations. This is mainly due to pointwise constraints on control functions that often take only discrete values as in typical problems of automatic control, a primary motivation for developing optimal control theory. Optimal control has always been a major source of inspiration as well as a fruitful territory for applications of advanced methods of variational analysis and generalized differentiation.

Key issues of variational analysis in finite-dimensional spaces have been addressed in the book "Variational Analysis" by Rockafellar and Wets [1165]. The development and applications of variational analysis in infinite dimensions require certain concepts and tools that cannot be found in the finite-dimensional theory. The *primary goals* of this book are to present basic concepts and principles of variational analysis unified in finite-dimensional and infinite-dimensional space settings, to develop a comprehensive generalized differential theory at the same level of perfection in both finite and infinite dimensions, and to provide valuable applications of variational theory to broad classes of problems in constrained optimization and equilibrium, sensitivity and stability analysis, control theory for ordinary, functional-differential and partial differential equations, and also to selected problems in mechanics and economic modeling.

Generalized differentiation lies at the heart of variational analysis and its applications. We systematically develop a *geometric dual-space approach* to generalized differentiation theory revolving around the *extremal principle*, which can be viewed as a local *variational* counterpart of the classical convex separation in nonconvex settings. This principle allows us to deal with *nonconvex* derivative-like constructions for sets (normal cones), set-valued mappings (coderivatives), and extended-real-valued functions (subdifferentials). These constructions are defined directly in dual spaces and, being nonconvex-valued, cannot be generated by any derivative-like constructions in primal spaces (like tangent cones and directional derivatives). Nevertheless, our basic nonconvex constructions enjoy comprehensive calculi, which happen to be significantly better than those available for their primal and/or convex-valued counterparts. Thus passing to *dual spaces*, we are able to achieve more beauty and harmony in comparison with primal world objects. In some sense, the dual viewpoint does indeed allow us to meet the perfection requirement in the fundamental statement by Euler quoted above.

Observe to this end that dual objects (multipliers, adjoint arcs, shadow prices, etc.) have always been at the center of variational theory and applications used, in particular, for formulating principal optimality conditions in the calculus of variations, mathematical programming, optimal control, and economic modeling. The usage of variations of optimal solutions in primal spaces can be considered just as a convenient tool for deriving necessary optimality conditions. There are no essential restrictions in such a "primal" approach in smooth and convex frameworks, since primal and dual derivative-like constructions are equivalent for these classical settings. It is not the case any more in the framework of modern variational analysis, where even *nonconvex primal space* local approximations (e.g., tangent cones) inevitably yield, *under duality, convex sets* of normals and subgradients. This convexity of dual objects leads to significant restrictions for the theory and applications. Moreover, there are many situations particularly identified in this book, where primal space approximations simply cannot be used for variational analysis, while the employment of dual space constructions provides comprehensive results. Nevertheless, tangentially generated/primal space constructions play an important role in some other aspects of variational analysis, especially in finite-dimensional spaces, where they recover in duality the nonconvex sets of our basic normals and subgradients at the point in question by *passing to the limit* from points nearby; see, for instance, the afore-mentioned book by Rockafellar and Wets [1165]

Among the abundant bibliography of this book, we refer the reader to the monographs by Aubin and Frankowska [54], Bardi and Capuzzo Dolcetta [85], Beer [92], Bonnans and Shapiro [133], Clarke [255], Clarke, Ledyaev, Stern and Wolenski [265], Facchinei and Pang [424], Klatte and Kummer [686], Vinter [1289], and to the comments given after each chapter for significant aspects of variational analysis and impressive applications of this rapidly growing area that are not considered in the book. We especially emphasize the concurrent and complementing monograph "Techniques of Variational Analysis" by Borwein and Zhu [164], which provides a nice introduction to some fundamental techniques of modern variational analysis covering important theoretical aspects and applications not included in this book.

The book presented to the reader's attention is self-contained and mostly collects results that have not been published in the monographical literature. It is split into two volumes and consists of eight chapters divided into sections and subsections. Extensive comments (that play a special role in this book discussing basic ideas, history, motivations, various interrelations, choice of terminology and notation, open problems, etc.) are given for each chapter. We present and discuss numerous references to the vast literature on many aspects of variational analysis (considered and not considered in the book) including early contributions and very recent developments. Although there are no formal exercises, the extensive remarks and examples provide grist for further thought and development. Proofs of the major results are complete, while there is plenty of room for furnishing details, considering special cases, and deriving generalizations for which guidelines are often given.

Volume I "Basic Theory" consists of four chapters mostly devoted to basic constructions of generalized differentiation, fundamental extremal and variational principles, comprehensive generalized differential calculus, and complete dual characterizations of fundamental properties in nonlinear study related to Lipschitzian stability and metric regularity with their applications to sensitivity analysis of constraint and variational systems.

Chapter 1 concerns the generalized differential theory in arbitrary *Banach spaces*. Our basic normals, subgradients, and coderivatives are directly defined in dual spaces via *sequential* weak* limits involving more primitive ε-normals and ε-subgradients of the Fréchet type. We show that these constructions have a variety of nice properties in the general Banach spaces setting, where the usage of ε-enlargements is crucial. Most such properties (including first-order and second-order calculus rules, efficient representations, variational descriptions, subgradient calculations for distance functions, necessary coderivative conditions for Lipschitzian stability and metric regularity, etc.) are collected in this chapter. Here we also define and start studying the so-called *sequential normal compactness* (SNC) properties of sets, set-valued mappings, and extended-real-valued functions that automatically hold in finite dimensions while being one of the most essential ingredients of variational analysis and its applications in infinite-dimensional spaces.

Chapter 2 contains a detailed study of the *extremal principle* in variational analysis, which is the main single tool of this book. First we give a direct variational proof of the extremal principle in finite-dimensional spaces based on a smoothing penalization procedure via the method of *metric approximations*. Then we proceed by infinite-dimensional variational techniques in Banach spaces with a Fréchet smooth norm and finally, by separable reduction, in the larger class of *Asplund spaces*. The latter class is well-investigated in the geometric theory of Banach spaces and contains, in particular, every reflexive space and every space with a separable dual. Asplund spaces play a prominent role in the theory and applications of variational analysis developed in this book. In Chap. 2 we also establish relationships between the (geometric) extremal principle and (analytic) variational principles in both conventional and enhanced forms. The results obtained are applied to the derivation of novel variational characterizations of Asplund spaces and useful representations of the basic generalized differential constructions in the Asplund space setting similar to those in finite dimensions. Finally, in this chapter we discuss abstract versions of the extremal principle formulated in terms of axiomatically defined normal and subdifferential structures on appropriate Banach spaces and also overview in more detail some specific constructions.

Chapter 3 is a cornerstone of the generalized differential theory developed in this book. It contains comprehensive *calculus rules* for basic normals, subgradients, and coderivatives in the framework of Asplund spaces. We pay most of our attention to *pointbased* rules via the limiting constructions *at* the points in question, for both assumptions and conclusions, having in mind that pointbased results indeed happen to be of crucial importance for applications. A number of the results presented in this chapter seem to be new even in the finite-dimensional setting, while overall we achieve the same level of perfection and generality in Asplund spaces as in finite dimensions. The main issue that distinguishes the finite-dimensional and infinite-dimensional settings is the necessity to invoke *sufficient amounts of compactness* in infinite dimensions that are not needed at all in finite-dimensional spaces. The required compactness is provided by the afore-mentioned SNC properties, which are included in the assumptions of calculus rules and call for their own calculus ensuring the preservation of SNC properties under various operations on sets and mappings. The absence of such a *SNC calculus* was a crucial obstacle for many successful applications of generalized differentiation in infinite-dimensional spaces to a range of infinite-dimensions problems including those in optimization, stability, and optimal control given in this book. Chapter 3 contains a broad spectrum of the SNC calculus results that are decisive for subsequent applications.

Chapter 4 is devoted to a thorough study of Lipschitzian, metric regularity, and linear openness/covering properties of set-valued mappings, and to their applications to sensitivity analysis of parametric constraint and variational systems. First we show, based on variational principles and the generalized differentiation theory developed above, that the necessary coderivative conditions for these fundamental properties derived in Chap. 1 in arbitrary Banach spaces happen to be *complete characterizations* of these properties in the Asplund space setting. Moreover, the employed variational approach allows us to obtain verifiable formulas for computing the *exact bounds* of the corresponding moduli. Then we present detailed applications of these results, supported by generalized differential and SNC calculi, to sensitivity and stability analysis of parametric constraint and variational systems governed by perturbed sets of feasible and optimal solutions in problems of optimization and equilibria, implicit multifunctions, complementarity conditions, variational and hemivariational inequalities as well as to some mechanical systems.

Volume II "Applications" also consists of four chapters mostly devoted to applications of basic principles in variational analysis and the developed generalized differential calculus to various topics in constrained optimization and equilibria, optimal control of ordinary and distributed-parameter systems, and models of welfare economics.

Chapter 5 concerns constrained optimization and equilibrium problems with possibly nonsmooth data. Advanced methods of variational analysis based on extremal/variational principles and generalized differentiation happen to be very useful for the study of constrained problems even with smooth initial data, since nonsmoothness naturally appears while applying penalization, approximation, and perturbation techniques. Our primary goal is to derive necessary optimality and suboptimality conditions for various constrained problems in both finite-dimensional and infinite-dimensional settings. Note that conditions of the latter – *suboptimality* – type, somehow underestimated in optimization theory, don't assume the existence of optimal solutions (which is especially significant in infinite dimensions) ensuring that "almost" optimal solutions "almost" satisfy necessary conditions for optimality. Besides considering problems with constraints of conventional types, we pay serious attention to rather new classes of problems, labeled as *mathematical problems with equilibrium constraints* (MPECs) and *equilibrium problems with equilibrium constraints* (EPECs), which are intrinsically nonsmooth while admitting a thorough analysis by using generalized differentiation. Finally, certain concepts of *linear subextremality* and *linear suboptimality* are formulated in such a way that the necessary optimality conditions derived above for conventional notions are seen to be *necessary and sufficient* in the new setting.

In *Chapter 6* we start studying problems of *dynamic optimization* and *optimal control* that, as mentioned, have been among the primary motivations for developing new forms of variational analysis. This chapter deals mostly with optimal control problems governed by *ordinary* dynamic systems whose state space may be infinite-dimensional. The main attention in the first part of the chapter is paid to the Bolza-type problem for evolution systems governed by constrained *differential inclusions*. Such models cover more conventional control systems governed by parameterized evolution equations with control regions generally dependent on state variables. The latter don't allow us to use control variations for deriving necessary optimality conditions. We develop the *method of discrete approximations*, which is certainly of numerical interest, while it is mainly used in this book as a direct vehicle to derive optimality conditions for continuous-time systems by passing to the limit from their discrete-time counterparts. In this way we obtain, strongly based on the generalized differential and SNC calculi, necessary optimality conditions in the extended Euler-Lagrange form for nonconvex differential inclusions in infinite dimensions expressed via our basic generalized differential constructions.

The second part of Chap. 6 deals with constrained optimal control systems governed by ordinary evolution equations of *smooth dynamics* in arbitrary Banach spaces. Such problems have essential specific features in comparison with the differential inclusion model considered above, and the results obtained (as well as the methods employed) in the two parts of this chapter are generally independent. Another major theme explored here concerns *stability* of the maximum principle under discrete approximations of nonconvex control systems. We establish rather surprising results on the *approximate maximum principle* for discrete approximations that shed new light upon both qualitative and quantitative relationships between continuous-time and discrete-time systems of optimal control.

In *Chapter 7* we continue the study of optimal control problems by applications of advanced methods of variational analysis, now considering systems with *distributed parameters*. First we examine a general class of *hereditary systems* whose dynamic constraints are described by both delay-differential inclusions and linear algebraic equations. On one hand, this is an interesting and not well-investigated class of control systems, which can be treated as a special type of variational problems for *neutral functional-differential inclusions* containing time delays not only in state but also in velocity variables. On the other hand, this class is related to differential-algebraic systems with a linear link between "slow" and "fast" variables. Employing the method of discrete approximations and the basic tools of generalized differentiation, we establish a strong variational convergence/stability of discrete approximations and derive extended optimality conditions for continuous-time systems in both Euler-Lagrange and Hamiltonian forms.

The rest of Chap. 7 is devoted to optimal control problems governed by *partial differential equations* with *pointwise* control and state constraints. We pay our primary attention to evolution systems described by *parabolic* and *hyperbolic* equations with controls functions acting in the Dirichlet and Neumann boundary conditions. It happens that such *boundary control* problems are the most challenging and the least investigated in PDE optimal control theory, especially in the presence of pointwise state constraints. Employing approximation and perturbation methods of modern variational analysis, we justify variational convergence and derive necessary optimality conditions for various control problems for such PDE systems including *minimax* control under *uncertain disturbances*.

The concluding *Chapter 8* is on applications of variational analysis to *economic modeling*. The major topic here is *welfare economics*, in the general nonconvex setting with infinite-dimensional commodity spaces. This important class of competitive equilibrium models has drawn much attention of economists and mathematicians, especially in recent years when nonconvexity has become a crucial issue for practical applications. We show that the methods of variational analysis developed in this book, particularly the extremal principle, provide adequate tools to study Pareto optimal allocations and associated price equilibria in such models. The tools of variational analysis and generalized differentiation allow us to obtain extended nonconvex versions of the so-called "second fundamental theorem of welfare economics" describing marginal equilibrium prices in terms of minimal collections of generalized normals to nonconvex sets. In particular, our approach and variational descriptions of generalized normals offer new economic interpretations of market equilibria via "nonlinear marginal prices" whose role in nonconvex models is similar to the one played by conventional linear prices in convex models of the Arrow-Debreu type.

The book includes a Glossary of Notation, common for both volumes, and an extensive Subject Index compiled separately for each volume. Using the Subject Index, the reader can easily find not only the page, where some notion and/or notation is introduced, but also various places providing more discussions and significant applications for the object in question.

Furthermore, it seems to be reasonable to title all the statements of the book (definitions, theorems, lemmas, propositions, corollaries, examples, and remarks) that are numbered in sequence within a chapter; thus, in Chap. 5 for instance, Example 5.3.3 precedes Theorem 5.3.4, which is followed by Corollary 5.3.5. For the reader's convenience, all these statements and numerated comments are indicated in the List of Statements presented at the end of each volume. It is worth mentioning that the list of acronyms is included (in alphabetic order) in the Subject Index and that the common principle adopted for the book notation is to use lower case Greek characters for numbers and (extended) real-valued functions, to use lower case Latin characters for vectors and single-valued mappings, and to use Greek and Latin upper case characters for sets and set-valued mappings.

Our notation and terminology are generally consistent with those in Rockafellar and Wets [1165]. Note that we try to distinguish everywhere the notions defined *at* the point and *around* the point in question. The latter indicates *robustness/stability* with respect to perturbations, which is critical for most of the major results developed in the book.

The book is accompanied by the abundant bibliography (with English sources if available), common for both volumes, which reflects a variety of topics and contributions of many researchers. The references included in the bibliography are discussed, at various degrees, mostly in the extensive commentaries to each chapter. The reader can find further information in the given references, directed by the author's comments.

We address this book mainly to researchers and graduate students in mathematical sciences; first of all to those interested in nonlinear analysis, optimization, equilibria, control theory, functional analysis, ordinary and partial differential equations, functional-differential equations, continuum mechanics, and mathematical economics. We also envision that the book will be useful to a broad range of researchers, practitioners, and graduate students involved in the study and applications of variational methods in operations research, statistics, mechanics, engineering, economics, and other applied sciences.

Parts of the book have been used by the author in teaching graduate classes on variational analysis, optimization, and optimal control at Wayne State University. Basic material has also been incorporated into many lectures and tutorials given by the author at various schools and scientific meetings during the recent years.

Acknowledgments

My first gratitude go to Terry Rockafellar who has encouraged me over the years to write such a book and who has advised and supported me at all the stages of this project.

Special thanks are addressed to Rafail Gabasov, my doctoral thesis adviser, from whom I learned optimal control and much more; to Alec Ioffe, Boris Polyak, and Vladimir Tikhomirov who recognized and strongly supported my first efforts in nonsmooth analysis and optimization; to Sasha Kruger, my first graduate student and collaborator in the beginning of our exciting journey to generalized differentiation; to Jon Borwein and Marián Fabian from whom I learned deep functional analysis and the beauty of Asplund spaces; to Ali Khan whose stimulating work and enthusiasm have encouraged my study of economic modeling; to Jiři Outrata who has motivated and influenced my growing interest in equilibrium problems and mechanics and who has intensely promoted the implementation of the basic generalized differential constructions of this book in various areas of optimization theory and applications; and to Jean-Pierre Raymond from whom I have greatly benefited on modern theory of partial differential equations.

During the work on this book, I have had the pleasure of discussing its various aspects and results with many colleagues and friends. Besides the individuals mentioned above, I'm particularly indebted to Zvi Artstein, Jim Burke, Tzanko Donchev, Asen Dontchev, Joydeep Dutta, Andrew Eberhard, Ivar Ekeland, Hector Fattorini, René Henrion, Jean-Baptiste Hiriart-Urruty, Alejandro Jofré, Abderrahim Jourani, Michal Kočvara, Irena Lasiecka, Claude Lemaréchal, Adam Levy, Adrian Lewis, Kazik Malanowski, Michael Overton, Jong-Shi Pang, Teemu Pennanen, Steve Robinson, Alex Rubinov, Andrzej Świech, Michel Théra, Lionel Thibault, Jay Treiman, Hector Sussmann, Roberto Triggiani, Richard Vinter, Nguyen Dong Yen, George Yin, Jack Warga, Roger Wets, and Jim Zhu for valuable suggestions and fruitful conversations throughout the years of the fulfillment of this project.

The continuous support of my research by the National Science Foundation is gratefully acknowledged.

As mentioned above, the material of this book has been used over the years for teaching advanced classes on variational analysis and optimization attended mostly by my doctoral students and collaborators. I highly appreciate their contributions, which particularly allowed me to improve my lecture notes and book manuscript. Especially valuable help was provided by Glenn Malcolm, Nguyen Mau Nam, Yongheng Shao, Ilya Shvartsman, and Bingwu Wang. Useful feedback and text corrections came also from Truong Bao, Wondi Geremew, Pankaj Gupta, Aychi Habte, Kahina Sid Idris, Dong Wang, Lianwen Wang, and Kaixia Zhang.

I'm very grateful to the nice people in Springer for their strong support during the preparation and publishing this book. My special thanks go to Catriona Byrne, Executive Editor in Mathematics, to Achi Dosajh, Senior Editor in Applied Mathematics, to Stefanie Zoeller, Assistant Editor in Mathematics, and to Frank Holzwarth from the Computer Science Editorial Department.

I thank my younger daughter Irina for her interest in my book and for her endless patience and tolerance in answering my numerous question on English. I would also like to thank my poodle Wuffy for his sharing with me the long days of work on this book. Above all, I don't have enough words to thank my wife Margaret for her sharing with me everything, starting with our high school years in Minsk.

Ann Arbor, Michigan *Boris Mordukhovich*
August 2005

Contents

Volume I Basic Theory

Volume II Applications

Volume I

Basic Theory

1

Generalized Differentiation in Banach Spaces

In this chapter we define and study basic concepts of *generalized differentiation* that lies at the heart of variational analysis and its applications considered in the book. Most properties presented in this chapter hold in *arbitrary* Banach spaces (some of them don't require completeness or even a normed structure, as one can see from the proofs). Developing a *geometric dual-space approach* to generalized differentiation, we start with *normals* to sets (Sect. 1.1), then proceed to *coderivatives* of set-valued mappings (Sect. 1.2), and then to *subdifferentials* of extended-real-valued functions (Sect. 1.3).

Unless otherwise stated, *all the spaces in question are Banach* whose norms are always denoted by $\|\cdot\|$. Given a space X, we denote by $I\!B_X$ its closed unit ball and by X^* its dual space equipped with the weak* topology w^*, where $\langle\cdot,\cdot\rangle$ means the canonical pairing. If there is no confusion, $I\!B$ and $I\!B^*$ stand for the closed unit balls of the space and dual space in question, while S and S^* are usually stand for the corresponding unit spheres ; also $B_r(x) := x + rI\!B$ with $r > 0$. The symbol * is used everywhere to indicate relations to *dual* spaces (dual elements, adjoint operators, etc.)

In what follows we often deal with set-valued mappings (multifunctions) $F\colon X \rightrightarrows X^*$ between a Banach space and its dual, for which the notation

$$\mathop{\mathrm{Lim\,sup}}_{x\to\bar{x}} F(x) := \Big\{x^* \in X^* \;\Big|\; \exists \text{ sequences } x_k \to \bar{x} \text{ and } x_k^* \stackrel{w^*}{\to} x^* \text{ with } x_k^* \in F(x_k) \text{ for all } k \in I\!N\Big\} \tag{1.1}$$

signifies the *sequential Painlevé-Kuratowski upper/outer limit* with respect to the norm topology of X and the weak* topology of X^*. Note that the symbol $:=$ means "equal by definition" and that $I\!N := \{1, 2, \ldots\}$ denotes the set of all natural numbers.

The linear combination of the two subsets Ω_1 and Ω_2 of X is defined by

$$\alpha_1\Omega_1 + \alpha_2\Omega_2 := \big\{\alpha_1 x_1 + \alpha_2 x_2 \;\big|\; x_1 \in \Omega_1,\; x_2 \in \Omega_2\big\}$$

with real numbers $\alpha_1, \alpha_2 \in I\!R := (-\infty, \infty)$, where we use the convention that $\Omega + \emptyset = \emptyset$, $\alpha\emptyset = \emptyset$ if $\alpha \in I\!R \setminus \{0\}$, and $\alpha\emptyset = \{0\}$ if $\alpha = 0$. Dealing with empty sets, we let $\inf \emptyset := \infty$, $\sup \emptyset := -\infty$, and $\|\emptyset\| := \infty$.

1.1 Generalized Normals to Nonconvex Sets

Throughout this section, Ω is a nonempty subset of a real Banach space X. Such a set is called *proper* if $\Omega \neq X$. In what follows the expressions

$$\mathrm{cl}\,\Omega,\quad \mathrm{co}\,\Omega,\quad \mathrm{clco}\,\Omega,\quad \mathrm{bd}\,\Omega,\quad \mathrm{int}\,\Omega$$

stand for the standard notions of *closure*, *convex hull* , *closed convex hull*, *boundary*, and *interior* of Ω, respectively. The *conic hull* of Ω is

$$\mathrm{cone}\;\Omega := \big\{\alpha x \in X \big|\; \alpha \geq 0,\; x \in \Omega\big\}\,.$$

The symbol cl^* signifies the *weak* topological closure* of a set in a dual space.

1.1.1 Basic Definitions and Some Properties

We begin the generalized differentiation theory with constructing generalized normals to arbitrary sets. To describe basic normals to a set Ω at a given point $\bar{x}$, we use a two-stage procedure: first define more primitive ε-normals (prenormals) to Ω at points x close to $\bar{x}$ and then pass to the sequential limit (1.1) as $x \to \bar{x}$ and $\varepsilon \downarrow 0$. Throughout the book we use the notation

$$x \xrightarrow{\Omega} \bar{x} \iff x \to \bar{x} \;\text{ with }\; x \in \Omega\,.$$

Definition 1.1 (generalized normals). *Let Ω be a nonempty subset of X.*

(i) *Given $x \in \Omega$ and $\varepsilon \geq 0$, define the* SET OF ε-NORMALS *to Ω at x by*

$$\widehat{N}_\varepsilon(x;\Omega) := \Big\{x^* \in X^* \;\Big|\; \limsup_{u \xrightarrow{\Omega} x} \frac{\langle x^*, u - x\rangle}{\|u - x\|} \leq \varepsilon\Big\}\,. \tag{1.2}$$

When $\varepsilon = 0$, elements of (1.2) are called FRÉCHET NORMALS *and their collection, denoted by $\widehat{N}(x;\Omega)$, is the* PRENORMAL CONE *to Ω at x. If $x \notin \Omega$, we put $\widehat{N}_\varepsilon(x;\Omega) := \emptyset$ for all $\varepsilon \geq 0$.*

(ii) *Let $\bar{x} \in \Omega$. Then $x^* \in X^*$ is a* BASIC/LIMITING NORMAL *to Ω at $\bar{x}$ if there are sequences $\varepsilon_k \downarrow 0$, $x_k \xrightarrow{\Omega} \bar{x}$, and $x_k^* \xrightarrow{w^*} x^*$ such that $x_k^* \in \widehat{N}_{\varepsilon_k}(x_k;\Omega)$ for all $k \in I\!N$. The collection of such normals*

$$N(\bar{x};\Omega) := \mathop{\mathrm{Lim\,sup}}_{\substack{x \to \bar{x} \\ \varepsilon \downarrow 0}} \widehat{N}_\varepsilon(x;\Omega) \tag{1.3}$$

is the (basic, limiting) NORMAL CONE *to Ω at $\bar{x}$. Put $N(\bar{x};\Omega) := \emptyset$ for $\bar{x} \notin \Omega$.*

It easily follows from the definitions that

$$\widehat{N}_\varepsilon(\bar{x};\Omega)=\widehat{N}_\varepsilon(\bar{x};\mathrm{cl}\,\Omega)\ \text{ and }\ N(\bar{x};\Omega)\subset N(\bar{x};\mathrm{cl}\,\Omega)$$

for every $\Omega\subset X$, $\bar{x}\in\Omega$, and $\varepsilon\ge 0$. Observe that both the prenormal cone $\widehat{N}(\cdot;\Omega)$ and the normal cone $N(\cdot;\Omega)$ are *invariant* with respect to equivalent norms on X while the ε-normal sets $\widehat{N}_\varepsilon(\cdot;\Omega)$ depend on a given norm $\|\cdot\|$ if $\varepsilon>0$. Note also that for each $\varepsilon\ge 0$ the sets (1.2) are obviously *convex and closed* in the norm topology of X^*; hence they are *weak* closed* in X^* when X is *reflexive.*

In contrast to (1.2), the basic normal cone (1.3) may be *nonconvex* in very simple situations as for $\Omega:=\big\{(x_1,x_2)\in I\!R^2|\ x_2\ge-|x_1|\big\}$, where

$$N((0,0);\Omega)=\big\{(v,v)\;\big|\;v\le 0\big\}\cup\big\{(v,-v)\;\big|\;v\ge 0\big\} \tag{1.4}$$

while $\widehat{N}((0,0);\Omega)=\{0\}$. This shows that $N(\bar{x};\Omega)$ *cannot be dual/polar* to any (even nonconvex) *tangential approximation* of Ω at $\bar{x}$ in the primal space X, since *polarity always implies convexity*; cf. Subsect. 1.1.2.

One can easily observe the following *monotonicity* properties of the ε-normal sets (1.2) with respect to ε as well as with respect to the set order:

$$\widehat{N}_\varepsilon(\bar{x};\Omega)\subset\widehat{N}_{\tilde{\varepsilon}}(\bar{x};\Omega)\ \text{ if }\ 0\le\varepsilon\le\tilde{\varepsilon}\,,$$

$$\widehat{N}_\varepsilon(\bar{x};\Omega)\subset\widehat{N}_\varepsilon(\bar{x};\widetilde{\Omega})\ \text{ if }\ \bar{x}\in\widetilde{\Omega}\subset\Omega\ \text{ and }\ \varepsilon\ge 0\,. \tag{1.5}$$

In particular, the decreasing property (1.5) holds for the prenormal cone $\widehat{N}(\bar{x};\cdot)$. Note however that neither (1.5) nor the opposite inclusion is valid for the basic normal cone (1.3). To illustrate this, we consider the two sets

$$\Omega:=\big\{(x_1,x_2)\in I\!R^2\;\big|\;x_2\ge-|x_1|\big\}\ \text{ and }\ \widetilde{\Omega}:=\big\{(x_1,x_2)\in I\!R^2\;\big|\;x_1\le x_2\big\}$$

with $\bar{x}=(0,0)\in\widetilde{\Omega}\subset\Omega$. Then

$$N(\bar{x};\widetilde{\Omega})=\big\{(v,-v)\;\big|\;v\ge 0\big\}\subset N(\bar{x};\Omega)\,,$$

where the latter cone is computed in (1.4). Furthermore, taking Ω as above and $\widetilde{\Omega}:=\big\{(x_1,x_2)\in I\!R^2|\ x_2\ge 0\big\}\subset\Omega$, we have

$$N(\bar{x};\Omega)\cap N(\bar{x};\widetilde{\Omega})=\{(0,0)\}\,,$$

which excludes any monotonicity relations.

The next property for representing normals to set products is common for both prenormal and normal cones.

Proposition 1.2 (normals to Cartesian products). *Consider an arbitrary point* $\bar{x} = (\bar{x}_1, \bar{x}_2) \in \Omega_1 \times \Omega_2 \subset X_1 \times X_2$. *Then*

$$\widehat{N}(\bar{x}; \Omega_1 \times \Omega_2) = \widehat{N}(\bar{x}_1; \Omega_1) \times \widehat{N}(\bar{x}_2; \Omega_2) ,$$

$$N(\bar{x}; \Omega_1 \times \Omega_2) = N(\bar{x}_1; \Omega_1) \times N(\bar{x}_2; \Omega_2) .$$

Proof. Since both prenormal and normal cones do not depend on equivalent norms on X_1 and X_2, we can fix any norms on these spaces and define a norm on the product $X_1 \times X_2$ by

$$\|(x_1, x_2)\| := \|x_1\| + \|x_2\| .$$

Given arbitrary $\varepsilon \geq 0$ and $x = (x_1, x_2) \in \Omega := \Omega_1 \times \Omega_2$, we easily check that

$$\widehat{N}_\varepsilon(x_1; \Omega_1) \times \widehat{N}_\varepsilon(x_2; \Omega_2) \subset \widehat{N}_{2\varepsilon}(x; \Omega) \subset \widehat{N}_{2\varepsilon}(x_1; \Omega_1) \times \widehat{N}_{2\varepsilon}(x_2; \Omega_2) ,$$

which implies both product formulas in the proposition. △

The prenormal cone $\widehat{N}(\cdot; \Omega)$ is obviously the smallest set among all the sets $\widehat{N}_\varepsilon(\cdot; \Omega)$. It follows from (1.2) that

$$\widehat{N}_\varepsilon(\bar{x}; \Omega) \supset \widehat{N}(\bar{x}; \Omega) + \varepsilon I\!B^*$$

for every $\varepsilon \geq 0$ and an arbitrary set Ω. If Ω is convex, then this inclusion holds as *equality* due to the following representation of ε-normals.

Proposition 1.3 (ε-normals to convex sets). *Let* Ω *be convex. Then*

$$\widehat{N}_\varepsilon(\bar{x}; \Omega) = \big\{x^* \in X^* \bigm| \langle x^*, x - \bar{x}\rangle \leq \varepsilon\|x - \bar{x}\| \ \textit{whenever} \ x \in \Omega\big\}$$

for any $\varepsilon \geq 0$ *and* $\bar{x} \in \Omega$. *In particular,* $\widehat{N}(\bar{x}; \Omega)$ *agrees with the normal cone of convex analysis.*

Proof. Note that the inclusion "$\supset$" in the above formula obviously holds for an arbitrary set Ω. Let us justify the opposite inclusion when Ω is convex. Consider any $x^* \in \widehat{N}_\varepsilon(\bar{x}; \Omega)$ and fix $x \in \Omega$. Then we have

$$x_\alpha := \bar{x} + \alpha(x - \bar{x}) \in \Omega \ \text{ for all } \ 0 \leq \alpha \leq 1$$

due to the convexity of Ω. Moreover, $x_\alpha \to \bar{x}$ as $\alpha \downarrow 0$. Taking an arbitrary $\gamma > 0$, we easily conclude from (1.2) that

$$\langle x^*, x_\alpha - \bar{x}\rangle \leq (\varepsilon + \gamma)\|x_\alpha - \bar{x}\| \ \text{ for small } \ \alpha > 0 ,$$

which completes the proof. △

It follows from Definition 1.1 that

$$\widehat{N}(\bar{x};\Omega) \subset N(\bar{x};\Omega) \quad \text{for any} \quad \Omega \subset X \quad \text{and} \quad \bar{x} \in \Omega\ . \tag{1.6}$$

This inclusion may be *strict* even for simple sets as the one in (1.4), where $\widehat{N}(\bar{x};\Omega) = \{0\}$ for $\bar{x} = 0 \in I\!R^2$. The equality in (1.6) singles out a class of sets that have certain "regular" behavior around $\bar{x}$ and unify good properties of both prenormal and normal cones at $\bar{x}$.

Definition 1.4 (normal regularity of sets). *A set $\Omega \subset X$ is (normally)* REGULAR *at $\bar{x} \in \Omega$ if*

$$N(\bar{x};\Omega) = \widehat{N}(\bar{x};\Omega)\ .$$

An important example of set regularity is given by sets Ω *locally convex* around $\bar{x}$, i.e., for which there is a neighborhood $U \subset X$ of $\bar{x}$ such that $\Omega \cap U$ is convex.

Proposition 1.5 (regularity of locally convex sets). *Let U be a neighborhood of $\bar{x} \in \Omega \subset X$ such that the set $\Omega \cap U$ is convex. Then Ω is regular at $\bar{x}$ with*

$$N(\bar{x};\Omega) = \big\{x^* \in X^* \;\big|\; \langle x^*, x - \bar{x}\rangle \le 0 \;\; \textit{for all} \;\; x \in \Omega \cap U\big\}\ .$$

Proof. The inclusion "$\supset$" follows from (1.6) and Proposition 1.3. To prove the opposite inclusion, we take any $x^* \in N(\bar{x};\Omega)$ and find the corresponding sequences of $(\varepsilon_k, x_k, x_k^*)$ from Definition 1.1(ii). Thus $x_k \in U$ for all $k \in I\!N$ sufficiently large. Then Proposition 1.3 ensures that, for such k,

$$\langle x_k^*, x - x_k\rangle \le \varepsilon_k \|x - x_k\| \quad \text{for all} \quad x \in \Omega \cap U\ .$$

Passing there to the limit as $k \to \infty$, we finish the proof. △

Further results and discussions on normal regularity of sets and related notions of regularity for functions and set-valued mappings will be presented later in this chapter and mainly in Chap. 3, where they are incorporated into *calculus rules*. We'll show that regularity is preserved under major calculus operations and ensure *equalities* in calculus rules for basic normal and subdifferential constructions. On the other hand, such regularity may fail in many situations important for the theory and applications. In particular, it *never holds* for sets in finite-dimensional spaces related to *graphs of nonsmooth locally Lipschitzian mappings*; see Theorem 1.46 below. However, the basic normal cone and associated subdifferentials and coderivatives enjoy desired properties in general "irregular" settings, in contrast to the prenormal cone $\widehat{N}(\bar{x};\Omega)$ and its counterparts for functions and mappings.

Next we establish two special representations of the basic normal cone to closed subsets of the finite-dimensional space $X = I\!R^n$. Since all the norms in finite dimensions are equivalent, we always select the *Euclidean norm*

$$\|x\| := \sqrt{x_1^2 + \ldots + x_n^2}$$

on $I\!R^n$, unless otherwise stated. In this case $X^* = X = I\!R^n$.

Given a nonempty set $\Omega \subset I\!R^n$, consider the associated *distance function*

$$\operatorname{dist}(x; \Omega) := \inf_{u \in \Omega} \|x - u\| \tag{1.7}$$

and define the *Euclidean projector* of x to Ω by

$$\Pi(x; \Omega) := \big\{ w \in \Omega \bigm| \|x - w\| = \operatorname{dist}(x; \Omega) \big\} .$$

If Ω is closed, the set $\Pi(x; \Omega)$ is nonempty for every $x \in I\!R^n$. The following theorem describes the basic normal cone to subsets $\Omega \subset I\!R^n$ that are *locally closed* around $\bar{x}$. The latter means that there is a neighborhood U of $\bar{x}$ for which $\Omega \cap U$ is closed.

Theorem 1.6 (basic normals in finite dimensions). *Let $\Omega \subset I\!R^n$ be locally closed around $\bar{x} \in \Omega$. Then the following representations hold:*

$$N(\bar{x}; \Omega) = \operatorname*{Lim\,sup}_{x \to \bar{x}} \widehat{N}(x; \Omega) , \tag{1.8}$$

$$N(\bar{x}; \Omega) = \operatorname*{Lim\,sup}_{x \to \bar{x}} \big[\operatorname{cone}(x - \Pi(x; \Omega))\big] . \tag{1.9}$$

Proof. First we prove (1.8), which means that one can equivalently put $\varepsilon = 0$ in definition (1.3) of basic normals to locally closed sets in finite-dimensions. The inclusion "$\supset$" in (1.8) is obvious; let us justify the opposite inclusion.

Fix $x^* \in N(\bar{x}; \Omega)$ and find, by Definition 1.1(ii), sequences $\varepsilon_k \downarrow 0$, $x_k \to \bar{x}$, and $x_k^* \to x^*$ such that $x_k \in \Omega$ and $x_k^* \in \widehat{N}_{\varepsilon_k}(x_k; \Omega)$ for all $k \in I\!N$. Taking into account that $X = X^* = I\!R^n$ and that Ω is locally closed around $\bar{x}$, for each $k = 1, 2, \ldots$ we form $x_k + \alpha x_k^*$ with some parameter $\alpha > 0$ and select $w_k \in \Pi(x_k + \alpha x_k^*; \Omega)$ from the Euclidean projector. Due to the choice of w_k one has the inequality

$$\|x_k + \alpha x_k^* - w_k\|^2 \le \alpha^2 \|x_k^*\|^2$$

and, since the norm is Euclidean,

$$\|x_k + \alpha x_k^* - w_k\|^2 = \|x_k - w_k\|^2 + 2\alpha \langle x_k^*, x_k - w_k \rangle + \alpha^2 \|x_k^*\|^2 .$$

This implies the estimate

$$\|x_k - w_k\|^2 \le 2\alpha \langle x_k^*, w_k - x_k \rangle \quad \text{for any} \quad \alpha > 0 . \tag{1.10}$$

Using the convergence $w_k \to x_k$ as $\alpha \downarrow 0$ and the definition of the ε_k-normals $x_k^* \in \widehat{N}_{\varepsilon_k}(x_k; \Omega)$, we find a sequence of positive numbers $\alpha = \alpha_k$ along which

$$\langle x_k^*, w_k - x_k \rangle \le 2\varepsilon_k \|w_k - x_k\| \quad \text{for every} \quad k \in I\!N .$$

This gives $\|x_k - w_k\| \le 4\alpha_k\varepsilon_k$ due to (1.10); hence $w_k \to \bar{x}$ as $k \to \infty$. Moreover, letting

$$w_k^* := x_k^* + \tfrac{1}{\alpha_k}(x_k - w_k) ,$$

we get $\|w_k^* - x_k^*\| \le 4\varepsilon_k$ and $w_k^* \to x^*$ as $k \to \infty$.

To justify (1.8), it remains to show that $w_k^* \in \widehat{N}(w_k;\Omega)$ for all k. Indeed, for every fixed $x \in \Omega$ we get

$$\begin{aligned} 0 &\le \|x_k + \alpha_k x_k^* - x\|^2 - \|x_k + \alpha_k x_k^* - w_k\|^2 \\ &= \langle \alpha_k x_k^* + x_k - x, \alpha_k x_k^* + x_k - w_k\rangle + \langle \alpha_k x_k^* + x_k - x, w_k - x\rangle \\ &- \langle \alpha_k x_k^* + x_k - w_k, x - w_k\rangle - \langle \alpha_k x_k^* + x_k - w_k, \alpha_k x_k^* + x_k - x\rangle \\ &= -2\alpha_k\langle w_k^*, x - w_k\rangle + \|x - w_k\|^2 , \end{aligned}$$

since the norm is Euclidean. The latter implies the estimate

$$\langle w_k^*, x - w_k\rangle \le \tfrac{1}{2\alpha_k}\|x - w_k\|^2 \quad \text{for all} \quad x \in \Omega ,$$

which obviously ensures that $w_k^* \in \widehat{N}(w_k;\Omega)$ by Definition 1.1(i). Thus we arrive at the first representation (1.8) of the basic normal cone.

To justify the second representation (1.9), it is sufficient to show that

$$\operatorname*{Lim\,sup}_{x\to\bar{x}} \widehat{N}(x;\Omega) = \operatorname*{Lim\,sup}_{x\to\bar{x}} \big[\mathrm{cone}(x - \Pi(x;\Omega))\big] .$$

Let us first prove the inclusion

$$\widehat{N}(x;\Omega) \subset \operatorname*{Lim\,sup}_{u\to x} \big[\mathrm{cone}(u - \Pi(u;\Omega))\big] \quad \text{for any} \quad x \in \Omega . \tag{1.11}$$

Given $x \in \Omega$ and $x^* \in \widehat{N}(x;\Omega)$, we put $x_k := x + \frac{1}{k}x^*$ and pick some $w_k \in \Pi(x_k;\Omega)$ for each $k \in \mathbb{N}$. The latter is clearly equivalent to

$$\begin{aligned} 0 &\le \|x_k - v\|^2 - \|x_k - w_k\|^2 = \langle x_k - v, x_k - w_k\rangle \\ &+ \langle x_k - v, w_k - v\rangle - \langle x_k - w_k, v - w_k\rangle - \langle x_k - w_k, x_k - v\rangle \\ &= -2\langle x_k - w_k, v - w_k\rangle + \|v - w_k\|^2 \quad \text{for all} \quad v \in \Omega , \end{aligned}$$

which *characterizes the Euclidean projector*: $w_k \in \Pi(x_k;\Omega)$ if and only if

$$\langle x_k - w_k, v - w_k\rangle \le \tfrac{1}{2}\|v - w_k\|^2 \quad \text{for all} \quad v \in \Omega .$$

Letting $v = x$ and using the definition of x_k, we get

$$\|x - w_k\|^2 + \tfrac{1}{k}\langle x^*, x - w_k\rangle \le \tfrac{1}{2}\|x - w_k\|^2 .$$

Since $x^* \in \widehat{N}(x;\Omega)$, the latter inequality gives

$$k\|x - w_k\| \le \frac{2\langle x^*, w_k - x\rangle}{\|x - w_k\|} \to 0 \quad \text{as} \quad k \to \infty$$

and therefore

$$k(x_k - w_k) = x^* + k(x - w_k) \to x^* \quad \text{as} \quad k \to \infty .$$

Thus we have (1.11) that implies the inclusion "$\subset$" in (1.9) by taking the Painlevé-Kuratowski upper limit as $x \to \bar{x}$ and using (1.8).

It remains to prove the opposite inclusion in (1.9). To furnish this, let us consider the *inverse* Euclidean projector

$$\Pi^{-1}(x;\Omega) := \big\{z \in X \;\big|\; x \in \Pi(z;\Omega)\big\}$$

to Ω at $x \in \Omega$. It follows from the above characterization of the Euclidean projector and the definition of $\widehat{N}(x;\Omega)$ that

$$\operatorname{cone}\big[\Pi^{-1}(x;\Omega) - x\big] \subset \widehat{N}(x;\Omega) \quad \text{for any} \quad x \in \Omega ,$$

which implies the inclusion "$\supset$" in (1.9) by taking the Painlevé-Kuratowski upper limit as $x \xrightarrow{\Omega} \bar{x}$ and using (1.8). $\triangle$

Note that, although the proof of representation (1.8) essentially employs properties of the Euclidean norm , the representation itself doesn't depend on a specific norm on $I\!R^n$ all of which are equivalent. In Chap. 2 we show, using variational arguments, that this representation of the basic normal cone holds in any *Asplund space*, i.e., in a Banach space where every convex continuous function is generically Fréchet differentiable (in particular, in any reflexive space). In fact, (1.8) is a characterization of Asplund spaces. Note however that $\varepsilon > 0$ *cannot be removed* from the definition of basic normals and the corresponding subdifferential and coderivative constructions without loss of important properties in the general Banach space setting; see below, in particular, the next subsection. Moreover, we'll see that *stability* with respect to ε-enlargements plays an essential role in the proof of some principal results in Asplund spaces and even in finite-dimensions.

On the contrary, representation (1.9) heavily depends on the Euclidean norm on $I\!R^n$ and is not valid even for convex sets if a norm in non-Euclidean. For example, we have

$$N((0,0);\Omega) = \big\{(0,v) \;\big|\; v \le 0\big\} \quad \text{for} \quad \Omega = \big\{x = (x_1,x_2) \in I\!R^2 \,|\, x_2 \ge 0\big\} ,$$

while the cone on the right-hand side of (1.9) equals to $\big\{(v_1,v_2) |\, v_2 + |v_1| \le 0\big\}$ when the norm is given by $\|x\| := \max\big\{|x_1|, |x_2|\big\}$.

We are not going to consider here special properties of the basic normal cone in finite-dimensional spaces referring the reader to the books by Mordukhovich [901] and Rockafellar and Wets [1165]. Let us just mention that this cone enjoys the following *robustness property*

$$N(\bar{x};\Omega) = \operatorname*{Lim\,sup}_{x\to\bar{x}} N(x;\Omega) \quad \text{for all } \bar{x}\in\Omega ,$$

which can be easily obtained via the standard diagonal process in finite dimensions. For closed sets $\Omega \subset I\!R^n$ this means that the *graph* of the set-valued mapping $N(\cdot;\Omega)$ is *closed*, which obviously implies that the values $N(x;\Omega)$ are closed for all $x \in \Omega$.

It happens that these properties don't hold in infinite dimensions, even in the case of the simplest *Hilbert space* of sequences $X = X^* = \ell^2$. The reason is that the basic normal cone is defined in terms of *sequential* limits but the weak* topology of X^* is not sequential, so the weak* sequential closure of a set may not be weak* sequentially closed. The following example, which is due to Fitzpatrick (1994, personal communication; see also [144]), shows that values of the basic normal cone may *not* be even *norm closed* in X^*, hence neither weak* closed nor weak* sequentially closed in the dual space.

Example 1.7 (nonclosedness of the basic normal cone in ℓ^2). *There are a closed subset Ω of the Hilbert space ℓ^2 and a boundary point $\bar{x} \in \Omega$ such that $N(\bar{x};\Omega)$ is not norm closed in ℓ^2.*

Proof. Consider a complete orthonormal basis $\{e_1, e_2, \ldots\}$ in the Hilbert space ℓ^2 and form a nonconvex subset of ℓ^2 by

$$\Omega := \big\{s(e_1 - je_j) + t(je_1 - e_m)\big|\ m > j > 1,\ \ s,t \geq 0\big\} \cup \big\{te_1\big|\ t \geq 0\big\} ,$$

which is obviously a cone. We can check that Ω is closed in ℓ^2. Let us show that the basic normal cone $N(0;\Omega)$ is not closed in the norm topology of ℓ^2. This follows from:

(i) $e_1^* + \frac{1}{j}e_j^* \in N(0;\Omega)$ for all $j = 2, 3, \ldots$,

(ii) $e_1^* + \frac{1}{j}e_j^* \to e_1^*$ as $j \to \infty$,

(iii) $e_1^* \notin N(0;\Omega)$,

where e_j^* are linear functionals generated by e_j. To justify (i), we define $e_{jm}^* := e_1^* + \frac{1}{j}e_j^* + je_m^*$ for $1 < j < m$ and observe that $e_{jm}^* \in \widehat{N}(\frac{1}{m}(je_1 - e_m);\Omega)$. For each j we have $\frac{1}{m}(je_1 - e_m) \to 0$ and $e_{jm}^* \xrightarrow{w} e_1^* + \frac{1}{j}e_j^*$ as $m \to \infty$, which gives (i). It is easy to check (ii), and so it remains to verify (iii).

Suppose that (iii) doesn't hold, i.e., $e_1^* \in N(0;\Omega)$. Then, by the definition of basic normals with $w^* = w$ (the weak convergence in $X^* = \ell^2$), there are sequences $x_k \xrightarrow{\Omega} 0$, $\varepsilon_k \downarrow 0$, and $x_k^* \xrightarrow{w} e_1^*$ such that $x_k^* \in \widehat{N}_{\varepsilon_k}(x_k;\Omega)$ for all $k \in I\!N$. Assume that some of x_k are of the form $x_k = t_k e_1$ with $t_k \geq 0$. Putting $u := x_k + re_1$ with $r > 0$, we get

$$\varepsilon_k \geq \limsup_{u \xrightarrow{\Omega} x_k} \left\langle x_k^*, \frac{u - x_k}{\|u - x_k\|} \right\rangle \geq \limsup_{r \downarrow 0} \left\langle x_k^*, \frac{re_1}{\|re_1\|} \right\rangle = \langle x_k^*, e_1 \rangle ,$$

and so the convergence $x_k^* \xrightarrow{w} e_1^*$ implies that all but finitely many of x_k are not of the form $x_k = t_k e_1$ for $t_k \geq 0$. Consequently, all but finitely many of x_k are of the form $s(e_1 - je_j) + t(je_1 - e_m)$, where $m > j > 1$ and $s, t \geq 0$.

Now consider a sequence of x_k in the form $s(e_1 - je_j) + t(je_1 - e_m)$ belonging to Ω for any choice of sequences $s = s(k) \geq 0$, $t = t(k) \geq 0$, $j = j(k) > 1$, and $m = m(k) > j(k)$. Taking $u := x_k + r(je_1 - e_m) \in \Omega$, we get

$$\begin{aligned} \varepsilon_k &\geq \limsup_{u \xrightarrow{\Omega} x_k} \left\langle x_k^*, \frac{u - x_k}{\|u - x_k\|} \right\rangle \geq \limsup_{r \downarrow 0} \left\langle x_k^*, \frac{r(je_1 - e_m)}{\|r(je_1 - e_m)\|} \right\rangle \\ &= \left\langle x_k^*, \frac{je_1 - e_m}{\|je_1 - e_m\|} \right\rangle, \end{aligned}$$

which gives the estimate

$$\langle x_k^*, e_1 - j^{-1} e_m \rangle \leq \varepsilon_k \sqrt{1 + j^{-2}} \tag{1.12}$$

On the other hand, considering $u := x_k + r(e_1 - je_j) \in \Omega$, we have

$$\begin{aligned} \varepsilon_k &\geq \limsup_{u \xrightarrow{\Omega} x_k} \left\langle x_k^*, \frac{u - x_k}{\|u - x_k\|} \right\rangle \geq \limsup_{r \downarrow 0} \left\langle x_k^*, \frac{r(e_1 - je_j)}{\|r(e_1 - je_j)\|} \right\rangle \\ &= \left\langle x_k^*, \frac{e_1 - je_j}{\|e_1 - je_j\|} \right\rangle , \end{aligned}$$

which implies

$$\langle x_k^*, e_1 \rangle \leq \langle x_k^*, je_j \rangle + \varepsilon_k \sqrt{1 + j^2} . \tag{1.13}$$

Letting $k \to \infty$ in (1.12), we get

$$1 \leq \liminf_{k \to \infty} \langle x_k^*, \tfrac{1}{j(k)} e_{m(k)} \rangle .$$

This shows that if the sequence of natural numbers $j(k)$ is unbounded, then the sequence of x_k^* is unbounded too. The later contradicts the weak convergence of x_k^* due to the classical Banach-Steinhaus theorem (uniform boundedness principle). Thus we have only finitely many $j(k)$, and then (1.13) contradicts the weak convergence $x_k^* \xrightarrow{w} e_1^*$ as $k \to \infty$. This justifies (iii). $\triangle$

1.1.2 Tangential Approximations

A conventional approach to the study of infinitesimal properties of sets at boundary points and related differential properties of functions and mappings involves *tangential* local approximations. As well known, the concept of tangents to the graph of a "smooth" function was in the very beginning of the classical differential calculus. Then tangential approximations/directional derivatives have been used as convenient tools of variational analysis, particularly for deriving necessary optimality conditions in constrained problems of the calculus of variations, mathematical programming, and optimal control with smooth and nonsmooth data.

In this subsection we present concepts of tangents most useful in variational analysis and its applications, discuss some of their properties, and establish relationships between them and generalized normals introduced in Subsect. 1.1.1. To define tangent vectors to a set, first recall two standard notions of limits for set-valued mappings. Unless otherwise stated, we always understand limits in the *sequential* sense, in contrast to topological/net limits for general non-metrizable topologies. Given a set-valued mapping $F\colon X \rightrightarrows Y$ between topological spaces, the Painlevé-Kuratowski *upper/outer* and *lower/inner* limits of F as $x \to \bar{x}$ is defined, respectively, by

$$\operatorname*{Lim\,sup}_{x\to\bar{x}} F(x) := \Big\{ y \in Y \Big|\ \exists \text{ sequences } x_k \to \bar{x} \text{ and } y_k \to y$$

$$\text{with } y_k \in F(x_k) \text{ for all } k \in I\!\!N \Big\} ,$$

$$\operatorname*{Lim\,inf}_{x\to\bar{x}} F(x) := \Big\{ y \in Y \Big|\ \forall \text{ sequence } x_k \to \bar{x}\ \exists\, y_k \in F(x_k) \text{ with } k \in I\!\!N$$

$$\text{such that } y_k \to y \text{ as } k \to \infty \Big\} .$$

Note that the above "Lim sup" has been defined in (1.1) for the case of mappings $F\colon X \rightrightarrows X^*$ acting into the dual space $Y = X^*$ equipped with the (sequential) weak* topology; this is the main setting considered in the book. The following constructions involve however "Lim sup" and "Lim inf" for set-valued mappings from a real line into a normed space X.

Definition 1.8 (tangents cones). *Let $\Omega \subset X$ with $\bar{x} \in \Omega$. Then:*

(i) *The set $T(\bar{x};\Omega) \subset X$ defined by*

$$T(\bar{x};\Omega) := \operatorname*{Lim\,sup}_{t\downarrow 0} \frac{\Omega - \bar{x}}{t} ,$$

where the "Lim sup" is taken with respect to the norm topology of X, is called the CONTINGENT CONE *to Ω at $\bar{x}$.*

(ii) *If the "Lim sup" in* (i) *is taken with respect to the weak topology of X, then the resulting construction, denoted by $T_W(\bar{x};\Omega)$, is called the* WEAK CONTINGENT CONE *to Ω at $\bar{x}$.*

(iii) *The set $T_C(\bar{x};\Omega) \subset X$ defined by*

$$T_C(\bar{x};\Omega) := \operatorname*{Lim\,inf}_{\substack{x \xrightarrow{\Omega} \bar{x} \\ t\downarrow 0}} \frac{\Omega - x}{t} ,$$

where the "Lim inf" is taken with respect to the norm topology of X, is called the CLARKE TANGENT CONE *to Ω at $\bar{x}$.*

The contingent cone $T(\bar{x};\Omega)$ is often called the *Bouligand tangent/contingent cone*, since it was introduced by Bouligand and independently by Severi; see Commentary to this chapter. This is a closed (but generally nonconvex) subcone of X that can be equivalently described as the collections of $v \in X$ such that there are sequences $\{x_k\} \subset \Omega$ and $\{\alpha_k\} \subset \mathbb{R}_+$ satisfying

$$x_k \to \bar{x} \;\text{ and }\; \alpha_k(x_k - \bar{x}) \to v \;\text{ as }\; k \to \infty\,.$$

Similarly, the weak contingent cone $T_W(\bar{x};\Omega)$ can be equivalently described as the collection of $v \in X$ such that there exist sequences $\{x_k\} \subset \Omega$ and $\{\alpha_k\} \subset \mathbb{R}_+$ satisfying the relations

$$x_k \to \bar{x} \;\text{ and }\; \alpha_k(x_k - \bar{x}) \xrightarrow{w} v \;\text{ as }\; k \to \infty\,.$$

The Clarke tangent cone (known also as the *regular tangent cone*) can be described in this way as the collection of $v \in X$ such that for every sequence $x_k \xrightarrow{\Omega} \bar{x}$ and every sequence $t_k \downarrow 0$ there is a sequence $v_k \to v$ satisfying

$$x_k + t_k v_k \in \Omega \;\text{ for all }\; k \in \mathbb{N}\,.$$

It follows immediately from the definitions that

$$T_C(\bar{x};\Omega) \subset T(\bar{x};\Omega) \subset T_W(\bar{x};\Omega)\,,$$

where the second inclusion holds as equality when X is finite-dimensional. In contrast to $T(\bar{x};\Omega)$ and $T_W(\bar{x};\Omega)$, the Clarke tangent cone is *always convex* (see [255, 1165]), although it may be essentially smaller than $T(\bar{x};\Omega)$ and $T_W(\bar{x};\Omega)$ even in finite dimensions.

The next theorem gives more precise relationships between the tangent cones from Definition 1.8. In its formulation we use the notion of a *Kadec norm* on a Banach space that is one for which the weak and norm topologies agree on the boundary of the unit sphere. It is well known in the geometric theory of Banach spaces that every reflexive space admits an equivalent Kadec norm that is also Fréchet differentiable off the origin.

Theorem 1.9 (relationships between tangent cones). *Let X be a Banach space, and let $\Omega \subset X$ be locally closed around $\bar{x}$. Then*

$$\operatorname*{Lim\,inf}_{x \xrightarrow{\Omega} \bar{x}} T(x;\Omega) \subset T_C(\bar{x};\Omega) \subset \operatorname*{Lim\,inf}_{x \xrightarrow{\Omega} \bar{x}} T_W(x;\Omega)\,,$$

where the second inclusion holds if X is reflexive. Moreover,

$$T_C(\bar{x};\Omega) = \operatorname*{Lim\,inf}_{x \xrightarrow{\Omega} \bar{x}} T_W(x;\Omega)$$

provided that the norm on X is Kadec and Fréchet differentiable off the origin.

Proof. To justify the first inclusion of the theorem, take arbitrary v from the set on the left-hand side. Then for any $\varepsilon > 0$ there is $\eta > 0$ such that

$$(v + \varepsilon I\!B) \cap T(x; \Omega) \neq \emptyset \quad \text{whenever} \quad x \in \Omega \cap (\bar{x} + \eta I\!B) .$$

Let $\nu := (\eta/2)(\|v\| + 2\varepsilon)^{-1}$ and show that

$$\big(x + t(v + 2\varepsilon\eta I\!B)\big) \cap \Omega \neq \emptyset \quad \text{for all} \quad x \in \Omega \cap (\bar{x} + \tfrac{\eta}{2} I\!B) \quad \text{and} \quad t \in (0, \nu) ,$$

which easily implies that $v \in T_C(\bar{x}; \Omega)$. To proceed, consider the set

$$T_\delta := \big\{ t \in (0, \nu) \big|\ \big(x + t(v + \delta I\!B)\big) \cap \Omega \neq \emptyset \big\}$$

that happens to be *dense* in $(0, \nu)$ whenever $\delta \in (\varepsilon, 2\varepsilon)$. Indeed, by the above choice of ν we find a sequence $t_k \downarrow 0$ such that

$$\big(x + t_k(v + \delta I\!B)\big) \cap \Omega \neq \emptyset \quad \text{as} \quad k \in I\!N, \qquad \text{and so} \quad T_\delta \neq \emptyset .$$

Pick arbitrarily $\tau \in (0, \nu) \setminus T_\delta$ and put $t_* := \sup \big[T_\delta \cap (0, \tau) \big]$, which obviously gives $\big(x + t_*(v + \delta I\!B)\big) \cap \Omega \neq \emptyset$. Taking into account the choice of ν and that

$$x + t_*(v + \delta I\!B) \subset \bar{x} + \tfrac{\eta}{2} I\!B + \nu(\|v\| + \delta) I\!B \subset \bar{x} + \eta I\!B ,$$

we find a sequence $t_k \downarrow 0$ such that

$$\big(x + (t_* + t_k)(v + \delta I\!B)\big) \cap \Omega \neq \emptyset \quad \text{for all} \quad k \in I\!N .$$

The latter means that $t_* = \tau$, and thus τ is a cluster point of the set T_δ. Due to $\delta \in (\varepsilon, 2\varepsilon)$ and an arbitrary choice of $\tau \in (0, \nu) \setminus T_\delta$, we get

$$\big(x + t(v + 2\varepsilon\eta I\!B)\big) \cap \Omega \neq \emptyset \quad \text{for all} \quad t \in (0, \nu) ,$$

which implies that $v \in T_C(\bar{x}; \Omega)$ and therefore justifies the first inclusion of the theorem in the general Banach space setting.

Suppose now that X is *reflexive* and justify the fulfillment of the second inclusion claimed in the theorem. Taking $v \in T_C(\bar{x}; \Omega)$ and $\varepsilon > 0$, select $\eta > 0$ so that for every $x \in (\bar{x} + \eta I\!B) \cap \Omega$ there is a sequence $t_k \downarrow 0$ and a sequence $\{v_k\} \subset v + \varepsilon I\!B$ with $x + t_k v_k \in \Omega$ whenever $k \in I\!N$. By the reflexivity of X we find $\bar{v} \in X$ satisfying

$$\bar{v} \in v + \varepsilon I\!B \quad \text{and} \quad v_k \xrightarrow{w} \bar{v} \quad \text{as} \quad k \to \infty .$$

It follows from the definition of the weak contingent cone that $\bar{v} \in T_W(x; \Omega)$. Since $\varepsilon > 0$ was chosen arbitrarily, we conclude that $v \in \operatorname{Lim\,inf} T_W(x; \Omega)$ as $x \to \bar{x}$ with $x \in \Omega$. This proves the second inclusion of the theorem.

As shown by Borwein and Strójwas [156, Theorem 3.2], the reflexivity of X is *necessary* for the validity of the second inclusion in the theorem. We refer the reader to Aubin and Frankowska [54, Theorem 4.1.13] and to Borwein and

Strójwas [156, Theorem 3.1] for the proofs of the equality formulated in the theorem under the additional assumptions made. △

Next we study connections between the above tangential approximations of sets and the generalized normals defined in Subsect. 1.1.1. The following theorem describes dual relations of Fréchet-type normals and ε-normals with elements of the contingent and weak contingent cones.

Theorem 1.10 (normal-tangent relations). *Let $\Omega \subset X$ be a subset of a Banach space, and let $\bar{x} \in \Omega$. Then*

$$\widehat{N}_\varepsilon(\bar{x};\Omega) \subset \big\{x^* \in X^* \big|\ \langle x^*, v\rangle \le \varepsilon\|v\| \ \ \textit{for all}\ \ v \in T(\bar{x};\Omega)\big\}$$

whenever $\varepsilon \ge 0$. Moreover,

$$\widehat{N}(\bar{x};\Omega) \subset \big\{x^* \in X^* \big|\ \langle x^*, v\rangle \le 0 \ \ \textit{for all}\ \ v \in T_W(\bar{x};\Omega)\big\}\ ,$$

where equality holds if X is reflexive. The first inclusion holds as equality if X is finite-dimensional.

Proof. To prove the first inclusion, fix $x^* \in \widehat{N}_\varepsilon(\bar{x};\Omega)$ with some $\varepsilon \ge 0$ and take an arbitrary tangent vector $v \in T(\bar{x};\Omega)$. It follows from Definition 1.8(i) that there are sequences $t_k \downarrow 0$ and $v_k \to v$ with $\bar{x} + t_k v_k \in \Omega$ for all $k \in I\!N$. Substituting the latter combination into definition (1.2) of ε-normals, we get

$$t_k\langle x^*, v_k\rangle \le \varepsilon\, t_k\|v_k\| \ \text{ for large }\ k \in I\!N\ ,$$

which yields by passing to the limit as $k \to \infty$ that $\langle x^*, v\rangle \le \varepsilon\|v\|$. This justifies the first inclusion of the theorem for an arbitrary number $\varepsilon \ge 0$.

If $\varepsilon = 0$, the above proof ensures the fulfillment of the second inclusion of the theorem, where the weak contingent cone replaces the contingent cone. Indeed, it is sufficient to apply the weak convergence of $v_k \xrightarrow{w} v$ for passing to the limit in $\langle x^*, v_k\rangle$ with zero on the right-hand side.

Assume now that X is reflexive and show that the second inclusion holds in this case as equality. To proceed, we fix $x^* \notin \widehat{N}(\bar{x};\Omega)$ and find by (1.2) a number $\widetilde{\varepsilon} > 0$ and a sequence $x_k \xrightarrow{\Omega} \bar{x}$ such that

$$\langle x^*, x_k - \bar{x}\rangle > \widetilde{\varepsilon}\|x_k - \bar{x}\| \ \text{ for large }\ k \in I\!N\ .$$

Put $\alpha_k := \|x_k - \bar{x}\|^{-1}$ for $k \in I\!N$ and suppose without loss of generality that

$$\frac{x_k - \bar{x}}{\|x_k - \bar{x}\|} \xrightarrow{w} v \ \text{ for some }\ v \in X$$

due to the weak sequential compactness of bounded sets in reflexive spaces. Thus $v \in T_W(\bar{x};\Omega)$ by Definition 1.8(ii). On the other hand, $\langle x^*, v\rangle \ge \widetilde{\varepsilon}$ by passing to the limit in the assumption above. This justifies the desired equality and completes the proof of the theorem. △

Corollary 1.11 (normal-tangent duality). *Let X be a reflexive space, and let $\Omega \subset X$ with $\bar{x} \in \Omega$. Then the prenormal/Fréchet normal cone to Ω at $\bar{x}$ is dual to the weak contingent cone to Ω at this point, i.e.,*

$$\widehat{N}(\bar{x};\Omega) = T_W^*(\bar{x};\Omega) := \big\{x^* \in X^* \big|\ \langle x^*, z\rangle \le 0 \ \ \textit{whenever} \ \ v \in T_W(\bar{x};\Omega)\big\}\,.$$

Thus one has the duality relationship

$$\widehat{N}(\bar{x};\Omega) = T^*(\bar{x};\Omega)$$

when X is finite-dimensional.

Proof. The first equality follows directly from Theorem 1.10. It obviously reduces to the second one if $\dim X < \infty$. △

Note that we don't have the converse duality relation $\widehat{N}^*(\bar{x};\Omega) = T(\bar{x};\Omega)$ between the Fréchet normal cone and the contingent cone, since the latter is typically nonconvex even for simple sets in finite dimensions, while *duality always generates convexity*. On the contrary, the *Clarke normal cone* to Ω at $\bar{x}$ defined by

$$N_C(\bar{x};\Omega) := T_C^*(\bar{x};\Omega)$$

enjoys the full duality

$$N_C^*(\bar{x};\Omega) = T_C(\bar{x};\Omega)$$

with the Clarke tangent cone from Definition 1.8(iii), being however substantially larger than the Fréchet normal cone and the basic normal cone. In particular, for the set $\Omega := \{(x_1,x_2) \in I\!R^2|\ x_2 \ge -|x_1|\}$, the basic normal cone is computed in (1.4), while $\widehat{N}((0,0);\Omega) = \{0\}$ and $N_C((0,0);\Omega) = \{(v_1,v_2) \in I\!R^2|\ v_2 \le -|v_1|\}$. A more striking example is provided by the graphical set $\Omega := \operatorname{gph}|x| \subset I\!R^2$, where

$$N((0,0);\Omega) = \big\{(v_1,v_2)\big|\ v_2 \le -|v_1|\big\} \cup \big\{(v_1,v_2)\big|\ v_2 = |v_1|\big\}$$

while $N_C((0,0);\Omega) = R^2$. The latter situation is typical for *graphical sets* generated by Lipschitzian single-valued mappings and the like: see Theorems 1.46 and 3.62 for the exact statements and also Subsect. 2.5.2 for equivalent representations of the Clarke normal cone.

As mentioned, the basic normal cone (1.3), which is generally nonconvex, cannot be dual to any tangential approximations. One has

$$\mathrm{cl}^*\mathrm{co}\, N(\bar{x};\Omega) \subset N_C(\bar{x};\Omega) \ \text{ and } \ T_C(\bar{x};\Omega) \subset N^*(\bar{x};\Omega)$$

in the general Banach space setting, where *equalities* hold in both inclusions above for closed subsets Ω of Asplund spaces; see Theorem 3.57.

Remark 1.12 (normal versus tangential approximations). The principal difference between tangential and normal approximations is that the former constructions provide local approximations of sets in *primal spaces*, while the latter ones are defined in *dual spaces* carrying "dual" information for the study of local behavior. Being applied to epigraphs of extended-real-valued functions and graphs of set-valued mappings, tangential approximations generate corresponding *directional derivatives/subderivatives* of functions and *graphical derivatives* of mappings, while normal approximations relate to *subdifferentials* and *coderivatives*, respectively; see below.

Conventional approaches to generalized differentiation start with tangential approximations and then proceed with dual-space constructions by polarity/duality correspondences. However, this way doesn't allow us to generate either the (nonconvex) basic normal cone or even the prenormal cone at reference points outside the settings discussed in Corollary 1.11. Nevertheless, as we'll see below, the basic normal cone and associated subdifferential and coderivative constructions for functions and mappings enjoy many useful properties in arbitrary Banach spaces and admit a comprehensive theory in the general Asplund space setting at the same level of perfection as in finite dimensions. It happens that the basic normal cone and associated subdifferential/coderivatives constructions enjoy much richer calculi in comparison with those available for tangential approximations and dual convex objects generated by them in finite and infinite dimensions.

It is worth mentioning that in our approach to calculus and related properties of basic normals, subgradients, and coderivatives one cannot see *any role of tangential approximations* in primal spaces. What becomes crucial, in both finite and – especially – infinite dimensions, is the focus on *perturbations* and their *stability* in *dual spaces*, which will be demonstrated throughout the book in various settings of calculus and applications. We can treat such a dual-space perturbation/approximation theory as a proper counterpart of classical variations and tangential approximations in general nonconvex frameworks of advanced variational analysis.

1.1.3 Calculus of Generalized Normals

This subsection contains some calculus results for generalized normals in Banach spaces that are important in what follows.

Let $f\colon X \to Y$ be a mapping between Banach spaces, and let Θ be a subset of Y. The *inverse image* of Θ under f is defined by

$$f^{-1}(\Theta) := \{x \in X \mid f(x) \in \Theta\}\,.$$

The main goal of this subsection is to establish calculus results for generalized normals from Definition 1.1 that provide relationships between normal vectors to nonempty sets Θ and their inverse images under differentiable mappings between arbitrary Banach spaces. These results play a significant role in many applications, in particular, those considered later in this chapter.

Recall that $f\colon X \to Y$ is *Fréchet differentiable* at $\bar{x}$ if there is a linear continuous operator $\nabla f(\bar{x})\colon X \to Y$, called the *Fréchet derivative* of f at $\bar{x}$, such that

$$\lim_{x\to\bar{x}} \frac{f(x)-f(\bar{x})-\nabla f(\bar{x})(x-\bar{x})}{\|x-\bar{x}\|} = 0 . \tag{1.14}$$

The most interesting applications require, however, the following stronger differentiability property.

Definition 1.13 (strict differentiability). *A mapping $f\colon X \to Y$ is* STRICTLY DIFFERENTIABLE *at $\bar{x}$ if*

$$\lim_{\substack{x\to\bar{x}\\ u\to\bar{x}}} \frac{f(x)-f(u)-\nabla f(\bar{x})(x-u)}{\|x-u\|} = 0 .$$

The RATE OF STRICT DIFFERENTIABILITY *of f at $\bar{x}$ is a function $r_f(\bar{x};\cdot)$ from $(0,\infty)$ into $[0,\infty]$ defined by*

$$r_f(\bar{x};\eta) := \sup_{\substack{x,u\in\bar{x}+\eta I\!B\\ x\neq u}} \frac{\|f(x)-f(u)-\nabla f(\bar{x})(x-u)\|}{\|x-u\|} .$$

It follows from Definition 1.13 that $r_f(\bar{x};\eta)\downarrow 0$ as $\eta\downarrow 0$ for strictly differentiable mappings. Observe that, in contrast to (1.14), strict differentiability involves some *uniformity* of the limit in the derivative definition with respect to variable *pairs* of points around $\bar{x}$. A simple example of a function $f\colon I\!R \to I\!R$ Fréchet differentiable but not strictly differentiable at $\bar{x}=0$ is given by

$$f(x) := \begin{cases} x^2 \text{ if } x \text{ is rational} , \\ 0 \quad \text{otherwise} . \end{cases}$$

If $f \in \mathcal{C}^1$ around $\bar{x}$, i.e., *continuously Fréchet differentiable* in a neighborhood of $\bar{x}$, then it is obviously strictly differentiable at this point but not vice versa. In fact it may not be even differentiable at points near $\bar{x}$ as in the following example of a continuous function $f : [-1,1] \to I\!R$, $\bar{x}=0$, defined by

$$f(x) := \begin{cases} x^2 & \text{if } x = 1/k,\ k\in I\!N , \\ 0 & \text{if } x = 0 , \\ \text{linear otherwise} . \end{cases}$$

Note that every mapping f strictly differentiable at $\bar{x}$ is *Lipschitz continuous* around $\bar{x}$, or *locally Lipschitzian* around this point, i.e., there is a neighborhood U of $\bar{x}$ and a constant $\ell \geq 0$ such that

$$\|f(x)-f(u)\| \leq \ell\|x-u\| \ \text{ for all } \ x,u\in U . \tag{1.15}$$

Let us establish relationships between ε-normals to sets and their inverse images under differentiable mappings at reference points. Recall that a linear operator $A\colon X \to Y$ is *surjective*, or *onto*, if $AX = Y$, i.e., the image of X under the operator A is the whole space Y.

Theorem 1.14 (ε-normals to inverse images under differentiable mappings). *Let $f\colon X \to Y$, $\Theta \subset Y$, and $\bar{y} := f(\bar{x}) \in \Theta$. The following assertions hold:*

(i) *If f is Fréchet differentiable at $\bar{x}$, then there is $c_1 > 0$ such that*

$$\widehat{N}_\varepsilon(\bar{x}; f^{-1}(\Theta)) \supset \nabla f(\bar{x})^* \widehat{N}_{c_1\varepsilon}(\bar{y}; \Theta) \quad \textit{for all} \;\; \varepsilon \ge 0\,.$$

(ii) *If f is strictly differentiable at $\bar{x}$ and $\nabla f(\bar{x})$ is surjective, then there is $c_2 > 0$ such that*

$$\widehat{N}_\varepsilon(\bar{x}; f^{-1}(\Theta)) \subset \nabla f(\bar{x})^* \widehat{N}_{c_2\varepsilon}(\bar{y}; \Theta) + \varepsilon I\!B^* \quad \textit{for all} \;\; \varepsilon \ge 0\,.$$

(iii) *If* $\dim Y < \infty$, *then the inclusion in* (ii) *holds provided that f is continuous around $\bar{x}$ and merely Fréchet differentiable at this point with the surjective derivative $\nabla f(\bar{x})$.*

Proof. To prove the inclusion in (i), we observe that (1.14) implies the existence of a number $\ell > 0$ and a neighborhood U of $\bar{x}$ such that

$$\|f(x) - f(\bar{x})\| \le \ell \|x - \bar{x}\| \quad \text{for all} \;\; x \in U\,.$$

Fix $y^* \in \widehat{N}_\varepsilon(\bar{y}; \Theta)$ and take an arbitrary sequence $x_k \to \bar{x}$ with $x_k \in f^{-1}(\Theta)$ for all $k \in I\!N$. Then we have $f(x_k) \to f(\bar{x}) = \bar{y}$ and

$$\begin{aligned}
\limsup_{x_k \to \bar{x}} \frac{\langle \nabla f(\bar{x})^* y^*, x_k - \bar{x}\rangle}{\|x_k - \bar{x}\|} &= \limsup_{x_k \to \bar{x}} \frac{\langle y^*, \nabla f(\bar{x})(x_k - \bar{x})\rangle}{\|x_k - \bar{x}\|} \\
&= \limsup_{x_k \to \bar{x}} \frac{\langle y^*, f(x_k) - f(\bar{x})\rangle}{\|x_k - \bar{x}\|} \\
&\le \limsup_{y \xrightarrow{\Theta} \bar{y}} \max\left\{0, \frac{\langle y^*, y - \bar{y}\rangle}{\ell^{-1}\|y - \bar{y}\|}\right\} \le \ell\varepsilon
\end{aligned}$$

due to the definitions of ε-normals, Fréchet differentiability, and adjoint linear operators. This ensures that $\nabla f(\bar{x})^* y^* \in \widehat{N}_{\ell\varepsilon}(\bar{x}; f^{-1}(\Theta))$ for any $\varepsilon \ge 0$. Thus we have (i) with $c_1 := \ell^{-1}$.

Next let us prove (ii). In the proof below we'll use the following property of *metric regularity* for f around $\bar{x}$ that holds under the assumptions in (ii): there are a constant $\mu > 0$ and neighborhoods U of $\bar{x}$ and V of $\bar{y}$ such that

$$\operatorname{dist}(x; f^{-1}(y)) \le \mu \|y - f(x)\| \quad \text{for any} \;\; x \in U, \quad y \in V\,. \tag{1.16}$$

This actually goes back to the classical results of Lyusternik [824] and Graves [522] and is known now as the Lyusternik-Graves theorem; cf. Theorem 1.57 in Subsect. 1.2.3 and the discussion therein.

Let us fix $x^* \in \widehat{N}_\varepsilon(\bar{x}; f^{-1}(\Theta))$ and show that

$$|\langle x^*, x\rangle| \le \varepsilon\|x\| \ \text{ for all } \ x \in \ker \nabla f(\bar{x})\,. \tag{1.17}$$

Taking any $x \in \ker \nabla f(\bar{x})$, one obviously has

$$\|f(\bar{x} + tx) - \bar{y}\| = o(t) \ \text{ for small } \ t > 0\,.$$

Then (1.16) implies that for any small $t > 0$ there is $x_t \in f^{-1}(\bar{y})$ with $\|\bar{x} + tx - x_t\| = o(t)$. Excluding the trivial case of $x = 0$, we get

$$\varepsilon \ge \limsup_{t\downarrow 0} \frac{\langle x^*, x_t - \bar{x}\rangle}{\|x_t - \bar{x}\|} = \limsup_{t\downarrow 0} \frac{\langle x^*, tx\rangle}{\|tx\|} = \frac{\langle x^*, x\rangle}{\|x\|}$$

for each $x \in \ker \nabla f(\bar{x})$. Since it is also true for $-x \in \ker \nabla f(\bar{x})$, we arrive at the desired estimate (1.17).

Note that (1.17) gives $\|x^*\|_L \le \varepsilon$ for the norm of the linear continuous functional x^* considered on the subspace $L := \ker \nabla f(\bar{x})$. Using the Hahn-Banach theorem, we extend $x^*|_L$ to some $\tilde{x}^* \in X^*$ with $\|\tilde{x}^*\| \le \varepsilon$. Now putting $\hat{x}^* := x^* - \tilde{x}^*$, we get $\hat{x}^* \in X^*$ such that

$$\|\hat{x}^* - x^*\| \le \varepsilon, \quad \langle \hat{x}^*, x\rangle = 0 \ \text{ for all } \ x \in \ker \nabla f(\bar{x})\,.$$

Taking into account that $\nabla f(\bar{x})X = Y$, this allows us to (uniquely) define a linear functional $\hat{y}^*$ on Y by

$$\langle \hat{y}^*, y\rangle := \langle \hat{x}^*, x\rangle \ \text{ with any } \ x \in \nabla f(\bar{x})^{-1}(y)\,.$$

Applying the metric regularity property (1.16) to the linear surjective operator $\nabla f(\bar{x})\colon X \to Y$ (which follows in this case from the classical open mapping theorem), we find a constant $\mu > 0$ such that for any $y \in Y$ there is $x \in \nabla f(\bar{x})^{-1}(y)$ satisfying $\|x\| \le \mu\|y\|$. This implies the boundedness of the linear functional $\hat{y}^*$ defined above, i.e., we have $\hat{y}^* \in Y^*$. Since $\nabla f(\bar{x})^*\hat{y}^* = \hat{x}^*$, it remains to prove that $\hat{y}^* \in \widehat{N}_{c_2\varepsilon}(\bar{y}; \Theta)$ with some constant $c_2 > 0$.

To furnish this, we use again the metric regularity property for the mapping f and its strict derivative. Picking any $y \in \Theta$ close to $\bar{y}$ and using (1.16) for f with some $\mu > 0$, we find $x_y \in f^{-1}(y)$ such that

$$\|x_y - \bar{x}\| \le \mu\|y - \bar{y}\|\,.$$

Further, taking into account that

$$\|y - \bar{y} - \nabla f(\bar{x})(x_y - \bar{x})\| = \|f(x_y) - f(\bar{x}) - \nabla f(\bar{x})(x_y - \bar{x})\| = o(\|x_y - \bar{x}\|)$$

and using (1.16) for the operator $\nabla f(\bar{x})$, we get $\hat{x}_y \in \nabla f(\bar{x})^{-1}(y - \bar{y})$ with

$$\|x_y - \bar{x} - \hat{x}_y\| = o(\|x_y - \bar{x}\|) .$$

Now putting all the above together, one has

$$\begin{aligned}\limsup_{y \xrightarrow{\Theta} \bar{y}} \frac{\langle \hat{y}^*, y - \bar{y} \rangle}{\|y - \bar{y}\|} &= \limsup_{y \xrightarrow{\Theta} \bar{y}} \frac{\langle \hat{x}^*, \hat{x}_y \rangle}{\|y - \bar{y}\|} \le \limsup_{y \xrightarrow{\Theta} \bar{y}} \max \left\{ 0, \frac{\langle \hat{x}^*, \hat{x}_y \rangle}{\mu^{-1}\|x_y - \bar{x}\|} \right\} \\ &= \limsup_{y \xrightarrow{\Theta} \bar{y}} \max \left\{ 0, \frac{\langle \hat{x}^*, x_y - \bar{x} \rangle}{\mu^{-1}\|x_y - \bar{x}\|} \right\} \\ &\le \mu \limsup_{x \xrightarrow{f^{-1}(\Theta)} \bar{x}} \max \left\{ 0, \left(\varepsilon + \frac{\langle x^*, x - \bar{x} \rangle}{\|x - \bar{x}\|} \right) \right\} \le 2\mu\varepsilon .\end{aligned}$$

This ensures that $\hat{y}^* \in \widehat{N}_{c_2\varepsilon}(\bar{y}; \Theta)$ with $c_2 := 2\mu$ and justifies (ii).

Observe that in the above proof we used the property of metric regularity only for $y = \bar{y}$ in (1.16). Such a weaker property also holds under the assumptions in (iii); this follows from the proofs of Theorem F in Halkin [543] and of Proposition 7 in Ioffe [594] based on the Brouwer fixed-point theorem; cf. also the proof of Theorem 6.37 in Subsect. 6.3.4. Thus we get (iii) and complete the proof of the theorem. △

Corollary 1.15 (Fréchet normals to inverse images under differentiable mappings). *Let $f: X \to Y$ be Fréchet differentiable at $\bar{x}$. Then*

$$\widehat{N}(\bar{x}; f^{-1}(\Theta)) \supset \nabla f(\bar{x})^* \widehat{N}(\bar{y}; \Theta) ,$$

where the equality holds when $\nabla f(\bar{x})$ is surjective and either $\dim Y < \infty$ *or f is strictly differentiable at $\bar{x}$.*

Proof. Follows from Theorem 1.14 for $\varepsilon = 0$. △

Our next goal is to obtain relationships between basic normals to sets and their inverse images at reference points. If f is continuously differentiable in a neighborhood of $\bar{x}$, we can employ the results of Theorem 1.14 for ε-normals at points x close to $\bar{x}$ and then pass to the limit as $x \to \bar{x}$ and $\varepsilon \downarrow 0$. The situation is more complicated when f is merely *strictly differentiable* at $\bar{x}$. Then one cannot use Theorem 1.14, since f may not be differentiable around $\bar{x}$. To proceed in the case of strict differentiability, we need to get more delicate *uniform* estimates of ε-normals to the sets under consideration at points nearby $\bar{x}$ and $f(\bar{x})$ that involve the (strict) derivative of f at $\bar{x}$ only. The following lemma provides the required estimates using the *rate* of strict differentiability of f at $\bar{x}$.

Lemma 1.16 (uniform estimates for ε-normals). *Let $f: X \to Y$ and $\Theta \subset Y$ with $\bar{y} = f(\bar{x}) \in \Theta$. Assume that f is strictly differentiable at $\bar{x}$. Then there are constants $c_1 > 0$ and $\bar{\eta} > 0$ such that for any $y^* \in \widehat{N}_\varepsilon(f(x); \Theta)$ with $\varepsilon \ge 0$, $x \in (\bar{x} + \eta I\!B) \cap f^{-1}(\Theta)$, and $\eta \in (0, \bar{\eta})$ one has*

$$\nabla f(\bar{x})^* y^* \in \widehat{N}_{\hat{\varepsilon}}(x; f^{-1}(\Theta)) \quad \text{with} \quad \hat{\varepsilon} := c_1\varepsilon + \|y^*\|\, r_f(\bar{x};\eta)\,.$$

If in addition $\nabla f(\bar{x})$ is surjective, then there are constants $c_2 > 0$ and $\bar{\eta} > 0$ such that for any $x^* \in \widehat{N}_\varepsilon(x; f^{-1}(\Theta))$ with $\varepsilon \geq 0$, $x \in (\bar{x} + \eta I\!\!B) \cap f^{-1}(\Theta)$, and $\eta \in (0, \bar{\eta})$ one has

$$x^* \in \nabla f(\bar{x})^* \widehat{N}_{\tilde{\varepsilon}}(f(x); \Theta) + \Big(\varepsilon + c_2(\varepsilon + \|x^*\|)\, r_f(\bar{x};\eta)\Big) I\!\!B^*\,,$$

where $\tilde{\varepsilon} := c_2\varepsilon + c_2\|x^*\|\, r_f(\bar{x};\eta)$.

Proof. Since f is strictly differentiable at $\bar{x}$, there is $\bar{\eta} > 0$ such that f is Lipschitz continuous on $\bar{x} + \bar{\eta} I\!\!B$ with some constant $\ell > 0$. Hence $r_f(\bar{x};\eta) < \infty$ for every $\eta \in (0,\bar{\eta})$. Now taking $y^* \in \widehat{N}_\varepsilon(f(x);\Theta)$ with $\varepsilon \geq 0$ and $x \in (\bar{x} + \eta I\!\!B) \cap f^{-1}(\Theta)$ for such η, we have

$$\begin{aligned}
\limsup_{u \xrightarrow{f^{-1}(\Theta)} x} \frac{\langle \nabla f(\bar{x})^* y^*, u - x\rangle}{\|u - x\|} &= \limsup_{u \xrightarrow{f^{-1}(\Theta)} x} \frac{\langle y^*, \nabla f(\bar{x})(u - x)\rangle}{\|u - x\|} \\
&\leq \limsup_{u \xrightarrow{f^{-1}(\Theta)} x} \frac{\langle y^*, f(u) - f(x)\rangle}{\|u - x\|} + \|y^*\|\, r_f(\bar{x};\eta) \\
&\leq \limsup_{v \xrightarrow{\Theta} y} \max\left\{0, \frac{\langle y^*, v - y\rangle}{\ell^{-1}\|v - y\|}\right\} + \|y^*\|\, r_f(\bar{x};\eta) \\
&\leq \ell\varepsilon + \|y^*\|\, r_f(\bar{x};\eta) = \hat{\varepsilon}\,,
\end{aligned}$$

which implies the first inclusion in the lemma with $c_1 := \ell$.

Let us justify the second inclusion assuming that $\nabla f(\bar{x})$ is surjective. The proof below is a modification of the proof of assertion (ii) in Theorem 1.14 with the full usage of the metric regularity property (1.16) not only for $y = \bar{y}$ but for all y from a neighborhood of $\bar{y}$.

Choose $\bar{\eta} > 0$ so that $r_f(\bar{x};\bar{\eta}) < \infty$ and for any $\eta \in (0,\bar{\eta})$ one has $\bar{x} + \eta I\!\!B \subset U$ with $f(\bar{x} + \eta I\!\!B) \subset V$ for the neighborhoods U and V in (1.16). Fix $\varepsilon \geq 0$, $\eta \in (0,\bar{\eta})$, $\hat{x} \in (\bar{x} + \eta I\!\!B) \cap f^{-1}(\Theta)$, $\hat{y} := f(\hat{x})$, and $x^* \in \widehat{N}_\varepsilon(\hat{x}; f^{-1}(\Theta))$. Let us show that (1.17) holds with ε replaced by

$$\varepsilon_0 := \varepsilon + \mu(\varepsilon + \|x^*\|)\, r_f(\bar{x};\eta)\,,$$

where $\mu > 0$ is a constant of metric regularity (1.16). This will obviously follow from

$$\langle x^*, x\rangle \leq \varepsilon_0\|x\| \quad \text{for any} \quad 0 \neq x \in \ker \nabla f(\bar{x})\,.$$

To prove the latter inequality, we pick an arbitrary $0 \neq x \in \ker \nabla f(\bar{x})$ and observe that

$$\|f(\hat{x} + tx) - \hat{y}\| \leq r_f(\bar{x};\eta)\, \|x\| t \quad \text{whenever} \quad t > 0\,.$$

Then the metric regularity of f around $\bar{x}$ implies the existence of $x_t \in f^{-1}(\hat{y})$ satisfying the estimate

$$\|\hat{x} + tx - x_t\| \le \mu\, r_f(\bar{x};\eta)\, \|x\| t \quad \text{for small } \; t > 0 \, .$$

If $\langle x^*, x_t - \hat{x}\rangle \le 0$ for some $t > 0$, then

$$\langle x^*, tx\rangle - \mu\|x^*\|\, r_f(\bar{x};\eta)\, \|x\| t \le 0, \quad x \in \ker \nabla f(\bar{x}) \, ,$$

and we get the required estimate. It remains to consider the case of

$$\langle x^*, x_{t_k} - \hat{x}\rangle > 0 \;\text{ for some }\; t_k \downarrow 0, \quad k \in I\!N \, .$$

In this case one has

$$\begin{aligned} \varepsilon \ge \limsup_{k\to\infty} \frac{\langle x^*, x_{t_k} - \hat{x}\rangle}{\|x_{t_k} - \hat{x}\|} &\ge \limsup_{k\to\infty} \frac{\langle x^*, t_k x\rangle - \mu\|x^*\|\, r_f(\bar{x};\eta)\, \|x\| t_k}{\|t_k x\| + \mu\, r_f(\bar{x};\eta)\, \|x\| t_k} \\ &= \frac{\langle x^*, x\rangle - \mu\|x^*\|\, r_f(\bar{x};\eta)\, \|x\|}{\|x\| + \mu\, r_f(\bar{x};\eta)\, \|x\|}, \quad x \in \ker \nabla f(\bar{x}) \, , \end{aligned}$$

which implies estimate (1.17) with $\varepsilon = \varepsilon_0$. Then similarly to the proof of Theorem 1.14(ii), we find $\hat{x}^* \in X^*$ such that

$$\|\hat{x}^* - x^*\| \le \varepsilon_0, \quad \langle \hat{x}^*, x\rangle = 0 \;\text{ for }\; x \in \ker \nabla f(\bar{x})$$

and define $\hat{y}^* \in Y^*$ by

$$\langle \hat{y}^*, y\rangle := \langle \hat{x}^*, x\rangle, \quad x \in \nabla f(\bar{x})^{-1}(y) \, .$$

Now let us show that there is a constant $c_2 > 0$ for which

$$\hat{y}^* \in \widehat{N}_{\tilde{\varepsilon}}(\hat{y};\Theta) \;\text{ with }\; \tilde{\varepsilon} = c_2\varepsilon + c_2\|x^*\|\, r_f(\bar{x};\eta) \, .$$

Applying (1.16) first to f with $x = \hat{x}$ and $y \in \Theta \cap V$ close to $\hat{y}$ and then to $\nabla f(\bar{x})$, we find $x_y \in f^{-1}(y)$ and $\hat{x}_y \in \nabla f(\bar{x})^{-1}(y - \hat{y})$ satisfying the estimates

$$\|x_y - \hat{x}\| \le \mu\|y - \hat{y}\|, \quad \|x_y - \hat{x} - \hat{x}_y\| \le \mu\, r_f(\bar{x};\eta)\, \|x_y - \hat{x}\| \, .$$

Putting the above constructions and estimates together, we get

$$\begin{aligned} \limsup_{y \xrightarrow{\Theta} \hat{y}} \frac{\langle \hat{y}^*, y - \hat{y}\rangle}{\|y - \hat{y}\|} &\le \limsup_{y \xrightarrow{\Theta} \hat{y}} \max\left\{0, \frac{\langle \hat{x}^*, \hat{x}_y\rangle}{\mu^{-1}\|x_y - \hat{x}\|}\right\} \\ &\le \limsup_{y \xrightarrow{\Theta} \hat{y}} \max\left\{0, \frac{\langle \hat{x}^*, x_y - \hat{x}\rangle}{\mu^{-1}\|x_y - \hat{x}\|} + \mu^2 r_f(\bar{x};\eta)\, \|\hat{x}^*\|\right\} \\ &\le \limsup_{y \xrightarrow{\Theta} \hat{y}} \max\left\{0, \mu\varepsilon_0 + \frac{\langle x^*, x_y - \hat{x}\rangle}{\mu^{-1}\|x_y - \hat{x}\|} + \mu^2 r_f(\bar{x};\eta)(\|x^*\| + \varepsilon_0)\right\} \\ &\le \mu\varepsilon_0 + \mu\varepsilon + \mu^2 r_f(\bar{x};\eta)(\|x^*\| + \varepsilon_0) \le c_2\varepsilon + c_2\|x^*\|\, r_f(\bar{x};\eta) \, , \end{aligned}$$

where $c_2 := \max\{\mu, 2\mu + 2\mu^2 r_f(\bar{x};\bar{\eta}) + \mu^3 r_f^2(\bar{x};\bar{\eta}), 2\mu^2 + \mu^3 r_f(\bar{x};\bar{\eta})\}$. To complete the proof, we observe that μ may be replaced with c_2 in the definition of ε_0; so we arrive at the second inclusion in the lemma. △

Theorem 1.17 (basic normals to inverse images under strictly differentiable mappings). *Let $f\colon X \to Y$ and $\Theta \subset Y$ with $\bar{y} = f(\bar{x}) \in \Theta$. Assume that f is strictly differentiable at $\bar{x}$ with the surjective derivative. Then one has*

$$N(\bar{x}; f^{-1}(\Theta)) = \nabla f(\bar{x})^* N(\bar{y}; \Theta) . \tag{1.18}$$

Proof. Pick any $y^* \in N(\bar{y};\Theta)$. Then using the definition of basic normals, the continuity of f around $\bar{x}$, and the metric regularity property (1.16) held due to the Lyusternik-Graves theorem, we find sequences $\varepsilon_k \downarrow 0$, $x_k \to \bar{x}$, and $y_k^* \xrightarrow{w^*} y^*$ satisfying

$$x_k \in f^{-1}(\Theta) \;\text{ and }\; y_k^* \in \widehat{N}_{\varepsilon_k}(f(x_k);\Theta) \;\text{ for all }\; k \in I\!N .$$

The above Lemma 1.16 implies that

$$\nabla f(\bar{x})^* y_k^* \in \widehat{N}_{\hat{\varepsilon}_k}(x_k; f^{-1}(\Theta)) \;\text{ with }\; \hat{\varepsilon}_k := c_1\varepsilon_k + \|y_k^*\|\, r_f\big(\bar{x}; \|x_k - \bar{x}\|\big)$$

for k sufficiently large. Since y_k^* are uniformly bounded and f is strictly differentiable at $\bar{x}$, we have $\hat{\varepsilon}_k \downarrow 0$ as $k \to \infty$. Thus $\nabla f(\bar{x})^* y^* \in N(\bar{x}; f^{-1}(\Theta))$, which proves the inclusion stated in the theorem.

To prove the opposite inclusion in (1.18) when the operator $\nabla f(\bar{x})$ is surjective, we take an arbitrary $x^* \in N(\bar{x}; f^{-1}(\Theta))$ and find sequences $\varepsilon_k \downarrow 0$, $x_k \to \bar{x}$, and $x_k^* \xrightarrow{w^*} x^*$ with $f(x_k) \in \Theta$ and $x_k^* \in \widehat{N}_{\varepsilon_k}(x_k; f^{-1}(\Theta))$ for $k \in I\!N$. Then Lemma 1.16 implies the existence of $c_2 > 0$ such that

$$x_k^* \in \nabla f(\bar{x})^* \widehat{N}_{\tilde{\varepsilon}_k}(f(x_k);\Theta) + \Big(\varepsilon_k + c_2(\varepsilon_k + \|x_k^*\|)\, r_f\big(\bar{x}; \|x_k - \bar{x}\|\big)\Big) I\!B^* ,$$

where $\tilde{\varepsilon}_k := c_2\varepsilon_k + c_2\|x_k^*\|\, r_f\big(\bar{x}; \|x_k - \bar{x}\|\big) \downarrow 0$ as $k \to \infty$. Now passing to the limit in the latter inclusion, we arrive at $x^* \in \nabla f(\bar{x})^* N(f(\bar{x});\Theta)$ and ends the proof of the theorem. △

Note that Theorem 1.17 ensures equality (1.18) for arbitrary sets Θ, which may not be normally regular at $\bar{y}$. Moreover, (1.18) and the equality in Corollary 1.15 allow us to show that the normal regularity of $f^{-1}(\Theta)$ at $\bar{x}$ is *equivalent* to the normal regularity of Θ at $\bar{x}$ provided that f is strictly differentiable at $\bar{x}$ with the surjective derivative. To proceed, we need the following fact from functional analysis that is useful also in the sequel.

Lemma 1.18 (properties of adjoint linear operators). *Let $A^*\colon Y^* \to X^*$ be the adjoint operator to a linear continuous operator $A\colon X \to Y$. Assume that A is surjective. Then for any $y^* \in Y^*$ one has*

$$\|A^*y^*\| \geq \kappa\|y^*\| \quad \textit{with} \quad \kappa = \inf\Big\{\|A^*y^*\| \;\Big|\; \|y^*\| = 1\Big\} \in (0, \infty)\,.$$

In particular, A^* is injective, i.e., $A^*y_1^* \neq A^*y_2^*$ if $y_1^* \neq y_2^*$.

Proof. Consider the canonical map $\pi: X \to X/\ker A$ between X and the quotient Banach space generated by $\ker A$, where the norm on $X/\ker A$ is defined by

$$\|x + \ker A\| := \inf_{u \in x + \ker A} \|u\|\,.$$

This clearly induces a linear isomorphism $\widetilde{A}: X/\ker A \to AX$ such that $A = \widetilde{A} \circ \pi$. Applying the classical *open mapping* theorem, we find a constant $\kappa > 0$ such that $\kappa B_Y \subset AB_X$. Then

$$\|A^*y^*\| = \sup_{x \in B_X} |\langle A^*y^*, x\rangle| = \sup_{x \in B_X} |\langle y^*, Ax\rangle| = \sup_{y \in AB_X} |\langle y^*, y\rangle|$$

$$\geq \sup_{y \in \kappa B_Y} |\langle y^*, y\rangle| = \kappa\|y^*\| \quad \text{for all} \quad y^* \in Y^*\,.$$

To complete the proof of the lemma, it remains to justify the above formula for κ. This follows from the relations

$$\|(\widetilde{A}^*)^{-1}\| = \Big(\inf_{\|y^*\|=1} \|\widetilde{A}^*y^*\|\Big)^{-1} = \Big(\inf_{\|y^*\|=1} \|A^*y^*\|\Big)^{-1}$$

by taking into account that $A^* = \pi^* \circ \widetilde{A}^*$ and $\|\pi^*z^*\| = \|z^*\|$. △

Theorem 1.19 (normal regularity of inverse images under strictly differentiable mappings). *Let $f: X \to Y$ be strictly differentiable at $\bar{x}$ with the surjective derivative $\nabla f(\bar{x})$. Then $f^{-1}(\Theta)$ is normally regular at $\bar{x}$ if and only if Θ is normally regular at $\bar{y} = f(\bar{x})$.*

Proof. Due to Theorem 1.17 and Corollary 1.15 we have (1.18) and

$$\widehat{N}(\bar{x}; f^{-1}(\Theta)) = \nabla f(\bar{x})^*\widehat{N}(\bar{y}; \Theta)\,.$$

Thus the normal regularity of Θ at $\bar{y}$ immediately implies the normal regularity of $f^{-1}(\Theta)$ at $\bar{x}$. To prove the opposite implication, we need to show that $N(\bar{y}; \Theta) \subset \widehat{N}(\bar{y}; \Theta)$ provided that $f^{-1}(\Theta)$ is normally regular at $\bar{x}$. Picking any $y_1^* \in N(\bar{y}; \Theta)$ and using the latter regularity, find $y_2^* \in \widehat{N}(\bar{y}; \Theta)$ such that $\nabla f(\bar{x})^*(y_1^* - y_2^*) = 0$. By Lemma 1.18 this implies that $y_1^* = y_2^*$, i.e., we have $y_1^* \in \widehat{N}(\bar{y}; \Theta)$ and complete the proof. △

More calculus and regularity results will be obtained in Chap. 3 in the Asplund space setting. In particular, we'll prove there far-going developments of Theorem 1.17 for nonsmooth and set-valued mappings, where the equality in (1.18) is replaced with the "right" inclusion "$\subset$". In general, nonsmooth calculus requires additional qualification conditions (which are automatic in the framework of Theorem 1.17) as well as some "sequential normal compactness" properties that always hold in finite-dimensional spaces. The latter properties are certainly of independent interest for general Banach spaces and occur to be an essential ingredient of the infinite-dimensional variational theory. We consider them next.

1.1.4 Sequential Normal Compactness of Sets

In this subsection we study some local properties of sets in Banach spaces that ensure the equivalence between the weak* and norm convergences to zero of ε-normals (1.2) in dual spaces. As mentioned above, such properties are very important for subsequent applications.

Definition 1.20 (sequential normal compactness). *A set $\Omega \subset X$ is* SEQUENTIALLY NORMALLY COMPACT (SNC) *at $\bar{x} \in \Omega$ if for any sequence $(\varepsilon_k, x_k, x_k^*) \in [0, \infty) \times \Omega \times X^*$ satisfying*

$$\varepsilon_k \downarrow 0, \ x_k \to \bar{x}, \ \ x_k^* \in \widehat{N}_{\varepsilon_k}(x_k; \Omega), \ \ \text{and} \ \ x_k^* \xrightarrow{w^*} 0$$

one has $\|x_k^*\| \to 0$ as $k \to \infty$.

It is easy to observe from the definition that Ω is SNC at $\bar{x} \in \Omega$ if its closure is SNC at this point. Note also that every nonempty set in a finite-dimensional space is SNC at each of its points. Our first result shows that the SNC property in infinite-dimensional spaces may hold only for sufficiently "large" sets.

Recall that the *affine hull* of Ω is defined as

$$\operatorname{aff} \Omega := \left\{ \sum_{i=1}^{l} \alpha_i x_i \ \middle| \ x_i \in \Omega, \ \alpha_i \in \mathbb{R}, \ \sum_{i=1}^{l} \alpha_i = 1, \ l \in \mathbb{N} \right\},$$

which is the smallest affine set containing Ω. It is clear that $\operatorname{aff} \Omega$ is a translation of a linear subspace of X. The closure of $\operatorname{aff} \Omega$ in X is called the *closed affine hull* of Ω and is denoted by $\overline{\operatorname{aff}}\, \Omega$. For any point $x \in \overline{\operatorname{aff}}\, \Omega$, the set $\overline{\operatorname{aff}}\, \Omega - x$ is a closed linear subspace of X that doesn't depend on the choice of x. The *codimension* of $\overline{\operatorname{aff}}\, \Omega$ is defined as the dimension of the quotient space $X/(\overline{\operatorname{aff}}\, \Omega - x)$. The *relative interior* $\operatorname{ri} \Omega$ of $\Omega \subset X$ is the interior of Ω with respect to $\overline{\operatorname{aff}}\, \Omega$.

Let us prove that any SNC set must be finite-codimensional, and this condition is a *characterization* of the SNC property for convex sets with nonempty relative interiors.

Theorem 1.21 (finite codimension of SNC sets). *A set $\Omega \subset X$ is sequentially normally compact at $\bar{x} \in \Omega$ only if*

$$\operatorname{codim} \overline{\operatorname{aff}}\, (\Omega \cap U) < \infty$$

for any neighborhood U of $\bar{x}$. In particular, a singleton in X is sequentially normally compact if and only if X is finite-dimensional. Moreover, when Ω is convex and $\operatorname{ri}\Omega \neq \emptyset$, *the sequential normal compactness of Ω at every $\bar{x} \in \Omega$ is equivalent to the finite codimension condition* $\operatorname{codim}\overline{\operatorname{aff}}\,\Omega < \infty$.

Proof. First we prove the necessity part for an arbitrary set $\Omega \subset X$. Since SNC is a local property, one may always assume that $\bar{x} = 0 \in \Omega$ and $U = X$. Then $L := \overline{\operatorname{aff}}\,\Omega$ is a closed linear subspace of X and its *annihilator*

$$L^{\perp} := \big\{x^* \in X^* \,\big|\, \langle x^*, x\rangle = 0 \;\text{ for all }\; x \in L\big\}$$

is obviously a subset of the prenormal cone $\widehat{N}(0;\Omega)$.

It is well known that $L^{\perp}$ is isometric to the dual quotient space $(X/L)^*$. Assuming that $\operatorname{codim}\Omega = \dim(X/L) = \infty$ and using the fundamental Josefson-Nissenzweig theorem (see, e.g., the book by Diestel [333, Chap. 12]), we find a sequence of vectors $x_k^* \in (X/L)^*$ such that

$$\|x_k^*\| = 1 \;\text{ for all }\; k \in I\!N \;\text{ and }\; x_k^* \xrightarrow{w^*} 0 \;\text{ as }\; k \to \infty \;\text{ in }\; (X/L)^* .$$

Invoking the mentioned isomorphism, we can treat $\{x_k^*\}$ as a sequence of norm-one vectors in $L^{\perp} \subset X^*$ converging to zero in the weak* topology of X^*. By the inclusions

$$L^{\perp} \subset \widehat{N}(0;\Omega) \subset \widehat{N}_{\varepsilon}(0;\Omega) \;\text{ for any }\; \varepsilon \geq 0 ,$$

we get a contradiction with the sequential normal compactness of Ω.

Let us prove the sufficiency part of theorem for convex sets with nonempty interiors. Without loss of generality, we assume that $0 \in \Omega$, hence $\overline{\operatorname{aff}}\,\Omega$ is a closed subspace of X. Since $\operatorname{codim}\overline{\operatorname{aff}}\,\Omega < \infty$, there is a finite-dimensional subspace $Z \subset X$ such that

$$X = \overline{\operatorname{aff}}\,\Omega \bigoplus Z, \;\text{ i.e., }\; X = \overline{\operatorname{aff}}\,\Omega + Z \;\text{ and }\; (\overline{\operatorname{aff}}\,\Omega) \cap Z = \{0\} .$$

One clearly has

$$\widehat{N}_{\varepsilon}(\bar{x};\Omega|_X) = \widehat{N}_{\varepsilon}(\bar{x};\Omega|_{\overline{\operatorname{aff}}\,\Omega}) \times Z^* \;\text{ for all }\; \bar{x} \in \Omega, \quad \varepsilon \geq 0 .$$

Taking into account that Z is finite-dimensional, it suffices to consider the case of $\overline{\operatorname{aff}}\,\Omega = X$ when $\operatorname{ri}\Omega = \operatorname{int}\Omega \neq \emptyset$.

Fix $\bar{x} \in \Omega$ and $x_0 \in \operatorname{int}\Omega$; then $x_0 + rI\!B \subset \Omega$ for some $r > 0$. Take arbitrary sequences of $x_k \in \Omega$ and $x_k^* \in \widehat{N}_{\varepsilon_k}(x_k;\Omega)$ with $x_k \to \bar{x}$, $\varepsilon_k \downarrow 0$, and $x_k^* \xrightarrow{w^*} 0$ as $k \to \infty$. We have $\|x_k^*\| \leq c$ for some constant $c > 0$ and all $k \in I\!N$. It follows from Proposition 1.3 that

$$\langle x_k^*, x - x_k\rangle \leq \varepsilon_k \|x - x_k\| \;\text{ for all }\; x \in \Omega, \quad k \in I\!N .$$

Since $x := x_0 + ru \in \Omega$ for any $u \in I\!B$, we get

$$\langle x_k^*, u \rangle \leq \tfrac{1}{r}\varepsilon_k \|x_0 + ru - x_k\| - \tfrac{1}{r}\langle x_k^*, x_0 - x_k \rangle \text{ for all } u \in I\!B ,$$

which gives

$$\|x_k^*\| \leq \alpha(\varepsilon_k + |\langle x_k^*, x_0 - x_k \rangle|), \quad k \in I\!N ,$$

with some $\alpha > 0$. Because of

$$|\langle x_k^*, x_0 - x_k \rangle| \leq |\langle x_k^*, x_0 - \bar{x} \rangle| + c\|\bar{x} - x_k\| ,$$

the latter clearly implies that $\|x_k^*\| \to 0$ as $k \to \infty$. △

Next we show that the SNC property of sets is *invariant* with respect to the inverse image operation defined by a strictly differentiable mapping whose derivative is surjective at the point of interest. This result is based on calculus rules established in the previous subsection.

Theorem 1.22 (SNC property for inverse images under strictly differentiable mappings). *Let $f: X \to Y$ be strictly differentiable at $\bar{x}$ with the surjective derivative $\nabla f(\bar{x})$, and let Θ be a subset of Y containing $\bar{y} := f(\bar{x})$. Then $f^{-1}(\Theta)$ is SNC at $\bar{x}$ if and only if Θ is SNC at $\bar{y}$.*

Proof. First assume that Θ is SNC at $\bar{y}$ and prove that $f^{-1}(\Theta)$ is SNC at $\bar{x}$. Take sequences $(\varepsilon_k, x_k, x_k^*)$ such that $f(x_k) \in \Theta$, $x_k^* \in \widehat{N}_{\varepsilon_k}(x_k; f^{-1}(\Theta))$ and $\varepsilon_k \downarrow 0$, $x_k \to \bar{x}$, $x_k^* \xrightarrow{w^*} 0$ as $k \to \infty$. Then x_k^* are uniformly bounded in X^*. By Lemma 1.16 we find sequences $\tilde{\varepsilon}_k \downarrow 0$, $\hat{\varepsilon}_k \downarrow 0$, and $y_k^* \in \widehat{N}_{\tilde{\varepsilon}_k}(f(x_k); \Theta)$ with

$$\|x_k^* - \nabla f(\bar{x})^* y_k^*\| \leq \hat{\varepsilon}_k, \quad k \in I\!N .$$

Now employing Lemma 1.18, we conclude that $y_k^* \xrightarrow{w^*} 0$. This implies $\|y_k^*\| \to 0$ due to the SNC property of Θ at $\bar{y}$ and the continuity of f at $\bar{x}$. Thus $\|x_k^*\| \to 0$ as well, which justifies the SNC property of $f^{-1}(\Theta)$ at $\bar{x}$.

To prove the opposite implication, we assume that $f^{-1}(\Theta)$ is SNC at $\bar{x}$ and pick arbitrary sequences $(\varepsilon_k, y_k, y_k^*)$ with $y_k^* \in \widehat{N}_{\varepsilon_k}(y_k; \Theta)$ and $\varepsilon_k \downarrow 0$, $y_k \xrightarrow{\Theta} \bar{y}$, $y_k^* \xrightarrow{w^*} 0$ as $k \to \infty$. The metric regularity property of f around $\bar{x}$ allows us to find $\mu > 0$ and $x_k \in f^{-1}(y_k)$ such that $\|x_k - \bar{x}\| \leq \mu\|y_k - \bar{y}\|$, i.e., $x_k \to \bar{x}$ with $y_k = f(x_k)$, $k \in I\!N$. Using again Lemma 1.16, we get a sequence $\hat{\varepsilon}_k \downarrow 0$ for which

$$x_k^* := \nabla f(\bar{x})^* y_k^* \in \widehat{N}_{\hat{\varepsilon}_k}(x_k; f^{-1}(\Theta)), \quad k \in I\!N .$$

Clearly $x_k^* \xrightarrow{w^*} 0$ and, since $f^{-1}(\Theta)$ is SNC at $\bar{x}$, we have $\|x_k^*\| \to 0$ as $k \to \infty$. Employing Lemma 1.18, we conclude that $\|y_k^*\| \to 0$, which completes the proof of the theorem. △

If $f(x) = Ax$ is a linear continuous operator between Banach spaces X and Y, then Theorem 1.22 ensures the equivalence between the SNC properties of

$\Theta \subset Y$ and the inverse image $A^{-1}(\Theta)$ at the corresponding points provided that A is surjective. Furthermore, in the linear case the surjectivity assumption can be relaxed as follows.

Proposition 1.23 (SNC property for inverse images under linear operators). *Let $A\colon X \to Y$ be a linear continuous operator whose range*

$$AX := \big\{ y \in Y \;\big|\; \exists x \in X \;\; \textit{with} \;\; y = Ax \big\}$$

is closed in Y. Take a set $\Theta \subset AX$ and assume that Θ is SNC at some point $\bar{y} := A\bar{x} \in \Theta$. Then its inverse image $A^{-1}(\Theta)$ is SNC at $\bar{x}$.

Proof. It is sufficient to show that any set $\Theta \subset AX$ sequentially normally compact at $\bar{y}$ (with respect to the whole space Y) is also SNC at $\bar{y}$ with respect to the smaller Banach space AX. Then we can use Theorem 1.22 for the surjective operator $A\colon X \to AX$.

To justify the mentioned claim, we use the necessity part of Theorem 1.21 ensuring that $\operatorname{codim} AX < \infty$ due to $\overline{\operatorname{aff}}\, \Theta \subset AX$. Hence the space AX is *complemented*, i.e., there is a closed subspace $Z \subset Y$ with $AX \bigoplus Z = Y$. Now denote by $\widehat{N}_\varepsilon(\cdot; \Theta|_{AX})$ the set of ε-normals to Θ with respect to AX and take arbitrary sequences $y_k \xrightarrow{\Theta} \bar{y}$, $\varepsilon_k \downarrow 0$, and $y_k^* \in \widehat{N}_{\varepsilon_k}(y_k; \Theta|_{AX})$ converging to zero in the weak* topology of $(AX)^*$. Since AX is complemented, we have $(y_k^*, 0) \in \widehat{N}_{\varepsilon_k}(y_k; \Theta)$, where $0 \in Z^*$ and $\widehat{N}_{\varepsilon_k}(\cdot; \Theta)$ is the set of ε_k-normals to Θ with respect to Y. Then the SNC property of Θ with respect to Y implies that $\|(y_k^*, 0)\|_{Y^*} \to 0$ and hence $\|y_k^*\|_{(AX)^*} \to 0$ as $k \to \infty$, i.e., Θ is SNC at $\bar{y}$ with respect to AX. $\triangle$

Next let us present some sufficient conditions for the SNC property of a set $\Omega \subset X$ that do not involve any normals to Ω, whereas they are expressed intrinsically in terms of the set Ω itself. Such conditions are related to a kind of Lipschitzian behavior of Ω *around* the point in question.

Definition 1.24 (epi-Lipschitzian and compactly epi-Lipschitzian sets). *Let $\Omega \subset X$ with $\bar{x} \in \operatorname{cl} \Omega$. Then:*

(i) *Ω is* COMPACTLY EPI-LIPSCHITZIAN (CEL) *around $\bar{x}$ if there are a compact set $C \subset X$, a neighborhood U of $\bar{x}$, a neighborhood O of the origin in X, and a number $\gamma > 0$ such that*

$$\Omega \cap U + tO \subset \Omega + tC \quad \textit{for all} \;\; t \in (0, \gamma)\,. \tag{1.19}$$

(ii) *Ω is* EPI-LIPSCHITZIAN *around $\bar{x}$ if the compact set C in (1.19) can be selected as a singleton.*

It is easy to see from the definition that if Ω is epi-Lipschitzian (compactly epi-Lipschitzian) around $\bar{x}$, then its closure has the same property around this point. When Ω is closed and C is a nonzero singleton in X, the epi-Lipschitzian property of Ω means that Ω is locally homeomorphic to the epigraph of a Lipschitz continuous function; hence the terminology.

If X is finite-dimensional, all subsets of X have the CEL property around all their points (with $C = I\!B$, the closed unit ball) . This is different from the epi-Lipschitzian property that may fail even for convex sets in $I\!R^n$. In fact, the epi-Lipschitzian property of convex sets admits the following simple characterization.

Proposition 1.25 (epi-Lipschitzian convex sets). *A convex set $\Omega \subset X$ is epi-Lipschitzian around any $\bar{x} \in \Omega$ if and only if* $\operatorname{int}\Omega \neq \emptyset$.

Proof. Let us show that a convex set $\Omega \subset X$ is epi-Lipschitzian around $\bar{x} \in \Omega$ if and only if there is $\upsilon \in X$ such that

$$\bar{x} + \gamma\upsilon \in \operatorname{int}\Omega \quad \text{for some} \quad \gamma > 0\,,$$

which clearly implies the result.

The necessity of the above condition is trivial. To prove the sufficiency, we take $\gamma > 0$ and a neighborhood V of the origin in X for which $\bar{x}+\gamma(\upsilon+V) \subset \Omega$. Choose another neighborhood $\widetilde{V}$ of $0 \in X$ such that $\frac{1}{\gamma}\widetilde{V} + \widetilde{V} \subset V$. Then we have the inclusions

$$x + \gamma(\upsilon + \widetilde{V}) \subset \bar{x} + \gamma(\upsilon + \tfrac{1}{\gamma}\widetilde{V} + \widetilde{V}) \subset \bar{x} + \gamma(\upsilon + V) \subset \Omega$$

for all $x \in \bar{x} + \widetilde{V}$. Since Ω is convex, it implies that

$$x + t(\upsilon + \widetilde{V}) \subset \Omega \quad \text{for all} \quad x \in \Omega \cap (\bar{x} + \widetilde{V}) \quad \text{and} \quad t \in (0, \gamma)\,.$$

Thus we get (1.19) with $U := \bar{x} + \widetilde{V}$, $O := \widetilde{V}$, and $C := \{-\upsilon\}$. △

Let us show that the CEL (and hence epi-Lipschitzian) property of Ω around $\bar{x} \in \Omega$ implies its SNC property at this point in any Banach space.

Theorem 1.26 (SNC property of CEL sets). *Let $\Omega \subset X$ be compactly epi-Lipschitzian around $\bar{x} \in \Omega$. Then it is sequentially normally compact at this point.*

Proof. Assuming that Ω is CEL around $\bar{x}$, we find a compact set $C \subset X$ and positive numbers γ and η such that

$$\Omega \cap (\bar{x} + \eta I\!B) + t\eta I\!B \subset \Omega + tC \quad \text{for all} \quad t \in (0, \gamma)\,.$$

Let us show that this implies the existence of a constant $\alpha > 0$ for which

$$\widehat{N}_\varepsilon(x; \Omega) \subset \Big\{x^* \in X^* \Big|\ \eta\|x^*\| \le \varepsilon(\alpha + \eta) + \max_{c \in C}\langle x^*, c\rangle\Big\} \tag{1.20}$$

whenever $x \in \Omega \cap (\bar{x} + \eta I\!B)$. Indeed, fixing $x \in \Omega \cap (\bar{x} + \eta I\!B)$ and employing the CEL property of Ω, for any $e \in I\!B$ and $t \in (0, \gamma)$ we pick a point $c_t \in C$

such that $x + t(\eta e - c_t) \in \Omega$. Due to the compactness of C, a subsequence of c_t converges to some point $\bar{c} \in C$ as $t \downarrow 0$. This easily implies, by definition (1.2), that

$$\langle x^*, \eta e - \bar{c} \rangle - \varepsilon \|\eta e - \bar{c}\| \le 0 \quad \text{for all} \quad x^* \in \widehat{N}_\varepsilon(x; \Omega) \, .$$

Since $e \in I\!B$ was chosen arbitrarily, the latter gives inclusion (1.20) with $\alpha := \max_{c \in C} \|c\|$.

Now take any sequences $\varepsilon_k \downarrow 0$, $x_k \xrightarrow{\Omega} \bar{x}$, and $x_k^* \xrightarrow{w^*} 0$ with $x_k^* \in \widehat{N}_{\varepsilon_k}(x_k; \Omega)$, Lucet $k \in I\!N$. The compactness of C implies that $\langle x_k^*, c \rangle \to 0$ *uniformly* in $c \in C$. Thus (1.20) ensures that $\|x_k^*\| \to 0$ as $k \to \infty$, i.e., Ω is SNC at $\bar{x}$. $\triangle$

Remark 1.27 (characterizations of CEL sets).

(i) The CEL property of *closed convex* sets $\Omega \subset X$ admits several explicit characterizations in the general framework of normed spaces X; we refer the reader to Borwein, Lucet and Mordukhovich [150] for more details. In particular, such a set Ω is CEL around every $\bar{x} \in \Omega$ *if and only if* its affine hull is a *closed finite-codimensional subspace* of X with $\operatorname{ri} \Omega \neq \emptyset$. Combining this characterization with the last part of Theorem 1.21, we conclude that the SNC and CEL properties *agree* in Banach spaces for any closed convex sets having closed affine hulls and nonempty *relative interiors.*

(ii) Characterizations of the CEL property for general closed sets are established by Ioffe [607] in terms of normal cones satisfying certain requirements in corresponding Banach spaces. When X is Asplund, the CEL property of Ω around $\bar{x} \in \Omega \subset X$ admits a *topological* limiting description in the form of Definition 1.20 with $\varepsilon_k = 0$, where sequences are replaced by *bounded nets.* We'll see in Chap. 2 that ε_k can be equivalently removed from the definition of the SNC property in the Asplund space setting. It is well known that for separable spaces X the weak* topology on $I\!B^* \subset X^*$ is metrizable, and there is no need to use nets in this case. Putting these facts together, we can conclude that the SNC property of Ω at $\bar{x} \in \Omega$ and CEL property of this set around $\bar{x}$ agree for closed subsets of *separable Asplund spaces.* Moreover, as proved in Fabian and Mordukhovich [422], these properties agree for a larger class of spaces including *weakly compactly generated* (WCG) Asplund spaces. This implies, in particular, that the SNC property of sets in such spaces is actually *around* $\bar{x} \in \Omega$. However, the SNC and CEL properties may *not* agree even for closed *convex cones* in nonseparable Asplund spaces admitting a C^∞-smooth renorm; see Example 3.6. Moreover, these properties *never* agree in Banach spaces whose dual unit ball is not weak* sequentially compact, in particular, in the standard spaces ℓ^∞ and $L^\infty[0, 1]$. We refer the reader to the aforementioned paper [422] for more results in this direction, where relationships between sequential and topological normal compactness properties are studied in detail in the framework of general Banach spaces. Let us emphasize that for most applications, in both Asplund and general Banach space settings, it suffices to use the SNC property without any separability assumptions; see the subsequent material of this book.

1.1.5 Variational Descriptions and Minimality

The very definition of basic normals to arbitrary sets allows us to study their properties by taking sequential limits of ε-normals (1.2) at neighboring points. The latter normals admit a useful variational description that follows directly from the definition of "lim sup" in (1.2).

Proposition 1.28 (variational description of ε-normals). *Given $\varepsilon \geq 0$ and $\bar{x} \in \Omega$, we have $x^* \in \widehat{N}_\varepsilon(\bar{x};\Omega)$ if and only if for any $\gamma > 0$ the function*

$$\psi(x) := \langle x^*, x - \bar{x}\rangle - (\varepsilon + \gamma)\|x - \bar{x}\|$$

attains a local maximum relative to Ω at $\bar{x}$.

This description characterizes ε-normals via *local* maximization of a *nonsmooth* function relative to the given set Ω. In particular, it holds for Fréchet normals ($\varepsilon = 0$) in arbitrary Banach spaces. In what follows we show that in the latter case one has more delicate variational descriptions that characterize Fréchet normals via *global* maximization over the set $\Omega \subset X$ of some "supporting" functions $s: X \to I\!R$ *smooth* in a certain sense. Theorem 1.30 bellow contains several results in this direction. If $s(\cdot)$ is required to be only Fréchet differentiable *at* $\bar{x}$, then such a variational description can be easily derived from Definition 1.1(i) in any Banach space. Using more involved arguments, we obtain significantly stronger results in Theorem 1.30 under additional geometric assumptions on the space in question. To proceed, let us first present the following lemma on smoothing real functions important in the proof of the theorem.

Lemma 1.29 (smoothing functions in $I\!R$). *Let $\rho: [0,\infty) \to [0,\infty)$ be a function having the right-hand derivative $\rho'_+(0)$ and satisfying the conditions:*

$$\rho(0) = \rho'_+(0) = 0 \quad \textit{and} \quad \rho(t) \leq \alpha + \beta t \quad \textit{for all} \quad t \geq 0$$

with positive constants α and β. Then there is a nondecreasing, convex, continuously differentiable function $\tau: [0,\infty) \to [0,\infty)$ such that

$$\tau(0) = \tau'_+(0) = 0 \quad \textit{and} \quad \tau(t) > \rho(t) \quad \textit{for all} \quad t > 0\,.$$

Proof. First let us prove that there exist $\gamma > 0$ and a nondecreasing, convex, continuously differentiable function $\sigma: [0, 2\gamma) \to [0,\infty)$ such that

$$\sigma(0) = \sigma'_+(0) = 0 \quad \text{and} \quad \sigma(t) > \rho(t) \quad \text{for} \quad t \in (0, 2\gamma)\,.$$

To construct such a function, we choose a sequence of positive numbers a_k such that $a_{k+1} < \frac{1}{2}a_k$ and

$$\rho(t) + t^2 < 2^{-(k+3)}t \ \text{ if } \ t \in [0, a_k]$$

for all $k \in I\!N$. Put $\gamma := \frac{1}{2}a_1$ and define a continuous function $r\colon [0, 2\gamma] \to [0, \infty)$ by $r(0) := 0$, $r(a_k) := 2^{-k}$, and r is linear on $[a_{k+1}, a_k]$ for all $k \in I\!N$. Then define a function $\sigma\colon [0, 2\gamma) \to [0, \infty)$ by

$$\sigma(t) := \int_0^t r(\xi)d\xi \ \text{ for } \ t \in [0, 2\gamma)$$

and show that it possesses the required properties. Its smoothness, monotonicity, convexity, and the equalities $\sigma(0) = \sigma'_+(0) = 0$ follow directly from the definition and standard facts of real analysis. To check the remaining properties, we fix $t \in (0, 2\gamma)$ and observe that $t \in [a_{k+1}, a_k)$ for some $k \in I\!N$. Then, by the construction of the functions σ and r, we get

$$\sigma(t) \geq \int_{a_{k+1}}^t r(\xi)d\xi + \int_{\frac{1}{2}a_{k+1}}^{a_{k+1}} r(\xi)d\xi \geq \int_{a_{k+1}}^t 2^{-(k+1)}d\xi + \int_{\frac{1}{2}a_{k+1}}^{a_{k+1}} 2^{-(k+2)}d\xi$$

$$= \frac{t - a_{k+1}}{2^{k+1}} + \frac{a_{k+1}}{2^{k+3}} \geq \frac{t}{2^{k+3}} > \rho(t) ,$$

which justifies the required properties of σ.

Next let us build a function $\tau\colon [0, \infty) \to (0, \infty)$ with the properties listed in the lemma. Given $\alpha, \beta > 0$, we choose $\lambda > 1$ such that $\lambda\sigma(\gamma) > \alpha + \beta\gamma$ and consider the following two cases.

First assume that $\lambda\sigma'(\gamma) \leq \beta$. In this case we find $\mu \geq \lambda$ such that $\mu\sigma'(\gamma) = \beta$ and define

$$\tau(t) := \begin{cases} \mu\sigma(t) & \text{if } \ 0 \leq t \leq \gamma , \\ \mu\sigma(\gamma) + \beta(t - \gamma) & \text{if } \ t > \gamma . \end{cases}$$

One can easily see that the function τ is nondecreasing, convex, and continuous everywhere on $[0, \infty)$ including $t = \gamma$. Moreover, $\tau'_-(\gamma) = \mu\sigma'(\gamma)$ and $\tau'_+(\gamma) = \beta = \mu\sigma'(\gamma)$ due to the choice of μ, which implies the continuous differentiability o τ on $[0, \infty)$. It follows from the definition that

$$\tau(0) = \tau'_+(0) = 0 \ \text{ and } \ \tau(t) \geq \sigma(t) > \rho(t) \ \text{ if } \ 0 < t \leq \gamma .$$

For $t > \gamma$ one has

$$\tau(t) = \mu\sigma(\gamma) + \beta(t - \gamma) > \alpha + \beta t \geq \rho(t)$$

due to the assumption on ρ. Thus we get the required properties of the above function τ in the case of $\lambda\sigma'(\gamma) \leq \beta$.

It remains to consider the other case when $\lambda\sigma'(\gamma) > \beta$. In this case we define a nondecreasing and convex function $\tau : [0, \infty) \to [0, \infty)$ by

$$\tau(t) := \begin{cases} \lambda\sigma(t) & \text{if} \quad 0 \le t \le \gamma\ , \\ \lambda\sigma(\gamma) - \lambda\gamma\sigma'(\gamma) + \lambda\sigma'(\gamma)t & \text{if} \quad t > \gamma\ . \end{cases}$$

Again, a straightforward verification yields that τ is a continuously differentiable function $[0, \infty)$ and satisfies all the requirements on $[0, \gamma]$. By the choice of λ we get

$$\tau(t) \ge \alpha + \beta\gamma + \lambda\sigma'(\gamma)(t - \gamma) > \alpha + \beta\gamma + \beta(t - \gamma) = \alpha + \beta t \ge \rho(t)$$

for $t > \gamma$, which completes the proof of the lemma. △

Recall that a Banach space X admits a *Fréchet smooth renorm* if there is an equivalent norm on X that is Fréchet differentiable at any nonzero point. In particular, every reflexive space admits a Fréchet smooth renorm. We'll also consider Banach spaces admitting an *$\mathcal{S}$-smooth bump function* with respect to a given class $\mathcal{S}$, i.e., a function $b: X \to I\!\!R$ such that $b(\cdot) \in \mathcal{S}$, $b(x_0) \ne 0$ for some $x_0 \in X$, and $b(x) = 0$ whenever x lies outside a ball in X. In what follows we deal with the three classes of $\mathcal{S}$-smooth functions on X: Fréchet smooth ($\mathcal{S} = \mathcal{F}$), Lipschitzian and Fréchet smooth ($\mathcal{S} = \mathcal{LF}$), and Lipschitzian and continuously differentiable ($\mathcal{S} = \mathcal{LC}^1$). It is well known that the class of spaces admitting a $\mathcal{LC}^1$-smooth bump function strictly includes the class of spaces with a Fréchet smooth renorm. Observe that all the spaces listed above belong to the class of Asplund spaces, where Fréchet normals play a role similar to ε-normals in the general Banach space setting; see Chap. 2.

Theorem 1.30 (smooth variational descriptions of Fréchet normals). *Let Ω be a nonempty subset of a Banach space X, and let $\bar{x} \in \Omega$. The following assertions hold:*

(i) *Given $x^* \in X^*$, we assume that there is a function $s: U \to I\!\!R$ defined on a neighborhood of $\bar{x}$ and Fréchet differentiable at $\bar{x}$ such that $\nabla s(\bar{x}) = x^*$ and $s(x)$ achieves a local maximum relative to Ω at $\bar{x}$. Then $x^* \in \widehat{N}(\bar{x}; \Omega)$. Conversely, for every $x^* \in \widehat{N}(\bar{x}; \Omega)$ there is a function $s: X \to I\!\!R$ such that $s(x) \le s(\bar{x}) = 0$ whenever $x \in \Omega$ and that $s(\cdot)$ is Fréchet differentiable at $\bar{x}$ with $\nabla s(\bar{x}) = x^*$.*

(ii) *Assume that X admits a Fréchet smooth renorm. Then for every $x^* \in \widehat{N}(\bar{x}; \Omega)$ there is a concave Fréchet smooth function $s: X \to I\!\!R$ that achieves its global maximum relative to Ω uniquely at $\bar{x}$ and such that $\nabla s(\bar{x}) = x^*$.*

(iii) *Assume that X admits an $\mathcal{S}$-smooth bump function, where $\mathcal{S}$ stands for one of the classes $\mathcal{F}$, $\mathcal{LF}$, or $\mathcal{LC}^1$. Then for every $x^* \in \widehat{N}(\bar{x}; \Omega)$ there is an $\mathcal{S}$-smooth function $s: X \to I\!\!R$ satisfying the conclusions in* (ii).

Proof. Under the assumptions in (i) we have

$$s(x) = s(\bar{x}) + \langle x^*, x - \bar{x}\rangle + o(\|x - \bar{x}\|) \le s(\bar{x})$$

for all $x \in \Omega$ near $\bar{x}$. Hence $\langle x^*, x - \bar{x}\rangle + o(\|x - \bar{x}\|) \le 0$ for such x, which implies that $x^* \in \widehat{N}(\bar{x};\Omega)$ due to Definition 1.1(i) with $\varepsilon = 0$. To justify the converse statement in (i), it is sufficient to check that the function

$$s(x) := \begin{cases} \min\big\{0, \langle x^*, x - \bar{x}\rangle\big\} & \text{if } x \in \Omega\,, \\ \langle x^*, x - \bar{x}\rangle & \text{otherwise} \end{cases}$$

is Fréchet differentiable at $\bar{x}$, which directly follows from the definitions.

Let us prove (ii). Fix an equivalent Fréchet smooth norm $\|\cdot\|$ on X and pick an arbitrary vector $x^* \in \widehat{N}(\bar{x};\Omega)$. Define the function

$$\rho(t) := \sup\big\{\langle x^*, x - \bar{x}\rangle \;\big|\; x \in \Omega,\; \|x - \bar{x}\| \le t\big\} \quad \text{for } t \ge 0\,, \tag{1.21}$$

which clearly satisfies all the assumptions of Lemma 1.29 due to the definition of Fréchet normals. Using this lemma, we get the corresponding function $\tau\colon [0,\infty) \to [0,\infty)$ and construct a function $s\colon X \to I\!R$ by

$$s(x) := -\tau(\|x - \bar{x}\|) - \|x - \bar{x}\|^2 + \langle x^*, x - \bar{x}\rangle, \quad x \in X\,.$$

Note that this function is concave on X with $s(\bar{x}) = 0$, since τ is convex and nondecreasing on $[0,\infty)$ with $\tau(0) = 0$. We also have

$$s(x) + \|x - \bar{x}\|^2 \le -\rho(\|x - \bar{x}\|) + \langle x^*, x - \bar{x}\rangle \le 0 = s(\bar{x}) \;\text{ for all } x \in \Omega\,,$$

which implies that $s(x)$ achieves its global maximum over Ω uniquely at $\bar{x}$. Observe that $s(x)$ is Fréchet differentiable at any $x \ne \bar{x}$ due the smoothness of the function τ and the norm $\|\cdot\|$ at nonzero point of X. To justify (ii), it remains to prove that $s(x)$ is Fréchet differentiable at $x = \bar{x}$ with $\nabla s(\bar{x}) = x^*$. The latter follows directly from the smoothness of τ with $\tau'_+(0) = 0$ by the classical chain rule.

Next let us prove (iii) simultaneously for all the three classes $\mathcal{S}$ listed in the theorem. Taking an $\mathcal{S}$-smooth bump function $b\colon X \to I\!R$, we can always assume that $0 \le b(x) \le 1$ for all $x \in X$, $b(0) = 1$, and $b(x) = 0$ if $\|x\| \ge 1$. Then consider a function $d\colon X \to [0,\infty)$ constructed in Lemma VIII.1.3 of the book by Deville, Godefroy and Zizler [331] as follows: $d(0) = 0$ and

$$d(x) := \frac{2}{h(x)} \quad \text{with } h(x) := \sum_{n=0}^{\infty} b(nx) \;\text{ for } x \ne 0\,.$$

It is proved in the mentioned lemma that

$$\|x\| \le d(x) \le \mu\|x\| \;\text{ if } \|x\| \le 1 \;\text{ and }\; d(x) = 2 \;\text{ if } \|x\| > 1$$

with some fixed $\mu > 1$, that d is Fréchet differentiable on $X \setminus \{0\}$, and it is Lipschitz continuous on X provided that the bump function b is Lipschitz continuous. Moreover, d is continuously differentiable on $X \setminus \{0\}$ if b has this property. We can easily check that the function d^2 as well as the composition $\tau \circ d$ of d with the function τ built above are Fréchet differentiable at 0 with

$$\nabla(d^2)(0) = \nabla(\tau \circ d)(0) = 0 .$$

Further, if d is Lipschitz continuous on X with modulus $l > 0$ and $0 \neq x \in X$ with $\|x\| \to 0$, then

$$\|\nabla(d^2)(x)\| = \|2d(x)\nabla d(x)\| \le l^2\|x\| \to 0 \quad \text{and}$$

$$\|\nabla(\tau \circ d)(x)\| = \|\tau'(d(x))\nabla d(x)\| \le l|\tau'(d(x))| \to 0 .$$

Putting these facts together, we conclude that the functions d^2 and $\tau \circ d$ are $\mathcal{S}$-smooth on X if the bump function b has this property, for each class $\mathcal{S}$ considered in the theorem.

Now we fix $x^* \in \widehat{N}(\bar{x}; \Omega)$ and take the function τ constructed in Lemma 1.29 for $\rho: [0, \infty) \to [0, \infty)$ defined in (1.21). Let $\psi: I\!R \to I\!R$ be an arbitrary $\mathcal{LC}^1$-function such that

$$\psi(t) = t \ \text{ for } \ t \ge 0 \ \text{ and } \ \psi(t) = -1 \ \text{ for } \ t \le -1 .$$

Choosing $\lambda > \max\{1, (\tau(\frac{1}{2}))^{-1}(1 + \|x^*\|)\}$, we form a function $\theta: X \to I\!R$ by

$$\theta(x) := \begin{cases} \psi\big(-\lambda\tau(d(x - v)) + \langle x^*, x - \bar{x}\rangle\big) & \text{if } \|x - \bar{x}\| \le 1 , \\ -1 & \text{otherwise} \end{cases}$$

and show that the combination

$$s(x) := \theta(x) - d^2(x - \bar{x}), \quad x \in X ,$$

has all the properties formulated in the theorem. It clearly follows from the facts that θ is $\mathcal{S}$-smooth on X and that $\theta(x) \le \theta(\bar{x}) = 0$ for all $x \in \Omega$.

We justify the required smoothness of θ by observing that

$$t(x) := -\lambda\tau(d(x - \bar{x})) + \langle x^*, x - \bar{x}\rangle \le \lambda\tau(\tfrac{1}{2}) + \|x^*\| < -1$$

if $\frac{1}{2} \le \|x - \bar{x}\| < 1$, and so $\theta(x) = \psi(t(x)) = -1$ for such x due to the choice of λ. To complete the proof of the theorem, it is sufficient to show that $\theta(x) \le 0$ if $x \in \Omega$ and $\|x - \bar{x}\| < \frac{1}{2}$, since $\theta(x) = -1 < 0$ for all other $x \in \Omega$.

Let us first consider the case when

$$-\lambda\tau(d(x - \bar{x})) + \langle x^*, x - \bar{x}\rangle \ge 0 .$$

Then, by properties of the functions involved in the construction of θ, we get

$$\theta(x) = -\lambda\tau(d(x - \bar{x})) + \langle x^*, x - \bar{x}\rangle \le -\rho(\|x - \bar{x}\|) + \langle x^*, x - \bar{x}\rangle \le 0 .$$

In the other case of

$$-\lambda\tau(d(x-\bar{x}))+\langle x^*,x-\bar{x}\rangle<0$$

we obviously have $\theta(x)\le\psi(0)=0$, which ends the proof. △

In the conclusion of this section we present a *minimality property* of the basic normal cone (1.3) among any normal structures satisfying natural requirements in Banach spaces. This property directly relates to Definition 1.1 and the variational description of ε-normals in Proposition 1.28.

Given a Banach space X, let us consider an abstract *prenormal structure* $\widehat{\mathcal{N}}$ on X that associates, with every nonempty subset $\Omega\subset X$, a set-valued mapping $\widehat{\mathcal{N}}(\cdot;\Omega)\colon X\rightrightarrows X^*$. We always assume that $\widehat{\mathcal{N}}(x;\Omega)=\emptyset$ for $x\notin\Omega$ and that $\widehat{\mathcal{N}}(x;\Omega)=\widehat{\mathcal{N}}(x;\widetilde{\Omega})$ if the sets Ω and $\widetilde{\Omega}$ coincide near $x\in\Omega$.

Of course, these assumptions are too broad and don't have any valuable consequences without additional requirements. To be useful, generalized normals should have some properties important for applications, particularly to optimization problems. From this viewpoint, a crucial requirement to generalized normals is their ability to describe necessary optimality conditions in problems of constrained optimization. The next result shows that the basic normal cone (1.3) is smaller than the *sequential* limit (1.1) of *any* prenormal structure supporting natural first-order optimality conditions.

Proposition 1.31 (minimality of the basic normal cone). *Given $\Omega\subset X$ and $\bar{x}\in\Omega$, we assume the following property of the prenormal structure $\widehat{\mathcal{N}}$ on X:*

(M) *For every $x^*\in X^*$, small $\varepsilon>0$, and $u\in\Omega\cap(\bar{x}+\varepsilon I\!B)$ providing a local minimum to the function*

$$\psi(x):=\langle x^*,x-u\rangle+\varepsilon\|x-u\|$$

over Ω, there is $v\in\Omega\cap(\bar{x}+\varepsilon B)$ such that

$$-x^*\in\eta I\!B^*+\widehat{\mathcal{N}}(v;\Omega)\quad\text{for all }\ \eta>\varepsilon\ .$$

Then one has the relationship

$$N(\bar{x};\Omega)\subset\mathcal{N}(\bar{x};\Omega):=\operatorname*{Lim\,sup}_{x\to\bar{x}}\widehat{\mathcal{N}}(x;\Omega)$$

between the basic normal cone (1.3) *and the sequential normal structure $\mathcal{N}$ generated by $\widehat{\mathcal{N}}$.*

Proof. Taking an arbitrary $x^*\in N(\bar{x};\Omega)$ in (1.3), we find sequences $\varepsilon_k\downarrow 0$, $x_k\to\bar{x}$, and $x_k^*\xrightarrow{w^*}x^*$ such that $x_k^*\in\widehat{N}_{\varepsilon_k}(x_k;\Omega)$ for all $k\in I\!N$. Due to Proposition 1.28 this implies that for any $k\in I\!N$ and any $\gamma>0$ one has

$$\langle x_k^*,x-x_k\rangle-(\varepsilon_k+\gamma)\|x-x_k\|\le 0\quad\text{for all }x\in\Omega\ \text{ near }\ x_k\ ,$$

and so x_k gives a local minimum to the function

$$\psi(x) := \langle -x_k^*, x - x_k \rangle + (\varepsilon_k + \gamma)\|x - x_k\|$$

belonging to the class specified in (M). Using this property with $\eta = 2\varepsilon_k + \gamma > \varepsilon_k + \gamma$, we get

$$x_k^* \in (2\varepsilon_k + \gamma)I\!B^* + \widehat{\mathcal{N}}(v_k; \Omega) \quad \text{with some } v_k \in \Omega \text{ near } x_k .$$

Since $\gamma > 0$ was chosen arbitrary, the latter ensures that $x^* \in \mathcal{N}(\bar{x}; \Omega)$ by passing to the limit as $k \to \infty$. △

The requirement on the prenormal structure $\widehat{\mathcal{N}}$ imposed in (M) means that $\widehat{\mathcal{N}}$ is adequate to describe "fuzzy" necessary optimality conditions in constrained optimization. It obviously holds when $v = u$ and $\eta = \varepsilon$ in (M), which corresponds to the "exact" necessary optimality condition (at the given minimum point) and is valid, in particular, for the sequential normal structure $\mathcal{N}$ generated by $\widehat{\mathcal{N}}$. Note that latter "exact" requirement on (pre)normal structure is more restrictive than the "fuzzy" one, but it is more convenient for applications. This requirement is fulfilled, in the case of closed subsets of arbitrary Banach spaces, for the normal cone of Clarke and for the "approximate" G-normal cone of Ioffe, which give constructive examples of broader *topological* normal structures and always contain the basic normal cone (1.3) due to Proposition 1.31; see Sect. 2.5.2 for more discussions. We'll show in Chap. 2 that the prenormal and normal cones from Definition 1.1 satisfy, respectively, the fuzzy and exact optimality conditions in property (M) for closed subsets of arbitrary Asplund spaces.

1.2 Coderivatives of Set-Valued Mappings

In this section we consider *set-valued mappings* (*multifunctions*) $F: X \rightrightarrows Y$ between Banach spaces, i.e., mappings from X into subsets of Y. When F happens to be *single-valued*, we usually use the notation $F = f: X \to Y$. We say that F is *closed-valued, convex-valued,* ... if all the values $F(x)$ are closed, convex, ..., respectively. Denote by

$$\operatorname{dom} F := \{x \in X \mid F(x) \neq \emptyset\}, \quad \operatorname{rge} F := \{y \in Y \mid \exists x \text{ with } y \in F(x)\}$$

the *domain* and the *range* of F. The *kernel* of F is

$$\ker F := \{x \in X \mid 0 \in F(x)\} .$$

Each set-valued mapping $F: X \rightrightarrows Y$ is uniquely associated with its *graph*

$$\operatorname{gph} F := \{(x, y) \in X \times Y \mid y \in F(x)\}$$

in the product space $X \times Y$. The space $X \times Y$ is Banach with respect to the *sum norm*

$$\|(x, y)\| := \|x\| + \|y\|$$

imposed on $X \times Y$ unless otherwise stated.

Given sets $\Omega \subset X$ and $\Theta \subset Y$, we define the *image* of Θ under F by

$$F(\Omega) := \{y \in Y \mid \exists x \in \Omega \text{ with } y \in F(x)\}$$

and the *inverse image* of Θ under F by

$$F^{-1}(\Theta) := \{x \in X \mid F(x) \cap \Theta \neq \emptyset\}.$$

The *inverse mapping* to $F: X \rightrightarrows Y$ is

$$F^{-1}: Y \rightrightarrows X \text{ with } F^{-1}(y) := \{x \in X \mid y \in F(x)\}.$$

It is clear that $\operatorname{dom} F^{-1} = \operatorname{rge} F$, $\operatorname{rge} F^{-1} = \operatorname{dom} F$, and

$$\operatorname{gph} F^{-1} = \{(y, x) \in Y \times X \mid (x, y) \in \operatorname{gph} F\}.$$

A set-valued mapping $F: X \rightrightarrows Y$ is *positively homogeneous* if $0 \in F(0)$ and $F(\alpha x) \supset \alpha F(x)$ for all $x \in X$ and $\alpha > 0$, or equivalently, when the graph of F is a cone in $X \times Y$. The *norm* of a positively homogeneous mapping F is defined by

$$\|F\| := \sup\{\|y\| \mid y \in F(x) \text{ and } \|x\| \leq 1\}. \tag{1.22}$$

1.2.1 Basic Definitions and Representations

Now let us describe the main derivative-like constructions for multifunctions we are going to study in this book. These objects are called *coderivatives*, since they provide a pointwise approximation of set-valued (in particular, single-valued) mappings between given spaces using elements of *dual spaces*. In the case of smooth single-valued mappings the coderivatives reduce to the classical *adjoint derivative* operator at the point in question. For general non-smooth and set-valued mappings they are constructed through normal vectors to graphs and are *not dual* to any derivative objects related to tangential approximations in initial spaces.

Following the pattern in constructing generalized normals, we first define preliminary coderivative objects at points nearby and then pass to the limit to construct coderivatives at the reference point. In this way we define two limiting coderivatives (different in infinite dimensions) depending on the convergence used on in the dual product space $X^* \times Y^*$.

Definition 1.32 (coderivatives). *Let* $F: X \rightrightarrows Y$ *with* $\operatorname{dom} F \neq \emptyset$.

(i) *Given* $(x, y) \in X \times Y$ *and* $\varepsilon \geq 0$, *we define the* ε-CODERIVATIVE *of* F *at* (x, y) *as a multifunction* $\widehat{D}^*_\varepsilon F(x, y): Y^* \rightrightarrows X^*$ *with the values*

$$\widehat{D}^*_\varepsilon F(x, y)(y^*) := \{x^* \in X^* \mid (x^*, -y^*) \in \widehat{N}_\varepsilon((x, y); \operatorname{gph} F)\}. \tag{1.23}$$

When $\varepsilon = 0$ *in* (1.23), *this construction is called the* PRECODERIVATIVE *or* FRÉCHET CODERIVATIVE *of* F *at* (x, y) *and is denoted by* $\widehat{D}^* F(x, y)$. *It follows from the definition that* $\widehat{D}^*_\varepsilon F(x, y)(y^*) = \emptyset$ *for all* $\varepsilon \geq 0$ *and* $y^* \in Y^*$ *if* $(x, y) \notin \operatorname{gph} F$.

(ii) *The* NORMAL CODERIVATIVE *of* F *at* $(\bar{x}, \bar{y}) \in \operatorname{gph} F$ *is a multifunction* $D^*_N F(\bar{x}, \bar{y}) \colon Y^* \rightrightarrows X^*$ *defined by*

$$D^*_N F(\bar{x}, \bar{y})(\bar{y}^*) := \mathop{\text{Lim sup}}_{\substack{(x,y)\to(\bar{x},\bar{y}) \\ y^* \xrightarrow{w^*} \bar{y}^* \\ \varepsilon\downarrow 0}} \widehat{D}^*_\varepsilon F(x, y)(y^*) . \tag{1.24}$$

That is, the normal coderivative (1.24) *is the collection of such* $\bar{x}^* \in X^*$ *for which there are sequences* $\varepsilon_k \downarrow 0$, $(x_k, y_k) \to (\bar{x}, \bar{y})$, *and* $(x^*_k, y^*_k) \xrightarrow{w^*} (\bar{x}^*, \bar{y}^*)$ *with* $(x_k, y_k) \in \operatorname{gph} F$ *and* $x^*_k \in \widehat{D}^*_{\varepsilon_k} F(x_k, y_k)(y^*_k)$. *We put* $D^*_N F(\bar{x}, \bar{y})(y^*) := \emptyset$ *for all* $y^* \in Y^*$ *if* $(\bar{x}, \bar{y}) \notin \operatorname{gph} F$.

(iii) *The* MIXED CODERIVATIVE *of* F *at* $(\bar{x}, \bar{y}) \in \operatorname{gph} F$ *is a multifunction* $D^*_M F(\bar{x}, \bar{y}) \colon Y^* \rightrightarrows X^*$ *defined by*

$$D^*_M F(\bar{x}, \bar{y})(\bar{y}^*) := \mathop{\text{Lim sup}}_{\substack{(x,y)\to(\bar{x},\bar{y}) \\ y^* \to \bar{y}^* \\ \varepsilon\downarrow 0}} \widehat{D}^*_\varepsilon F(x, y)(y^*) . \tag{1.25}$$

That is, the mixed coderivative (1.25) *is the collection of such* $\bar{x}^* \in X^*$ *for which there are sequences* $\varepsilon_k \downarrow 0$, $(x_k, y_k, y^*_k) \to (\bar{x}, \bar{y}, \bar{y}^*)$, *and* $x^*_k \xrightarrow{w^*} \bar{x}^*$ *with* $(x_k, y_k) \in \operatorname{gph} F$ *and* $x^*_k \in \widehat{D}^*_{\varepsilon_k} F(x_k, y_k)(y^*_k)$. *We put* $D^*_M F(\bar{x}, \bar{y})(y^*) := \emptyset$ *for all* $y^* \in Y^*$ *if* $(\bar{x}, \bar{y}) \notin \operatorname{gph} F$.

We always omit $\bar{y}$ in the coderivative notation if $F(\bar{x}) = \{\bar{y}\}$. Note that

$$D^*_N F(\bar{x}, \bar{y})(y^*) = \big\{x^* \in X^* \bigm| (x^*, -y^*) \in N((\bar{x}, \bar{y}); \operatorname{gph} F)\big\} , \tag{1.26}$$

i.e., the normal coderivative (1.24) is uniquely determined by the basic normal cone (1.3) to the graph of F; hence the name. The only difference in the construction of the mixed coderivative (1.25) in comparison with (1.24) is that the weak* convergence is used in (1.24) for both sequences x^*_k and y^*_k, while the convergence in (1.25) is *mixed*: the norm convergence of $y^*_k \to \bar{y}^*$ and the weak* convergence of $x^*_k \xrightarrow{w^*} \bar{x}^*$.

Observe that generalized normals to arbitrary sets in Definition 1.1 can be expressed in terms of the corresponding coderivatives for set indicator mappings useful in the sequel.

Proposition 1.33 (coderivatives of indicator mappings). *Given spaces* X *and* Y, *we consider a nonempty subset* $\Omega \subset X$ *and define the* INDICATOR MAPPING $\Delta \colon X \to Y$ *of* Ω *relative to* Y *by*

$$\Delta(x;\Omega) := \begin{cases} 0 \in Y \; if \;\; x \in \Omega\,, \\ \emptyset \qquad if \;\; x \notin \Omega\,. \end{cases}$$

Then for any $\bar{x} \in \Omega$ and $y^* \in Y^*$ one has

$$\widehat{D}^*_\varepsilon \Delta(\bar{x};\Omega)(y^*) = \widehat{N}_\varepsilon(\bar{x};\Omega), \quad \varepsilon \ge 0\,;$$

$$D^*_N \Delta(\bar{x};\Omega)(y^*) = D^*_M \Delta(\bar{x};\Omega)(y^*) = N(\bar{x};\Omega)\,.$$

Proof. Immediately follows from the definitions due to $\text{gph}\, \Delta = \Omega \times \{0\}$. $\triangle$

Clearly $D^*_N F(\bar{x},\bar{y}) = D^*_M F(\bar{x},\bar{y}) := D^* F(\bar{x},\bar{y})$ if $\dim Y < \infty$. Observe that these coderivatives often have *nonconvex* values; so they *cannot be dual* to a tangentially generated derivative. For example, consider the simplest nonsmooth convex function $\varphi(x) = |x|$, $x \in I\!R$. By Theorem 1.6 we can easily compute the basic normal cone to $\text{gph}\,|x| \subset I\!R^2$ at (0,0). Then (1.26) gives

$$D^*\varphi(0,0)(\lambda) = \begin{cases} [-\lambda,\lambda] \;\; \text{if } \lambda \ge 0\,, \\ \{-\lambda,\lambda\} \; \text{if } \lambda < 0\,. \end{cases}$$

Note also that coderivative values may be *empty* at points of the mapping graph for simple continuous functions. It happens, e.g., for $\varphi(x) = |x|^\alpha$ with $x \in I\!R$ and $0 < \alpha < 1$, where

$$D^*\varphi(0,0)(\lambda) = \begin{cases} I\!R \; \text{if } \lambda \ge 0\,, \\ \emptyset \;\; \text{if } \lambda < 0\,. \end{cases}$$

Moreover, for the class of *convex-valued* and inner/lower semicontinuous multifunctions, points of the coderivative domain induce a certain *extremal property* important for various applications, especially in optimal control.

Recall that $F\colon X \rightrightarrows Y$ is *inner semicontinuous* at $\bar{x} \in \text{dom}\, F$ if for every $y \in F(\bar{x})$ and every sequence $x_k \to \bar{x}$ with $x_k \in \text{dom}\, F$ there are $y_k \in F(x_k)$ such that $y_k \to y$ as $k \to \infty$.

Theorem 1.34 (extremal property of convex-valued multifunctions). *Let $F\colon X \rightrightarrows Y$ be inner semicontinuous at $\bar{x} \in \text{dom}\, F$ and convex-valued around this point. Assume that $y^* \in \text{dom}\, D^*_N F(\bar{x},\bar{y})$ for some $\bar{y} \in F(\bar{x})$. Then one has*

$$\langle y^*, \bar{y}\rangle = \min_{y \in F(\bar{x})} \langle y^*, y\rangle\,.$$

Proof. Due to $D^*_N F(\bar{x},\bar{y})(y^*) \ne \emptyset$ and (1.26) there is $x^* \in X^*$ with $(x^*, -y^*) \in N((\bar{x},\bar{y}); \text{gph}\, F)$. Using Definition 1.1, we find sequences $\varepsilon_k \downarrow 0$, $(x_k, y_k) \to (\bar{x},\bar{y})$ with $y_k \in F(x_k)$, and $(x^*_k, y^*_k) \xrightarrow{w^*} (x^*, y^*)$ such that

$$\limsup_{(x,y)\to(x_k,y_k),\ y\in F(x)} \frac{\langle x_k^*, x - x_k\rangle - \langle y_k^*, y - y_k\rangle}{\|(x,y) - (x_k, y_k)\|} \leq \varepsilon_k \quad \text{for each} \quad k \in \mathbb{N} .$$

When $x = x_k$, this implies that $-y_k^* \in \widehat{N}_{\varepsilon_k}(y_k; F(x_k))$. Since all the sets $F(x_k)$ are convex, we get from Proposition 1.3 that

$$\langle y_k^*, y - y_k\rangle \geq -\varepsilon_k\|y - y_k\| \quad \text{for all} \quad y \in F(x_k), \quad k \in \mathbb{N} .$$

Now assume that there is $\tilde{y} \in F(\bar{x})$ such that

$$\langle y^*, \tilde{y}\rangle < \langle y^*, \bar{y}\rangle .$$

Using the inner semicontinuity property of F at $\bar{x}$, we find a sequence of $\tilde{y}_k \to \tilde{y}$ with $\tilde{y}_k \in F(x_k)$ for all $k \in \mathbb{N}$. Then we easily deduce from the convergences involved that

$$\langle y_k^*, \tilde{y}_k - y_k\rangle < -\varepsilon_k\|\tilde{y}_k - y_k\| \quad \text{for large} \quad k \in \mathbb{N} .$$

This contradiction completes the proof. △

It follows from the definitions for general mappings $F: X \rightrightarrows Y$ that

$$\widehat{D}^*F(\bar{x},\bar{y})(y^*) \subset D_M^*F(\bar{x},\bar{y})(y^*) \subset D_N^*F(\bar{x},\bar{y})(y^*) \tag{1.27}$$

for any $y^* \in Y^*$, and that all the three multifunctions are positively homogeneous in y^* containing $x^* = 0$ when $y^* = 0$ and $(\bar{x},\bar{y}) \in \operatorname{gph} F$. We can easily see that the first inclusion in (1.27) is often strict. It happens, in particular, for the above function $\varphi(x) = |x|$, where

$$\widehat{D}^*\varphi(0,0)(\lambda) = \begin{cases} [-\lambda, \lambda] \text{ if } \lambda \geq 0 , \\ \emptyset \qquad\quad \text{ if } \lambda < 0 . \end{cases}$$

The second inclusion in (1.27) obviously holds as equality if $\dim Y < \infty$. Let us show that this inclusion may be *strict* even for single-valued and Lipschitz continuous mappings from the real line into Hilbert spaces.

Example 1.35 (difference between mixed and normal coderivatives). *Let H be an arbitrary Hilbert space. Then there is a mapping $f: \mathbb{R} \to H$, which is Lipschitz continuous on $[-1,1]$ and such that $\widehat{D}^*f(0) = D_M^*f(0)$ while*

$$D_M^*f(0)(y^*) \neq D_N^*f(0)(y^*) \quad \textit{whenever } y^* \in H .$$

Proof. Take a sequence of orthonormal vectors $\{e_1, e_2, \ldots\}$ in a Hilbert space and define a mapping $f: [-1,1] \to H$ by

$$f(x) := \begin{cases} 2^{-k}e_k \text{ if } |x| = 2^{-k}, \\ 0 \qquad\quad \text{if } x = 0 , \\ \text{linear otherwise.} \end{cases}$$

It is easy to check that f is Lipschitz continuous on $[-1,1]$. Taking into account that $\langle y^*, e_k\rangle \to 0$ as $k \to \infty$, we compute

$$\widehat{D}^* f(x)(y^*) = \langle y^*, 2e_k - e_{k+1}\rangle \cdot \operatorname{sign} x \ \text{ if } \ 2^{-(k+1)} < |x| < 2^{-k};$$

$$\widehat{D}^* f(0)(y^*) = D^*_M f(0)(y^*) = \{0\} \ \text{ for all } \ y^* \in H \ .$$

It remains to show that $D^*_N f(0)(y^*)$ contains nonzero elements whenever $y^* \in H$. Picking $y^* \in H$, we choose a sequence of positive numbers x_k such that $x_k \to 0$ and $x_k \neq 2^{-j}$ for all $k, j \in I\!N$. Then put

$$y^*_k := -y^* - \upsilon_k \ \text{ and } \ \lambda_k := \langle y^*_k, 2e_{j_k} - e_{j_k+1}\rangle \ ,$$

where $\upsilon_k := (2e_{j_k} - e_{j_k+1})/\|2e_{j_k} - e_{j_k+1}\|$ and the index j_k is such that $2^{-(j_k+1)} < x_k < 2^{-j_k}$. We can check that $\upsilon_k \stackrel{w}{\to} 0$ with $\|\upsilon_k\| = 1$ and that

$$(\lambda_k, y^*_k) \in \widehat{N}((x_k, f(x_k)); \operatorname{gph} f), \quad y^*_k \stackrel{w}{\to} -y^*, \ \text{ and } \ \lambda_k \to -1 \ \text{ as } \ k \to \infty \ .$$

Thus $(-1, -y^*) \in N((0,0); \operatorname{gph} f)$ and $-1 \in D^*_N f(0)(y^*)$. △

Observe that f in Example 1.35 is *not* Fréchet differentiable at $\bar{x} = 0$, since the latter would easily yield $\nabla f(0) = 0$, which doesn't hold due to

$$\frac{\|f(x_k)\|}{|x_k|} = 1 \not\to 0 \ \text{ for } \ x_k = 2^{-k} \to 0 \ \text{ as } \ k \to \infty \ .$$

On the other hand, this mapping is *weakly* Fréchet differentiable at $\bar{x}$ (even *strictly-weakly* $\mathcal{F}$-differentiable at this point) in the sense of Definition 3.63; see Subsect. 3.2.4 for more discussions.

Similarly to the case of set regularity in Definition 1.4, we can consider a "regular" behavior of set-valued mappings at points of their graphs, which corresponds to equalities in (1.27). In this way we introduce two notions of graphical regularity for set-valued mappings based on properties of their normal and mixed coderivatives, respectively.

Definition 1.36 (graphical regularity of multifunctions). *Let* $F\colon X \rightrightarrows Y$ *and* $(\bar{x}, \bar{y}) \in \operatorname{gph} F$. *Then:*

(i) *F is N-REGULAR at $(\bar{x}, \bar{y})$ if $D^*_N F(\bar{x}, \bar{y}) = \widehat{D}^* F(\bar{x}, \bar{y})$.*

(ii) *F is M-REGULAR at $(\bar{x}, \bar{y})$ if $D^*_M F(\bar{x}, \bar{y}) = \widehat{D}^* F(\bar{x}, \bar{y})$.*

It follows from (1.23) and (1.26) with $\varepsilon = 0$ that F is N-regular at $(\bar{x}, \bar{y})$ if and only if the graph of F is normally regular at this point. Obviously N-regularity always implies M-regularity of F at $(\bar{x}, \bar{y})$ but not vice versa, as Example 1.35 shows. Let us present some sufficient conditions that ensure both regularities in Definition 1.36.

First we consider *convex-graph* multifunctions, i.e., such $F\colon X \rightrightarrows Y$ whose graphs are convex subsets of $X\times Y$. In this case we have a special representation of the coderivatives that follows from the form of the normal cone to convex sets.

Proposition 1.37 (coderivatives of convex-graph multifunctions). *Let $F\colon X \rightrightarrows Y$ be convex-graph. Then F is N-regular at every point $(\bar{x},\bar{y})\in \operatorname{gph} F$ and one has the coderivative representations*

$$D_N^*F(\bar{x},\bar{y})(y^*)=D_M^*F(\bar{x},\bar{y})(y^*)$$

$$=\Big\{x^*\in X^*\;\Big|\;\langle x^*,\bar{x}\rangle-\langle y^*,\bar{y}\rangle=\max_{(x,y)\in \operatorname{gph} F}\big[\langle x^*,x\rangle-\langle y^*,y\rangle\big]\Big\}\,.$$

Proof. Due to (1.23) and (1.26) it follows from Proposition 1.3 and Proposition 1.5 as $\varepsilon=0$. $\triangle$

Next we establish relationships between coderivatives and derivatives of single-valued differentiable mappings that imply the graphical regularity of $f\colon X\to Y$ if f is strictly differentiable at $\bar{x}$.

Theorem 1.38 (coderivatives of differentiable mappings). *Let $f\colon X\to Y$ be Fréchet differentiable at $\bar{x}$. Then*

$$\widehat{D}^*f(\bar{x})(y^*)=\big\{\nabla f(\bar{x})^*y^*\big\}\;\;\textit{for all}\;\;y^*\in Y^*\,.$$

If, moreover, f is strictly differentiable at $\bar{x}$, then

$$D_N^*f(\bar{x})(y^*)=D_M^*f(\bar{x})(y^*)=\big\{\nabla f(\bar{x})^*y^*\big\}\;\;\textit{for all}\;\;y^*\in Y^*\,,$$

and thus f is N-regular at this point.

Proof. Observe that for any $f\colon X\to Y$ the inclusion $x^*\in \widehat{D}^*f(\bar{x})(y^*)$ means that, taking an arbitrary $\gamma>0$, one has

$$\langle x^*,x-\bar{x}\rangle-\langle y^*,f(x)-f(\bar{x})\rangle\le\gamma\big(\|x-\bar{x}\|+\|f(x)-f(\bar{x})\|\big)$$

when x sufficiently close to $\bar{x}$. If f is Fréchet differentiable at $\bar{x}$, we easily get from (1.14) and the definition of adjoint linear operators that $\nabla f(\bar{x})^*y^*\in \widehat{D}^*f(\bar{x})(y^*)$ for every $y^*\in Y^*$. Conversely, picking any $x^*\in\widehat{D}^*f(\bar{x})(y^*)$ and using the Fréchet differentiability of f at $\bar{x}$, we have

$$\langle x^*-\nabla f(\bar{x})^*y^*,x-\bar{x}\rangle\le\gamma\,\|x-\bar{x}\|\;\;\text{for all}\;\;x\in U\,,$$

where the neighborhood U of $\bar{x}$ depends on γ, (x^*,y^*), and $\|\nabla f(\bar{x})\|$. Since $\gamma>0$ was chosen arbitrarily, the latter implies that $x^*=\nabla f(\bar{x})^*y^*$, which justifies the first equality in the theorem.

Now assume that f is strictly differentiable at $\bar{x}$ and prove the second part of the theorem. It is sufficient to show that $x^* = \nabla f(\bar{x})^* y^*$ for any $x^* \in D_N^* f(\bar{x})(y^*)$ and $y^* \in Y^*$. Due to (1.24) and (1.3) we have sequences $\varepsilon_k \downarrow 0$, $x_k \to \bar{x}$, and $(x_k^*, y_k^*) \stackrel{w^*}{\to} (x^*, y^*)$ such that

$$\langle x_k^*, x - x_k\rangle - \langle y_k^*, f(x) - f(x_k)\rangle \le \varepsilon_k\big(\|x - x_k\| + \|f(x) - f(x_k)\|\big)$$

for all x close enough to x_k and all $k \in I\!N$. It follows from Definition 1.13 of strict differentiability that for any sequence $\gamma_j \downarrow 0$ as $j \to \infty$ there is a sequence of neighborhoods U_j of $\bar{x}$ with

$$\|f(u) - f(x) - \nabla f(\bar{x})(u - x)\| \le \gamma_j \|u - x\| \quad \text{for all} \;\; x, u \in U_j, \quad j \in I\!N\,.$$

This allows us to select a subsequence $\{k_j\}$ of natural numbers such that

$$\langle x_{k_j}^* - \nabla f(\bar{x})^* y_{k_j}^*, x - x_{k_j}\rangle \le \widetilde{\varepsilon}_j \|x - x_{k_j}\| \quad \text{for all} \;\; x \in U_{k_j}, \quad j \in I\!N\,,$$

where U_{k_j} is a neighborhood of x_{k_j} and where $\widetilde{\varepsilon}_j := (\ell + 1)(\varepsilon_{k_j} + \gamma_j \|y_{k_j}^*\|)$ with a Lipschitz constant $\ell > 0$ of f around $\bar{x}$. The latter implies that

$$\|x_{k_j}^* - \nabla f(\bar{x})^* y_{k_j}^*\| \le \widetilde{\varepsilon}_j \quad \text{for large} \;\; j \in I\!N\,,$$

which gives $x^* = \nabla f(\bar{x})^* y^*$ due to

$$\widetilde{\varepsilon}_j \downarrow 0, \qquad x_{k_j}^* - \nabla f(\bar{x})^* y_{k_j}^* \stackrel{w^*}{\to} x^* - \nabla f(\bar{x})^* y^* \;\; \text{as} \;\; j \to \infty$$

and the weak* lower semicontinuity of the norm on X^*. △

Theorem 1.38 shows that the coderivatives under consideration can be viewed as proper set-valued generalizations of the *adjoint linear operator* to the classical derivative at the point in question. Note that, in the case of nonsmooth mappings and multifunctions, coderivative values do not depend linearly on the variable y^* but exhibit a *positively homogeneous* dependence. If f itself is a linear continuous operator, then its coderivatives reduce to the classical adjoint linear operator.

Corollary 1.39 (coderivatives of linear operators). *Let $A\colon X \to Y$ be linear and continuous. Then it is N-regular at every point $\bar{x} \in X$ with*

$$D_N^* A(\bar{x})(y^*) = D_M^* A(\bar{x})(y^*) = \big\{A^* y^*\big\} \;\; \textit{for all} \;\; \bar{x} \in X, \quad y^* \in Y^*\,.$$

Proof. Follows immediately from Theorem 1.38 with $f(x) = Ax$. △

We'll see in Subsect. 1.2.4 and then in Chap. 3 that both properties of N-regularity and M-regularity enjoy *rich calculi*, i.e., they are preserved under various compositions of single-valued and set-valued mappings, being incorporated into coderivative calculus.

Note that the strict differentiability assumption in Theorem 1.38 is sufficient but *not necessary* for graphical regularity of single-valued mappings. A simple example is provided by the function $\varphi(x) = |x|^\alpha$ with $0 < \alpha < 1$ considered above, which is clearly N-regular at $\bar{x} = 0$. Observe that this function is *not locally Lipschitzian* around the point in question, and it is crucial for the regularity property; cf. Theorem 1.46 in the next subsection.

1.2.2 Lipschitzian Properties

Lipschitzian properties of single-valued and set-valued mappings play a principal role in many aspects of variational analysis and its applications. They are often decisive from both viewpoints of reasonable assumptions ensuring the validity of important results and favorable conclusions, especially related to stability of solutions with respect to perturbations, rates of convergence in approximating and numerical procedures, etc. A crucial feature of the classical Lipschitz continuity (1.15) in comparison with the general continuity concept for single-valued mappings is a *linear rate* of continuity quantified by some modulus (Lipschitz constant) ℓ. In what follows we study natural extensions of Lipschitz continuity to set-valued mappings and show that the coderivative constructions defined above are helpful in both single-valued and set-valued cases. The necessary coderivative conditions for Lipschitzian properties obtained in this subsection are widely used in subsequent applications considered in this book, particularly to generalized differential calculus, optimization, and optimal control.

Definition 1.40 (Lipschitzian properties of set-valued mappings). *Let* $F\colon X \rightrightarrows Y$ *with* $\operatorname{dom} F \neq \emptyset$.

(i) *Given nonempty subsets* $U \subset X$ *and* $V \subset Y$, *we say that* F *is* LIPSCHITZ-LIKE *on* U *relative to* V *if there is* $\ell \geq 0$ *such that*

$$F(x) \cap V \subset F(u) + \ell\|x - u\| I\!B \quad \text{for all } x, u \in U\,. \tag{1.28}$$

(ii) *Given* $(\bar{x}, \bar{y}) \in \operatorname{gph} F$, *we say that* F *is* LOCALLY LIPSCHITZ-LIKE *around* $(\bar{x}, \bar{y})$ *with modulus* $\ell \geq 0$ *if there are neighborhoods* U *of* $\bar{x}$ *and* V *of* $\bar{y}$ *such that* (1.28) *holds. The infimum of all such moduli* $\{\ell\}$ *is called the* EXACT LIPSCHITZIAN BOUND *of* F *around* $(\bar{x}, \bar{y})$ *and is denoted by* $\operatorname{lip} F(\bar{x}, \bar{y})$.

(iii) F *is* LIPSCHITZ CONTINUOUS *on* U *if* (1.28) *holds as* $V = Y$. *Furthermore,* F *is* LOCALLY LIPSCHITZIAN *around* $\bar{x}$ *with the exact bound* $\operatorname{lip} F(\bar{x})$ *if* $V = Y$ *in* (ii).

The local Lipschitz-like property is also known as the *pseudo-Lipschitzian property* or the *Aubin property* of multifunctions. Note that the local properties in the above definition are *stable/robust* with respect to small perturbations of the reference points and hold for F if and only if they hold for the mapping $\overline{F}\colon X \rightrightarrows Y$ with $\overline{F}(x) := \operatorname{cl}(F(x))$.

It follows from the definition that the Lipschitz continuity of F on U is equivalent to

$$\text{haus}(F(x), F(u)) \le \ell\|x-u\| \quad \text{for all} \;\; x, u \in U \; ,$$

where $\text{haus}(\Omega_1, \Omega_2)$ is the *Pompieu-Hausdorff distance* (often referred to as simply the Hausdorff distance) between two subsets of Y that is defined by

$$\text{haus}(\Omega_1, \Omega_2) := \inf\big\{\eta \ge 0 \;\big|\; \Omega_1 \subset \Omega_2 + \eta I\!B,\; \Omega_2 \subset \Omega_1 + \eta I\!B\big\} \; .$$

Note that the Pompieu-Hausdorff distance furnishes a *metric* on the space of all nonempty and *compact* subsets of Y. Thus, if a multifunction $F\colon X \rightrightarrows Y$ is compact-valued, its Lipschitz continuity in Definition 1.40(iii) is equivalent to the classical Lipschitz continuity of a single-valued mapping $x \to F(x)$ from X to the space of all nonempty, compact subsets of Y equipped with the Pompieu-Hausdorff metric.

Of course, for single-valued mappings $f\colon X \to Y$ all the properties in Definition 1.40 reduce to the classical Lipschitz continuity. For general set-valued mappings $F\colon X \rightrightarrows Y$ the local Lipschitz-like property can be viewed as a localization of Lipschitzian behavior not only relative to a point of the domain but also relative to a particular point of the *image* $\bar{y} \in F(\bar{x})$. It admits the following useful characterization in terms of the local Lipschitz continuity of the (scalar) *distance function* (1.7) to the moving set $F(x)$ with respect to *both* variables (x, y).

Theorem 1.41 (scalarization of the Lipschitz-like property). *For any multifunction $F\colon X \rightrightarrows Y$ with $(\bar{x}, \bar{y}) \in \text{gph}\, F$ the following properties are equivalent:*

(a) *F is locally Lipschitz-like around $(\bar{x}, \bar{y})$.*
(b) *A scalar function $\rho\colon X \times Y \to I\!R$ defined by*

$$\rho(x, y) := \text{dist}(y; F(x)) = \inf_{v \in F(x)} \|y - v\|$$

is locally Lipschitzian around $(\bar{x}, \bar{y})$.

Proof. Due to the nature of the distance function we can easily observe that the local Lipschitz continuity of ρ around $(\bar{x}, \bar{y})$ is equivalent to the existence of neighborhoods U of $\bar{x}$, V of $\bar{y}$, and a constant $\ell \ge 0$ such that ρ is finite on $U \times V$ and

$$\rho(u, y) \le \rho(x, y) + \ell\|x - u\| \quad \text{for all} \;\; x, u \in U, \quad y \in V \; . \tag{1.29}$$

To have (a)$\Rightarrow$(b), it suffices to show that (1.28) with some neighborhoods U, V implies (1.29) with generally different neighborhoods $\widetilde{U}$, $\widetilde{V}$. It follows from (1.28) that

$$\text{dist}(y; F(u) + \ell\|x - u\| I\!B) \le \text{dist}(y; F(x) \cap V) \quad \text{for all} \;\; x, u \in U, \quad y \in Y \; .$$

Since $\operatorname{dist}(y; F(u)) - \eta \le \operatorname{dist}(y; F(u) + \eta I\!B)$ for any $\eta \ge 0$, this gives

$$\operatorname{dist}(y; F(u)) - \ell\|x - u\| \le \operatorname{dist}(y; F(x) \cap V) \ \text{ for all } \ x, u \in U, \quad y \in Y .$$

The latter obviously implies (1.29) with some neighborhoods $\widetilde{U}$ of $\bar{x}$ and $\widetilde{V}$ of $\bar{y}$ for which

$$\operatorname{dist}(y; F(x) \cap V) = \operatorname{dist}(y; F(x)) \ \text{ if } \ x \in \widetilde{U}, \quad y \in \widetilde{V} . \tag{1.30}$$

We need to prove the existence of such neighborhoods $\widetilde{U}$ and $\widetilde{V}$. To furnish this, we choose $\gamma > 0$ with $\bar{y} + \gamma I\!B \subset V$ and put $\widetilde{V} := \bar{y} + \frac{1}{3}\gamma I\!B$. Then for any $y \in \widetilde{V}$ one has $y + \frac{2}{3}\gamma I\!B \subset V$, and so

$$\operatorname{dist}(y; F(x) \cap V) = \operatorname{dist}(y; F(x)) \ \text{ if } \ \operatorname{dist}(y; F(x)) \le \tfrac{2}{3}\gamma .$$

Furthermore, since $\operatorname{dist}(y; F(x)) \le \operatorname{dist}(\bar{y}; F(x)) + \|y - \bar{y}\|$, we get

$$\operatorname{dist}(y; F(x)) \le \tfrac{2}{3}\gamma \ \text{ when } \ \operatorname{dist}(\bar{y}; F(x)) \le \tfrac{1}{3}\gamma, \quad y \in \widetilde{V} .$$

To ensure (1.30) with the specified $\widetilde{V}$, we need to find a neighborhood $\widetilde{U}$ of $\bar{x}$ satisfying the property

$$\operatorname{dist}(\bar{y}; F(x)) \le \tfrac{1}{3}\gamma \ \text{ for all } \ x \in \widetilde{U} .$$

The existence of such $\widetilde{U}$ follows from (1.28) that obviously implies

$$\operatorname{dist}(\bar{y}; F(x)) \le \ell\|x - \bar{x}\| \ \text{ for all } \ x \in U .$$

Hence we can take $\widetilde{U} := \bar{x} + \eta I\!B$, where $\eta > 0$ satisfies $\ell\eta \le \frac{1}{3}\gamma$ and $\bar{x} + \eta I\!B \subset U$. This gives (a)$\Rightarrow$(b).

Conversely, let F be closed-valued and (1.29) hold. Picking $x, u \in U$ and $y \in F(x) \cap V$ in (1.29), we have $\operatorname{dist}(y; F(x)) = 0$ and

$$\operatorname{dist}(y; F(u)) \le \operatorname{dist}(y; F(x)) + \ell\|x - u\| = \ell\|u - x\| ,$$

which gives (1.28) with ℓ replaced by $\ell + \varepsilon$ for some $\varepsilon > 0$. Since the local Lipschitz-like property of F is invariant with respect to taking the closure of its values, we get (b)$\Rightarrow$(a) in the general case. △

Let us discuss more about relationships between the local Lipschitzian and Lipschitz-like properties of multifunctions. It follows directly from the definitions that if F is locally Lipschitzian around $\bar{x} \in \operatorname{dom} F$, then it is locally Lipschitz-like around $(\bar{x}, \bar{y})$ for every $\bar{y} \in F(\bar{x})$ with

$$\operatorname{lip} F(\bar{x}) \ge \sup\big\{\operatorname{lip} F(\bar{x}, \bar{y}) \ \big| \ \bar{y} \in F(\bar{x})\big\} . \tag{1.31}$$

The next result shows that the converse holds with the equality in (1.31) when F satisfies some additional assumptions.

Recall that $F\colon X \rightrightarrows Y$ is *locally compact* around $\bar{x} \in \operatorname{dom} F$ if there exist a neighborhood O of $\bar{x}$ and a compact set $C \subset Y$ such that $F(O) \subset C$. Furthermore, F is said to be *closed at* $\bar{x}$ if for every $y \notin F(\bar{x})$ there are neighborhoods U of $\bar{x}$ and V of y such that $F(x) \cap V = \emptyset$ for all $x \in U$. The latter obviously implies that F is closed-valued at $\bar{x}$. It is easy to see that F is closed at $\bar{x}$ if, for every $\bar{y} \in F(\bar{x})$, the graph of F is a closed subset of $X \times Y$ for all $(x, y) \in \operatorname{gph} F$ near $(\bar{x}, \bar{y})$.

Theorem 1.42 (Lipschitz continuity of locally compact multifunctions). *Let $F\colon X \rightrightarrows Y$ be closed at some point $\bar{x} \in \operatorname{dom} F$ and locally compact around this point. Then F is locally Lipschitzian around $\bar{x}$ if and only if it is locally Lipschitz-like around $(\bar{x}, \bar{y})$ for every $\bar{y} \in F(\bar{x})$. In this case*

$$\operatorname{lip} F(\bar{x}) = \max\big\{\operatorname{lip} F(\bar{x}, \bar{y}) \;\big|\; \bar{y} \in F(\bar{x})\big\} < \infty\,.$$

Proof. Taking a compact set $C \subset Y$ and a neighborhood O of $\bar{x}$ from the local compactness assumption, we have

$$F(x) \cap C = F(x) \;\text{ for all }\; x \in O\,.$$

Suppose without loss of generality that all the neighborhoods of $\bar{x}$ considered below are subsets of O. We need to show that the local Lipschitz-like property of F around $(\bar{x}, \bar{y})$, for all $\bar{y} \in F(\bar{x})$, implies that F is locally Lipschitzian around $\bar{x}$ with the equality in (1.31). On the contrary, assume that the inequality is strict in (1.31), i.e.,

$$\operatorname{lip} F(\bar{x}) > \operatorname{lip} F(\bar{x}, \bar{y}) \;\text{ for all }\; \bar{y} \in F(\bar{x})\,.$$

Then for each $\bar{y} \in F(\bar{x})$ we find a number $0 \le \ell_{\bar{y}} < \operatorname{lip} F(\bar{x})$ and neighborhoods $U_{\bar{y}}$ of $\bar{x}$ and $V_{\bar{y}}$ of $\bar{y}$ such that

$$F(x) \cap V_{\bar{y}} \subset F(u) + \ell_{\bar{y}}\|x - u\|I\!B \;\text{ for all }\; x, u \in U_{\bar{y}}, \quad \bar{y} \in F(\bar{x})\,.$$

Since $F(\bar{x})$ is a compact subset of Y, we can select from $\{V_{\bar{y}}\}$ a *finite* covering $\{V_i\}$, $i = 1, \ldots, n$, of the set $F(\bar{x})$. Taking the corresponding numbers ℓ_i and neighborhoods U_i, $i = 1, \ldots, n$, let us denote

$$\widehat{V} := \bigcup_{i=1}^{n} V_i, \quad \widehat{U} := \bigcap_{i=1}^{n} U_i, \quad \widehat{\ell} := \max_{i=1,\ldots,n} \ell_i\,.$$

Thus we have

$$F(x) \cap \widehat{V} \subset F(u) + \widehat{\ell}\|x - u\|I\!B \;\text{ for all }\; x, u \in \widehat{U}\,.$$

Consider now the relative complement $C \setminus \widehat{V}$, which is a compact set with $F(\bar{x}) \cap (C \setminus \widehat{V}) = \emptyset$. Because F is closed at $\bar{x}$, for any $y \in C \setminus \widehat{V}$ there are neighborhoods $\widetilde{U}_y$ of $\bar{x}$ and $\widetilde{V}_y$ of y such that

$$F(x) \cap \widetilde{V}_y = \emptyset \ \text{ when } \ x \in \widetilde{U}_y, \quad y \in C \setminus \widehat{V} .$$

Again, using the compactness of $C \setminus \widehat{V}$, we extract from $\{\widetilde{V}_y\}$ a finite covering $\{\widetilde{V}_j\}$, $j = 1, \ldots, m$, of the set $C \setminus \widehat{V}$. Letting

$$\widetilde{V} := \bigcup_{j=1}^{m} \widetilde{V}_j \ \text{ and } \ \widetilde{U} := \bigcap_{j=1}^{m} \widetilde{U}_j ,$$

one clearly has

$$F(x) \cap \widetilde{V} = \emptyset \ \text{ for all } \ x \in \widetilde{U} .$$

Putting all the above together, we arrive at

$$F(x) \subset F(u) + \widehat{\ell}\|x - u\|I\!B \ \text{ for all } \ x, u \in \widehat{U} \cap \widetilde{U} ,$$

which means that $\widehat{\ell} < \operatorname{lip} F(\bar{x})$, a contradiction. This proves that F is locally Lipschitzian around $\bar{x}$ with the equality in (1.31). Moreover, the maximum is realized due to the upper semicontinuity of $\operatorname{lip} F(\cdot,\cdot)$ on the graph of F. $\triangle$

Next let us derive important *necessary coderivative conditions* for the local properties in Definition 1.40 in the case of arbitrary Banach spaces. We start with *neighborhood* conditions expressed in terms of ε-coderivatives (1.23) at points near the reference one. Let us emphasize that for the validity of these necessary conditions, as well as the point conditions in the following Theorem 1.44, it is very essential that the Lipschitzian properties under consideration are *around* the reference points, i.e., both x and u vary in (1.28). We'll see in Chap. 4 that such conditions, even with $\varepsilon = 0$, turn out to be also *sufficient* for these and related properties of multifunctions with *equalities* in the exact bound formulas in the case of *Asplund spaces.*

Theorem 1.43 (ε-coderivatives of Lipschitzian mappings). *Let $F\colon X \rightrightarrows Y$, $\bar{x} \in \operatorname{dom} F$, and $\varepsilon \geq 0$. The following hold:*

(i) *If F is locally Lipschitz-like around some $(\bar{x}, \bar{y}) \in \operatorname{gph} F$ with modulus $\ell \geq 0$, then there is $\eta > 0$ such that*

$$\sup\Big\{\|x^*\| \ \Big|\ x^* \in \widehat{D}^*_\varepsilon F(x,y)(y^*)\Big\} \leq \ell\|y^*\| + \varepsilon(1+\ell) \tag{1.32}$$

whenever $x \in \bar{x} + \eta I\!B$, $y \in F(x) \cap (\bar{y} + \eta I\!B)$, and $y^* \in Y^*$. Therefore

$$\operatorname{lip} F(\bar{x},\bar{y}) \geq \inf_{\eta>0} \sup\Big\{\|\widehat{D}^* F(x,y)\| \ \Big|\ x \in B_\eta(\bar{x}),\ y \in F(x) \cap B_\eta(\bar{y})\Big\} .$$

(ii) *If F is locally Lipschitzian around $\bar{x}$, then there is $\eta > 0$ such that* (1.32) *holds whenever $x \in \bar{x} + \eta I\!B$, $y \in F(x)$, and $y^* \in Y^*$. Therefore*

$$\operatorname{lip} F(\bar{x}) \geq \inf_{\eta>0} \sup\Big\{\|\widehat{D}^* F(x,y)\| \ \Big|\ x \in B_\eta(\bar{x}),\ y \in F(x)\Big\} .$$

Proof. Let us prove (i) assuming that $\ell > 0$ (the case of $\ell = 0$ is trivial). The local Lipschitz-like property ensures the existence of $\eta > 0$ for which

$$F(x) \cap (\bar{y} + \eta I\!B) \subset F(u) + \ell\|x - u\| I\!B \text{ if } x, u \in \bar{x} + 2\eta I\!B \,.$$

We are going to show that (1.32) holds with the numbers η and ℓ selected above. Pick arbitrary elements $(x, y) \in (\operatorname{gph} F) \cap [(\bar{x} + \eta I\!B) \times (\bar{y} + \eta I\!B)]$, $x^* \in \widehat{D}_\varepsilon^* F(x, y)(y^*)$, and $\gamma > 0$. Employing definitions (1.23) and (1.2), we find a positive number $\alpha \leq \{\eta, \ell\eta\}$ such that

$$\langle x^*, u - x\rangle - \langle y^*, v - y\rangle \leq (\varepsilon + \gamma)\big(\|u - x\| + \|v - y\|\big) \tag{1.33}$$

for all $(u, v) \in \operatorname{gph} F$ with $\|u - x\| \leq \alpha$ and $\|v - y\| \leq \alpha$. Now choose $u \in x + \alpha\ell^{-1} I\!B$ and observe that

$$\|u - \bar{x}\| \leq \|u - x\| + \|x - \bar{x}\| \leq 2\eta \,.$$

Thus one can apply the local Lipschitz-like property with $y \in F(x) \cap (\bar{y} + \eta I\!B)$ and the chosen u. In this way we find $v \in F(u)$ such that

$$\|v - y\| \leq \ell\|x - u\| \leq \ell \cdot \ell^{-1}\alpha = \alpha \,.$$

Substituting these u and v into (1.33), we get

$$\langle x^*, u - x\rangle \leq \alpha\|y^*\| + (\varepsilon + \gamma)(\alpha\ell^{-1} + \alpha)$$

holding for every $u \in x + \alpha\ell^{-1} I\!B$. Therefore

$$\alpha\ell^{-1}\|x^*\| \leq \alpha\|y^*\| + \alpha(\varepsilon + \gamma)(\ell^{-1} + 1) \,,$$

which yields (1.32), since $\gamma > 0$ was chosen arbitrarily. In turn, (1.32) implies

$$\operatorname{lip} F(\bar{x}, \bar{y}) \geq \inf_{\eta > 0} \sup\Big\{(\|x^*\| - \varepsilon)/(\varepsilon + 1) \;\Big|\; x^* \in \widehat{D}_\varepsilon^* F(x, y)(y^*),\ x \in B_\eta(\bar{x}) \,,$$
$$y \in F(x) \cap B_\eta(\bar{y}),\ \|y^*\| \leq 1,\ \varepsilon \geq 0\Big\} \,,$$

which surely gives the exact bound estimate in (i) as $\varepsilon = 0$. Assertion (ii) easily follows from (i) and Definition 1.40. △

Passing to the limit in the neighborhood conditions of Theorem 1.43, we can derive *point conditions* valid for local Lipschitzian mappings in terms of the *mixed coderivative* (1.25) computed only at reference points. The next theorem shows that the local properties in Definition 1.40 imply the *norm-boundedness* of the mixed coderivative and provides relationships between the coderivative norm (1.22) and the corresponding exact Lipschitzian bounds.

Theorem 1.44 (mixed coderivatives of Lipschitzian mappings). *Let* $F\colon X \rightrightarrows Y$ *with* $\bar{x} \in \operatorname{dom} F$. *The following hold:*

(i) *If* F *is locally Lipschitz-like around some* $(\bar{x}, \bar{y}) \in \operatorname{gph} F$, *then*

$$\|D_M^* F(\bar{x}, \bar{y})\| \le \operatorname{lip} F(\bar{x}, \bar{y}) < \infty \tag{1.34}$$

and therefore

$$D_M^* F(\bar{x}, \bar{y})(0) = \{0\} . \tag{1.35}$$

(ii) *If* F *is locally Lipschitzian around* $\bar{x}$, *then*

$$\sup_{\bar{y} \in F(\bar{x})} \|D_M^* F(\bar{x}, \bar{y})\| \le \operatorname{lip} F(\bar{x})$$

and therefore

$$D_M^* F(\bar{x}, \bar{y})(0) = \{0\} \quad \textit{for all} \quad \bar{y} \in F(\bar{x}) .$$

Proof. Clearly (ii) follows from (i) due to (1.31). Furthermore, (1.34) implies (1.35), since

$$\|x^*\| \le \|D_M^* F(\bar{x}, \bar{y})\| \cdot \|y^*\| \quad \text{for all} \quad x^* \in D_M^* F(\bar{x}, \bar{y})(y^*), \qquad y^* \in Y^* .$$

To establish (1.34), we need to show that if F is locally Lipschitz-like around $(\bar{x}, \bar{y})$ with modulus $\ell \ge 0$, then

$$\|D_M^* F(\bar{x}, \bar{y})\| \le \ell .$$

Take any $(x^*, y^*) \in X^* \times Y^*$ with $x^* \in D_M^* F(\bar{x}, \bar{y})(y^*)$. Using Definition 1.32(iii) of the mixed coderivative, we find sequences $\varepsilon_k \downarrow 0$, $(x_k, y_k, y_k^*) \to (\bar{x}, \bar{y}, y^*)$, and $x_k^* \xrightarrow{w^*} x^*$ such that

$$y_k \in F(x_k) \quad \text{and} \quad x_k^* \in \widehat{D}_{\varepsilon_k}^* F(x_k, y_k)(y_k^*)$$

for all $k \in I\!N$. Due to (1.32) we have

$$\|x_k^*\| \le \ell \|y_k^*\| + \varepsilon_k (1 + \ell)$$

for all k sufficiently large. Remember that $\|y_k^* - y^*\| \to 0$ as $k \to \infty$ (which is crucial in the construction of the mixed coderivative) and that the norm function is weak* lower semicontinuous on X^*. Then passing to the limit in the latter inequality, we get

$$\|x^*\| \le \ell \|y^*\| \quad \text{for any} \quad x^* \in D_M^* F(\bar{x}, \bar{y})(y^*) .$$

This implies $\|D_M^* F(\bar{x}, \bar{y})\| \le \ell$ due to the norm definition (1.22) for positively homogeneous multifunctions. △

Let us emphasize that in Theorem 1.44 one *cannot replace the mixed coderivative* D_M^* *with the normal coderivative* D_N^* if $\dim Y = \infty$. Indeed, the function f from Example 1.35 is single-valued and locally Lipschitzian around $\bar{x} = 0$ with $D_N^* f(0)(0) \neq \{0\}$ and $\|D_N^* f(0)\| = \infty$.

Theorem 1.44 is useful in many applications, in particular, to coderivative calculus and related questions fully considered in Chap. 3. Moreover, we'll prove in Chap. 4 that each of the conditions (1.34) and (1.35) is not only *necessary* but also *sufficient* for the local Lipschitz-like property of set-valued mappings between Asplund spaces, together with some "partial normal compactness" assumptions that are automatic in finite-dimensions when the first inequality in (1.34) holds as equality.

Next let us consider another type of Lipschitzian behavior of multifunctions that is also a generalization of the classical local Lipschitz continuity to the case of set-valued mappings. We'll see that Theorem 1.44 and calculus rules in Subsect. 1.1.2 are useful for the study of this kind of behavior.

Recall that a linear continuous operator $A: X \to Y$ is *invertible* if it is surjective and injective (one-to-one) simultaneously, i.e., A is a linear isomorphism between X and Y.

Definition 1.45 (graphically hemi-Lipschitzian and hemismooth mappings). *Let $F: X \rightrightarrows Y$ with $(\bar{x}, \bar{y}) \in \operatorname{gph} F$.*

(i) *F is* GRAPHICALLY HEMI-LIPSCHITZIAN *around $(\bar{x}, \bar{y})$ if there is a mapping $g: X \times Y \to Z$ from $X \times Y$ into another Banach space Z such that g is strictly differentiable at $(\bar{x}, \bar{y})$ with the surjective derivative $\nabla g(\bar{x}, \bar{y})$, and*

$$(\operatorname{gph} F) \cap O = g^{-1}\big((\operatorname{gph} f) \cap O_1\big)$$

for some neighborhoods O of $(\bar{x}, \bar{y})$, O_1 of $\bar{z} := g(\bar{x}, \bar{y})$ and a locally Lipschitzian mapping $f: X_1 \to Y_1$ with $X_1 \times Y_1 = Z$. If in addition $\nabla g(\bar{x}, \bar{y})$ is invertible, then F is said to be GRAPHICALLY LIPSCHITZIAN *around $(\bar{x}, \bar{y})$.*

(ii) *F is* GRAPHICALLY HEMISMOOTH *at $(\bar{x}, \bar{y})$ if it is graphically hemi-Lipschitzian around this point and the mapping f in* (i) *can be chosen as strictly differentiable at $\bar{u} \in X_1$ with $(\bar{u}, f(\bar{u})) = \bar{z}$. If, moreover, $\nabla g(\bar{x}, \bar{y})$ is invertible, then F is said to be* GRAPHICALLY SMOOTH *at $(\bar{x}, \bar{y})$.*

Roughly speaking, the graphical hemi-Lipschitzian (resp. hemismooth) property of multifunctions means that the graph of $F: X \rightrightarrows Y$ is locally represented, up to a smooth local transformation of $X \times Y$ with the surjective derivative, as the graph of a single-valued Lipschitz continuous (resp. strictly differentiable) mapping. If $\nabla g(\bar{x}, \bar{y})$ happens to be invertible in Definition 1.45, then the inverse mapping g^{-1} is locally single-valued and strictly differentiable at $\bar{z}$. This follows from Leach's inverse mapping theorem; see Theorem 1.60 below. In finite dimensions such a one-to-one transformation $g: X \times Y \to X \times Y$ is actually a *change of coordinates* around $(\bar{x}, \bar{y})$ under which a graphically Lipschitzian (resp. graphically smooth) multifunction can be locally identified with the graph of some single-valued Lipschitz continuous (resp. strictly differentiable) mapping.

Of course, every single-valued locally Lipschitzian mapping $f: X \to Y$ is graphically Lipschitzian, and f is graphically smooth if and only if it is strictly differentiable at the point in question. The *inverse* multifunction $f^{-1}: Y \rightrightarrows X$ is also graphically Lipschitzian around $(f(\bar{x}), \bar{x})$ if f is Lipschitz continuous around $\bar{x}$. A less obvious and highly important for applications class of graphically Lipschitzian multifunctions is formed by *maximal monotone* mappings $F: X \rightrightarrows X$ in Hilbert spaces, i.e., those for which

$$\langle x_1 - x_2, y_1 - y_2 \rangle \geq 0 \text{ for all } x_i \in X, \quad y_i \in F(x_i), \quad i = 1, 2,$$

and no enlargement of the graph of F is possible in $X \times X$ without destroying monotonicity. This class includes, in particular, *subdifferential mappings* for convex and saddle functions. Moreover, the graphical Lipschitzian property holds for subdifferential mappings associated with a vast class of so-called "prox-regular" functions typically encountered in finite-dimensional optimization. We refer the reader to Rockafellar [1153] and to the book by Rockafellar and Wets [1165] for more details and discussions.

It occurs that graphically hemi-Lipschitzian (graphically Lipschitzian) mappings between finite-dimensional spaces are graphically regular *if and only if* they are graphically hemismooth (resp. graphically smooth) at points in question. We'll prove this in the next theorem, where D^*F stands for the common coderivative of F in finite dimensions defined by (1.26). Analogs of these results in infinite dimensions will be presented in Subsect. 3.2.4.

Theorem 1.46 (graphical regularity for graphically hemi-Lipschitzian multifunctions). *Let F be a multifunction between finite-dimensional spaces, and let $(\bar{x}, \bar{y}) \in \operatorname{gph} F$. The following hold:*

(i) *Assume that F is graphically hemi-Lipschitzian around $(\bar{x}, \bar{y})$. Then F is graphically regular at $(\bar{x}, \bar{y})$ if and only if it is graphically hemismooth at this point.*

(ii) *Assume that F is graphically Lipschitzian around $(\bar{x}, \bar{y})$. Then F is graphically regular at $(\bar{x}, \bar{y})$ if and only if it is graphically smooth at this point.*

Proof. Assertion (ii) clearly follows from (i) and the definitions. To justify (i), let us first establish its counterpart for single-valued mappings.

Claim. *If $f: I\!R^n \to I\!R^m$ is locally Lipschitzian around $\bar{x}$, then its graphical regularity at $\bar{x}$ is equivalent to its strict differentiability at this point.*

The graphical regularity of strictly differentiable mappings is proved in Theorem 1.38. It remains to prove the converse implication for locally Lipschitzian mappings between finite-dimensional spaces. Applying Theorem 1.44, we immediately conclude that

$$D^* f(\bar{x})(0) := \{x^* \in I\!R^n \mid (x^*, 0) \in N((\bar{x}, f(\bar{x})); \operatorname{gph} f)\} = \{0\}$$

when f is Lipschitz continuous around $\bar{x}$. Further, it follows from Theorem 3.5 in Rockafellar [1153] that, for every locally Lipschitzian function $f: I\!R^n \to I\!R^m$, the *convexified* (*Clarke*) *normal cone*

$$N_C((\bar{x}, f(\bar{x})); \text{gph}\, f) := \text{clco}\, N((\bar{x}, f(\bar{x})); \text{gph}\, f)$$

is actually a *linear subspace* of dimension $q \geq m$, where $q = m$ *if and only if* f is *strictly differentiable* at $\bar{x}$; cf. Theorem 3.62 and Corollary 3.67 in Subsect. 3.2.4. Assuming the graphically regularity of f at $\bar{x}$ and taking into account that the basic normal cone is convex-valued in this case and always closed-valued in finite dimensions, we have $N((\bar{x}, f(\bar{x})); \text{gph}\, f) = N_C((\bar{x}, f(\bar{x})); \text{gph}\, f)$. Hence there is a matrix $A \in I\!R^{(n+m-q)\times n}$ such that

$$D^* f(\bar{x})(0) = \{x^* \in I\!R^n \mid Ax^* = 0\} = \{0\} .$$

This implies that $n + m - q = n$. Thus f is strictly differentiable at $\bar{x}$, which proves the claim.

Now let us consider the general case of a mapping $F\colon I\!R^n \rightrightarrows I\!R^m$ that is graphically hemi-Lipschitzian around $(\bar{x}, \bar{y})$. Without loss of generality we can assume that

$$\text{gph}\, F = g^{-1}(\text{gph}\, f) ,$$

where g is strictly differentiable at $(\bar{x}, \bar{y})$ with the surjective derivative and where f is locally Lipschitzian around $\bar{u}$ with $(\bar{u}, f(\bar{u})) = g(\bar{x}, \bar{y})$. It follows from Theorem 1.19 that the normal regularity of gph F at $(\bar{x}, \bar{y})$ is equivalent to the normal regularity of $g^{-1}(\text{gph}\, f)$ at $(\bar{u}, f(\bar{u}))$. The above claim implies that f is strictly differentiable at $\bar{u}$. Thus F is graphically hemismooth at $(\bar{x}, \bar{y})$, which completes the proof of the theorem. △

1.2.3 Metric Regularity and Covering

In this subsection we consider important properties of multifunctions, known as metric regularity and covering/linear openness, that occur to be closely related to Lipschitzian properties of *inverse* mappings. In the classical cases of linear and smooth operators these properties go back to basic principles of functional analysis given by the Banach-Schauder open mapping theorem and its nonlinear Lyusternik-Graves generalization that we have already used in Subsect. 1.1.2. Appropriate extensions of metric regularity and covering properties to nonsmooth and set-valued mappings play a fundamental role in variational analysis and optimization. In what follows we study these properties and their relationships (actually equivalence) to the Lipschitzian properties of inverse mappings considered in the previous subsection. In this way we get necessary conditions for covering and metric regularity of multifunctions in terms of coderivatives. The results obtained are significant for subsequent applications in this book and imply, in particular, that the classical surjectivity assumption on strict derivatives is *not only sufficient but also necessary* for openness and metric regularity in the Lyusternik-Graves theorem proved below; see Theorem 1.57.

Let us start with the definition of metric regularity for arbitrary multifunctions. Remember that $\text{dist}(x; \emptyset) = \infty$ due to (1.7) and $\inf \emptyset := \infty$.

Definition 1.47 (metric regularity). *Let $F: X \rightrightarrows Y$ with* $\operatorname{dom} F \neq \emptyset$.

(i) *Given nonempty subsets $U \subset X$ and $V \subset Y$, we say that F is* METRICALLY REGULAR *on U relative to V if there are numbers $\mu > 0$ and $\gamma > 0$ such that*

$$\operatorname{dist}(x; F^{-1}(y)) \leq \mu \operatorname{dist}(y; F(x)) \tag{1.36}$$

for all $x \in U$ and $y \in V$ satisfying $\operatorname{dist}(y; F(x)) \leq \gamma$.

(ii) *Given $(\bar{x}, \bar{y}) \in \operatorname{gph} F$, we say that F is* LOCALLY METRICALLY REGULAR *around $(\bar{x}, \bar{y})$ with modulus $\mu > 0$ if* (i) *holds with some neighborhoods U of $\bar{x}$ and V of $\bar{y}$. The infimum of all such moduli $\{\mu\}$, denoted by* $\operatorname{reg} F(\bar{x}, \bar{y})$, *is called the* EXACT REGULARITY BOUND *of F around $(\bar{x}, \bar{y})$.*

(iii) *F is* SEMI-LOCALLY METRICALLY REGULAR *around $\bar{x} \in \operatorname{dom} F$ (resp. around $\bar{y} \in \operatorname{rge} F$) with modulus $\mu > 0$ if* (i) *holds with a neighborhood U of $\bar{x}$ and $V = Y$ (resp. with a neighborhood V of $\bar{y}$ and $U = X$). The infimum of all such moduli is denoted by* $\operatorname{reg} F(\bar{x})$ *(resp. by* $\operatorname{reg} F(\bar{y})$*).*

Metric regularity (1.36) provides, for given points (x, y), a *linear estimate* of the distance between x and the solution map to the (generalized) equation $y \in F(u)$ through the distance between y and $F(x)$, which is easier to compute. Modifications (i)–(iii) in Definition 1.47 describe different conditions imposed on (x, y) that are typical for applications. The next proposition shows that in the case of local metric regularity the condition $\operatorname{dist}(y; F(x)) \leq \gamma$ can be equivalently dismissed.

Proposition 1.48 (equivalent descriptions of local metric regularity). *For any multifunction $F: X \rightrightarrows Y$ with* $\operatorname{dom} F \neq \emptyset$, *any* $(\bar{x}, \bar{y}) \in \operatorname{gph} F$, *and any $\mu > 0$ the following properties are equivalent:*

(a) *F is locally metrically regular around $(\bar{x}, \bar{y})$ with modulus μ;*

(b) *there are neighborhoods U of $\bar{x}$ and V of $\bar{y}$ such that* (1.36) *holds for all $x \in U$ and $y \in V$;*

(c) *there are neighborhoods U of $\bar{x}$ and V of $\bar{y}$ such that* (1.36) *holds for all $x \in U$ and $y \in V$ with $F(x) \cap V \neq \emptyset$.*

Proof. Obviously (b)⇒(a) and (b)⇒(c). Let us prove that (a)⇒(b). To perform this, it suffices to show that for any numbers $\eta > 0$ and $\gamma > 0$ there is $\nu > 0$ such that (1.36) holds for all $x \in \bar{x} + \nu I\!B$ and $y \in \bar{y} + \nu I\!B$ provided that it holds for every $x \in \bar{x} + \eta I\!B$ and $y \in \bar{y} + \eta I\!B$ with $\operatorname{dist}(y; F(x)) \leq \gamma$. Given (μ, η, γ), we put

$$\nu := \min\{\eta,\ \gamma\mu/(\mu + 1)\}\ .$$

Taking $x \in \bar{x} + \nu I\!B$ and $y \in \bar{y} + \nu I\!B$, we only need to consider the case when $\operatorname{dist}(y; F(x)) > \gamma$. Note that $\operatorname{dist}(\bar{x}; F^{-1}(y)) \leq \mu \operatorname{dist}(y; F(\bar{x}))$ due to (a) and

$$\operatorname{dist}(y; F(\bar{x})) \leq \|y - \bar{y}\| \leq \nu \leq \gamma\ .$$

Thus we have

$$\operatorname{dist}(x; F^{-1}(y)) \le \operatorname{dist}(\bar{x}; F^{-1}(y)) + \|x - \bar{x}\| \le \mu \operatorname{dist}(y; F(\bar{x})) + \|x - \bar{x}\|$$

$$\le \mu \|y - \bar{y}\| + \|x - \bar{x}\| \le \nu(\mu + 1) \le \gamma \mu$$

$$< \mu \operatorname{dist}(y; F(x))$$

due to the choice of ν. This proves that properties (a) and (b) are equivalent with the same modulus μ.

It remains to show that (c)$\Rightarrow$(a). Fix U and $\eta > 0$ such that (1.36) holds for all $x \in U$ and $y \in V := \operatorname{int}(\bar{y} + \eta I\!B)$ satisfying $F(x) \cap V \neq \emptyset$. Then take $\gamma := \frac{\eta}{3}$, $\widetilde{V} := \operatorname{int}(\bar{y} + \frac{\eta}{3} I\!B)$ and consider $y \in \widetilde{V}$ with $\operatorname{dist}(y; F(x)) \le \gamma$. For every such y we select $v \in F(x)$ satisfying $\|y - v\| \le \operatorname{dist}(y; F(x)) + \frac{\eta}{3}$ and get

$$\|v - \bar{y}\| \le \|v - y\| + \|y - \bar{y}\| < \operatorname{dist}(y; F(x)) + \frac{\eta}{3} + \frac{\eta}{3} \le \gamma + \frac{2\eta}{3} = \eta\,,$$

i.e., $v \in \operatorname{int}(\bar{y} + \eta I\!B)$. Thus $F(x) \cap \operatorname{int}(\bar{y} + \eta I\!B) \neq \emptyset$, which implies (a). $\triangle$

We see that each of the properties (b) and (c) in Proposition 1.48 can be chosen as an equivalent definition of local metric regularity with the same exact regularity bound $\operatorname{reg} F(\bar{x}, \bar{y})$. Note that an analog of the equivalence (a)$\Leftrightarrow$(c) holds also for semi-local metric regularity from Definition 1.47(iii). We'll justify and use this fact in the proof of the next theorem that establishes the equivalence between the corresponding Lipschitzian and metric regularity properties of arbitrary multifunctions.

Theorem 1.49 (relationships between Lipschitzian and metric regularity properties). *Let $F\colon X \rightrightarrows Y$ with* $\operatorname{dom} F \neq \emptyset$, *and let $\ell > 0$. Then the following hold:*

(i) *F is locally Lipschitz-like around $(\bar{x}, \bar{y}) \in \operatorname{gph} F$ if and only if its inverse $F^{-1}\colon Y \rightrightarrows X$ is locally metrically regular around $(\bar{y}, \bar{x}) \in \operatorname{gph} F^{-1}$ with the same modulus. Moreover, the latter is equivalent to the existence of neighborhoods U of $\bar{x}$, V of $\bar{y}$ and a number $\ell \ge 0$ such that*

$$F(x) \cap V \subset F(u) + \ell\|x - u\| I\!B \quad \textit{for all} \;\; u \in U,\; x \in X\,. \tag{1.37}$$

In this case one has the equality $\operatorname{lip} F(\bar{x}, \bar{y}) = \operatorname{reg} F^{-1}(\bar{y}, \bar{x})$.

(ii) *F is locally Lipschitzian around $\bar{x} \in \operatorname{dom} F$ if and only if F^{-1} is semi-locally metrically regular around $\bar{x} \in \operatorname{rge} F^{-1}$. In this case one has the equality* $\operatorname{lip} F(\bar{x}) = \operatorname{reg} F^{-1}(\bar{x})$.

Proof. We just prove assertion (ii). The proof of (i) is similar with taking into account the equivalence between properties (a) and (b) in Proposition 1.48. Note that (1.37) doesn't contain any restriction on x, in contrast to (1.28), which is due to the localization in both domain and range spaces.

To prove (ii), we first assume that F is locally Lipschitzian around $\bar{x}$ and denote $\ell := \operatorname{lip} F(\bar{x}) < \infty$. Then for any $\varepsilon > 0$ one has

$$F(x) \subset F(u) + (\ell + \varepsilon)\|x - u\| I\!B \quad \text{whenever} \quad x, u \in U\ ,$$

which immediately implies that

$$\text{dist}(y; F(u)) \le (\ell + \varepsilon)\|x - u\| \quad \text{if} \quad y \in F(x) \quad \text{and} \quad x, u \in U\ .$$

Choosing $r > 0$ with $\bar{x} + r I\!B \subset U$, it is easy to see from the above that

$$\text{dist}(y; F(u)) \le (\ell + \varepsilon)\,\text{dist}(u; F^{-1}(y)) \tag{1.38}$$

whenever $u \in \bar{x} + r I\!B$ and $F^{-1}(y) \cap (\bar{x} + r I\!B) \neq \emptyset$. Denote now $\widetilde{U} := \bar{x} + (r/3) I\!B$ and show that (1.38) holds for any $u \in \widetilde{U}$ and $y \in Y$ with $\text{dist}(u, F^{-1}) \le \gamma := r$. Indeed, for such u and y one gets $\|\widetilde{x} \in F^{-1}(y)$ with $\widetilde{x} - u\| \le r/3$ which yields $\|\widetilde{x} - \bar{x}\| \le r$ and hence $F^{-1}(y) \cap (\bar{x} + r I\!B) \neq \emptyset$. The latter means that F^{-1} is semi-locally metrically regular around $\bar{x}$ with modulus $\ell + \varepsilon$. Since $\varepsilon > 0$ was chosen arbitrarily, we have $\text{reg}\, F^{-1}(\bar{x}) \le \ell = \text{lip}\, F(\bar{x})$.

Conversely, let F^{-1} be semi-locally metrically regular around $\bar{x} \in \text{rge}\, F^{-1}$ with $\text{reg}\, F^{-1}(\bar{x}) := \mu$. Then for any $\varepsilon > 0$ we find positive numbers r and $\gamma < 3r$ such that

$$\text{dist}(y; F(u)) \le (\mu + \varepsilon)\text{dist}(u, F^{-1}(y))$$

whenever $u \in \bar{x} + r I\!B$ and $y \in Y$ satisfy $\text{dist}(u; F^{-1}(y)) \le \gamma$. Since

$$\text{dist}(u; F^{-1}(y)) \le \|u - \widetilde{x}\| \le \|u - \bar{x}\| + \|\widetilde{x} - \bar{x}\| < \gamma$$

if $\widetilde{x} \in F^{-1}(y) \cap (\bar{x} + (\gamma/3) I\!B)$, one has

$$\text{dist}(y; F(u)) \le (\mu + \varepsilon)\text{dist}(u; F^{-1}(y))$$

whenever $u \in \bar{x} + (\gamma/3) I\!B$ and $y \in Y$ with $F^{-1}(y) \cap (\bar{x} + (\gamma/3) I\!B) \neq \emptyset$. Shrinking the latter ball if necessary, we find a neighborhood U of $\bar{x}$ such that

$$F(x) \subset F(u) + (\mu + 2\varepsilon)\|u - x\| I\!B \quad \text{for} \quad x, u \in U,\ \ y \in Y\ ,$$

which implies the local Lipschitzian property of F around $\bar{x}$ with modulus $\mu + 2\varepsilon$. Since $\varepsilon > 0$ was chosen arbitrarily, we get $\text{lip}\, F(\bar{x}) \le \mu = \text{reg}\, F^{-1}(\bar{x})$ and complete the proof of the theorem. △

Now let us consider relationships between the notions of local and semi-local metric regularity in Definition 1.47. Obviously that semi-local metric regularity of F around $\bar{x} \in \text{dom}\, F$ (resp. around $\bar{y} \in \text{rge}\, F$) implies its local metric regularity around $(\bar{x}, \bar{y})$ for every $\bar{y} \in F(\bar{x})$ (resp. for every $\bar{x} \in F^{-1}(\bar{y})$), and one has

$$\text{reg}\, F(\bar{x}) \ge \sup_{\bar{y} \in F(\bar{x})} \big\{\text{reg}\, F(\bar{x}, \bar{y})\big\}, \quad \text{reg}\, F(\bar{y}) \ge \sup_{\bar{x} \in F^{-1}(\bar{y})} \big\{\text{reg}\, F(\bar{x}, \bar{y})\big\}\ .$$

Let us present conditions under which the converse implications take place and the latter inequalities become equalities. Note that the properties of multifunctions used in the next proposition are discussed right before Theorem 1.42.

Proposition 1.50 (relationships between local and semi-local metric regularity). *For any multifunction* $F\colon X \rightrightarrows Y$ *with* $\operatorname{dom} F \neq \emptyset$ *the following assertions hold:*

(i) *Given* $\bar{x} \in \operatorname{dom} F$*, assume that* F *is closed at* $\bar{x}$ *and locally compact around this point. Then* F *is semi-locally metrically regular around* $\bar{x}$ *if and only if it is locally metrically regular around* $(\bar{x}, \bar{y})$ *for every* $\bar{y} \in F(\bar{x})$*. In this case one has*

$$\operatorname{reg} F(\bar{x}) = \max\big\{\operatorname{reg} F(\bar{x}, \bar{y}) \;\big|\; \bar{y} \in F(\bar{x})\big\} < \infty\,.$$

(ii) *Given* $\bar{y} \in \operatorname{rge} F$*, assume that* F^{-1} *is closed at* $\bar{y}$ *and locally compact around this point. Then* F *is semi-locally metrically regular around* $\bar{y}$ *if and only if it is locally metrically regular around* $(\bar{x}, \bar{y})$ *for every* $\bar{x} \in F^{-1}(\bar{y})$*. In this case one has*

$$\operatorname{reg} F(\bar{y}) = \max\big\{\operatorname{reg} F(\bar{x}, \bar{y}) \;\big|\; \bar{x} \in F^{-1}(\bar{y})\big\} < \infty\,.$$

Proof. Assertion (ii) follows from Theorems 1.42 and 1.49. Assertion (i) is independent but can be justified similarly to the proof of Theorem 1.42; see the proof of Theorem 4.2(c) in Mordukhovich [909] for more details. △

As shown above, the properties of local and semi-local (global relative to *domain* spaces) metric regularity of arbitrary multifunctions are equivalent, correspondingly, to the local Lipschitz-like and local Lipschitzian properties of their inverses. It also happens that metric regularity of a multifunction F is closely related to the so-called *covering* properties of F we consider next. In this respect, the other notion of semi-local metric regularity of F in Definition 1.47 (global relative to *image* spaces) plays a major role.

Definition 1.51 (covering properties). *Let* $F\colon X \rightrightarrows Y$ *with* $\operatorname{dom} F \neq \emptyset$.

(i) *Given nonempty subsets* $U \subset X$ *and* $V \subset Y$*, we say that* F *has the* COVERING PROPERTY *on* U *relative to* V *if there is* $\kappa > 0$ *such that*

$$F(x) \cap V + \kappa r I\!B \subset F(x + r I\!B) \quad \text{whenever} \quad x + r I\!B \subset U \quad \text{as } r > 0\,. \tag{1.39}$$

(ii) *Given* $(\bar{x}, \bar{y}) \in \operatorname{gph} F$*, we say that* F *has the* LOCAL COVERING PROPERTY *around* $(\bar{x}, \bar{y})$ *with modulus* $\kappa > 0$ *if there are neighborhoods* U *of* $\bar{x}$ *and* V *of* $\bar{y}$ *such that* (1.39) *holds. The supremum of all such moduli* $\{\kappa\}$*, denoted by* $\operatorname{cov} F(\bar{x}, \bar{y})$*, is called the* EXACT COVERING BOUND *of* F *around* $(\bar{x}, \bar{y})$.

(iii) F *has the* SEMI-LOCAL COVERING PROPERTY *around* $\bar{x} \in \operatorname{dom} F$ *with modulus* $\kappa > 0$ *if there is a neighborhood* U *of* $\bar{x}$ *such that* (1.39) *holds as* $V = Y$*. The supremum of all such moduli is denoted by* $\operatorname{cov} F(\bar{x})$.

The local covering property in Definition 1.51(ii) is also known as *openness at a linear rate* or *linear openness* of F around $(\bar{x}, \bar{y})$. For single-valued mappings $f\colon X \to Y$ it relates to a conventional openness property of f at $\bar{x}$

generated by the mixed coderivative of inverse mappings. Observe that (1.40) corresponds to taking the *reversed* convergence (strong in X^* and weak* in Y^*) in definition (1.25) of the mixed coderivative. Of course, $\widetilde{D}^*_M F(\bar{x},\bar{y}) = D^*_N F(\bar{x},\bar{y})$ if $\dim X < \infty$, and $\widetilde{D}^*_M F(\bar{x},\bar{y}) = D^*_M F(\bar{x},\bar{y})$ if both X and Y are finite-dimensional. Note also that there is no difference between these three coderivatives if F is N-regular at $(\bar{x},\bar{y})$. However, in the general setting the reversed coderivative (1.40) *doesn't enjoy a satisfactory calculus* developed for the normal and mixed coderivatives in Subsects. 1.2.4 and 3.1.2. This restricts the range of its applications in comparison with D^*_N and D^*_M.

Theorem 1.54 (coderivative conditions from local metric regularity and covering). *Let $F\colon X \rightrightarrows Y$ with $(\bar{x},\bar{y}) \in \operatorname{gph} F$. Assume that F is locally metrically regular around $(\bar{x},\bar{y})$ with modulus $\mu > 0$ or, equivalently, F has the local covering property around $(\bar{x},\bar{y})$ with modulus μ^{-1}. Then the following assertions hold:*

(i) *There is $\eta > 0$ such that*

$$\inf\Big\{\|x^*\| \ \Big|\ x^* \in \widehat{D}^* F(x,y)(y^*)\Big\} \ge \mu^{-1}\|y^*\| \tag{1.41}$$

whenever $x \in \bar{x} + \eta I\!B$, $y \in F(x) \cap (\bar{y} + \eta I\!B)$, and $y^* \in Y^*$. In this case

$$\operatorname{reg} F(\bar{x},\bar{y}) \ge \inf_{\eta>0} \sup\Big\{\|\widehat{D}^* F(x,y)^{-1}\| \ \Big|\ x \in B_\eta(\bar{x}),\ y \in F(x) \cap B_\eta(\bar{y})\Big\},$$

$$\operatorname{cov} F(\bar{x},\bar{y}) \le \sup_{\eta>0} \inf\Big\{\|x^*\| \ \Big|\ x^* \in \widehat{D}^* F(x,y)(y^*),\ x \in B_\eta(\bar{x}),\ y \in F(x) \cap B_\eta(\bar{y}),\ \|y^*\| = 1\Big\}.$$

(ii) *One has the equivalent conditions*

$$D^*_M F^{-1}(\bar{y},\bar{x})(0) = \{0\} \Longleftrightarrow \ker \widetilde{D}^*_M F(\bar{x},\bar{y}) = \{0\} \tag{1.42}$$

and the exact bounds estimates

$$\operatorname{reg} F(\bar{x},\bar{y}) \ge \|D^*_M F^{-1}(\bar{y},\bar{x})\| = \|\widetilde{D}^*_M F(\bar{x},\bar{y})^{-1}\|,$$

$$\operatorname{cov} F(\bar{x},\bar{y}) \le \inf\Big\{\|x^*\| \ \Big|\ x^* \in \widetilde{D}^*_M F(\bar{x},\bar{y})(y^*),\ \|y^*\| = 1\Big\}.$$

Proof. To prove (i), we observe that one always has

$$y^* \in \widehat{D}^* F^{-1}(y,x)(x^*) \Longleftrightarrow -x^* \in \widehat{D}^* F(x,y)(-y^*).$$

From here we get $\|\widehat{D}^* F^{-1}(x,y)\| = \|\widehat{D}^* F(x,y)^{-1}\|$ and then derive all the conclusions in (i) from Theorem 1.43(i) due to the equivalence results of Theorems 1.49(i) and 1.52(ii). These equivalences also imply both conditions (1.42) and the estimate for the regularity bound in (ii) due to condition (1.35) in Theorem 1.44 and definition (1.40).

It remains to justify the estimate for the covering bound in (ii). This follows from the above and the observation that

$$1\Big/\|H^{-1}\| = \inf\big\{\|y\| \;\big|\; y \in H(x),\ \|x\| = 1\big\}$$

for any positively homogeneous multifunction $H\colon X \rightrightarrows Y$. △

The results obtained easily imply the corresponding necessary coderivative conditions with the exact bounds estimates for semi-local covering and metric regularity properties. For brevity we present only the necessary conditions.

Corollary 1.55 (coderivative conditions from semi-local metric regularity and covering). *Let* $F\colon X \rightrightarrows Y$ *with* $\operatorname{dom} F \neq \emptyset$. *The following assertions hold:*

(i) *Assume that* F *is semi-locally metrically regular around* $\bar{x} \in \operatorname{dom} F$ *with modulus* $\mu > 0$ *or, equivalently,* F *has the semi-local covering property around* $\bar{x}$ *with modulus* μ^{-1}. *Then there is* $\eta > 0$ *such that* (1.41) *is fulfilled for any* $x \in \bar{x} + \eta I\!B$, $y \in F(x)$, *and* $y^* \in Y^*$, *and also the equivalent conditions* (1.42) *hold for every* $\bar{y} \in F(\bar{x})$.

(ii) *Assume that* F *is semi-locally metrically regular around* $\bar{y} \in \operatorname{rge} F$ *with modulus* $\mu > 0$. *Then there is* $\eta > 0$ *such that* (1.41) *is fulfilled for any* $y \in F(x) \cap (\bar{y} + \eta I\!B)$ *with* $x \in X$ *and any* $y^* \in Y^*$. *Also the equivalent conditions* (1.42) *hold for every* $\bar{x} \in F^{-1}(\bar{y})$ *in this case.*

Proof. Follows directly from the definitions and Theorem 1.54. △

If $F = f\colon X \to Y$ is single-valued, there is no difference between the local and semi-local metric regularity and covering properties of f around the reference point $\bar{x}$ with $\bar{y} = f(\bar{x})$. Let us consider the case when f is *strictly differentiable* at $\bar{x}$ and present a *complete characterization* of metric regularity and covering with *precise formulas* for computing the corresponding exact bounds. The *necessity part* of this characterization with a lower (resp. upper) estimate for the exact bound of metric regularity (resp. covering) is a special case of the general coderivative results from Theorem 1.54 and the following Lemma 1.56 on the automatic *closedness* of the derivative image for metrically regular mappings. The *sufficiency part* of Theorem 1.57 with the opposite side estimates is the essence of the celebrated Lyusternik-Graves theorem – in fact of its proof – that is reproduced in the arguments below.

Let us start with the afore-mentioned lemma that holds, as well as Theorem 1.57, in arbitrary Banach spaces.

Lemma 1.56 (closed derivative images of metrically regular mappings). *Let* $f\colon X \to Y$ *be metrically regular around* $\bar{x}$ *and Fréchet differentiable at this point. Then the linear image space* $\nabla f(\bar{x})X$ *is closed in* Y.

Proof. Choose $\eta > 0$ such that for some $\mu > 0$ we have

$$\operatorname{dist}\big(x; f^{-1}(\bar{x})\big) \le \mu \|f(x) - f(\bar{x})\| \quad \text{whenever} \quad x \in \bar{x} + \eta I\!B\ ;$$

this is a consequence of metric regularity. Denote $A := \nabla f(\bar{x})$ and fix an arbitrary point $y_0 \in \operatorname{cl}(AX)$. Then there is a sequence of $y_k \to y_0$ with $y_k \in AX$ and $\|y_{k+1} - y_k\| \le 2^{-k}$ as $k \in I\!N$. To proceed, we construct a sequence of $x_k \in X$ satisfying the estimates

$$\|x_{k+1} - x_k\| \le \frac{3\mu}{2^k} \quad \text{and} \quad \|y_k - Ax_k\| \le \frac{1}{2^k} \quad \text{for all} \quad k \in I\!N\ .$$

Define x_k iteratively. First let x_1 be any point with $Ax_1 = y_1$. Then having $x_1, \ldots, x_k$ satisfying the above estimates, construct x_{k+1} as follows. Fix $u \in f^{-1}(y_{k+1}) - x_k$ and choose $t > 0$ satisfying $t\|u\| \le \eta$ and

$$\Big\|\frac{f(\bar{x} + tz) - f(\bar{x})}{t} - Az\Big\| \le \frac{1}{2^{k+2}} \quad \text{whenever} \quad z \in \Big\{\|u\|, \frac{3\mu}{2^k}\Big\} I\!B\ ,$$

which implies the relationships

$$\begin{aligned}
\|f(\bar{x} + tu) - f(\bar{x})\| &\le t\Big(\|Au\| + \frac{1}{2^{k+2}}\Big) = t\Big(\|y_{k+1} - Ax_k\| + \frac{1}{2^{k+2}}\Big) \\
&\le t\Big(\|y_{k+1} - y_k\| + \|y_k - Ax_k\| + \frac{1}{2^{k+2}}\Big) \\
&\le t\Big(\frac{1}{2^k} + \frac{1}{2^k} + \frac{1}{2^{k+2}}\Big) \le \frac{3t}{2^k}\ .
\end{aligned}$$

Now using the metric regularity of f around $\bar{x}$, find $\widetilde{x}$ with $f(\widetilde{x}) = f(\bar{x} + tu)$ and $\|\widetilde{x} - \bar{x}\| \le 3\mu t/2^k$. Putting $v := (\widetilde{x} - \bar{x})/t$ and $x_{k+1} := x_k + v$, we get $\|x_{j+1} - x_j\| \le 3\mu t/2^j$ for $j = k, k+1$. It remains to show that

$$\|y_{k+1} - Ax_{k+1}\| \le \frac{1}{2^{k+1}}\ .$$

To justify this, observe from the above constructions that

$$\Big\|\frac{f(\bar{x} + tv) - f(\bar{x})}{t} - Av\Big\| \le \frac{1}{2^{k+2}}, \qquad \Big\|\frac{f(\bar{x} + tu) - f(\bar{x})}{t} - Av\Big\| \le \frac{1}{2^{k+2}}\ ,$$

and hence $\|Au - Av\| = \|y_{k+1} - ax_{k+1}\| \le 1/2^{k+1}$. Thus $\{x_k\}$ is a Cauchy sequence in X that converges to some point x_0. Furthermore, $Ax_k = y_k \to y_0$, which gives $Ax_0 = y_0$ and completes the proof of the lemma. △

Now we are ready to prove the mentioned fundamental characterization of metric regularity and covering for strictly differentiable mappings between general Banach spaces.

Theorem 1.57 (metric regularity and covering for strictly differentiable mappings). *Let $f: X \to Y$ be strictly differentiable at $\bar{x}$. Then f is metrically regular around $\bar{x}$ (equivalently, f has the covering property around this point) if and only if the derivative operator $\nabla f(\bar{x}): X \to Y$ is surjective. In this case one has the exact formulas*

$$\operatorname{reg} f(\bar{x}) = \big\|\big(\nabla f(\bar{x})^*\big)^{-1}\big\|, \qquad \operatorname{cov} f(\bar{x}) = \inf\Big\{\|\nabla f(\bar{x})^* y^*\| \;\Big|\; \|y^*\| = 1\Big\}.$$

Proof. First we justify the *necessity* of the surjectivity of the derivative operator $\nabla f(\bar{x})$ for the metric regularity of f around $\bar{x}$. It follows from Theorem 1.38 and the definitions that

$$\widetilde{D}_M^* f(\bar{x})(y^*) = \big\{\nabla f(\bar{x})^* y^*\big\} \text{ for all } y^* \in Y^*$$

when f is strictly differentiable at $\bar{x}$. Hence the metric regularity of f around $\bar{x}$ gives by (1.42) that

$$\ker \nabla f(\bar{x})^* = \{0\}, \text{ i.e., } \nabla f(\bar{x})^* y^* = 0 \Longrightarrow y^* = 0.$$

The latter easily implies, since the image space $\nabla f(\bar{x})X$ is *closed* in Y by Lemma 1.56, that the operator $\nabla f(\bar{x})$ is *surjective.* Indeed, the opposite assumption immediately contradicts the separation (or, equivalently, Hahn-Banach) theorem. Observe furthermore that the surjectivity of $\nabla f(\bar{x})$ implies by Lemma 1.18 that the inverse operator to $\nabla f(\bar{x})^*$ is single-valued. Thus we get the relationships

$$\operatorname{reg} f(\bar{x}) \ge \|(\nabla f(\bar{x})^*)^{-1}\|, \qquad \operatorname{cov} f(\bar{x}) \le \inf\Big\{\|\nabla f(\bar{x})^* y^*\| \;\Big|\; \|y^*\| = 1\Big\}$$

from the general coderivative estimates of Theorem 1.54(ii).

Next let us prove that the surjectivity of $\nabla f(\bar{x})$ is also *sufficient* for the metric regularity (covering) of f around $\bar{x}$, in which case the above estimates hold as equalities. For definiteness we'll proceed with the covering property.

Put $A := \nabla f(\bar{x})$. It follows from the surjectivity of A (see the proof of Lemma 1.18) that for any $y \in Y$ there is $x \in A^{-1}(y)$ satisfying

$$\|x\| \le \mu\|y\| \text{ with } \mu^{-1} = \inf\Big\{\|A^* y^*\| \;\Big|\; \|y^*\| = 1\Big\}. \tag{1.43}$$

Using the strict differentiability of f at $\bar{x}$, for every $\gamma \in (0, \mu^{-1})$ we find a neighborhood U of $\bar{x}$ such that

$$\|f(x_1) - f(x_2) - A(x_1 - x_2)\| \le \gamma\,\|x_1 - x_2\| \text{ for all } x_1, x_2 \in U.$$

Let us show

$$f(\hat{x}) + (\mu^{-1} - \gamma) r I\!B \subset f(\hat{x} + r I\!B) \text{ whenever } \hat{x} + r I\!B \subset U,\ r > 0.$$

By definition this means that f has the covering property around $\bar{x}$ with modulus $\kappa = \mu^{-1} - \gamma$. Since $\gamma > 0$ can be taken arbitrarily small, we get

$$\operatorname{cov} f(\bar{x}) \geq \mu^{-1} = \inf\left\{ \|\nabla f(\bar{x})^* y^*\| \;\Big|\; \|y^*\| = 1 \right\},$$

which will end the proof of the theorem.

It remains to prove the above inclusion for f, where one can obviously take $\hat{x} = 0$ and $f(\hat{x}) = 0$ without loss of generality. The latter means that for every $y \in (\mu^{-1} - \gamma) r I\!B$ the equation $y = f(x)$ has a solution $x \in r I\!B \subset U$. This is actually the main result (Theorem 1) in Graves [522].

Fix $y \in Y$ with $\|y\| \leq (\mu^{-1} - \gamma) r$ and construct the desired solution x as the limit of a sequence $\{x_k\}$, $k = 1, 2, \ldots$, recurrently defined in the following way. Starting with $x_0 := 0$, we use (1.43) to construct x_k by the iterative procedure of Newton's type:

$$Ax_k = y - f(x_{k-1}) + Ax_{k-1} \quad \text{with} \quad \|x_k - x_{k-1}\| \leq \mu \, \|y - f(x_{k-1})\|$$

for all $k \in I\!N$. It follows from the above construction that

$$\|x_{k+1} - x_k\| \leq \mu(\mu\gamma)^k \, \|y\| \quad \text{and}$$

$$\|x_k\| \leq \sum_{j=1}^{k} \|x_j - x_{j-1}\| \leq \mu \, \|y\| \sum_{j=1}^{k} (\mu\gamma)^{j-1}$$

$$\leq \mu \, \|y\|/(1 - \mu\gamma) = \|y\|/(\mu^{-1} - \gamma) \leq r$$

for every $k \in I\!N$. Thus $\{x_k\}$ is a Cauchy sequence that converges to some $x \in X$ with $\|x\| \leq r$. Passing to the limit in the iterations as $k \to \infty$, we obtain $y = f(x)$ and complete the proof of the theorem. △

The following corollary of Theorem 1.57 for linear operators gives a refinement of the classical Banach-Schauder open mapping theorem.

Corollary 1.58 (metric regularity and covering for linear operators). *A linear and continuous operator $A\colon X \to Y$ is metrically regular around every point $\bar{x} \in X$ (equivalently, it has the covering property around $\bar{x}$) if and only if A is surjective. In this case one has*

$$\operatorname{reg} A(\bar{x}) = \|(A^*)^{-1}\|, \quad \operatorname{cov} A(\bar{x}) = \inf\left\{ \|A^* y^*\| \;\Big|\; \|y^*\| = 1 \right\} \quad \textit{for all } \bar{x} \in X .$$

Proof. Follows immediately from Theorem 1.57 with $f(x) = Ax$. △

Throughout this subsection we have considered relationships between properties of mappings and their inverses that may be set-valued even for simple smooth functions. Another direct corollary of Theorem 1.57 provides the following characterization of the local Lipschitz-like property of inverses to strictly differentiable mappings.

Corollary 1.59 (Lipschitz-like inverses to strictly differentiable mappings). *Let $f\colon X \to Y$ be strictly differentiable at $\bar{x}$, and let $\bar{y} = f(\bar{x})$. Then the inverse mapping $f^{-1}\colon Y \rightrightarrows X$ is locally Lipschitz-like around $(\bar{y}, \bar{x})$ if and only if $\nabla f(\bar{x})$ is surjective. In this case one has*

$$\operatorname{lip} f^{-1}(\bar{y}, \bar{x}) = \left\| \left(\nabla f(\bar{x})^*\right)^{-1} \right\| .$$

Proof. Follows from Theorem 1.57 and the equivalence in Theorem 1.49(i). △

The result in Corollary 1.59 can be interpreted as a kind of "set-valued inverse mapping theorem", since it infers good (Lipschitz-like) behavior of inverse multifunctions. However, the main objective of conventional inverse mapping theorems, as well as implicit mapping theorems implied by them, is to find efficient conditions ensuring that f^{-1} is locally *single-valued* and inherits the same analytic/differential properties as the given mapping f.

The classical inverse mapping theorem concerns the case of $f \in \mathcal{C}^1$ around $\bar{x}$ and proves that $f^{-1} \in \mathcal{C}^1$ around $\bar{y} = f(\bar{x})$ if $\nabla f(\bar{x})$ is invertible. Leach [748] extended this result to the case of mappings f strictly differentiable at $\bar{x}$. He formally introduced the notion of strict differentiability for this purpose although the corresponding construction actually appeared in Graves' proof of his seminal result; cf. the proof of Theorem 1.57. Let us show, based on Theorem 1.57, that the invertibility of the strict derivative $\nabla f(\bar{x})$ is *necessary and sufficient* for f^{-1} to be strictly differentiable at $\bar{y}$. Moreover, we give precise formulas for computing the exact metric regularity, covering, and Lipschitzian bounds of f^{-1} in this case.

Theorem 1.60 (strictly differentiable inverses). *Let $f\colon X \to Y$ be strictly differentiable at $\bar{x}$, and let $\bar{y} = f(\bar{x})$. Then f^{-1} is locally single-valued around $\bar{y}$ and strictly differentiable at this point if and only if $\nabla f(\bar{x})$ is invertible. In this case one has*

$$\nabla f^{-1}(\bar{y}) = \nabla f(\bar{x})^{-1}, \quad \operatorname{lip} f^{-1}(\bar{y}) = \left\| \left(\nabla f(\bar{x})^*\right)^{-1} \right\| ,$$

$$\operatorname{reg} f^{-1}(\bar{y}) = \|\nabla f(\bar{x})^*\| ,$$

$$\operatorname{cov} f^{-1}(\bar{y}) = \inf \left\{ \left\| \nabla \left(f(\bar{x})^{-1}\right)^* x^* \right\| \,\Big|\, \|x^*\| = 1 \right\} .$$

Proof. Assume that $\nabla f(\bar{x})$ is invertible and show first that f^{-1} is locally single-valued around $\bar{y}$. If it is not the case, for any neighborhood U of $\bar{x}$ we find $x_1, x_2 \in U$ such that $f(x_1) = f(x_2)$. Then

$$\frac{\|\nabla f(\bar{x})(x_1 - x_2)\|}{\|x_1 - x_2\|} = \frac{\|f(x_1) - f(x_2) - \nabla f(\bar{x})(x_1 - x_2)\|}{\|x_1 - x_2\|} .$$

This clearly contradicts the strict differentiability of f at $\bar{x}$ and the existence of $\alpha > 0$ with $\|\nabla f(\bar{x})x\| \ge \alpha \|x\|$ for all $x \in X$, which follows from the invertibility of $\nabla f(\bar{x})$.

Next let us prove that f^{-1} is strictly differentiable at $\bar{y}$ with $\nabla f^{-1}(\bar{y}) = \nabla f(\bar{x})^{-1}$. Taking arbitrary $y_i = f(x_i)$, $i = 1, 2$, near $\bar{y}$ and denoting $\gamma(x_1, x_2) := f(x_1) - f(x_2) - \nabla f(\bar{x})(x_1 - x_2)$, we have

$$\|f^{-1}(y_1) - f^{-1}(y_2) - \nabla f(\bar{x})^{-1}(y_1 - y_2)\|$$

$$= \|x_1 - x_2 - \nabla f(\bar{x})^{-1}(f(x_1) - f(x_2))\|$$

$$= \|x_1 - x_2 - \nabla f(\bar{x})^{-1}(\nabla f(\bar{x})(x_1 - x_2) + \gamma(x_1, x_2))\|$$

$$= \|\nabla f(\bar{x})^{-1}(\gamma(x_1, x_2))\| \le \|\nabla f(\bar{x})^{-1}\| \cdot \|\gamma(x_1, x_2)\| .$$

By Theorem 1.57 the function f is metrically regular around $\bar{x}$, which gives $\mu > 0$ such that $\|x_1 - x_2\| \le \mu\|y_1 - y_2\|$. This implies

$$\|\gamma(x_1, x_2)\| \Big/ \|y_1 - y_2\| \le \|\gamma(x_1, x_2)\| \Big/ \mu^{-1}\|x_1 - x_2\| \to 0 \;\text{ as }\; y_1, y_2 \to \bar{y} ,$$

which proves the claim and the sufficiency part of the theorem.

In this case f^{-1} is locally Lipschitzian around $\bar{y}$, and thus $\operatorname{lip} f^{-1}(\bar{y}) = \|\nabla f(\bar{x})^{-1}\|$ due to Corollary 1.59. The formulas for $\operatorname{reg} f^{-1}(\bar{y})$ and $\operatorname{cov} f^{-1}(\bar{y})$ follow directly from Theorem 1.57.

Conversely, if f^{-1} is locally single-valued and strictly differentiable at $\bar{y}$, then both f and f^{-1} are metrically regular around $\bar{x}$ and $\bar{y}$, respectively. Hence both $\nabla f(\bar{x})$ and $\nabla f^{-1}(\bar{y})$ are surjective due to the necessity in Theorem 1.57, which implies the invertibility of $\nabla f(\bar{x})$. △

Remark 1.61 (restrictive metric regularity). Observe that Definition 1.47 of metric regularity doesn't depend on the linear structure of the spaces in question and applies to arbitrary metric spaces. In this way, given a mapping $f\colon X \to Y$ between Banach spaces, we can consider the metric regularity of the *restricted mapping* $f\colon X \to f(X)$, where the image space Y is replaced by the metric space $f(X)$. This notion is naturally to call the *restrictive metric regularity* (RMR) of f around $\bar{x}$.

If f is strictly differentiable at $\bar{x}$ with the surjective derivative $\nabla f(\bar{x})$, then the classical Lyusternik-Graves theorem ensures the metric regularity of $f\colon X \to Y$ around $\bar{x}$, and the surjectivity of $\nabla f(\bar{x})$ is also *necessary* for the latter property; see Theorem 1.57. What could we say about the *restrictive* metric regularity of f when $\nabla f(\bar{x})$ *is not surjective*? This issue is addressed in the paper by Mordukhovich and B. Wang [967, 968], where the notion of restrictive metric regularity is studied in depth with applications to the first-order and second-order generalized differential calculus and to the sequential normal compactness of set and mappings. In particular, the following generalization of the Lyusternik-Graves theorem involving the *paratingent cone*

$$\widetilde{T}(\bar{x};\Omega) := \big\{ v \in X \big|\, \exists v_k \to v,\ t_k \downarrow 0,\ x_k \xrightarrow{\Omega} \bar{x} \ \text{ with }\ x_k + t_k v_k \in \Omega \big\}$$

to Ω at $\bar{x}$ is obtained (note that the image space $\nabla f(\bar{x})X$ is closed in Y under the RMR property of f around $\bar{x}$; this follows from the proof of Lemma 1.56):

Let $f\colon X \to Y$ be a mapping between Banach spaces that is strictly differentiable at $\bar{x}$. Then the restrictive metric regularity of f around $\bar{x}$ implies that $\widetilde{T}(f(\bar{x}); f(X)) = \nabla f(\bar{x})X$, and the converse implication holds when codim $\nabla f(\bar{x})X < \infty$.

Applications of the restrictive metric regularity to the generalized differential calculus and SNC properties of sets and mappings are similar to those presented in this book, but *without surjectivity* assumption on $\nabla f(\bar{x})$. In particular, a counterpart of Theorem 1.17 is formulated as follows:

Let $f\colon X \to Y$ be strictly differentiable at $\bar{x}$, and let the space $\nabla f(\bar{x})X$ be complemented in Y. Then one has the two generally independent equalities:

$$N\big(\bar{x}; f^{-1}(\Theta)\big) = \nabla f(\bar{x})^* N\big(f(\bar{x}); \Theta \cap f(X)\big)\ ,$$

$$\big(\nabla f(\bar{x})^*\big)^{-1} N\big(\bar{x}; \Theta \cap f(X)\big) = N\big(f(\bar{x}); \Theta \cap f(X)\big)$$

provided that f has the RMR property around $\bar{x}$.

Note that the complementarity requirement on $\nabla f(\bar{x})X$ above may be replaced by the more general *w^*-extensibility* property of $\nabla f(\bar{x})X$ in the sense of Definition 1.122, which always holds if $I\!B^*$ is weak* sequentially compact; see Proposition 1.123. We refer the reader to the afore-mentioned papers [967, 968] for more results, applications, and discussions in this direction.

1.2.4 Calculus of Coderivatives in Banach Spaces

This subsection contains calculus results for coderivatives of set-valued mappings between arbitrary *Banach* spaces. We pay the main attention to normal and mixed coderivatives from Definition 1.32 that are the most important for applications. The results obtained concern sum and chain rules for coderivatives and incorporate the corresponding calculus for graphical regularity of multifunctions. We'll come back to this subject in Chap. 3, where much more calculus rules (*full calculus*) will be developed for set-valued mappings between *Asplund* spaces.

Let us start with *sum rules* for coderivatives of two mappings, one of which is single-valued and differentiable. The following theorem ensures sum rules with *equalities.*

Theorem 1.62 (coderivative sum rules with equalities). *Let $f\colon X \to Y$ be Fréchet differentiable at $\bar{x}$, and let $F\colon X \rightrightarrows Y$ be an arbitrary set-valued mapping such that $\bar{y} - f(\bar{x}) \in F(\bar{x})$ for some $\bar{y} \in Y$. The following hold:*

(i) *For all $y^* \in Y^*$ one has*

$$\widehat{D}^*(f+F)(\bar{x},\bar{y})(y^*) = \nabla f(\bar{x})^* y^* + \widehat{D}^* F(\bar{x},\bar{y}-f(\bar{x}))(y^*)\,.$$

(ii) *If f is strictly differentiable at $\bar{x}$, then*

$$D^*(f+F)(\bar{x},\bar{y})(y^*) = \nabla f(\bar{x})^* y^* + D^* F(\bar{x},\bar{y}-f(\bar{x}))(y^*)$$

for all $y^* \in Y^*$, where D^* stands either for the normal coderivative (1.24) *or for the mixed coderivative* (1.25). *Moreover, the mapping $f+F$ is N-regular (resp. M-regular) at $(\bar{x},\bar{y})$ if and only if F is N-regular (resp. M-regular) at the point $(\bar{x},\bar{y}-f(\bar{x}))$.*

Proof. The inclusions "$\subset$" in both formulas can be proved similarly to Theorem 1.38. Applying them to the sum $(f+F)+(-f)$, we get the opposite inclusions and thus establish the equalities. The regularity statements follow from the combination of (i), (ii), and the definitions. △

Next let us derive formulas for computing coderivatives of *compositions*

$$(F\circ G)(x) := F(G(x)) = \bigcup\Big\{F(y)\;\Big|\; y\in G(x)\Big\}$$

for mappings between Banach spaces. To proceed, we need to define some notions used in what follows.

Definition 1.63 (inner semicontinuous and inner semicompact multifunctions). *Let $S\colon X \rightrightarrows Y$ with $\bar{x}\in \operatorname{dom} S$.*

(i) *Given $\bar{y}\in S(\bar{x})$, we say that the mapping S is* INNER SEMICONTINUOUS *at $(\bar{x},\bar{y})$ if for every sequence $x_k\to\bar{x}$ there is a sequence $y_k\in S(x_k)$ converging to $\bar{y}$ as $k\to\infty$.*

(ii) *S is* INNER SEMICOMPACT *at $\bar{x}$ if for every sequence $x_k\to\bar{x}$ there is a sequence $y_k\in S(x_k)$ that contains a convergent subsequence as $k\to\infty$.*

The inner semicontinuity of S at $(\bar{x},\bar{y})$ for *every* $\bar{y}\in S(\bar{x})$ goes back to the standard notion of inner/lower semicontinuity of S at $\bar{x}$ recalled and used in Subsect. 1.2.1; see Theorem 1.34. The latter notion clearly implies the inner semicompactness of S at $\bar{x}$, which may be substantially weaker than the inner semicontinuity. In particular, any nonempty-valued mapping that is locally compact around $\bar{x}$ (locally bounded when $\dim Y<\infty$) is obviously inner semicompact *around* $\bar{x}$, i.e., at each x from some neighborhood of $\bar{x}$. Under additional assumptions imposed in the results below, the inner semicompactness of mappings S at $\bar{x}$ implies that S is *closed-graph at* $\bar{x}$ (but not around this point), i.e., $\bar{y}\in S(\bar{x})$ whenever $x_k\to\bar{x}$ and $y_k\to\bar{y}$ with $y_k\in S(x_k)$. Note that, in contrast to the inner semicontinuity property (i), the inner semicompactness property (ii) in Definition 1.63 cannot be equivalently formulated via the convergence of the whole sequence $\{y_k\}$, $k\in I\!N$, and requires passing to a *subsequence*.

To formulate the first theorem on coderivatives of compositions, let us consider the multifunction

$$\Phi(x,y) := F(y) + \Delta((x,y);\operatorname{gph} G)$$

involving the indicator mapping Δ defined in Proposition 1.33. This multifunction plays a significant role in the proof of various chain rules considered below; see also Chap. 3.

Theorem 1.64 (coderivatives of compositions). *Let $G\colon X \rightrightarrows Y$, $F\colon Y \rightrightarrows Z$, $\bar{z} \in (F \circ G)(\bar{x})$, and*

$$S(x,z) := G(x) \cap F^{-1}(z) = \big\{ y \in G(x) \;\big|\; z \in F(y) \big\} .$$

The following hold for both coderivatives $D^* = D^*_N$ and $D^* = D^*_M$ for all $z^* \in Z^*$:

(i) *Given $\bar{y} \in S(\bar{x},\bar{z})$, assume that S is inner semicontinuous at $(\bar{x},\bar{z},\bar{y})$. Then one has*

$$D^*(F \circ G)(\bar{x},\bar{z})(z^*) \subset \big\{ x^* \in X^* \;\big|\; (x^*,0) \in D^*\Phi(\bar{x},\bar{y},\bar{z})(z^*) \big\} .$$

(ii) *Assume that S is inner semicompact at $(\bar{x},\bar{z})$, where G is closed-graph at $\bar{x}$ and F^{-1} is closed-graph at $\bar{z}$. Then one has*

$$D^*(F \circ G)(\bar{x},\bar{z})(z^*) \subset \Big\{ x^* \in X^* \;\Big|\; (x^*,0) \in \bigcup_{\bar{y} \in S(\bar{x},\bar{z})} D^*\Phi(\bar{x},\bar{y},\bar{z})(z^*) \Big\} .$$

(iii) *Let $G = g$ be single-valued around $\bar{x}$. Then one has*

$$D^*(F \circ g)(\bar{x},\bar{z})(z^*) = \big\{ x^* \in X^* \;\big|\; (x^*,0) \in D^*\Phi(\bar{x},g(\bar{x}),\bar{z})(z^*) \big\}$$

if either g is Lipschitz continuous around $\bar{x}$ and $\dim Y < \infty$, or g is strictly differentiable at $\bar{x}$. In each of these cases $F \circ g$ is N-regular (M-regular) at $(\bar{x},\bar{z})$ if Φ has the corresponding property at $(\bar{x},g(\bar{x}),\bar{z})$.

Proof. We prove the theorem for the case of $D^* = D^*_N$; for $D^* = D^*_M$ the proof is similar. Let us start with (i). Take arbitrary (x^*,z^*) with $x^* \in D^*(F \circ G)(\bar{x},\bar{z})(z^*)$ and find sequences $\varepsilon_k \downarrow 0$, $(x_k,z_k) \to (\bar{x},\bar{z})$, and $(x_k^*,z_k^*) \xrightarrow{w^*} (x^*,z^*)$ such that

$$z_k \in (F \circ G)(x_k) \;\text{ and }\; (x_k^*,-z_k^*) \in \widehat{N}_{\varepsilon_k}((x_k,z_k);\operatorname{gph} F \circ G), \;\; k \in I\!N .$$

Using the inner semicontinuity of S at $(\bar{x},\bar{z},\bar{y})$, one gets $y_k \in S(x_k,z_k)$ with $y_k \to \bar{y}$ as $k \to \infty$. For each $k \in I\!N$ we have

$$\begin{aligned}
&\limsup_{\substack{(x,y,z)\to(x_k,y_k,z_k)\\ z\in\Phi(x,y)}} \frac{\langle (x_k^*,0,-z_k^*),(x,y,z)-(x_k,y_k,z_k)\rangle}{\|(x,y,z)-(x_k,y_k,z_k)\|}\\
&= \limsup_{\substack{(x,y,z)\to(x_k,y_k,z_k)\\ y\in G(x),\; z\in F(y)}} \frac{\langle x_k^*,x-x_k\rangle - \langle z_k^*,z-z_k\rangle}{\|(x,y,z)-(x_k,y_k,z_k)\|}\\
&\le \max\Big\{0,\; \limsup_{\substack{(x,z)\to(x_k,z_k)\\ z\in(F\circ G)(x)}} \frac{\langle x_k^*,x-x_k\rangle - \langle z_k^*,z-z_k\rangle}{\|(x,z)-(x_k,z_k)\|}\Big\} \le \varepsilon_k .
\end{aligned}$$

This gives $(x_k^*, 0, -z_k^*) \in \widehat{N}_{\varepsilon_k}((x_k, y_k, z_k); \operatorname{gph} \Phi)$ and justifies (i) by passing to the limit as $k \to \infty$.

To justify (ii), we proceed similarly to (i) and find, by the inner semicompactness of S at $(\bar{x}, \bar{z})$, a subsequence of $y_k \in S(x_k, z_k)$ that converges to some point $\bar{y}$. Since $y_k \in G(x_k) \cap F^{-1}(z_k)$ and the graphs of G and F^{-1} are closed at the corresponding points, we obtain that $\bar{y} \in G(\bar{x}) \cap F^{-1}(\bar{z}) = S(\bar{x}, \bar{z})$. Then the proof of (i) leads to the conclusion in (ii).

Let us finally prove (iii). In both cases there g is Lipschitz continuous around $\bar{x}$ with some modulus $\ell \geq 0$. Taking any (x^*, z^*) with $(x^*, 0) \in D^*\Phi(\bar{x}, g(\bar{x}), \bar{z})(z^*)$, we find sequences $\varepsilon_k \downarrow 0$, $(x_k, z_k) \to (\bar{x}, \bar{z})$, and $(x_k^*, y_k^*, z_k^*) \xrightarrow{w^*} (x^*, 0, z^*)$ such that $z_k \in F(g(x_k))$ and

$$\limsup_{\substack{x \to x_k,\, z \to z_k \\ z \in F(g(x))}} \frac{\langle (x_k^*, y_k^*, -z_k^*),\ (x, g(x), z) - (x_k, g(x_k), z_k) \rangle}{\|(x, g(x), z) - (x_k, g(x_k), z_k)\|} \leq \varepsilon_k$$

for all $k \in I\!N$. The latter implies

$$\limsup_{\substack{x \to x_k,\, z \to z_k \\ z \in F(g(x))}} \frac{\langle x_k^*, x - x_k \rangle - \langle z_k^*, z - z_k \rangle}{\|(x, z) - (x_k, z_k)\|} \leq \widetilde{\varepsilon}_k := (\ell + 1)(\varepsilon_k + \|y_k^*\|) \ .$$

If $\dim Y < \infty$, then $\widetilde{\varepsilon}_k \downarrow 0$ as $k \to \infty$, which proves (iii) in this case.

Assume now that g is strictly differentiable at $\bar{x}$. Following the proof of Theorem 1.38, we take an arbitrary sequence $\gamma_j \downarrow 0$ as $j \to \infty$ and derive from above that

$$\limsup_{\substack{x \to x_k,\, z \to z_k \\ z \in F(g(x))}} \frac{\langle x_{k_j}^* + \nabla g(\bar{x})^* y_{k_j}^*, x - x_{k_j} \rangle - \langle z_{k_j}^*, z - z_{k_j} \rangle}{\|(x, z) - (x_{k_j}, z_{k_j})\|} \leq \widetilde{\varepsilon}_j \ ,$$

where $\widetilde{\varepsilon}_j := (\ell + 1)(\varepsilon_{k_j} + \gamma_j \|y_{k_j}^*\|) \downarrow 0$ as $j \to \infty$. This implies

$$x_{k_j}^* + \nabla g(\bar{x})^* y_{k_j}^* \in \widehat{D}_{\widetilde{\varepsilon}_j}^*(F \circ g)(x_{k_j}, z_{k_j})(z_{k_j}^*)$$

and then $x^* \in D^*(F \circ g)(\bar{x}, \bar{z})(z^*)$, since $x_{k_j}^* + \nabla g(\bar{x})^* y_{k_j}^* \xrightarrow{w^*} x^*$ as $j \to \infty$.

It remains to justify the regularity statement in (iii). This easily follows from the equality proved in (iii) and the observation that

$$\widehat{D}^*(F \circ g)(\bar{x}, \bar{z})(z^*) = \big\{ x^* \in X^* \ \big|\ (x^*, 0) \in \widehat{D}^*\Phi(\bar{x}, g(\bar{x}), \bar{z})(z^*) \big\}$$

if g is locally Lipschitzian around $\bar{x}$. △

Note that the results of Theorem 1.64 provide the "right" inclusions and equalities for representing the coderivatives of compositions but not in a *chain rule* form, since they involve the coderivatives of the auxiliary multifunction Φ instead of the ones for F and G. To derive coderivative chain rules in this way, it suffices to employ a sum rule for representing the coderivatives of Φ. For now let us use the sum rule of Theorem 1.62(ii) available in arbitrary Banach spaces. Further results in this direction will be obtained in Chap. 3, where coderivative sum rules (and hence chain rules) will be established for general multifunctions in the Asplund space setting.

The following theorem gives parallel chain rules for the normal and mixed coderivatives of compositions. Observe, however, that just the *normal* coderivative of the inner mapping G is used in both cases. To simplify the notation, we omit the coderivative argument $z^* \in Z^*$ in chain rules.

Theorem 1.65 (coderivative chain rules with strictly differentiable outer mappings). *Let $G: X \rightrightarrows Y$, $f: Y \to Z$, and $\bar{z} \in (f \circ G)(\bar{x})$. The following hold for both coderivatives $D^* = D^*_N$ and $D^* = D^*_M$:*

(i) *Assume that $G \cap f^{-1}$ is inner semicontinuous at $(\bar{x}, \bar{z}, \bar{y})$ for some given $\bar{y} \in G(\bar{x})$ with $f(\bar{y}) = \bar{z}$ and that f is strictly differentiable at $\bar{y}$. Then*

$$D^*(f \circ G)(\bar{x}, \bar{z}) \subset D^*_N G(\bar{x}, \bar{y}) \circ \nabla f(\bar{y})^* .$$

(ii) *Assume that $G \cap f^{-1}$ is inner semicompact at $(\bar{x}, \bar{z})$, where G and f^{-1} are closed-graph at the corresponding points. Assume also that f is strictly differentiable at every $\bar{y} \in G(\bar{x}) \cap f^{-1}(\bar{z})$. Then*

$$D^*(f \circ G)(\bar{x}, \bar{z}) \subset \bigcup_{\bar{y} \in G(\bar{x}) \cap f^{-1}(\bar{z})} D^*_N G(\bar{x}, \bar{y}) \circ \nabla f(\bar{y})^* .$$

(iii) *Let $G = g$ be single-valued and either Lipschitz continuous around $\bar{x}$ with $\dim Y < \infty$ or strictly differentiable at this point. Then*

$$D^*_M(f \circ g)(\bar{x}) = D^*_N(f \circ g)(\bar{x}) = D^* g(\bar{x}) \circ \nabla f(g(\bar{x}))^* .$$

Moreover, $f \circ g$ is N-regular at $\bar{x}$ if g is N-regular at this point.

Proof. Follows from Theorem 1.64 by computing the coderivatives of Φ via the sum rule of Theorem 1.62(ii) and Proposition 1.33. △

Note that assertion (iii) of Theorem 1.65 ensures an *equality* chain rule for both normal and mixed coderivatives (which agree in this case) with *no regularity* assumptions on g unless g is strictly differentiable at $\bar{x}$. In the latter case this result reduces to the classical chain rule for compositions of strictly differentiable mappings between Banach spaces.

Next let us consider the case when the inner mapping g in the composition $F \circ g$ is strictly differentiable at the reference point. In this case we derive coderivative chain rules with *equalities* from the calculus results for normal cones in Subsect. 1.1.2. Similarly to Theorem 1.65, we don't impose any regularity assumptions on F but relate its graphical (normal and mixed) regularity with the corresponding regularity of the composition $F \circ g$.

Theorem 1.66 (coderivative chain rules with surjective derivatives of inner mappings). *Let $g: X \to Y$, $F: Y \rightrightarrows Z$, and $\bar{z} \in (F \circ g)(\bar{x})$. Assume that g is strictly differentiable at $\bar{x}$ with the surjective derivative $\nabla g(\bar{x})$. Then the following hold:*

$$\widehat{D}^*(F \circ g)(\bar{x}, \bar{z}) = \nabla g(\bar{x})^* \widehat{D}^* F(g(\bar{x}), \bar{z}) ,$$

$$D^*(F \circ g)(\bar{x}, \bar{z}) = \nabla g(\bar{x})^* D^* F(g(\bar{x}), \bar{z}) ,$$

where D^* stands either for D_N^* or for D_M^*. Moreover, $F \circ g$ is N-regular (resp. M-regular) at $(\bar{x}, \bar{z})$ if and only if F has the corresponding regularity property at $(g(\bar{x}), \bar{z})$.

Proof. Let I be the identity operator on Z. Then $(g, I): X \times Z \to Y \times Z$ is strictly differentiable at $(\bar{x}, \bar{z})$ with the surjective derivative $\nabla(g, I)(\bar{x}, \bar{z})$. One can easily observe that $(g, I)^{-1}(\text{gph}\, F) = \text{gph}(F \circ g)$. Thus the chain rules in the theorem for $\widehat{D}^*$ and $D^* = D_N^*$ follow from Corollary 1.15 and Theorem 1.17, respectively. To prove the chain rule for the case of $D^* = D_M^*$, we apply Lemma 1.16 to the set $(g, I)^{-1}(\text{gph}\, F)$ and then pass to the limit similarly to the proof of Theorem 1.17 using the strong convergence of $z_k^* \to z^*$ in the construction of mixed coderivatives for F and $F \circ g$. The regularity statements of the theorem follow from the chain rules obtained and the injectivity of $\nabla g(\bar{x})^*$; see Lemma 1.18. △

1.2.5 Sequential Normal Compactness of Mappings

In this subsection we consider sequential normal compactness properties of general multifunctions between Banach spaces. These properties, which are automatic in finite dimensions, play a crucial role in many aspects of infinite-dimensional variational analysis particularly related to furnishing limiting procedures and deriving efficient *pointbased* conditions for Lipschitzian behavior, metric regularity, generalized differential calculus, optimization, etc.; see the subsequent chapters of this book. In Subsect. 1.1.3 we have introduced and studied the sequential normal compactness property of arbitrary sets in Banach spaces. This naturally induces the corresponding property of set-valued mappings when applied to their graphs. However, the case of mappings allows us to consider also a weaker (less restrictive) property that exploits different convergences in domain and range spaces. The latter property, called "*partial* sequential normal compactness", is especially important for various results involving coderivatives. Here we study both properties of multifunctions in the framework of arbitrary Banach spaces and obtain efficient conditions for their fulfillment and preservation under some operations. A much richer calculus of sequential normal compactness is developed in Chap. 3 for mappings between Asplund spaces.

Definition 1.67 (sequential normal compactness of multifunctions). *Let $F: X \rightrightarrows Y$ with $(\bar{x}, \bar{y}) \in \text{gph}\, F$. Then:*

(i) *F is* SEQUENTIALLY NORMALLY COMPACT (SNC) *at $(\bar{x}, \bar{y})$ if for any sequence $(\varepsilon_k, x_k, y_k, x_k^*, y_k^*) \in [0, \infty) \times (\operatorname{gph} F) \times X^* \times Y^*$ satisfying*

$$\varepsilon_k \downarrow 0,\ (x_k, y_k) \to (\bar{x}, \bar{y}),\ x_k^* \in \widehat{D}^*_{\varepsilon_k} F(x_k, y_k)(y_k^*),\quad \text{and}\ (x_k^*, y_k^*) \xrightarrow{w^*} (0, 0)$$

one has $\|(x_k^*, y_k^*)\| \to 0$ as $k \to \infty$.

(ii) *F is* PARTIALLY SEQUENTIALLY NORMALLY COMPACT (PSNC) *at $(\bar{x}, \bar{y})$ if for any sequence $(\varepsilon_k, x_k, y_k, x_k^*, y_k^*) \in [0, \infty) \times (\operatorname{gph} F) \times X^* \times Y^*$ satisfying*

$$\varepsilon_k \downarrow 0,\ (x_k, y_k) \to (\bar{x}, \bar{y}),\ x_k^* \in \widehat{D}^*_{\varepsilon_k} F(x_k, y_k)(y_k^*),\ x_k^* \xrightarrow{w^*} 0,\quad \text{and}\ \|y_k^*\| \to 0$$

one has $\|x_k^*\| \to 0$ as $k \to \infty$.

We may omit $\bar{y}$ in the above definition if F is single-valued. Observe that the SNC property of a set-valued mapping agrees with the SNC property of its graph in the sense of Definition 1.20. Note also that the PSNC property always holds when $\dim X < \infty$. There is no difference between the two properties in Definition 1.67 if $\dim Y < \infty$, but otherwise the PSNC property is implied by the SNC one and may be strictly weaker even for linear continuous operators. The following proposition shows that the PSNC (but not SNC) property always holds for the important class of Lipschitz-like multifunctions, thanks to the necessary condition for such mappings in terms of ε-coderivatives obtained in Theorem 1.43. Moreover, in this case the PSNC property holds *around* $(\bar{x}, \bar{y})$, i.e., at any point (x, y) sufficiently close to $(\bar{x}, \bar{y})$.

Proposition 1.68 (PSNC property of Lipschitz-like multifunctions). *Let $F\colon X \rightrightarrows Y$ be locally Lipschitz-like around $(\bar{x}, \bar{y}) \in \operatorname{gph} F$. Then it is partially sequentially normally compact at this point.*

Proof. If follows from Theorem 1.43(i) and Definition 1.67(ii). △

Corollary 1.69 (SNC properties of single-valued mappings and their inverses). *Let $f\colon X \to Y$ be Lipschitz continuous around $\bar{x}$. Then:*

(i) *f is PSNC at $(\bar{x}, f(\bar{x}))$. Moreover, it is SNC at this point if* $\dim Y < \infty$.

(ii) *If f is strictly differentiable at $\bar{x}$ with the surjective derivative $\nabla f(\bar{x})$, then f^{-1} has the PSNC property around $(f(\bar{x}), \bar{x})$.*

Proof. Assertion (i) follows directly from Proposition 1.68. To prove (ii), we conclude from Corollary 1.59 that f^{-1} is Lipschitz-like around $(f(\bar{x}), \bar{x})$, and again apply the proposition. △

It will be proved in Subsect. 3.1.3 that the finite dimensionality condition $\dim Y < \infty$ is not only sufficient but also *necessary* for the SNC property of the so-called *w^*-strictly Lipschitzian* (in particular, strictly differentiable) mappings $f\colon X \to Y$ defined in Asplund spaces.

Another essential fact related to sequential normal compactness that will be established in Subsect. 3.1.3 is the PSNC property of inversions to *generalized Fredholm* operators important in applications to optimization problems with operator constraints and particularly to optimal control. Such generalized Fredholm operators are built upon some *compactly strictly Lipschitzian* mappings, which form a remarkable subclass of strictly Lipschitzian ones.

Next we establish some results on "calculus of sequential normal compactness" for mappings between Banach spaces. In what follows we obtain conditions ensuring that these properties are preserved under certain additions and compositions. Such results are naturally related to calculus rules for normal cones and coderivatives.

Theorem 1.70 (SNC properties under additions with strictly differentiable mappings). *Let $f\colon X \to Y$ be strictly differentiable at $\bar{x}$, and let $F\colon X \rightrightarrows Y$ be an arbitrary multifunction such that $\bar{y} - f(\bar{x}) \in F(\bar{x})$ for some $\bar{y} \in Y$. Then $f + F$ is SNC (resp. PSNC) at $(\bar{x}, \bar{y})$ if and only if F has the corresponding property at $(\bar{x}, \bar{y} - f(\bar{x}))$.*

Proof. Let us prove the "if" part of the theorem in a parallel way for both SNC and PSNC properties. Taking $x_k^* \in \widehat{D}^*_{\varepsilon_k}(f + F)(x_k, y_k)(y_k^*)$ for each $k \in I\!N$, one has from the definitions that

$$\langle x_k^*, x - x_k\rangle - \langle y_k^*, y - y_k\rangle \le 2\varepsilon_k(\|x - x_k\| + \|y - y_k\|)$$

for all $(x, y) \in \operatorname{gph}(f + F)$ sufficiently close to (x_k, y_k). Denote $\widetilde{y}_k := y_k - f(x_k)$. Now using the strict differentiability of f at $\bar{x}$ similarly to the proof of Theorem 1.38, we pick an arbitrary sequence $\gamma_j \downarrow 0$ as $j \to \infty$ and get

$$\langle x_{k_j}^* - \nabla f(\bar{x})^* y_{k_j}^*, x - x_{k_j}\rangle - \langle y_{k_j}^*, y - \widetilde{y}_{k_j}\rangle \le \widetilde{\varepsilon}_j(\|x - x_{k_j}\| + \|y - \widetilde{y}_{k_j}\|)$$

$$\text{with } \widetilde{\varepsilon}_j := (\ell + 1)(2\varepsilon_{k_j} + \gamma_j\|y_{k_j}^*\|)$$

for all $(x, y) \in \operatorname{gph} F$ sufficiently close to $(x_{k_j}, \widetilde{y}_{k_j})$ and $j \in I\!N$ sufficiently large, where ℓ is a Lipschitz constant of f around $\bar{x}$. This gives

$$x_{k_j}^* - \nabla f(\bar{x})^* y_{k_j}^* \in \widehat{D}^*_{\widetilde{\varepsilon}_j} F(x_{k_j}, \widetilde{y}_{k_j})(y_{k_j}^*)\ .$$

One can see that $\widetilde{\varepsilon}_j \downarrow 0$, $\widetilde{y}_{k_j} \to \bar{y} - f(\bar{x})$, and $x_{k_j}^* - \nabla f(\bar{x})^* y_{k_j}^* \overset{w^*}{\to} 0$ as $j \to \infty$ provided that $\varepsilon_k \downarrow 0$, $(x_k, y_k) \to (\bar{x}, \bar{y})$, and $(x_k^*, y_k^*) \overset{w^*}{\to} (0, 0)$ as $k \to \infty$. From here we easily conclude that the SNC (resp. PSNC) property of F at $(\bar{x}, \bar{y} - f(\bar{x}))$ implies the corresponding property of $f + F$ at $(\bar{x}, \bar{y})$. The opposite implication follows from the "if" part applied to $(f + F) + (-f)$. $\triangle$

Next let us consider the composition $F \circ G$ of set-valued mappings between Banach spaces. First we relate the sequential normal compactness properties of $F \circ G$ with the ones for the auxiliary multifunction $\Phi(x, y) = F(y) + \Delta((x, y); \operatorname{gph} G)$ with the indicator mapping $\Delta\colon X \times Y \to Z$ defined in Proposition 1.33.

Proposition 1.71 (SNC properties under compositions). *Let $G: X \rightrightarrows Y$, $F: Y \rightrightarrows Z$, and $\bar{z} \in (F \circ G)(\bar{x})$. Assume that the multifunction $G(x) \cap F^{-1}(z)$ is inner semicontinuous at $(\bar{x}, \bar{z}, \bar{y})$ for some $\bar{y} \in G(\bar{x}) \cap F^{-1}(\bar{z})$. Then $F \circ G$ is SNC (resp. PSNC) at $(\bar{x}, \bar{z})$ if Φ has the corresponding property at $(\bar{x}, \bar{y}, \bar{z})$.*

Proof. Take sequences $(\varepsilon_k, x_k, z_k, x_k^*, z_k^*) \in [0, \infty) \times X \times Z \times X^* \times Z^*$ with $\varepsilon_k \downarrow 0$, $(x_k, z_k) \to (\bar{x}, \bar{z})$, $(x_k^*, z_k^*) \xrightarrow{w^*} (0, 0)$,

$$z_k \in (F \circ G)(x_k), \quad \text{and} \quad x_k^* \in \widehat{D}^*_{\varepsilon_k}(F \circ G)(x_k, z_k)(z_k^*), \ k \in I\!N .$$

Using the inner semicontinuity of $G \cap F^{-1}$ at $(\bar{x}, \bar{z}, \bar{y})$ for the given $\bar{y}$, we find $y_k \in G(x_k) \cap F^{-1}(z_k)$ converging to $\bar{y}$. It was actually shown in the proof of Theorem 1.64(i) that

$$(x_k^*, 0) \in \widehat{D}^*_{\varepsilon_k} \Phi(x_k, y_k, z_k)(z_k^*) \ \text{ for all } \ k \in I\!N . \tag{1.44}$$

From here we can easily conclude that the SNC (resp. PSNC) property of Φ at $(\bar{x}, \bar{y}, \bar{z})$ implies the corresponding property of $F \circ G$ at $(\bar{x}, \bar{z})$. △

To obtain the SNC properties of $F \circ G$ in terms of the ones for F and G, one can proceed similarly to the proof of Theorem 1.65 employing a sum rule for Φ. However, this way is limited for the SNC calculus. The reason is that, due to Proposition 1.33, the indicator mapping $\Delta(\cdot; \Omega)$ is PSNC at $\bar{x} \in \Omega$ at $\bar{x}$ *if and only if* Ω is SNC at this point, and Δ is *never* SNC at $\bar{x}$ unless the image space is finite-dimensional. Combining therefore Proposition 1.71 and Theorem 1.70, we can only conclude that $f \circ G$ is PSNC if G is SNC and f is strictly differentiable at the corresponding points but cannot get any conclusions on the SNC property of $f \circ G$ when $\dim Z = \infty$. Better results are given in the next theorem based on a chain rule for ε-coderivatives.

Theorem 1.72 (SNC properties under compositions with strictly differentiable outer mappings). *Consider $G: X \rightrightarrows Y$, $f: Y \to Z$, and $\bar{z} \in (f \circ G)(\bar{x})$. Assume that $G \cap f^{-1}$ is inner semicontinuous at $(\bar{x}, \bar{z}, \bar{y})$ for some $\bar{y} \in G(\bar{x}) \cap f^{-1}(\bar{z})$, and that f is strictly differentiable at $\bar{y}$. The following assertions hold:*

(i) *If G is PSNC at $(\bar{x}, \bar{y})$, then the composition $f \circ G$ is PSNC at $(\bar{x}, \bar{z})$.*

(ii) *If G is SNC at $(\bar{x}, \bar{y})$ and $\nabla f(\bar{y})$ is surjective, then the composition $f \circ G$ is SNC at $(\bar{x}, \bar{z})$.*

Proof. Taking sequences $(\varepsilon_k, x_k, z_k, x_k^*, z_k^*)$ as in the proof of Proposition 1.71, we find $y_k \to \bar{y}$ such that $y_k \in G(x_k) \cap f^{-1}(z_k)$ and (1.44) holds with $\Phi(x, y) = f(y) + \Delta((x, y); \text{gph}\, G)$. Then we use the strict differentiability of f at $\bar{y}$ and, following the proof of Theorem 1.70, derive from (1.44) that

$$x_{k_j}^* \in \widehat{D}^*_{\widetilde{\varepsilon}_j} G(x_{k_j}, y_{k_j})(\nabla f(\bar{y})^* z_{k_j}^*) \ \text{ for all } \ j \in I\!N ,$$

where $\widetilde{\varepsilon}_j := (\ell + 1)(2\varepsilon_{k_j} + \gamma_j \|\nabla f(\bar{y})^* z_{k_j}^*\|)$, ℓ is a Lipschitz constant of f around $\bar{y}$, and $\gamma_j \downarrow 0$ as $j \to \infty$. The latter clearly implies that $\|x_{k_j}^*\| \to 0$ if

G is assumed to be PSNC at $(\bar{x}, \bar{y})$. If G is SNC at this point, then we have in addition that $\|\nabla f(\bar{y})^* z_{k_j}^*\| \to 0$. By Lemma 1.18 this yields $\|z_{k_j}^*\| \to 0$ as $j \to \infty$ provided that $\nabla f(\bar{y})$ is surjective.

We have proved both assertions (i) and (ii) of the theorem along a subsequence $\{k_j\}$ of the original sequence. This doesn't restrict the generality, since the original sequence was chosen arbitrarily. △

Note that the surjectivity assumption on $\nabla f(\bar{y})$ is essential for the validity of assertion (ii) in the theorem. Indeed, consider $G(x) \equiv X$ and $f(x) \equiv 0$. Then $(f \circ G)(x) \equiv 0$ is never SNC unless $\dim X < \infty$, although G is obviously SNC at every point.

Let us present an efficient corollary of Theorem 1.72 that ensures the SNC properties of compositions with Lipschitz-like inner mappings G.

Corollary 1.73 (SNC compositions with Lipschitz-like inner mappings). *Let $\bar{z} \in (f \circ G)(\bar{x})$. Fix $\bar{y} \in G(\bar{x}) \cap f^{-1}(\bar{z})$ and assume the following: G is locally Lipschitz-like around $(\bar{x}, \bar{y})$, f is strictly differentiable at $\bar{y}$, and $G \cap f^{-1}$ is inner semicontinuous at $(\bar{x}, \bar{z}, \bar{y})$. Then $f \circ G$ is PSNC at $(\bar{x}, \bar{z})$. Moreover, $f \circ G$ is SNC at this point if* $\dim Y < \infty$ *and $\nabla f(\bar{y})$ is surjective.*

Proof. Follows from the theorem due to Proposition 1.68. △

The next result concerns the SNC properties of compositions in which outer mappings are arbitrary but inner mappings are strictly differentiable with surjective derivatives. It turns out that both properties in Definition 1.67 are *invariant* under such compositions.

Theorem 1.74 (SNC properties under compositions with strictly differentiable inner mappings). *Let $g\colon X \to Y$, $F\colon Y \rightrightarrows Z$, and $\bar{z} \in (F \circ g)(\bar{x})$. Assume that g is strictly differentiable at $\bar{x}$ with the surjective derivative $\nabla g(\bar{x})$. Then $F \circ g$ is SNC (resp. PSNC) at $(\bar{x}, \bar{z})$ if and only if F has the corresponding property at $(g(\bar{x}), \bar{x})$.*

Proof. We have observed in the proof of Theorem 1.66 that

$$\operatorname{gph}(F \circ g) = (g, I)^{-1}(\operatorname{gph} F)\,,$$

where I is the identity operator on Z. Since $\nabla(g, I)(\bar{x}, \bar{z})$ is surjective, the equivalence between the SNC property of $F \circ g$ and the one for F follows directly from Theorem 1.22. The proof of the equivalence in the case of PSNC is similar based on Lemma 1.16. △

The calculus results obtained above allow us to establish the sequential normal compactness properties of set-valued mappings built upon "basic" SNC and PSNC mappings via various compositions. We know from Theorem 1.26 and Proposition 1.68 that the SNC and PSNC properties are inherent in sets and mappings possessing a kind of local Lipschitzian behavior. Let us present a PSNC analog of Theorem 1.26 for the case of mappings that are just "partial" CEL.

A set-mapping $F\colon X \rightrightarrows Y$ is said to be *partially compactly epi-Lipschitzian* around $(\bar{x}, \bar{y}) \in \operatorname{gph} F$ (relative to X) if there are neighborhoods U of $(\bar{x}, \bar{y})$ and O of the origin in X, as well as a number $\gamma > 0$ and a compact set $C \subset X \times Y$ such that

$$(\operatorname{gph} F) \cap U + t(O \times \{0\}) \subset \operatorname{gph} F + tC \tag{1.45}$$

for all $t \in (0, \gamma)$. Note that this property is intrinsically defined in terms of the given mapping F with no use of generalized differential constructions.

One can see that (1.45), which is a partial counterpart of the CEL property in Definition 1.24, always holds when $\dim X < \infty$. Observe also that the partial CEL property is different from the Lipschitz-like property of set-valued mappings in Definition 1.40. Let us show, similarly to Theorem 1.26, that the partial CEL property always implies the PSNC property (even a *stronger* version of it; see Definition 3.3 and the subsequent discussion) for general multifunctions between Banach spaces.

Theorem 1.75 (PSNC property of partial CEL mappings). *Let $F\colon X \rightrightarrows Y$ be partially compactly epi-Lipschitzian around $(\bar{x}, \bar{y}) \in \operatorname{gph} F$. Then for any sequence $(\varepsilon_k, x_k, y_k, x_k^*, y_k^*) \in [0, \infty) \times (\operatorname{gph} F) \times X^* \times Y^*$ satisfying*

$$\varepsilon_k \downarrow 0,\ (x_k, y_k) \to (\bar{x}, \bar{y}),\ x_k^* \in \widehat{D}^*_{\varepsilon_k} F(x_k, y_k)(y_k^*),\ \text{ and } \ (x_k^*, y_k^*) \xrightarrow{w^*} (0, 0)$$

one has $\|x_k^*\| \to 0$ as $k \to \infty$. In particular, F has the PSNC property at the reference point $(\bar{x}, \bar{y})$.

Proof. Fix $\eta > 0$ such that $B_\eta(\bar{x}, \bar{y}) \subset U$ and $\eta I\!B \subset O$ for the neighborhoods in (1.45). Taking any sequence $(\varepsilon_k, x_k, y_k, x_k^*, y_k^*)$ in the theorem, we have

$$(x_k^*, -y_k^*) \in \widehat{N}_{\varepsilon_k}((x_k, y_k); \operatorname{gph} F) \ \text{ with } \ (x_k, y_k) \in (\operatorname{gph} F) \cap B_\eta(\bar{x}, \bar{y})$$

for big $k \in I\!N$. Now using (1.45) for each fixed k, we find sequences $t_j \downarrow 0$ and $c_j \in C$ such that

$$(x_k, y_k) + t_j \eta(e, 0) - t_j c_j \in \operatorname{gph} F \ \text{ for all } \ e \in I\!B, \quad j \in I\!N\ .$$

Since C is compact, we may assume that c_j converges to some $\bar{c} \in C$ as $j \to \infty$. It is easy to conclude from the construction of ε_k-normals that

$$\big\langle (x_k^*, y_k^*),\ (\eta e, 0) - \bar{c} \big\rangle \le \varepsilon_k \big(\|(\eta e, 0) - \bar{c}\| \big)\ .$$

This gives

$$\eta\|x_k^*\| \le \max_{c\in C} \langle (x_k^*, y_k^*), c\rangle + \varepsilon_k(\alpha+\eta) ,$$

where $\alpha := \max_{c\in C} \|c\|$. The latter implies that $\|x_k^*\| \to 0$ as $k \to \infty$, since $\varepsilon_k \downarrow 0$ and $\langle (x_k^*, y_k^*), c\rangle \to 0$ uniformly in $c \in C$ due to $(x_k^*, y_k^*) \xrightarrow{w^*} (0,0)$ and the compactness of C. △

1.3 Subdifferentials of Nonsmooth Functions

This section is devoted to generalized differential properties of *extended-real-valued functions* $\varphi\colon X \to \overline{I\!R} := [-\infty, \infty]$ defined on arbitrary Banach spaces. Given a point $\bar{x} \in X$ at which the function φ is finite but may not admit a classical derivative/gradient $\varphi'(\bar{x}) = \nabla\varphi(\bar{x}) \in X^*$, we consider *subgradient sets*, called usually "subdifferentials", for φ at $\bar{x}$ that provide set-valued extensions of derivative operators for nondifferentiable functions.

Extended-real-valued functions are particularly convenient for applications to constrained optimization problems and allow one to incorporate constraints into cost functionals. Dealing with *minimization* problems, we mostly concern *lower* generalized differential properties of nonsmooth functions described by sets of lower subgradients called (lower) subdifferentials. For some significant applications (including those to minimization problems) we also need to consider upper generalized differential properties of nonsmooth functions in the framework of unilateral/one-sided variational analysis. Such upper properties for φ, related to lower ones for $-\varphi$, can be conveniently described via collections of upper subgradients for φ at $\bar{x}$ that are sometimes called "superdifferentials." In what follows we employ the terminology of *subgradients* and *subdifferentials* (omitting, as a rule, the adjective "lower") in the case of lower generalized differential constructions, while *upper subgradients* and *upper subdifferentials* are used for their upper counterparts. We'll pay the main attention to the study of lower subdifferential constructions whose properties symmetrically induce the ones for upper subgradients. As already mentioned, there are important issues in variational analysis and optimization that require both lower and upper subgradients; see, e.g., mean value results in Chap. 3 and applications to nonsmooth minimization problems in Chap. 5.

Having in mind lower properties of $\varphi\colon X \to \overline{I\!R}$, we say that φ is *proper* if $\varphi(x) > -\infty$ for all $x \in X$ and its *domain*

$$\operatorname{dom}\varphi := \{x \in X \,|\, \varphi(x) < \infty\}$$

is nonempty. With any φ we associate its *epigraph* and *hypergraph*

$$\operatorname{epi}\varphi := \{(x,\alpha) \in X\times I\!R \,|\, \alpha \ge \varphi(x)\},\quad \operatorname{hypo}\varphi := \{(x,\alpha) \in X\times I\!R \,|\, \alpha \le \varphi(x)\} .$$

Obviously $\operatorname{gph}\varphi = \operatorname{epi}\varphi \cap \operatorname{hypo}\varphi$. One can easily see that local closedness of the epigraph, hypergraph, and graph around $(\bar{x}, \varphi(\bar{x}))$ corresponds to the local lower semicontinuity, upper semicontinuity, and continuity of φ around $\bar{x}$, respectively. Recall that φ is *lower semicontinuous* (*l.s.c.*) *at* a point $\bar{x}$ with $|\varphi(\bar{x})| < \infty$ if

$$\varphi(\bar{x}) \le \liminf_{x \to \bar{x}} \varphi(x) .$$

We say that φ is l.s.c. *around* $\bar{x}$ when it is l.s.c. at any point of some neighborhood of $\bar{x}$. The *upper semicontinuity* (*u.s.c.*) of φ is defined symmetrically from the lower semicontinuity of $-\varphi$. The *continuity* of φ at $\bar{x}$ means that φ is l.s.c. and u.s.c. at this point simultaneously. Throughout the book we use the notation

$$x \xrightarrow{\varphi} \bar{x} \Longleftrightarrow x \to \bar{x} \;\text{ with }\; \varphi(x) \to \varphi(\bar{x}) ,$$

where $\varphi(x) \to \varphi(\bar{x})$ is superfluous if φ is continuous at $\bar{x}$.

1.3.1 Basic Definitions and Relationships

Developing a geometric approach to the generalized differentiation of extended-real-valued functions, we define our main subdifferential constructions through basic normals to epigraphs. Then we study their relationships with coderivatives and discuss some important properties obtained in this way. First let us describe basic normals to epigraphical sets.

Proposition 1.76 (basic normals to epigraphs). *Let* $\varphi\colon X \to \overline{I\!R}$ *with* $(\bar{x}, \bar{\alpha}) \in \operatorname{epi}\varphi$. *Then* $\lambda \ge 0$ *for every* $(x^*, -\lambda) \in N((\bar{x}, \bar{\alpha}); \operatorname{epi}\varphi)$, *and so there are uniquely defined subsets* D *and* D^∞ *of* X^* *such that*

$$N((\bar{x}, \varphi(\bar{x})); \operatorname{epi}\varphi) = \big\{(\lambda(x^*, -1)\big|\, x^* \in D,\ \lambda > 0\big\} \cup \big\{(x^*, 0)\big|\, x^* \in D^\infty\big\} .$$

Proof. Taking any $(x^*, -\lambda) \in N((\bar{x}, \bar{\alpha}); \operatorname{epi}\varphi)$ and using Definition 1.1, we find sequences $\varepsilon_k \downarrow 0$, $(x_k, \alpha_k) \overset{\operatorname{epi}\varphi}{\to} (\bar{x}, \bar{\alpha})$, $x_k^* \overset{w^*}{\to} x^*$, and $\lambda_k \to \lambda$ such that

$$\limsup_{(x,\alpha) \overset{\operatorname{epi}\varphi}{\to} (x_k, \alpha_k)} \frac{\langle x_k^*, x - x_k\rangle - \lambda_k(\alpha - \alpha_k)}{\|(x, \alpha) - (x_k, \alpha_k)\|} \le \varepsilon_k$$

for all $k \in I\!N$. Letting $x = x_k$ and then $k \to \infty$, we get $\lambda \ge 0$, which implies the above representation. $\triangle$

The set D in Proposition 1.76 characterizes "sloping" normals to the epigraph, while D^∞ is the collection of "horizontal" normals. We take these sets as the definitions of the (lower) basic and singular subdifferentials of φ at $\bar{x}$, respectively.

Definition 1.77 (basic and singular subdifferentials). *Consider a function* $\varphi\colon X \to \overline{I\!R}$ *and a point* $\bar{x} \in X$ *with* $|\varphi(\bar{x})| < \infty$.

(i) *The set*

$$\partial\varphi(\bar{x}) := \big\{x^* \in X^*\big|\ (x^*, -1) \in N((\bar{x}, \varphi(\bar{x})); \operatorname{epi}\varphi)\big\}$$

is the (basic, limiting) SUBDIFFERENTIAL *of φ at $\bar{x}$, and its elements are* BASIC SUBGRADIENTS *of φ at this point. We put $\partial\varphi(\bar{x}) := \emptyset$ if $|\varphi(\bar{x})| = \infty$.*

(ii) *The set*

$$\partial^\infty\varphi(\bar{x}) := \left\{x^* \in X^* \middle|\ (x^*, 0) \in N((\bar{x}, \varphi(\bar{x})); \operatorname{epi}\varphi)\right\}$$

is the SINGULAR SUBDIFFERENTIAL *of φ at $\bar{x}$, and its elements are* SINGULAR SUBGRADIENTS *of φ at this point. We put $\partial^\infty\varphi(\bar{x}) := \emptyset$ if $|\varphi(\bar{x})| = \infty$.*

Thus we define the basic and singular subdifferentials of an extended-real-valued function through basic normals to its epigraph. Below we show that the basic subdifferential agrees with the classical gradient for strictly differentiable functions as well as with the subdifferential of convex analysis when φ is convex. The singular subdifferential occurs to be useful for the study of non-Lipschitzian functions. As we'll see below, both subdifferential constructions in Definition 1.77 enjoy rich calculi and valuable applications for general classes of nonsmooth functions reflecting their lower generalized differentiability properties. Following the tradition in convex analysis, we skip here the minus sign in the lower subdifferential notation $\partial = \partial^-$ (in contrast to some previous work, e.g., Mordukhovich [901, 909]) but keep the plus sign for the corresponding upper subdifferentials, which are defined through basic normals to hypergraphs and reflect upper generalized differential properties of nonsmooth functions.

Definition 1.78 (upper subgradients). *Given $\varphi: X \to \overline{I\!R}$ and $\bar{x} \in X$ with $|\varphi(\bar{x})| < \infty$, we define the (basic, limiting)* UPPER SUBDIFFERENTIAL *of φ at $\bar{x}$ and the* SINGULAR UPPER SUBDIFFERENTIAL *of φ at $\bar{x}$ by*

$$\partial^+\varphi(\bar{x}) := \left\{x^* \in X^* \middle|\ (-x^*, 1) \in N((\bar{x}, \varphi(\bar{x})); \operatorname{hypo}\varphi)\right\},$$

$$\partial^{\infty,+}\varphi(\bar{x}) := \left\{x^* \in X^* \middle|\ (-x^*, 0) \in N((\bar{x}, \varphi(\bar{x})); \operatorname{hypo}\varphi)\right\},$$

respectively. We put $\partial^+\varphi(\bar{x}) = \partial^{\infty,+}\varphi(\bar{x}) = \emptyset$ if $|\varphi(\bar{x})| = \infty$.

If φ is concave, $\partial^+\varphi(\bar{x})$ reduces to the classical upper subdifferential of convex analysis. Note that $\partial\varphi$ and $\partial^+\varphi$ may be considerably different even in the case of convex and concave functions. The simplest example is given by $\varphi(x) = -|x|$ at $\bar{x} = 0 \in I\!R$, where

$$\partial\varphi(0) = \{-1, 1\} \quad \text{while} \quad \partial^+\varphi(0) = [-1, 1].$$

Note that the first set in *nonconvex*, which is typical for both lower and upper subdifferential constructions introduced.

One can easily observe that

$$\partial^+\varphi(\bar{x}) = -\partial(-\varphi)(\bar{x}) \quad \text{and} \quad \partial^{\infty,+}\varphi(\bar{x}) = -\partial^\infty(-\varphi)(\bar{x}).$$

In some cases (in particular, for mean value results involving nonsmooth functions) one needs to consider the union of the corresponding lower and upper subdifferentials

$$\partial^0\varphi(\bar{x}) := \partial\varphi(\bar{x}) \cup \partial^+\varphi(\bar{x}), \quad \partial^{\infty,0}\varphi(\bar{x}) := \partial^\infty\varphi(\bar{x}) \cup \partial^{\infty,+}\varphi(\bar{x}) \tag{1.46}$$

called the *symmetric subdifferential* and the *singular symmetric subdifferential* of φ at $\bar{x}$, respectively. Note that

$$\partial^0(-\varphi)(\bar{x}) = -\partial^0\varphi(\bar{x}) \text{ and } \partial^{\infty,0}(-\varphi)(\bar{x}) = -\partial^{\infty,0}\varphi(\bar{x}) ,$$

which means that, in contrast to the one-sided lower and upper subdifferential constructions from Definitions 1.77 and 1.78, the symmetric subdifferential and singular symmetric subdifferential in (1.46) possess the classical two-sided symmetry. In what follows we mostly confine ourselves to the study of (lower) subdifferential properties that obviously induce the corresponding results for the upper and symmetric subdifferentials.

Let us start with computing subgradients for *indicator functions* of arbitrary sets. For this class of extended-real-valued functions both subdifferentials in Definition 1.77 reduce to the basic normal cone.

Proposition 1.79 (subdifferentials of indicator functions). *Consider a nonempty set $\Omega \subset X$ and its indicator function $\delta(\cdot;\Omega)\colon X \to \overline{I\!R}$ defined by*

$$\delta(x;\Omega) := 0 \text{ if } x \in \Omega \text{ and } \delta(x;\Omega) := \infty \text{ if } x \notin \Omega .$$

Than for any $\bar{x} \in \Omega$ one has

$$\partial\delta(\bar{x};\Omega) = \partial^\infty\delta(\bar{x};\Omega) = N(\bar{x};\Omega) .$$

Proof. This follows from the definitions and Proposition 1.2 applied to $\operatorname{epi}\delta(\cdot;\Omega) = \Omega \times [0,\infty)$. △

Next let us consider relationships between subgradients and coderivatives. Given $\varphi\colon X \to \overline{I\!R}$, we associate with it the *epigraphical multifunction* E_φ from X into $I\!R$ defined by

$$E_\varphi(x) := \big\{\alpha \in I\!R \big|\ \alpha \ge \varphi(x)\big\} .$$

Since E_φ takes values in $I\!R$, there is no difference between its normal and mixed coderivatives in Definition 1.32; as usual, we denote this common (basic) coderivative by D^*. Note that $\operatorname{gph} E_\varphi = \operatorname{epi}\varphi$. Thus, for every $\bar{x}$ where φ is finite, we can equivalently define the basic and singular subdifferentials of φ at $\bar{x}$ through the coderivative of E_φ:

$$\partial\varphi(\bar{x}) = D^*E_\varphi(\bar{x},\varphi(\bar{x}))(1) \text{ and } \partial^\infty\varphi(\bar{x}) = D^*E_\varphi(\bar{x},\varphi(\bar{x}))(0) . \tag{1.47}$$

This allows us to derive some results for subdifferentials of extended-real-valued functions from those obtained for coderivatives of set-valued mappings. On the other hand, we can consider the coderivative $D^*\varphi(\bar{x})$ of a single-valued mapping $\varphi\colon X \to \overline{I\!R}$ provided that φ is finite around $\bar{x}$. The following theorem establishes links between this coderivative and (basic and singular) subgradients of continuous functions.

Theorem 1.80 (subdifferentials from coderivatives of continuous functions). *Let* $\varphi\colon X \to \overline{I\!R}$ *be continuous around* $\bar{x}$. *Then*

$$\partial\varphi(\bar{x}) = D^*\varphi(\bar{x})(1) \quad\textit{and}\quad \partial^\infty\varphi(\bar{x}) \subset D^*\varphi(\bar{x})(0) .$$

Proof. Observe that the continuity of φ around $\bar{x}$ implies that the set $\operatorname{epi}\varphi$ is closed and $\operatorname{gph}\varphi = \operatorname{bd}(\operatorname{epi}\varphi)$ near $(\bar{x}, \varphi(\bar{x}))$. Thus the inclusions

$$\partial\varphi(\bar{x}) \subset D^*\varphi(\bar{x})(1) \quad\text{and}\quad \partial^\infty\varphi(\bar{x}) \subset D^*\varphi(\bar{x})(0)$$

follow from the fact that for any closed set $\Omega \subset X$ in a Banach space one has

$$N(\bar{x}; \Omega) \subset N(\bar{x}; \operatorname{bd}\Omega) \quad\text{at every}\quad \bar{x} \in \operatorname{bd}\Omega .$$

To prove this, we take $0 \neq x^* \in N(\bar{x}; \Omega)$ and find sequences $\varepsilon_k \downarrow 0$, $x_k \xrightarrow{\Omega} \bar{x}$, and $x_k^* \xrightarrow{w^*} x^*$ such that $x_k^* \in \widehat{N}_{\varepsilon_k}(x_k; \Omega)$ for all $k \in I\!N$. Since the norm $\|\cdot\|$ on X^* is weak* lower semicontinuous, we have

$$\liminf_{k\to\infty} \|x_k^*\| \geq \|x^*\| > 0 ,$$

which implies that $x_k \notin \operatorname{int}\Omega$ for large k due to the construction (1.2). Thus $x_k \in \operatorname{bd}\Omega$ for such $k \in I\!N$. Now using (1.5), we conclude that $x_k^* \in \widehat{N}_{\varepsilon_k}(x_k; \operatorname{bd}\Omega)$, and hence $x^* \in N(\bar{x}; \operatorname{bd}\Omega)$.

To complete the proof of the theorem, it remains to show that

$$(x^*, -1) \in N((\bar{x}, \varphi(\bar{x})); \operatorname{gph}\varphi) \Longrightarrow (x^*, -1) \in N((\bar{x}, \varphi(\bar{x})); \operatorname{epi}\varphi) .$$

Take $(x^*, -1) \in N((\bar{x}, \varphi(\bar{x})); \operatorname{gph}\varphi)$ and find by definition sequences $\varepsilon_k \downarrow 0$, $x_k \to \bar{x}$, $x_k^* \xrightarrow{w^*} x^*$, and $\lambda_k \to -1$ such that $(x^*, \lambda_k) \in \widehat{N}_{\varepsilon_k}((x_k, \varphi(x_k)); \operatorname{gph}\varphi)$ for all $k \in I\!N$. Without loss of generality we let $\lambda_k = -1$. Our goal is to show that $(x_k^*, -1) \in \widehat{N}_{\varepsilon_k}((x_k, \varphi(x_k)); \operatorname{epi}\varphi)$.

Suppose that the latter doesn't hold for some $k \in I\!N$ fixed in what follows. Then there is $0 < \gamma < 1 - \varepsilon_k$ and sequences $(u_j, \alpha_j) \xrightarrow{\operatorname{epi}\varphi} (x_k, \varphi(x_k))$ as $j \to \infty$ satisfying the relation

$$\langle x_k^*, u_j - x_k\rangle + (\varphi(x_k) - \alpha_j) > (\varepsilon_k + \gamma)\|(u_j, \alpha_j) - (x_k, \varphi(x_k))\|, \quad j \in I\!N .$$

Since $\alpha_j \geq \varphi(u_j)$ and $\varphi(u_j) \to \varphi(x_k)$ as $j \to \infty$, we have

$$\|(u_j - x_k, \varphi(u_j) - \varphi(x_k))\| \leq \|(u_j - x_k, \alpha_j - \varphi(x_k))\| + \alpha_j - \varphi(u_j)$$

and therefore

$$\langle x_k^*, u_j - x_k\rangle + \varphi(x_k) - \varphi(u_j) > (\varepsilon_k + \gamma)\|(u_j, \varphi(u_j)) - (x_k, \varphi(x_k))\|$$

for all $j \in I\!N$, which means that $(x_k^*, -1) \notin \widehat{N}_{\varepsilon_k}((x_k, \varphi(x_k)); \operatorname{gph}\varphi)$. Thus we arrive at a contradiction and complete the proof of the theorem. △

Note that the inclusion $\partial^\infty\varphi(\bar{x}) \subset D^*\varphi(\bar{x})(0)$ may be *strict* for continuous functions. An example is provided by the function

$$\varphi(x) := \begin{cases} -x^{1/3} \text{ if } x \geq 0 , \\ 0 \qquad \text{otherwise} . \end{cases} \tag{1.48}$$

Employing representation (1.9) from Theorem 1.6, we compute

$$N((0,0); \operatorname{epi}\varphi) = \big\{(v,0)\big|\ v \leq 0\big\} \cup \big\{(0,v)\big|\ v \leq 0\big\}$$

and $N((0,0); \operatorname{gph}\varphi) = N((0,0); \operatorname{epi}\varphi) \cup I\!R^2_+$. Thus $\partial^\infty\varphi(0) = (-\infty, 0]$ and $D^*\varphi(0)(0) = (-\infty, \infty)$.

Corollary 1.81 (subdifferentials of Lipschitzian functions). *Let φ be Lipschitz continuous around $\bar{x}$ with modulus $\ell \geq 0$. Then*

$$\partial^\infty\varphi(\bar{x}) = \{0\} \ \ \textit{and} \ \ \|x^*\| \leq \ell \ \ \textit{for all} \ \ x^* \in \partial\varphi(\bar{x}) .$$

Proof. Using Theorem 1.44 for the locally Lipschitzian mapping $F = \varphi\colon X \to I\!R$, we have $D^*\varphi(\bar{x})(0) = \{0\}$ and $\|D^*\varphi(\bar{x})\| \leq \ell$. This directly implies the results of the corollary due to Theorem 1.80. △

Note that $\partial\varphi(0) = \{0\}$ in the case of function (1.48), which is continuous but not locally Lipschitzian around $\bar{x} = 0$. This shows that the *local Lipschitz continuity is not necessary for the boundedness of the basic subdifferential.*

It is easy to check that locally Lipschitzian functions on finite-dimensional spaces have *at least one* basic subgradient at the point in question. Indeed, it follows from Theorem 1.6 that $N(\bar{x}; \Omega) \neq \{0\}$ if $\bar{x} \in \operatorname{bd}\Omega$ for closed sets $\Omega \subset I\!R^n$, in particular, for $\Omega = \operatorname{epi}\varphi$ at graphical points of continuous functions. This implies by Proposition 1.76 that in finite dimensions the nontriviality condition $\partial^\infty\varphi(\bar{x}) = \{0\}$ yields $\partial\varphi(\bar{x}) \neq \emptyset$, which is always the case for locally Lipschitzian functions due to Corollary 1.81. The Lipschitz condition is essential here; cf. the continuous function $\varphi(x) = x^{1/3}$ on $I\!R$ with $\partial\varphi(0) = \partial^+\varphi(0) = \emptyset$. In arbitrary Banach spaces one may have $\partial\varphi(\bar{x}) = \emptyset$ for locally Lipschitzian functions, but it never happens in the case of *Asplund spaces*; see Corollary 2.25 in Subsect. 2.2.3. We'll also see that in Asplund spaces the condition $\partial^\infty\varphi(\bar{x}) = \{0\}$ is not only necessary but also *sufficient* for the local Lipschitzian property of l.s.c. functions satisfying a certain sequential normal compactness assumption, which is automatics in finite dimensions.

It follows from (1.46) and Corollary 1.81 that

$$\partial^{\infty,0}\varphi(\bar{x}) = \{0\} \ \text{ and } \ \|x^*\| \leq \ell \ \text{ for all } \ x^* \in \partial^0\varphi(\bar{x})$$

if φ is Lipschitz continuous around $\bar{x}$. Another useful corollary of Theorem 1.80 concerns strictly differentiable functions.

Corollary 1.82 (subdifferentials of strictly differentiable functions). *Let φ be strictly differentiable at $\bar{x}$. Then*

$$\partial\varphi(\bar{x}) = \partial^+\varphi(\bar{x}) = \partial^0\varphi(\bar{x}) = \{\nabla\varphi(\bar{x})\}\ .$$

Proof. Follows from Theorem 1.80 and Theorem 1.38 applied to the mapping $f = \varphi\colon X \to I\!R$, and the constructions of $\partial^+\varphi(\bar{x})$ and $\partial^0\varphi(\bar{x})$. △

Note that $\partial\varphi(\bar{x})$ may be a singleton for continuous functions that are not strictly differentiable at $\bar{x}$ as, e.g., in (1.48). The latter is not possible for locally Lipschitzian functions on Asplund spaces; see Chap. 3. On the other hand, $\varphi\colon I\!R \to I\!R$ may be Lipschitz continuous and *differentiable* at $\bar{x}$, but *not strictly* differentiable at this point, while both $\partial\varphi(\bar{x})$ and $\partial^+\varphi(\bar{x})$ are not singletons. Such an example is given by the function

$$\varphi(x) := \begin{cases} x^2\sin(1/x) & \text{if } x \neq 0\ , \\ 0 & \text{if } x = 0\ , \end{cases} \tag{1.49}$$

where $\nabla\varphi(0) = 0$ and $\partial\varphi(0) = \partial^+\varphi(0) = [-1, 1]$.

1.3.2 Fréchet-Like ε-Subgradients and Limiting Representations

Now we consider two kinds of (Fréchet-like) ε-subdifferentials of extended-real-valued functions that provide convenient approximating tools for the study of our basic subdifferential constructions in Banach spaces.

Definition 1.83 (ε-subgradients). *Let $\varphi\colon X \to \overline{I\!R}$ be finite at a point $\bar{x}$, and let $\varepsilon \geq 0$.*

(i) *The set*

$$\widehat{\partial}_{g\varepsilon}\varphi(\bar{x}) := \big\{x^* \in X^*\big|\ (x^*, -1) \in \widehat{N}_\varepsilon((\bar{x}, \varphi(\bar{x})); \operatorname{epi}\varphi)\big\}$$

is the GEOMETRIC ε-SUBDIFFERENTIAL *of φ at $\bar{x}$ with elements called* GEOMETRIC ε-SUBGRADIENTS *of φ at $\bar{x}$. We put $\widehat{\partial}_{g\varepsilon}\varphi(\bar{x}) := \emptyset$ if $|\varphi(\bar{x})| = \infty$.*

(ii) *The set*

$$\widehat{\partial}_{a\varepsilon}\varphi(\bar{x}) := \Big\{x^* \in X^*\Big|\ \liminf_{x\to\bar{x}} \frac{\varphi(x) - \varphi(\bar{x}) - \langle x^*, x - \bar{x}\rangle}{\|x - \bar{x}\|} \geq -\varepsilon\Big\}\ ,$$

also denoted by $\widehat{\partial}_\varepsilon\varphi(\bar{x})$, is the ANALYTIC ε-SUBDIFFERENTIAL *of φ at $\bar{x}$ with elements called* ANALYTIC ε-SUBGRADIENTS *of φ at $\bar{x}$. We put $\widehat{\partial}_{a\varepsilon}\varphi(\bar{x}) := \emptyset$ if $|\varphi(\bar{x})| = \infty$.*

One can easily see that both ε-subdifferentials are *convex* for an arbitrary function $\varphi\colon X \to \overline{I\!R}$ whenever $\varepsilon \geq 0$. However, these sets may be *empty*, when ε is sufficiently small, even for simple Lipschitzian functions on $I\!R$ as, e.g.,

$\varphi(x) = -|x|$ at $\bar{x} = 0$. As for ε-normals in Subsect. 1.1.1, we observe that both ε-subdifferentials are *norm-closed* in X^*; hence they are *weakly closed* if the space X is reflexive.

Directly from the definitions we get the following descriptions of geometric ε-subgradients of φ via ε-coderivatives of the epigraphical multifunction E_φ and analytic ε-subgradients of φ via minimization of an auxiliary function.

Proposition 1.84 (descriptions of ε-subgradients). *For any $\varphi: X \to \overline{I\!R}$ finite at $\bar{x}$ and any $\varepsilon \geq 0$ one has:*

(i) $\widehat{\partial}_{g\varepsilon}\varphi(\bar{x}) = \widehat{D}^*_\varepsilon E_\varphi(\bar{x}, \varphi(\bar{x}))(1)$.

(ii) $x^* \in \widehat{\partial}_{a\varepsilon}\varphi(\bar{x})$ *if and only if for every $\gamma > 0$ the function*

$$\psi(x) := \varphi(x) - \varphi(\bar{x}) - \langle x^*, x - \bar{x}\rangle + (\varepsilon + \gamma)\|x - \bar{x}\|$$

attains a local minimum at $\bar{x}$.

This implies useful estimates for ε-subgradients as well as for horizontal ε-normals to epigraphs of locally Lipschitzian functions.

Proposition 1.85 (ε-subgradients of locally Lipschitzian functions). *Let $\varphi: X \to \overline{I\!R}$ be finite around $\bar{x}$, and let $\varepsilon \geq 0$. The following hold:*

(i) *φ is Lipschitz continuous around $\bar{x}$ if and only if E_φ is Lipschitz-like around $(\bar{x}, \varphi(\bar{x}))$.*

(ii) *If φ is Lipschitz continuous around $\bar{x}$ with modulus $\ell \geq 0$, then there is $\eta > 0$ such that*

$$\|x^*\| \leq \varepsilon(1+\ell) \ \text{ whenever } \ (x^*, 0) \in \widehat{N}_\varepsilon((x, \varphi(x)); \operatorname{epi}\varphi), \quad x \in \bar{x} + \eta I\!B\,,$$

$$\|x^*\| \leq \ell + \varepsilon(1+\ell) \ \text{ whenever } \ x^* \in \widehat{\partial}_{g\varepsilon}\varphi(x), \quad x \in \bar{x} + \eta I\!B\,,$$

$$\|x^*\| \leq \ell + \varepsilon \ \text{ whenever } \ x^* \in \widehat{\partial}_{a\varepsilon}\varphi(x), \quad x \in \bar{x} + \eta I\!B\,.$$

Proof. Assertion (i) is derived from the definitions. To justify the first two estimates in (ii), we apply Theorem 1.43(i) for ε-coderivatives of epigraphical multifunctions. The last estimate in (ii) follows directly from Proposition 1.84(ii) and the local Lipschitz continuity of φ around $\bar{x}$. △

One can check that for the indicator functions $\varphi(x) = \delta(x; \Omega)$ both geometric and analytic ε-subdifferentials at $\bar{x} \in \Omega$ reduce to the set of ε-normals to Ω at this point:

$$\widehat{\partial}_{g\varepsilon}\delta(\bar{x}; \Omega) = \widehat{\partial}_{a\varepsilon}\delta(\bar{x}; \Omega) = \widehat{N}_\varepsilon(\bar{x}; \Omega) \ \text{ for all } \ \varepsilon \geq 0\,. \tag{1.50}$$

The following theorem establishes relationships between geometric and analytic ε-subgradients in the general case of extended-real-valued functions.

Theorem 1.86 (relationships between ε-subgradients). *Let $\varphi\colon X \to \overline{I\!R}$ with $|\varphi(\bar{x})| < \infty$. Then*

$$\widehat{\partial}_{a\varepsilon}\varphi(\bar{x}) \subset \widehat{\partial}_{g\varepsilon}\varphi(\bar{x}) \quad \textit{for all} \quad \varepsilon \geq 0 \,.$$

Conversely, if $x^* \in \widehat{\partial}_{g\varepsilon}\varphi(\bar{x})$ for some $0 \leq \varepsilon < 1$, then

$$x^* \in \widehat{\partial}_{a\tilde{\varepsilon}}\varphi(\bar{x}) \quad \textit{with} \quad \tilde{\varepsilon} := \varepsilon(1 + \|x^*\|)/(1-\varepsilon) \,.$$

Proof. Pick $x^* \in \widehat{\partial}_{a\varepsilon}\varphi(\bar{x})$ and show that $(x^*, -1) \in \widehat{N}_\varepsilon((\bar{x}, \varphi(\bar{x})); \operatorname{epi}\varphi)$ for each $\varepsilon \geq 0$. Using Proposition 1.84(ii), for any $\gamma > 0$ we find a neighborhood U of $\bar{x}$ such that

$$\varphi(x) - \varphi(\bar{x}) - \langle x^*, x - \bar{x}\rangle \geq -(\varepsilon + \gamma)\|x - \bar{x}\| \quad \text{for all} \quad x \in U \,.$$

This immediately implies that

$$\langle x^*, x - \bar{x}\rangle + \varphi(\bar{x}) - \alpha \leq (\varepsilon + \gamma)\|(x, \alpha) - (\bar{x}, \varphi(\bar{x}))\|$$

if $x \in U$ and $\alpha \geq \varphi(x)$, which means that the function

$$\psi(x, \alpha) := \langle x^*, x - \bar{x}\rangle - (\alpha - \varphi(\bar{x})) - (\varepsilon + \gamma)\|(x, \alpha) - (\bar{x}, \varphi(\bar{x}))\|$$

attains a local maximum relative to the set $\Omega := \operatorname{epi}\varphi$ at $(\bar{x}, \varphi(\bar{x}))$. Employing Proposition 1.28, we conclude that $x^* \in \widehat{\partial}_{g\varepsilon}\varphi(\bar{x})$.

To prove the converse inclusion in the theorem, fix $\varepsilon \geq 0$ and assume on the contrary that $x^* \notin \widehat{\partial}_{a\tilde{\varepsilon}}\varphi(\bar{x})$ with the specified $\tilde{\varepsilon}$. Then there are $\gamma > 0$ and a sequence $x_k \to \bar{x}$ such that

$$\varphi(x_k) - \varphi(\bar{x}) - \langle x^*, x_k - \bar{x}\rangle + (\tilde{\varepsilon} + \gamma)\|x_k - \bar{x}\| < 0 \quad \text{for all} \quad k \in I\!N \,.$$

Letting $\alpha_k := \varphi(\bar{x}) + \langle x^*, x_k - \bar{x}\rangle - (\tilde{\varepsilon} + \gamma)\|x_k - \bar{x}\|$, we observe that $\alpha_k \to \varphi(\bar{x})$ as $k \to \infty$ and that $(x_k, \alpha_k) \in \operatorname{epi}\varphi$ for all $k \in I\!N$. This yields

$$\begin{aligned}\frac{\langle x^*, x_k - \bar{x}\rangle - (\alpha_k - \varphi(\bar{x}))}{\|(x_k, \alpha_k) - (\bar{x}, \varphi(\bar{x}))\|} &= \frac{(\tilde{\varepsilon} + \gamma)\|x_k - \bar{x}\|}{\|(x_k - \bar{x}), \langle x^*, x_k - \bar{x}\rangle - (\tilde{\varepsilon} + \gamma)\|x_k - \bar{x}\|)\|} \\ &\geq \frac{\tilde{\varepsilon} + \gamma}{1 + \|x^*\| + (\tilde{\varepsilon} + \gamma)} > \frac{\tilde{\varepsilon}}{1 + \|x^*\| + \tilde{\varepsilon}} = \varepsilon\end{aligned}$$

for all $k \in I\!N$ due to $\gamma > 0$ and the choice of $\tilde{\varepsilon}$. The latter clearly implies that $(x^*, -1) \notin \widehat{N}_\varepsilon((\bar{x}, \varphi(\bar{x})); \operatorname{epi}\varphi)$, which means that $x^* \notin \widehat{\partial}_{g\varepsilon}\varphi(\bar{x})$ and completes the proof of the theorem. △

It follows from Theorem 1.86 that for $\varepsilon = 0$ both sets of geometric and analytic subgradient in Definition 1.83 reduce to the same set of *Fréchet* (*lower*) *subgradients* $\widehat{\partial}\varphi(\bar{x}) := \widehat{\partial}_0\varphi(\bar{x})$ expressed (when $|\varphi(\bar{x})| < \infty$) either in the geometric form $(x^*, -1) \in \widehat{N}((\bar{x}, \varphi(\bar{x})); \operatorname{epi} \varphi)$ via the prenormal cone $\widehat{N}$ or analytically by

$$\widehat{\partial}\varphi(\bar{x}) = \Big\{x^* \in X^* \Big|\ \liminf_{x \to \bar{x}} \frac{\varphi(x) - \varphi(\bar{x}) - \langle x^*, x - \bar{x}\rangle}{\|x - \bar{x}\|} \geq 0\Big\}. \tag{1.51}$$

This set is called the *presubdifferential* or *Fréchet subdifferential* of φ at $\bar{x}$.

Symmetrically to Definition 1.83 we can define the corresponding upper constructions, which reduce for $\varepsilon = 0$ to the *Fréchet upper subdifferential* $\widehat{\partial}^+\varphi(\bar{x}) := -\widehat{\partial}(-\varphi)(\bar{x})$ of φ at $\bar{x}$ with $|\varphi(\bar{x})| < \infty$ described by

$$\widehat{\partial}^+\varphi(\bar{x}) = \Big\{x^* \in X^* \Big|\ \limsup_{x \to \bar{x}} \frac{\varphi(x) - \varphi(\bar{x}) - \langle x^*, x - \bar{x}\rangle}{\|x - \bar{x}\|} \leq 0\Big\}. \tag{1.52}$$

Note that the sets $\widehat{\partial}\varphi(\bar{x})$ and $\widehat{\partial}^+\varphi(\bar{x})$ may be empty simultaneously for continuous functions on $I\!R$, e.g., for $\varphi(x) = x^{1/3}$ at $\bar{x} = 0$. Furthermore, the following useful observation holds as a direct consequence of definitions (1.51), (1.52), and (1.14).

Proposition 1.87 (subgradient description of Fréchet differentiability). *Let $\varphi\colon X \to \overline{I\!R}$ with $|\varphi(\bar{x})| < \infty$. Then $\widehat{\partial}\varphi(\bar{x}) \neq \emptyset$ and $\widehat{\partial}^+\varphi(\bar{x}) \neq \emptyset$ if and only if φ is Fréchet differentiable at $\bar{x}$, in which case $\widehat{\partial}\varphi(\bar{x}) = \widehat{\partial}^+\varphi(\bar{x}) = \{\nabla\varphi(\bar{x})\}$.*

Therefore, when one of the sets $\widehat{\partial}\varphi(\bar{x})$ and $\widehat{\partial}^+\varphi(\bar{x})$ is not a singleton, the other is empty. This distinguishes the latter constructions from the basic ones $\partial\varphi(\bar{x})$ and $\partial^+\varphi(\bar{x})$, which are nonempty simultaneously for every locally Lipschitzian functions on $I\!R^n$ (actually on any Asplund spaces). In contrast to the symmetric subdifferential $\partial^0\varphi(\bar{x})$ in (1.46), the union $\widehat{\partial}\varphi(\bar{x}) \cup \widehat{\partial}^+\varphi(\bar{x})$ always reduces to either $\widehat{\partial}\varphi(\bar{x})$ or $\widehat{\partial}^+\varphi(\bar{x})$. Note that φ may *not* be Fréchet differentiable at $\bar{x}$ while $\widehat{\partial}\varphi(\bar{x})$ is a *singleton*. A simple example is provided by the function

$$\varphi(x) := \begin{cases} \max\{0, x\sin(1/x)\} & \text{if } x \neq 0, \\ 0 & \text{if } x = 0, \end{cases}$$

where $\widehat{\partial}\varphi(0) = \{0\}$ and $\widehat{\partial}^+\varphi(0) = \emptyset$.

The next theorem, which is a subdifferential counterpart of Theorem 1.30, provides important variational descriptions of Fréchet subgradients of nonsmooth functions in terms of smooth supports. The corresponding notation and terminology are introduced at the beginning of Subsect. 1.1.4.

Theorem 1.88 (variational descriptions of Fréchet subgradients). *For every proper function $\varphi\colon X \to \overline{I\!R}$ finite at $\bar{x}$ the following hold:*

(i) *Given* $x^* \in X^*$, *we assume that there is a function* $s: U \to I\!R$ *defined on a neighborhood of* $\bar{x}$ *and Fréchet differentiable at* $\bar{x}$ *such that* $\nabla s(\bar{x}) = x^*$ *and* $\varphi(x) - s(x)$ *achieves a local minimum at* $\bar{x}$. *Then* $x^* \in \widehat{\partial}\varphi(\bar{x})$. *Conversely, for every* $x^* \in \widehat{\partial}\varphi(\bar{x})$ *there is a function* $s: X \to I\!R$ *with* $s(\bar{x}) = \varphi(\bar{x})$ *and* $s(x) \le \varphi(x)$ *whenever* $x \in X$ *such that* $s(\cdot)$ *is Fréchet differentiable at* $\bar{x}$ *with* $\nabla s(\bar{x}) = x^*$.

(ii) *Assume that* X *admits an* $\mathcal{S}$*-smooth bump function, where* $\mathcal{S}$ *stands for one of the classes* $\mathcal{F}$, $\mathcal{LF}$, *or* $\mathcal{LC}^1$. *Then for every* $x^* \in \widehat{\partial}\varphi(\bar{x})$ *there is a function* $s: U \to I\!R$ *defined and* $\mathcal{S}$*-smooth on a neighborhood of* $\bar{x}$ *such that* $\nabla s(\bar{x}) = x^*$ *and*

$$\varphi(x) - s(x) - \|x - \bar{x}\|^2 \ge \varphi(\bar{x}) - s(\bar{x}) \quad \text{for all } x \in U \,, \tag{1.53}$$

where $s(\cdot)$ *can be chosen to be concave if* X *admits a Fréchet smooth renorm. In the latter case we can take* $U = X$ *if* φ *is bounded from below.*

(iii) *Let* $x^* \in \widehat{\partial}\varphi(\bar{x})$, *where* φ *is bounded from below on the space* X *admitting an* $\mathcal{S}$*-smooth bump function of one the types listed above. Then there is a bump function* $b: X \to I\!R$ *such that* $\nabla b(\bar{x}) = x^*$ *and*

$$\varphi(x) - b(x) \ge \varphi(\bar{x}) - b(\bar{x}) \quad \text{for all } x \in X \,.$$

Furthermore, under the assumptions made there are $\mathcal{S}$*-smooth functions* $s: X \to I\!R$ *and* $\theta: X \to [0, \infty)$ *such that* $\nabla s(\bar{x}) = x^*$, $\theta(x) = 0$ *only for* $x = 0$, $\theta(x) \le \|x\|^2$ *for* $\|x\| \le 1$, *and*

$$\varphi(x) - s(x) - \theta(x - \bar{x}) \ge \varphi(\bar{x}) - s(\bar{x}) \quad \text{for all } x \in X \,. \tag{1.54}$$

Proof. Assertion (i) follows from Theorem 1.30(i) due to the above geometric description of Fréchet subgradients.

To prove (ii) in the case of smooth bumps, we observe that the condition $x^* \in \widehat{\partial}\varphi(\bar{x})$ implies the existence of $r \in (0, 1)$ such that φ is bounded from below on the ball $B_{2r}(\bar{x})$. Letting

$$\rho(t) := \sup\big\{\varphi(\bar{x}) - \varphi(x) + \langle x^*, x - \bar{x}\rangle \big|\, x \in X,\ \|x - \bar{x}\| \le t\big\}, \quad t \ge 0\,,$$

we observe that $\rho(t) < \infty$ for $t \in [0, r]$. Then $\widetilde{\rho}(t) := \min\{\rho(t), \rho(r)\}$ satisfies the assumptions of Lemma 1.29 due to the definition of Fréchet subgradients. Let τ and d be the functions built, respectively, in this lemma from $\rho := \widetilde{\rho}$ and in the proof of Theorem 1.30 from the given $\mathcal{S}$-smooth bump on X. Putting

$$s(x) := -\tau(d(x - \bar{x})) - d^2(x - \bar{x}) + \varphi(\bar{x}) + \langle x^*, x - \bar{x}\rangle \,,$$

one can check that it has the properties listed in (ii) with $U := \operatorname{int} B_r(\bar{x})$. If X admits a Fréchet smooth renorm $\|\cdot\|$, we get $d(x) = \|x\|$, which implies the concavity of $s(x)$ and that the support inequality (1.53) holds globally if φ is bounded from below on X.

The proof of (iii) is similar to the one in the last part of Theorem 1.30; we refer the reader to the proof of Theorem 4.6 in Fabian and Mordukhovich [419] for more details. △

Note that estimates (1.53) and (1.54) imply that $\varphi(x)-s(x)$ achieves its minimum (local and global, respectively) *uniquely* at $\bar{x}$ with the following *well-posedness* property:

$$\|x_k-\bar{x}\|\to 0 \text{ whenever } \varphi(x_k)-s(x_k)\to\varphi(\bar{x})-s(\bar{x}) \text{ as } k\to\infty\,.$$

Representations of basic subgradients via ε-subgradients and Fréchet subgradients of extended-real-valued functions are given by the following theorem.

Theorem 1.89 (limiting representations of basic subgradients). *Let* $\varphi\colon X\to\overline{I\!R}$ *with* $|\varphi(\bar{x})|<\infty$. *Then*

$$\partial\varphi(\bar{x})=\mathop{\mathrm{Lim\,sup}}_{\substack{x\xrightarrow{\varphi}\bar{x}\\ \varepsilon\downarrow 0}}\widehat{\partial}_{g\varepsilon}\varphi(x)=\mathop{\mathrm{Lim\,sup}}_{\substack{x\xrightarrow{\varphi}\bar{x}\\ \varepsilon\downarrow 0}}\widehat{\partial}_{a\varepsilon}\varphi(x)\,. \tag{1.55}$$

Moreover, when φ *is l.s.c. around* $\bar{x}$ *and* $\dim X<\infty$ *one has*

$$\partial\varphi(\bar{x})=\mathop{\mathrm{Lim\,sup}}_{x\xrightarrow{\varphi}\bar{x}}\widehat{\partial}\varphi(x)\,. \tag{1.56}$$

Proof. The first representation in (1.55) follows from Definition 1.1 and 1.83. This immediately implies the inclusion "$\supset$" in the second representation of (1.55) due to $\widehat{\partial}_{a\varepsilon}\varphi(x)\subset\widehat{\partial}_{g\varepsilon}\varphi(x)$ in Theorem 1.86. To prove the opposite inclusion, we pick $x^*\in\partial\varphi(\bar{x})$ and find $\varepsilon_k\downarrow 0$, $x_k\xrightarrow{\varphi}\bar{x}$, and $x_k^*\xrightarrow{w^*}x^*$ with $x_k^*\in\widehat{\partial}_{g\varepsilon_k}\varphi(x_k)$ for all $k\in I\!N$. It follows from the second part of Theorem 1.86 that $x_k^*\in\widehat{\partial}_{a\tilde{\varepsilon}_k}\varphi(x_k)$ with $\tilde{\varepsilon}_k:=\varepsilon_k(1+\|x_k^*\|)/(1-\varepsilon_k)$. Since the sequence $\{x_k^*\}$ is bounded in X^*, we have $\tilde{\varepsilon}_k\downarrow 0$ as $k\to\infty$, which justifies the second representation in (1.55). Representation (1.56) follows, under the assumptions made, from the normal cone representation (1.8) in Theorem 1.6. △

We'll see in Subsect. 2.4.1 that the subdifferential representation (1.56) holds in any Asplund spaces and, moreover, it characterizes this class of Banach spaces. Since Fréchet subgradients are usually easier to compute for typical nonsmooth functions, representation (1.56) is convenient for calculating basic subgradients. For example, let us consider the function

$$\varphi(x):=|x_1|-|x_2|,\quad x=(x_1,x_2)\in I\!R^2\,, \tag{1.57}$$

which is Lipschitz continuous on $I\!R^2$ and differentiable at every $x\in I\!R^2$ with $x_1x_2\neq 0$. One has $\nabla\varphi(x)\in\big\{(1,1),\ (1,-1),\ (-1,1),\ (-1,-1)\big\}$ for any such x. It is easy to calculate Fréchet subgradients from their analytic description given in (1.51):

$$\widehat{\partial}\varphi(x) = \begin{cases} (1,-1) & \text{if } x_1 > 0,\ x_2 > 0\ , \\ (-1,-1) & \text{if } x_1 < 0,\ x_2 > 0\ , \\ (-1,1) & \text{if } x_1 < 0,\ x_2 < 0\ , \\ (1,1) & \text{if } x_1 > 0,\ x_2 < 0\ , \\ \{(v,-1)|\ -1 \le v \le 1\} & \text{if } x_1 = 0,\ x_2 > 0\ , \\ \{(v,1)|\ -1 \le v \le 1\} & \text{if } x_1 = 0,\ x_2 < 0\ , \\ \emptyset & \text{if } x_2 = 0\ . \end{cases}$$

By Theorem 1.89 we get

$$\partial\varphi(0) = \big\{(v,1)\big|\ -1 \le v \le 1\big\} \cup \big\{(v,-1)\big|\ -1 \le v \le 1\big\}\ .$$

Similarly one can calculate Fréchet upper subgradients from (1.52) and, using the upper counterpart of (1.56), compute the basic upper subdifferential as

$$\partial^+\varphi(0) = \big\{(-1,v)\big|\ -1 \le v \le 1\big\} \cup \big\{(1,v)\big|\ -1 \le v \le 1\big\}\ .$$

Hence the symmetric subdifferential $\partial^0\varphi(0) = \partial\varphi(0) \cup \partial^+\varphi(0)$ is this case is the boundary of the unit square in $I\!R^2$.

In general Banach space setting one cannot removed $\varepsilon > 0$ from the subdifferential representations (1.55), which are crucial for the validity of many important results. To illustrate this, let us use (1.55) for establishing links between the *mixed* coderivative (1.25) of single-valued mappings $f\colon X \to Y$ between arbitrary Banach spaces and basic subgradients of their *scalarization*

$$\langle y^*, f\rangle(x) := \langle y^*, f(x)\rangle, \quad y^* \in Y^*\ . \tag{1.58}$$

Theorem 1.90 (scalarization of the mixed coderivative). *Let $f\colon X \to Y$ be continuous around $\bar{x}$. Then*

$$\partial\langle y^*, f\rangle(\bar{x}) \subset D^*_M f(\bar{x})(y^*) \ \ \textit{for all}\ \ y^* \in Y^*\ .$$

If in addition f is Lipschitz continuous around $\bar{x}$, then

$$D^*_M f(\bar{x})(y^*) = \partial\langle y^*, f\rangle(\bar{x}) \ \ \textit{for all}\ \ y^* \in Y^*\ .$$

Proof. Let $x^* \in \partial\langle y^*, f\rangle(\bar{x})$. Using (1.55), we find sequences $\varepsilon_k \downarrow 0$, $x_k \to \bar{x}$, and $x_k^* \stackrel{w^*}{\to} x^*$ with $x_k^* \in \widehat{\partial}_{a\varepsilon_k}\langle y^*, f\rangle(x_k)$ for $k \in I\!N$. Due to Definition 1.83(ii) for each k there is a neighborhood U_k of x_k such that

$$\langle y^*, f\rangle(x) - \langle y^*, f\rangle(x_k) - \langle x_k^*, x - x_k\rangle \ge -2\varepsilon_k\|x - x_k\| \ \ \text{when}\ \ x \in U_k\ .$$

The latter implies that

$$\limsup_{x\to x_k}\frac{\langle x_k^*, x-x_k\rangle-\langle y^*, f(x)-f(x_k)\rangle}{\|(x-x_k, f(x)-f(x_k))\|}\le 2\varepsilon_k\ ,$$

and hence $(x_k^*,-y^*)\in\widehat{N}_{2\varepsilon_k}((x_k, f(x_k));\operatorname{gph} f)$ for each $k\in I\!N$. This gives $x^*\in D_M^*f(\bar{x})(y^*)$ due to the coderivative definitions in (1.23) and (1.25), which completes the proof of the theorem.

To prove the opposite inclusion, we pick $x^*\in D_M^*f(\bar{x})(y^*)$ and find sequences $\varepsilon_k\downarrow 0$, $x_k\to\bar{x}$, $x_k^*\stackrel{w^*}{\to}x^*$, and $y_k^*\to y^*$ such that $(x_k^*,-y_k^*)\in\widehat{N}_{\varepsilon_k}((x_k, f(x_k));\operatorname{gph} f)$ for $k\in I\!N$. Hence

$$\langle x_k^*, x-x_k\rangle-\langle y_k^*, f(x)-f(x_k)\rangle\le 2\varepsilon_k(1+\ell)\|x-x_k\|\ \text{ for all }\ x\in x_k+\eta_k I\!B$$

with some sequence $\eta_k\downarrow 0$, where $\ell>0$ is a Lipschitz constant of f around $\bar{x}$. The latter yields

$$x_k^*\in\widehat{\partial}_{a\tilde{\varepsilon}_k}\langle y^*, f\rangle(x_k)\ \text{ with }\ \tilde{\varepsilon}_k:=2\varepsilon_k(1+\ell)+\ell\|y_k^*-y^*\|\ .$$

Since $\|y_k^*-y^*\|\to 0$, we have $\tilde{\varepsilon}_k\downarrow 0$ as $k\to\infty$, and hence $x^*\in\partial\langle y^*, f\rangle(\bar{x})$ due to (1.55). $\triangle$

Example 1.35 shows that a similar scalarization formula *doesn't* hold for the *normal* coderivative (1.24) of Lipschitzian mappings with values in Hilbert spaces. In Subsect. 3.1.3 we obtain such a normal scalarization under additional assumptions on Lipschitzian mappings defined on Asplund spaces.

It immediately follows from Theorem 1.89 that $\widehat{\partial}\varphi(\bar{x})\subset\partial\varphi(\bar{x})$ for every function $\varphi\colon X\to\overline{I\!R}$ on a Banach space X. This inclusion is often *strict*, which may happen even for Fréchet differentiable functions on $I\!R$; see, e.g., (1.49) with $\widehat{\partial}\varphi(0)=\{0\}$ and $\partial\varphi(0)=[-1,1]$. The case of equality in the latter inclusion signifies some "lower regularity" of φ at $\bar{x}$ expressed in terms of subdifferentials. The next definition describes two modifications of lower subdifferential regularity for extended-real-valued functions.

Definition 1.91 (lower regularity of functions). *Let $\varphi\colon X\to\overline{I\!R}$ be finite at $\bar{x}$. Then:*

(i) *φ is* LOWER REGULAR *at $\bar{x}$ if $\partial\varphi(\bar{x})=\widehat{\partial}\varphi(\bar{x})$.*

(ii) *φ is* EPIGRAPHICALLY REGULAR *at $\bar{x}$ if the set* $\operatorname{epi}\varphi\subset X\times I\!R$ *is normally regular at $(\bar{x},\varphi(\bar{x}))$.*

Similarly we define *upper regularity* of φ at $\bar{x}$ by $\partial^+\varphi(\bar{x})=\widehat{\partial}^+\varphi(\bar{x})$ and *hypergraphical regularity* of φ at this point via normal regularity from Definition 1.4 applied to the hypergraph of φ at $(\bar{x},\varphi(\bar{x}))$. As usual, we mainly deal with lower regularity properties that symmetrically induce the corresponding upper ones.

Proposition 1.92 (lower regularity relationships).

(i) *Let* $\Omega \subset X$ *with* $\bar{x} \in \Omega$. *Then both lower regularity and epigraphical regularity of the indicator function* $\delta(\cdot;\Omega)$ *at* $\bar{x}$ *are equivalent to the normal regularity of* Ω *at this point.*

(ii) *Let* $\varphi\colon X \to \overline{I\!R}$ *with* $|\varphi(\bar{x})| < \infty$. *Then* φ *is epigraphically regular at* $\bar{x}$ *if and only if it is lower regular at* $\bar{x}$ *and*

$$\partial^\infty \varphi(\bar{x}) = \widehat{\partial}^\infty \varphi(\bar{x}) := \big\{x^* \in X^* \big|\, (x^*,0) \in \widehat{N}((\bar{x},\varphi(\bar{x}));\operatorname{epi}\varphi)\big\} .$$

Thus epigraphical regularity and lower regularity of φ *at* $\bar{x}$ *are equivalent if* φ *is Lipschitz continuous around* $\bar{x}$.

Proof. Assertion (i) follows directly from the definitions, Proposition 1.79, and formulas (1.50) as $\varepsilon = 0$. To prove assertion (ii), observe similarly to Proposition 1.76 that

$$\widehat{N}((\bar{x},\varphi(\bar{x}));\operatorname{epi}\varphi) = \big\{\lambda(x^*,-1)\big|\, x^* \in \widehat{\partial}\varphi(\bar{x}),\ \lambda > 0\big\} \cup \big\{(x^*,0)\big|\, x^* \in \widehat{\partial}^\infty\varphi(\bar{x})\big\} .$$

This clearly implies the first part of (ii). The second part of (ii) follows from Corollary 1.81, which ensures that $\partial^\infty\varphi(\bar{x}) = \widehat{\partial}^\infty\varphi(\bar{x}) = \{0\}$ for locally Lipschitzian functions. △

Note that lower regularity of φ at $\bar{x}$ may be less restrictive than its epigraphical regularity as for the function $\varphi\colon I\!R \to I\!R$ given by

$$\varphi(x) := \begin{cases} -\sqrt{x - 1/n} & \text{if } 1/n \le x < 1/n + 1/n^4,\ n \in I\!N\,, \\ 0 & \text{otherwise}\,. \end{cases}$$

One can check that this function is Fréchet differentiable at $\bar{x} = 0$ with $\partial\varphi(0) = \widehat{\partial}\varphi(0) = \widehat{\partial}^\infty\varphi(0) = \{0\}$ and $\partial^\infty\varphi(0) = (-\infty,0]$.

If $\varphi\colon X \to \overline{I\!R}$ is *convex*, its epigraphical regularity follows directly from Proposition 1.5 applied to the convex set $\Omega := \operatorname{epi}\varphi$. The next theorem gives more detailed descriptions of ε-subgradients and basic (lower and upper) subgradients for convex functions.

Theorem 1.93 (subgradients of convex functions). *Let* $\varphi\colon X \to \overline{I\!R}$ *be convex and finite at* $\bar{x}$. *Then for every* $\varepsilon \ge 0$ *one has the following representations of the* ε*-subdifferentials:*

$$\widehat{\partial}_{g\varepsilon}\varphi(\bar{x}) = \Big\{x^* \in X^* \Big|\, \langle x^*, x - \bar{x}\rangle \le \varphi(x) - \varphi(\bar{x}) + \varepsilon\big(\|x - \bar{x}\| + |\varphi(x) - \varphi(\bar{x})|\big) \quad \textit{whenever } x \in X\Big\},$$

$$\widehat{\partial}_{a\varepsilon}\varphi(\bar{x}) = \Big\{x^* \in X^* \Big|\, \langle x^*, x - \bar{x}\rangle \le \varphi(x) - \varphi(\bar{x}) + \varepsilon\|x - \bar{x}\| \quad \textit{whenever } x \in X\Big\} \tag{1.59}$$

Furthermore, φ is epigraphically regular at $\bar{x}$ and

$$\partial^0\varphi(\bar{x}) = \partial\varphi(\bar{x}) = \big\{x^* \in X^* \big|\ \langle x^*, x - \bar{x}\rangle \le \varphi(x) - \varphi(\bar{x}) \ \textit{ for all } \ x \in X\big\}\,.$$

Proof. The representation of geometric ε-subgradients follows from Proposition 1.3 with $\Omega = \operatorname{epi}\varphi$ and representation (1.59) of analytic ones due to $\widehat{\partial}_{a\varepsilon}\varphi(\bar{x}) \subset \widehat{\partial}_{g\varepsilon}\varphi(\bar{x})$. The inclusion "$\supset$" in (1.59) is obvious. To justify the opposite inclusion, pick an arbitrary subgradient $x^* \in \widehat{\partial}_{a\varepsilon}\varphi(\bar{x})$ and, employing the local variational description of analytic ε-subgradients from Proposition 1.84(ii), conclude that for any given $\eta > 0$ the function

$$\psi(x) := \varphi(x) - \varphi(\bar{x}) - \langle x^*, x - \bar{x}\rangle + (\varepsilon + \eta)\|x - \bar{x}\|$$

attains a local minimum at $\bar{x}$. Since ψ is convex, $\bar{x}$ happens to be its *global* minimizer. Hence

$$\psi(x) = \varphi(x) - \varphi(\bar{x}) - \langle x^*, x - \bar{x}\rangle + (\varepsilon + \eta)\|x - \bar{x}\| \ge \psi(\bar{x}) = 0$$

for all $x \in X$. Taking into account that $\eta > 0$ was chosen arbitrarily, we get (1.59). Using now (1.55) and then representation (1.59) at points $x_k \xrightarrow{\varphi} \bar{x}$ with $\varepsilon_k \downarrow 0$, we arrive at

$$\partial\varphi(\bar{x}) = \big\{x^* \in X^* \big|\ \langle x^*, x - \bar{x}\rangle \le \varphi(x) - \varphi(\bar{x}) \ \text{ whenever } \ x \in X\big\}\,.$$

It remains to show that $\partial^+\varphi(\bar{x}) \subset \partial\varphi(\bar{x})$ for any convex function finite at $\bar{x}$. To furnish this, we observe that if $\widehat{\partial}^+_{a\varepsilon}\varphi(x) := -\widehat{\partial}_{a\varepsilon}(-\varphi)(x) \ne \emptyset$ for some $x \in X$ and $\varepsilon > 0$, then φ is bounded from above around x. It implies, for convex functions, that φ is continuous and subdifferentiable at this point in the sense of convex analysis, which gives $\widehat{\partial}\varphi(x) \ne \emptyset$ due to (1.59). Since $\widehat{\partial}^+_{a\varepsilon}\varphi(x) \subset \widehat{\partial}\varphi(x) + \varepsilon I\!B^*$, the inclusion $\partial^+\varphi(\bar{x}) \subset \partial\varphi(\bar{x})$ follows now from (1.55) and its upper counterpart. △

Note that the set on the right-hand side of (1.59) is the subdifferential of the convex function $\varphi(x) + \varepsilon\|x - \bar{x}\|$ at $\bar{x}$. By the classical Moreau-Rockafellar theorem this set is equal to $\partial\varphi(\bar{x}) + \varepsilon I\!B^*$ for any proper convex function $\varphi\colon X \to \overline{I\!R}$. Observe that for $\varepsilon > 0$ the latter set is *different* from the standard ε-subdifferential/approximate subdifferential of convex analysis defined as the collection of $x^* \in X^*$ satisfying

$$\langle x^*, x - \bar{x}\rangle \le \varphi(x) - \varphi(\bar{x}) + \varepsilon \ \text{ for all } \ x \in X\,;$$

see, e.g., Hiriart-Urruty and Lemaréchal [575].

Symmetrically, *concave* functions $\varphi\colon X \to \overline{I\!R}$ are hypergraphically (hence upper) regular at every point where they are finite, and their upper subgradients satisfy an upper counterpart of Theorem 1.93. Note that the lower and upper regularity under consideration are clearly notions of *unilateral analysis.*

In particular, a locally Lipschitzian function φ on a finite-dimensional space (actually on any Asplund space) *cannot* be simultaneously lower and upper regular at the reference point $\bar{x}$ unless it is Fréchet differentiable at $\bar{x}$. It easily follows from Proposition 1.87 and from the fact that both $\partial\varphi(\bar{x})$ and $\partial^+\varphi(\bar{x})$ are nonempty in this case; see the discussion after Corollary 1.81. On the other hand, example (1.49) shows that there are Lipschitz continuous functions, which are Fréchet differentiable at $\bar{x}$ but neither lower nor upper regular at this point. Of course, it never happens for *strictly differentiable* functions $\varphi\colon X \to I\!R$ that exhibit even *graphical regularity* in the sense of Definition 1.36 (there is no difference between N-regularity and M-regularity in this case).

Proposition 1.94 (two-sided regularity relationships). *Let $\varphi\colon X \to \overline{I\!R}$ be continuous around $\bar{x}$. Consider the following properties:*

(a) *φ is graphically regular at $\bar{x}$;*

(b) *φ is lower regular and upper regular at $\bar{x}$ simultaneously;*

(c) *φ is strictly differentiable at $\bar{x}$.*

Then (c)$\Rightarrow$(a)$\Rightarrow$(b). *Conversely,* (b)$\Rightarrow$(a) *if φ is locally Lipschitzian around $\bar{x}$, and* (a)$\Rightarrow$(c) *if φ is locally Lipschitzian and* $\dim X < \infty$.

Proof. Implication (c)$\Rightarrow$(a) follows from Theorem 1.38. To get (a)$\Rightarrow$(b), we first note that $\partial\varphi(\bar{x}) = D^*\varphi(\bar{x})(1)$ due to Theorem 1.80. Moreover, it follows from the proof of this theorem that $\widehat{\partial}\varphi(\bar{x}) = \widehat{D}^*\varphi(\bar{x})(1)$. Similarly we have $\partial^+\varphi(\bar{x}) = -D^*\varphi(\bar{x})(-1)$ and $\widehat{\partial}^+\varphi(\bar{x}) = -\widehat{D}^*\varphi(\bar{x})(-1)$. This gives (a)$\Rightarrow$(b) for any continuous function. If φ is Lipschitz continuous around $\bar{x}$, then $D^*\varphi(\bar{x})(0) = \widehat{D}^*\varphi(\bar{x})(0) = \{0\}$ due to Theorem 1.44, which yields the converse implication (b)$\Rightarrow$(a). Finally, (a)$\Rightarrow$(c) follows from Theorem 1.46 under the assumptions made. △

More results on lower regularity and related properties will be obtained in Subsect. 1.3.4 and then in Chap. 3, where they are incorporated into subdifferential calculus. We'll see, in particular, that lower regularity is preserved under various *unilateral operations* like sums, maxima, etc. and ensures *equalities* in the corresponding calculus rules. In the next subsection we consider subdifferentiation and lower regularity issues for an important class of Lipschitzian functions.

1.3.3 Subdifferentiation of Distance Functions

Given an nonempty subset $\Omega \subset X$ of a Banach space, we consider the *distance function* $d_\Omega\colon X \to I\!R$ associated with the set by

$$d_\Omega(x) := \operatorname{dist}(x;\Omega) = \inf_{u\in\Omega} \|x-u\| \,.$$

This class of functions plays an important role in optimization and variational analysis. One can see that d_Ω is *nonsmooth* and *Lipschitz continuous* globally on X with modulus $\ell = 1$. In what follows we compute subgradients and of the distance function d_Ω to at a point $\bar{x}$ in terms of the corresponding generalized normals to considering the two distinct cases: $\bar{x} \in \Omega$ and $\bar{x} \notin \Omega$. This allows us, in particular, to establish relationships between the properties of lower regularity for d_Ω and normal regularity for Ω. We start with deriving two-sided estimates for analytic ε-subgradients of d_Ω at $\bar{x} \in \Omega$, which induce the corresponding estimates for geometric ε-subgradients due to Theorem 1.86.

In this subsection and in the rest of the book the *notation* $\widehat{\partial}_\varepsilon \varphi(\bar{x})$ *stands for the analytic ε-subdifferential* of φ at $\bar{x}$ from Definition 1.83(ii).

Proposition 1.95 (ε-subgradients of distance functions at in-set points). *Let $\Omega \subset X$ with $\bar{x} \in \Omega$, and let $\varepsilon \geq 0$. Then*

$$\widehat{\partial}_\varepsilon d_\Omega(\bar{x}) \subset \left\{x^* \in \widehat{N}_\varepsilon(\bar{x}; \Omega) \middle| \; \|x^*\| \leq 1 + \varepsilon\right\},$$

$$\widehat{\partial}_\varepsilon d_\Omega(\bar{x}) \supset \left\{x^* \in \widehat{N}_{\varepsilon/4}(\bar{x}; \Omega) \middle| \; \|x^*\| \leq 1 + \varepsilon/4\right\}.$$

Proof. It follows from the definitions that

$$x^* \in \widehat{\partial}_\varepsilon d_\Omega(\bar{x}) \Longrightarrow x^* \in \widehat{N}_\varepsilon(\bar{x}; \Omega) \;\text{ and }\; \langle x^*, x\rangle \leq (1+\varepsilon)\|x\| \;\; \forall x \in X.$$

The latter gives $\|x^*\| \leq 1+\varepsilon$ and justifies the first inclusion in the proposition.

To establish the second inclusion, let us pick any $x^* \in \widehat{N}_{\varepsilon/4}(\bar{x}; \Omega)$ satisfying $\|x^*\| \leq 1 + \varepsilon/4$ and, given $x \notin \Omega$, find $u \in \Omega$ with

$$\|x - u\| \leq \text{dist}(x; \Omega) + \|x - \bar{x}\|^2 .$$

Taking into account that $\|u - \bar{x}\| \leq 3\|x - \bar{x}\|$ for x close to $\bar{x}$, we have

$$\liminf_{\substack{x \to \bar{x} \\ x \notin \Omega}} \frac{d_\Omega(x) - d_\Omega(\bar{x}) - \langle x^*, x - \bar{x}\rangle}{\|x - \bar{x}\|} \geq \liminf_{\substack{x \to \bar{x} \\ x \notin \Omega}} \frac{(1 - \|x^*\|)\|x - u\| - \langle x^*, u - \bar{x}\rangle}{\|x - \bar{x}\|}$$

$$\geq \min\left\{0, 1 - \|x^*\| - \limsup_{\substack{x \to \bar{x} \\ x \notin \Omega}} \frac{\langle x^*, u - \bar{x}\rangle}{\|x - \bar{x}\|}\right\} \geq -\frac{\varepsilon}{4} - \frac{3\varepsilon}{4} = -\varepsilon .$$

It remains to observe that

$$\liminf_{\substack{x \to \bar{x} \\ x \in \Omega}} \frac{d_\Omega(x) - d_\Omega(\bar{x}) - \langle x^*, x - \bar{x}\rangle}{\|x - \bar{x}\|} \geq -\varepsilon$$

if $x^* \in \widehat{N}_{\varepsilon/4}(\bar{x}; \Omega)$. Thus $x^* \in \widehat{\partial}_\varepsilon d_\Omega(\bar{x})$. △

Corollary 1.96 (Fréchet subgradients of distance functions at in-set points). *For any set $\Omega \subset X$ with $\bar{x} \in \Omega$ one has the representations*

$$\widehat{\partial} d_\Omega(\bar{x}) = \widehat{N}(\bar{x}; \Omega) \cap I\!B^*, \qquad \widehat{N}(\bar{x}; \Omega) = \bigcup_{\lambda > 0} \lambda \widehat{\partial} d_\Omega(\bar{x}) .$$

Proof. The second representation immediately follows from the first one, which is the case of $\varepsilon = 0$ in Proposition 1.95. $\triangle$

Thus we have an equivalent description of the prenormal cone to a arbitrary set in terms of the presubdifferential of the (Lipschitzian) distance function. Let us obtain a similar description of the basic normal cone to closed subsets of Banach spaces.

Theorem 1.97 (basic normals via subgradients of distance functions at in-set points). *Let $\Omega \subset X$ be nonempty and closed. Then*

$$N(\bar{x};\Omega) = \bigcup_{\lambda>0} \lambda \partial d_\Omega(\bar{x}) \quad \textit{for any } \bar{x} \in \Omega .$$

Proof. Picking $x^* \in N(\bar{x};\Omega)$ and using the definition of basic normals, we find sequences $\varepsilon_k \downarrow 0$, $x_k \xrightarrow{\Omega} \bar{x}$, and $x_k^* \xrightarrow{w^*} x^*$ with $x_k^* \in \widehat{N}_{\varepsilon_k}(x_k;\Omega)$ for $k \in I\!N$. Since $\{x_k^*\}$ is bounded, there is a bounded sequence of $\lambda_k > 0$ such that $\|x_k^*\|/\lambda_k \le 1+\varepsilon_k$. Then the second inclusion in Proposition 1.95 gives $x_k^* \in \lambda_k \widehat{\partial}_{\tilde{\varepsilon}_k} d_\Omega(x_k)$ with $\tilde{\varepsilon}_k := 4\varepsilon_k$. Employing representation (1.55), we get $x^* \in \lambda \partial d_\Omega(\bar{x})$ with some $\lambda > 0$, which justifies the inclusion "$\subset$" in the theorem for an arbitrary set Ω.

Let us prove the opposite inclusion when Ω is closed. Take $x^* \in \partial d_\Omega(\bar{x})$ and find sequences $\varepsilon_k \downarrow 0$, $x_k \to \bar{x}$, and $x^* \xrightarrow{w^*} x^*$ with $x_k^* \in \widehat{\partial}_{\varepsilon_k} d_\Omega(x_k)$. If $x_k \in \Omega$ along a subsequence of k, we end the proof by passing to the limit in the first inclusion of Proposition 1.95. Assume that $x_k \notin \Omega$ for all $k \in I\!N$. In this case there are $\eta_k \downarrow 0$ with

$$\langle x_k^*, x - x_k\rangle \le 2\varepsilon_k \|x - x_k\| \quad \text{whenever } x \in B_{\eta_k}(x_k) \cap \Omega, \quad k \in I\!N .$$

Choose $\rho_k \downarrow 0$ with $\rho_k < \min\{\eta_k^2,\ \frac{1}{k} d_\Omega(x_k)\}$ and take $\nu_k \downarrow 1$ such that $(\nu_k - 1) d_\Omega(x_k) < \rho_k^2$. Then we pick $\tilde{x}_k \in \Omega$ satisfying $\|\tilde{x}_k - x_k\| \le \nu_k d_\Omega(x_k)$ and observe that

$$\langle x_k^*, u\rangle \le d_\Omega(x_k+u) - \nu_k^{-1}\|x_k - \tilde{x}_k\| + \varepsilon_k\|u\|$$

$$\le d_\Omega(\tilde{x}_k+u) + (1-\nu_k^{-1})\|x_k - \tilde{x}_k\| + 2\varepsilon_k\|u\|$$

if $\|u\| \le \eta_k$. Then

$$\langle x_k^*, x - \tilde{x}_k\rangle \le (1-\nu_k^{-1})\|x_k - \tilde{x}_k\| + 2\varepsilon_k\|x - \tilde{x}_k\|$$

for all $x \in \Omega \cap B_{\eta_k}(\tilde{x}_k)$, and hence

$$0 \le \varphi_k(x) := -\langle x_k^*, x - \tilde{x}_k\rangle + 2\varepsilon_k\|x - \tilde{x}_k\| + \gamma_k^2, \quad x \in \Omega \cap B_{\eta_k}(\tilde{x}_k) ,$$

where $\gamma_k^2 := (1-\nu_k^{-1})\|x_k - \tilde{x}_k\|$. The latter gives

$$\gamma_k^2 = \varphi_k(\tilde{x}_k) \le \inf_{x \in \Omega \cap B_{\eta_k}(\tilde{x}_k)} \varphi_k(x) + \gamma_k^2$$

for each $k \in I\!N$, and we can apply the Ekeland variational principle (see Theorem 2.26 in Subsect. 2.3.1) to the continuous function φ_k on the complete metric space $\Omega \cap B_{\eta_k}(\tilde{x}_k)$. According to this result, there is $\hat{x}_k \in \Omega \cap B_{\eta_k}(\tilde{x}_k)$ such that $\|\hat{x}_k - \tilde{x}_k\| \le \gamma_k$ and

$$\langle -x_k^*, \hat{x}_k - \tilde{x}_k \rangle + 2\varepsilon_k \|\hat{x}_k - \tilde{x}_k\| \le \langle -x_k^*, x - \tilde{x}_k \rangle + 2\varepsilon_k \|x - \tilde{x}_k\| + \gamma_k \|x - \hat{x}_k\| .$$

Taking into account that $\gamma_k^2 \le \nu_k(1 - \nu_k^{-1}) d_\Omega(x_k) < \rho_k^2$ and then letting $r_k := \rho_k - \gamma_k > 0$, we get

$$\|x - \hat{x}_k\| \le r_k \Longrightarrow \|x - \tilde{x}_k\| \le \|x - \hat{x}_k\| + \gamma_k \le \rho_k \le \eta_k .$$

It follows from the above estimates that

$$\langle x_k^*, x - \hat{x}_k \rangle \le (2\varepsilon_k + \gamma_k)\|x - \hat{x}_k\| \quad \text{whenever} \quad x \in \Omega \cap B_{r_k}(\hat{x}_k) ,$$

and hence $x_k^* \in \widehat{N}_{2\varepsilon_k + \gamma_k}(\hat{x}_k; \Omega)$ for all $k \in I\!N$. Passing to the limit as $k \to \infty$ and taking into account that $\gamma_k \downarrow 0$ and $\hat{x}_k \to \bar{x}$, we finally get $x^* \in N(\bar{x}; \Omega)$, which ends the proof of the theorem. △

The results obtained allow us to show that, for any point $\bar{x} \in \Omega$, the lower regularity of d_Ω at $\bar{x} \in \Omega$ is completely determined by the normal regularity of Ω at this point.

Corollary 1.98 (regularity of sets and distance functions at in-set points). *Let $\Omega \subset X$ be a closed set with $\bar{x} \in \Omega$. Then Ω is normally regular at $\bar{x}$ if and only if the distance function d_Ω is lower regular at this point.*

Proof. Follows from the definitions and the normal cone representations in Corollary 1.96 and Theorem 1.97. △

Next let us consider the case of $\bar{x} \notin \Omega$ and derive the relationship between Fréchet subgradients of the distance function $d_\Omega(\cdot)$ and Fréchet normals of the ρ-*enlargement* of Ω *relative to* $\bar{x}$ defined by

$$\Omega(\rho) := \big\{ x \in X \big|\, d_\Omega(x) \le \rho \big\} \quad \text{with} \quad \rho := d_\Omega(\bar{x}) .$$

Note that the ρ-enlargement of Ω is always closed for any $\rho \ge 0$, even when Ω is not. Furthermore, $\Omega(\rho) = \Omega + \rho I\!B$ if Ω is either compact in Banach spaces or closed in finite dimensions.

Theorem 1.99 (ε-subgradients of distance functions at out-of-set points). *For any $\emptyset \ne \Omega \subset X$, any $\bar{x} \notin \Omega$, and any $\varepsilon \ge 0$ sufficiently small the following inclusions hold:*

$$\big\{x^*\in\widehat N_{\varepsilon/4}\big(\bar x;\Omega(\rho)\big)\big|\;1-\varepsilon/4\le\|x^*\|\le 1+\varepsilon/4\big\}\subset\widehat\partial_\varepsilon d_\Omega(\bar x)$$

$$\subset\big\{x^*\in\widehat N_\varepsilon\big(\bar x;\Omega(\rho)\big)\big|\;1-\varepsilon\le\|x^*\|\le 1+\varepsilon\big\}\;\;\textit{with}\;\;\rho=d_\Omega(\bar x)\;.$$

In particular, for $\varepsilon=0$ *one has*

$$\widehat\partial d_\Omega(\bar x)=\widehat N\big(\bar x;\Omega(\rho)\big)\cap\big\{x^*\in X^*\big|\;\|x^*\|=1\big\}\;.$$

Proof. For simplicity we consider only the case of $\varepsilon=0$; the proof for $\varepsilon>0$ is similar. First let us check the representation

$$d_{\Omega(\rho)}(x)=d_\Omega(x)-\rho\;\;\text{for any}\;\;x\notin\Omega(\rho)\;\;\text{and}\;\;\rho>0\;.$$

To proceed, we fix $x\notin\Omega(\rho)$ and take any $u\in\Omega(x)$ with $d_\Omega(u)\le\rho$. Then for every $\varepsilon>0$ there is $u_\varepsilon\in\Omega$ satisfying

$$\|u-u_\varepsilon\|\le d_\Omega(u)+\varepsilon\le\rho+\varepsilon\;,$$

which obviously yields

$$\|u-x\|\ge\|u_\varepsilon-x\|-\|u_\varepsilon-u\|\ge d_\Omega(x)-\|u_\varepsilon-u\|\ge d_\Omega(x)=\rho-\varepsilon\;.$$

Since the estimate $\|u-x\|\ge d_\Omega(x)-\rho-\varepsilon$ holds for all $u\in\Omega(\rho)$ and all $\varepsilon>0$, we get the inequality

$$d_{\Omega(\rho)}(x)\ge d_\Omega(x)-\rho\;.$$

To prove the opposite inequality, let us fix $u\in\Omega$ and define the continuous function $\varphi\colon I\!R_+\to I\!R$ by

$$\varphi(t):=d_\Omega(tx+(1-t)u)\;.$$

Since $\varphi(0)=0$ and $\varphi(1)>\rho$, there is $t_0\in(0,1)$ with $\varphi(t_0)=\rho$ by the classical intermediate value theorem. Putting now $v:=t_0x+(t-t_0)u$, we have $d_\Omega(v)=\rho$ and $\|x-u\|=\|x-v\|+\|v-u\|$. Hence

$$\|x-u\|\ge\|x-v\|+d_\Omega(v)=\|x-v\|+\rho$$

by $u\in\Omega$ and $v\in\Omega(\rho)$, which implies $\|x-u\|\ge d_{\Omega(\rho)}(x)+\rho$ and the desired equality $d_{\Omega(\rho)}(x)=d_\Omega(x)-\rho$.

Using this representation of $d_{\Omega(\rho)}$, let us prove the equality claimed in the theorem starting with the inclusion"$\subset$" therein. From now we fix $\rho=d_\Omega(\bar x)$. Take any $x^*\in\widehat\partial d_\Omega(\bar x)$ and fix $\varepsilon>0$. Then, by the definition of Fréchet subgradients, there is $\nu>0$ such that

$$\langle x^*,x-\bar x\rangle\le d_\Omega(x)-d_\Omega(\bar x)+\varepsilon\|x-\bar x\|\;\;\text{whenever}\;\;x\in\bar x+\nu I\!B\;,$$

which implies $\langle x^*,x-\bar x\rangle\le\varepsilon\|x-\bar x\|$ for all $x\in(\bar x+\nu I\!B)\cap\Omega(\rho)$ by virtue of $d_\Omega(x)-d_\Omega(\bar x)\le 0$ when $x\in\Omega(\rho)$. The latter gives $x^*\in\widehat N(\bar x;\Omega(\rho))$.

Let us show that $\|x^*\| = 1$ whenever $x^* \in \widehat{\partial} d_\Omega(\bar{x})$. Using again the definition of Fréchet subgradients of d_Ω at $\bar{x}$ with ε and ν therein, we put

$$r := \min\Big\{1, \varepsilon, \frac{\nu}{1 + d_\Omega(\bar{x})}\Big\}$$

and choose $x_r \in \Omega$ so that $\|\bar{x} - x_r\| \le d_\Omega(\bar{x}) + r^2$. For $x := \bar{x} + r(x_r - \bar{x})$ one obviously has the estimates

$$\|x - \bar{x}\| \le r\|\bar{x} - x_r\| \le r d_\Omega(\bar{x}) + r^2 \le r\big(1 + d_\Omega(\bar{x})\big) \le \nu \, ,$$

and therefore

$$\begin{aligned}\langle x^*, x - \bar{x}\rangle &\le \|x - \bar{x}\| - \|\bar{x} - x_r\| + r^2 + \varepsilon r\|\bar{x} - x_r\| \\ &= -r\|\bar{x} - x_r\| + r^2 + \varepsilon r\|\bar{x} - x_r\| \, .\end{aligned}$$

Taking into account the above choice of x, we get

$$\langle x^*, x_r - \bar{x}\rangle \le -\|\bar{x} - x_r\| + \varepsilon(1 + \|\bar{x} - x_r\|) \, ,$$

which readily gives

$$\frac{\langle x^*, \bar{x} - x_r\rangle}{\|\bar{x} - x_r\|} \ge 1 - \varepsilon\Big(1 + \frac{1}{\|\bar{x} - x_r\|}\Big) \ge 1 - \varepsilon\Big(1 + \frac{1}{d_\Omega(\bar{x})}\Big) \, ,$$

and thus $\|x^*\| \ge 1$. Since $\|x^*\| \le 1$ by the Lipschitz continuity of d_Ω with modulus $\ell = 1$, we conclude that $\|x^*\| = 1$ and complete the proof of the inclusion "$\subset$" in the theorem.

To justify the opposite inclusion, fix $x^* \in \widehat{N}(\bar{x}; \Omega(\rho))$ with $\|x^*\| = 1$ and take arbitrary $\varepsilon > 0$ and $\eta \in (0, 1)$. By the first equality in Corollary 1.96 we get $x^* \in \widehat{\partial} d_{\Omega(\rho)}(\bar{x})$, and hence there is $\nu_1 > 0$ such that

$$\langle x^*, x - \bar{x}\rangle \le d_{\Omega(\rho)}(x) - d_{\Omega(\rho)}(\bar{x}) + \varepsilon\|x - \bar{x}\| \ \text{ whenever } \ x \in \bar{x} + \nu_1 I\!B \, .$$

It follows from the representation of $d_{\Omega(\rho)}$ established above that

$$\langle x^*, x - \bar{x}\rangle \le d_\Omega(x) - d_\Omega(\bar{x}) + \varepsilon\|x - \bar{x}\| \ \text{ whenever } \ x \in \big(\bar{x} + \nu_1 I\!B\big) \setminus \Omega(\rho) \, .$$

On the other hand, the inclusion $x^* \in \widehat{N}(\bar{x}; \Omega(\rho))$ implies the existence of $\nu_2 > 0$ ensuring the estimate

$$\langle x^*, x - \bar{x}\rangle \le (\varepsilon/2)\|x - \bar{x}\| \ \text{ for all } \ x \in \big(\bar{x} + \nu_2 I\!B\big) \cap \Omega(\rho) \, .$$

Since $\|x^*\| = 1$, we choose $u \in X$ such that $\|u\| = 1$ and $\langle x^*, u\rangle \ge 1 - \eta$. Fix $\nu_3 \in (0, \nu_2/2)$ and $x \in (\bar{x} + \nu_3 I\!B) \cap \Omega(\rho)$ and put $\gamma_x := d_\Omega(\bar{x}) - d_\Omega(x) \ge 0$. Then $x + \gamma_x u \in \Omega(\rho) \cap (\bar{x} + \nu I\!B)$ due to

$$d_\Omega(x + \gamma_x u) \le d_\Omega(x) + \gamma_x = d_\Omega(\bar{x}) = \rho \ \text{ and}$$

$$\|x + \gamma_x u - \bar{x}\| \le \|x - \bar{x}\| + \gamma_x \le 2\|x - \bar{x}\| \le 2\nu_3 \le \nu_2 ,$$

which implies that $\langle x^*, x + \gamma_x u - \bar{x}\rangle \le \varepsilon\|x - \bar{x}\|$ and hence

$$\langle x^*, x - \bar{x}\rangle = \langle x^*, x + \gamma_x u - \bar{x}\rangle - \langle x^*, \gamma_x u\rangle \le \varepsilon\|x - \bar{x}\| - \gamma_x(1-\eta)$$

$$\le \varepsilon\|x - \bar{x}\| + \big(d_\Omega(x) - d_\Omega(\bar{x})\big)(1-\eta) .$$

Since $\eta > 0$ was chosen arbitrary, one has

$$\langle x^*, x - \bar{x}\rangle \le \varepsilon\|x - \bar{x}\| + d_\Omega(x) - d_\Omega(\bar{x}) \ \text{ whenever } \ x \in (\bar{x} + \nu_3 I\!B) \cap \Omega(\rho) ,$$

and therefore the latter holds for all $x \in \bar{x} + \nu I\!B$ with $\nu := \min\{\nu_1, \nu_3\}$. Thus we get $x^* \in \widehat{\partial} d_\Omega(\bar{x})$ and complete the proof of the theorem. △

Do we have analogs of the inclusions in Theorem 1.99 for basic normals and subgradients? It happens that the answer is *negative* for the crucial inclusion

$$\partial d_\Omega(\bar{x}) \subset N\big(\bar{x}; \Omega(\rho)\big) \cap I\!B^* \ \text{ with } \ \rho = d_\Omega(\bar{x})$$

even in *finite dimensions*. A simple *counterexample* is provided by the set

$$\Omega := \big\{(x_1, x_2) \in I\!R^2 \big|\ x_1^2 + x_2^2 \ge 1\big\}$$

with $\bar{x} = (0,0)$. Indeed, in this case $d_\Omega(\bar{x}) = 1$ and $\Omega(\rho) = \Omega + \rho I\!B = I\!R^2$ for $\rho = 1$, hence $N\big(\bar{x}; \Omega(\rho)\big) = \{0\}$. On the other hand, it is easy to compute the distance function

$$d_\Omega(x_1, x_2) = 1 - \sqrt{x_1^2 + x_2^2}$$

in this case, and so to see that $\partial d_\Omega(\bar{x})$ is the unit sphere of $I\!R^2$.

To derive a correct inclusion important for subsequent applications, we need to change a bit the construction of the subdifferential $\partial d_\Omega(\cdot)$, which seems to be appropriate for describing generalized differential properties of distance functions at out-of-set points. The idea behind this modification is that, in the limiting procedure from ε-subgradients, we consider only those points $x_k \to \bar{x}$, where the function values are *to the right* of the one at $\bar{x}$. In this way we can define other "sided" subdifferential modifications that are not used in what follows.

Definition 1.100 (right-sided subdifferential). *Given $\varphi\colon X \to \overline{I\!R}$ finite at $\bar{x}$, define the* RIGHT-SIDED SUBDIFFERENTIAL *of φ at $\bar{x}$ by*

$$\partial_{\ge}\varphi(\bar{x}) := \mathop{\mathrm{Lim\,sup}}_{\substack{x \overset{\varphi+}{\to} \bar{x} \\ \varepsilon \downarrow 0}} \widehat{\partial}_\varepsilon \varphi(x) ,$$

where $x \overset{\varphi+}{\to} \bar{x}$ means that $x \to \bar{x}$ with $\varphi(x) \to \varphi(\bar{x})$ and $\varphi(x) \ge \varphi(\bar{x})$.

We obviously have the inclusions

$$\widehat{\partial}\varphi(\bar{x}) \subset \partial_{\geq}\varphi(\bar{x}) \subset \partial\varphi(\bar{x}) ,$$

i.e., $\partial_{\geq}\varphi(\bar{x}) = \partial\varphi(\bar{x})$ for functions φ lower regular at $\bar{x}$, in particular, for strictly differentiable and convex functions. On the other hand, the right-sided subdifferential may be empty for Lipschitzian functions in finite dimensions as for the one in the example above, where

$$\widehat{\partial}\varphi(x) = \emptyset \ \text{ whenever } \ \varphi(x) \geq \varphi(\bar{x}), \quad \text{so } \ \partial_{\geq}\varphi(\bar{x}) = \emptyset .$$

It is important to emphasize that

$$\partial_{\geq}\varphi(\bar{x}) = \partial\varphi(\bar{x}), \ \text{ and thus } \ 0 \in \partial_{\geq}\varphi(\bar{x})$$

when φ attains its *local minimum* at $\bar{x}$. In particular, one has

$$\partial_{\geq}d_{\Omega}(\bar{x}) = \partial d_{\Omega}(\bar{x}) \ \text{ whenever } \ \bar{x} \in \Omega .$$

The next theorem gives the required relationships between subgradients of the distance function at out-of set points and basic normals to the enlargement of Ω in terms of the *right-sided subdifferential* from Definition 1.100. Moreover, the latter construction allows us to derive the out-of-set counterpart of the *equality* in Theorem 1.97.

Theorem 1.101 (right-sided subgradients of distance functions and basic normals at out-of-set points). *Let $\Omega \subset X$ be a nonempty closed subset of a Banach space, and let $\bar{x} \notin \Omega$. The following assertions hold:*

(i) *One has the inclusion*

$$\partial_{\geq}d_{\Omega}(\bar{x}) \subset N\big(\bar{x}; \Omega(\rho)\big) \cap I\!B^* \ \ \textit{with} \ \ \rho = d_{\Omega}(\bar{x}) .$$

If in addition the latter enlargement $\Omega(\rho)$ is SNC at $\bar{x}$, then

$$\partial_{\geq}d_{\Omega}(\bar{x}) \subset \big[N\big(\bar{x}; \Omega(\rho)\big) \cap I\!B^*\big] \setminus \{0\} .$$

(ii) *One always has the equality*

$$N(\bar{x}; \Omega_{\rho}) = \bigcup_{\lambda \geq 0} \lambda \partial_{\geq} d_{\Omega}(\bar{x}) \ \ \textit{with} \ \ \rho = d_{\Omega}(\bar{x}) .$$

Proof. To prove the first inclusion in (i), we take any $x^* \in \partial_{\geq}d_{\Omega}(\bar{x})$ and find $\varepsilon_k \downarrow 0$, $x_k \to \bar{x}$ with $d_{\Omega}(x_k) \geq d_{\Omega}(\bar{x})$, and $x_k^* \xrightarrow{w^*} x^*$ such that

$$x_k^* \in \widehat{\partial}_{\varepsilon_k} d_{\Omega}(x_k) \ \text{ for all } \ k \in I\!N .$$

It follows from Theorem 1.99 that $1-\varepsilon_k \le \|x_k^*\| \le 1+\varepsilon_k$ for all $k \in I\!N$ sufficiently large. Denote for convenience $\Omega(\bar{x}) := \Omega(\rho)$ with $\rho = d_\Omega(\bar{x})$ and consider the following two cases:

(a) There is a subsequence of $\{x_k\}$ such that $d_\Omega(x_k) = d_\Omega(\bar{x})$ along this subsequence.

(b) Otherwise. Since $d_\Omega(x_k) > d_\Omega(\bar{x})$, we have in this case that $x_k \notin \Omega(\bar{x})$ for all $k \in I\!N$.

In case (a) we get from the second inclusion in Theorem 1.99 that

$$x_k^* \in \widehat{N}_{\varepsilon_k}\big(x_k; \Omega(\bar{x})\big)$$

along the subsequence of x_k under consideration. Then passing to the limit as $k \to \infty$ with taking into account the lower semicontinuity of the norm functions in the weak* topology of X^*, we arrive at

$$x^* \in N\big(\bar{x}; \Omega(\bar{x})\big) \cap I\!B^* ,$$

which justifies the first inclusion from (i) in case (a). The second inclusion in this case follows directly from the definition of the SNC property for the fixed enlargement set $\Omega(\bar{x})$.

Now consider the remaining case (b) when $x_k \notin \Omega(\bar{x})$ for all $k \in I\!N$. As established in the proof of the first part of Theorem 1.99,

$$d_\Omega(x) = d_\Omega(\bar{x}) + d_{\Omega(\bar{x})}(x) \quad \text{whenever} \quad x \notin \Omega(\bar{x}) .$$

Hence for every $k \in I\!N$ one has the relations

$$x_k^* \in \widehat{\partial}_{\varepsilon_k} d_\Omega(x_k) = \widehat{\partial}_{\varepsilon_k}\big[d_\Omega(\bar{x}) + d_{\Omega(\bar{x})}\big](x_k) = \widehat{\partial}_{\varepsilon_k} d_{\Omega(\bar{x})}(x_k) .$$

Let $\widetilde{\varepsilon}_k := \|x_k - \bar{x}\|$. Following the proof of Theorem 1.97 for the set $\Omega(\bar{x})$, with the usage of Ekeland's variational principle, we find $\widetilde{x}_k \in \Omega(\bar{x})$ such that

$$\|\widetilde{x}_k - x_k\| \le d_{\Omega(\bar{x})}(x_k) + \varepsilon_k \le \widetilde{\varepsilon}_k + \varepsilon_k \quad \text{and} \quad x_k^* \in \widehat{N}\big(\widetilde{x}_k; \Omega(\bar{x})\big)$$

whenever $k \in I\!N$. Since $\widetilde{\varepsilon}_k + \varepsilon_k \downarrow 0$ as $k \to \infty$, it gives $x^* \in N\big(\bar{x}; \Omega(\bar{x})\big)$. The facts that $x^* \in I\!B^*$ and that $x^* \ne 0$ if $\Omega(\bar{x})$ is SNC at $\bar{x}$ are justified similarly to case (a). Thus we complete the proof of assertion (i) of the theorem.

It follows directly from the first inclusion in (i) that

$$\bigcup_{\lambda > \ge 0} \lambda \partial_{\ge} d(\bar{x}; \Omega) \subset N\big(\bar{x}; \Omega(\bar{x})\big) .$$

For proving assertion (ii), it remains therefore to justify the opposite inclusion. Take $x^* \in N\big(\bar{x}; \Omega(\bar{x})\big)$ and suppose that $x^* \ne 0$; the other case is trivial. Then there are $\varepsilon_k \downarrow 0$, $x_k \to \bar{x}$ with $x_k \in \Omega(\bar{x})$, and $x_k^* \xrightarrow{w^*} x^*$ such that

$$x_k^* \in \widehat{N}_{\varepsilon_k}\big(x_k; \Omega(\bar{x})\big) \ \text{ for all } \ k \in I\!N .$$

By the norm weak* lower semicontinuity we have

$$\liminf_{k\to\infty} \|x_k^*\| \ge \|x^*\| > 0$$

Thus there exist subsequences of (x_k, x_k^*), without relabeling, and a sequence $\epsilon_k \downarrow 0$ satisfying

$$\frac{x_k^*}{\|x_k^*\|} \in \widehat{N}_{\epsilon_k/4}\big(x_k; \Omega(\bar{x})\big), \quad k \in I\!N .$$

Employing the first inclusion in Theorem 1.99, we get

$$x_k^* \in \|x_k^*\| \widehat{\partial}_{\epsilon_k} d_\Omega(x_k) \ \text{ as } \ k \to \infty .$$

Note that $d_\Omega(x_k) \le 0$ by the choice of $x_k \in \Omega(\bar{x})$. At the same time the strict inequality $d_\Omega(x_k) < 0$ is not possible for k sufficiently large due to $0 \ne x_k^* \in \widehat{N}_{\varepsilon_k}\big(x_k; \Omega(\bar{x})\big)$. Selecting now a convergent subsequence of $\|x_k^*\|$ and using Definition 1.100 of the right-sided subdifferential, we find $\lambda > 0$ such that $x^* \in \lambda \partial_{\ge} d_\Omega(\bar{x})$, which completes the proof of the theorem. △

Observe that we may unify the statements of Theorem 1.97 and of assertion (ii) in Theorem 1.101, since $\partial_{\ge} d_\Omega(\bar{x}) = \partial d_\Omega(\bar{x})$ if $\bar{x} \in \Omega$. Note also that some sufficient conditions for the SNC property of the set enlargement $\Omega(\rho) = \Omega(\bar{x})$ used in Theorem 1.101(i) are given subsequently in Theorem 3.83 in the framework of Asplund spaces.

Finally in this subsection, we derive results of the *projection type* that allow us to estimate subgradients of the distance function $d_\Omega(\bar{x})$ at *out-of-set* points $\bar{x} \notin \Omega$ via normals to Ω at projection or perturbed projection points of Ω. Let us start with estimating ε-subgradients of $d_\Omega(\bar{x})$ at $\bar{x} \notin \Omega$ in the case when the *projection set*

$$\Pi(\bar{x}; \Omega) := \big\{ w \in \Omega \big| \ \|w - \bar{x}\| = d_\Omega(\bar{x}) \big\}$$

in *nonempty.* In this case we get the following useful inclusion.

Proposition 1.102 (ε-subgradients of distance functions and ε-normals at projection points). *Let $\Omega \subset X$ be a nonempty subset of a Banach space, let $\bar{x} \notin \Omega$, and let $\Pi(\bar{x}; \Omega) \ne \emptyset$. Then for any $\varepsilon \in [0, 1]$ one has*

$$\widehat{\partial}_\varepsilon d_\Omega(\bar{x}; \Omega) \subset \bigcap_{w \in \Pi(\bar{x}; \Omega)} \widehat{N}_\varepsilon(w; \Omega) \cap \big[1 - \varepsilon, 1 + \varepsilon\big] S^* .$$

Proof. Pick $x^* \in \widehat{\partial}_\varepsilon d_\Omega(\bar{x})$ and, by definition of ε-subgradients, for any $\gamma > 0$ find $\delta > 0$ such that

$$\langle x^*, x - \bar{x}\rangle \le (\varepsilon + \gamma)\|x - \bar{x}\| + d_\Omega(x) - d_\Omega(\bar{x}) \quad \text{whenever} \quad \|x - \bar{x}\| \le \delta .$$

Now given *any* projection element $w \in \Pi(\bar{x};\Omega)$ and any $x \in \bar{x} + \delta I\!B$, we have

$$\begin{aligned}\langle x^*, x - w\rangle &\le (\varepsilon + \gamma)\|x - w\| + d_\Omega(x - w + \bar{x}) - \|\bar{x} - w\| \\ &\le (\varepsilon + \gamma)\|x - w\| ,\end{aligned}$$

and hence $x^* \in \widehat{N}_\varepsilon(w;\Omega)$.

It remains to show that for any $x^* \in \widehat{\partial}_\varepsilon d_\Omega(\bar{x})$ with $\bar{x} \notin \Omega$ and $\varepsilon \in [0,1]$ one has the estimates

$$1 - \varepsilon \le \|x^*\| \le 1 + \varepsilon .$$

Observe that the upper estimate above follows directly from the definition of ε-subgradients and the Lipschitz continuity of $d_\Omega(\cdot)$ with modulus $\ell = 1$.

Taking an arbitrary $x^* \in \widehat{\partial}_\varepsilon d_\Omega(\bar{x})$, let us justify the lower estimate $\|x^*\| \ge 1 - \varepsilon$ for it assuming that $\varepsilon \in (0,1)$ without loss of generality. By definition of ε-subgradients, for each $\nu \in (\varepsilon, 1]$ there is $\delta > 0$ such that

$$\langle x^*, x - \bar{x}\rangle \le \nu\|x - \bar{x}\| + d_\Omega(x) - d_\Omega(\bar{x}) \quad \text{whenever} \quad x \in \bar{x} + \delta I\!B .$$

Fixing $t \in (0,1)$, select $x_t \in \Omega$ satisfying

$$\|x_t - \bar{x}\| \le (1 + t^2) d_\Omega(\bar{x})$$

and then $z_t \in (x_t, \bar{x}) := \{(1-\alpha)x_t + \alpha\bar{x} |\ \alpha \in (0,1)\}$ satisfying

$$\|\bar{x} - z_t\| = t\|x_t - \bar{x}\| .$$

One clearly has $z_t \in \bar{x} + \delta I\!B$ for all t sufficiently small. Thus substituting z_t into the above inequality for x^* and taking into account that $d_\Omega(z_t) \le \|x_t - z_t\|$ by the choice of x_t, we get

$$\langle x^*, z_t - \bar{x}\rangle \le \nu\|\bar{x} - z_t\| + \|x_t - z_t\| - (1 + t^2)^{-1}\|x_t - \bar{x}\| .$$

This gives by the choice of z_t that

$$\langle x^*, t(x_t - \bar{x})\rangle \le \nu t\|x_t - \bar{x}\| + (1 - t)\|x_t - \bar{x}\| - (1 + t^2)^{-1}\|x_t - \bar{x}\| ,$$

which implies the estimate

$$\langle x^*, \bar{x} - x_t\rangle \ge (\gamma_t - \nu)\|x_t - \bar{x}\| \quad \text{with} \quad \gamma_t := t^{-1}[(1 - t^2)^{-1} + t - 1] ,$$

and therefore $\|x^*\| \ge \gamma_t - \nu$. Since the latter holds for any $\nu \downarrow \varepsilon$ with $\gamma_t \to 1$ as $t \uparrow 1$, we finally get $\|x^*\| \ge 1 - \varepsilon$ and complete the proof. $\triangle$

Next let us consider the case when the projection set $\Pi(\bar{x};\Omega)$ may be *empty* and, given $\eta > 0$, define the *perturbed projection* set by

$$\Pi_\eta(\bar{x};\Omega) := \big\{ w \in \Omega \big|\ \|w - \bar{x}\| \le d_\Omega(\bar{x}) + \eta \big\} .$$

Theorem 1.103 (ε-subgradients of distance functions and ε-normals to perturbed projections). *Let $\Omega \subset X$ be a closed subset of a Banach space, and let $\bar{x} \notin \Omega$. Then for every $\varepsilon \in [0,1]$ one has the upper estimate*

$$\widehat{\partial}_\varepsilon d_\Omega(\bar{x};\Omega) \subset \bigcap_{\eta>0} \bigcup_{w\in\Pi_\eta(\bar{x};\Omega)} \Big\{\widehat{N}_{\varepsilon+\eta}(w;\Omega) \cap \big[1-\varepsilon, 1+\varepsilon\big]S^*\Big\} .$$

Proof. Fixed $x^* \in \widehat{\partial}_\varepsilon d_\Omega(\bar{x})$ and $\eta > 0$, for any $\gamma \in (0, \eta/2)$ find $\delta > 0$ with

$$\langle x^*, x - \bar{x}\rangle \le d_\Omega(x) - d_\Omega(\bar{x}) + (\varepsilon + \gamma)\|x - \bar{x}\|$$

whenever $\|x - \bar{x}\| \le \delta$. Take $0 < \widetilde{\eta} < \min\{\gamma, \delta/4\}$ and choose $z \in \Omega$ satisfying

$$\|z - \bar{x}\| \le d_\Omega(\bar{x}) + \widetilde{\eta}^2 .$$

Then for any $x \in \Omega \cap (z + \delta I\!B)$ we have the estimates

$$\begin{aligned}\langle x^*, x - z\rangle &\le d_\Omega(x - z + \bar{x}) - \|\bar{x} - z\| + \widetilde{\eta}^2 + (\varepsilon + \gamma)\|x - z\| \\ &\le (\varepsilon + \gamma)\|x - z\| + \widetilde{\eta}^2 .\end{aligned}$$

Consider the real-valued function

$$\varphi(x) := -\langle x^*, x - z\rangle + (\varepsilon + \gamma)\|x - z\| + \widetilde{\eta}^2 ,$$

which is obviously continuous on the complete metric space $W := \Omega \cap (\bar{x} + \delta I\!B)$. It follows from the above constructions that

$$\varphi(z) \le \inf_W \varphi(x) + \widetilde{\eta}^2 .$$

Employing Ekeland's variational principle from Theorem 2.26, we find $w \in W$ satisfying $\|w - z\| < \widetilde{\eta}$ and

$$\begin{aligned}-\langle x^*, w - z\rangle + (\varepsilon + \gamma)\|w - z\| + \widetilde{\eta}^2 \le -\langle x^*, x - z\rangle + (\varepsilon + \gamma)\|x - z\| \\ +\widetilde{\eta}^2 + \widetilde{\eta}\|w - x\|\end{aligned}$$

for all $x \in W$. This implies the estimates

$$\langle x^*, x - z\rangle \le (\varepsilon + \gamma + \widetilde{\eta})\|x - z\| \le (\varepsilon + 2\gamma)\|x - w\| \le (\varepsilon + \eta)\|x - w\|$$

whenever $x \in W$. Furthermore, by the choice of $\widetilde{\eta}$ we have $w + \widetilde{\eta} I\!B \subset z + \delta I\!B$ and therefore

$$\langle x^*, x - w\rangle \le (\varepsilon + \eta)\|x - w\| \ \text{ for all } \ x \in \Omega \cap (w + \widetilde{\eta} I\!B) ,$$

which justifies the inclusion $x^* \in \widehat{N}_{\varepsilon+\eta}(w;\Omega)$. Note that

$$\|w - \bar{x}\| \le \|w - z\| + \|z - \bar{x}\| \le \widetilde{\eta} + d_\Omega(\bar{x}) + \widetilde{\eta} \le d_\Omega(\bar{x}) + \eta ,$$

and hence $w \in \Pi_\eta(\bar{x}; \Omega)$. Observe finally that the estimates

$$1 - \varepsilon \le \|x^*\| \le 1 + \varepsilon$$

follow from the proof of Proposition 1.102. △

The concluding results of this subsection provide *upper estimates* of the whole *basic subdifferential* of the distance function $d_\Omega(\cdot)$ at out-of-set points via the basic normal cone to Ω at the corresponding projections. To establish the principal theorem in this direction, we impose a certain *well-posedness* of the best approximation problem for Ω, which automatically holds under some natural geometric assumptions; see below.

Definition 1.104 (well-posedness of best approximations). *Let $\Omega \subset X$ be an nonempty subset of a Banach space, and let $\bar{x} \notin \Omega$. We say that the best approximation problem to Ω from $\bar{x}$ is* WELL POSED *if either one of the following properties holds:*

(a) *every sequence of $x_k \in \Omega$ satisfying $x_k \to \bar{x}$ and*

$$\|x_k - \bar{x}\| \to d_\Omega(\bar{x}) \quad \text{as} \quad k \to \infty$$

contains a convergent subsequence;

(b) *for every sequence of $x_k \to \bar{x}$ with $\widehat{\partial}_{\varepsilon_k} d_\Omega(x_k) \ne \emptyset$ as $\varepsilon_k \downarrow 0$ there is a sequence of $w_k \in \Pi(x_k; \Omega)$ that contains a convergent subsequence.*

Observe that the main difference between properties (a) and (b) in Definition 1.104 is that instead of the compactness requirement on *minimizing* sequences of *in-set* points $x_k \in \Omega$ in (a), a similar compactness is imposed in (b) on some *projection* sequence to $x_k \notin \Omega$ satisfying the subdifferential condition $\widehat{\partial}_{\varepsilon_k} d_\Omega(x_k) \ne \emptyset$ with $\varepsilon_k \downarrow 0$. Note that one can equivalently put $\varepsilon_k = 0$ in the latter condition for locally closed subsets Ω of Asplund spaces.

Theorem 1.105 (projection formulas for basic subgradients of distance functions at out-of-set points). *Let $\Omega \subset X$ be a closed subset of a Banach space, and let $\bar{x} \notin \Omega$. Assume that the best approximation problem to Ω from $\bar{x}$ is well posed. Then*

$$\partial d_\Omega(\bar{x}) \subset \bigcup_{w \in \Pi(\bar{x};\Omega)} \big[N(w; \Omega) \cap I\!B^*\big] .$$

The stronger inclusion

$$\partial d_\Omega(\bar{x}) \subset \bigcup_{w \in \Pi(\bar{x};\Omega)} \big[N(w; \Omega) \cap I\!B^*\big] \setminus \{0\}$$

holds when Ω is SNC at every projection point $w \in \Pi(\bar{x}; \Omega)$. Furthermore,

$$\partial d_\Omega(\bar{x}) \subset \bigcup_{w \in \Pi(\bar{x};\Omega)} \big[N(w;\Omega) \cap S^*\big]$$

if the space X is finite-dimensional.

Proof. Assuming without loss of generality that $\partial d_\Omega(\bar{x}) \neq \emptyset$, we take an arbitrary subgradient $x^* \in \partial d_\Omega(\bar{x})$ and find by definition sequences $\varepsilon_k \downarrow 0$, $x_k \to \bar{x}$, and $x_k^* \xrightarrow{w^*} x^*$ such that

$$x_k^* \in \widehat{\partial}_{\varepsilon_k} d_\Omega(x_k) \ \text{ for all } \ k \in \mathbb{N}\ .$$

Suppose first that the well-posedness property in (b) holds and find a sequence of $w_k \in \Pi(x_k;\Omega)$ converging to some w that clearly belongs to $\Pi(\bar{x};\Omega)$. Moreover, $x_k \notin \Omega$ for all large $k \in \mathbb{N}$. Employing Proposition 1.102, we get a sequence of x_k^* satisfying

$$x_k^* \in \widehat{N}_{\varepsilon_k}(w_k;\Omega) \ \text{ with } \ 1-\varepsilon_k \le \|x_k^*\| \le 1+\varepsilon_k, \quad k \in \mathbb{N}\ .$$

Passing to the limit as $k \to \infty$, we arrive at $x^* \in N(w;\Omega)$, which justifies the first inclusion of the theorem in case (b). The two other inclusions easily follow from the above constructions under the additional assumptions made.

It remains to justify the first inclusion of the theorem under the well-posedness property in (a). Taking $x^* \in \partial d_\Omega(\bar{x})$ and having sequences $(\varepsilon_k, x_k, x_k^*)$ as above, we employ now Theorem 1.103 and get $w_k \in \Omega$ such that

$$x_k^* \in \widehat{N}_{\varepsilon_k}(w_k;\Omega), \quad 1-\varepsilon_k \le \|x_k^*\| \le 1+\varepsilon_k, \quad \text{and}$$

$$d_\Omega(x_k) \le \|x_k - w_k\| \le d_\Omega(x_k) + 2\varepsilon_k\ .$$

This gives the estimates

$$\big|\, \|w_k - \bar{x}\| - d_\Omega(\bar{x})\big| \le \big|\|w_k - \bar{x}\| - \|w_k - x_k\|\,\big| + \big|\, \|w_k - x_k\| - d_\Omega(x_k)\big|$$

$$+\big|d_\Omega(x_k) - d_\Omega(\bar{x})\big| \le 2\|x_k - \bar{x}\| + \big|\, \|w_k - x_k\| - d_\Omega(x_k)\big| \to 0\,,$$

which imply that $\|w_k - \bar{x}\| \to d_\Omega(\bar{x})$ as $k \to \infty$. It follows from the well-posedness property (a) that there is $w \in \Pi(\bar{x};\Omega)$ such that $w_k \to w$ along some subsequence as $k \to \infty$. Thus $x^* \in N(w;\Omega)$ with $\|x^*\| \le 1$. △

Observe that the well-posedness requirement of the theorem is clearly satisfied, via property (b), if the projection sets $\Pi(\cdot;\Omega)$ are nonempty and *uniformly compact* around $\bar{x}$. The latter assumptions are *not needed* under some geometric properties of the space X and the set Ω in question. Recall again (cf. Subsect. 1.1.2) that the norm $\|\cdot\|$ on a Banach space X is *Kadec* if the strong and weak convergence agree on the boundary of its unit sphere. It is well known that every locally uniformly convex space (in particular, every *reflexive* space) admits an *equivalent Kadec norm.*

Corollary 1.106 (basic subgradients of distance functions in spaces with Kadec norms). *Let X be a reflexive Banach space with an equivalent Kadec norm. Given an nonempty set $\Omega \subset X$ and $\bar{x} \notin \Omega$, assume that:*
- *either Ω is weakly closed,*
- *or Ω is closed and $\widehat{\partial} d_\Omega(\bar{x}) \neq \emptyset$.*

Then the best approximation problem to Ω from $\bar{x}$ is well posed. This implies that $\Pi(\bar{x}; \Omega) \neq \emptyset$ and that the first inclusion of Theorem 1.105 holds, while the second one is also fulfilled under the additional SNC assumption made.

Proof. Let Ω be weakly closed. To justify the well-posedness of the best approximation problem via property (a) in Definition 1.104, take any sequence of $x_k \in \Omega$ with $\|x_k - \bar{x}\| \to d_\Omega(\bar{x})$ as $k \to \infty$. Since X is reflexive, we may assume without loss of generality that x_k weakly converge to some $w \in X$. Thus $w \in \Omega$ by the weak closedness of Ω. Observe that

$$\|w - \bar{x}\| \le \liminf_{k \to \infty} \|x_k - \bar{x}\| = d_\Omega(\bar{x}) ,$$

which implies that $w \in \Pi(\bar{x}; \Omega)$ and that $\|x_k - \bar{x}\| \to \|w - \bar{x}\|$. Since the norm on X is Kadec, we get $\|x_k - w\| \to 0$ as $k \to \infty$. The latter justifies the well-posedness property of Theorem 1.105 and thus the inclusions therein provided that Ω is weakly closed. If $\widehat{\partial} d_\Omega(\bar{x}) \neq \emptyset$, then the well-posedness property of the theorem follows from Lemma 6 in Borwein and Giles [146] provided that Ω is just closed in the norm topology of X. △

Note that the inclusions of Theorem 1.105 are generally *strict* even for convex sets in finite dimensions, as in the case of $\Omega := \text{epi}\,(\|\cdot\|) \subset I\!R^2$ with $\bar{x} = (-1, 0)$. On the other hand, both the basic subdifferential and the Fréchet subdifferential of the distance function for any closed set $\Omega \subset I\!R^n$ at $\bar{x} \notin \Omega$ can be *computed* via the *Euclidean projector* $\Pi(\cdot; \Omega)$ by

$$\partial d_\Omega(\bar{x}) = \frac{\bar{x} - \Pi(\bar{x}; \Omega)}{d_\Omega(\bar{x})}, \quad \widehat{\partial} d_\Omega(\bar{x}) = \begin{cases} (\bar{x} - \bar{w})/\|\bar{x} - \bar{w}\| & \text{if } \Pi(\bar{x}; \Omega) = \{\bar{w}\} , \\ \emptyset & \text{otherwise ;} \end{cases}$$

cf. Mordukhovich [901, Proposition 2.7] and Rockafellar and Wets [1165, Example 8.53]. This particularly provides an interesting observation that the *distance function d_Ω is lower regular at $\bar{x} \notin \Omega \subset I\!R^n$ if and only if the Euclidean projector $\Pi(\bar{x}; \Omega)$ is a singleton.* Thus we have a broad class of Lipschitzian functions, which *fail to be lower regular* at intrinsic points. Note that the above formula for computing the basic subdifferential of the distance functions *does't hold in infinite dimensions*, while the inclusion "$\subset$" is valid. Indeed, the equality is violated in any Hilbert space for the orthonormal basis $\Omega := \{e_1, e_2, \ldots\}$ at $\bar{x} = 0 \notin \Omega$.

We refer the reader to the papers by Mordukhovich and Nam [935, 936] for more details and discussions on the above material and also to extended subdifferential results for the distance function to varying/moving sets

$$\rho(x,y) := \inf_{v \in F(x)} \|y - v\| = d\big(y; F(x)\big)$$

useful in many aspects of variational analysis and optimization; see, in particular, Theorem 1.41.

1.3.4 Subdifferential Calculus in Banach Spaces

Here we present a part of subdifferential calculus for extended-real-valued functions valid in arbitrary Banach spaces. We obtain calculus rules describing behavior of basic and singular subgradients from Definition 1.77 (and hence the corresponding upper subgradients) under various operations important for applications. Some of these results follow directly from the coderivative calculus of Subsect. 1.2.4; the others take into account specific features of (extended) real-valued functions. We incorporate regularity statements into calculus rules and also discuss related calculus results for "sequential normal epi-compactness" of functions induced by those in Subsect. 1.2.5.

Dealing with functions that may take infinite values, we adopt the natural conventions on extended arithmetic described in Sect. 1E of the book by Rockafellar and Wets [1165]. One obviously has

$$\partial(\lambda\varphi)(\bar{x}) = \begin{cases} \lambda\partial\varphi(\bar{x}) & \text{if } \lambda \geq 0\,, \\ \lambda\partial^+\varphi(\bar{x}) & \text{otherwise} \end{cases}$$

and similarly for ∂^∞, $\widehat{\partial}$, and the corresponding upper subdifferentials. The next proposition gives subdifferential sum rules ensuring equalities with no regularity assumptions.

Proposition 1.107 (subdifferential sum rules with equalities). *Given an arbitrary function $\psi\colon X \to \overline{I\!R}$ finite at $\bar{x}$, the following hold:*

(i) *For any $\varphi\colon X \to \overline{I\!R}$ Fréchet differentiable at $\bar{x}$ one has*

$$\widehat{\partial}(\varphi + \psi)(\bar{x}) = \nabla\varphi(\bar{x}) + \widehat{\partial}\psi(\bar{x})\,.$$

(ii) *For any $\varphi\colon X \to \overline{I\!R}$ strictly differentiable at $\bar{x}$ one has*

$$\partial(\varphi + \psi)(\bar{x}) = \nabla\varphi(\bar{x}) + \partial\psi(\bar{x})\,.$$

Moreover, $\varphi + \psi$ is lower (resp. epigraphically) regular at $\bar{x}$ if and only if ψ has the corresponding property at this point.

(iii) *For any $\varphi\colon X \to \overline{I\!R}$ Lipschitz continuous around $\bar{x}$ one has*

$$\partial^\infty(\varphi + \psi)(\bar{x}) = \partial^\infty\psi(\bar{x})\,.$$

Proof. Assertions (i) and (ii) follow from Theorem 1.62 and Proposition 1.92. Let us prove the inclusion "$\subset$" in (iii). Given $x^* \in \partial^\infty(\varphi + \psi)(\bar{x})$, we find sequences $\varepsilon_k \downarrow 0$, $(x_k, \alpha_k) \overset{\text{epi}(\varphi+\psi)}{\to} (\bar{x}, (\varphi+\psi)(\bar{x}))$, $x_k^* \overset{w^*}{\to} x^*$, $\nu_k \to 0$, and $\eta_k \downarrow 0$ such that

$$\langle x_k^*, x - x_k\rangle + \nu_k(\alpha - \alpha_k) \le 2\varepsilon_k(\|x - x_k\| + |\alpha - \alpha_k|)$$

for all $(x, \alpha) \in \text{epi}\,(\varphi + \psi)$ with $x \in x_k + \eta_k I\!B$ and $|\alpha - \alpha_k| \le \eta_k$, $k \in I\!N$. Let $\ell > 0$ be a Lipschitz modulus of φ around $\bar{x}$, let $\tilde{\eta}_k := \eta_k/2(\ell+1)$, and let $\tilde{\alpha}_k := \alpha_k - \varphi(x_k)$. We have $(x_k, \tilde{\alpha}_k) \overset{\text{epi}\,\psi}{\to} (\bar{x}, \psi(\bar{x}))$ and check that

$$(x, \alpha + \varphi(x)) \in \text{epi}\,(\varphi + \psi), \quad |(\alpha + \varphi(x)) - \alpha_k| \le \eta_k$$

whenever $(x, \alpha) \in \text{epi}\,\psi$, $x \in x_k + \tilde{\eta}_k I\!B$, and $|\alpha - \tilde{\alpha}_k| \le \tilde{\eta}_k$. Hence

$$\langle x^*, x - x_k\rangle + \nu_k(\alpha - \tilde{\alpha}_k) \le \tilde{\varepsilon}_k(\|x - x_k\| + |\alpha - \tilde{\alpha}_k|) \text{ with } \tilde{\varepsilon}_k := 2\varepsilon_k(1+\ell) + |\nu_k|\ell$$

for any $(x, \alpha) \in \text{epi}\,\psi$ with $x \in x_k + \tilde{\eta}_k I\!B$ and $|\alpha - \tilde{\alpha}_k| \le \tilde{\eta}_k$. This implies $(x_k^*, \nu_k) \in \widehat{N}_{\tilde{\varepsilon}_k}((x_k, \tilde{\alpha}_k); \text{epi}\,\psi)$ for all $k \in I\!N$, and hence $(x^*, 0) \in N((\bar{x}, \psi(\bar{x})); \text{epi}\,\psi)$ due to $\tilde{\varepsilon}_k \downarrow 0$ as $k \to \infty$. Thus we get the inclusion "$\subset$" in (iii). Applying it to the sum $\psi = (\psi + \varphi) + (-\varphi)$, one has $\partial^\infty \psi(\bar{x}) \subset \partial^\infty(\varphi + \psi)(\bar{x})$, which gives the equality in (iii). △

Next we consider subdifferentiation of the so-called *marginal functions* generally defined by

$$\mu(x) := \inf\left\{\varphi(x, y)\big|\ y \in G(x)\right\}, \tag{1.60}$$

where $\varphi\colon X \times Y \to \overline{I\!R}$ is an extended-real-valued cost function and $G\colon X \rightrightarrows Y$ is a set-valued constraint mapping between Banach spaces. Marginal functions (1.60) can be interpreted as *value functions* in parametric optimization problems of the form

$$\text{minimize } \varphi(x, y) \text{ subject to } y \in G(x).$$

They play an important role in variational analysis, optimization, control theory, and various applications. It is well known that marginal functions (1.60) don't usually admit a classical derivative even for smooth and simple initial data φ and G. In what follows we calculate basic and singular subgradients of (1.60) and present applications of the obtained results to subdifferential chain rules and related calculus.

The next theorem gives upper estimates of the subdifferentials $\partial\mu(\bar{x})$ and $\partial^\infty\mu(\bar{x})$ in terms of the corresponding subdifferentials of the extended function

$$\vartheta(x, y) := \varphi(x, y) + \delta((x, y); \text{gph}\,G).$$

The results involve the *argminimum mapping* $M\colon X \to Y$ defined by

$$M(x) := \left\{y \in G(x)\big|\ \varphi(x, y) = \mu(x)\right\}$$

and depend on inner semicontinuous/semicompact properties of M formulated in Definition 1.63. Recall that G is closed-graph *at* $\bar{x}$ if $\bar{y} \in G(\bar{x})$ whenever $x_k \to \bar{x}$ and $y_k \to \bar{y}$ with $y_k \in G(x_k)$ as $k \to \infty$.

Theorem 1.108 (subdifferentiation of marginal functions). *Let the marginal function* (1.60) *is finite at* $\bar{x}$ *with* $M(\bar{x}) \neq \emptyset$. *The following hold:*

(i) *Given* $\bar{y} \in M(\bar{x})$, *assume that* M *is inner semicontinuous at* $(\bar{x}, \bar{y})$. *Then one has*

$$\partial\mu(\bar{x}) \subset \{x^* \in X^* \big| \, (x^*, 0) \in \partial\vartheta(\bar{x}, \bar{y})\} ,$$

$$\partial^\infty\mu(\bar{x}) \subset \{x^* \in X^* \big| \, (x^*, 0) \in \partial^\infty\vartheta(\bar{x}, \bar{y})\} .$$

(ii) *Assume that* M *is inner semicompact at* $\bar{x}$, *that* G *is closed-graph at* $\bar{x}$, *and that* φ *is l.s.c. on* $\operatorname{gph} G$ *when* $x = \bar{x}$. *Then one has*

$$\partial\mu(\bar{x}) \subset \Big\{x^* \in X^* \Big| \, (x^*, 0) \in \bigcup_{\bar{y} \in M(\bar{x})} \partial\vartheta(\bar{x}, \bar{y})\Big\} ,$$

$$\partial^\infty\mu(\bar{x}) \subset \Big\{x^* \in X^* \Big| \, (x^*, 0) \in \bigcup_{\bar{y} \in M(\bar{x})} \partial^\infty\vartheta(\bar{x}, \bar{y})\Big\} .$$

Proof. To justify (i), we first prove the estimate for $\partial\mu(\bar{x})$. Picking $x^* \in \partial\mu(\bar{x})$ and using (1.55), we find sequences $\varepsilon_k \downarrow 0$, $x_k \xrightarrow{\mu} \bar{x}$, and $x_k^* \xrightarrow{w^*} x^*$ with $x_k^* \in \widehat{\partial}_{\varepsilon_k}\mu(x_k)$ for all $k \in I\!N$. Hence there is $\eta_k \downarrow 0$ such that

$$\langle x_k^*, x - x_k\rangle \le \mu(x) - \mu(x_k) + 2\varepsilon_k\|x - x_k\| \quad \text{whenever} \quad x \in x_k + \eta_k I\!B .$$

By constructions of μ, ϑ, and M one has

$$\langle (x_k^*, 0), (x, y) - (x_k, y_k)\rangle \le \vartheta(x, y) - \vartheta(x_k, y_k) + 2\varepsilon_k(\|x - x_k\| + \|y - y_k\|)$$

for all $y_k \in M(x_k)$ and $(x, y) \in (x_k, y_k) + \eta_k I\!B$, $k \in I\!N$. This gives $(x_k^*, 0) \in \widehat{\partial}_{\tilde{\varepsilon}_k}\vartheta(x_k, y_k)$ with $\tilde{\varepsilon}_k := 2\varepsilon_k$. Since M is inner semicontinuous at $(\bar{x}, \bar{y})$, we select a sequence of $y_k \in M(x_k)$ converging to $\bar{y}$. Observe that $\vartheta(x_k, y_k) \to \vartheta(\bar{x}, \bar{y})$ due to $\mu(x_k) \to \mu(\bar{x})$. Thus $(x^*, 0) \in \partial\vartheta(\bar{x}, \bar{y})$, which justifies the first inclusion in (i).

To prove the second inclusion in (i), we take $x^* \in \partial^\infty\mu(\bar{x})$ and get sequences $\varepsilon_k \downarrow 0$, $x_k \xrightarrow{\mu} \bar{x}$, $(x_k^*, \nu_k) \xrightarrow{w^*} (x^*, 0)$, and $\eta_k \downarrow 0$ such that

$$\langle x_k^*, x - x_k\rangle + \nu_k(\alpha - \alpha_k) \le 2\varepsilon_k(\|x - x_k\| + |\alpha - \alpha_k|)$$

if $(x, \alpha) \in \operatorname{epi}\mu$, $x \in x_k + \eta_k I\!B$, and $|\alpha - \alpha_k| \le \eta_k$ for $k \in I\!N$. Similarly to the proof of (i) we select $y_k \to \bar{y}$ with $y_k \in M(x_k)$, $\alpha_k \downarrow \vartheta(\bar{x})$, and

$$(x_k^*, 0, \nu_k) \in \widehat{N}_{2\varepsilon_k}((x_k, y_k, \alpha_k); \operatorname{epi}\vartheta), \quad k \in I\!N .$$

Passing to the limit as $k \to \infty$, one has $(x^*, 0) \in \partial^\infty\vartheta(\bar{x})$, which completes the proof of (i).

Let us justify assertion (ii) of the theorem under the assumptions made. Proceeding as in the proof of (i), we get the corresponding sequences $\{x_k\}$ and $\{y_k\}$ satisfying

$$x_k \to \bar{x}, \quad \mu(x_k) \to \mu(\bar{x}), \quad y_k \in G(x_k), \quad \varphi(x_k, y_k) = \mu(x_k) .$$

By the inner semicompactness of M at $\bar{x}$ there is a subsequence of y_k that converges to some $\bar{y}$ (without relabeling). It follows from the closed-graph assumption on G that $\bar{y} \in G(\bar{x})$. Similarly to the proof of (i), it remains to show that $\varphi(\bar{x}, \bar{y}) = \mu(\bar{x})$, which then implies both inclusions in (ii). Involving the lower semicontinuity of φ on $\operatorname{gph} G$ and the above choice of x_k and y_k, one therefore has

$$\varphi(\bar{x}, \bar{y}) \le \liminf_{k\to\infty} \varphi(x_k, y_k) = \liminf_{k\to\infty} \mu(x_k) = \mu(\bar{x}) ,$$

which ends the proof of the theorem. $\triangle$

When the cost function φ in (1.60) is strictly differentiable at points in question, we get the following corollary of Theorem 1.108 that gives upper estimates of $\partial\mu(\bar{x})$ and $\partial^\infty\mu(\bar{x})$ in terms of partial gradients of φ and the normal coderivative of G. For simplicity we consider only case (i) of the theorem.

Corollary 1.109 (marginal functions with smooth costs). *Given $\bar{y} \in M(\bar{x})$ in (1.60), we assume that M is inner semicontinuous at $(\bar{x}, \bar{y})$ and that φ is strictly differentiable at this point. Then*

$$\partial\mu(\bar{x}) \subset \nabla_x\varphi(\bar{x}, \bar{y}) + D_N^*G(\bar{x}, \bar{y})(\nabla_y\varphi(\bar{x}, \bar{y})), \quad \partial^\infty\mu(\bar{x}) \subset D_N^*G(\bar{x}, \bar{y})(0) .$$

Proof. Follows from Theorem 1.108(i) by applying the sum rules of Proposition 1.107 to the function ϑ with the usage of Proposition 1.79 and representation (1.26) of the normal coderivative. $\triangle$

Now let us consider a special case of (1.60) when the constraint mapping $G := g\colon X \to Y$ is single-valued. Then the marginal function $\mu(x)$ reduces to the *composition*

$$\mu(x) = (\varphi \circ g)(x) := \varphi(x, g(x)) , \tag{1.61}$$

which is the standard one $\varphi(g(x))$ when φ doesn't depend on x. The next theorem provides the *exact calculation* (equalities) for the basic and singular subdifferentials of compositions (1.61) in the case of locally Lipschitzian mappings g. Its first assertion ensures that the inclusions of Theorem 1.108 become equalities in this case. The second assertion gives precise formulas for computing the basic subdifferential of (1.61) in terms of the mixed coderivative of g and the subdifferential of its scalarization, which improve the result of Corollary 1.109. Both assertions also contain additional regularity statements.

Theorem 1.110 (subdifferentiation of compositions: equalities). *Let $\varphi: X \times Y \to \overline{I\!R}$ be finite at $(\bar{x}, \bar{y})$ with $\bar{y} := g(\bar{x})$, and let $g: X \to Y$ be Lipschitz continuous around $\bar{x}$. Then the following hold for composition* (1.61)*:*

(i) *One has*

$$\partial(\varphi \circ g)(\bar{x}) = \big\{x^* \in X^* \big|\ (x^*, 0) \in \partial\vartheta(\bar{x}, g(\bar{x}))\big\}\ ,$$

$$\partial^\infty(\varphi \circ g)(\bar{x}) = \big\{x^* \in X^* \big|\ (x^*, 0) \in \partial^\infty\vartheta(\bar{x}, g(\bar{x}))\big\}$$

if either g is strictly differentiable at $\bar{x}$ or $\dim Y < \infty$. In the latter case $\varphi \circ g$ is lower (resp. epigraphically) regular at $\bar{x}$ if $\vartheta := \varphi + \delta(\cdot; \operatorname{gph} g)$ has the corresponding property at $(\bar{x}, \bar{y})$.

(ii) *Assume that φ is strictly differentiable at $(\bar{x}, \bar{y})$. Then*

$$\partial(\varphi \circ g)(\bar{x}) = \nabla_x\varphi(\bar{x}, \bar{y}) + D^*_M g(\bar{x})(\nabla_y\varphi(\bar{x}, \bar{y}))$$

$$= \nabla_x\varphi(\bar{x}, \bar{y}) + \partial\langle\nabla_y\varphi(\bar{x}, \bar{y}), g\rangle(\bar{x})\ .$$

Moreover, $\varphi \circ g$ at $\bar{x}$ is lower regular at $\bar{x}$ if g is M-regular at this point.

Proof. One can check, using (1.47), that (i) is a special case of Theorem 1.64(iii) with $G(x) := (x, g(x))$ and $F := E_\varphi$, the epigraphical multifunction. Then observe that both representations in (ii) are equivalent due to Theorem 1.90 and that the regularity statement follows directly from the first equality in (ii). It remains to prove the second representation in (ii).

Take an arbitrary sequence $\gamma_j \downarrow 0$ and, by the strict differentiability of φ at $(\bar{x}, \bar{y})$, find $\eta_j \downarrow 0$ such that

$$|\varphi(u, g(u)) - \varphi(x, g(x)) - \langle\nabla_x\varphi(\bar{x}, \bar{y}), u - x\rangle - \langle\nabla_y\varphi(\bar{x}, \bar{y}), g(u) - g(x)\rangle|$$

$$\le \gamma_j(\|u - x\| + \|g(u) - g(x)\|) \ \text{ for all } \ x, u \in B_{\eta_j}(\bar{x}), \quad j \in I\!N\ .$$

Then pick $x^* \in \partial(\varphi \circ g)(\bar{x})$ and get $\varepsilon_k \downarrow 0$, $x_k \to \bar{x}$, and $x_k^* \xrightarrow{w^*} x^*$ with $x_k^* \in \widehat{\partial}_{\varepsilon_k}(\varphi \circ g)(x_k)$, $k \in I\!N$. This allows us to select a sequence $k_j \to \infty$ as $j \to \infty$ such that $\|x_{k_j} - \bar{x}\| \le \eta_j/2$ and

$$\varphi(x, g(x)) - \varphi(x_{k_j}, g(x_{k_j})) - \langle x^*_{k_j}, x - x_{k_j}\rangle \ge -2\varepsilon_{k_j}\|x - x_{k_j}\|$$

for all $x \in x_{k_j} + (\eta_j/2)I\!B$, $j \in I\!N$. Combining this with the above inequality from strict differentiability, one gets

$$\langle\nabla_y\varphi(\bar{x}, \bar{y}), g(x) - g(x_{k_j})\rangle - \langle x^*_{k_j} - \nabla_x\varphi(\bar{x}, \bar{y}), x - x_{k_j}\rangle$$

$$\ge -\big[2\varepsilon_{k_j} + \gamma_j(\ell + 1)\big]\|x - x_{k_j}\| \ \text{ for } \ x \in x_{k_j} + (\eta_j/2)I\!B, \quad j \in I\!N\ ,$$

where ℓ is a Lipschitz modulus of g around $\bar{x}$. Thus

$$x^*_{k_j} - \nabla_x\varphi(\bar{x},\bar{y}) \in \widehat{\partial}_{\tilde{\varepsilon}_j}\langle\nabla_y\varphi(\bar{x},\bar{y}), g\rangle(x_{k_j}) \;\text{ with }\; \tilde{\varepsilon}_j := 2\varepsilon_{k_j} + \gamma_j(\ell+1)\,.$$

Passing to the limit as $j \to \infty$, we arrive at $x^* - \nabla_x\varphi(\bar{x},\bar{y}) \in \partial\langle\nabla_y\varphi(\bar{x},\bar{y}), g\rangle(\bar{x})$. To verify the opposite inclusion, we employ similar arguments starting with a point $x^* \in \partial\langle\nabla_y\varphi(\bar{x},\bar{y}), g\rangle(\bar{x})$. △

The second representation in Theorem 1.110(ii) can be treated as a *subdifferential chain rule* for compositions with strictly differentiable outer functions. It easily implies the corresponding formulas for subgradients of products and quotients involving Lipschitz continuous functions that generalize the classical *product and quotient rules.*

Corollary 1.111 (subdifferentiation of products and quotients). *Let $\varphi\colon X \to \overline{I\!R}$, $i = 1, 2$, be Lipschitz continuous around $\bar{x}$. The following hold:*

(i) *One always has*

$$\partial(\varphi_1\cdot\varphi_2)(\bar{x}) = \partial\big(\varphi_2(\bar{x})\varphi_1 + \varphi_1(\bar{x})\varphi_2\big)(\bar{x})\,.$$

If in addition φ_1 is strictly differentiable at $\bar{x}$, then

$$\partial(\varphi_1\cdot\varphi_2)(\bar{x}) = \nabla\varphi_1(\bar{x})\varphi_2(x) + \partial\big(\varphi_1(\bar{x})\varphi_2\big)(\bar{x})\,.$$

In the latter case $\varphi_1\cdot\varphi_2$ is lower regular at $\bar{x}$ if and only if the function $x \to \varphi_1(\bar{x})\varphi_2(x)$ is lower regular at this point.

(ii) *Assume that $\varphi_2(\bar{x}) \neq 0$. Then*

$$\partial(\varphi_1/\varphi_2)(\bar{x}) = \frac{\partial\big(\varphi_2(\bar{x})\varphi_1 - \varphi_1(\bar{x})\varphi_2\big)(\bar{x})}{[\varphi_2(\bar{x})]^2}\,.$$

If in addition φ_1 is strictly differentiable at $\bar{x}$, then

$$\partial(\varphi_1/\varphi_2)(\bar{x}) = \frac{\nabla\varphi_1(\bar{x})\varphi_2(\bar{x}) + \partial\big(-\varphi_1(\bar{x})\varphi_2\big)(\bar{x})}{[\varphi_2(\bar{x})]^2}\,.$$

In the latter case φ_1/φ_2 is lower regular at $\bar{x}$ if and only if the function $x \to \varphi_1(\bar{x})\varphi_2(x)$ is upper regular at this point.

(iii) *Let $\varphi\colon X \to \overline{I\!R}$ be Lipschitz continuous around $\bar{x}$ with $\varphi(\bar{x}) \neq 0$. Then*

$$\partial(1/\varphi)(\bar{x}) = -\frac{\partial^+\varphi(\bar{x})}{\varphi^2(\bar{x})}\,.$$

Moreover, $1/\varphi$ is lower regular at φ if and only if φ is upper regular at this point.

Proof. To prove (i), represent $\varphi_1\cdot\varphi_2$ as composition (1.61) with $\varphi\colon I\!R^2 \to I\!R$ and $g\colon X \to I\!R^2$ defined by

$$\varphi(y_1, y_2) := y_1\cdot y_2 \;\text{ and }\; g(x) := \big(\varphi_1(x), \varphi_2(x)\big)\,.$$

Then Theorem 1.110(ii) gives the first equality in (i), which implies the second one and the regularity statement due to Proposition 1.107(ii). The proof of (ii) is similar with $\varphi(y_1, y_2) := y_1/y_2$ and the same mapping g in composition (1.61). Assertion (iii) is a special case of (ii) with $\varphi_1 = 1$ and $\varphi_2 = \varphi$. △

Let us consider another important class of compositions (1.61) with strictly differentiable inner mappings. The next proposition contains *equality-type subdifferential chain rules* in the case of *surjective* derivatives. It follows from the corresponding results for coderivatives based on the normal cone calculus from Subsect. 1.1.2.

Proposition 1.112 (subdifferentiation of compositions with surjective derivatives of inner mappings). *Consider composition* (1.61), *where* $g\colon X \to Y$ *is strictly differentiable at* $\bar{x}$ *with the surjective derivative* $\nabla g(\bar{x})$ *and where* $\varphi(x, y) = \varphi_1(x) + \varphi_2(y)$ *with* $\varphi_2\colon Y \to \overline{I\!R}$ *finite at* $\bar{y} = g(\bar{x})$. *The following assertions hold:*

(i) *If* φ_1 *is strictly differentiable at* $\bar{x}$, *then*

$$\partial(\varphi \circ g)(\bar{x}) = \nabla\varphi_1(\bar{x}) + \nabla g(\bar{x})^* \partial\varphi_2(\bar{y}) .$$

In this case $\varphi \circ g$ *is lower (resp. epigraphically) regular at* $\bar{x}$ *if and only if* φ_2 *has the corresponding property at* $\bar{y}$.

(ii) *If* φ_1 *is Lipschitz continuous around* $\bar{x}$, *then*

$$\partial^\infty(\varphi \circ g)(\bar{x}) = \nabla g(\bar{x})^* \partial^\infty \varphi_2(\bar{y}) .$$

Proof. The subdifferential chain rules and regularity conclusions for the composition $\varphi_2 \circ g$ follow from Theorem 1.66 with $F := E_{\varphi_2}$. To get the whole statement, we then need to apply Proposition 1.107 to $\varphi_1 + \varphi_2 \circ g$. △

Next let us consider *minimum functions* of the form

$$\big(\min \varphi_i\big)(x) := \min\big\{\varphi_i(x)\big|\ i = 1, \ldots, n\big\} ,$$

where $\varphi_i\colon X \to \overline{I\!R}$ and $n \geq 2$. Note that such functions are nonsmooth (even when all φ_i are smooth) and belong to the class of marginal functions (1.60). However, its argminimum mapping

$$M(x) = \big\{i \in \{1, \ldots, n\}\big|\ \varphi_i(x) = \big(\min \varphi_i\big)(x)\big\}$$

doesn't satisfy the assumptions of Theorem 1.108 at nontrivial points. In the following proposition we directly derive an efficient upper estimate of $\partial\big(\min \varphi_i\big)(\bar{x})$ in terms of basic subgradients of the involved functions φ_i.

Proposition 1.113 (subdifferentiation of minimum functions). *Let* φ_i *be finite at* $\bar{x}$ *for all* $i = 1, \ldots, n$ *and l.s.c. at* $\bar{x}$ *for* $i \notin M(\bar{x})$. *Then*

$$\partial\big(\min \varphi_i\big)(\bar{x}) \subset \bigcup\big\{\partial\varphi_i(\bar{x})\big|\ i \in M(\bar{x})\big\} .$$

Proof. Consider a sequence of $x_k \in X$ such that $x_k \to \bar{x}$ and $\varphi_i(x_k) \to (\min \varphi_i)(\bar{x})$ for $i \notin M(\bar{x})$. Using the lower semicontinuity of φ_i at $\bar{x}$ for $i \notin M(\bar{x})$, we get $M(x_k) \subset M(\bar{x})$. It follows from the construction of analytic ε-subgradients that

$$\widehat{\partial}_\varepsilon \big(\min \varphi_i \big)(x_k) \subset \bigcup \big\{ \widehat{\partial}_\varepsilon \varphi_i(x_k) \big|\ i \in M(\bar{x}) \big\}$$

for any $\varepsilon \geq 0$ and $k \in I\!N$. The latter implies the inclusion in the proposition due to representation (1.55) of basic subgradients. △

It is well known that one of the most fundamental principles of classical analysis is the *Fermat rule* (or *stationary principle*) discovered in 1636 for polynomials [442], according to which gradients of differentiable functions must vanish at points of local minima and maxima. The following proposition contains nonsmooth counterparts of this rule for the case of arbitrary extended-real-valued functions in terms of their lower and upper subgradients, which naturally distinguish between minima and maxima.

Proposition 1.114 (nonsmooth versions of Fermat's rule). *Let $\varphi\colon X \to \overline{I\!R}$ be finite at $\bar{x}$. Then $0 \in \widehat{\partial}\varphi(\bar{x}) \subset \partial\varphi(\bar{x})$ if φ has a local minimum at $\bar{x}$, and $0 \in \widehat{\partial}^+\varphi(\bar{x}) \subset \partial^+\varphi(\bar{x})$ if φ has a local maximum at $\bar{x}$. Thus*

$$0 \in \widehat{\partial}\varphi(\bar{x}) \cup \widehat{\partial}^+\varphi(\bar{x}) \subset \partial^0\varphi(\bar{x})$$

if $\bar{x}$ is either a local minimum or a local maximum point of φ.

Proof. The inclusion $0 \in \widehat{\partial}\varphi(\bar{x})$ at points of local minimum follows directly from the definition of Fréchet subgradients in (1.51). This implies the other statements, since we always have $\widehat{\partial}\varphi(\bar{x}) \subset \partial\varphi(\bar{x})$ as well as $\widehat{\partial}^+\varphi(\bar{x}) = -\widehat{\partial}(-\varphi)(\bar{x}) \subset \partial^+\varphi(\bar{x})$. △

As we have mentioned above, the union $\widehat{\partial}\varphi(\bar{x}) \cup \widehat{\partial}^+\varphi(\bar{x})$ always reduces to one of the sets $\widehat{\partial}\varphi(\bar{x})$ and $\widehat{\partial}^+\varphi(\bar{x})$, while the symmetric subdifferential $\partial^0\varphi(\bar{x})$ in (1.46) has an independent meaning; see, e.g., the calculation in (1.57). The main difference between the Fréchet-like constructions $\widehat{\partial}$ and our basic ones is that the latter have much better calculus, which is crucial for applications.

Following the line in standard calculus, we obtain a nonsmooth version of the *Lagrange mean value* theorem in Banach spaces, which is based on the generalized Fermat rule from Proposition 1.114.

Proposition 1.115 (mean values). *Let $a, b \in X$ and let $\varphi\colon X \to \overline{I\!R}$ be continuous on $[a, b] := \big\{ a + t(b-a) \big|\ 0 \leq t \leq 1 \big\}$. Then there is a number $\theta \in (0, 1)$ such that*

$$\varphi(b) - \varphi(a) \in \partial_t^0 \varphi(a + \theta(b-a))\ ,$$

where the set on the right-hand side stands for the symmetric subdifferential of the function $t \to \varphi(a + t(b-a))$ at $t = \theta$.

Proof. Consider a function $\phi\colon [0,1] \to I\!R$ defined by

$$\phi(t) := \varphi(a + t(b-a)) + t(\varphi(a) - \varphi(b)), \quad 0 \le t \le 1 .$$

This function is continuous on $[0,1]$ with $\phi(0) = \phi(1) = \varphi(a)$. Thus, by the classical Weierstrass theorem, it attains both global minimum and maximum on $[0,1]$. Excluding the trivial case when ϕ is constant on $[0,1]$, we conclude that there is an interior point $\theta \in (0,1)$ at which ϕ attains either its minimal or maximal value over $[0,1]$. Employing Proposition 1.114, one has $0 \in \partial^0\phi(\theta)$. Observe that ϕ is the sum of two functions one of which is smooth. We end the proof by using Proposition 1.107(ii). △

Note that ∂^0 cannot be replaced with ∂ in Theorem 1.115 as follows from the example of $\varphi(x) = -|x|$ on $[-1,1]$. If φ is strictly differentiable at every point of the interval $(a,b) \subset X$, we can apply the chain rule to the composition

$$\varphi(a + t(b-a)) = (\varphi \circ g)(t) \;\text{ with }\; g(t) := a + t(b-a)$$

(cf. Theorem 1.110) and get the classical mean value theorem in Banach spaces. However, the chain rules obtained above don't allow us to proceed in this way without the strict differentiability assumption on φ. Observe that the chain rule from Proposition 1.112 is not applicable in this setting, since the derivative of $g\colon I\!R \to X$ is not surjective. In Chap. 3 we develop more involved calculus in Asplund spaces that contains, in particular, extended coderivative and subdifferential chain rules with no surjectivity assumptions and also there counterparts for nonsmooth and set-valued mappings. Such an enhanced (full) calculus is based on the *extremal principle* and related variational results of Chap. 2.

To conclude this subsection, we consider an epigraphical version of the sequential normal compactness (SNC) property for extended-real-valued functions. This property is needed in what follows, particularly for the enhanced subdifferential calculus in Chap. 3.

Definition 1.116 (sequential normal epi-compactness of functions). *Let $\varphi\colon X \to \overline{I\!R}$ be finite at $\bar{x}$. We say that φ is* SEQUENTIALLY NORMALLY EPI-COMPACT (SNEC) *at $\bar{x}$ if its epigraph is sequentially normally compact at $(\bar{x}, \varphi(\bar{x}))$.*

Due to relationships between subdifferentials and coderivatives of epigraphical multifunctions, this can be equivalently described in terms of ε-subgradients of φ and their singular counterparts. In the case of Asplund spaces, a convenient description of the SNEC property via Fréchet subgradients is given in Subsect. 2.4.2.

We need to distinguish between the SNEC and SNC properties of real-valued functions; cf. Definition 1.67 for $\varphi\colon X \to I\!R$. The latter is equivalent to the SNC property of $\operatorname{gph}\varphi$ at $(\bar{x}, \varphi(\bar{x}))$, being more restrictive than the SNEC one due to the decreasing relation (1.5) for ε-normals. Note that there is no difference between the SNC and PSNC properties for real-valued functions.

It follows from Theorem 1.26 that φ is SNEC at $\bar{x}$ if its epigraph is compactly epi-Lipschitzian around $(\bar{x}, \varphi(\bar{x}))$. This happens, in particular, when either $\dim X < \infty$ or φ is *directionally Lipschitzian* around $\bar{x}$, which corresponds to the epi-Lipschitzian property of $\operatorname{epi}\varphi$ around $(\bar{x}, \varphi(\bar{x}))$; see Rockafellar [1147] for more details on directionally Lipschitzian functions. Hence every function φ Lipschitz continuous around $\bar{x}$ is SNEC at this point; moreover, it has the SNC property by Corollary 1.69(i).

For efficient applications of the SNEC property it is important to have calculus results that ensure its preservation under various operations. Due to Definition 1.116 such a calculus is induced by the corresponding results for general multifunctions applied to the case of epigraphical ones. The next proposition gives a useful necessary and sufficient condition in this direction for arbitrary Banach spaces.

Proposition 1.117 (SNEC property under compositions with strictly differentiable inner mappings). *Let $g: X \to Y$ be strictly differentiable at $\bar{x}$ with the surjective derivative $\nabla g(\bar{x})$ and let $\varphi: Y \to \overline{I\!R}$ be finite at $\bar{y} = g(\bar{x})$. Then $\varphi \circ g$ is SNEC at $\bar{x}$ if and only if φ has this property at $\bar{y}$.*

Proof. Follows from Theorem 1.74 with $F = E_\varphi$. △

Note that other results of Subsect. 1.2.5 dealing with the SNC and PSNC properties under additions and compositions provide sufficient conditions for the SNEC property of real-valued functions generated in this way. In Chap. 3 we present more developed calculus for all these properties in the case of Asplund spaces.

1.3.5 Second-Order Subdifferentials

All the previous material was related to the *first-order* generalized differentiation. Now let us describe some *second-order* generalized differential constructions for extended-real-valued functions. We adopt the classical "derivative-of-derivative" approach to the second-order differentiation that regards second derivatives as first derivatives of gradient mappings. Developing such an approach to the second-order subdifferentiation of nonsmooth functions, one faces the fact that first-order subgradient mappings are *multifunctions*. Therefore, to describe "second-order subgradients" of extended-real-valued functions, certain derivative-like constructions for set-valued mappings should be employed. In this way we define second-order subdifferentials of functions $\varphi: X \to \overline{I\!R}$ on Banach spaces via *coderivatives* of the basic subgradient mapping $\partial\varphi: X \rightrightarrows X^*$ that provide *dual-space* approximations of $\partial\varphi(\cdot)$. Such constructions possess a good calculus and turn out to be useful for the study of a range of problems in optimization and variational analysis, especially those related to robust stability of variational systems; see below.

The general scheme of defining second-order subdifferentials of φ at $\bar{x}$ relative to $\bar{y} \in \partial\varphi(\bar{x})$ is as follows:

$$\partial^2\varphi(\bar{x},\bar{y})(u) = (D^*\partial\varphi)(\bar{x},\bar{y})(u) , \tag{1.62}$$

where $\partial\varphi(\cdot)$ stands for some first-order subdifferential mapping and where D^* stands for its coderivative. Considering for definiteness only *lower* subdifferential constructions, apply this scheme to the basic subdifferential ∂ from Definition 1.77(i) and the two limiting coderivatives ($D^* = D^*_N$ and $D^* = D^*_M$) defined in (1.24) and (1.25), respectively.

Definition 1.118 (second-order subdifferentials). *Let $\varphi\colon X \to \overline{I\!R}$ be finite at $\bar{x}$, and let $\bar{y} \in \partial\varphi(\bar{x})$. Then:*

(i) *The mapping $\partial^2_N\varphi(\bar{x},\bar{y})\colon X^{**} \rightrightarrows X^*$ with the values*

$$\partial^2_N\varphi(\bar{x},\bar{y})(u) := (D^*_N\partial\varphi)(\bar{x},\bar{y})(u), \quad u \in X^{**} ,$$

is the NORMAL SECOND-ORDER SUBDIFFERENTIAL *of φ at $\bar{x}$ relative to $\bar{y}$.*

(ii) *The mapping $\partial^2_M\varphi(\bar{x},\bar{y})\colon X^{**} \rightrightarrows X^*$ with the values*

$$\partial^2_M\varphi(\bar{x},\bar{y})(u) := (D^*_M\partial\varphi)(\bar{x},\bar{y})(u), \quad u \in X^{**} ,$$

is the MIXED SECOND-ORDER SUBDIFFERENTIAL *of φ at $\bar{x}$ relative to $\bar{y}$.*

Using the coderivatives of the first-order upper subdifferential from Definition 1.78, we can define the corresponding *second-order upper subdifferentials* of φ at $\bar{x}$ relative to $\bar{y} \in \partial^+\varphi(\bar{x})$, which symmetrically reduce to the second-order lower subdifferentials of $-\varphi$ and are not considered in what follows.

There is no difference between $\partial^2_N\varphi(\bar{x},\bar{y})$ and $\partial^2_M\varphi(\bar{x},\bar{y})$ if the normal and mixed coderivatives agree for $\partial\varphi$ at $(\bar{x},\bar{y})$; then we use the symbol $\partial^2\varphi(\bar{x},\bar{y})$ in Definition 1.118. It happens, in particular, if X is finite-dimensional and also if $\partial\varphi$ is N-regular at $(\bar{x},\bar{y})$. The latter always holds for $\mathcal{C}^2$ (and for slightly more general) functions when, moreover, the values of the second-order subdifferential mappings are singletons and coincide with images of the adjoint operator to the classical second-order derivative.

Proposition 1.119 (second-order subdifferentials of twice differentiable functions). *Let $\varphi \in \mathcal{C}^1$ around $\bar{x}$, and let its derivative operator $\nabla\varphi\colon X \to X^*$ be strictly differentiable at $\bar{x}$ with the strict derivative denoted by $\nabla^2\varphi(\bar{x})$. Then*

$$\partial^2_N\varphi(\bar{x})(u) = \partial^2_M\varphi(\bar{x})(u) = \big\{\nabla^2\varphi(\bar{x})^*u\big\} \text{ for all } u \in X^{**} .$$

Proof. If $\varphi \in \mathcal{C}^1$ around $\bar{x}$, then $\partial\varphi(x) = \{\nabla\varphi(x)\}$ for all x near $\bar{x}$. Applying the coderivative representation of Theorem 1.38 to the mapping $f\colon X \to X^*$ with $f(x) := \nabla\varphi(x)$, we arrive at the result. △

When $\varphi \in \mathcal{C}^2$ around $\bar{x}$ and X is finite-dimensional, $\nabla^2\varphi(\bar{x})$ reduces to the classical Hessian matrix for which $\nabla^2\varphi(\bar{x})^* = \nabla^2\varphi(\bar{x})$.

In general, both $\partial_N^2\varphi(\bar{x}, \bar{y})$ and $\partial_M^2\varphi(\bar{x}, \bar{y})$ are positively homogeneous mappings from X^{**} into X^* whose calculation involves evaluations of generalized normals to $\operatorname{gph}\partial\varphi$. In finite dimensions it is convenient to use the representations of basic normals from Theorem 1.6. For illustration we consider $\varphi(x) := |x|$ on $I\!R$ and compute $\partial^2\varphi(0,1)$. In this case

$$\partial\varphi(x) = \begin{cases} 1 & \text{if } x > 0\,, \\ [-1,1] & \text{if } x = 0\,, \\ -1 & \text{if } x < 0; \end{cases} \qquad \partial^2\varphi(0,1)(u) = \begin{cases} 0 & \text{if } u > 0\,, \\ (-\infty,\infty) & \text{if } u = 0\,, \\ (-\infty,0] & \text{if } u < 0\,, \end{cases}$$

since one easily has from representation (1.8) that

$$N((0,1); \operatorname{gph}\partial\varphi) = \{(v_1,v_2)\big|\ v_1 \le 0,\ v_2 \ge 0\} \\ \cup\{(v,0)\big|\ v < 0\} \cup \{(0,v)\big|\ v < 0\}\,.$$

For another example let us consider $\varphi(x) := \frac{1}{2}x^2 \operatorname{sign} x$ that is differentiable on $I\!R$ with $\nabla\varphi(x) = |x|$. Based on the calculation of the coderivative of $|x|$ in Subsect. 1.2.1 (right after Proposition 1.33), we have

$$\partial^2\varphi(0)(u) = \begin{cases} [-u,u] & \text{if } u \ge 0\,, \\ \{u,-u\} & \text{if } u < 0\,. \end{cases}$$

The function from the latter example belongs to the so-called $\mathcal{C}^{1,1}$*-class* around the reference point $\bar{x}$. This class consists of functions φ that are continuously differentiable around $\bar{x}$ with the gradient $\nabla\varphi$ locally Lipschitzian around this point. The calculation of the mixed second-order subdifferential for such functions can be essentially simplified due to the following representation. Similar result for the normal second-order subdifferential holds under additional assumptions on functions φ and spaces X; see Subsect. 3.1.3.

Proposition 1.120 (mixed second-order subdifferentials of $\mathcal{C}^{1,1}$ functions). *Let $\varphi \in \mathcal{C}^{1,1}$ around $\bar{x}$. Then*

$$\partial_M^2\varphi(\bar{x})(u) = \partial\langle u, \nabla\varphi\rangle(\bar{x}) \quad \textit{for all} \quad u \in X^{**}\,.$$

Proof. This follows from the scalarization formula in Theorem 1.90. △

We refer the reader to the papers by Dontchev and Rockafellar [364] and by Mordukhovich and Outrata [939] that contain efficient computations of the second-order subdifferentials for attractive classes of nonsmooth functions in finite dimensions. In the first paper it is done for the class of indicator functions of *polyhedral convex sets* that naturally appear in many important applications of variational analysis and optimization, in particular, to stability and sensitivity issues. The second paper covers the class of so-called *separable piecewise* $\mathcal{C}^2$ functions that are especially important for applications to mathematical programs with equilibrium constraints and frequently arise, e.g., in the modeling of mechanical equilibria; see the above papers and their references for more details. Using calculus rules, one can extend these and related results to other classes of functions via various compositions.

Our primary goal in the second-order theory is to develop principal calculus (sum and chain) rules for the second-order subdifferentials defined above. In this subsection we present results obtained in general Banach spaces; other results are given in Subsect. 3.2.5, where some spaces in question are assumed to be Asplund.

To derive second-order sum and chain rules for ∂_N^2 and ∂_M^2, we proceed via Definition 1.118 applying calculus rules for the normal and mixed coderivatives to set-valued mappings generated by the basic first-order subdifferential. In this way we have to restrict ourselves to favorable classes of functions for which the corresponding first-order subdifferential calculus rules hold as *equalities*, since neither normal nor mixed coderivative enjoys monotonicity properties that may allow one to use an inclusion-type subdifferential calculus. We begin with a simple *sum rule* for the second-order subdifferentials.

Proposition 1.121 (equality sum rule for second-order subdifferentials). *Let $\bar{y} \in \partial(\varphi_1 + \varphi_2)(\bar{x})$, where $\varphi_1 \in \mathcal{C}^1$ around $\bar{x}$ with $\nabla\varphi_1$ strictly differentiable at $\bar{x}$ while $\varphi_2\colon X \to \overline{I\!R}$ is finite at $\bar{x}$ with $\bar{y}_2 := \bar{y} - \nabla\varphi_1(\bar{x}) \in \partial\varphi_2(\bar{x})$. Then one has*

$$\partial^2(\varphi_1 + \varphi_2)(\bar{x}, \bar{y})(u) = \nabla^2\varphi_1(\bar{x})^* u + \partial^2\varphi_2(\bar{x}, \bar{y}_2)(u), \quad u \in X^{**},$$

for both normal ($\partial^2 = \partial_N^2$) and mixed ($\partial^2 = \partial_M^2$) second-order subdifferentials.

Proof. If $\varphi_1 \in \mathcal{C}^1$ around $\bar{x}$, then there is a neighborhood U of $\bar{x}$ such that the equality

$$\partial(\varphi_1 + \varphi_2)(x) = \nabla\varphi_1(x) + \partial\varphi_2(x), \quad x \in U,$$

holds whenever $\varphi_2\colon X \to \overline{I\!R}$; see Proposition 1.107(ii). Applying to the latter equality the coderivative sum rule from Theorem 1.62(ii) for $D^* = D_N^*$ and $D^* = D_M^*$, we conclude the proof of the proposition. △

Next we consider *chain rules* for the second-order subdifferentials of compositions $(\varphi \circ g)(x) := \varphi(g(x))$ involving inner mappings $g\colon X \to Z$ between Banach spaces and extended-real-valued outer functions $\varphi\colon Z \to \overline{I\!R}$. To obtain the central result in this direction, we need to introduce first the following extensibility property, which is related to but somewhat different from the so-called *Banach extensibility property* (see, e.g., Diestel [333]) and plays an essential role in proving the second-order chain rule.

Definition 1.122 (weak* extensibility). *Let V be a closed linear subspace of a Banach space X. Then V is w^*-EXTENSIBLE in X if every sequence $\{v_k^*\} \subset V^*$ with $v_k^* \xrightarrow{w^*} 0$ in V^* as $k \to \infty$ contains a subsequence $\{v_{k_j}^*\}$ such that each $v_{k_j}^*$ can be extended to a linear bounded functional $x_j^* \in X^*$ with $x_j^* \xrightarrow{w^*} 0$ in X^* as $j \to \infty$.*

The w^*-extensibility property always holds in the following two broad settings of Banach spaces.

Proposition 1.123 (sufficient conditions for weak* extensibility). *Let V be a closed linear subspace of a Banach space X. Then V is w^*-extensible in X if one of the following conditions holds:*

(a) *V is complemented in X, i.e., there is a closed linear subspace $L \subset X$ such that $V \bigoplus L = X$.*

(b) *The closed unit ball of X^* is weak* sequentially compact (in particular, if X is either Asplund or WCG).*

Proof. Let V be complemented in X, and let $\Pi\colon X \to V$ be a projection operator. Putting $x_k^* := \langle v_k^*, \Pi(x)\rangle$ on X, we conclude that x_k^* is an extension of v_k^* with $x_k^* \xrightarrow{w^*} 0$, i.e., V is w^*-extensible in X in case (a).

To justify this property in case (b) for every $V \subset X$, we take an arbitrary sequence v_k^* from Definition 1.122 and observe that it is bounded in V^* due to the weak* convergence. By the Hahn-Banach theorem we extend each v_k^* to $\tilde{x}_k^* \in X^*$ such that the sequence $\{\tilde{x}_k^*\}$ is still bounded in X^*. Since $I\!B_{X^*}$ is assumed to be weak* sequentially compact, there exist $x^* \in X^*$ and a weak* convergent subsequence $\tilde{x}_{k_j}^* \xrightarrow{w^*} x^*$ as $j \to \infty$. Observe that $x^* = 0$ on V due to the weak* convergence $v_k^* \xrightarrow{w^*} 0$ in V^*. Putting $x_j^* := \tilde{x}_{k_j}^* - x^*$, we complete the proof of the proposition. △

Let us demonstrate that the weak* extensibility property may not hold even in some classical Banach spaces.

Example 1.124 (violation of weak* extensibility). *The subspace $V = c_0$ is not w^*-extensible in $X = \ell^\infty$.*

Proof. Recall that c_0 is a Banach space of all real sequences converging to zero that is endowed with the supremum norm. Let $v_k^* := \xi_k^* \in c_0^*$, where ξ_k^* maps every vector from c_0 to its k-th component. Assume that there is an increasing sequence of $k_j \in I\!N$ such that $v_{k_j}^*$ can be extended to $x_j^* \in (\ell^\infty)^*$ with $x_j^* \xrightarrow{w^*} 0$. Define a closed linear subspace of ℓ^∞ by

$$Z := \big\{(\alpha_1, \alpha_2, \ldots) \in \ell^\infty \big|\ \alpha_k = 0 \ \text{ if } \ k \notin \{k_1, k_2, \ldots\}\big\}$$

and a linear bounded operator $A\colon \ell^\infty \to Z$ by

$$A(\alpha_1, \alpha_2, \ldots) := (\beta_1, \beta_2, \ldots) \text{ for all } (\alpha_1, \alpha_2, \ldots) \in \ell^\infty ,$$

where one has

$$\beta_k = \begin{cases} \alpha_i & \text{if } k = k_j,\ j \in I\!N , \\ 0 & \text{otherwise} . \end{cases}$$

Taking the above sequence $\{x_j^*\}$, we denote $z_j^* := x_j^*|_Z$ and form a linear bounded operator $T\colon Z \to c_0$ by

$$T(z) := \big(\langle z_1^*, z\rangle, \langle z_2^*, z\rangle, \ldots\big) \in c_0 \quad \text{for all} \quad z \in Z .$$

Then the operator $(T \circ A)\colon \ell^\infty \to c_0$ is bounded and its restriction $(T \circ A)|c_0$ is the identity operator on c_0. Therefore $(T \circ A)$ is a projection of ℓ^∞ to c_0, which means that c_0 is complemented in ℓ^∞. It is well known that the latter is not true, and hence we get a contradiction. This proves that c_0 is not w^*-extensible in ℓ^∞. $\triangle$

Next we show that linear operators with w^*-extensible ranges enjoy a certain stability property, which is crucial for the subsequent application to the second-order chain rule.

Proposition 1.125 (stability property for linear operators with weak* extensible ranges). *Let $A\colon X \to Y$ be a linear bounded operator between Banach spaces. Assume that the range of A is closed and w^*-extensible in Y and take $x_k^* \in \operatorname{rge} A^*$ with $x_k^* \xrightarrow{w^*} x^*$. Then $(A^*)^{-1}(x^*) \neq \emptyset$, and for every $y^* \in (A^*)^{-1}(x^*)$ there is a sequence $y_k^* \in (A^*)^{-1}(x_k^*)$ that contains a subsequence weak* converging to y^*.*

Proof. It is well known that the range A^*Y^* of the adjoint operator to A is weak* closed in X^* if $V := AX$ is closed in Y. Thus $x^* \in A^*Y^*$, i.e., $(A^*)^{-1}(x^*) \neq \emptyset$. Take any $y^* \in (A^*)^{-1}(x^*)$, arbitrarily choose $\hat{y}_k^* \in (A^*)^{-1}(x_k^*)$, and let $v_k^* := \hat{y}_k^*|_V$. Then $v_k^* \xrightarrow{w^*} y^*|_V$ in V^*. Since the space V is closed and w^*-extensible in Y, we find an extension $\tilde{y}_k^*$ of $v_k^* - y^*|_V$ for each $k \in I\!N$ such that $\{\tilde{y}_k^*\}$ contains a subsequence weak* converging to zero. Now letting $y_k^* := y^* + \tilde{y}_k^*$, we check that $A^*y_k^* = x_k^*$ and that $\{y_k^*\}$ contains a subsequence weak* converging to y^*. $\triangle$

To establish chain rules for second-order subdifferentials, we need the following basic lemma giving chain rules for coderivatives of special compositions whose structure as well as imposed assumptions correspond to the second-order setting. These special structure and assumptions allow us to obtain more precise results that are *not* implied by chain rules for general compositions (except the inclusion for normal coderivatives); see below.

Lemma 1.126 (special chain rules for coderivatives). *Let $G: X \rightrightarrows Y$ and $f: X \times Y \to Z$ be mappings between Banach spaces, and let*

$$(f \circ G)(x) := f(x, G(x)) = \bigcup \Big\{ f(x, y) \Big| \ y \in G(x) \Big\} . \tag{1.63}$$

Given $\bar{x} \in \operatorname{dom} G$, we assume that:

(a) *$f(x, \cdot) \in \mathcal{L}(Y, Z)$ around $\bar{x}$, i.e., it is a linear bounded operator from Y into Z. Moreover, $f(\bar{x}, \cdot)$ is injective and its range is closed in Z.*

(b) *The mapping $x \to f(x, \cdot)$ from X into the operator space $\mathcal{L}(Y, Z)$ is strictly differentiable at $\bar{x}$.*

Take any $\bar{y} \in G(\bar{x})$ and denote $\bar{z} := f(\bar{x}, \bar{y})$. Then one has

$$D^*_M(f \circ G)(\bar{x}, \bar{z})(z^*) = \nabla_x f(\bar{x}, \bar{y})^* z^* + D^*_M G(\bar{x}, \bar{y})\big(f(\bar{x}, \cdot)^* z^*\big) , \tag{1.64}$$

$$D^*_N(f \circ G)(\bar{x}, \bar{z})(z^*) \subset \nabla_x f(\bar{x}, \bar{y})^* z^* + D^*_N G(\bar{x}, \bar{y})\big(f(\bar{x}, \cdot)^* z^*\big) \tag{1.65}$$

for all $z^* \in Z^*$. If in addition the range of $f(\bar{x}, \cdot)$ is w^*-extensible in Z, then (1.65) holds as equality.

Proof. Consider the mapping $h(x) := f(x, \cdot)$ from X into $\mathcal{L}(Y, Z)$ and denote by $A: X \to \mathcal{L}(Y, Z)$ its strict derivative at $\bar{x}$. Let $\ell > 0$ be a Lipschitz modulus of h around $\bar{x}$. For any $y \in Y$ we define a linear operator $A_y: X \to Z$ by $A_y(x) := A(x)y$ and easily check that it is bounded. Moreover, the operator $y \to A_y$ from Y into $\mathcal{L}(X, Z)$ is linear and bounded as well. By enlarging ℓ if necessary, we assume that the norm of this operator is less than ℓ. Also it is clear that $A_y = \nabla_x f(\bar{x}, y)$ for all $y \in Y$.

Our first step is to prove the inclusions "$\subset$" in (1.64) and (1.65) simultaneously. Proceeding by definitions of these coderivatives, we start with ε-normals

$$(x^*, -z^*) \in \widehat{N}_\varepsilon((\hat{x}, \hat{z}); \operatorname{gph}(f \circ G)) ,$$

where $\hat{z} := f(\hat{x}, \hat{y})$, $(\hat{x}, \hat{y}) \in \operatorname{gph} G$ with $\|\hat{x} - \bar{x}\| < \eta$ for some small $\eta > 0$.

Using the definition of ε-normals and involving the rate of strict differentiability $r_h(\bar{x}; \eta)$ for the above mapping h at $\bar{x}$ (see Definition 1.13), we get the estimate

$$\limsup_{(x,y) \xrightarrow{\operatorname{gph} G} (\hat{x}, \hat{y})} \frac{\langle x^* - A^*_{\bar{y}} z^*, x - \hat{x} \rangle - \langle f(\bar{x}, \cdot)^* z^*, y - \hat{y} \rangle}{\|x - \hat{x}\| + \|y - \hat{y}\|} \le \hat{\varepsilon} ,$$

where $\hat{\varepsilon} := c\varepsilon + c\|z^*\|\big(r_h(\bar{x}; \eta) + \|\hat{x} - \bar{x}\| + \|\hat{y} - \bar{y}\|\big)$ with some constant $c > 0$. Thus one has

$$\big(x^* - A^*_{\bar{y}} z^*, -f(\bar{x}, \cdot)^* z^*\big) \in \widehat{N}_{\hat{\varepsilon}}((\hat{x}, \hat{y}); \operatorname{gph} G) . \tag{1.66}$$

To justify the inclusions "$\subset$" in (1.64) and (1.65) simultaneously, we take $x^* \in D^*(f \circ G)(\bar{x}, \bar{z})(z^*)$ and find sequences $\varepsilon_k \downarrow 0$, $x_k \to \bar{x}$, $y_k \in G(x_k)$, $(x^*_k, -z^*_k) \in \widehat{N}((x_k, z_k); \operatorname{gph}(f \circ G))$ with $z_k := f(x_k, y_k)$ such that $z_k \to \bar{z}$,

$x_k^* \xrightarrow{w^*} x^*$, and that $\|z_k^* - z^*\| \to 0$ for $D^* = D_M^*$ and $z_k^* \xrightarrow{w^*} z^*$ for $D^* = D_N^*$. Then we get the inclusions in (1.64) and (1.65) by passing to the limit in (1.66) provided that $y_k \to \bar{y}$. To prove the latter convergence, we observe that the open mapping theorem and the injectivity of $f(\bar{x}, \cdot)$ ensure the existence of a constant $\mu > 0$ such that

$$\|f(\bar{x}, u) - f(\bar{x}, v)\| \ge \mu\|u - v\| \quad \text{whenever} \quad u, v \in Y \,.$$

Therefore, involving the above Lipschitz modulus ℓ, one has

$$\|z_k - \bar{z}\| = \big\|[f(\bar{x}, y_k) - f(\bar{x}, \bar{y})] + [f(x_k, y_k - \bar{y}) - f(\bar{x}, y_k - \bar{y})]$$

$$+[f(x_k, \bar{y}) - f(\bar{x}, \bar{y})]\big\| \ge \|y_k - \bar{y}\|\big(\mu - \ell\|x_k - \bar{x}\|\big) - \ell\|x_k - \bar{x}\| \cdot \|\bar{y}\| \,,$$

which implies that $y_k \to \bar{y}$ as $k \to \infty$.

Next let us show that the opposite inclusions hold in (1.64) and (1.65) under the assumptions made; in fact, there are no additional assumptions in the case of mixed coderivatives (1.64). To proceed simultaneously in both cases, we take $(\hat{x}, \hat{y})$ as above and pick arbitrary (x^*, z^*) satisfying

$$\big(x^*, -f(\bar{x}, \cdot)^* z^*\big) \in \widehat{N}_\varepsilon((\hat{x}, \hat{y}); \operatorname{gph} G) \,.$$

Thus for any given $\gamma > 0$ one has

$$\theta := \langle x^*, x - \hat{x}\rangle - \langle f(\bar{x}, \cdot)^* z^*, y - \hat{y}\rangle \le (\varepsilon + \gamma)\big(\|x - \hat{x}\| + \|y - \hat{y}\|\big) \tag{1.67}$$

whenever $(x, y) \in \operatorname{gph} G$ are sufficiently close to $(\hat{x}, \hat{y})$. Let us obtain a lower estimate for θ in (1.67) using the strict differentiability of the above mapping $h\colon X \to \mathcal{L}(Y, Z)$ at $\bar{x}$ with the rate $r_h(\bar{x}; \eta)$ and elementary transformations. In this way we get:

$$\theta = \langle x^*, x - \hat{x}\rangle - \langle z^*, f(\bar{x}, y) - f(\bar{x}, \hat{y})\rangle$$

$$= \langle x^* + A_{\bar{y}}^* z^*, x - \hat{x}\rangle - \langle z^*, A_{\bar{y}}(x - \hat{x})\rangle - \langle z^*, f(\bar{x}, y) - f(\bar{x}, \hat{y})\rangle$$

$$\ge \langle x^* + A_{\bar{y}}^* z^*, x - \hat{x}\rangle - \langle z^*, A_y(x - \hat{x})\rangle - \langle z^*, f(\hat{x}, y) - f(\hat{x}, \hat{y})\rangle$$

$$-\ell\|z^*\| \cdot \|y - \bar{y}\| \cdot \|x - \hat{x}\| - \ell\|z^*\| \cdot \|\hat{x} - \bar{x}\| \cdot \|y - \hat{y}\|$$

$$\ge \langle x^* + A_{\bar{y}}^* z^*, x - \hat{x}\rangle - \langle z^*, f(x, y) - f(\hat{x}, y)\rangle - r_h(\bar{x}; \eta)\|z^*\| \cdot \|y\| \cdot \|x - \hat{x}\|$$

$$-\langle z^*, f(\hat{x}, y) - f(\hat{x}, \hat{y})\rangle - \ell\|z^*\|\big(\|y - \bar{y}\| \cdot \|x - \hat{x}\| + \|\hat{x} - \bar{x}\| \cdot \|y - \hat{y}\|\big)$$

$$= \langle x^* + A_{\bar{y}}^* z^*, x - \hat{x}\rangle - \langle z^*, f(x, y) - f(\hat{x}, \hat{y})\rangle - r_h(\bar{x}; \eta)\|z^*\| \cdot \|y\| \cdot \|x - \hat{x}\|$$

$$-\ell\|z^*\|\big(\|y - \bar{y}\| \cdot \|x - \hat{x}\| + \|\hat{x} - \bar{x}\| \cdot \|y - \hat{y}\|\big) \,.$$

Now we are going to give an upper estimate of the number on the right-hand side of (1.67). To proceed, we first observe that, by the open mapping theorem and the injectivity of $f(\bar{x},\cdot)$, there is $\mu > 0$ such that

$$\mu\|y\| \le \|f(\bar{x},y)\| \quad \text{for all} \quad y \in Y\,.$$

Then taking any $T \in \mathcal{L}(Y,Z)$, we get

$$\|Ty\| = \|(f(\bar{x},\cdot)-T)y - f(\bar{x},y)\| \ge \|f(\bar{x},y)\| - \|(f(\bar{x},\cdot)-T)y\|$$

$$\ge \big(\mu - \|f(\bar{x},\cdot)-T\|\big)\cdot\|y\|\,.$$

This implies the existence of a constant $\mu_1 > 0$ with the uniform estimate $\mu_1\|y\| \le \|Ty\|$ for all $y \in Y$ and all T sufficiently close to $f(\bar{x},\cdot)$. It gives therefore that

$$\|f(x,y) - f(\hat{x},\hat{y})\| = \|f(x,y) - f(\hat{x},y) + f(\hat{x},y-\hat{y})\|$$

$$\ge \|f(\hat{x},y-\hat{y})\| - \|f(x,y)-f(\hat{x},y)\| \ge \mu_1\|y-\hat{y}\| - L\|x-\hat{x}\|\cdot\|y\|$$

for $(x,y) \in \operatorname{gph} G$ close to $(\hat{x},\hat{y})$ while $(\hat{x},\hat{y})$ is close to $(\bar{x},\bar{y})$. Thus we obtain the estimate

$$\|y-\hat{y}\| \le \mu_2\big(\|x-\hat{x}\| + \|f(x,y)-f(\hat{x},\hat{y})\|\big)$$

for all such (x,y) and $(\hat{x},\hat{y})$, with some constant $\mu_2 > 0$. Putting these estimates together, one has

$$(x^* + A^*_{\bar{y}}z^*, -z^*) \in \widehat{N}_{\hat{\varepsilon}}((\hat{x},\hat{z}); \operatorname{gph}(f\circ G))\,, \tag{1.68}$$

where $\hat{z} := f(\hat{x},\hat{y})$ and $\hat{\varepsilon}$ is defined as above with a different constant $c > 0$.

To prove the opposite inclusions in (1.64) and (1.65), we need passing to the limit in (1.68) as $(\hat{x},\hat{y}) \to (\bar{x},\bar{y})$ along some sequence. Pick arbitrary (x^*,z^*) with $x^* \in D^*G(\bar{x},\bar{y})(f(\bar{x},\cdot)^*z^*)$, where D^* stands for either mixed or normal coderivative. Then there are sequences $\varepsilon_k \downarrow 0$, $(x_k,y_k) \to (\bar{x},\bar{y})$ with $(x_k,y_k) \in \operatorname{gph} G$, and $x_k^* \in D^*_{\varepsilon_k}G(x_k,y_k)(y_k^*)$ such that $x_k^* \xrightarrow{w^*} x^*$ and either $\|y_k^* - f(\bar{x},\cdot)^*z^*\| \to 0$ when $D^* = D^*_M$, or $y_k^* \xrightarrow{w^*} f(\bar{x},\cdot)^*z^*$ when $D^* = D^*_N$. Note that $\hat{\varepsilon}_k \downarrow 0$ for the corresponding $\hat{\varepsilon}_k$ in (1.68). To complete the proof of the lemma, it is sufficient to show that there are $z_k^* \in Z^*$ such that $f(\bar{x},\cdot)^*z_k^* = y_k^*$ for all $k \in I\!N$, and that either $\|z_k^* - z^*\| \to 0$ for $D^* = D^*_M$ or $z_k^* \xrightarrow{w^*} z^*$ for $D^* = D^*_N$ along a subsequence. We consider the cases of mixed and normal coderivatives separately.

(i) Let $D^* = D^*_M$. Since $f(\bar{x},\cdot)$ is injective with the closed range, it is easy to see that the adjoint operator $f(\bar{x},\cdot)^*$ is surjective and hence metrically regular. This ensures the existence of $\mu > 0$ and $\hat{z}_k^* \in (f(\bar{x},\cdot)^*)^{-1}(y_k^* - f(\bar{x},\cdot)^* z^*)$ satisfying the estimate

$$\|\hat{z}_k^*\| \le \mu \|y_k^* - f(\bar{x},\cdot)^* z^*\| .$$

Putting $z_k^* := \hat{z}_k^* + z^*$, we get $f(\bar{x},\cdot)^* z_k^* = y_k^*$ and $\|z_k^* - z^*\| \to 0$ as $k \to \infty$.

(ii) Let $D^* = D_N^*$. In this case the subspace $f(\bar{x}, Y)$ is assumed to be w^*-extensible in Z. Then the existence of the desired sequence $\{z_k^*\}$ follows from Proposition 1.125. △

Note that inclusion (1.65) for the normal coderivative can be derived from the chain rule of Theorem 1.65(i) applied to (1.63) represented as the standard composition

$$f(x, G(x)) = f(\widetilde{G}(x)) \quad \text{with} \quad \widetilde{G}(x) := (x, G(x)) .$$

Indeed, under the injectivity assumption on $f(\bar{x},\cdot)$ the corresponding mapping $\widetilde{G} \cap f^{-1}$ in Theorem 1.65 is single-valued and continuous. The equality in (1.65) and the entire case (1.64) for the mixed coderivative are due to the special setting of Lemma 1.126.

Now we are ready to derive the central result of the second-order subdifferential calculus in general Banach spaces.

Theorem 1.127 (second-order chain rules with surjective derivatives of inner mappings). *Let $\bar{y} \in \partial(\varphi \circ g)(\bar{x})$ with $g\colon X \to Z$ and $\varphi\colon Z \to \overline{\mathbb{R}}$, where X and Z are Banach. Assume that $g \in \mathcal{C}^1$ around $\bar{x}$ with the surjective derivative $\nabla g(\bar{x})\colon X \to Z$ and that the mapping $\nabla g\colon X \to \mathcal{L}(X, Z)$ is strictly differentiable at $\bar{x}$. Let $\bar{v} \in Z^*$ be a unique functional satisfying*

$$\bar{y} = \nabla g(\bar{x})^* \bar{v} \quad \text{and} \quad \bar{v} \in \partial\varphi(\bar{z}) \quad \text{with} \quad \bar{z} := g(\bar{x}) .$$

Then for all $u \in X^{**}$ one has

$$\partial_M^2(\varphi \circ g)(\bar{x}, \bar{y})(u) = \nabla^2\langle \bar{v}, g\rangle(\bar{x})^* u + \nabla g(\bar{x})^* \partial_M^2 \varphi(\bar{z}, \bar{v})(\nabla g(\bar{x})^{**} u) ,$$

$$\partial_N^2(\varphi \circ g)(\bar{x}, \bar{y})(u) \subset \nabla^2\langle \bar{v}, g\rangle(\bar{x})^* u + \nabla g(\bar{x})^* \partial_N^2 \varphi(\bar{z}, \bar{v})(\nabla g(\bar{x})^{**} u) .$$

Moreover, the latter inclusion becomes an equality if the range of $\nabla g(\bar{x})^*$ is w^*-extensible in X^*. This is true under one of the following conditions:

(a) *The range of $\nabla g(\bar{x})^*$ is complemented in X^*, which holds, in particular, when the kernel of $\nabla g(\bar{x})$ is complemented in X.*

(b) *The closed unit ball of X^{**} is weak* sequentially compact, which holds, in particular, when either X is reflexive or X^* is separable.*

Proof. Using the first-order subdifferential sum rule from Proposition 1.112(i), we have the equality

$$\partial(\varphi \circ g)(x) = \nabla g(x)^* \partial\varphi(g(x)) := (f \circ G)(x)$$

for all x around $\bar{x}$, where the mappings $f: X \times Z^* \to X^*$ and $G: X \rightrightarrows Z^*$ in the latter representation are defined by

$$f(x, v) := \nabla g(x)^* v, \quad G(x) := \partial\varphi(g(x)) .$$

Thus we represent $\partial(\varphi \circ g)$ as composition (1.63) and apply Lemma 1.126 to this composition. Let us check that its assumptions hold under the assumptions made in the theorem. Actually the only assumption needed to be checked is the injectivity of the operator $\nabla g(\bar{x})^*: Z^* \to X^*$, which follows from the assumed surjectivity of $\nabla g(\bar{x})$ due to Lemma 1.18. $\triangle$

Note that the normal coderivative inclusion in Theorem 1.127 may be also obtained by applying the coderivative chain rule from Theorem 1.65 to the standard composition

$$f \circ \widetilde{G} \text{ with } f(x, v) = \nabla g(x)^* v \text{ and } \widetilde{G}(x) := \Big(x, \partial\varphi(g(x))\Big)$$

and then the coderivative chain rule from Theorem 1.66 to the composition $\partial\varphi \circ g$. Moreover, this inclusion becomes an equality if $\nabla g(\bar{x})$ is invertible. Indeed, in this case g^{-1} is locally single-valued and strictly differentiable at $\bar{z}$ by Theorem 1.60, and one gets the opposite inclusion considering the composition $\varphi = \psi \circ g^{-1}$ with $\psi := \varphi \circ g$. Moreover, it is possible to show that the case when $\nabla g(\bar{x})$ is surjective and has the complemented kernel in X can be reduced to the one with $\nabla g(\bar{x})$ invertible. However, the general equality case for normal coderivatives in Theorem 1.127 and the entire case for mixed coderivatives don't seem to be derivable from the results of Subsect. 1.2.4.

The last result of this subsection provides equalities for both second-order subdifferentials of compositions $\varphi \circ g$ in general Banach spaces, where φ but not g is assumed to be twice differentiable. Given a Lipschitz continuous mapping $g: X \to Z$, we define the following *second-order coderivative* sets for g at $(\bar{x}, \bar{v}, \bar{y}) \in X \times Z^* \times X^*$ with $\bar{y} \in \partial\langle\bar{v}, g\rangle(\bar{x})$

$$D^2 g(\bar{x}, \bar{v}, \bar{y})(u) := \Big(D^* \partial\langle\cdot, g\rangle\Big)(\bar{x}, \bar{v}, \bar{y})(u), \quad u \in X^{**} , \tag{1.69}$$

used in formulations of the next theorem and related results of Chap. 3. In (1.63), D^* stands for either *normal* ($D^* = D^*_N$, then $D^2 = D^2_N$) or *mixed* ($D^* = D^*_M$, then $D^2 = D^2_M$) coderivative of the mapping $(x, v) \to \partial\langle v, g\rangle(x)$. If g is strictly differentiable at $\bar{x}$, then $\partial\langle\bar{v}, g\rangle(\bar{x}) = \nabla g(\bar{x})^* \bar{v}$ and we omit $\bar{y}$ in the arguments of $D^2 g$.

Theorem 1.128 (second-order chain rules with twice differentiable outer mappings). *Let g be strictly differentiable at $\bar{x}$, let $\varphi \in \mathcal{C}^1$ around $\bar{z} := g(\bar{x})$ with $\nabla\varphi$ strictly differentiable at this point, and let $\bar{v} := \nabla\varphi(\bar{z})$. Assume that the operator $\nabla^2\varphi(\bar{z})\nabla g(\bar{x}): X \to Z^*$ is surjective. Then*

$$\partial^2(\varphi \circ g)(\bar{x})(u) = \bigcup_{(x^*,v^*)\in D^2 g(\bar{x},\bar{v})(u)} \Big[x^* + \nabla g(\bar{x})^* \nabla^2 \varphi(\bar{z})^* v^*\Big]$$

for all $u \in X^{**}$, *where* ∂^2 *and* D^2 *stand for the corresponding normal and mixed second-order constructions. These chain rules hold without the above surjectivity assumption if* ∇g *is strictly differentiable at* $\bar{x}$. *In the latter case*

$$D^2_N g(\bar{x},\bar{v})(u) = D^2_M g(\bar{x},\bar{v})(u) = \Big(\nabla^2\langle \bar{v}, g\rangle(\bar{x})^* u, \nabla g(\bar{x})^{**} u\Big).$$

Proof. Since $\varphi \in \mathcal{C}^1$ and g is locally Lipschitzian, Theorem 1.110(ii) ensures the existence of a neighborhood U of $\bar{x}$ such that

$$\partial(\varphi \circ g)(x) = \partial\langle \nabla\varphi(g(x)), g\rangle(x) := (F \circ h)(x), \quad x \in U,$$

where the mappings $F\colon X \times Z^* \rightrightarrows X^*$ and $h\colon X \to X \times Z^*$ are defined by

$$F(x,v) := \partial\langle v, g\rangle(x), \quad h(x) := \Big(x, \nabla\varphi(g(x))\Big).$$

If h is strictly differentiable at $\bar{x}$ with the surjective derivative operator, then one has by Theorem 1.66 that

$$D^*(F \circ h)(\bar{x},\bar{y})(u) = \nabla h(\bar{x})^* D^* F(\bar{x},\bar{v},\bar{y})(u), \quad u \in X^{**},$$

for both normal and mixed coderivatives, where $\bar{y} = \nabla g(\bar{x})^* \bar{v}$ if g is strictly differentiable at $\bar{x}$. Note that $\nabla^2(\varphi \circ g)(\bar{x}) = \nabla^2\varphi(\bar{z})\nabla g(\bar{x})$ in the framework of theorem, and that the surjectivity of the latter operator implies the surjectivity of $\nabla h(\bar{x})$. This proves the theorem under the surjectivity assumption made. The last claim in theorem easily follows from the above procedure due to Theorem 1.65(iii); this is actually a classical second-order chain rule for strict derivatives. △

In Subsect. 3.2.5 we obtain second-order subdifferential sum and chain rules in the form of inclusions under less restrictive assumptions on functions and mappings in Asplund space settings.

1.4 Commentary to Chap. 1

1.4.1. Motivations and Early Developments in Nonsmooth Analysis. Nonsmooth phenomena have been known for a long time in mathematics and applied sciences. To deal with nonsmoothness, various kinds of generalized derivatives were introduced in the classical theory of real functions and in the theory of distributions; see, e.g., Bruckner [182], Saks [1186], Schwartz [1197], and Sobolev [1218]. However, those generalized derivatives, which "ignore sets of density zero," are of little help for optimization theory and variational analysis, where the main interest is in behavior of functions at *individual* points of maxima, minima, equilibria, and other optimization-related notions.

The concepts of generalized differentiability appropriate for applications to optimization were defined in *convex analysis*: first geometrically as the *normal cone* to a convex set that goes back to Minkowski [882], and then – much later – analytically as the *subdifferential* of an extended-real-valued convex function. The latter notion, inspired by the work of Fenchel [441], was explicitly introduced by Moreau [981] and Rockafellar [1140] who emphasized the *set-valuedness* of the new generalized derivative with values in dual spaces and the decisive role of subdifferential *calculus rules*. The central result in this direction, called now the Moreau-Rockafellar theorem on subdifferential sums, is based on the *separation principle* for convex sets around which the whole convex analysis actually revolves.

Convex analysis and separation theorems play a crucial role not only in studying convex sets, functions, and convex optimization problems but also in more general nonconvex settings via *convex approximations*. This idea, largely motivated by applications to optimal control, has been much explored in nonsmooth analysis and optimization starting with the early 1960s. The initial inspiration came from the Pontryagin maximum principle and its proof given by Boltyanskii; see [124, 1102]. Note that a similar approach to abnormal problems in the calculus of variation was developed by McShane [860] whose work didn't receive a proper attention till the formulation and proof of the maximum principle; compare, e.g., Bliss [119] and Hestenes [565]. Roughly speaking, the underlying idea was to construct, by using special *needle-type* control variations, a *convex tangent cone* approximating the reachable set of system endpoints so that the optimal endpoint lies at its boundary and thus can be separated by a supporting hyperplane. Such a convex approximation approach was strongly developed and applied to new classes of extremal problems by Dubovitskii and Milyutin [369, 370] (see also the book by Girsanov [507]) and then by Gamkrelidze [496, 497], Halkin [539, 541], Hestenes [565], Neustadt [1001, 1002], Ioffe and Tikhomirov [618], and others.

1.4.2. Tangents and Directional Derivatives. Observe that among tangent cones to arbitrary sets successfully used in nonsmooth analysis and optimization from the early 1960s and onwards we can find the so-called "contingent cone" introduced in 1930 independently by Bouligand [167] and by Severi [1202] in the framework of contingent equations and differential geometry. It is interesting to observe that the mentioned seminal papers by Bouligand and Severi were published (in French and Italian, respectively) in the *same issue* (!) of Annales de la Société Polonaise de Mathématique; see also Bouligand [168] and Verchenko and Kolmogorov [1285] for further developments at that time related to differential geometry and real analysis. Then this cone was rediscovered and applied to optimization theory by Dubovitskii and Milyutin [369, 370] under the name "cone of variations admissible by equality constraints." The reader can find more discussions on these and related tangential constructions in Aubin and Frankowska [54] and Ursescu [1276].

Analytically tangent cone approximations of sets correspond to *directional derivatives* of functions, while convex subcones of tangents correspond to sublinear majorants of directional derivatives. It is well known that every *convex* function $\varphi\colon X \to (-\infty, \infty]$ on a Banach space admits the *classical directional derivative*

$$\varphi'(\bar{x}; v) := \lim_{t \downarrow 0} \frac{\varphi(\bar{x} + tv) - \varphi(\bar{x})}{t} \tag{1.70}$$

in all direction $v \in X$ at any point of its *efficient domain*

$$\operatorname{dom} \varphi := \{x \in X \mid \varphi(x) < \infty\} .$$

Moreover, the function of directions $v \mapsto \varphi'(\bar{x}; v)$ is convex as well. These properties of the existence of the directional derivative (1.70) and its convexity with respect to directions hold not only for convex functions and, obviously, for classical differentiable functions, but also for a broader class of functions called *locally convex* by Ioffe and Tikhomirov [618] and closely related to them *quasidifferentiable* functions in the sense of Pshenichnyi [1106]. The latter class contains, in particular, *maximum functions* of the type

$$\varphi(x) := \max_{u \in U} \vartheta(x, u)$$

generated by smooth functions $\vartheta(\cdot, u)$ and compact sets U; (cf. Danskin [307] and Demyanov and Malozemov [319]); this class is closed under taking linear combinations with nonnegative coefficients. In [320], Demyanov and Rubinov extended the notion of quasidifferentiability to the class of functions for which the classical directional derivative exists and admits a special representation via maxima and minima over pairs of compact convex sets; see also Demyanov and Rubinov [321, 322], Gorokhovik [515, 516], and Pallaschke and Urbański [1041] for more references, recent developments, related geometric aspects, and applications.

Since even simple continuous functions on real line may not be directionally differentiable as, e.g.,

$$\varphi(x) := \begin{cases} x \sin(1/x) & \text{if } x \neq 0 , \\ 0 & \text{if } x = 0 , \end{cases}$$

an important issue in nonsmooth analysis has been to define generalized directional derivatives that automatically exist and have some useful properties. Among the most attractive constructions of this type appeared in the 1970s and 1980s is

$$d^- \varphi(\bar{x}; v) := \liminf_{\substack{z \to v \\ t \downarrow 0}} \frac{\varphi(\bar{x} + tz) - \varphi(\bar{x})}{t} \tag{1.71}$$

called "lower semiderivative" by Penot [1064], "contingent derivative/epi-derivative" by Aubin [48], "lower Dini (or Dini-Hadamard) directional derivative" by Ioffe [594, 607], and "subderivative" by Rockafellar and Wets [1165]. This directional derivative goes back, for the case of real functions, to the classical (1878) "derivate numbers" by Dini [335], while in the general case they can be equivalently described geometrically via the *contingent cone* from Definition 1.8(i) by

$$d^-\varphi(\bar{x};v) = \inf\left\{\nu \in I\!R \,\middle|\, (v,\nu) \in T\big((\bar{x},\varphi(\bar{x}));\operatorname{epi}\varphi\big)\right\}. \tag{1.72}$$

Note that one can put $z = v$ in (1.71) if φ is locally Lipschitzian around $\bar{x}$.

The key disadvantage of the generalized directional derivative $d^-\varphi(\bar{x};v)$ is its *nonconvexity* with respect to directions v that takes place in many common situations. This nonconvexity doesn't allow one to employ tools of convex analysis (based on separation) and eventually leads to a poor calculus available for (1.71). A standard procedure to overcome these difficulties is to build a positively homogeneous *convex upper approximation* (majorant) of (1.71) that corresponds by (1.72) to forming a convex subcone of the contingent cone and thus brings us back to the *realm of convex analysis.* We refer the reader to [54, 52, 89, 313, 337, 464, 569, 588, 733, 763, 764, 852, 870, 871, 1002, 1040, 1072, 1109, 1264, 1265, 1266, 1311] for various constructions of this type, which are not always uniquely and efficiently defined. Another approach to introduce directional derivatives with good properties is to *postulate* the existence of some limits and thus to deal with classes of functions that satisfy such assumptions; see, e.g., [44, 54, 1135, 1156, 1165, 1204, 1248] for constructions and results in this vein particularly related to notions of *epi-convergence.*

1.4.3. Constructions by Clarke and Related Developments. A refined *generalized directional derivative* of locally *Lipschitzian* functions that is *automatically convex* in directions was introduced in the 1973 dissertation by Clarke [243], conducted under supervision of Rockafellar, and then was published in [244]. The crucial role of this pioneering contribution to the development and applications of nonsmooth analysis (the term coined by Clarke) is difficult to overstate.

It seems that the original motivation came from the intention to derive necessary optimality conditions for variational and optimal control problems, with no convexity assumptions on state variables, using "Rockafellar's convex theory [1143, 1145] as a starting point" (see [245, p. 80]). Clarke's generalized derivative defined by

$$\varphi^\circ(\bar{x};v) := \limsup_{\substack{x\to\bar{x}\\ t\downarrow 0}} \frac{\varphi(x+tv)-\varphi(x)}{t} \tag{1.73}$$

made it possible to reduce the variational problem

$$\text{minimize } \left\{l(x(0,x(1)) + \int_0^1 L(t,x(t),\dot{x}(t))\,dt\right\}$$

with a *Lipschitzian* integrand $L(t,\cdot,\cdot)$ and an extended-real-valued endpoint function l to a *convex* problem of this type considered by Rockafellar, i.e., where both l and $L(t,\cdot,\cdot)$ are convex functions; see [245] for all the details in deriving the generalized Euler-Lagrange inclusion in Clarke's terms.

Observe that the generalized directional derivative (1.73) is different not only from the Dini-like directional derivative (1.71) but also from the classical directional derivative (1.70). The key issue is that in (1.73), contrary to (1.70) and (1.71), the initial point $\bar{x}$ is *perturbed*, which provides some *uniformity* (and hence robustness) with respect to the initial data. By definition, Clarke's directional derivative is a majorant of both lower Dini directional derivative (1.71) and its *upper* counterpart

$$d^+\varphi(\bar{x};v) := \limsup_{t\downarrow 0}\frac{\varphi(\bar{x}+tv)-\varphi(\bar{x})}{t}$$

for locally Lipschitzian functions, i.e.,

$$d^-\varphi(\bar{x};v) \le d^+\varphi(\bar{x};v) \le \varphi^\circ(\bar{x};v) \quad \text{for all } \; v\in X\,.$$

As mentioned, the generalized directional derivative $\varphi^\circ(\bar{x};v)$ may not reduce to the classical one $\varphi'(\bar{x};v)$ when the latter exists, even for simple real functions like $\varphi(x)=-|x|$ at $\bar{x}=0$. The case of

$$\varphi^\circ(\bar{x};v)=\varphi'(\bar{x};v) \quad \text{for all } \; v\in X$$

postulates *Clarke regularity* of φ at $\bar{x}$, which is equivalent to

$$d^-\varphi(\bar{x};v)=d^+\varphi(\bar{x};v)=\varphi^\circ(\bar{x};v), \qquad v\in X\,,$$

and corresponds geometrically to the equality

$$T(\bar{x};v)=T_C(\bar{x};v) \quad \text{whenever } \; v\in X \tag{1.74}$$

between the contingent cone and Clarke's tangent cone considered in Subsect. 1.1.2; cf. Clarke [255] and Rockafellar and Wets [1165]. It is well known that Clarke's directional derivative is usually far from the best (and even adequate) local approximation of a function in the absence of regularity.

Having any positively homogeneous (in directions v) function $\varphi^\bullet(\bar{x};v)$, which can be considered as a local approximation of $\varphi\colon X\to\overline{I\!R}$ finite at $\bar{x}$ (in particular, the directional derivatives mentioned above), the corresponding *subdifferential* of φ at $\bar{x}$ is defined by the *duality* correspondence

$$\partial^\bullet\varphi(\bar{x}) := \big\{x^*\in X^*\big|\; \langle x^*,v\rangle \le \varphi^\bullet(\bar{x};v) \;\text{ for all }\; v\in X\big\}\,. \tag{1.75}$$

This is a standard way to introduce subgradients via directional derivatives. For convex functions it gives the classical subdifferential of convex analysis:

$$\begin{aligned}\partial\varphi(\bar{x}) &= \big\{x^*\in X^*\big|\; \langle x^*,v\rangle \le \varphi'(\bar{x};v) \;\text{ for all }\; v\in X\big\}\\ &= \big\{x^*\in X^*\big|\; \langle x^*,x-\bar{x}\rangle \le \varphi(x)-\varphi(\bar{x}) \;\text{ for all }\; x\in X\big\}\,,\end{aligned}$$

where the second representation is due to the global nature of convexity, while the first one defines the subdifferential of locally convex functions and the like. Clarke's subdifferential (or *generalized gradient* [243, 244]) of locally Lipschitzian functions is defined in this way by

$$\partial_C\varphi(\bar{x}) = \big\{x^* \in X^* \big|\ \langle x^*, v\rangle \le \varphi^\circ(\bar{x}; v)\ \text{ for all }\ v \in X\big\}\,. \tag{1.76}$$

In finite dimensions the generalized gradient admits the equivalent representation

$$\partial_C\varphi(\bar{x}) = \text{co}\Big\{\lim_{x_k\to\bar{x}} \nabla\varphi(x_k)\Big\}\,, \tag{1.77}$$

where the set under the convex hull in (1.77) is nonempty and compact by the classical Rademacher theorem [1114] ensuring that a Lipschitz continuous function on an open subset of $I\!R^n$ is a.e. differentiable. The latter set was introduced by Shor [1207], under the name of the "set of almost-gradients," from the viewpoint of numerical optimization of nonsmooth functions. Note that Shor also considered the convexified set in (1.77), under the name of the "set of generalized almost-gradients," however, no calculus rules were obtained; see also [1208, 683, 1111] for more details and references. Observe that the nonconvex set of almost-gradients in (1.77) doesn't reduce to the subdifferential even for simple convex functions (e.g., $\varphi(x) = |x|$), so the *convexification* operation in (1.77) is *crucial.* Being convexified, the generalized gradient $\partial_C\varphi(\cdot)$ possesses a reasonably good calculus on the class of Lipschitz continuous function; in particular, it satisfies the inclusion sum rule

$$\partial_C(\varphi_1 + \varphi_2)(\bar{x}) \subset \partial_C\varphi_1(\bar{x}) + \partial_C\varphi_2(\bar{x})$$

the proof of which is based on the *convex separation* theorem similarly to most other results of Clarke's nonsmooth analysis [255].

Definition 1.8(iii) of the Clarke tangent cone $T_C(\bar{x}; \Omega)$ is different from the original one [243, 244] given via the generalized directional derivative (1.73) of the (Lipschitzian) *distance function* $\text{dist}(\cdot; \Omega)$; the equivalence between the two definitions follows from the proof of [244, Proposition 3.7] and was first observed by Thibault [1244]; see also [1248]. As discussed above, $T_C(\bar{x}; \Omega)$ is a geometric counterpart of the directional derivative $\varphi^\circ(\bar{x}; v)$, while Clarke's normal cone to Ω at $\bar{x}$ is a *dual* object defined by

$$N_C(\bar{x}; \Omega) := \big\{x^* \in X^* \big|\ \langle x^*, v\rangle \le 0\ \text{ for all }\ v \in T_C(\bar{x}; \Omega)\big\}\,. \tag{1.78}$$

It can always be described via the *weak* closure* of the cone spanned on the generalized gradient of the distance function

$$N_C(\bar{x}; \Omega) = \text{cl}^*\Big\{\bigcup_{\lambda\ge 0} \lambda\partial_C\text{dist}(\bar{x}; \Omega)\Big\}\,.$$

This implies, by [244, Proposition 3.2] and [255, Theorem 2.5.6] established for closed subsets $\Omega \subset I\!R^n$, the following representation:

$$N_C(\bar{x};\Omega) = \text{clco}\left\{0, \lim \frac{u_k}{\|u_k\|}\,\Big|\, u_k \perp \Omega \text{ at } x_k \to \bar{x},\ u_k \to 0\right\}, \tag{1.79}$$

where the notation $u \perp \Omega$ at x signifies that u is a *perpendicular* to Ω at $x \in \Omega$, i.e., there is z such that $u = z - x$ and x is the unique closest point to z in Ω.

Using the route well understood in convex analysis, Clarke's generalized gradient of *lower semicontinuous* (l.s.c.) functions $\varphi\colon X \to \overline{I\!R}$ was originally defined via the normal cone to the epigraph of φ by

$$\partial_C\varphi(\bar{x}) := \left\{x^* \in X^* \,\big|\, (x^*, -1) \in N_C\big((\bar{x}, \varphi(\bar{x})); \text{epi}\,\varphi\big)\right\},$$

and then it was equivalently described by Rockafellar [1147, 1149] in the analytic duality way (1.75) via his generalized directional derivative (*upper subderivative*) $\varphi^\bullet = \varphi^\uparrow$ given by

$$\varphi^\uparrow(\bar{x}; v) := \sup_{\gamma > 0}\left\{\limsup_{\substack{x \xrightarrow{\varphi} \bar{x} \\ t \downarrow 0}}\left[\inf_{\|z - v\| \le \gamma} \frac{\varphi(x + tz) - \varphi(x)}{t}\right]\right\}.$$

Rockafellar's subderivative $\varphi^\uparrow(\bar{x}; v)$ is convex in directions, reduces to $\varphi^\circ(\bar{x}; v)$ for locally Lipschitzian functions φ, and happens to be the *support function* for the generalized gradient of arbitrary l.s.c. functions $\varphi\colon X \to \overline{I\!R}$ finite at $\bar{x}$:

$$\varphi^\uparrow(\bar{x}; v) = \sup\left\{\langle x^*, v\rangle \,\big|\, x^* \in \partial_C\varphi(\bar{x})\right\}.$$

The achieved duality relationships between $\partial_C\varphi(\bar{x})$ and $\varphi^\uparrow(\bar{x}; v)$ allowed Rockafellar [1146, 1147, 1148, 1149], based mainly on the *machinery of convex analysis*, to develop calculus rules and related results for the Clarke generalized gradient of l.s.c. functions; see also Aubin [48] and Hiriart-Urruty [570, 571, 572]. However, some important properties have been lost in the non-Lipschitzian case; in particular, the so-called *robustness property*

$$\partial_C\varphi(\bar{x}) = \mathop{\text{Lim}\,\text{sup}}_{x \xrightarrow{\varphi} \bar{x}} \partial_C\varphi(x)$$

doesn't hold true for l.s.c. functions, e.g., when φ is the indicator function of the set

$$\Omega := \left\{(x_1, x_2, x_3) \in I\!R^3 \,\big|\, x_3 = x_1 x_2\right\}$$

with $\bar{x} = 0 \in I\!R^3$; see more details on this example in Rockafellar [1147, 1149].

The full and beautiful duality between directional derivatives/tangents and subgradients/normals achieved in the Clarke-Rockafellar theory and related calculus rules for these constructions made the fundamental ground for many important, breakthrough applications to optimization, calculus of variations, optimal control, and other areas of nonlinear and variational analysis. The *convexity* of the generalized gradient and normal cone seemed to be crucial for the theory and applications involving the eventual usage of separation theorems. Note to this end that any subdifferential/normal cone constructions in dual spaces generated by *polarity* relations like (1.75) are *automatically convex* regardless of the convexity of the generating directional derivatives and sets of tangents.

1.4.4. Motivations to Avoid Convexity. It is well known that Clarke's generalized gradient of Lipschitzian functions is *unimprovable* (minimal in size) among any *convex-valued and robust* extensions of the subdifferential of convex analysis with some properties desired for applications. This statement has been first proved by Lebourg [749], where the desired property is a nonsmooth version of the classical mean value theorem. Furthermore, it follows from the results by Ioffe [599, Theorem 8.1] (cf. also Mordukhovich [901, Section 4.6] and Mordukhovich and Shao [949, Theorem 9.7]) that $\partial_C\varphi(\bar{x})$ is the smallest among any robust and convex-valued subdifferentials $\partial^\bullet\varphi(\bar{x})$ satisfying the inclusion sum rule mentioned above and a nonsmooth counterpart of the Fermat stationary principle: $0 \in \partial^\bullet\varphi(\bar{x})$ whenever $\bar{x}$ provides a local minimum to φ.

On the other hand, it has been well recognized that the generalized gradient may be *too large* for many important applications, in particular, to necessary optimality conditions. It is easy to give simple examples (as the trivial ones: minimize $-|x|$ over $I\!R$; also minimize $|x_1| - |x_2|$ over $I\!R^2$), where $0 \in \partial_C\varphi(\bar{x})$ while $\bar{x}$ is far removed from the minimum that can be directly detected by other necessary conditions for minimization. Another serious drawback of these convex constructions concerns deficient conditions obtained in their terms for some fundamental properties in nonlinear analysis related to covering of nonsmooth operators, metric regularity, open mapping theorems, Lipschitzian stability, and the like; see, e.g., the corresponding results and discussions in Dmitruk, Milyutin and Osmolovskii [337], Warga [1320], Rockafellar [1154], etc. In basic calculus [255, Sect. 2.3], the weakest point concerns chain rules that either require smoothness of some mappings in compositions or involve unsatisfactory convexification.

But probably the most striking undesirable phenomenon arises in geometric considerations, where the normal cone (1.78) to *graphical sets* with nonsmooth boundaries often happens to be the *whole space* or at least a linear subspace of big dimension. Consider, for instance, the graph of the simplest nonsmooth function $\varphi(x) = |x|,\ x \in I\!R$. Then one can easily check that $N_C((0,0); \operatorname{gph}\varphi) = I\!R^2$. The same picture comes into view at the "complementarity corner," i.e., for the boundary of the nonnegative orthant in $I\!R^n$ appearing in complementarity conditions. Indeed, we have on the plane

$$N_C((0,0);\Omega) = I\!R^2 \quad \text{for} \quad \Omega := \big\{(x_1,x_2) \in I\!R^2 \big|\ x_1x_2 = 0,\ x_1 \ge 0,\ x_2 \ge 0\big\}\ .$$

which of course was observed by people working on complementarity problems and variational inequalities.

Comprehensive results in this direction were obtained by Rockafellar [1153] for the tangent cone $T_C(\bar{x};\Omega)$ in finite dimensions; they imply by polarity the corresponding conclusions for Clarke normals. It has been proved in [1153, Theorem 3.2] that for every mapping $f\colon I\!R^n \to I\!R^m$ Lipschitz continuous around $\bar{x}$, the normal cone $N_C((\bar{x}, f(\bar{x})); \text{gph}\, f)$ is actually a *linear subspace* of dimension $q \geq m$, where $q = m$ *if and only if* f is *strictly differentiable* at $\bar{x}$. Furthermore, this result was extended in [1153, Theorem 3.5] to the so-called "Lipschitzian manifolds," which are locally homeomorphic to the graph of a locally Lipschitzian vector function. It has been shown in [1153] that the class of Lipschitzian manifolds (called *graphically Lipschitzian* sets in [1165]) includes graphs of *maximal monotone* set-valued mappings , in particular, graphs of *subdifferential mappings* for convex and saddle functions. Such subdifferential mappings have been long recognized in variational analysis as convenient tools for describing *variational inequalities* and *complementarity conditions*; see Robinson [1130, 1131]. More recently, it has been proved by Poliquin and Rockafellar [1090] that subdifferential mappings for the so-called "prox-regular" functions, that are typically encountered in finite-dimensional optimization, also belong to the class of graphically Lipschitzian mappings, for which therefore Clarke's normal cone has the mentioned subspace property. To this end, let us refer the reader to a recent result by Dontchev and Rockafellar [365] showing that the graphical Lipschitzian property is preserved under "ample parameterizations" important for sensitivity analysis of variational inclusions/generalized equations and related problems.

It is worth mentioning that the set counterpart of prox-regular functions, called "prox-regular sets" by Poliquin and Rockafellar [1090] has been already introduced and studied by Federer [437] in geometric measure theory under the name "sets of positive reach." Such sets are also called "sets with property ρ" by Plaskacz [1081] and by "proximally smooth sets" by Clarke, Stern and Wolenski [271].

1.4.5. Basic Normals and Subgradients. Due to the *unimprovability* of Clarke's generalized differential constructions among any *convex-valued* ones with reasonable properties including robustness, the only way to avoid the drawbacks discussed above is to *give up the convexity* of the normal cone and subdifferential. This inevitably presumes that one should abandon the conventional scheme of convex and nonsmooth analysis generating normals and subgradients via polarity correspondences from tangents and directional derivatives that automatically yields the convexity of polar/dual objects; cf. (1.75) and (1.78). Furthermore, the theory of such nonconvex dual-space constructions (optimality conditions, calculus rules, etc.) *cannot make any appeal* to the traditional techniques of convex analysis based on separation theorems.

The nonconvex *basic/limiting normal cone* to closed sets and the corresponding subdifferential of l.s.c. extended-real-valued functions satisfying these requirements were introduced by Mordukhovich in the beginning of 1975, who was not familiar with Clarke's constructions at that time. The initial motivation came from the intention to derive necessary optimality conditions for optimal control problems with endpoint geometric constraints by passing to the limit from free endpoint control problems, which are much easier to handle. This was published in [887] (first in Russian and then translated into English), where the original normal cone definition was given in finite-dimensional spaces by

$$N(\bar{x};\Omega) = \operatorname*{Lim\,sup}_{x \to \bar{x}} \big[\text{cone}(x - \Pi(x;\Omega))\big] \tag{1.80}$$

via the Euclidean projector $\Pi(\cdot;\Omega)$, while the *basic subdifferential* $\partial\varphi(\bar{x})$ was defined geometrically via the normal cone to the epigraph of φ; see Definition 1.77. It is written in the final version of [887], after discussions with Ioffe, that Clarke's normal cone is the closed convex closure of (1.80) in finite-dimensional spaces. We see, by Theorem 1.6, that the normal cone (1.80) is equivalent in finite dimensions to the basic normal cone used in this book.

It is worth mentioning that the basic normal cone (1.80) appeared in [887] as a *by-product* of the *method of metric approximations* introduced in that paper, which allowed us to reduce nonsmooth constrained problems to *smooth* problems of unconstrained optimization; see also [889, 717, 892], where this method was applied to general classes of extremal problems containing mathematical programs with equality, inequality and geometric constraints, minimax and vector optimization problems, optimal control problems for systems with smooth dynamics and also for dynamical systems governed by discrete-time and continuous-time differential inclusions. Moreover, this method directly leads to studying the general concept of *local extremal points* and establishing the *extremal principle*; see the proof of Theorem 2.8 in Chap. 2 and Commentary to that chapter.

Note that the method of metric approximations shares some similarities with the penalty function method, which was employed for deriving necessary optimality conditions in smooth constrained problems; compare, e.g., McShane [864], Berkovitz [106], and Polyak [1097]. We also used a modified penalty method for nonsmooth constrained problems of optimization and optimal control [893], but the results obtained in this vein impose more requirements on the (scalar) cost functional in comparison with the method of metric approximations, which treats cost and constraint functions fully *symmetrically* and thus allows us to cover multiobjective and equilibrium problems as well as general extremal points of set systems.

1.4.6. Fréchet-like representations. It was realized after a while (at the end of the 1970s) that the basic normal cone (1.80) and the corresponding basic subdifferential from Definition 1.77(i) can be represented via limits of Fréchet-like constructions in finite-dimensional spaces (which are dual geometrically to the contingent cone $T(\bar{x};\Omega)$ and analytically to the lower Dini directional derivative $d^-\varphi(\bar{x};v)$ in finite dimensions), while the infinite-dimensional setting requires the usage of *sequential* limits of

ε-enlargements; thus we came up to the basic definitions used in this book. Besides the afore-mentioned papers, we refer the reader to the joint work by Kruger and Mordukhovich [718, 719] and to Kruger's dissertation [706] conducted under supervision of Mordukhovich. It has been also realized around the same time that the metric approximation method is useful not only for deriving necessary optimality conditions in terms of the nonconvex generalized differential constructions but also for normal and subgradient *calculus rules* in finite-dimensional spaces and in Banach spaces with *Fréchet smooth* renorms under certain Lipschitzian assumptions.

First calculus results in the fully non-Lipschitzian setting were obtained by Mordukhovich [894] in finite-dimensional spaces. In particular, it was proved there by the method of metric approximations that the intersection rule for basic normals

$$N(\bar{x};\Omega_1\cap\Omega_2)\subset N(\bar{x};\Omega_1)+N(\bar{x};\Omega_2) \tag{1.81}$$

holds provided that the sets Ω_i are locally closed around $\bar{x}\in\Omega_2\cap\Omega_2$ and that the *basic qualification condition*

$$N(\bar{x};\Omega_1)\cap\big(-N(\bar{x};\Omega_2)\big)=\{0\} \tag{1.82}$$

is satisfied. Moreover, (1.81) holds as *equality* if both sets Ω_i are *normally regular* at $\bar{x}$ in the sense of [894], i.e., when

$$N(\bar{x};\Omega)=\widehat{N}(\bar{x};\Omega). \tag{1.83}$$

Note that in finite-dimensional spaces the normal regularity (1.83) happens to agree with Clarke's tangential regularity (1.74) due to the convexity of $\widehat{N}(\bar{x};\Omega)$ (and hence of $N(\bar{x};\Omega)$ in this case) and by the duality relations between tangents and normals in finite dimensions discussed in Subsect. 1.1.2. It is not the case however in infinite-dimensional spaces; see Bounkhel and Thibault [172] for a comprehensive study of various regularity notions in nonsmooth analysis and the comparison between them.

We refer the reader to the book by Mordukhovich [901] and the bibliography therein for a unified theory, mostly in finite dimensions but with full discussions of infinite-dimensional extensions, based on his generalized differential constructions and their applications to problems of optimization, optimal control for discrete-time and continuous-time systems, and related topics developed up to the end of 1986.

In *infinite-dimensional* Banach spaces, as adopted in this book, we build our *basic normals* from Definition 1.1 as *sequential limits of ε-normals* belonging to

$$\widehat{N}_\varepsilon(\bar{x};\Omega)=\Big\{x^*\in X^*\Big|\ \limsup_{x\xrightarrow{\Omega}\bar{x}}\frac{\langle x^*,x-\bar{x}\rangle}{\|x-\bar{x}\|}\le\varepsilon\Big\},\quad \varepsilon\ge 0\,.$$

The latter set first appeared in Kruger and Mordukhovich [718]. Note its relationship with the *local ε-support* by Ekeland and Lebourg [400] defined by

$$S_\varepsilon(\bar{x};\Omega) := \Big\{x^* \in X^* \Big|\ \exists \nu > 0 \text{ with } \langle x^*, x-\bar{x}\rangle \le \varepsilon\|x-\bar{x}\| \text{ whenever } x \in \Omega \text{ and } \|x-\bar{x}\| < \nu\Big\}, \quad \varepsilon > 0\,.$$

One can easily see that

$$\widehat{N}_\varepsilon(\bar{x};\Omega) = \bigcup_{\gamma>\varepsilon} S_{\varepsilon+\gamma}(\bar{x};\Omega) \text{ for any } \varepsilon \ge 0$$

and observe that the "0-support" set $S_0(\bar{x};\Omega)$ carries little information even in finite dimensions, while the cone of "0-normals" $\widehat{N}_0(\bar{x};\Omega) = \widehat{N}(\bar{x};\Omega)$ plays a very important role in our considerations, in both finite-dimensional and infinite-dimensional settings. Similar observations can be made about the ε-subdifferentials $\widehat{\partial}_{g\varepsilon}\varphi(\bar{x})$ and $\widehat{\partial}_{a\varepsilon}\varphi(\bar{x})$ defined in Subsect. 1.3.2 following the pattern of [718, 719, 706], which are functional counterparts of ε-normals. Note that the construction $\widehat{\partial}\varphi(\bar{x}) := \partial_0\varphi(\bar{x})$ from (1.51), which we call "Fréchet subdifferential" or "presubdifferential," is labeled as "regular subdifferential" in Rockafellar and Wets [1165]); an equivalent construction in finite dimensions appeared in Bazaraa, Goode and Nashed [89] under the name "the set of $\ge$ gradients."

Of course, Fréchet had nothing to do with such normals and subgradients; we keep this name to emphasize parallels with the classical differentiation, where the Fréchet derivative is the basic tool of nonlinear analysis. It is worth mentioning that Fréchet, a student of Hadamard, introduced his derivative [473] in infinite-dimensional spaces not being familiar with the fact that the same definition, for functions of finitely many variables, had been already used by Weierstrass in his lectures at the University of Berlin in the end of the 1870s and the beginning of 1880s, which were published only in 1927 [1326] although partly incorporated in some German and English textbooks (e.g., by Scholtz and by Young) written in the beginning of the 20th century under the influence of Weierstrass; see Tikhomirov [1257] and Brinkhuis and Tikhomirov [178] for more information. We also refer the reader to the survey paper by Averbukh and Smolyanov [68] for various classical (and neoclassical) derivatives in analysis, with thorough discussions of the history and relationships between them in the general setting of linear topological spaces.

Thus starting with the late 1970s, the Fréchet-like normals and subgradients have played a prominent role in optimization and nonsmooth analysis; we refer the reader to [156, 146, 157, 163, 164, 172, 329, 413, 415, 420, 419, 593, 600, 634, 654, 657, 707, 708, 713, 718, 800, 801, 802, 901, 935, 946, 949, 952, 960, 1007, 1249, 1263, 1311, 1345] for more discussions. The Fréchet subdifferential $\widehat{\partial}\varphi(\bar{x})$ is also known as "subdifferential in the sense of *viscosity solutions*" and has been broadly used, starting with the 1983 paper by Crandall and Lions [297], in partial differential equations of the Hamilton-Jacobi type with many applications to optimal control, stochastic control, differential games, etc.; the reader can find more information in

[85, 86, 215, 265, 295, 296, 330, 331, 425, 458, 471, 688, 702, 701, 721, 793, 818, 819, 869, 1230, 1231, 1240, 1241, 1359]. Note also that constructions of this type have long traditions in the Italian school of variational inequalities and related topics; see, e.g., the papers by Marino and Tosques [851], Degiovanni, Marino and Tosques [313], and the references therein.

1.4.7. Approximate Subdifferentials. The other line of extensions of Mordukhovich's generalized differential constructions to infinite-dimensional spaces was strongly developed by Ioffe in the series of many publications starting from 1981. He began [589] with the subdifferential construction

$$\partial_M \varphi(\bar{x}) := \overline{\operatorname{Lim\,sup}}_{\substack{x \xrightarrow{\varphi} \bar{x} \\ \varepsilon \downarrow 0}} \partial_\varepsilon^- \varphi(x)\,, \tag{1.84}$$

called him by the *M-subdifferential*, where $\overline{\operatorname{Lim\,sup}}$ signifies the *topological* counterpart of the Painlevé-Kuratowski upper limit (1.1) with sequences in X^* replaced by nets, and where the ε-subdifferential construction

$$\partial_\varepsilon^- \varphi(x) := \big\{x^* \in X^* \big|\ \langle x^*, v\rangle \le d^- \varphi(x; v) + \varepsilon\|v\|\big\} \tag{1.85}$$

is a polar/dual object generated by the "ε-shifted" lower Dini derivative (1.71). It is not hard to check (cf. the proof of Theorem 1.10) that one has the relationship

$$\widehat{\partial}_\varepsilon \varphi(\bar{x}) \subset \partial_\varepsilon^- \varphi(\bar{x})$$

between the Fréchet ε-subdifferential $\widehat{\partial}_\varepsilon \varphi(\bar{x})$ from Definition 1.83(ii) and the Dini one (1.85), where equality holds in finite dimensions; in the latter case ε may be omitted in both limiting constructions of the basic subdifferential $\partial\varphi(\bar{x})$ (see Theorem 1.89) and the Dini-generated M-subdifferential (1.84), which both reduce to the original construction by Mordukhovich; cf. Kruger and Mordukhovich [718, 719] and Ioffe [596]. In general the M-subdifferential, which has useful properties in spaces with *Gâteaux smooth* renorms, may be essentially *larger* than our basic one (it may be even larger than Clarke's generalized gradient for non-Lipschitzian function; see Treiman [1262, 1263]).

Further infinite-dimensional improvements of the M-subdifferential and the corresponding M-normal cone reduced to (1.80) in finite dimensions, have been developed by Ioffe [590, 591, 592, 597, 599, 607] under the common name of "approximate normals and subdifferentials" including "analytic" (A) and "geometric" (G) ones as well as their "nuclei"; see Subsect. 2.5.2B for more details and discussions. Note that the adjective "approximate" indicates the relation to the original approximation technique [887] generating and/or inspiring these kinds of nonconvex constructions. Indeed, Ioffe wrote in [591, p. 3]: "It all essentially arises from thinking over Mordukhovich's approximate approach to necessary conditions for an extremum [887]"; see also [594, p.

518] and [596, p. 389]. Observe that the best of these constructions, the so-called "nuclei of the G-subdifferential and the G-normal cone" may be still larger than our basic constructions out of WCG (weakly compactly generated) spaces, even in those admitting a Fréchet smooth renorm; see Borwein and Fitzpatrick [141], Mordukhovich and Shao [949, Sect. 9], and Subsect. 3.2.3 of this book. On the other hand, they have essentially better (actually those needed for the majority of applications) calculus properties than our basic constructions in non-Asplund settings, being however significantly more complicated.

1.4.8. Further Historical Remarks. Coming back to finite dimensions, observe that the unconvexified limiting set in the braces $\{\cdots\}$ in representation (1.79) of Clarke's normal cone agrees with the basic normal cone by Mordukhovich. To the best of our knowledge, this set was first designated for its own sake in the Western literature, under the name of "limiting proximal normal cone," in the 1985 paper by Rockafellar [1155], where it was used as an auxiliary tool to derive extended calculus formulas and necessary optimality conditions in terms of Clarke's normals and subgradients via certain perturbation techniques. Some amount of calculus, particularly related to subdifferentiation of marginal functions and inf-convolutions, was developed in [1155] for limiting proximal normals and associated limiting sets of "proximal subgradients" introduced by Rockafellar in [1150] to recover Clarke's generalized gradient via the closed convex hull of such limits in finite dimensions; see Treiman [1262, 1263], Borwein and Strójwas [156, 157], and Loewen [798, 799] for infinite-dimensional extensions. However, the major calculus results and necessary optimality conditions were obtained by employing the *convexification* procedure, i.e., in terms of Clarke's constructions. In particular, the basic intersection formula (1.81) and related calculus results were derived by Rockafellar [1155] in Clarke's terms with qualifications conditions of type (1.82) expressed via Clarke's normals and subgradients. But, as discussed above, these formulas and many other results of this type have already been available *without any convexification*!

This clear gap between Western and Russian developments was definitely due to the lack of communication and personal contacts between Eastern and Western researchers during the Cold War. The situation has been dramatically changed after Mordukhovich's first talk at a scientific meeting in the West, which happened at the International Workshop in Quantitative Analysis in Sensitivity Analysis and Optimization organized by Clarke, Rockafellar, and Wets and held near Montreal in February 1989, just about a month following his immigration to the United States. Indeed, after learning Mordukhovich's results presented in his talk (which "...came as a surprise..."[1157]) and reading his book on the flight back from Montreal, Rockafellar was able to prove the main calculus results without any convexification on the basis of his own methods developed in [1150, 1155]. As he wrote in his letter to Mordukhovich [1157] accompanied his note [1158] shortly after the Montreal meeting: "... Oddly, as soon as the formulas you had established... had sunk in, I had no trouble at all proving them on the basis of other facts already familiar. But it had never occurred to me to push in such a direction!"

It seems that Clarke designated and utilized the nonconvex normal cone and subdifferential in question for the first time in his 1989 book [257], with the reference to Mordukhovich. He used the names of "prenormal cone" and "presubdifferential" for these nonconvex constructions reserving the terms "normal cone" and "subdifferential" for his convexified normal cone and generalized gradient. In [257, Sect. 1.4], Clarke provided another proof of the basic intersection rule (1.81) and related subdifferential results obtained earlier by Mordukhovich, using for these purposes a perturbation technique similar to that in "fuzzy calculus" developed by Ioffe [594]. Recognizing advantages of the latter calculus results in comparison with those in terms of the convexified objects $N_C(\bar{x};\Omega)$ and $\partial_C\varphi(\bar{x})$, Clarke nevertheless emphasized in the discussion of [257, p. 15] his preference to work in terms of $N_C(\bar{x};\Omega)$ and $\partial_C\varphi(\bar{x})$ for certain reasons related, first of all, to the polarity with the tangent cone and directional derivative. At the same time he indicated, in the footnote comments to the major necessary optimality conditions for variational and control problems considered in [257], that transversality conditions therein can be given in more precise terms of the "prenormal cone" and "presubdifferential" referring to the original work by Mordukhovich.

It is worth mentioning to this end that even in many papers after 1989 (and of course in earlier Western publications in this direction, with probably one essential exception of Warga's work employed his derivate containers [1316, 1317, 1319, 1321]), transversality conditions in nonsmooth optimal control and the calculus of variations were written in terms of Clarke's normal cone and generalized gradient, with no comments about possible refinements; see, e.g., [255, 256, 267, 268, 272, 273, 274, 276, 595, 666, 667, 803, 804, 808, 1178, 1291, 1292]. The recognition of the possibility of using the nonconvex normal cone and subdifferential to obtain refined Euler-Lagrange and Hamiltonian conditions for optimality came to the West even later in the 1990s, although results of this type have been developed in the Russian literature since 1980; see Mordukhovich [892, 897, 901, 902, 908], Smirnov [1215, 1216], and Commentary to Chap. 6 for more details and discussions.

1.4.9. Some Advantages of Nonconvexity. Eventually it has been recognized that the *nonconvexity* of the basic/limiting normal cone (1.80) and its infinite-dimensional extensions, as well as the corresponding subdifferentials, *is not a disadvantage* but, in most cases, *just the opposite*: it provides an opportunity to develop a much better calculus, to derive more precise results in variational theory, and to enlarge essentially a spectrum of applications in comparison with the convexified constructions. Furthermore, it allows us to define and efficiently apply the *basic coderivative* construction

$$D^*F(\bar{x},\bar{y})(y^*) := \big\{x^* \in X^* \big|\, (x^*,-y^*) \in N\big((\bar{x},\bar{y}); \operatorname{gph} F\big)\big\}\,. \tag{1.86}$$

for a set-valued mapping $F: X \rightrightarrows Y$ between Banach spaces at a graph point $(\bar{x}, \bar{y}) \in \operatorname{gph} F$ via the nonconvex normal cone (1.80) and its infinite-dimensional extensions. It was first done in the 1980 paper of Mordukhovich [892] motivated by applications to adjoint systems in optimal control systems but then it happened to be useful in many fundamental aspects of variational analysis and its applications (e.g., characterizations of metric regularity and Lipschitzian stability, sensitivity analysis for constraint and variational systems, optimality conditions for variational and equilibrium problems with equilibrium constraints, etc.; see numerous results, discussions, and comments in this book). It is important to emphasize that, by Rockafellar's theorem [1153] discussed above, the usage of Clarke's normal in scheme (1.86) with *graphical* sets therein doesn't lead to satisfactory constructions and results, since the *subspace property* holds for the latter cone *due to its convexity.*

Another opportunity provided by the nonconvex normal cone (1.80) and its infinite-dimensional generalizations is to define the *second-order subdifferential* of an extended-real-valued function $\varphi: X \to \overline{I\!R}$ at a point $(\bar{x}, \bar{y}) \in \operatorname{gph} \partial\varphi$ by

$$\partial^2\varphi(\bar{x}, \bar{y})(u) := (D^*\partial\varphi)(\bar{x}, \bar{y})(u), \quad u \in X^{**}, \tag{1.87}$$

i.e., as the coderivative of the first-order subdifferential. It was first done in the 1992 paper of Mordukhovich [907] motivated by applications to sensitivity analysis for systems described via (first-order) subdifferentials or normal cones in Robinson's framework of generalized equations, which covers variational inequalities, complementarity conditions, etc.; see [1130, 1131]. Again, the usage of Clarke's convexified normal cone in this scheme doesn't lead to valuable results, particularly for the case of convex functions φ corresponding to the classical variational inequalities and complementarity problems, where φ is the indicator function of a convex set. Indeed, by the afore-mentioned Rockafellar's results [1153], the graph of the subdifferential of a convex function is a Lipschitzian manifold (as for any maximal monotone relation), and hence the subspace property of Clarke's normal cone always holds in this case; see more discussions in Rockafellar [1154, Remark 3.13] and Mordukhovich [912, Sect. 3]. On the other hand, the coderivative and second-order subdifferential constructions (1.86) and (1.87) enjoy rich calculi in finite-dimensional and infinite-dimensional spaces being useful for many applications; see the corresponding parts of this book, with subsequent comments and references.

1.4.10. List of Major Topics and Contributors. Great progress has been made, particularly in recent years, in the study and applications of the basic/limiting generalized differential constructions under consideration and associated variational techniques in both finite-dimensional and infinite-dimensional settings. Let us present a *partial list* of the major topics in variational analysis and its applications, where the usage of these constructions happens to be crucial while leading to essentially new results and perspectives. The list is accompanied by the names of the main contributors/users and their publications (in alphabetical order), being definitely incomplete in these rapidly growing areas and reflecting of course the author's knowledge and understanding. More comments will be made while discussing specific results later in the book. Note that the list below mostly contains publications that employ *limiting* procedures involving Fréchet-like and similar normals and subgradients (or, equivalently, proximal ones in finite-dimensional and Hilbert space settings), with *no mandatary convexification*:

Calculus Rules for Nonconvex Normal Cone, First-Order Subdifferentials, and Coderivatives: Allali and Thibault [15], Borwein and Ioffe [147], Borwein, Mordukhovich and Shao [151], Borwein, Treiman and Zhu [158], Borwein and Zhu [162, 163, 164], Eberhard and Nyblom [382], Fabian and Mordukhovich [419], Geremew, Mordukhovich and Nam [503], Ioffe [590, 590, 596, 597, 599, 600, 603, 604, 607], Ioffe and Penot [614], Ivanov [622], Jourani [643, 644, 646], Jourani and Théra [650] Jourani and Thibault [652, 653, 654, 657, 658, 659, 660], Kruger [706, 708, 708, 709], Kruger and Mordukhovich [718, 719], Ledyaev and Zhu [754], Lee, Tam and Yen [755], Minchenko [879], Mordukhovich [892, 894, 901, 907, 908, 910, 917], Mordukhovich and Nam [935, 936, 934], Mordukhovich, Nam and Yen [937], Mordukhovich and Shao [949, 950, 952, 953], Mordukhovich, Shao and Zhu [954], Mordukhovich and B. Wang [963, 967, 968], Ngai, Luc and Théra [1007], Ngai and Théra [1008], Penot [1070], Rockafellar [1155, 1158, 1160, 1161, 1162], Rockafellar and Wets [1165], Thibault [1249, 1252], and Treiman [1267, 1269].

Second-Order Subdifferential Calculus: Dutta and Dempe [377], Dontchev and Rockafellar [364], Eberhard, Nyblom and Ralph [383], Eberhard and Pearce [384], Eberhard and Wenczel [387], Ioffe and Penot [615], Levy and Mordukhovich [769], Levy, Poliquin and Rockafellar [771], Mordukhovich [910, 912, 923], Mordukhovich and Outrata [939], Mordukhovich and B. Wang [967, 968], Poliquin and Rockafellar [1090, 1092], Rockafellar (personal communication; see [769, 923, 939]), Rockafellar and Zagrodny [1168], and Ward [1307].

Metric Regularity, Openness/Covering at Linear Rate, and Robust Lipschitzian Properties for Nonsmooth and Set-Valued Mappings: Azé, Corvellec and Lucchetti [70], Borwein and Zhu [163, 164], Galbraith [491], Geremew, Mordukhovich and Nam [503], Glover and Ralph [510], Ioffe [589, 596, 598, 607, 608], Jourani and Thibault [651, 655, 656, 657, 661], Kruger [709, 711, 714, 715], Kummer [727, 728], Ledyaev and Zhu [751], Levy and Poliquin [770], Mordukhovich [894, 901, 907, 909, 917, 924], Mordukhovich and Shao [946, 951, 953], Mordukhovich and B. Wang [967, 968], Ngai and Théra [1008], Penot [1068, 1071], Rockafellar and Wets [1165], Zhang and Treiman [1363], and Zheng and Ng [1365].

Regularity Perturbation, Distance to Infeasibility, and Conditioning in Variational Analysis and Optimization: Cánovas, Dontchev, Lopez and Parra [219], Dontchev and Lewis [360], Dontchev, Lewis and

Rockafellar [361], Dontchev and Rockafellar [366], Ioffe [609, 610], and Mordukhovich [924].

Studies of Structural, Generic, and Compactness-Like Properties of Sets, Functions, and Set-Valued Mappings: Aussel, Corvellec and Lassonde [61, 62], Aussel, Daniilidis and Thibault [63], Bernard and Thibault [108, 109, 110], Borwein, Borwein and Wang [136], Borwein and Fitzpatrick [141, 142], Borwein, Fitzpatrick and Girgensohn [144], Borwein, Lucet and Mordukhovich [150], Bounkhel [170], Borwein, Moors and Wang [152], Bounkhel and Thibault [172, 173], Clarke, Ledyaev, Stern and Wolenski [265], Clarke, Stern and Wolenski [271], Colombo and Goncharov [277, 278], Colombo and Marigonda [279], Cornet and Czarnecki [289], Correa, Gajardo and Thibault [291], Correa, Jofré and Thibault [292], Eberhard [381], Edmond and Thibault [389], Fabian and Mordukhovich [422], Henrion [555, 556], Guillaume [525], Ioffe [607], Jofré, Luc and Théra [634], Jourani [648, 645, 649], Jourani and Thibault [661], Lewis [778], Loewen [800, 802], Marcellin [848], Mifflin and Sagastizábal [873, 874], Mordukhovich and Shao [949, 950, 951, 953], Mordukhovich and B. Wang [961, 964, 965, 967], Penot [1071], Poliquin and Rockafellar [1089, 1090, 1091], Poliquin, Rockafellar and Thibault [1093], Rockafellar and Wets [1165], and Wang [1303].

Variational Convergence, Approximation, and Regularization in Generalized Differentiation and Related Topics: Benoist [99], Cornet and Czarnecki [289, 290], Czarnecki and Rifford [304], Eberhard [381], Eberhard and Nyblom [382], Eberhard, Nyblom and Ralph [383], Eberhard, Sivakumaran and Wenczel [386], Eberhard and Wenczel [387], Geoffroy and Lassonde [501], Ioffe [596], Jourani [646], Kruger [705, 713], Kruger and Mordukhovich [719], Levy, Poliquin and Thibault [772], Mordukhovich [901], Poliquin [1088], Poliquin and Rockafellar [1090, 1091], Poliquin, Rockafellar and Thibault [1093], Rockafellar and Wets [1165], and Rockafellar and Zagrodny [1168].

Efficient Conditions for Error Bounds, Calmness, and Sharp Minima: Azé and Corvellec [69], Azé and Hiriart-Urruty [71], Bosch, Jourani and Henrion [166], Burke [189], Henrion and Jourani [559], Henrion, Jourani and Outrata [560], Henrion and Outrata [561, 562], Jourani [647], Jourani and Ye [662], Li and Singer [784], Mordukhovich, Nam and Yen [937], Ng and Zheng [1005], Ngai and Théra [1010], Papi and Sbaraglia [1050, 1051], Studniarski and Ward [1229], Wu and Ye [1334, 1335], Zhang [1362], and Zheng and Ng [1365].

Computational Algorithms in Nonsmooth Analysis: Bolte, Daniilidis and Lewis [122, 122], Burke, Lewis and Overton [196, 197, 199], Flegel [454], Hare and Lewis [549], Klatte and Kummer [686, 687], Kočvara, Kružik and Outrata [689], Kočvara and Outrata [690, 691], Kummer [726, 727, 728], Lewis [778], Mifflin and Sagastizábal [873, 874], Outrata [1030], and Papi and Sbaraglia [1052].

Applications to Stability and Sensitivity Analysis for Constraint and Variational Systems: Azé, Corvellec and Lucchetti [70], Azé and Hiriart-Urruty [71], Bosch, Jourani and Henrion [166], Burke, Lewis and Overton [195], Dontchev and Rockafellar [364], Geremew, Mordukhovich and Nam [503], Henrion and Jourani [559], Henrion, Jourani and Outrata [560], Henrion and Outrata [561, 562], Jeyakumar and Yen [631], Jourani [647], Jourani and Ye [662], Klatte and Henrion [685], Klatte and Kummer [686, 687], Kummer [725, 726, 728], Ledyaev and Zhu [751], Levy [767, 768], Lee, Tam and Yen [755], Levy and Mordukhovich [769], Levy, Poliquin and Rockafellar [771], Lucet and Ye [816], Mordukhovich [907, 910, 911, 912, 913, 924, 927, 929], Mordukhovich and Nam [935, 934], Mordukhovich and Outrata [939], Mordukhovich and Shao [951], Outrata [1030], Papi and Sbaraglia [1050], Poliquin and Rockafellar [1092], Robinson [1137, 1138, 1139], Rockafellar and Wets [1165], Rückmann [1183], Zhang [1362], Zhang and Treiman [1363], and Zheng and Ng [1365].

First-Order Optimality/Suboptimality and Qualification Conditions in Nondifferentiable Programming and Related Problems: Arutyunov and Pereira [37], Bector, Chandra and Dutta [90], Bertsekas and Ozdaglar [112, 1035], Borwein, Treiman and Zhu [158], Borwein and Zhu [163, 164], Dutta [374, 375, 376], Glover and Craven [508], Glover, Craven and Flåm [509], Ioffe [589, 596, 603, 611], Kruger [706, 705, 714, 715], Kruger and Mordukhovich [718, 719], Lassonde [747], Ledyaev and Zhu [754], Mordukhovich [892, 893, 897, 901, 922, 925], Mordukhovich, Nam and Yen [937, 938], Mordukhovich and B. Wang [962], Mureşan [988], Rockafellar [1158, 1160], Ralph [1115], Rockafellar and Wets [1165], Thibault [1250], Treiman [1267, 1268], and Ye [1339, 1340].

Optimality Conditions for Multiobjective Problems: Amahroq and Gadhi [16], Bellaassali and Jourani [93], Borwein and Zhu [164], Craven and Luu [300], Eisenhart [395], Dutta [376], Dutta and Tammer [378], El Abdouni and Thibault [402], Gadhi [489], Govil and Mehra [518], Ha [531, 532], Jahn, Khan and Zeilinger [628], Jourani [645], Kruger and Mordukhovich [718, 719], Mordukhovich [892, 897, 901, 926, 928], Mordukhovich, Treiman and Zhu [958], Mordukhovich, Outrata and Červinka [940], Thibault [1250], Ye and Zhu [1345], Ward and Lee [1312], Zheng and Ng [1364], and Zhu [1372].

Second-Order Optimality Conditions: Arutyunov and Pereira [37], Eberhard and Pearce [384], Eberhard, Pearce and Ralph [385], Eberhard and Wenczel [387], Jahn, Khan and Zeilinger [628], Levy, Poliquin and Rockafellar [771], Mordukhovich [925, 926], Poliquin and Rockafellar [1092], and Ward [1308, 1310].

Optimization and Equilibrium Problems with Equilibrium Constraints: Anitescu [20], Dutta and Dempe [377], Flegel [454], Flegel and Kanzow [455, 456], Flegel, Kanzow and Outrata [457], Hu and Ralph [584], Jiang and Ralph [632], Kočvara, Kružik and Outrata [689], Kočvara and Outrata [690], Lucet and Ye [816], Mordukhovich [925, 926, 928], Mordukhovich, Outrata and Červinka [940], Outrata [1024, 1025, 1027, 1026, 1028, 1029, 1030], Ralph [1116], Scheel and Scholtes [1191], Scholtes [1192], Treiman [1268], Ye [1338, 1339, 1342], Ye and Ye [1343], Ye and Zhu [1345], and Zhang [1360, 1361].

Eigenvalue Analysis and Optimization: Borwein and Zhu [164], Burke, Lewis and Overton [194, 195, 198, 200], Burke and Overton [202, 203, 204], Ciligot-Travain and Traore [242], Dontchev and Lewis [360], Jourani and Ye [662], Ledyaev and Zhu [752, 753, 754], Lewis [775, 779], Lewis and Sendov [782, 783], and Sendov [1200]; cf. also Overton [1033] and Overton and Womersley [1034] for earlier results in this direction concerning eigenvalues of symmetric matrices.

Stochastic Programming and Related Topics: Dentcheva and Römisch [324], Glover, Craven and Flåm [509], Henrion [557, 558], Henrion and Outrata [562], Henrion and Römisch [563, 564], Outrata and Römisch [1032], and Papi and Sbaraglia [1051, 1052]. Note that there are many other problems of stochastic optimization and related areas, which are intrinsically nonsmooth and potentially cover a large territory for applying the generalized differential tools of variational analysis developed in this book; see, e.g., Birge and Qi [115], Dentcheva and Ruszczyński [325], Pennanen [1061], Schultz [1196], Wets [1327], and the references therein.

Necessary Conditions in the Calculus of Variations and Optimal Control for Ordinary Discrete and Differential Systems: Arutyunov and Aseev [33], Aseev [39, 40, 41], Bellaassali and Jourani [93], Bessis, Ledyaev and Vinter [113], Clarke [257, 258, 260, 261], Clarke, Ledyaev, Stern and Wolenski [264, 265], Eisenhart [395], Ferreira, Fontes and Vinter [443], Ferreira and Vinter [444], Ginsburg and Ioffe [506], Ioffe [605], Ioffe and Rockafellar [616], Kruger and Mordukhovich [717], Loewen [801], Loewen and Rockafellar [805, 806, 807], Marcelli [845], Marcelli, Outkine and Sytchev [847], Mordukhovich [887, 889, 893, 897, 901, 902, 904, 914, 915, 916, 921], Mordukhovich and Shvartsman [955], de Pinho [1074], de Pinho, Ferreira and Fontes [1075, 1076], de Pinho and Ilchmann [1077], de Pinho and Vinter [1078, 1079], de Pinho, Vinter and Zheng [1080], Rampazzo and Vinter [1118], Rockafellar [1161, 1162], Rowland and Vinter [1179], Silva and Vinter [1211], Smirnov [1215, 1216], Vinter [1289], Vinter and Woodford [1293], Vinter and Zheng [1294, 1295, 1296], Woodford [1331], and Zhu [1372].

Qualitative Analysis of Ordinary Control Systems, Sensitivity, Stability, and Controllability: Borwein and Zhu [161], Clarke [261], Clarke, Ledyaev, Stern and Wolenski [264, 265], Galbraith [491, 492], Galbraith and Vinter [493], Ioffe [605], Jourani [647], Ledyaev and Zhu [754], Loewen and Rockafellar [807], Mordukhovich [901, 915], Rockafellar and Wolenski [1166,

1167], Shvartsman and Vinter [1210], Smirnov [1216], Vinter [1289], Vinter and Wolenski [1292], and Wolenski and Zhuang [1330].

Optimal Control of Time-Delay and Functional-Differential Systems: Clarke and Wolenski [275], Ginsburg and Ioffe [506], Minchenko [878], Minchenko and Sirotko [880], Minchenko and Volosevich [881], Mordukhovich [921], Mordukhovich and Trubnik [959], Mordukhovich and L. Wang [973, 974, 975, 976, 977], Ortiz [1021], and Ortiz and Wolenski [1022].

Generalized Solutions to Hamilton-Jacobi Equations, Stabilization, and Feedback Synthesis of Control Systems: Clarke, Ledyaev, Sontag and Subbotin [263], Clarke, Ledyaev, Stern and Wolenski [264, 265], Clarke and Stern [269], Luo and Eberhard [819], Freeman and Kokotović [474], Galbraith [490, 491, 492], Goebel [511], Ledyaev and Zhu [754], Malisoff, Rifford and Sontag [837], Rifford [1124], Rockafellar [1164], Rockafellar and Wolenski [1166, 1167], Sontag [1220], and Wolenski and Zhuang [1330].

Analysis, Control, and Optimization of Evolution and Partial Differential Systems: Bounkhel and Thibault [173], Colombo and Goncharov [277], Colombo and Wolenski [280], Edmond and Thibault [390], Gavrilov and Sumin [500], Guillaume [525], Ioffe [611], Marcellin [848], Mordukhovich [932], Mordukhovich and D. Wang [970, 971], Rossi and Savaré [1176], and Sumin [1233].

Variational Analysis and Generalized Differentiation on Smooth and Riemannian Manifolds: This area of research has been recently started in the work by Borwein and Zhu [164], Dontchev and Lewis [360], Ledyaev and Zhu [752, 753, 754], and Rolewicz [1172]; cf. also Chryssochoos and Vinter [240].

Applications to the Qualitative Theory of Dynamical Systems, Geometry of Banach Spaces, Real and Complex Analysis: Avelin [66, 67], Benabdellah [96], Benabdellah, Castaing, Salvadori and Syam [97], Bolte, Daniilidis and Lewis [122, 122], Bounkhel and Thibault [173], Borwein, Borwein and Wang [136], Borwein, Fabian, Kortezov and Loewen [139], Borwein, Fabian and Loewen [140], Borwein and Fitzpatrick [141, 143], Borwein, Fitzpatrick and Girgensohn [144], Borwein and Jofré [148], Borwein, Moors and Wang [152], Borwein, Treiman and Zhu [158], Borwein and Zhu [163, 164], Fabian and Mordukhovich [419, 422], Ha [530, 531], Ioffe [607], Jourani [649], Jourani and Thibault [661], Mordukhovich and Shao [949], Mordukhovich and B. Wang [960], Rolewicz [1171, 1172], Rossi and Savaré [1176], and Wang [1303, 1304].

Applications to Mechanical, Physical, and Engineering Problems: Anitescu [20], Benabdellah [96], Benabdellah, Castaing, Salvadori and Syam [97], Bounkhel and Thibault [173], Burke, Lewis and Overton [194, 195, 197], Burke and Luke [201], Luke, Burke and Lyon [817], Colombo and Goncharov [277], Edmond and Thibault [390], Freeman and Kokotović [474], Kočvara, Kružik and Outrata [689], Kočvara and Outrata [690, 691], Mordukhovich and Outrata [939], Outrata [1024, 1027, 1028, 1030], Rossi and Savaré [1176], and Vinter [1289].

Applications to Economics and Finance: Bellaassali and Jourani [93], Borwein and Zhu [164], Bounkhel and Jofré [171], Cornet [288], Cornet and Czarnecki [290], Flåm [452], Flåm and Jourani [453], Florenzano, Gourdel and Jofré [460], Jofré [633], Jofré and Rivera [635], Habte [533], Khan [669, 670, 671], Kočvara and Outrata [690], Malcolm and Mordukhovich [836], Mordukhovich [920, 922, 930], Mordukhovich, Outrata and Červinka [940], Outrata [1029, 1030], Papi and Sbaraglia [1051, 1052], Villar [1288], and Zhu [1375].

1.4.11. Generalized Normals in Banach Spaces. Now let us comment on the major results presented in Sect. 1.1, which is mainly devoted to the study of our basic *geometric constructions* in the framework of arbitrary Banach spaces. Theorem 1.6 was first formulated in Kruger and Mordukhovich [718] and Mordukhovich [892], where relations with tangent/contingent approximations were established as well. Complete proofs of these results were given in [719, 901]; cf. also Ioffe [596] for an equivalent representation of the basic normal cone in finite dimensions via limits of dual vectors to the contingent cone. Note that representation (1.8) of the basic normal cone in Theorem 1.6 was adopted by Rockafellar and Wets [1165] as the basic definition of the (general) normal cone in finite-dimensional spaces.

Polarity relationships between *tangents* and *normals* of the type discussed in Subsect. 1.1.2 were considered in many publications; see particularly [89, 156, 600, 705, 719, 1165]. Both inclusion relations involving Clarke's tangent cone and the contingent/weak contingent ones in Theorem 1.9 were established by Kruger [705] in the infinite-dimensional settings of the theorem; cf. also Cornet [285] and Penot [1065] for the finite-dimensional equality

$$T_C(\bar{x};\Omega) = \operatorname*{Lim\,inf}_{x \xrightarrow{\Omega} \bar{x}} T(x;\Omega)$$

that follows from Theorem 1.9. The first inclusion of this theorem was also proved by Treiman [1262] in Banach spaces, while the second one was given by Penot [1065] in reflexive spaces. The equality formula of Theorem 1.9 under the additional Kadec and Fréchet smooth assumptions was established by Borwein and Strójwas [156].

The results of Subsect. 1.1.3 are mostly based on the paper by Mordukhovich and B. Wang [967]. Note that the notion of *strict differentiability* largely used in this subsection was formally introduced by Leach [748], while it was already known to Peano [1054] and was actually used by Graves [522] in his proof of the celebrated Lyusternik-Graves theorem; see Theorem 1.57 and the paper by Dontchev [352]. Observe also that the *uniform estimates* for ε-normals derived in Lemma 1.16 (considered here and everywhere in the book as preliminary results versus pointwise assertions in terms of the basic/limiting constructions) should be distinguished from "fuzzy calculus" rules initiated by Ioffe [591, 594] in somewhat different settings, since the former provide more precise estimates *uniformly* on the *entire* neighborhoods of the points in question with computing the corresponding constants. A finite-dimensional version of Theorem 1.17 with the full rank assumption on the Jacobian was proved, in a different way, by Rockafellar and Wets [1165].

The *sequential normal compactness* (SNC) property of sets from Subsect. 1.1.4 was introduced by Mordukhovich and Shao in [951] (preprint of 1994) and then named "SNC" in [950]. Note that arguments involving an interplay between the weak* and norm convergences of normal elements to zero in dual spaces have been often used (explicitly or implicitly) in different aspects of infinite-dimensional variational analysis to avoid triviality conclusions; see, e.g., Borwein and Strójwas [155, 156], Ginsburg and Ioffe [506], Ioffe [595, 598, 607], Jourani and Thibault [655, 656, 661], Kruger [707, 709], Loewen [800, 801], Mordukhovich [901, 917], Mordukhovich and Shao [949], and Penot [1068, 1071]. Theorems 1.21 and 1.22 were established by Mordukhovich and B. Wang [967].

The *compactly epi-Lipschitzian* (CEL) property of sets was introduced by Borwein and Strójwas [155] as an extension of the *epi-Lipschitzian* property by Rockafellar [1147]. In contrast to the epi-Lipschitzian property largely related to *nonempty interiors* (see Proposition 1.25 for convex sets), the CEL property holds for *every* set in finite dimensions. Comprehensive characterizations of the CEL property for closed and *convex* sets in normed spaces were given by Borwein, Lucet and Mordukhovich [150]; see Remark 1.27(i). Further elaborations and deep developments of these results, in the framework of separation theorems in Hilbert spaces, were obtained by Ernst and Théra [409]. The proof of Theorem 1.26 is based on Loewen's arguments from [800]; cf. also Mordukhovich and Shao [949].

Complete characterizations of CEL sets in Banach spaces via the *topological/net* convergence of normal elements in dual spaces were obtained in the fundamental study by Ioffe [607] with the usage of *variational principles*; see Remark 1.27(ii). These characterizations show that the CEL property is actually a proper *topological counterpart* of the SNC one. Comprehensive relationships between the CEL and SNC properties of sets in general Banach spaces were established by Fabian and Mordukhovich [422] and discussed in Remark 1.27(ii).

A *smooth variational description* of Fréchet normals in general Banach spaces from Theorem 1.30(i) of Subsect. 1.1.5 was observed by Mordukhovich [925]. The much *more delicate* descriptions from assertions (ii) and (iii) of this theorem under the additional geometric assumptions on the space in question are geometric/normal counterparts of the corresponding subgradient descriptions established by Fabian and Mordukhovich [419]; see Theorem 1.88 in

Subsect. 1.3.2. Note that assertion (iii) of Theorem 1.30 for $\mathcal{S} = \mathcal{LF}$ follows from the variational description of Fréchet subgradients derived by Deville, Godefroy and Zizler [330, 331]. It was also proved by Rockafellar and Wets [1165] in finite-dimensional spaces. Let us emphasize that the *Fréchet-like normal/subgradient structure is crucial for such smooth variational descriptions* important in many applications including those in this book.

It is worth mentioning that a generalized normal concept of the *variational type* given in Theorem 1.30(iii) goes back, in finite dimensions, to Hörmander [581, 582] who applied it to partial differential equations and complex analysis; see also Avelin [66, 67]. Subdifferential concepts of this type were initiated and strongly developed by Crandall and Lions [297], Crandall, Evans and Lions [295] in the theory of *viscosity solutions* to Hamilton-Jacobi and related equations, which then became one of the most active and flourishing areas in nonlinear analysis and partial differential equations with various applications to optimal control, differential games, stochastic equations, etc.; see, e.g., [85, 296, 458, 1230] and the references therein. Such subdifferential concepts have been adopted and applied to problems in nonsmooth and variational analysis by Deville et al. [328, 329, 330, 331] and especially by Borwein and Zhu [160, 163, 164] under the name of "viscosity" or "smooth" subdifferentials. Note that smooth normals and subgradients of this kind are equivalent to the Fréchet ones from Definition 1.1(i) and Subsect. 1.3.2 under some *smoothness assumptions* on the space in question, which are always imposed in the aforementioned publications and which are not only sufficient but also *necessary* for such descriptions of Fréchet-like constructions; see Fabian and Mordukhovich [419]. On the other hand, any smoothness restrictions can be *avoided* while using the constructions adopted in this book, in both prelimiting and limiting frameworks.

The *minimality property* of the basic normal cone from Proposition 1.31 observed by Mordukhovich [920] is strongly related to the corresponding subdifferential result obtained by Mordukhovich and Shao [949]. Previous minimality results in this direction, under more restrictive requirements, were first observed by Ioffe [596] and then developed by Ioffe [599] and Mordukhovich [894, 901].

1.4.12. Derivatives and Coderivatives of Set-Valued Mappings. In Sect. 1.2 we start studying generalized differentiation of set-valued (in particular, single-valued) mappings employing the *graphical/geometric approach* to generalized differentiation that relates derivative-like constructions for mappings with infinitesimal approximations of their graphs. Such a graphical approach goes back to the very beginning of classical differentiation when Fermat (1636) defined the original derivative notion for a polynomial function at a given point via the *tangent slope* to its graph. Fermat's geometric approach was strongly developed in the modern framework by Aubin who defined, in his 1981 paper [48], a derivative notion for a set-valued mapping via the *contingent cone* to its graph at the point in question; cf. also Pshenichnyi

[1107, 1109] for earlier developments. Various *tangentially generated derivatives* of this type for nonsmooth functions and mappings were introduced and studied in many publications employing different tangential approximations of graphs; see, e.g., [28, 29, 52, 54, 58, 60, 91, 133, 186, 465, 469, 517, 594, 630, 686, 774, 1068, 1060, 879, 1094, 1159, 1165, 1168, 1247, 1278].

The other line of the graphical approach to generalized differentiation was developed by Mordukhovich who introduced, in his 1980 paper [892], the *coderivative* notion for general set-valued mappings via the basic normal cone (1.80) to their graphs. This is conceptually different from tangentially generated derivatives in the line of Aubin and Pshenichnyi due to the *absence of duality* between tangent and normal cones in general nonconvex settings; of course, for smooth and convex-graph mappings the two approaches are equivalent. Observe that coderivatives provide extensions of the *adjoint* derivative operator to nonsmooth and set-valued mappings, while tangentially generated derivatives extend the classical derivative concept to arbitrary mappings.

As mentioned, the first coderivative was defined in [892] by formula (1.86) via the nonconvex normal cone (1.80) in finite dimensions. It was motivated by applications to optimal control of differential inclusions $\dot{x} \in F(x,t)$, and D^*F was employed in [892] (under the name of "adjoint mapping") to describe the *adjoint system* in necessary optimality conditions of the Euler-Lagrange type for differential inclusions; for convex-graph mappings this agrees with "locally conjugate/adjoint" operations used by Pshenichnyi. The very appropriate term "coderivative" for constructions of type (1.86) for set-valued mappings was later suggested by Ioffe [594, 596]. The notions of graphical N-regularity and M-regularity from Definition 1.36 appeared in Mordukhovich [917], while in finite dimensions they both go back to his earlier publications [892, 901].

In infinite-dimensional settings, we distinguish between two limiting coderivatives that both play a basic role in our analysis: the normal coderivative and the mixed coderivative from Definition 1.32. The normal coderivative described by (1.26) via the basic normal cone (1.3) is not actually different from the original definition of [892] in finite dimensions depending only on the normal cone in question, while the mixed coderivative is a pure infinite-dimensional construction. It first appeared in Mordukhovich [917] (see also Mordukhovich and Shao [953]), although the idea of using a mixed convergence on the product of dual spaces was earlier explored by Penot [1071] (preprint of 1995). However, the construction of [1071] (defined in terms of convergent nets, not sequences) is different from the mixed coderivative of Definition 1.32(iii) by the *reserved order* of mixed convergence: weak* in the domain variable and strong in the image one. The main disadvantage of the latter construction is the *lack of calculus*, even in the case of real-valued functions; cf. Remark 3.22. In contrast, our limiting coderivatives from Definition 1.32, both normal and mixed, enjoy comprehensive calculi and thus various applications being fully independent and irreplaceable in infinite dimensions.

The difference between the normal and mixed coderivative in Example 1.35 was demonstrated by Mordukhovich and Shao [953], while the mapping in this example was taken from Ioffe [598]. The extremal property of convex-valued multifunctions from Theorem 1.34 and the coderivative representations for differentiable mappings from Theorem 1.38 go back to the early work of Mordukhovich [892, 901].

1.4.13. Lipschitzian Properties. In Subsect. 1.2.2 we begin a comprehensive study of Lipschitzian properties for (generally) set-valued mappings, which play a central role in many aspects of variational analysis and its applications, particularly those considered in this book. The Lipschitz continuity of functions (introduced in the 19th century by Lipschitz [796] in the framework of differential equations) has been well recognized in the classical analysis (probably starting with Peano) as a *linear rate* counterpart of the standard continuity that, due to its linear rate, is very convenient from both theoretical/qualitative and numerical/quantitative viewpoints. The classical Lipschitz property plays a significant role in convex analysis, where it is actually indistinguishable from the standard continuity of convex functions, and especially in Clarke's nonsmooth analysis that is largely revolves around locally Lipschitzian functions.

Set-valued mappings are of special interest in variational analysis and optimization due, in particular, to the necessity of analyzing the behavior of (moving) sets of feasible and optimal solutions to constraint and variational systems with respect to parameter perturbations. This is mainly a subject of sensitivity and/or stability analysis, where notions of *Lipschitzian stability* play a crucial role. Appropriate extensions of the Lipschitz continuity to set-valued mappings are therefore heavily required. The standard notion of the (Hausdorff) Lipschitz continuity for a multifunction $F\colon X \rightrightarrows Y$, corresponding actually to the classical Lipschitz property of a *single-valued* mapping with values in the space of *compact* subsets of Y endowed with the *Pompieu-Hausdorff distance* (see [552, 1101, 1165]), may be *restrictive* for the needs of variational analysis. A significant restriction comes from the compactness requirement (boundedness in finite dimensions) on the set values. This is not often the case for solution maps to parametric variational inequalities and other optimization-related problems. A simple while very important example of unbounded sets is provided by epigraphs of real-valued functions significant in the theory and many applications.

An appropriate version of Lipschitzian behavior for set-valued mappings, with no compactness restriction, was discovered by Aubin [49] who was motivated by applications to sensitivity analysis for convex optimization problems. Aubin's property is a localization of Lipschitzian behavior in a neighborhood of a given point from the *graph* of F, being indeed *the most natural* counterpart of the classical local Lipschitz continuity in the case of set-valued mappings. Furthermore, Aubin's property happens to be *equivalent* to the standard local Lipschitz continuity of the corresponding (scalar) distance function due to Theorem 1.41 established by Rockafellar [1154]. Thus the term "pseudo-Lipschitz" suggested by Aubin for this property seems to be rather misleading, since "pseudo" means "false." In [364, 1165] this property was called the "Aubin property," without specifying its Lipschitzian nature. Other names for this behavior were suggested, e.g., in [686, 728]. In our opinion, the term "*Lipschitz-like*" accepted in this book better reflects the nature and the sense of Aubin's extension of the classical Lipschitz property to set-valued mappings.

Observe that, in accordance with the classical local Lipschitz continuity, both Hausdorff and Aubin local Lipschitzian properties involve the comparison between *all pairs* of points from a neighborhood of the reference point in question. This implies the *robustness* of both Hausdorff and Aubin set-valued extensions with respect to perturbations of the reference point, i.e., these Lipschitzian properties, as well as the classical one, are properties *around* the given point. Throughout the book we distinguish such properties from those *at* the given point that are usually not robust.

Other robust Lipschitzian properties for set-valued mappings, which seem to be essentially finite-dimensional in nature, were defined and studied by Rockafellar [1154], Loewen and Rockafellar [805], Rockafellar and Wets [1165], and Galbraith [491]. Theorem 1.42 is an infinite-dimensional version of Rockafellar's results established in [1154]. More discussions on such properties can be found in [1165].

The study of "non-robust" properties of set-valued mappings, corresponding to the fixed $u = \bar{x}$ in the basic inclusion (1.28) of Definition 1.40, was initiated by Robinson [1130] under the name of the "upper Lipschitzian" property, where $V = I\!R^m$ in (1.28); note that such behavior doesn't go back to the classical Lipschitz continuity in the case of single-valued mappings. In [1132], Robinson established the upper-Lipschitzian property for the so-called *piecewise polyhedral* mappings important in applications to sensitivity analysis for some classes of optimization problems particularly including linear programming; cf. Walkup and Wets [1299] and Robinson [1126, 1127] for previous results in this direction. The upper Lipschitzian property and its modifications were called later "calmness" properties by Rockafellar and Wets [1165]. These and related Lipschitzian properties of set-valued mappings were studied and applied in many publications; see, e.g., [91, 424, 482, 519, 550, 559, 560, 561, 562, 641, 768, 773, 686, 687, 1339, 1362].

One of the *strongest advantages* of the coderivative constructions from Definition 1.32 is the possibility to provide in their terms complete *dual characterizations* for robust Lipschitzian behavior of set-valued and single-valued mappings and for the corresponding properties of metric regularity and covering. Subsection 1.2.2 contains *necessary coderivative conditions* for robust Lipschitzian behavior in arbitrary Banach spaces. Theorems 1.43 and 1.44 were established in Mordukhovich [917] and Mordukhovich and Shao [953], while in finite dimensions the results of Theorem 1.44 go back to the earlier work by Mordukhovich: to [892, 901] for the local Lipschitzian property and to

[907] for the Lipschitz-like one. Estimate (1.32) in general Banach spaces was first obtained by Mordukhovich and Shao [946] for $\varepsilon = 0$; the given simplified proof follows the ideas from Jourani and Thibault [661].

The concepts of *graphically Lipschitzian* and *graphically smooth* mappings from Definition 1.45 go back to Rockafellar [1153] who introduced them under the names of "Lipschitzian manifolds" and "strictly smooth sets" for their graphs; the "graphical" terminology was first adopted by Rockafellar and Wets [1165]. The hemi-Lipschitzian and hemismooth versions of Definition 1.45 appeared in Mordukhovich and B. Wang [965]. Due to the results by Rockafellar [1153] in their extensions in Poliquin and Rockafellar [1090] and Dontchev and Rockafellar [365], the graphical Lipschitzian property holds for broad collections of greatly important mappings typically encountered in finite-dimensional variational analysis and optimization. They particularly include *subdifferential mappings* for convex, saddle, and (essentially more general) prox-regular functions being invariant under the so-called "ample parametrization."

Theorem 1.46 on the equivalence between the graphical regularity and the graphical smooth (resp. hemismooth) properties was established by Mordukhovich [912] for graphically Lipschitzian mappings and by Mordukhovich and B. Wang [965] for graphically hemi-Lipschitzian ones based on Rockafellar's results [1153] on the *subspace property* of Clarke normals in finite dimensions and on the normal cone (equality type) calculus from Subsect. 1.1.3. We refer the reader to Subsect. 3.2.4 and the corresponding comments to Chap. 3 given in Sect. 3.4 for infinite-dimensional extensions of these and related results.

1.4.14. Metric Regularity and Linear Openness. Metric regularity and covering/linear openness properties we begin to study in Subsect. 1.2.3 have been long recognized among the most fundamental in nonlinear analysis. Their origin goes back to the classical Banach-Schauder *open mapping* theorem for linear operators [76, 1190] established in the early 1930s. A celebrated nonlinear extension of the Banach-Schauder result was obtained in 1934 by Lyusternik [824] and independently (in a different but largely equivalent form) in the 1950 paper by Graves [522]. This result, called now the *Lyusternik-Graves theorem*, and the methods developed for its proof reproduced in the arguments of Theorem 1.57 play a crucial role in many aspects of the classical nonlinear analysis as well as of modern variational analysis and their numerous applications; see, e.g., [337, 352, 355, 361, 587, 608, 676, 677, 1100, 1110, 1129] for more results, discussions, references, and applications.

The key estimate (1.36) in the definition of metric regularity with $y = \bar{y} = f(\bar{x})$ for $\mathcal{C}^1$ functions $F = f\colon X \to Y$ appeared in the original Lyusternik's proof [824] of his result regarding the description of the tangent space to a smooth manifold; it is worth mentioning that his theorem was motivated by applications to *Lagrange multipliers* in a variational problem with the equality/operator constraint $f(x) = 0$ given by a smooth mapping between Banach spaces. Graves established in his proof, which was actually applied to mappings f *strictly differentiable* at $\bar{x}$ though the latter notion was not explicitly defined, the covering/openness part (1.39) of the theorem; both regularity and covering parts are now known to be *equivalent.* The equivalence between these properties for Lipschitz continuous mappings was first observed probably by Dmitruk, Milyutin and Osmolovskii [337, Introduction], with no proof given; cf. also Ioffe [589, 598]. Note that Graves' original version of the covering/openness theorem was definitely underestimated in [337]; see more discussions in Dontchev [352].

The next step in obtaining distance estimates of type (1.36) for set-valued mappings given by *inequalities*, which probably reflect the main feature of modern (after linear programming) optimization in contrast to the classical one, was the 1952 paper by Hoffman [579] who derived estimates for the distance to sets of solutions given by linear equality and inequality systems in finite dimensions. Hoffman's type estimates, known now as *error bounds*, has become an important part of modern optimization theory developed in many publications; see, e.g., [59, 60, 71, 88, 188, 190, 191, 205, 424, 445, 639, 647, 686, 716, 692, 784, 842, 1003, 1004, 1005, 1045, 1126, 1334, 1353] and the references therein.

Seminal contributions to the study of metric regularity and openness properties of set-valued mappings governed by nonlinear smooth equality and inequality systems as well as convex processes, were made by Robinson in the series of publications in the 1970s; see [1125, 1127, 1128, 1129]. His fundamental theorem on metric regularity and covering/openness for *convex processes*, discovered independently by Ursescu [1275] (cf. Theorem 4.21 in this book and its "closed graph" version in Aubin and Ekeland [52, Theorem 3.3.1]), has been of great importance and influence for the development and applications of variational analysis.

Early extensions of the Lyusternik-Graves theorem to *nonsmooth* and *nonconvex* systems were obtained, for single-valued Lipschitzian mappings $f\colon X \to Y$ between Banach spaces in terms of Clarke subgradients, by Ioffe [587] and by Milyutin in [337, Sect. 5]. In fact, Ioffe considered not the full metric regularity property as defined in (1.36) for *all* y around $\bar{y}$ but its weaker *one-point* counterpart with $y = \bar{y} = f(\bar{x})$ in (1.36). The latter regularity *at a point* called recently "subregularity" by Dontchev and Rockafellar [366] is useful for certain important applications, e.g., to the theory of necessary optimality and controllability conditions. Its covering counterpart was investigated by Warga (see, e.g., [1318, 1320, 1322]), under the name of "fat homeomorphism," in terms of his derivate containers. However, such one-point properties are *not robust*, which creates difficulties for their comprehensive study and implementation, especially in infinite dimensions.

Milyutin was probably the first who strongly emphasized (in his talks and personal communications, long before publishing [337]) the importance to consider regularity and covering properties of operators in *entire neighborhoods* (or *around* reference points – the terminology adopted in this book), with *uniform* estimates. He also realized from the very beginning that his sufficient condition for covering of Lipschitzian operators in terms of Clarke subgradients, as well as the related implicit function theorem by Magaril-Il'yaev [826], were incomplete and *far removed from the necessity*, while the classical Lyusternik regularity condition $\nabla f(\bar{x})X = Y$ was an equivalent to covering for smooth mappings.

The "regularity" terminology was originally employed by Lyusternik to indicate the fulfillment of his surjectivity condition $\nabla f(\bar{x})X = Y$. In the same sense it has been later used in most of the Russian literature; see, e.g., Ioffe and Tikhomirov [618]. Robinson's usage of the word "regularity" in [1128, 1129] related actually to the openness property of type (1.39), which was called "covering" by Milyutin et al. (see, e.g., [337]). Ioffe [589, 596, 598] used the term "surjection" for a similar property defined at a point; he reserved "regularity" [587] for the distance estimate (1.36) with $y = \bar{y} = f(\bar{x})$. The term "metric regularity" for the distance estimate, which seems to be very appropriate and is widely accepted nowadays, was first employed by Borwein [137]. The "openness at a linear rate" terminology goes back to Dolecki [339]; Rockafellar and Wets [1165] called this property "linear openness."

The *equivalences* between the local properties of metric regularity, covering/linear openness for set-valued mappings, and Lipschitzian behavior of Aubin's type for their inverses were proved by Borwein and Zhuang [165] and by Penot [1066]. They didn't however include the correspondences between modulus/exact bounds into their theorems. The equivalence results and terminology of Subsect. 1.2.3, including local and nonlocal concepts, were developed by Mordukhovich [909].

Note that *nonlocal* (global, semi-local) metric regularity and related properties of set-valued and single-valued mappings happened to be important in many applications, in particular, to optimal control (see, e.g., Dmitruk [336]) and numerical methods in optimization and equilibria (see, e.g., Ralph [1116]). Observe that the nonlocal properties studied in Subsect. 1.2.3 are different from those in the recent paper by Ioffe [608] who developed the metric regularity theory for mappings between *metric spaces*. Mordukhovich and B. Wang [967, 968] introduced and studied the property of "restrictive metric regularity" for mappings $f\colon X \to Y$ between Banach spaces that reduced to the standard metric estimate of type (1.36) for the restrictive mapping $f\colon X \to f(X)$ between X and the *metric* space $f(X) \subset Y$ while taking into account the *Banach* space nature of both spaces X and Y; see Remark 1.61 for more discussions. Another notion of nonlocal *directional* metric regularity has been recently introduced and studied by Arutyunov and Izmailov [36] motivated by applications to sensitivity analysis in optimization.

Necessary coderivative conditions for the metric regularity and covering properties, with the exact bound estimates, presented in Theorem 1.54 and Corollary 1.55 follow from the corresponding Lipschitzian results of Subsect. 1.2.2 due to the obtained equivalence relationships; cf. Mordukhovich [894, 901, 917], Kruger [709], and Mordukhovich and Shao [946, 953]. These necessary conditions are important in the subsequent applications, especially to coderivative calculus rules in Chap. 3. The *sufficiency* of these conditions and their applications will be discussed in Chap. 4, with full commentaries and references given in Sect. 4.5.

Theorem 1.57 gives *complete characterizations* of the covering and metric regularity properties for single-valued mappings between Banach space that are strictly differentiable at the point in question. Its *sufficiency* part is the essence of (the proof of) the classical Lyusternik-Graves theorem. As mentioned, Lyusternik [824] formally established the tangent space result for $\mathcal{C}^1$ mappings, while his proof contained in fact the metric regularity estimate (1.36). Graves [522] obtained the covering property, actually for strictly differentiable mappings; his arguments are exactly reproduced in the proof of the sufficient part of Theorem 1.57. Note that both proofs by Lyusternik and Graves were based on an *iterative process*, which happened to be a certain – essential – modification of the classical Newton's tangent method, called "Lyusternik's iterative process" in [337].

It seems that the *necessity* part of Theorem 1.57 and the *precise formulas* for the exact regularity and covering bounds were first established in finite-dimensions by Mordukhovich [894, 901, 909] as a simple corollary of general coderivative characterizations of the metric regularity and covering properties for set-valued mappings. It was later observed that these results for $\mathcal{C}^1$ (as well as for strictly differentiable) mappings could be derived by conventional arguments of functional analysis; cf. Cominetti [282], Ioffe [607], and Dontchev, Lewis and Rockafellar [361]. Note that a rigorous proof of Theorem 1.57 requires the closedness of derivative images for metrically regular mappings; this fact presented in Lemma 1.56 was established by Mordukhovich and B. Wang [967]. Of course, the possibility to obtain the *necessity* and *exact bound formulas* in terms of the first-order differential constructions are due to the *linear rate* in the properties under consideration; this was probably not realized in the classical Lyusternik-Graves theorem. Higher-order versions of these properties were studied, e.g., in [165, 466, 467, 469, 521, 608].

The *inverse mapping* results of Theorem 1.60 are established in this book is a consequence of the covering characterization of Theorem 1.57. The sufficient part of this theorem is Leach's extension [748] of the classical ($\mathcal{C}^1$) inverse function theorem to the then-new class of strictly differentiable mappings; see also the corresponding extension of the related *implicit function* theorem by Nijenhuis [1011] and the recent book by Krantz and Parks [699] on implicit function theorems with many historical details. The necessity of the *invertibility* assumption on $\nabla f(\bar{x})$ for the existence of a locally single-valued and strictly differentiable inverse was probably first observed by Dontchev [351] as a consequence of his general results on the preservation of certain Lipschitzian and differentiability properties for solution maps to "generalized equations" under *strong approximations* in the sense of Robinson [1136]. We refer the reader to Clarke [252, 255], Dontchev [350], Dontchev and Hager [356], Hiriart-Urruty [570], Ioffe [589], Jongen, Klatte and Tammer [639] Kummer [725, 726], Levy [767], Robinson [1136], Rockafellar and Wets [1165], Warga [1318, 1320, 1322], and the bibliographies therein for *nonsmooth versions* of the implicit and inverse function theorems with various applications.

1.4.15. Coderivative Calculus in Banach Spaces. Subsection 1.2.4 contains calculus rules of the "right" inclusion and *equality* types for Fréchet, normal, and mixed coderivatives in arbitrary Banach spaces, with the corresponding regularity statements. The sum and chain rules from Theorems 1.62, 1.64, and 1.65 were derived by Mordukhovich and Shao [950, 953] extending the finite-dimensional results and arguments of Mordukhovich [910]. Note that the *ε-enlargements* in the construction of both normal and mixed limiting coderivatives are *crucial* for the validity of the sum and chain rules even in finite dimensions, being indeed *unavoidable* in general Banach space settings.

The reader recognizes from Definition 1.63(i) that the notion introduced therein is actually the classical notion of lower semicontinuity for set-valued mappings; the appropriate name of *inner semicontinuity* was suggested by Rockafellar and Wets [1165] to distinguish it from the lower semicontinuity of real-valued functions. The property of inner/lower *semicompactness* from Definition 1.63(ii) was defined by Mordukhovich and Shao [949]. The chain rules from Theorem 1.66 were established by Mordukhovich and B. Wang [967].

The *SNC property* of set-valued mappings from Definition 1.67(i) is directly induced by the SNC property of sets defined in Subsect. 1.1.4, while the PSNC (i.e., *partial* SNC) property essentially takes into account the natural *product structure* of the graph space for set-valued mappings $F\colon X \rightrightarrows Y$ exploring *different convergences* of sequences in X^* and Y^*. The latter property was formulated by Mordukhovich and Shao [950, 951]; it versions and modifications can be found, under various names, in Ioffe [604, 607], Jourani and Thibault [659, 661], and Penot [1071].

The *automatic PSNC* property of *Lipschitz-like* (Aubin's "pseudo-Lipschitzian") mappings in Proposition 1.68 was first observed by Mordukhovich [917]; it directly follows from the necessary coderivative condition for the Lipschitz-like behavior established in Theorem 1.43. The SNC calculus results from Theorems 1.70, 1.71, 1.72, and 1.74 were established by Mordukhovich and B. Wang [967].

The *partial CEL* property defined in (1.45) was introduced by Jourani and Thibault [655] who actually established the implication in Theorem 1.75, although not explicitly formulated therein.

1.4.16. Subgradients of Extended-Real-Valued Functions. In Sect. 1.3 we start a comprehensive study of generalized differential/subdifferential properties for extended-real-valued functions on Banach spaces. The comments on the history and genesis of generalized differential concepts were given above in Subsects. 1.4.1–1.4.9. We pay the main attention to the *basic/limiting subdifferential* of Definition 1.77 introduced by Mordukhovich

[887] via the basic normal cone (1.80) in finite dimensions. *Singular subgradients* were introduced by Rockafellar [1150] as "singular limiting proximal subgradients" (the name and ∞-notation appeared later in [1155]) via the limits of proximal subgradients of the type considered in Theorem 2.38 with the replacement of Fréchet subgradients by proximal subgradients, which is possible in finite dimensions. Rockafellar's singular subdifferential construction was motivated by seeking an analytic representation of Clarke's generalized gradient for *non-Lipschitzian* functions. The equivalent (in $I\!R^n$) definition of the singular subdifferential $\partial^\infty\varphi(\bar{x})$ via basic *horizontal normals* to the epigraph of φ was independently given by Mordukhovich [894] motivated by establishing appropriate/minimal *qualification conditions* for subdifferential calculus rules involving non-Lipschitzian functions. These conditions, particularly

$$\partial^\infty\varphi_1(\bar{x}) \cap \big(- \partial^\infty\varphi_2(\bar{x})\big) = \{0\}$$

for the sum rule and the induced one for the chain rule, are automatic in the Lipschitzian case. Note that Rockafellar and Wets [1165] used the terms "subgradient" (or "general subgradient") and "horizontal subgradient" for elements of the sets $\partial\varphi(\bar{x})$ and $\partial^\infty\varphi(\bar{x})$, respectively.

The framework of *extended* (by infinite values) real-valued functions, very convenient in variational analysis and optimization, was originated independently in the early 1960s by Moreau [980] and Rockafellar [1140], under the influence of the 1951 lecture notes by Fenchel [441]; see Commentary to Chap. 1 in Rockafellar and Wets [1165] for more details.

Although basic and singular subgradients are defined for arbitrary extended-real-valued functions finite at the point in question, the most useful properties and applications of them concern *lower semicontinuous* functions introduced by Baire in 1899; see [72]. The importance of l.s.c. functions (versus continuous ones) has been well realized in the classical calculus of variations, first probably by Tonelli who established the existence of minimizers for integral functional of the calculus of variations under the *convexity* of integrals with respect to derivative variables. The latter ensures the *lower semicontinuity* of integral functionals in *weak* topologies of the Lebesgue spaces , while *continuity* corresponds to *linearity* in that framework; see Tonelli [1260], Cesari [235], and Olech [1020] for more details and references.

The *upper* subdifferential from Definition 1.78 and the *symmetric* subdifferential defined in (1.42), which may be essentially different from the lower one (in contrast to the case of Clarke's generalized gradients) were first considered by Kruger and Mordukhovich [718, 719, 892] motivated by applications to optimization; the symmetric subdifferential (called "generalized differential" [718, 892]) happened to be especially useful for the mean value theorems for nonsmooth functions established in [706, 708, 894, 901, 949].

A useful result of Theorem 1.80 seems to be derived here for the first time, while its corollaries are well known. Note that the *equality* for the basic subdifferential in Theorem 1.80 *doesn't* generally hold for *l.s.c.* functions as claimed in [708].

Epsilon-subgradients in Definition 1.83 were introduced and studied in the early work by Kruger and Mordukhovich motivated by seeking convenient representations of basic subgradients in *infinite dimensions*; see [706, 708, 718, 719]. Theorem 1.86 was proved by Kruger [706, 708] and then by Ioffe [600]. *Smooth variational descriptions* in assertions (ii) and (iii) of Theorem 1.88 were established by Fabian and Mordukhovich [419]; see also the above comments in Subsect. 1.4.11 related to the corresponding descriptions of Fréchet normals from Theorem 1.30.

The *scalarization* formula for the *mixed* coderivative in Theorem 1.90 was obtained by Mordukhovich and Shao [953]; another proof is given in this book. In finite dimensions, this formula goes back to Ioffe [596] and Mordukhovich [894] following in fact from the "generalized epigraph" results established by Kruger in his dissertation [706]; see also [707, 901].

The *lower/subdifferential regularity* notion from Definition 1.91(i) goes back to Mordukhovich [894]. It is generally different from the epigraphical regularity (ii) of that definition, which is induced by normal regularity of sets from Definition 1.4 applied to epigraphs and hence involving also singular subgradients. Note that lower regularity of locally Lipschitzian functions reduces to Clarke regularity in finite dimensions (see Subsect. 1.4.3), but it is *no longer* the case in (even Hilbert) *infinite-dimensional* spaces; see Bounkhel and Thibault [172] for a detailed study.

As follows from Theorem 1.93, Fréchet-like ε-subgradients of *convex* functions in the sense of Definition 1.83, which reduce to classical subgradients of convex analysis for $\varepsilon = 0$, are *different* for $\varepsilon > 0$ from conventional ε-subgradients of convex functions introduced by Brøndsted and Rockafellar [179] and used in a number of applications under various names including "ε-subgradients" [683, 733, 853, 1017, 1142, 1353], "approximate subgradients" [575, 987, 1199], "ε-enlargements" [186, 187], "ε-Fenchel subgradients" [849], etc. We don't consider such ε-constructions in this book.

1.4.17. Subgradients of Distance Functions. Subdifferential properties of the *distance functions* considered in Subsect. 1.3.3 are highly important in many aspects of variational analysis and its applications due to a special role played by such functions in *variational principles* and *variational techniques.* We pay the main attention to studying the *standard distance* from a variable point to a fixed set in Banach spaces, while most of the results obtained in Subsect. 1.3.3 can also be derived in the case of the *extended distance function*

$$\rho(x, y) = \operatorname{dist}(y; F(x)) := \inf_{v \in F(x)} \|y - v\| \tag{1.88}$$

generated by set-valued mappings (or moving sets); see the comments given below. However, there are principal differences between subdifferential results for distance functions at *in-set* and *out-of-set* points.

Relations for ε-subgradients of the standard distance function at set points from Proposition 1.95 were established by Kruger [705]; Corollary 1.96 on Fréchet subgradients can be also found in Ioffe [600]. Theorem 1.97 on computing basic normals to a set via basic subgradients of the distance function is due to Thibault [1249] who actually derived it for the extended distance function (1.88). Theorem 1.99 on ε-subgradients of the distance function at out-of-set points via ε-normals to set enlargement was obtained by Kruger [705]; however, his proof didn't contain all the necessary details. The complete proof presented in the book is taken from the paper by Bounkhel and Thibault [172].

It has been recently observed by Mordukhovich and Nam [935, 936] that counterparts of Thibault's relationships (as in Theorem 1.97) between basic subgradients of distance functions at in-set points and basic normals to the corresponding sets *don't hold* at *out-of-set* points, even in finite dimensions. Motivated by this observation, they introduced the new *sided* modifications of the basic subdifferential (see Definition 1.100) and established Theorem 1.101 on evaluating right-sided subgradients of the standard distance function via *set enlargements*, as well its analog for the extended distance function (1.88). Note that a different sided subdifferential of the standard distance function, involving limits of Clarke normals, was introduced by Cornet and Czarnecki [290] motivated by applications to existence theorems for generalized equilibria.

The afore-mentioned papers [935, 936] contain also various *projection inclusions* for ε-subgradients and basic subgradients of the distance function, particularly those presented in Subsect. 1.3.3, while the estimates $1 - \varepsilon \leq \|x^*\| \leq 1 + \varepsilon$ in Proposition 1.102 and Theorem 1.103 were proved by Jourani and Thibault [657]. Previous results of the projection type were established by Borwein, Fitzpatrick and Giles [145], Borwein and Giles [146] and Burke, Ferris and Quian [193] via Clarke's constructions. Other results on differentiability and subdifferentiability of distance functions, with some remarkable specifications in finite-dimensional and Hilbert space settings, can be found in Borwein and Ioffe [147], Bounkhel [170], Clarke [255], Clarke et al. [146, 271], Fitzpatrick [451], Ioffe [596, 599, 600], Mordukhovich [901], Mordukhovich and Nam [935, 936], Poliquin, Rockafellar and Thibault [1093], Rockafellar [1142], Rockafellar and Wets [1165], Thibault [1253], Wu and Ye [1336], etc.

1.4.18. Subdifferential Calculus in Banach Spaces. Most of the subdifferential calculus rules presented in Subsect. 1.3.4 for functions on arbitrary Banach spaces are taken from Mordukhovich and Shao [947]; see also Mordukhovich [901, 907] and Rockafellar and Wets [1165] for preceding results in finite-dimensional spaces. The subdifferential inclusions for marginal functions from Theorem 1.108 go back to Rockafellar [1155] in finite dimensions.

Various results on subdifferentiation of the *marginal functions* (1.60) in general Banach spaces have been recently obtained by Mordukhovich, Nam and Yen [937] using both *lower* and *upper Fréchet subgradients.* It was shown, in particular, that

$$\widehat{\partial}\mu(\bar{x}) \subset \bigcap_{(x^*,y^*)\in\widehat{\partial}^+\varphi(\bar{x},\bar{y})} \Big[x^* + \widehat{D}^*G(\bar{x},\bar{y})(y^*)\Big] \tag{1.89}$$

provided that $\widehat{\partial}^+\varphi(\bar{x},\bar{y}) \neq \emptyset$, which is the case, e.g., for rather broad classes of *semiconcave* and other *upper regular* functions φ; see more discussions in Subsects. 5.1.1 and 5.5.4. Moreover, the upper estimate (1.89) is *exact* (i.e., holds as *equality*) in many important situations. The results obtained in this way imply new calculus rules and optimality conditions involving Fréchet-like constructions in arbitrary Banach spaces; see also another paper of the same authors [938].

Observe that the subdifferential sum and chain rules of the *equality* type presented in Subsect. 1.3.4, as well as the related product and quotient rules, don't require *any* regularity assumptions. On the other hand, the corresponding *calculus* for both lower and epigraphical *regularity notions* are incorporated into these results.

The *SNEC property* of extended-real-valued functions was defined by Mordukhovich and Shao [950]; it is *automatic* when either the space in question is finite-dimensional or the function considered is *directionally Lipschitzian* in the sense of Rockafellar [1147]. The SNEC calculus result of Proposition 1.117 was derived by Mordukhovich and B. Wang [967] as a consequence of the more general SNC calculus for sets and set-valued mappings.

1.4.19. Second-Order Generalized Differentiation. The study of second-order generalized differential properties of real-valued functions started with Alexandrov's theorem [8] (1939) who, being motivated by applications to differential geometry, established the *almost everywhere* twice differentiability of convex functions in finite dimensions. Note that Alexandrov didn't introduce any generalized derivative; it came later in the framework of nonsmooth analysis motivated mostly by applications to optimization. Observe also that *no* special theory of second-order generalized differentiation had been created in *convex analysis*; it is probably due to the fact that first-order necessary optimality conditions for convex functions happen to be sufficient as well; see Chap. 13 in Rockafellar and Wets [1165] and the subsequent paper by Rockafellar [1163] for more discussions.

There are definitely much more possibilities to construct second-order generalized derivatives in comparison with first-order ones. Even in classical analysis on finite-dimensional spaces there exist at least *two ways* to do so, which are not equivalent unless a function is $\mathcal{C}^2$: via *Taylor's expansion* and via the "*derivative-of-derivative*" approach. When a function is nonsmooth (of either first or second order), one can explore a variety of different directional derivatives; this indeed has been done in many publications. We are not going to discuss here numerous second-order generalized differential constructions introduced and applied in the framework of variational analysis and beyond, referring the reader to the books by Aubin and Frankowska [54], Bonnans and Shapiro [133], Hiriart-Urruty and Lemaréchal [575], Rockafellar and Wets [1165], to the survey paper by Crandall, Ishii and Lions [296], and to many other publications, e.g., [8, 56, 102, 153, 236, 282, 283, 301, 328, 381, 384, 387, 466, 469, 502, 577, 601, 613, 615, 628, 765, 771, 772, 939, 1037, 1038, 1067, 1091, 1092, 1156, 1163, 1198, 1306, 1307, 1308, 1337, 1358].

The *dual* derivative-of-derivative approach to second-order generalized differentiation was developed by Mordukhovich who introduced in [907] the *second-order subdifferential* $\partial^2\varphi(\bar{x},\bar{y})$ in form (1.87) for extended-real-valued functions $\varphi\colon X \to \overline{I\!R}$. The original definition was given in finite dimensions being motivated by applications to sensitivity analysis for variational systems. In this approach the set of basic subgradients $\partial\varphi(\bar{x}) \subset X^*$ stands for a first-order generalized derivative of φ at $\bar{x}$, while the coderivative D^* plays a role of an adjoint derivative operator for the set-valued mapping $\partial\varphi\colon X \rightrightarrows X^*$ at $\bar{y} \in \partial\varphi(\bar{x})$. The distinction between the *normal* and *mixed* second-order subdifferentials from Definition 1.118, depending on the coderivative type employed via (1.87), was first made in [917].

Note that one can use of course another first-order subdifferential ∂ in (1.87) to define the corresponding second-order construction, as it was done by Mordukhovich and Outrata [939] with the Clarke subdifferential $\partial = \partial_C\varphi(\bar{x})$ and by Eberhard and Wenczel [387] with the proximal one $\partial = \partial_P\varphi(\bar{x})$. The type of *coderivatives* in (1.87), or normal cones to the *graph* of $\partial\varphi(\cdot)$, is however much more essential. In particular, the replacement of the basic normal cone $N(\cdot;\Omega)$ by its Clarke counterpart for $\Omega = \operatorname{gph}\partial\varphi$ in scheme (1.87) *doesn't* lead to an adequate second-order construction in view of the *subspace property* of the Clarke normal cone to Lipschitzian manifolds, which is the case of *any* reasonable first-order subdifferential operator $\partial\varphi(\cdot)$, already for convex functions φ on $I\!R^n$! We refer the reader to the above discussions in Subsects. 1.4.9 and 1.4.13 and to the references therein for more details.

The second-order subdifferential constructions of type (1.87) were studied and applied, sometimes under the names of "generalized Hessians" or "coderivative Hessians," to a large spectrum of problems in variational analysis and its applications including second-order necessary and sufficient optimality conditions; stability of solution maps to problems in constrained optimization, complementarity conditions, variational and hemivariational inequalities along with their generalizations; optimization and equilibrium problems with equilibrium constraints; optimal control of evolution systems; various mechanical equilibria, etc. The interested reader can find the corresponding results and discussions in Dontchev and Rockafellar [364], Eberhard, Pearce and Ralph [385], Eberhard, Pearce and Sivakumaran [384], Eberhard and Wenczel [387], Kočvara and Outrata [690], Levy and Mordukhovich [769], Levy, Poliquin and Rockafellar [771], Lucet and Ye [816], Mordukhovich [907, 910, 911, 912, 913, 921, 923, 925, 926, 928], Mordukhovich and Outrata [939], Mordukhovich and B. Wang [967], Outrata [1024, 1027, 1028, 1030], Poliquin and Rockafellar [1092], Rockafellar and Wets [1165], Rockafellar and Zagrodny [1168], Treiman [1268], Ye [1338, 1339], Ye and Ye [1343], Ye and Zhu [1345], Zhang [1360, 1361, 1362], and in other publications.

1.4.20. Second-Order Subdifferential Calculus in Banach Spaces. Subsection 1.3.5 collects some properties and calculus results for both normal and mixed second-order subdifferentials from Definition 1.118 that hold in general Banach space settings. The properties presented in the beginning of this subsection simply follow from the subdifferential definitions and the corresponding coderivative properties; they demonstrate that the second-order subdifferentials under consideration are natural extensions of the *adjoint Hessian* to the case of extended-real-valued functions that are not $\mathcal{C}^2$. Recall that no adjoint/transposition operation is needed for the classical Hessian matrix in finite dimensions.

Regarding second-order calculus results, let us emphasize that they can be developed only for those classes of functions, which enjoy the first-order subdifferential calculus in the form of *equalities*. This is due to the *absence of monotonicity* with respect to inclusions for either normal or mixed coderivative.

The inclusion chain rule in Theorem 1.127 was obtained by Mordukhovich and Outrata [939] in finite dimensions and then was extended by Mordukhovich [923] to arbitrary Banach spaces. Furthermore, based on the idea suggested by Rockafellar in finite dimensions (cf. [1165, Exercises 6.7 and 10.7] for the first-order constructions), the latter chain rule for the *normal* second-order subdifferential was proved in [923] to hold as *equality* provided that the subspace $\ker \nabla g(\bar{x})$ is *complemented* in X.

Another approach to second-order chain rules was developed by Mordukhovich and B. Wang [967] based on deriving in Lemma 1.126 certain coderivative chain rules for compositions whose specific structure is appropriate for applications to generalized second-order subdifferentiation. Observe particularly that the afore-mentioned specific structure allows us to obtain the notable chain rule (1.64), where the *mixed coderivative* is used for the *inner* mapping. This is significantly different from the general coderivative chain rules presented in Subsects. 1.2.4 and 3.1.2 in both Banach and Asplund space settings; cf. the arguments and discussions therein.

Employing this approach, the new chain rules presented in Theorem 1.127 were established in [967] for both mixed and normal second-order subdifferentials. It is remarkable to observe that the "mixed" chain rule of this theorem holds as *equality* in *arbitrary* Banach spaces! The equality statement in the corresponding "normal" result requires the *weak* extensibility* property of the Banach space in question (see Definition 1.122) introduced and studied by Mordukhovich and B. Wang [967]. The fairly general sufficient conditions obtained in [968] for this property ensure the *equality-type* chain rule for the normal second-order subdifferential in Theorem 1.127 that essentially extends the previous result of [923].

The *second-order coderivative* (1.69) of Lipschitzian mappings was introduced by Mordukhovich [923] who employed it therein to establish the second-order chain rules of Theorem 1.128 for compositions with nonsmooth inner mappings. Let us finally mention that efficient formulas to compute the second-order constructions under consideration were derived by Dontchev and Rockafellar [364] and Mordukhovich and Outrata [939] for rather general classes of functions in finite-dimensional spaces, while more specific calculations and applications can be found in Flegel [454], Flegel and Kanzow [456], Flegel, Kanzow and Outrata [457], Henrion, Jourani and Outrata [560], Kočvara and Outrata [690], Mordukhovich [911, 912], Outrata [1024, 1025, 1027, 1026, 1028, 1030], Poliquin and Rockafellar [1090], Ye [1338, 1339, 1342], Ye and Ye [1343], Zhang [1360, 1361], etc.

2

Extremal Principle in Variational Analysis

It is well known that the convex separation principle plays a fundamental role in many aspects of nonlinear analysis, optimization, and their applications. Actually the whole convex analysis revolves around using separation theorems for convex sets. In problems with nonconvex data separation theorems are applied to convex approximations. This is a conventional way to derive necessary optimality conditions in constrained optimization: first build *tangential convex* approximations of the problem data around an optimal solution in *primal* spaces and then apply convex separation theorems to get supporting elements in dual spaces (Lagrange multipliers, adjoint arcs, prices, etc.). For problems of nonsmooth optimization this approach inevitably leads to the usage of *convex* sets of normals and subgradients, whose calculus is also based on convex separation theorems.

This chapter is devoted to another principle in variational analysis, called the *extremal principle*, which can be viewed as a variational counterpart of the convex separation principle in nonconvex settings. The extremal principle provides necessary conditions for local extremal points of set systems in terms of generalized normals to nonconvex sets with *no use* of tangential approximations and convex separation. It is the base for subsequent applications in this book to nonconvex calculus, optimization, and related topics.

We mainly consider three versions of the extremal principle in Banach spaces formulated, respectively, in terms of ε-normals, Fréchet normals, and basic normals from Chap. 1. It will be shown, by direct variational arguments and the method of separable reduction, that the class of *Asplund spaces* is the most suitable framework for the validity and applications of these results. We also establish relationships between the extremal principle and other basic results in variational analysis, obtain a number of variational characterizations of Asplund spaces in terms of the normal and subgradient constructions studied above, and derive their simplified representations important in what follows. Finally, we discuss some abstract versions of the extremal principle in terms of axiomatically defined normal and subdifferential structures in appropriate Banach spaces.

2.1 Set Extremality and Nonconvex Separation

In this section we introduce a general concept of set extremality and study its relationships with conventional notions of optimal solutions in constrained optimization and separation of sets. We formulate three basic versions of the extremal principle and prove the strongest one in finite-dimensional spaces. As usual, our standard framework is Banach spaces unless otherwise stated.

2.1.1 Extremal Systems of Sets

We start with the definition of extremal systems of sets that may belong to linear topological spaces.

Definition 2.1 (local extremality of set systems). *Let $\Omega_1, \ldots, \Omega_n$ be nonempty subsets of a space X for $n \geq 2$, and let $\bar{x}$ be a common point of these sets. We say that $\bar{x}$ is a* LOCAL EXTREMAL POINT *of the set system $\{\Omega_1, \ldots, \Omega_2\}$ if there are sequences $\{a_{ik}\} \subset X$, $i = 1, \ldots, n$, and a neighborhood U of $\bar{x}$ such that $a_{ik} \to 0$ as $k \to \infty$ and*

$$\bigcap_{i=1}^{n} \left(\Omega_i - a_{ik}\right) \cap U = \emptyset \quad \textit{for all large } k \in I\!N \,.$$

In this case $\{\Omega_1, \ \ldots, \Omega_n, \bar{x}\}$ is said to be an EXTREMAL SYSTEM *in X.*

Loosely speaking, the local extremality of sets at a common point means that they can be locally "pushed apart" by a small perturbation (translation) of even one of them. For $n = 2$ the local extremality of $\{\Omega_1, \Omega_2, \bar{x}\}$ can be equivalently described as follows: there exists a neighborhood U of $\bar{x}$ such that for any $\varepsilon > 0$ there is $a \in \varepsilon I\!B$ with $(\Omega_1 + a) \cap \Omega_2 \cap U = \emptyset$. Note that the condition $\Omega_1 \cap \Omega_2 = \{\bar{x}\}$ doesn't necessary imply that $\bar{x}$ is a local extremal point of $\{\Omega_1, \Omega_2\}$. A simple example is given by $\Omega_1 := \{(v, v) |\ v \in I\!R\}$ and $\Omega_2 := \{(v, -v) |\ v \in I\!R\}$.

It is clear that every *boundary point* $\bar{x}$ of a closed set Ω is a local extremal point of the pair $\{\Omega, \bar{x}\}$. In general, this geometric concept of extremality covers conventional notions of optimal solutions to various problems of scalar and vector optimization. In particular, let $\bar{x}$ be a local solution to the following problem of *constrained optimization*:

$$\text{minimize } \varphi(x) \text{ subject to } x \in \Omega \subset X \,.$$

Then one can easily check that $(\bar{x}, \varphi(\bar{x}))$ is a local extremal point of the set system $\{\Omega_1, \Omega_2\}$ in $X \times I\!R$ with $\Omega_1 = \operatorname{epi} \varphi$ and $\Omega_2 = \Omega \times \{\varphi(\bar{x})\}$. Indeed, we satisfy the requirements of Definition 2.1 with $a_{1k} = (0, \nu_k)$, $a_{2k} = 0$, and $U = O \times I\!R$, where $\nu_k \uparrow 0$ and where O is a neighborhood of the local minimizer $\bar{x}$. In the subsequent parts of the book the reader will find many other examples of extremal systems in problems related to optimization, variational principles, generalized differential calculus, and applications to welfare economics.

The next simple property of extremal systems is useful in what follows.

Proposition 2.2 (interiors of sets in extremal systems). *For every extremal system $\{\Omega_1, \ldots, \Omega_n, \bar{x}\}$ in X one has*

$$(\text{int}\, \Omega_1) \cap \ldots \cap (\text{int}\, \Omega_{n-1}) \cap \Omega_n \cap U = \emptyset\,, \tag{2.1}$$

where U is a neighborhood of the local extremal point $\bar{x}$.

Proof. Assuming the contrary, pick any point x from the intersection in (2.1) and take arbitrary sequences $a_{ik} \to 0$, $i = 1, \ldots, n$, in X. Since $x \in \text{int}\, \Omega_i \cap U$ for $i = 1, \ldots, n-1$, we have $x - a_{nk} \in U$ and $x + a_{ik} - a_{nk} \in \Omega_i$ for $i = 1, \ldots, n-1$ and $k \in I\!N$ large enough. Thus $x - a_{nk} \in (\Omega_i - a_{ik}) \cap U$ for all $i = 1, \ldots, n$ and large k, which contradicts the set extremality. △

Now we establish relationships between the concept of set extremality from Definition 2.1 and the conventional separation property for a finite number of sets that may be nonconvex. Recall that sets $\Omega_i \subset X$, $i = 1, \ldots, n$, are said to be *separated* if there exist vectors $x_i^* \in X^*$, not equal to zero simultaneously, and numbers α_i such that

$$\langle x_i^*, x\rangle \le \alpha_i \quad \text{for all} \quad x \in \Omega_i, \qquad i = 1, \ldots, n\,,$$

$$x_1^* + \ldots + x_n^* = 0, \qquad \alpha_1 + \ldots + \alpha_n \le 0\,.$$

Note that if the sets Ω_i are separated and have a common point, then the last condition must hold as equality.

Proposition 2.3 (extremality and separation). *Let $\Omega_1, \ldots, \Omega_n$ $(n \ge 2)$ be subsets of X that have at least one common point. The following hold:*

(i) *If these sets are separated, then the system $\{\Omega_1, \ldots, \Omega_n, \bar{x}\}$ is extremal for every common point $\bar{x}$ of these sets.*

(ii) *The converse is true if all Ω_i are convex and* $\text{int}\, \Omega_i \neq \emptyset$ *for* $i = 1, \ldots, n-1$.

Proof. Assume that Ω_i are separated with $x_n^* \neq 0$, which doesn't restrict the generality. Pick any $a \in X$ with $\langle x_n^*, a\rangle > 0$ and put $a_k := a/k$ for all $k \in I\!N$. Let us show that

$$\Omega_1 \cap \ldots \cap \Omega_{n-1} \cap (\Omega_n - a_k) = \emptyset, \quad k \in I\!N\,,$$

which obviously implies the extremality of $\{\Omega_1, \ldots, \Omega_n, \bar{x}\}$ for every common point $\bar{x}$. Assuming the contrary and taking any x from the latter intersection, one has by the separation property that

$$\langle x_i^*, x\rangle \le \alpha_i, \quad i = 1, \ldots, n-1, \quad \text{and} \quad \langle x_n^*, x + a_k\rangle \le \alpha_n, \quad k \in I\!N\,.$$

Summing up, we arrive at $\alpha_1 + \ldots + \alpha_n \ge \frac{1}{k}\langle x_n^*, a\rangle > 0$, a contradiction. Thus (i) holds. The converse assertion (ii) follows from Proposition 2.2 and the separation theorem for convex sets. △

Note that, for *convex sets in finite dimensions*, Proposition 2.3(ii) holds with *no interiority* assumption on Ω_i, $i = 1, \ldots, n-1$. This follows from the extremal principle established below in Theorem 2.8. Hence for $\dim X < \infty$ the *extremality and separation of convex sets are unconditionally equivalent.* One will also see that the extremal principle allows us to relax interiority assumptions on convex sets Ω_i, $i = 1, \ldots, n-1$, ensuring the validity of Proposition 2.3(ii) in infinite dimensions.

Corollary 2.4 (extremality criterion for convex sets). *Let Ω_i, $i = 1, \ldots, n$, be convex sets in X having at least one point in common. Assume that $\operatorname{int} \Omega_i \neq \emptyset$ for $i = 1, \ldots, n-1$. Then condition* (2.1) *with $U = X$ is necessary and sufficient for extremality of the system $\{\Omega_1, \ldots, \Omega_n, \bar{x}\}$, where $\bar{x}$ is any common point of these sets.*

Proof. Follows from Propositions 2.2 and 2.3(i), since condition (2.1) ensures the separation (and hence extremality) property of n convex sets with nonempty interiors of all but one of them. △

Note that the convexity of Ω_i is essential for the extremality criterion in Corollary 2.4. A counterexample is provided by the sets

$$\Omega_1 := I\!R^2_+ \cup I\!R^2_-, \; \Omega_2 := \big\{(x_1, x_2)\big|\ x_1 \le 0, x_2 \ge 0\big\} \cup \big\{(x_1, x_2)\big|\ x_1 \ge 0, x_2 \le 0\big\} .$$

2.1.2 Versions of the Extremal Principle and Supporting Properties

In this subsection we define three basic versions of the extremal principle in Banach spaces and show that they can be treated as a kind of local separation of nonconvex sets around extremal points. We also discuss their relationships with supporting properties of nonconvex sets expressed in terms of generalized normals from Definition 1.1.

Definition 2.5 (versions of the extremal principle). *Let $\{\Omega_1, \ldots, \Omega_n, \bar{x}\}$ be an extremal system in X. We say that:*

(i) *$\{\Omega_1, \ldots, \Omega_n, \bar{x}\}$ satisfies the* ε-EXTREMAL PRINCIPLE *if for every $\varepsilon > 0$ there are $x_i \in \Omega_i \cap (\bar{x} + \varepsilon I\!B)$ and $x_i^* \in X^*$ such that*

$$x_i^* \in \widehat{N}_\varepsilon(x_i; \Omega_i), \quad i = 1, \ldots, n , \tag{2.2}$$

$$x_1^* + \ldots + x_n^* = 0, \quad \|x_1^*\| + \ldots + \|x_n^*\| = 1 . \tag{2.3}$$

(ii) *$\{\Omega_1, \ldots, \Omega_n, \bar{x}\}$ satisfies the* APPROXIMATE EXTREMAL PRINCIPLE *if for every $\varepsilon > 0$ there are $x_i \in \Omega_i \cap (\bar{x} + \varepsilon I\!B)$ and*

$$x_i^* \in \widehat{N}(x_i; \Omega_i) + \varepsilon I\!B^*, \quad i = 1, \ldots, n , \tag{2.4}$$

such that (2.3) *holds.*

(iii) $\{\Omega_1, \ldots, \Omega_n, \bar{x}\}$ *satisfies the* EXACT EXTREMAL PRINCIPLE *if there are basic normals*

$$x_i^* \in N(\bar{x}; \Omega_i), \quad i = 1, \ldots, n, \tag{2.5}$$

such that (2.3) *holds.*

We say that the corresponding version of the extremal principle holds IN *the space* X *if it holds for every extremal system* $\{\Omega_1, \ldots, \Omega_n, \bar{x}\}$ *in* X, *where all the sets* Ω_i *are* (*locally*) *closed around* $\bar{x}$.

It is clear that the number 1 in the nontriviality condition of (2.3) can be replaced with any other positive number, which should be independent of ε in versions (i) and (ii). Note that ε in "ε-extremal principle" is just a part of the notation (and not a subject to change unlike anywhere else), which emphasizes the difference between (2.2) and (2.4). Since one always has $\widehat{N}(x; \Omega) + \varepsilon I\!B^* \subset \widehat{N}_\varepsilon(x; \Omega)$, the ε-extremal principle follows from the approximate extremal principle for any extremal system in a Banach space X. We'll see below that these two versions of the extremal principle are actually equivalent if they apply to *every* extremal system in X.

Thus the relations of the extremal principle provide necessary conditions for local extremal points of set systems and can be viewed as *generalized Euler equations* in an abstract geometric setting. They also can be treated as proper variational counterparts of local separation for nonconvex sets. To see this, we first consider the exact extremal principle for two sets. Then (2.3) and (2.5) reduce to: there is $x^* \in X^*$ with

$$0 \neq x^* \in N(\bar{x}; \Omega_1) \cap \big(-N(\bar{x}; \Omega_2)\big). \tag{2.6}$$

When both Ω_1 and Ω_2 are *convex*, (2.6) means

$$\langle x^*, u_1 \rangle \leq \langle x^*, u_2 \rangle \quad \text{for all} \quad u_1 \in \Omega_1 \quad \text{and} \quad u_2 \in \Omega_2,$$

which is exactly the classical separation property for two convex sets. Similarly, relations (2.3) and (2.5) for n convex sets ($n > 2$) give the conventional separation property considered in the preceding subsection.

Note that, in contrast to the classical separation, the extremal principle applies only to *local extremal points* of set systems. As shown in Proposition 2.3, it is *always the case* for every common point of sets separated in the classical sense. Therefore, any sufficient condition for convex separation implies set extremality. The above discussion allows us to view the extremal principle as a local *variational* extension of the classical separation to nonconvex sets. It is important to emphasize that in many situations occurring in applications, even in the case of convex sets, the local extremality of points in question can be checked *automatically* from the problem statement, and we don't need to care about any interiority-like conditions, etc. This supports a variational approach to such problems (which may be not of a variational nature) based on the extremal principle; see below.

Considering "fuzzy" versions (i) and (ii) of the extremal principle for systems of two sets, we reduce them to the following relations: for every $\varepsilon > 0$ there are $x_i \in \Omega_i \cap (\bar{x} + \varepsilon I\!B)$, $i = 1, 2$, and $x^* \in X^*$ with $\|x^*\| = 1$ such that, respectively,

$$x^* \in \widehat{N}_\varepsilon(x_1; \Omega_1) \cap \big(- \widehat{N}_\varepsilon(x_2; \Omega_2)\big) ,$$

$$x^* \in \big(\widehat{N}(x_1; \Omega_1) + \varepsilon I\!B^*\big) \cap \big(- \widehat{N}(x_2; \Omega_2) + \varepsilon I\!B^*\big) .$$

For convex sets they coincide, due to Proposition 1.3, and provide an *approximate separation* of Ω_1 and Ω_2 near $\bar{x}$. Likewise, relations (2.2)–(2.4) of the extremal principle in the general case under consideration can be viewed as a local variational counterpart of the approximate local separation for nonconvex sets.

Next let us consider a special case of extremal systems generated by *boundary points* $\bar{x}$ of locally closed sets $\Omega \subset X$, i.e., extremal systems of the type $\{\Omega, \{\bar{x}\}, \bar{x}\}$ in the notation of Definition 2.1. Then the exact extremal principle gives the *nontriviality property* for the basic normal cone:

$$N(\bar{x}; \Omega) \neq \{0\} \text{ if and only if } \bar{x} \in \operatorname{bd} \Omega . \tag{2.7}$$

Note that the "only if" part follows immediately from Definition 1.1 for any closed set $\Omega \subset X$, and the "if" part is an easy consequence of the exact extremal principle whenever it holds in X. When Ω is convex, condition (2.7) reduces to the classical *supporting hyperplane* theorem; so in general (2.7) can be viewed as a local extension of this result to nonconvex sets. Applying the other versions of the extremal principle, we get some approximate supporting properties of nonconvex sets in terms of ε-normals and Fréchet normals at points near $\bar{x}$.

Proposition 2.6 (approximate supporting properties of nonconvex sets). *Given a proper closed set $\Omega \subset X$ and a point $\bar{x} \in \operatorname{bd} \Omega$, one has the following:*

(i) *If the ε-extremal principle holds for $\{\Omega, \{\bar{x}\}, \bar{x}\}$, then whenever $\varepsilon > 0$ and $M > \varepsilon$ there is $x \in B_\varepsilon(\bar{x}) \cap \operatorname{bd} \Omega$ such that $\widehat{N}_\varepsilon(x; \Omega) \setminus M I\!B^* \neq \emptyset$.*

(ii) *If the approximate extremal principle holds for $\{\Omega, \{\bar{x}\}, \bar{x}\}$, then for every $\varepsilon > 0$ there is $x \in B_\varepsilon(\bar{x}) \cap \operatorname{bd} \Omega$ such that $\widehat{N}(x; \Omega) \neq \{0\}$.*

Therefore, the validity of the approximate extremal principle (the ε-extremal principle) in X implies, respectively, the density of the set

$$\Big\{x \in \operatorname{bd} \Omega \Big|\ \widehat{N}(x; \Omega) \neq \{0\}\Big\} \tag{2.8}$$

for every proper closed subset $\Omega \subset X$, and the set

$$\Big\{x \in \operatorname{bd} \Omega \Big|\ \widehat{N}_\varepsilon(x; \Omega) \setminus M I\!B^* \neq \emptyset\Big\} \tag{2.9}$$

for every proper closed subset $\Omega \subset X$, every $\varepsilon > 0$, and every $M > \varepsilon$.

Proof. Assertion (i) for $0 < M < 1/2$ follows immediately from Definition 2.5(i) with $n = 2$, $\Omega_1 = \Omega$, and $\Omega_2 = \{\bar{x}\}$. Let us prove it for any $M > \varepsilon$. Fix arbitrary $\varepsilon > 0$ and $M \geq 1/2$ and employ the relations of the ε-extremal principle to $\{\Omega, \{\bar{x}\}, \bar{x}\}$ with $\tilde{\varepsilon} := \varepsilon/(2M+1)$. We find $x \in \Omega$ and $\tilde{x}^* \in X^*$ satisfying

$$\|x - \bar{x}\| \leq \tilde{\varepsilon} < \varepsilon, \; \tilde{x}^* \in \widehat{N}_{\tilde{\varepsilon}}(x; \Omega), \quad \text{and} \quad \|\tilde{x}^*\| = 1/2 \; ,$$

which implies that $x \in \operatorname{bd} \Omega$. Then putting $x^* := (2M + 1)\tilde{x}^*$ and using the definition of ε-normals (1.2), we get

$$\limsup_{u \xrightarrow{\Omega} x} \frac{\langle x^*, u - x \rangle}{\|u - x\|} = (2M + 1) \limsup_{u \xrightarrow{\Omega} x} \frac{\langle \tilde{x}^*, u - x \rangle}{\|u - x\|} \leq (2M + 1)\tilde{\varepsilon} = \varepsilon \; ,$$

i.e., $x^* \in \widehat{N}_\varepsilon(x; \Omega)$ with $\|x^*\| = (2M + 1)/2 > M$. This gives (i).

To prove (ii), we use the approximate extremal principle for $\{\Omega, \{\bar{x}\}, \bar{x}\}$ with $\varepsilon \in (0, 1/2)$. In this way we find $x \in B_\varepsilon(x) \cap \Omega$ and $x^* \in \widehat{N}(x; \Omega) + \varepsilon I\!B^*$ with $\|x^*\| = 1/2$. The latter yields $x \in \operatorname{bd} \Omega$ and $\widehat{N}(x; \Omega) \neq \{0\}$. △

If Ω is *convex*, then (2.8) describes the set of *support points* to Ω. Hence the approximate extremal principle in a Banach space X implies the density of support points to every closed convex subset of X, which is the contents of the celebrated Bishop-Phelps theorem (see Theorem 3.18 in Phelps [1073]).

A natural question arises about the reverse implications in Proposition 2.6, i.e., about the possibility to derive relations of the approximate extremal principle (resp. the ε-extremal principle) from the density of sets (2.8) and (2.9) for *every* proper closed subset of X. To explore this way, let us fix an extremal system $\{\Omega_1, \Omega_2, \bar{x}\}$ and observe that the local extremality of $\bar{x} \in \Omega_1 \cap \Omega_2$ implies that $0 \in \operatorname{bd}(\Omega_1 - \Omega_2)$. Hence one can apply the mentioned density results to the set $\Omega_1 - \Omega_2$ around the origin if $\Omega_1 - \Omega_2$ is *assumed to be closed.* For simplicity let us consider the case of (2.8) and find $x_i \in \Omega_i$, $i = 1, 2$, such that

$$\widehat{N}(x_1 - x_2; \Omega_1 - \Omega_2) \neq \{0\} \quad \text{and} \quad \|x_1 - x_2\| \leq \varepsilon \; .$$

Taking $x^* \in \widehat{N}(x_1 - x_2; \Omega_1 - \Omega_2)$ with $\|x^*\| = 1/2$, we have from (1.2) that

$$\limsup_{u \xrightarrow{\Omega_1 - \Omega_2} x_1 - x_2} \frac{\langle x^*, u - (x_1 - x_2) \rangle}{\|u - (x_1 - x_2)\|} \leq 0 \; .$$

Now putting $u = v - x_2$, $v \in \Omega_1$ and then $u = x_1 - v$, $v \in \Omega_2$, one gets $x^* \in \widehat{N}(x_1 : \Omega_1)$ and $-x^* \in \widehat{N}(x_2; \Omega_2)$. In this way we arrive at all the relations of the approximate extremal principle *except* that $x_i \in \bar{x} + \varepsilon I\!B^*$, $i = 1, 2$. Thus we cannot obtain the reverse statements in Proposition 2.6 using the reduction of local extremal points to the boundary of $\Omega_1 - \Omega_2$. Moreover, the above arguments actually provide *characterizations* of the supporting properties $\widehat{N}_\varepsilon(x; \Omega) \setminus M I\!B^* \neq \emptyset$ and $\widehat{N}(x; \Omega) \neq \{0\}$ in terms of relations (2.2)–(2.4), which *don't involve* extremal points and their small perturbations.

Proposition 2.7 (characterizations of supporting properties). *Given a Banach space X and numbers $\varepsilon \ge 0$ and $M \ge \varepsilon$, the following properties are equivalent:*

(a) *For every proper closed set $\Omega \subset X$ there exists $x \in \operatorname{bd}\Omega$ satisfying $\widehat{N}_\varepsilon(x;\Omega) \setminus M I\!B^* \ne \emptyset$, which corresponds to $\widehat{N}(x;\Omega) \ne \{0\}$ if $\varepsilon = 0$.*

(b) *Let Ω_1 and Ω_2 be arbitrary subsets of X such that $\Omega_1 - \Omega_2$ is proper and closed around the origin. Then there are $x_1 \in \Omega_1$ and $x_2 \in \Omega_2$ satisfying*

$$0 \in \big(\widehat{N}_\varepsilon(x_1;\Omega_1) \setminus M I\!B^*\big) + \widehat{N}_\varepsilon(x_2;\Omega_2)\,.$$

Proof. To establish (a)⇒(b), we take $\Omega := \Omega_1 - \Omega_2$ in (a) and use the above arguments for $x_1 - x_2 \in \Omega_1 - \Omega_2$ and $x^* \in \widehat{N}_\varepsilon(x_1 - x_2;\Omega_1 - \Omega_2)$ with $\|x^*\| > M > \varepsilon \ge 0$. Implication (b)⇒(a) is proved similarly to Proposition 2.6 putting $\Omega_1 := \Omega$ and $\Omega_2 := \{\bar{x}\}$, where $\bar{x}$ is a fixed boundary point of Ω. △

2.1.3 Extremal Principle in Finite Dimensions

In this subsection we give a direct proof of the exact extremal principle in finite-dimensional spaces. The proof is based on the *method of metric approximations*, which provides an efficient approximation of extremal set systems by families of *smooth* problems of *unconstrained optimization.* Without loss of generality we use the Euclidean norm on X.

Theorem 2.8 (exact extremal principle in finite dimensions). *The exact extremal principle holds in any space X with* $\dim X < \infty$.

Proof. Let $\bar{x}$ be a local extremal point of the set system $\{\Omega_1,\ldots,\Omega_n\}$, where all the sets Ω_i are closed around $\bar{x}$. Take sequences $\{a_{ik}\}$ and a neighborhood U from Definition 2.1 and assume without loss of generality that $U = X$. For each $k = 1,2,\ldots$ we consider the following problem of unconstrained minimization:

$$\text{minimize}\;\; d_k(x) := \left[\sum_{i=1}^n \operatorname{dist}^2(x + a_{ik};\Omega_{ik})\right]^{1/2} + \|x - \bar{x}\|^2, \quad x \in X\,. \tag{2.10}$$

Since the function d_k is continuous and its level sets are bounded, there is an optimal solution x_k to (2.10) by the classical Weierstrass theorem. Due to the local extremality of $\bar{x}$ one has

$$\alpha_k := \left[\sum_{i=1}^n \operatorname{dist}^2(x_k + a_{ik};\Omega_i)\right]^{1/2} > 0\,.$$

Taking into account that x_k is an optimal solution to (2.10), we get

$$d_k(x_k) = \alpha_k + \|x_k - \bar{x}\|^2 \le \left[\sum_{i=1}^n \|a_{ik}\|^2\right]^{1/2} \downarrow 0 ,$$

which implies that $x_k \to \bar{x}$ and $\alpha_k \downarrow 0$ as $k \to \infty$.

Now let us arbitrarily pick $w_{ik} \in \Pi(x_k + a_{ik}; \Omega_i)$ for $i = 1, \ldots, n$ (the *best approximations* to $x_k + a_{ik}$ in the closed set Ω_i) and consider the problem:

$$\text{minimize } \rho_k(x) := \left[\sum_{i=1}^n \|x + a_{ik} - w_{ik}\|^2\right]^{1/2} + \|x - \bar{x}\|^2 \tag{2.11}$$

that obviously has the same optimal solution x_k as (2.10). Since $\alpha_k > 0$ and the norm $\|\cdot\|$ is Euclidean, $\rho_k(x)$ is continuously differentiable around x_k. Thus (2.11) is a *smooth* problem of unconstrained minimization. Employing the classical Fermat rule in (2.11), we get

$$\nabla \rho_k(x_k) = \sum_{i=1}^n x_{ik}^* + 2(x_k - \bar{x}) = 0 , \tag{2.12}$$

where $x_{ik}^* = (x_k + a_{ik} - w_{ik})/\alpha_k$, $i = 1, \ldots, n$, with

$$\|x_{1k}^*\|^2 + \ldots + \|x_{nk}^*\|^2 = 1 .$$

Taking into account the compactness of the unit sphere in finite dimensions, we find vectors $x_i^* \in X = X^*$, $i = 1, \ldots, n$, satisfying the normalization condition in (2.3) and such that $x_{ik}^* \to x_i^*$ as $k \to \infty$. Passing to the limit in (2.12), one gets the first condition in (2.3) as well. It follows from representation (1.9) of basic normals in Theorem 1.6 that $x_i^* \in N(\bar{x}; \Omega_i)$ for all $i = 1, \ldots, n$. This completes the proof of the exact extremal principle in finite-dimensional spaces. △

Corollary 2.9 (nontriviality of basic normals in finite dimensions). *Let* $\dim X < \infty$. *Then the nontriviality property* (2.7) *holds for basic normals to every proper closed set* $\Omega \subset X$.

Proof. Follows from the extremal principle as discussed above. It can also be proved directly by using the definition of boundary points and representation (1.9) in Theorem 1.6. △

The proof of the exact extremal principle given in Theorem 2.8 is essentially based on the *geometry of finite-dimensional spaces.* Namely, it uses the *compactness* of the closed unit ball and the unit sphere as well as *variational properties* of the *Euclidean norm* that have been also exploited above for representation (1.9) of the basic normal cone. An important feature of finite-dimensional spaces is that they always admit a *smooth renorm* (by the Euclidean norm) differentiable away from the origin.

In the next section we justify, based on *variational arguments*, all the three versions of the extremal principle formulated above for a broad class of infinite-dimensional spaces that possess remarkable geometric properties *not related* to the Euclidean norm.

2.2 Extremal Principle in Asplund Spaces

The results of this section play a *crucial role* for the whole subsequent material of the book. We start with a direct variational proof of the approximate extremal principle in spaces admitting a *Fréchet smooth renorm*, which form a special subclass of Asplund spaces. Then we develop the method of *separable reduction* for Fréchet-like normals and subgradients that allows us to reduce certain problems involving such constructions in *nonseparable* Banach spaces to separable ones. This method is particularly helpful for the class of *Asplund spaces*, where every separable subspace admits a Fréchet smooth renorm. In such a way we prove the *extremal principle* in Asplund spaces (in both approximate and exact forms) and then establish *variational characterizations* of this class of Banach spaces.

2.2.1 Approximate Extremal Principle in Smooth Banach Spaces

In this subsection we pay the main attention to the proof of the approximate extremal principle in Banach spaces that admit *Fréchet smooth renorming*, i.e., an equivalent norm Fréchet differentiable at any nonzero point. It is well known that this class includes every reflexive Banach space; see, e.g., Diestel [332]. Since the prenormal cone $\widehat{N}$ is invariant with respect to equivalent norms on X, we don't restrict the generality by assuming that $\|\cdot\|$ is such a smooth norm on X.

Theorem 2.10 (approximate extremal principle in Fréchet smooth spaces). *The approximate extremal principle holds in any space X admitting a Fréchet smooth renorm.*

Proof. We first prove the theorem for the case of two sets and then obtain the general statement by induction. Let $\bar{x} \in \Omega_1 \cap \Omega_2$ be a local extremal point of some sets Ω_i closed around $\bar{x}$. We have a neighborhood U of $\bar{x}$ such that for any $\varepsilon > 0$ there is $a \in X$ with $\|a\| \le \varepsilon^3/2$ and $(\Omega_1 + a) \cap \Omega_2 \cap U = \emptyset$. Assume for simplicity that $U = X$ and also that $\varepsilon < 1/2$. Then considering the function

$$\varphi(z) := \|x_1 - x_2 + a\| \quad \text{for} \quad z = (x_1, x_2) \in X \times X \; ,$$

we conclude that $\varphi(z) > 0$ on $\Omega_1 \times \Omega_2$, and hence φ is Fréchet differentiable at any point $z \in \Omega_1 \times \Omega_2$. In what follows we use the product norm $\|z\| := (\|x_1\|^2 + \|x_2\|^2)^{1/2}$ that is obviously Fréchet differentiable away from the origin in $X \times X$. Observe the link between the above function φ and the distance function (2.10) used in the proof of the extremal principle in finite dimensions. In contrast to the finite-dimensional proof of Theorem 2.8, now we cannot use the compactness of the unit ball and the Weierstrass existence theorem, which are replaced below by variational arguments based on the completeness of X and then on the smoothness of the norm.

To proceed, we take $z_0 := (\bar{x}, \bar{x})$ and form the set

$$W(z_0) := \big\{z \in \Omega_1 \times \Omega_2 \big|\ \varphi(z) + \varepsilon\|z - z_0\|^2/2 \le \varphi(z_0)\big\}$$

that is nonempty and closed. Moreover, for each $z \in W(z_0)$ one has

$$\|x_1 - \bar{x}\|^2 + \|x_2 - \bar{x}\|^2 \le 2\varphi(z_0)/\varepsilon = 2\|a\|/\varepsilon \le \varepsilon^2 ,$$

which implies that $W(z_0) \subset B_\varepsilon(\bar{x}) \times B_\varepsilon(\bar{x})$. Next let us inductively define sequences of vectors $z_k \in \Omega_1 \times \Omega_2$ and nonempty closed sets $W(z_k)$, $k \in I\!N$, as follows. Given z_k and $W(z_k)$, $k = 0, 1, \ldots$, we select $z_{k+1} \in W(z_k)$ satisfying

$$\varphi(z_{k+1}) + \varepsilon\sum_{j=0}^{k}\frac{\|z_{k+1} - z_j\|^2}{2^{j+1}} < \inf_{z \in W(z_k)}\Big\{\varphi(z) + \varepsilon\sum_{j=0}^{k}\frac{\|z - z_j\|^2}{2^{j+1}}\Big\} + \frac{\varepsilon^3}{2^{3k+2}} .$$

Then we form the set

$$W(z_{k+1}) := \Big\{z \in \Omega_1 \times \Omega_2 \Big|\ \varphi(z) + \varepsilon\sum_{j=0}^{k+1}\frac{\|z - z_j\|^2}{2^{j+1}} \le \varphi(z_{k+1}) + \varepsilon\sum_{j=0}^{k}\frac{\|z_{k+1} - z_j\|^2}{2^{j+1}}\Big\} .$$

It is easy to check that $\{W(z_k)\}$ is a *nested* sequence of nonempty closed subsets of $\Omega_1 \times \Omega_2$. Let us show that $\operatorname{diam} W(z_k) := \sup\big\{\|z - w\| \ \big|\ z, w \in W(z_k)\big\} \to 0$ as $k \to \infty$. Indeed, for each $z \in W(z_{k+1})$ and $k \in I\!N$ we have

$$\frac{\varepsilon\|z - z_{k+1}\|^2}{2^{k+2}} \le \varphi(z_{k+1}) + \varepsilon\sum_{j=0}^{k}\frac{\|z_{k+1} - z_j\|^2}{2^{j+1}} - \Big(\varphi(z) + \varepsilon\sum_{j=0}^{k}\frac{\|z - z_j\|^2}{2^{j+1}}\Big)$$

$$\le \varphi(z_{k+1}) + \varepsilon\sum_{j=0}^{k}\frac{\|z_{k+1} - z_j\|^2}{2^{j+1}} - \inf_{z \in W(z_k)}\Big\{\varphi(z) + \varepsilon\sum_{j=0}^{k}\frac{\|z - z_j\|^2}{2^{j+1}}\Big\} < \frac{\varepsilon^3}{2^{3k+2}} ,$$

which implies that $\operatorname{diam} W(z_k) \le \varepsilon/2^{k-1} \to 0$. Thus (due to the completeness of X) $\cap_{k=0}^{\infty} W(z_k) = \{\bar{z}\}$ with $z_k \to \bar{z} = (\bar{x}_1, \bar{x}_2) \in \Omega_1 \times \Omega_2$ as $k \to \infty$. By $\bar{z} \in W(z_0)$ one has $\bar{z} \in B_\varepsilon(\bar{x}) \times B_\varepsilon(\bar{x})$. Let us show that $\bar{z}$ is a minimum point of the function

$$\phi(z) := \varphi(z) + \varepsilon\sum_{j=0}^{\infty}\frac{\|z - z_j\|^2}{2^{j+1}}$$

over the set $\Omega_1 \times \Omega_2$. Indeed, taking $\bar{z} \neq z \in \Omega_1 \times \Omega_2$ and using the construction of $W(z_k)$, we find $k \in I\!N$ such that

$$\varphi(z) + \varepsilon \sum_{j=0}^{k} \frac{\|z - z_j\|^2}{2^{j+1}} > \varphi(z_k) + \varepsilon \sum_{j=0}^{k-1} \frac{\|z_k - z_j\|^2}{2^{j+1}} . \tag{2.13}$$

This implies that $\bar{z}$ is a minimum point of ϕ over $\Omega_1 \times \Omega_2$, since the sequence on the right-hand side of (2.13) is nonincreasing as $k \to \infty$. Therefore the function $\psi(z) := \phi(z) + \delta(z; \Omega_1 \times \Omega_2)$ achieves at $\bar{z}$ its minimum over $X \times X$. Thus $0 \in \widehat{\partial}\psi(\bar{z})$ by the generalized Fermat rule of Proposition 1.114. Note that ϕ is Fréchet differentiable at $\bar{z}$ due to $\varphi(\bar{z}) \neq 0$ and the smoothness of $\|\cdot\|^2$. Now applying the sum rule of Proposition 1.107(i) and then (1.50) as $\varepsilon = 0$ and the product formula of Proposition 1.2, we get

$$-\nabla\phi(\bar{z}) \in \widehat{N}(\bar{z}; \Omega_1 \times \Omega_2) = \widehat{N}(\bar{x}_1; \Omega_1) \times \widehat{N}(\bar{x}_2; \Omega_2) .$$

It follows from the construction of ϕ that $\nabla\phi(\bar{z}) = (u_1^*, u_2^*) \in X^* \times X^*$, where

$$u_1^* = x^* + \varepsilon \sum_{j=0}^{\infty} w_{1j}^* \frac{\|\bar{x}_1 - x_{1j}\|}{2^j}, \quad u_2^* = -x^* + \varepsilon \sum_{j=0}^{\infty} w_{2j}^* \frac{\|\bar{x}_2 - x_{2j}\|}{2^j}$$

with $(x_{1j}, x_{2j}) = z_j$, $x^* = \nabla(\|\cdot\|)(\bar{x}_1 - \bar{x}_2 + a)$, and

$$w_{ij}^* = \begin{cases} \nabla(\|\cdot\|)(\bar{x}_i - x_{ij}) \text{ if } \bar{x}_i - x_{ij} \neq 0 , \\ 0 \qquad\qquad\qquad\qquad \text{otherwise} \end{cases}$$

for $j = 0, 1, \ldots$ and $i = 1, 2$. One clearly has $\sum_{j=0}^{\infty} \|w_{ij}^*\| \cdot \|\bar{x}_i - x_{ij}\|/2^j \leq 1$, $i = 1, 2$, and $\|x^*\| = 1$. Thus putting $x_i := \bar{x}_i$ and $x_i^* := (-1)^i x^*/2$ for $i = 1, 2$, we arrive at relations (2.3) and (2.4) of the approximate extremal principle in the case of two sets.

Now let us consider the general case of n sets $\{\Omega_1, \ldots, \Omega_n\}$ in X and prove the approximate extremal principle by induction when $n > 2$. It is easy to see that if $\bar{x}$ is a local extremal point of $\{\Omega_1, \ldots, \Omega_n\}$, then the point $\bar{z} = (\bar{x}, \ldots, \bar{x}) \in X^{n-1}$ is locally extremal for the system of two sets

$$\Lambda_1 := \Omega_1 \times \ldots \times \Omega_{n-1} \text{ and } \Lambda_2 := \big\{(x, \ldots, x) \in X^{n-1} \big|\, x \in \Omega_n\big\} ,$$

which are closed around $\bar{z}$ if all Ω_i are assumed to be closed around $\bar{x}$. It is obvious that X^{n-1} admits a Fréchet smooth renorm if X does. Hence we can employ the previous consideration with $n = 2$ and get the approximate extremal principle for $\{\Lambda_1, \Lambda_2, \bar{z}\}$. In this way, taking into account Proposition 1.2 and the representation

$$\widehat{N}(\bar{z}; \Lambda_2) = \big\{(x_1^*, \ldots, x_{n-1}^*) \in (X^*)^{n-1} \big|\, x_1^* + \ldots + x_{n-1}^* \in \widehat{N}(\bar{x}; \Omega_n)\big\} ,$$

we finish the proof of the theorem. △

Remark 2.11 (bornologically smooth spaces). The arguments used in the proof of Theorem 2.10 for $n = 2$ are now typical in the area of *variational principles*; cf. Li and Shi [785] and discussions in the next section. In particular, they can be modified to prove the smooth variational principle of Borwein and Preiss [154] in spaces admitting a smooth renorm with respect to any given bornology on X. Recall that a *bornology* β on X is a family of bounded and centrally symmetric subsets of X whose union is X, which is closed under multiplication by positive numbers and such that the union of any two members of β is contained in some member of β. The *Fréchet bornology* considered above is the strongest one, where β consists of all bounded symmetric subsets of X. The weakest one is the *Gâteaux bornology*, where β consists of all finite subsets of X. It is well known that every separable Banach space admits a Gâteaux smooth renorm. There are useful bornologies in-between; particularly the *Hadamard bornology*, where β consists of all compact symmetric subsets of X.

One can check that the way of proving Theorem 2.10 allows us to justify the approximate extremal principle (under a suitable modification of generalized normals to nonconvex sets) in Banach spaces admitting a *smooth renorm of any kind.* Actually the corresponding versions of the approximate extremal principle and the smooth variational principle are *equivalent* in Banach spaces with smooth renorms; see Borwein, Mordukhovich and Shao [151] for more details. It will be shown in Section 2.3 that a smoothness of the space in question is not only sufficient and but also *necessary* for the validity of smooth variational principles. On the other hand, the version of the extremal principle in Definition 2.5 will be justified in arbitrary Asplund spaces, which may not admit even a Gâteaux smooth renorm. This is due to the possibility of *separable reduction* for Fréchet-like normals and subgradients considered next.

2.2.2 Separable Reduction

In this subsection we develop the method of separable reduction that allows us to reduce certain problems involving Fréchet-like constructions from an arbitrary Banach space to the case of *separable subspaces.* The main goal is to obtain separable reduction results valuable for applications to the extremal principle in the approximate form of Definition 2.5(ii). A suitable assertion for this purpose can be formulated as follows.

Given proper functions $f_i: X \to \overline{I\!R}$, $i = 1, \ldots, N$, a separable subspace Y_0 of X, and a number $M > 0$, there is a closed separable subspace Y of X such that $Y_0 \subset Y$ and

$$0 \in \big(\widehat{\partial} f_1(x_1) \setminus M I\!B^*\big) + \widehat{\partial} f_2(x_2) + \ldots + \widehat{\partial} f_N(x_N) \tag{2.14}$$

whenever $x_1, x_2, \ldots, x_N \in Y$ and

$$0 \in \big(\widehat{\partial} f_{1|Y}(x_1) \setminus M\mathbb{B}^*\big) + \widehat{\partial} f_{2|Y}(x_2) + \ldots + \widehat{\partial} f_{N|Y}(x_N)\,, \tag{2.15}$$

where $f_{|Y}$ denotes the restriction of f to Y and where $\mathbb{B}^* = \mathbb{B}_{X^*}$.

This result, being applied to the indicator functions $f_i(x) := \delta(x;\Omega_i)$, $i = 1,\ldots,n$, with $f_{n+1}(x) := \varepsilon\|x\|$, ensures the desired separable reduction of the approximate extremal principle for n sets from a nonseparable space X to its separable subspace Y, provided that the initial subspace Y_0 is properly selected; see below. Note that it is *crucial* to have $M > 0$ in (2.14) and (2.15) independently from the other data; otherwise we don't get the *nontriviality* condition in the extremal principle.

To justify the desired separable reduction, we have to overcome essential technical difficulties in constructing a separable subspace $Y_0 \subset Y \subset X$ for the given data. This requires working only with elements of the primal Banach space X. However, formulations of the extremal principle and the assertion needed for its separable reduction involve elements of the dual space X^*. Thus an important part of the separable reduction procedure is to translate the required assertion into the language of the *space X only.* We'll do it first for convex functions, based on the fundamental duality in convex analysis, and then apply to general extended-real-valued functions using some *convexification* via *infimal convolution*, which is possible due to the very definition of Fréchet subgradients.

Lemma 2.12 (primal characterization of convex subgradients). *Let $\varphi\colon X \to \overline{\mathbb{R}}$ be a proper convex function with $0 \in \operatorname{dom}\varphi$. Then for any given $M > 0$ one has*

$$\partial\varphi(0) \setminus M\mathbb{B}^* \neq \emptyset \tag{2.16}$$

if and only if there are $c \ge 0$, $\gamma > 0$, and a nonempty open set $U \subset X$ such that the following properties hold:

(a) *$\varphi(h) \ge \varphi(0) - c\|h\|$ for all $h \in X$;*

(b) *$\varphi(th) \ge \varphi(0) + (M+\gamma)t\|h\|$ whenever $h \in U$ and $t \in [0,1]$.*

In this case for every $0 \neq h \in U$ there is $x^* \in \partial\varphi(0)$ with $\langle x^*, h\rangle > M\|h\|$.

Proof. To prove the necessity, we pick any $x^* \in \partial\varphi(0) \setminus M\mathbb{B}^*$ and observe that (a) holds with $c = \|x^*\|$. Then choose $\gamma > 0$ with $\|x^*\| > M + \gamma$ and find a nonempty open set $U \subset X$ such that $\langle x^*, h\rangle > (M+\gamma)\|x^*\|$ for every $h \in U$. This implies (b).

Let us prove the sufficiency, which includes the last statement of the lemma. Take (c,γ,U) satisfying (a) and (b) and then fix $0 \neq h \in U$. By (b) we find nonempty open convex sets $U_0 \subset U$ and $U_1 \subset \mathbb{R}$ such that $0 \notin U_0$, $h \in U_0$, $0 \notin U_1$, and

$$M < \tau/\|u\| < M + \gamma \quad \text{whenever} \quad (u,\tau) \in U_0 \times U_1\,.$$

Since φ is convex, we get from (b) that $\varphi'_+(0)(u) \ge (M+\gamma)\|u\|$ whenever $u \in U_0$. Consider the nonempty convex sets

$$C_1 := \{(u,t) \in X \times I\!R \,|\, \varphi(u) \le t\}, \quad C_2 := \bigcup_{\lambda>0} \lambda(U_0 \times U_1)$$

and observe that $C_1 \cap C_2 = \emptyset$. Indeed, if $\lambda(u,\tau) \in C_1 \cap C_2$ for some $\lambda > 0$, then one has

$$\lambda\tau \ge \varphi(\lambda u) \ge \varphi'_+(0)(\lambda u) = \lambda\varphi'_+(0)u \ge (M+\gamma)\lambda\|u\| > \lambda\tau$$

due to the choice of τ, a contradiction. Since C_2 is open, we apply the classical separation theorem and find $(0,0) \ne (\widetilde{x}^*, \widetilde{\nu}) \in (X \times I\!R)^* = X^* \times I\!R$ such that

$$l := \inf \big\langle (\widetilde{x}^*, \widetilde{\nu}), C_1 \big\rangle \ge \sup \big\langle (\widetilde{x}^*, \widetilde{\nu}), C_2 \big\rangle =: r\,.$$

Note that $l \le 0$ due to $(0,0) \in C_1$ and that $r \ge 0$ due to the structure of C_2. Thus $l = r = 0$, and we have

$$\begin{aligned} &\inf \big\{ \langle \widetilde{x}^*, u\rangle + \widetilde{\nu}t \,\big|\, (u,t) \in X \times I\!R,\ \varphi(u) \le \varphi(0) + t \big\} \\ &= \sup \big\{ \lambda\langle \widetilde{x}^*, u\rangle + \lambda\tau\widetilde{\nu} \,\big|\, (u,\tau) \in U_0 \times U_1,\ \lambda > 0 \big\} = 0\,. \end{aligned} \tag{2.17}$$

Since $\widetilde{\nu}t = \langle \widetilde{x}^*, 0\rangle + \widetilde{\nu}t \ge 0$ for all $t \ge 0$, we get $\widetilde{\nu} \ge 0$. To proceed, we first assume that $\widetilde{\nu} > 0$. Then putting $t = \varphi(u)$ in (2.17), we have $\langle -\widetilde{x}^*/\widetilde{\nu}, u\rangle \le \varphi(u) = \varphi(u) - \varphi(0)$ if $u \in \operatorname{dom}\varphi$. This also obviously holds if $\varphi(u) = \infty$, and so we conclude that $-\widetilde{x}^*/\widetilde{\nu} \in \partial\varphi(0)$.

On the other hand, it follows from (2.17) for $\tau \in U_1$ and $u = h$ that $\langle \widetilde{x}^*, h\rangle + \tau\widetilde{\nu} \le 0$, and hence

$$\| -\widetilde{x}^*/\widetilde{\nu}\| \ge \big\langle -\widetilde{x}^*/\widetilde{\nu}, h/\|h\| \big\rangle \ge \tau/\|h\| > M$$

due to the choice of τ. Thus we obtain

$$\langle -\widetilde{x}^*/\widetilde{\nu}, h\rangle > M\|h\| \quad \text{and} \quad -\widetilde{x}^*/\widetilde{\nu} \in \partial\varphi(0) \setminus M I\!B^*\,,$$

which justifies (2.16) in the case of $\widetilde{\nu} > 0$. We haven't used (a) so far.

Next let us consider the remaining case of $\widetilde{\nu} = 0$ in (2.17) and justify (2.16) using (a). In this case we necessarily have $\widetilde{x}^* \ne 0$ and get from (2.17) that $\langle \widetilde{x}^*, u\rangle \ge 0$ for all $u \in \operatorname{dom}\varphi$ and $\langle \widetilde{x}^*, u\rangle \le 0$ for all $u \in U_0$. Since U_0 is a neighborhood of h, the latter yields $\langle \widetilde{x}^*, h\rangle < 0$. Form the closed convex set

$$C_3 := \{(u,t) \in X \times I\!R \,|\, t < -c\|u\|\}$$

and observe that $C_1 \cap C_3 = \emptyset$ due to (a). Employing again the separation theorem, we find $(0,0) \ne (\widehat{x}^*, \widehat{\nu}) \in X^* \times I\!R$ such that

$$l := \inf \big\langle (\widehat{x}^*, \widehat{\nu}), C_1 \big\rangle \ge \sup \big\langle (\widehat{x}^*, \widehat{\nu}), C_3 \big\rangle =: r\,.$$

It is easy to check that $l = r = 0$, and thus

$$\inf\big\{\langle\widehat{x}^*,u\rangle+\widehat{\nu}t\big|\ (u,t)\in X\times I\!R,\ \varphi(u)\le\varphi(0)+t\big\}$$
$$=\sup\big\{\langle\widehat{x}^*,u\rangle+\widehat{\nu}t\big|\ (u,t)\in X\times I\!R,\ t<-c\|u\|\big\}=0\,, \tag{2.18}$$

which implies that $\widehat{\nu}\ge 0$. In fact we have $\widehat{\nu}>0$, since otherwise (2.18) yields $\langle\widehat{x}^*,u\rangle\le 0$ whenever $u\in X$, which contradicts the nontriviality of $(\widehat{x}^*,\widehat{\nu})$. Thus (2.18) gives $-\widehat{x}^*/\widehat{\nu}\in\partial\varphi(0)$ similarly to the case of (2.17). Now put

$$x^*:=-\widehat{x}^*/\widehat{\nu}-K\widetilde{x}^*\ \text{ with }\ K>\max\left\{0,\ -\frac{M\|h\|+\langle\widehat{x}^*/\widehat{\nu},h\rangle}{\langle\widetilde{x}^*,h\rangle}\right\} \tag{2.19}$$

and observe that, by the definition of $\partial\varphi(0)$ and the condition $\langle\widetilde{x}^*,u\rangle\ge 0$ for all $u\in\operatorname{dom}\varphi$, we have

$$\varphi(u)-\varphi(0)\ge\langle-\widehat{x}^*/\widehat{\nu},u\rangle\ge\langle x^*,u\rangle\quad\text{if}\quad u\in\operatorname{dom}\varphi;$$

so $x^*\in\partial\varphi(0)$. Moreover, using (2.19) and $\langle\widetilde{x}^*,h\rangle<0$, we conclude that

$$\langle x^*,h\rangle=\langle-\widehat{x}^*/\widehat{\nu},h\rangle-K\langle\widetilde{x}^*,h\rangle>M\|h\|\,,$$

which yields $\|x^*\|>M$ and hence (2.16). △

The next lemma provides a primal characterization of subdifferential sums for convex functions with a nontriviality condition crucial for subsequent applications to the extremal principle.

Lemma 2.13 (primal characterization of subdifferential sums for convex functions). *Let $\varphi_i\colon X\to\overline{I\!R}$, $j=1,\ldots,N$, be proper convex functions with $0\in\operatorname{dom}\varphi_1\cap\ldots\cap\operatorname{dom}\varphi_N$ and $N>1$. Given any $M>0$, one has*

$$0\in\big(\partial\varphi_1(0)\setminus M I\!B^*\big)+\partial\varphi_2(0)+\ldots+\partial\varphi_N(0) \tag{2.20}$$

if and only if there are $c\ge 0$, $\gamma>0$ and a nonempty open set $U\subset X$ such that the following hold:

(a) $\sum_{j=1}^N\varphi_j(h_j)\ge\sum_{j=1}^N\varphi_j(0)-c\max\big\{\|h_j-h_1\|\ \big|\ j=2,\ldots,N\big\}$ *for all* $h_1,\ldots,h_N\in X$;

(b) $\sum_{j=1}^N\varphi_j(th_j)\ge\sum_{j=1}^N\varphi_j(0)+(M+\gamma)t\max\big\{\|h_j-h_1\|\ \big|\ j=2,\ldots,N\big\}$ *for all* $h_1,\ldots,h_N\in X$ *with* $h_j-h_1\in U$, $j=2,\ldots,N$, *and for all* $t\in[0,1]$.

Proof. Assume that (2.20) holds and find $x_j^*\in\partial\varphi_j(0)$, $j=1,\ldots,N$, such that $\|x_1^*\|>M$ and $x_1^*+\ldots+x_N^*=0$. Then

$$\sum_{j=1}^N\varphi_j(h_j)-\sum_{j=1}^N\varphi_j(0)\ge\sum_{j=1}^N\langle x_j^*,h_j\rangle=\sum_{j=2}^N\langle x_j^*,h_j-h_1\rangle$$

$$\ge-\sum_{j=2}^N\|x_j^*\|\max\big\{\|h_j-h_1\|\ \big|\ j=2,\ldots,N\big\}$$

for all $h_1,\ldots,h_N \in X$, which gives (a) with $c := \sum_{j=2}^N \|x_j^*\|$. To justify (b), we take $\gamma > 0$ and an open set $\emptyset \neq U \subset X$ such that

$$\sum_{j=2}^N \langle x_j^*, h\rangle = -\langle x_1^*, h\rangle > (M+\gamma)\|h\| \ \text{ for all } \ h \in U\ .$$

By diminishing U if necessary, we may assume that

$$\sum_{j=2}^N \langle x_j^*, h_j\rangle > (M+\gamma)\max\big\{\|h_j\|\ \big|\ j=2,\ldots,N\big\}$$

whenever $h_2,\ldots,h_n \in U^{N-1}$. Then

$$\varphi_1(th_1) + \sum_{j=2}^N \varphi_j(th_j) - \sum_{j=1}^N \varphi_j(0) \ge t\sum_{j=2}^N \langle x_j^*, h_j - h_1\rangle$$

$$\ge (M+\gamma)t\max\big\{\|h_j - h_1\|\ \big|\ j=2,\ldots,N\big\}$$

whenever $h_1,\ldots,h_N \in X$ with $h_j - h_1 \in U,\ j=2,\ldots,N$, and $t \in [0,1]$. This gives (b) and proves the necessity in the lemma.

To prove the sufficiency, we assume that $c,\ \gamma$, and U are such that (a) and (b) hold. Define the *inf-convolution*

$$\varphi(h_2,\ldots,h_N) := \inf\Big\{\varphi_1(x) + \sum_{j=2}^N \varphi_j(x+h_j)\Big|\ x \in X\Big\}$$

for $(h_2,\ldots,h_N) \in X^{N-1}$ and observe that φ is a proper convex function on X^{N-1} with $0 \in \operatorname{dom}\varphi$. It is easy to check that properties (a) and (b) of this lemma implies that φ satisfies properties (a) and (b) of Lemma 2.12 on the product space X^{N-1} with the norm $\|(h_2,\ldots,h_N)\| := \max\big\{\|h_j\|\ \big|\ j = 2,\ldots,N\big\}$. Thus for fixed $0 \neq h \in U$ we find $z^* := (x_2^*,\ldots,x_N^*) \in (X^{N-1})^*$ such that $z^* \in \partial\varphi(0,\ldots,0)$ and $\langle z^*, (h,\ldots,h)\rangle > M\max\big\{\|h\|,\ldots,\|h\|\big\}$, i.e.,

$$\Big\langle \sum_{j=2}^N x_j^*, h\Big\rangle > M\|h\|\ . \tag{2.21}$$

Since $z^* \in \partial\varphi(0)$, the definition of φ gives

$$\varphi_1(x)+\sum_{j=2}^N \varphi_j(x+h_j) \ge \sum_{j=1}^N \varphi_j(0)+\langle z^*, (h_2,\ldots,h_N)\rangle = \sum_{j=1}^N \varphi_j(0)+\sum_{j=2}^N \langle x_j^*, h_j\rangle$$

for all $x \in X$ and all $(h_2,\ldots,h_N) \in X^{N-1}$. If we fix here one j and put $h_i = x = 0$ for all $i \neq j$, we get $x_j^* \in \partial\varphi_j(0),\ j = 2,\ldots,N$. If we put $h_j = -x,\ j = 2,\ldots,N$, we get $x^* := -(x_2^* + \ldots + x_N^*) \in \partial\varphi_1(0)$. Hence

$$0 \in \partial\varphi_1(0) + \ldots + \partial\varphi_N(0) \quad \text{and} \quad x^* \in \partial\varphi_1(0)\setminus M I\!B_{X^*}$$

due to (2.21), which completes the proof of the lemma. △

Now let us consider a general proper function $f\colon X \to \overline{I\!R}$, a point $x \in \operatorname{dom} f$ and associated with them two *convex* functions of the inf-convolution type. First, given positive numbers δ and ϵ, we define $\varphi_{f,x,\delta,\epsilon} : X \to [-\infty,\infty]$ by

$$\varphi_{f,x,\delta,\epsilon}(h) := \inf\Big\{\sum_{i=1}^m \alpha_i\Big[f(x+h_i)+\epsilon\|h_i\|\Big]\Big|\, m\in I\!N,\ h_i\in X,$$
$$\|h_i\|<\delta,\ \alpha_i\ge 0,\ i=1,\ldots,m,\ \sum_{i=1}^m\alpha_i=1,\ \sum_{i=1}^m\alpha_i h_i=h\Big\} \tag{2.22}$$

if $\|h\| < \delta$ and $\varphi_{f,x,\delta,\epsilon}(h) := \infty$ otherwise. Then, given a sequence $\Delta := (\delta_i)_{i=1}^\infty$ with $\delta_1 > \delta_2 > \cdots > 0$ and $\delta_i \downarrow 0$, we define $\varphi_{f,x,\Delta}\colon X \to \overline{I\!R}$ by

$$\varphi_{f,x,\Delta}(h) := \inf\Big\{\sum_{i=1}^m \alpha_i\varphi_{f,x,\delta_i,1/i}(h_i)\Big|\, m\in I\!N,\ h_i\in X,$$
$$\alpha_i\ge 0,\ i=1,\ldots,m,\ \sum_{i=1}^m\alpha_i=1,\ \sum_{i=1}^m\alpha_i h_i=h\Big\}, \tag{2.23}$$

where each $\varphi_{f,x,\delta_i,1/i}$, $i \in I\!N$, is constructed in (2.22). It follows from the definitions that both functions (2.22) and (2.23) are convex and not greater than $f(x)$ at $h = 0$. Moreover, the Fréchet subdifferential of f at x is closely related to the subdifferential of $\varphi_{f,x,\Delta}$ at zero. One can easily check that if $\widehat{\partial} f(x) \ne \emptyset$, then $\varphi_{f,x,\Delta}(0) = f(x)$ and $\widehat{\partial} f(x) \supset \partial\varphi_{f,x,\Delta}(0) \ne \emptyset$ for some Δ. On the other hand, if $\partial\varphi_{f,x,\Delta}(0) \ne \emptyset$ for some Δ and $\varphi_{f,x,\Delta}(0) = f(x)$, then $\partial\varphi_{f,x,\Delta}(0) \subset \widehat{\partial} f(x)$ as well.

The following corollary of Lemma 2.13 provides an equivalent translation of the basic assertion (2.14) into the language of the primal space X.

Corollary 2.14 (primal characterization for sums of Fréchet subdifferentials). *Let $f_j\colon X \to \overline{I\!R}$ be arbitrary proper functions, let $x_j \in \operatorname{dom} f_j$ as $= 1,\ldots,N$ and $N > 1$. Then for any given $M > 0$ one has (2.14) if and only if there are $c \ge 0$, $\gamma > 0$, a sequence $\Delta = (\delta_i)_{i=1}^\infty \subset (0,\infty)$ with $\delta_i \downarrow 0$, and a nonempty open set $U \subset X$ such that the following hold:*

(a) $\sum_{j=1}^N \varphi_{f_j,x_j,\Delta}(h_j) \ge \sum_{j=1}^N f_j(x_j) - c\max\big\{\|h_j - h_1\| \;\big|\; j = 2,\ldots,N\big\}$ *for all $h_1,\ldots,h_N \in X$;*

(b) $\sum_{j=1}^N \varphi_{f_j,x_j,\Delta}(th_j) \ge \sum_{j=1}^N f_j(x_j) + (M+\gamma)t\max\big\{\|h_j - h_1\| \;\big|\; j = 2,\ldots,N\big\}$ *for all $h_1,\ldots,h_N \in X$ with $h_j - h_1 \in U$, $j = 2,\ldots,N$, and for all numbers $t \in [0,1]$.*

Proof. If (2.14) holds, then $\widehat{\partial} f_j(x_j) \neq \emptyset$, and hence $\varphi_{f,x,\Delta}(0) = f_j(x_j)$, $j = 1, \ldots, N$, for some sequence Δ. Then conditions (a) and (b) of the corollary immediately follow from the corresponding conditions of Lemma 2.13. In the other direction, if conditions (a) and (b) of the corollary hold, then $\varphi_{f_j,x_j,\Delta}(0) = f_j(x_j)$ by (a), and so (2.14) follows from the sufficiency in Lemma 2.13 for the convex functions $\varphi_j = \varphi_{f_j,x_j,\Delta}$, $j = 1, \ldots, N$, and the mentioned relationships between $\widehat{\partial} f(x)$ and $\partial\varphi_{f,x,\Delta}(0)$. △

Next we establish the basic separable reduction result for assertion (2.14) that lies at the ground of the whole separable reduction technique for the extremal principle.

Theorem 2.15 (basic separable reduction). *Let $f_1, \ldots, f_N\colon X \to \overline{I\!R}$, $N > 1$, be proper functions bounded from below, and let Y_0 be a separable subspace of X. Then there is a closed separable subspace $Y \subset X$ such that $Y_0 \subset Y$ and, given any $M > 0$, assertion* (2.14) *holds whenever $x_1, x_2, \ldots, x_N \in Y$ and one has* (2.15).

Proof. Our strategy is to build Y inductively starting with Y_0 and then to derive (2.14) from (2.15) and $(x_1, \ldots, x_N) \in Y^N$ based on the primal characterization of (2.14) in Corollary 2.14.

Let $\mathcal{A}$ be the countable set of all matrices $(\alpha_i^j \mid i \in I\!N,\ j = 1, \ldots N)$ with rational nonnegative entries such that $\alpha_i^j > 0$ only for finitely many pairs $(i, j) \in I\!N \times \{1, \ldots, N\}$ and that $\sum_{i=1}^\infty \alpha_i^j = 1$ for all $j = 1, \ldots, N$. Let $\mathcal{B}$ be the countable set of all matrices $(\beta_{il}^j \mid i, l \in I\!N,\ j = 1, \ldots N)$ with rational nonnegative entries such that $\beta_{il}^j > 0$ only for finitely many triples $(i, l, j) \in I\!N^2 \times \{1, \ldots, N\}$ and that $\sum_{l=1}^\infty \beta_{il}^j = 1$ for all $i \in I\!N$ and $j = 1, \ldots, N$. Let $\mathcal{D}$ be the countable set of all sequences $(\delta_i)_{i=1}^\infty$ with rational entries for which $0 < \delta_1 \ge \delta_2 \ge \cdots \ge 0$ and $\delta_i = 0$ if $i \in I\!N$ is sufficiently large. Given $j = 1, \ldots, N$ and $x \in \operatorname{dom} f_j$, let $\eta_j(x) > 0$ be such that f_j is bounded from below on the ball around x with radius $\eta_j(x)$.

For $\bar{x} := (x_1, \ldots, x_N) \in X^N$, for $a := (\alpha_i^j) \in \mathcal{A}$, for $b := (\beta_{il}^j) \in \mathcal{B}$, for $\bar{r} := (r_2, \ldots, r_N) \in (0, \infty)^{N-1}$, for $\Delta := (\delta_i) \in \mathcal{D}$ satisfying $\delta_i > 0$ whenever $\max\{\alpha_i^1, \ldots, \alpha_i^N\} > 0$ and $\delta_1 < \min\{\eta_1(x_1), \ldots, \eta_N(x_N)\}$, and for $k \in I\!N$ we find $u_{il}^j(\bar{x}, a, b, \bar{r}, \Delta, k) \in X$, $i, l \in I\!N$, $j = 1, \ldots, N$, such that $\|u_{il}^j(\bar{x}, a, b, \bar{r}, \Delta, k)\| < \delta_i$ if $\delta_i > 0$ and $u_{il}^j(\ldots) = 0$ if $\delta_i = 0$ for all $i, l \in I\!N$ and $j = 1, \ldots, N$, that

$$\Big\|\sum_{i=1}^\infty \alpha_i^j \sum_{l=1}^\infty \beta_{il}^j u_{il}^j(\bar{x}, a, b, \bar{r}, \Delta, k) - \sum_{i=1}^\infty \alpha_i^1 \sum_{l=1}^\infty \beta_{il}^1 u_{il}^1(\ldots)\Big\| < r_j, \quad j = 2, \ldots, N\,,$$

and that

$$\sum_{j=1}^{N}\sum_{i=1}^{\infty}\alpha_i^j\sum_{l=1}^{\infty}\beta_{il}^j\Big[f_j(x_j+u_{il}^j(\overline{x},a,b,\overline{r},\varDelta,k))+\tfrac{1}{i}\|u_{il}^j(\overline{x},a,b,\overline{r},\varDelta,k)\|\Big]$$

$$<\frac{1}{k}+\sum_{j=1}^{N}\sum_{i=1}^{\infty}\alpha_i^j\sum_{l=1}^{\infty}\beta_{il}^j\Big[f_j(x_j+h_{il}^j)+\tfrac{1}{i}\|h_{il}^j\|\Big]$$

whenever $h_{il}^j\in X$, $\|h_{il}^j\|<\delta_i$ if $\delta_i>0$ and $h_{il}^j=0$ if $\delta_i=0$, and

$$\Big\|\sum_{i=1}^{\infty}\alpha_i^j\sum_{l=1}^{\infty}\beta_{il}^jh_{il}^j-\sum_{i=1}^{\infty}\alpha_i^1\sum_{l=1}^{\infty}\beta_{il}^1h_{il}^1\Big\|<r_j,\ \ j=2,\ldots,N\,.$$

Further, for $\overline{x},a,b,\overline{r},\varDelta,k$ as above and for $h\in X$ with $\|h\|<\delta_1$ we find $g_{il}^j(\overline{x},h,a,b,\overline{r},\varDelta,k)\in X,\ i,l\in I\!N,\ j=1,\ldots,N$, such that

$$\|g_{il}^j(\overline{x},h,a,b,\overline{r},\varDelta,k)\|<\delta_i\ \text{ if }\ \delta_i>0\ \text{ and }\ g_{il}^j(\ldots)=0\text{ if }\ \delta_i=0\,,$$

$$\Big\|\sum_{i=1}^{\infty}\alpha_i^j\sum_{l=1}^{\infty}\beta_{il}^jg_{il}^j(\overline{x},h,a,b,\overline{r},\varDelta,k)-\sum_{i=1}^{\infty}\alpha_i^1\sum_{l=1}^{\infty}\beta_{il}^1g_{il}^1(\ldots)-h\Big\|<r_j$$

if $j=2,\ldots,N$, and that

$$\sum_{j=1}^{N}\sum_{i=1}^{\infty}\alpha_i^j\sum_{l=1}^{\infty}\beta_{il}^j\Big[f_j(x_j+g_{il}^j(\overline{x},h,a,b,\overline{r},\varDelta,k))+\tfrac{1}{i}\|g_{il}^j(\overline{x},h,a,b,\overline{r},\varDelta,k)\|\Big]$$

$$<\frac{1}{k}+\sum_{j=1}^{N}\sum_{i=1}^{\infty}\alpha_i^j\sum_{l=1}^{\infty}\beta_{il}^j\Big[f_j(x_j+h_{il}^j)+\tfrac{1}{i}\|h_{il}^j\|\Big]$$

whenever $h_{il}^j\in X$, $\|h_{il}^j\|<\delta_i$ if $\delta_i>0$ and $h_{il}^j=0$ if $\delta_i=0$, and

$$\Big\|\sum_{i=1}^{\infty}\alpha_i^j\sum_{l=1}^{\infty}\beta_{il}^jh_{il}^j-\sum_{i=1}^{\infty}\alpha_i^1\sum_{l=1}^{\infty}\beta_{il}^1h_{il}^1-h\Big\|<r_j,\ \ j=2,\ldots,N\,.$$

Now we are ready to construct the required separable subspace $Y\subset X$. By induction we build separable subspaces $Y_0\subset Y_1\subset\ldots\subset X$ as follows. If Y_n was already constructed for some $n\in I\!N\cup\{0\}$ (Y_0 is given), take any countable subset $C_n\subset Y_n$ dense in Y_n. Then let Y_{n+1} be the closed linear span of Y_n and the points

$$u_{il}^j(\overline{x},a,b,\overline{r},\varDelta,k),\quad g_{il}^j(\overline{x},h,a,b,\overline{r},\varDelta,k)\,,$$

where $\overline{x}=(x_1,\ldots,x_N)\in C_n^N$, $h\in C_n$, $\|h\|<\delta_1$, $\overline{r}\in(0,\infty)^{N-1}$ with rational entries, $\varDelta=(\delta_i)\in\mathcal{D}$ with $\delta_1<\min\{\eta_1(x_1),\ldots,\eta_N(x_N)\}$, $a\in\mathcal{A}$, $b\in\mathcal{B}$, $j=1,\ldots,N$, and $i,l,k\in I\!N$. Denoting $Y:=\operatorname{cl}\big[\bigcup\{Y_n|\ n\in I\!N\}\big]$ and

$C := \bigcup\{C_n \mid n \in I\!N\}$, we see that $\mathrm{cl}\, C = Y$ and Y is a separable subspace of X containing Y_0.

Fix any $M > 0$. We need to prove that for every given $\bar{x} = (x_1, \ldots, x_N) \in Y^N$ satisfying (2.15) one has (2.14). According to Corollary 2.14 the latter is equivalent to the fulfillment of conditions (a) and (b) therein. Using (2.15), we find $x_j^* \in \widehat{\partial}(f_{j|Y})(x_j),\ j = 1, \ldots, N$, such that $\|x_1^*\| > M$ and $x_1^* + \ldots + x_N^* = 0$. Due to the definition of Fréchet subgradients there is a sequence of rational numbers $\delta_1 > \delta_2 > \ldots > 0$ with

$$f_j(x_j + h) + \tfrac{1}{i}\|h\| \geq f_j(x_j) + \langle x_j^*, h\rangle \quad \text{whenever} \quad h \in Y,\ \|h\| < 2\delta_i\ , \tag{2.24}$$

$i \in I\!N$, and $j = 1, \ldots, N$. We always take $\delta_1 < \min\{\eta_1(x_1), \ldots, \eta_N(x_N)\}$ and show that conditions (a) and (b) of Corollary 2.14 hold along the chosen sequence $\Delta = \{\delta_1, \delta_2, \ldots\}$. Since $\bar{x} \in Y^N$, for any $n \in I\!N$ and $j = 1, \ldots, N$ we find $x_n^j \in C_n \subset Y$ and rational numbers γ_n^j satisfying

$$\|x_j - x_n^j\| \leq \gamma_n^j \leq 2\|x_j - x_n^j\| \quad \text{and} \quad \|x_j = x_n^j\| \to 0 \quad \text{as} \quad n \to \infty\ .$$

First we verify condition (a) of Corollary 2.14 with $c := \sum_{j=2}^N \|x_j^*\|$. Fix any $h_1, \ldots, h_N \in X$ and assume without loss of generality that $\|h_j\| < \delta_1$ for all $j = 1, \ldots, N$. Consider any $a = (\alpha_i^j) \in \mathcal{A}$, any $b = (\beta_{il}^j) \in \mathcal{B}$, any $h_{il}^j \in X$ with $\|h_{il}^j\| < \delta_i,\ i, l \in I\!N,\ j = 1, \ldots, N$, such that

$$\sum_{i=1}^\infty \alpha_i^j \sum_{l=1}^\infty \beta_{il}^j h_{il}^j = h_j \quad \text{for all} \quad j = 1, \ldots, N\ . \tag{2.25}$$

Find $i_0 \in I\!N$ so large that $\alpha_i^j = 0$ for all $i \geq i_0$ and $j = 1, \ldots, N$. Then we put $h_{il}^j = 0$ whenever $i \geq i_0$. Taking any rational numbers $r_j > \|h_j - h_1\|,\ j = 2, \ldots, N$, we observe that

$$\begin{aligned} &\|h_{il}^j\| + \gamma_n^j < \delta_i,\ i < i_0,\ l \in I\!N,\ j = 1, \ldots, N\ , \\ &\text{and} \quad \|h_j - h_1\| + \gamma_n^j + \gamma_n^1 < r_j,\ j = 2, \ldots, N \end{aligned} \tag{2.26}$$

for all $n \in I\!N$ sufficiently large. Denote $\overline{x_n} := (x_n^1, \ldots, x_n^N),\ n \in I\!N$, and

$$h_{il}^{j,n} := h_{il}^j + x_j - x_n^j, \quad i, l \in I\!N,\ j = 1, \ldots, N\ . \tag{2.27}$$

Finally, putting $\overline{\Delta} := (\delta_1, \delta_2, \ldots, \delta_{i_0}, 0, 0, \ldots)$ and using the u_{il}^j-part in the construction of Y, we get the following chain of inequalities valid for all large numbers $n \in I\!N$:

$$\sum_{j=1}^N \sum_{i=1}^\infty \alpha_i^j \sum_{l=1}^\infty \beta_{il}^j \Big[f_j(x_j + h_{il}^j) + \tfrac{1}{i}\|h_{il}^j\|\Big] = \sum_{j=1}^N \sum_{i=1}^\infty \alpha_i^j \sum_{l=1}^\infty \beta_{il}^j \Big[f_j(x_n^j + h_{il}^{j,n})$$

$$+\tfrac{1}{i}\|h_{il}^j\|\Big] \ge \sum_{j=1}^N \sum_{i=1}^\infty \alpha_i^j \sum_{l=1}^\infty \beta_{il}^j \Big[f_j(x_n^j + h_{il}^{j,n}) + \tfrac{1}{i}\|h_{il}^{j,n}\|\Big] - \tfrac{1}{i}\sum_{j=1}^N \gamma_n^j$$

$$> -\frac{1}{n} - \sum_{j=1}^N \gamma_n^j + \sum_{j=1}^N \sum_{i=1}^\infty \alpha_i^j \sum_{l=1}^\infty \beta_{il}^j \Big[f_j(x_n^j + u_{il}^j(\overline{x_n}, a, b, \overline{r}, \overline{\Delta}, n)$$

$$+\tfrac{1}{i}\|u_{il}^j(\ldots)\|\Big] \quad \Big(\text{as } \|h_{il}^{j,n}\| \le \|h_{il}^j\| + \gamma_n^j < \delta_i, \text{ if } \ i \le i_0, \quad \text{and}$$

$$\Big\|\sum_{i=1}^\infty \alpha_i^j \sum_{l=1}^\infty \beta_{il}^j h_{il}^{n,j} - \sum_{i=1}^\infty \alpha_i^1 \sum_{l=1}^\infty \beta_{il}^1 h_{il}^{1,n}\Big\| \le \|h_j - h_1\| + \gamma_n^j + \gamma_n^1 < r_j\Big)$$

$$\ge -\frac{1}{n} - 2\sum_{j=1}^N \gamma_n^j + \sum_{j=1}^N \sum_{i=1}^\infty \alpha_i^j \sum_{l=1}^\infty \beta_{il}^j \Big[f_j\big(x_j + x_n^j - x_j$$

$$+u_{il}^j(\overline{x_n}, a, b, \overline{r}, \overline{\Delta}, n)\big) + \tfrac{1}{i}\|x_n^j - x_j + u_{il}^j(\ldots)\|\Big]$$

$$\ge -\frac{1}{n} - 2\sum_{j=1}^N \gamma_n^j + \sum_{j=1}^N f_j(x_j) + \sum_{j=1}^N \Big\langle \xi_j, x_n^j - x_j$$

$$+\sum_{i=1}^\infty \alpha_i^j \sum_{l=1}^\infty \beta_{il}^j u_{il}^j(\overline{x_n}, a, b, \overline{r}, \overline{\Delta}, n)\Big\rangle$$

$$\Big(\text{as } x_n^j - x_j + u_{il}^j(\ldots) \in Y \ \text{ and } \ \|x_n^j - x_j + u_{il}^j(\ldots)\| < \gamma_n^j + \delta_i < 2\delta_i\Big)$$

$$= -\frac{1}{n} - 2\sum_{j=1}^N \gamma_n^j + \sum_{j=1}^N f_j(x_j) + \sum_{j=1}^N \langle x_j^*, x_n^j - x_j\rangle$$

$$+\sum_{j=2}^N \Big\langle x_j^*, \sum_{i=1}^\infty \alpha_i^j \sum_{l=1}^\infty \beta_{il}^j u_{il}^j(\overline{x_n}, a, b, \overline{r}, \overline{\Delta}, n) - \sum_{i=1}^\infty \alpha_i^1 \sum_{l=1}^\infty \beta_{il}^1 u_{il}^1(\ldots)\Big\rangle$$

$$\Big(\text{as } x_1^* + x_2^* + \ldots + x_N^* = 0\Big)$$

$$\ge -\frac{1}{n} - 2\sum_{j=1}^N \gamma_n^j + \sum_{j=1}^N f_j(x_j) - \sum_{j=1}^N \|x_j^*\|\gamma_n^j$$

$$-\sum_{j=2}^N \|x_j^*\| \Big\|\sum_{i=1}^\infty \alpha_i^j \sum_{l=1}^\infty \beta_{il}^j u_{il}^j(\overline{x_n}, a, b, \overline{r}, \overline{\Delta}, n) - \sum_{i=1}^\infty \alpha_i^1 \sum_{l=1}^\infty \beta_{il}^1 u_{il}^1(\ldots)\Big\|$$

$$\ge -\frac{1}{n} - 2\sum_{j=1}^N \gamma_n^j + \sum_{j=1}^N f_j(x_j) - \sum_{j=1}^N \|x_j^*\|\gamma_n^j - \sum_{j=2}^N \|x_j^*\| r_j\,.$$

Letting $n \to \infty$, we get the estimate

$$\sum_{j=1}^N \sum_{i=1}^\infty \alpha_i^j \sum_{l=1}^\infty \beta_{il}^j \big[f_j(x_j + h_{il}^j) + \tfrac{1}{i}\|h_{il}^j\|\big] \ge \sum_{j=1}^N f_j(x_j) - \sum_{j=2}^N \|x_j^*\| r_j.$$

Then letting $r_j \to \tilde{r}_j := \|h_j - h_1\|$ for $j = 2, \ldots, N$, we arrive at

$$\sum_{j=1}^{N}\sum_{i=1}^{\infty}\alpha_i^j\sum_{l=1}^{\infty}\beta_{il}^j\Big[f_j(x_j+h_{il}^j)+\tfrac{1}{i}\|h_{il}^j\|\Big] \ge \sum_{j=1}^{N} f_j(x_j)-c\max\big\{\tilde{r}_j\big|\ j=2,\ldots,N\big\},$$

which ensures condition (a) of Corollary 2.14 with $c := \sum_{j=2}^{N}\|x_j^*\|$ due to the definition of $\varphi_{f_j,x_j,\Delta}$ in (2.23) along the sequence Δ selected in (2.24).

To complete the proof of the theorem, it remains to verify condition (b) in Corollary 2.14 along the sequence Δ, some number $\gamma > 0$, and an open set $U \subset X$. Since $\|x_1^*\| > M$, we find $y \in Y$ with $\|y\| \le \delta_1$ and $\gamma \in (0,1)$ so that

$$\langle x_1^*, y\rangle > (M + 3\gamma)\|y\|\ . \tag{2.28}$$

Choose a number ζ satisfying

$$0 < \zeta < \min\Big\{\delta_1 - \|y\|, \gamma\,\|y\|\Big(\sum_{j=1}^{N}\|x_j^*\|\Big)^{-1}, \gamma\,\|y\|\Big[2(M+\gamma)\Big]^{-1}\Big\} \tag{2.29}$$

and put $U := \big\{h \in X\big|\ \|h - y\| < \zeta\big\}$. Now fix any $t \in (0,1]$ and any $h_1, \ldots, h_N \in X$ with $h_j - h_1 \in U$; then $\|h_j - h_1\| < \delta,\ j = 2, \ldots, N$. We may assume without loss of generality that $\|th_j\| \le \delta_1$ for all $j = 1, \ldots, N$. Since $\|h_j - h_1 - y\| < \zeta$, there is a rational number η with $\|th_j - th_1 - ty\| < \eta < t\zeta$ for all $j = 2, \ldots, N$. This allows us to find $h_0 \in C$ such that

$$\|th_j - th_1 - h_0\| < \eta,\ \ j = 2, \ldots, N, \quad \text{and} \quad \|h_0 - ty\| < t\zeta\ . \tag{2.30}$$

As in the proof of the first part of the theorem, we pick any $a = (\alpha_i^j) \in \mathcal{A}$, any $b = (\beta_{il}^j) \in \mathcal{B}$, and any $h_{il}^j \in X$, with $\|h_{il}^j\| < \delta_i,\ i, l \in I\!N,\ j = 1, \ldots, N$, and such that (2.25) holds. Find $i_0 \in I\!N$ so large that $\alpha_i^j = 0$ whenever $i \ge i_0$ and $j = 1, \ldots, N$. We may choose $h_{il}^j = 0$ whenever $i \ge i_0$. Thus we have (2.26) for all large $n \in I\!N$. Take $\overline{\Delta} = (\delta_1, \delta_2, \ldots, \delta_{i_0}, 0, 0, \ldots)$, define $\overline{x_n}$ and $h_{il}^{j,n}$ as in (2.27), and put $\overline{r_n} := (\eta + \gamma_n^2 + \gamma_n^1, \ldots, \eta + \gamma_n^N + \gamma_n^1)$. Now using the g_{il}^j-part in the construction of Y, we perform the following chain of inequalities for all $n \in I\!N$ sufficiently large:

$$\begin{aligned}
&\sum_{j=1}^{N}\sum_{i=1}^{\infty}\alpha_i^j\sum_{l=1}^{\infty}\beta_{il}^j\Big[f_j(x_j + h_{il}^j) + \tfrac{1}{i}\|h_{il}^j\|\Big]\\
&\ge \sum_{j=1}^{N}\sum_{i=1}^{\infty}\alpha_i^j\sum_{l=1}^{\infty}\beta_{il}^j\Big[f_j(x_n^j + h_{il}^{j,n}) + \tfrac{1}{i}\|h_{il}^{j,n}\|\Big] - \tfrac{1}{i}\sum_{j=1}^{N}\gamma_n^j\\
&> -\frac{1}{n} - \sum_{j=1}^{N}\gamma_n^j + \sum_{j=1}^{N}\sum_{i=1}^{\infty}\alpha_i^j\sum_{l=1}^{\infty}\beta_{il}^j\Big[f_j(x_n^j + g_{il}^j(\overline{x_n}, h_0, a, b, \overline{r_n}, \overline{\Delta}, n)
\end{aligned}$$

$$+\tfrac{1}{i}\|g_{il}^j(\ldots)\|\Big]\ \Bigg(\text{as } \|h_{il}^{j,n}\| \le \|h_{il}^j\| + \gamma_n^j < \delta_i,\ i \le i_0, \text{ and } \Bigg\|\sum_{i=1}^{\infty}\alpha_i^j\sum_{l=1}^{\infty}\beta_{il}^j h_{il}^{n,j}$$

$$-\sum_{i=1}^{\infty}\alpha_i^1\sum_{l=1}^{\infty}\beta_{il}^1 h_{il}^{1,n} - h_0\Bigg\| \le \|th_j - th_1 - h_0\| + \gamma_n^j + \gamma_n^1 < \eta + \gamma_n^j + \gamma_n^1\Bigg)$$

$$\ge -\frac{1}{n} - 2\sum_{j=1}^{N}\gamma_n^j + \sum_{j=1}^{N}\sum_{i=1}^{\infty}\alpha_i^j\sum_{l=1}^{\infty}\beta_{il}^j\Big[f_j\big(x_j + x_n^j - x_j$$

$$+g_{il}^j(\overline{x_n}, h_0, a, b, \overline{r_n}, \overline{\Delta}, n)\big) + \tfrac{1}{i}\|x_n^j - x_j + g_{il}^j(\ldots)\|\Big]$$

$$\ge -\frac{1}{n} - 2\sum_{j=1}^{N}\gamma_n^j + \sum_{j=1}^{N} f_j(x_j)$$

$$+\sum_{j=1}^{N}\Big\langle x_j^*, x_n^j - x_j + \sum_{i=1}^{\infty}\alpha_i^j\sum_{l=1}^{\infty}\beta_{il}^j g_{il}^j(\overline{x_n}, h_0, a, b, \overline{r_n}, \overline{\Delta}, n)\Big\rangle$$

$$\Big(\text{as } x_n^j - x_j + g_{il}^j(\ldots) \in Y \text{ and } \|x_n^j - x_j + g_{il}^j(\ldots)\| < \gamma_n^j + \delta_i < 2\delta_i\Big)$$

$$= -\frac{1}{n} - 2\sum_{j=1}^{N}\gamma_n^j + \sum_{j=1}^{N} f_j(x_j) + \sum_{j=1}^{N}\langle x_j^*, x_n^j - x_j\rangle - \langle x_1^*, h_0\rangle$$

$$+\sum_{j=2}^{N}\Big\langle x_j^*, \sum_{i=1}^{\infty}\alpha_i^j\sum_{l=1}^{\infty}\beta_{il}^j g_{il}^j(\overline{x_n}, h_0, a, b, \overline{r_n}, \overline{\Delta}, n)$$

$$-\sum_{i=1}^{\infty}\alpha_i^1\sum_{l=1}^{\infty}\beta_{il}^1 g_{il}^1(\ldots) - h_0\Big\rangle\quad\Big(\text{as } x_1^* + x_2^* + \ldots + x_N^* = 0\Big)$$

$$\ge -\frac{1}{n} - 2\sum_{j=1}^{N}\gamma_n^j + \sum_{j=1}^{N} f_j(x_j) - \sum_{j=1}^{N}\|x_j^*\|\gamma_n^j - \langle x_1^*, h_0\rangle$$

$$-\sum_{j=2}^{N}\|x_j^*\|\Bigg\|\sum_{i=1}^{\infty}\alpha_i^j\sum_{l=1}^{\infty}\beta_{il}^j g_{il}^j(\overline{x_n}, h_0, a, b, \overline{r_n}, \overline{\Delta}, n)$$

$$-\sum_{i=1}^{\infty}\alpha_i^1\sum_{l=1}^{\infty}\beta_{il}^1 g_{il}^1(\ldots) - h_0\Bigg\| \ge -\frac{1}{n} - 2\sum_{j=1}^{N}\gamma_n^j + \sum_{j=1}^{N} f_j(x_j) - \sum_{j=1}^{N}\|x_j^*\|\gamma_n^j$$

$$-\langle x_1^*, h_0\rangle - \sum_{j=2}^{N}\|x_j^*\|(\eta + \gamma_n^j + \gamma_n^1).$$

Letting $n \to \infty$, we get

$$\sum_{j=1}^{N}\sum_{i=1}^{\infty}\alpha_i^j\sum_{l=1}^{\infty}\beta_{il}^j\Big[f_j(x_j + h_{il}^j) + \tfrac{1}{i}\|h_{il}^j\|\Big] \ge \sum_{j=1}^{N} f_j(x_j) - \langle x_1^*, h_0\rangle - \sum_{j=2}^{N}\|x_j^*\|\eta\,.$$

Now using (2.28)–(2.30), we finally have

$$\sum_{j=1}^{N}\sum_{i=1}^{\infty}\alpha_i^j\sum_{l=1}^{\infty}\beta_{il}^j\Big[f_j(x_j+h_{il}^j)+\tfrac{1}{i}\|h_{il}^j\|\Big]-\sum_{j=1}^{N}f_j(x_j)$$

$$\ge -\langle x_1^*,h_0\rangle-\sum_{j=2}^{N}\|x_j^*\|t\zeta\ge -\langle x_1^*,ty\rangle-\|x_1^*\|\cdot\|ty-h_0\|-\sum_{j=2}^{N}\|x_j^*\|t\zeta$$

$$>(M+3\gamma)\|ty\|-\sum_{j=1}^{N}\|x_j^*\|t\zeta>(M+2\gamma)\|ty\|>(M+\gamma)(\|h_0\|-t\zeta)$$

$$+\gamma t\|y\|>(M+\gamma)(t\|h_j-h_1\|-2t\zeta)+\gamma t\|y\|>(M+\gamma)t\|h_j-h_1\|$$

for all $j=2,\ldots,N$ and $t\in[0,1]$. Due to the definition of $\varphi_{f_j,x_j,\Delta}$ in (2.23) we get condition (b) in Corollary 2.14 and end the proof of the theorem. △

Note that the boundedness from below assumption on the functions $f_1,\ldots,f_N$ in Theorem 2.15 can be dropped by an additional separable reduction. As a consequence of Theorem 2.15, we arrive at the following result needed for the separable reduction of the extremal principle.

Corollary 2.16 (separable reduction for the extremal principle). *Let Y_0 be a separable subspace of a (nonseparable) Banach space X, and let $\varepsilon>0$. Given nonempty subsets $\Omega_1,\ldots,\Omega_n$ of X, $n\ge 2$, there is a closed separable subspace $Y\subset X$ such that $Y_0\subset Y$ and, for any fixed $M>0$, one has*

$$0\in\big(\widehat{N}(x_1;\Omega_1)\setminus M\mathbb{B}_{X^*}\big)+\widehat{N}(x_2;\Omega_2)+\ldots+\widehat{N}(x_n;\Omega_n)+\varepsilon\mathbb{B}_{X^*} \tag{2.31}$$

whenever $x_1,x_2,\ldots,x_N\in Y$ and

$$0\in\big(\widehat{N}(x_1;\Omega_1\cap Y)\setminus M\mathbb{B}_{X^*}\big)+\widehat{N}(x_2;\Omega_2\cap Y)+\ldots+\widehat{N}(x_n;\Omega_n\cap Y)+\varepsilon\mathbb{B}_{Y^*}\,.$$

Proof. This follows from Theorem 2.15 applied to $n+1$ functions

$$f_i(x):=\delta(x;\Omega_i),\ i=1,\ldots,n,\quad\text{and}\quad f_{n+1}(x):=\varepsilon\|x\|$$

with $x_1,\ldots,x_n\in Y$ and $x_{n+1}=0$. △

2.2.3 Extremal Characterizations of Asplund Spaces

In this subsection we consider a general class of Banach spaces, called Asplund spaces, which plays a prominent role in the subsequent variational analysis. We show, based on separable reduction, that the approximate extremal principle unconditionally holds in Asplund spaces, is equivalent to the version of the extremal principle in terms of ε-normals, and provides a *characterization* of this class of Banach spaces. Furthermore, we justify the validity of the exact extremal principle in Asplund spaces under the sequential normal compactness condition imposed on all but one of the sets involved in the extremal system. We also obtain related characterizations of Asplund spaces in terms of supporting properties of Fréchet normals and ε-normals at boundary points of closed sets.

Definition 2.17 (Asplund spaces). *A Banach space X is* ASPLUND, *or it has the* ASPLUND PROPERTY, *if every convex continuous function $\varphi: U \to I\!R$ defined on an open convex subset U of X is Fréchet differentiable on a dense subset of U.*

Note that Definition 2.17 is equivalent to the standard definition of Asplund spaces, which requires the *generic* Fréchet differentiability of φ on U, i.e., its Fréchet differentiability on a dense G_δ subset of U. This follows from the well-known fact that the collection of points where a convex continuous function is Fréchet differentiable is automatically a G_δ set. For simplicity we always put $U = X$ in Definition 2.17 that doesn't restrict the generality.

The class of Asplund spaces is well investigated in the geometric theory of Banach spaces. We refer the reader to the books of Deville, Godefroy and Zizler [331], Fabian [416], Phelps [1073], and to the survey paper of Yost [1348] for various characterizations, classifications, properties, and examples of Asplund spaces. Note that this class includes all Banach spaces having Fréchet smooth bump functions (in particular, spaces with Fréchet smooth renorms, hence every reflexive space); spaces with separable duals; spaces of continuous functions $\mathcal{C}(K)$ on a scattered compact Hausdorff space K (i.e., such that every subset of K has an isolated point); the classical space of sequences c_0 with the supremum norm and its generalization $c_0(\Gamma)$ to an arbitrary set Γ, etc. Although Asplund spaces are generally related to the Fréchet type of differentiability and subdifferentiability, they may fail to have even an equivalent norm Gâteaux differentiable off the origin.

Asplund spaces possess many useful properties some of them are employed in what follows. Let us mention that every closed subspace of an Asplund space is Asplund itself; moreover, *every separable Asplund space admits a Fréchet differentiable renorm*, which is especially important for the method of separable reduction. It is also important that the class of Asplund spaces is stable under Cartesian products and linear isomorphisms. A crucial topological property of duals to Asplund spaces is that the dual unit ball $I\!B^*$ is weak* *sequentially* compact.

There is a number of nice geometric characterizations of Asplund spaces. One of the most striking characterizations is that *X is Asplund if and only if every separable closed subspace of X has a separable dual.* In the sequel we often use another characterization of Banach spaces not having the Asplund property: they admit a "rough" equivalent norm *nowhere Fréchet differentiable.* The exact formulation is as follows.

Proposition 2.18 (Banach spaces with no Asplund property). *Let X be a Banach space with the norm $\|\cdot\|$. Then X is not Asplund if and only if there exist a number $\vartheta > 0$ and an equivalent norm $|\cdot|$ on X satisfying $|\cdot| \le \|\cdot\|$ and*

$$\limsup_{h\to 0}\left[\frac{|x+h|+|x-h|-2|x|}{\|h\|}\right] > \vartheta \quad \text{for all} \;\; x \in X\,. \tag{2.32}$$

Proof. It is not difficult to show (cf. Proposition 1.23 in Phelps [1073]) that condition (2.32) implies that the convex function $\varphi(x) = |x|$ is nowhere Fréchet differentiable on X. Thus (2.32) doesn't hold if X is Asplund.

To prove the converse statement, we recall that a *weak* slice* of $\Lambda \subset X^*$ is a set of the form

$$S(x,\Lambda,\alpha) := \big\{x^* \in \Lambda \big|\; \langle x^*, x\rangle > \sigma_\Lambda(x) - \alpha\big\}\,,$$

where $x \in X$, $\alpha > 0$, and $\sigma_\Lambda(x) := \sup\big\{\langle x^*, x\rangle \big|\; x^* \in \Lambda\big\}$. Assuming that X is not Asplund and applying Theorem 2.32 from Phelps [1073], we find a convex symmetric subset $\Lambda \subset I\!\!B^*$ with nonempty interior in X^* and a number $\vartheta > 0$ such that Λ doesn't admit a weak* slice of diameter less than 2ϑ. Observe that $|x| := \sigma_\Lambda(x)$ defines an equivalent norm on X with $|\cdot| \le \|\cdot\|$. For any fixed $0 \ne x \in X$ we take an arbitrary small $t > 0$ and select $x_1^*, x_2^* \in S(x,\Lambda,t\vartheta/2)$ such that $\|x_1^* - x_2^*\| > 2\vartheta$. Then we find $h \in X$, $\|h\| = 1$, with $\langle x_1^* - x_2^*, h\rangle > 2\vartheta$. This yields the estimates

$$\left[\frac{|x+th|+|x-th|-2|x|}{\|th\|}\right] \ge \left[\frac{\langle x_1^*, x+th\rangle + \langle x_2^*, x-th\rangle - 2|x|}{t}\right]$$

$$> \frac{1}{t}\left[|x| - \frac{t\vartheta}{2} + |x| - \frac{t\vartheta}{2} - 2|x|\right] + \langle x_1^* - x_2^*, h\rangle > -\vartheta + 2\vartheta = \vartheta$$

and implies the required inequality (2.32). △

Based on Proposition 2.18, we now construct an important example showing that in any non-Asplund space there are simple sets with pathological behavior of normals to every boundary point.

Example 2.19 (degeneracy of normals in non-Asplund spaces). *Let X be a Banach space with no Asplund property. Then there exists a closed epi-Lipschitzian set $\Omega \subset X$ for which the following hold:*

(a) *There is $K > 1$ such that*

$$\|x^*\| \le K\varepsilon \;\; \text{for all} \;\; x^* \in \widehat{N}_\varepsilon(x;\Omega), \quad \text{all} \;\; x \in \operatorname{bd}\Omega, \quad \text{and all} \;\; \varepsilon > 0\,.$$

(b) *Ω is normally regular at every boundary point with*

$$N(\bar{x};\Omega) = \widehat{N}(\bar{x};\Omega) = \{0\} \;\; \text{for all} \;\; \bar{x} \in \operatorname{bd}\Omega\,.$$

Proof. Take an arbitrary non-Asplund space X and represent it in the form $X = Z \times I\!R$ with the norm $\|(z,\alpha)\| := \|z\| + |\alpha|$ for $(z,\alpha) \in X$. Then Z is non-Asplund as well, since the opposite implies the Asplund property of X. By Proposition 2.18 we find a number $\vartheta > 0$ and a norm $|\cdot|$ on Z, which is equivalent to the original norm $\|\cdot\|$, so that $|\cdot| \le \|\cdot\|$ and one has (2.32) with $X = Z$ and $x = z$. Based on the norm $|\cdot|$, we construct a set $\Omega \subset X$ in the epigraphical form

$$\Omega := \big\{(z,\alpha) \in X \big|\ \alpha \ge \varphi(z)\big\} \ \text{ with } \ \varphi := -|\cdot| \ \text{ and } \ \mathrm{bd}\,\Omega = \mathrm{gph}\,\varphi\,. \tag{2.33}$$

Since φ in (2.33) is Lipschitz continuous on X, the set Ω is epi-Lipschitzian at every boundary point. To justify (a), we need to find a constant $K > 1$ providing the estimate

$$\|(z^*,\lambda)\| \le K\varepsilon \ \text{ if } \ (z^*,\lambda) \in \widehat{N}_\varepsilon\big((z,\varphi(z));\Omega\big),\ z \in Z,\ \varepsilon > 0\,, \tag{2.34}$$

where $\|(z^*,\lambda)\| := \max\big\{\|z^*\|,\ |\lambda|\big\}$ is the dual norm to $\|(z,\alpha)\| = \|z\| + |\alpha|$. Fix arbitrary $\bar z \in Z$ and $(z^*,\lambda) \in \widehat{N}_\varepsilon((\bar z,\varphi(\bar z));\Omega)$. It follows directly from the definition of $\widehat{N}_\varepsilon$ that

$$\langle z^*, z - \bar z\rangle + \lambda(\alpha - \varphi(\bar z)) \le 2\varepsilon(\|z - \bar z\| + |\alpha - \varphi(\bar z)|)$$

for all $(z,\alpha) \in \mathrm{epi}\,\varphi$ around $(\bar z,\varphi(\bar z))$. Putting here $z = \bar z$, one gets $\lambda \le 2\varepsilon$. Since $|\cdot| \le \|\cdot\|$ and $|\varphi(z) - \varphi(\bar z)| \le |z - \bar z|$, we conclude that

$$\langle z^*, z - \bar z\rangle + \lambda(\varphi(z) - \varphi(\bar z)) \le 4\varepsilon\|z - \bar z\|$$

and further that

$$\langle z^*, z - \bar z\rangle \le (4\varepsilon + |\lambda|)\|z - \bar z\|$$

for all z around $\bar z$. The latter gives

$$\|z^*\| \le 4\varepsilon + |\lambda| \ \text{ for any } \ (z^*,\lambda) \in \widehat{N}_\varepsilon\big((\bar z,\varphi(\bar z));\Omega\big)\,. \tag{2.35}$$

Let us show that (2.35) ensures (2.34) with $K := \max\big\{6, 4 + 8/\vartheta\big\}$. Indeed, for $\lambda \ge 0$ we get from (2.35) that $\|(z^*,\lambda)\| \le 6\varepsilon$ and arrive at (2.34) with $K = 6$. For $\lambda < 0$ we have from the above definition of $\widehat{N}_\varepsilon((\bar z,\varphi(\bar z));\Omega$ with $\varphi = -|\cdot|$ that

$$|z| - |\bar z| - \Big\langle \frac{z^*}{\lambda}, z - \bar z\Big\rangle \le -\frac{4\varepsilon}{\lambda}\|z - \bar z\|$$

for all z around $\bar z$. Putting there $2\bar z - z$ instead of z, we get

$$|2\bar z - z| - |\bar z| + \Big\langle \frac{z^*}{\lambda}, z - \bar z\Big\rangle \le -\frac{4\varepsilon}{\lambda}\|z - \bar z\|\,.$$

Adding the two previous inequalities together, we arrive at

$$|\bar{z}+(z-\bar{z})|+|\bar{z}-(z-\bar{z})|-2|\bar{z}| \le -\frac{8\varepsilon}{\lambda}\|z-\bar{z}\| .$$

The latter implies, according to Proposition 2.18 with $x=\bar{z}$ and $h=z-\bar{z}$, that $|\lambda|<8\varepsilon/\vartheta$, where ϑ is the fixed positive number from (2.32). Thus (2.35) gives $\|z^*\|\le 4\varepsilon+(8\varepsilon/\vartheta)$ for $\lambda<0$, and we arrive at (2.34) with $K=4+8/\vartheta$, which justifies (a).

Property (b) follows from (a) due to Definitions 1.1 and 1.4 by passing to the limit as $\varepsilon\downarrow 0$ and $x\to\bar{x}$. △

Now we are ready to establish the main result of this section ensuring that the first two versions of the extremal principle in Definition 2.5, being applied to *every* extremal system in a Banach space X, are equivalent to the Asplund property of X.

Theorem 2.20 (extremal characterizations of Asplund spaces). *Let X be a Banach space. The following are equivalent:*

(a) *X is Asplund.*

(b) *The approximate extremal principle holds in X.*

(c) *The ε-extremal principle holds in X.*

Proof. First we prove (a)$\Rightarrow$(b). Let X be an Asplund space, and let $\bar{x}$ be a local extremal point of some sets $\Omega_1,\ldots,\Omega_n$ closed around $\bar{x}$. By Definition 2.1 we take sequences $\{a_{ik}\}\subset X$, $i=1,\ldots,n$, and then consider a separable subspace Y_0 of X defined as

$$Y_0:=\operatorname{span}\big\{\bar{x},a_{ik}\big|\ i=1,\ldots,n,\quad k\in I\!N\big\} .$$

Applying the separable reduction result of Corollary 2.16, for every fixed $\varepsilon>0$ we find a closed separable subspace $Y_0\subset Y\subset X$ that ensures the fulfillment of (2.31) under the conditions imposed in the corollary. Observe that

$$\big\{\Omega_1\cap Y,\ldots,\Omega_n\cap Y,\bar{x}\big\} \tag{2.36}$$

is an extremal system in the space Y. Indeed, $\bar{x}$ is obviously a common point of the sets $\Omega_i\cap Y$, $i=1,\ldots,n$, since $\bar{x}\in Y_0\subset Y$. On the other hand, these sets shifted by the corresponding sequences a_{ik}, $i=1,\ldots,n$, don't have any common points in the neighborhood $U\cap Y$ of $\bar{x}$ in Y for all large $k\in I\!N$. Since $a_{ik}\in Y_0\subset Y$, this means that $\bar{x}$ is a local extremal point of the set system $\{\Omega_1\cap Y,\ldots,\Omega_n\cap Y\}$ in the space Y.

Since Y is a separable Asplund space, it admits an equivalent Fréchet smooth (re)norm denoted again by $\|\cdot\|$. Thus one can apply Theorem 2.10 ensuring the fulfillment of the approximate extremal principle for the extremal system (2.36) in Y. Without loss of generality we assume that $\varepsilon<1/4$ and use relations (2.3) and (2.4) of the extremal principle with ε/n. In this way we find $x_i\in\Omega_i\cap\big(\bar{x}+(\varepsilon/n)I\!B_Y\big)$ and

$$y_i^*\in\widehat{N}(x_i;\Omega_i\cap Y)+(\varepsilon/n)I\!B_{Y^*}$$

satisfying (2.3) for y_i^*. Hence $\|y_i^*\| > 1/2n$ for at least one $i \in \{1,\ldots,n\}$; let it hold for $i = 1$. Thus we have $y_i^* = \widetilde{y}_i^* + u_i^*$ with $\widetilde{y}_i^* \in \widehat{N}(x_i;\Omega_i \cap Y)$ and $\|u_i^*\| \le \varepsilon/n$ for $i = 1,\ldots,n$ and with

$$\|\widetilde{y}_1^*\| \ge \|y_1^*\| - \frac{\varepsilon}{n} > \frac{1-2\varepsilon}{2n} > \frac{1}{4n} := M > 0\,.$$

This implies the relation

$$0 \in \big(\widehat{N}(x_1;\Omega_1 \cap Y) \setminus \tfrac{1}{4n} I\!B_{X^*}\big) + \widehat{N}(x_2;\Omega_2 \cap Y) + \ldots + \widehat{N}(x_n;\Omega_n \cap Y) + \varepsilon I\!B_{Y^*}\,.$$

Due to Corollary 2.16 we get (2.31) with $M = 1/4n$. The latter means that there are $\widetilde{x}_i^* \in \widehat{N}(x_i;\Omega_i)$, $i = 1,\ldots,n$, and $v^* \in X^*$ with $\|v^*\| \le \varepsilon$ satisfying $\|\widetilde{x}_1^*\| > 1/4n$ and $\widetilde{x}_1^* + \ldots + \widetilde{x}_n^* + v^* = 0$. Now denoting $x_i^* := \widetilde{x}_i^*$ for $i = 1,\ldots,n-1$ and $x_n^* := \widetilde{x}_n^* + v^*$, we have all the relations in (2.3) and (2.4) except the normalization condition $\|x_1^*\| + \ldots + \|x_n^*\| = 1$. Since $\gamma := \|\widetilde{x}_1^*\| + \ldots + \|\widetilde{x}_n^*\| > 1/4n$ independently of ε, we can easily obtain the normalization condition for x_i^*/γ by adjusting ε in (2.4). This gives (a)$\Rightarrow$(b).

As mentioned above, (b)$\Rightarrow$(c) always holds. It remains to justify (c)$\Rightarrow$(a). Assuming that X is not Asplund, we have the closed set Ω from Example 2.19. Then the ε-extremal principle is not valid for $\big\{\Omega, \{\bar{x}\}, \bar{x}\big\}$ with any $\bar{x} \in \operatorname{bd}\Omega$, since the opposite contradicts Proposition 2.6(i) with $M = K\varepsilon > \varepsilon$. △

As a consequence of the results obtained, we arrive at the following characterizations of Asplund spaces via supporting properties of closed sets expressed in terms of Fréchet normals and ε-normals at boundary points.

Corollary 2.21 (boundary characterizations of Asplund spaces). *Let X be a Banach space. The following are equivalent:*

(a) *X is Asplund.*

(b) *For every proper closed subset Ω of X the set of points $x \in \operatorname{bd}\Omega$ with $\widehat{N}(x;\Omega) \ne \{0\}$ is dense in the boundary of Ω.*

(c) *For every proper closed subset Ω of X there is $x \in \operatorname{bd}\Omega$ such that $\widehat{N}(x;\Omega) \ne \{0\}$.*

(d) *For every proper closed subset Ω of X, every $\varepsilon > 0$, and every $M > \varepsilon$ the set of points $x \in \operatorname{bd}\Omega$ with $\widehat{N}_\varepsilon(x;\Omega) \setminus M I\!B^* \ne \emptyset$ is dense in the boundary of Ω.*

(e) *For every proper closed subset Ω of X, every $\varepsilon > 0$, and every $M > \varepsilon$ there is $x \in \operatorname{bd}\Omega$ such that $\widehat{N}_\varepsilon(x;\Omega) \setminus M I\!B^* \ne \emptyset$.*

Proof. Implication (a)$\Rightarrow$(b) follows from Theorem 2.20 and Proposition 2.6(ii). Implications (b)$\Rightarrow$(c)$\Rightarrow$(e) and (b)$\Rightarrow$(d)$\Rightarrow$(e) are trivial. Implication (e)$\Rightarrow$(a) follows from Example 2.19; see the end of the proof of Theorem 2.20. △

As follows from the above proof, an arbitrary number $M > \varepsilon$ in (d) and (e) can be equivalently replaced with $K\varepsilon$, $K > 1$. Related characterizations of Asplund spaces in terms of ε-normals can be written in the form: *for every proper closed subset $\Omega \subset X$ there is $\lambda > 0$ such that for each $\varepsilon > 0$ the set*

$$\Big\{x \in \operatorname{bd}\Omega \Big| \; \exists\, x^* \in \widehat{N}_\varepsilon(x;\Omega) \;\text{ with }\; \|x^*\| = \lambda\Big\}$$

is dense in the boundary of Ω, or is just nonempty; see Mordukhovich and B. Wang [960] for the proof and discussions.

We can see from the above results that the supporting properties (b)–(e) in Corollary 2.21 applied to *every* closed subset of X are equivalent to the "fuzzy" versions of the extremal principle in Theorem 2.20, since each of them characterizes Asplund spaces. This is essentially based on properties of Fréchet normals and ε-normals in Asplund spaces: cf. the related discussions in Subsect. 2.1.2. It follows from the proofs that for the equivalencies in Corollary 2.21 one can consider only *epigraphical sets* of type (2.33).

Next let us obtain conditions ensuring the fulfillment of the *exact* extremal principle in Definition 2.5(iii). For this purpose we employ the sequential normal compactness (SNC) property of sets introduced in Subsect. 1.1.3.

Theorem 2.22 (exact extremal principle in Asplund spaces).

(i) *Let X be an Asplund space, and let $\{\Omega_1,\ldots,\Omega_n,\bar{x}\}$ be an extremal system in X such that all Ω_i are locally closed around $\bar{x}$ and all but one of Ω_i are sequentially normally compact at $\bar{x}$. Then the exact extremal principle holds for $\{\Omega_1,\ldots,\Omega_n,\bar{x}\}$.*

(ii) *Conversely, let the exact extremal principle hold for every extremal system $\{\Omega_1,\Omega_2,\bar{x}\}$ in X, where both sets Ω_i are closed and one of them is sequentially normally compact at $\bar{x}$. Then X is Asplund.*

Proof. To justify (i), we use the ε-extremal principle that holds in any Asplund space by Theorem 2.20. Take a sequence of $\varepsilon_k \downarrow 0$ as $k \to \infty$ and consider the corresponding sequence of x_{ik} and x^*_{ik}, $i = 1,\ldots,n$, satisfying (2.2) and (2.3) with $\varepsilon = \varepsilon_k$. Then $x_{ik} \to \bar{x}$ for all $i = 1,\ldots,n$. Since the sequences $\{x^*_{ik}\}$ are bounded in X^* and since bounded sets in duals to Asplund spaces are weak* sequentially compact, we find $x^*_i \in X^*$ such that $x^*_{ik} \xrightarrow{w^*} x^*_i$ for $i = 1,\ldots,n$. Passing to the limit in (2.2) as $k \to \infty$ and using the definition of basic normals, we get (2.5). Also one obviously has $x^*_1 + \ldots + x^*_n = 0$. It remains to show that $(x^*_1,\ldots,x^*_n) \neq 0$ under the SNC assumptions of the theorem. On the contrary, assume that $x^*_i = 0$ while Ω_i are SNC at $\bar{x}$ for $i = 1,\ldots,n-1$. By Definition 1.20 the latter implies that $\|x^*_{ik}\| \to 0$ as $k \to \infty$ for $i = 1,\ldots,n-1$. Hence

$$\|x^*_{nk}\| \le \|x^*_{1k}\| + \ldots + \|x^*_{n-1k}\| \to 0 \;\text{ as }\; k \to \infty\,,$$

which contradicts the nontriviality condition $\|x^*_{1k}\| + \ldots + \|x^*_{nk}\| = 1$ for all $k \in I\!N$ and ends the proof of (i).

To prove (ii), we assume that X is not an Asplund space and represent it as $X = Z \times I\!R$, where Z must be non-Asplund as well. Then consider

$\Omega_1 := \{0\} \times (-\infty, 0] \in Z \times I\!R$ and $\Omega_2 := \Omega$ defined in (2.33). One can easily check that $\bar{x} = (0,0)$ is a local extremal point of these closed sets in X. Since Ω_2 is epi-Lipschitzian at $\bar{x}$, it is SNC at this point due to Theorem 1.26. However, the exact extremal principle doesn't hold for $\{\Omega_1, \Omega_2, \bar{x}\}$. Indeed, we have $N((0,0);\Omega_2) = \{(0,0)\}$ from property (b) in Example 2.19, while $N((0,0);\Omega_1) = Z^* \times [0,\infty)$. That is, $N(\bar{x};\Omega_1) \cap \big(-N(\bar{x};\Omega_2)\big) = \{(0,0)\}$, which justifies (ii) and ends the proof of the theorem. △

Let us show that the SNC assumption in Theorem 2.22 is *essential* for the fulfillment of the exact extremal principle in infinite-dimensional spaces.

Example 2.23 (violation of the exact extremal principle in the absence of SNC). *Every infinite-dimensional separable Banach space contains an extremal system $\{\Omega_1, \Omega_2, \bar{x}\}$ that doesn't satisfy the relations of the exact extremal principle.*

Proof. Let X be a separable Banach space, and let $\{e_k\}_1^\infty$ be unit independent vectors that densely span X. Consider the sets

$$\Omega_1 := \operatorname{clco}\Big\{\frac{e_n}{2^n},\ -\frac{e_n}{2^n}\ \Big|\ n \in I\!N\Big\},$$

and $\Omega_2 = \{0\}$, which are convex and compact in the norm topology of X. Note that Ω_1 and Ω_2 are *not SNC* unless X is finite-dimensional; see Theorem 1.21. Let us check that $0 \in \Omega_1 \cap \Omega_2$ is a local extremal point of the set system $\{\Omega_1, \Omega_2\}$. Indeed, taking

$$a := \sum_{n=1}^{\infty} \frac{e_n}{n^2} \in X,$$

we observe that for any sequence of $\nu_k \downarrow 0$ one has

$$\Omega_1 \cap (\nu_k a + \Omega_2) = \Omega_1 \cap \{\nu_k a\} = \emptyset .$$

It follows from the structure of Ω_1 that $N(0;\Omega_1) = \{0\}$, and thus $\{\Omega_1, \Omega_2, 0\}$ doesn't satisfy the exact extremal principle. △

Next we consider some properties of the basic normal cone $N(\cdot;\Omega)$ on *boundaries* of closed sets. It immediately follows from Corollary 2.21 that in Asplund spaces the sets of point $x \in \operatorname{bd}\Omega$ with $N(x;\Omega) \neq \{0\}$ is *dense* in the boundary of any proper closed subset $\Omega \subset X$. Moreover, Example 2.19 shows that even nonemptiness of this set for any Ω of type (2.33) implies that X in Asplund. Theorem 2.22 gives conditions under which this nontriviality property of basic normals holds at *every* boundary point of closed sets.

Corollary 2.24 (nontriviality of basic normals in Asplund spaces). *Let X be an Asplund space, and let Ω be a proper closed subset of X. Then $N(\bar{x};\Omega) \neq \{0\}$ at every point $\bar{x} \in \operatorname{bd}\Omega$ where the set Ω is sequentially normally compact.*

Proof. Follows from Theorem 2.22 applied to the system $\{\Omega, \{\bar{x}\}, \bar{x}\}$. △

Note that the result of Corollary 2.24 gives a new condition for the *supporting hyperplane* property even for *closed convex cones* in Asplund spaces, where the SNC assumption may be strictly weaker than the CEL one; see Remark 1.27 with its references and Example 3.6 in Subsect. 3.1.1.

In conclusion of this section we present a consequence of the results above that characterizes Asplund spaces via the existence of basic subgradients for every locally Lipschitzian function.

Corollary 2.25 (subdifferentiability of Lipschitzian functions on Asplund spaces). *Let X be a Banach space. Then $\partial\varphi(\bar{x}) \neq \emptyset$ for every function $\varphi\colon X \to \overline{I\!R}$ locally Lipschitzian around $\bar{x}$ if and only if X is Asplund.*

Proof. Consider any function φ on an Asplund space X that is Lipschitz continuous around $\bar{x}$. Then $N((\bar{x}, \varphi(\bar{x})); \operatorname{epi}\varphi) \neq \{(0,0)\}$ due to Corollary 2.24. By Corollary 1.81 we have $\partial\varphi(\bar{x}) \neq \emptyset$. Conversely, if X is not Asplund, then $\partial\varphi(x) \equiv \emptyset$ on X for the Lipschitz continuous function φ in (2.33). △

2.3 Relations with Variational Principles

By *variational principles*, in the conventional terminology of variational analysis, one means a group of results stating that for any lower semicontinuous (l.s.c.) and bounded from below function $\varphi\colon X \to \overline{I\!R}$ and a point x_0 close to its minimum there is an arbitrary small perturbation $\theta(\cdot)$ such that the resulting function $\varphi + \theta$ achieves its minimum at some point $\bar{x}$ near x_0. A variational principle is said to be *smooth* when the perturbation function may be chosen as smooth in some sense. The first general variational principle was established by Ekeland [396, 397] in complete metric spaces. Among smooth variational principles the most powerful are those by Borwein and Preiss [154] in Banach spaces with smooth renorms and by Deville, Godefroy and Zizler [331] in Banach spaces with smooth bump functions. Variational principles play a prominent role in many aspects of nonlinear analysis, optimization, and numerous applications.

For $\dim X < \infty$ such principles easily follow from the classical Weierstrass existence theorem and the compactness of the unit ball $I\!B \subset X$. In the case of infinite-dimensional spaces they ensure the existence of optimal solutions to perturbed problems and hence lead, by employing some calculus, to "almost" minimal points of the original function φ that "almost" satisfy necessary optimality conditions in terms of corresponding subgradients of φ. If X admits a smooth variational principle, such conditions can be obtained in terms of Fréchet subgradients by using the simple rule of Proposition 1.107(i). However, as we'll see below, smooth variational principles may be applied *only if* X has some *smoothness* properties, while the required subgradient conditions can be derived from the approximate extremal principle in *any Asplund space.* In this way we establish relationships between the extremal principle and appropriate versions of variational principles in X and obtain variational characterizations of Asplund spaces in terms of Fréchet subgradients and ε-subgradients of lower semicontinuous functions.

2.3.1 Ekeland Variational Principle

Let us start with the fundamental variational principle of Ekeland that turns out to be a *characterization* of complete metric spaces (X, d).

Theorem 2.26 (Ekeland's variational principle). *Let (X, d) be a metric space. The following hold:*

(i) *Assume that X is complete and that $\varphi\colon X \to \overline{\mathbb{R}}$ is a proper l.s.c. function bounded from below. Let $\varepsilon > 0$ and $x_0 \in X$ be given such that $\varphi(x_0) \le \inf_X \varphi + \varepsilon$. Then for any $\lambda > 0$ there is $\bar{x} \in X$ satisfying*

(a) $\varphi(\bar{x}) \le \varphi(x_0)$,
(b) $d(\bar{x}, x_0) \le \lambda$,
(c) $\varphi(x) + (\varepsilon/\lambda) d(x, \bar{x}) > \varphi(\bar{x})$ *for all* $x \ne \bar{x}$.

(ii) *Conversely, X is complete if for every Lipschitz continuous function $\varphi\colon X \to \mathbb{R}$ bounded from below and every $\varepsilon > 0$ there is $\bar{x} \in X$ satisfying*

(a′) $\varphi(\bar{x}) \le \inf_X \varphi + \varepsilon$ *and property* **(c)** *above with* $\lambda = 1$.

Proof. Let us justify (i) observing that it is sufficient to consider the case of $\varepsilon = \lambda = 1$. Indeed, the general case in (i) can be easily reduced to this special case applied to the function $\tilde{\varphi}(x) := \varepsilon^{-1}\varphi(x)$ on the metric space $(X, \tilde{d})$ with $\tilde{d}(x, y) := \lambda^{-1} d(x, y)$. Putting $\varepsilon = \lambda = 1$ in what follows, we first prove that there always exists $\bar{x} \in X$ satisfying (c) under the assumptions in (i). Define a mapping $T\colon X \rightrightarrows X$ by

$$T(x) := \big\{ u \in X \big|\ \varphi(u) + d(x, u) \le \varphi(x) \big\} .$$

Starting with an arbitrary point $x_1 \in \operatorname{dom} \varphi$, we inductively construct a sequence $\{x_k\}$, $k \in \mathbb{N}$, as follows. Assume that x_k is known and select the next iteration x_{k+1} so that

$$x_{k+1} \in T(x_k) \ \text{ and } \ \varphi(x_{k+1}) < \inf_{x \in T(x_k)} \varphi(x) + \frac{1}{k}, \quad k \in \mathbb{N} .$$

Observe that all $T(x_k)$ are nonempty and closed. Moreover, $T(x_{k+1}) \subset T(x_k)$ due to the triangle inequality. This gives

$$d(u, x_{k+1}) \le \varphi(x_{k+1}) - \varphi(u) \le \inf_{x \in T(x_k)} \varphi(x) + \frac{1}{k} - \varphi(u)$$

$$\le \inf_{x \in T(x_{k+1})} \varphi(x) + \frac{1}{k} - \varphi(u) \le \frac{1}{k}$$

for all $u \in T(x_{k+1})$, $k \in \mathbb{N}$. Therefore

$$\operatorname{diam} T(x_k) := \sup_{x,u \in T(x_k)} d(x,u) \to 0 \text{ as } k \to \infty .$$

Due to the completeness of X we conclude that the sets $T(x_k)$ shrink to a single point:

$$\bigcap_{k=1}^{\infty} T(x_k) = \{\bar{x}\} \text{ for some } \bar{x} \in X .$$

The latter implies (c) by the construction of $T(x_k)$.

Now given $x_0 \in X$ with $\varphi(x_0) \leq \inf_X \varphi + 1$, we consider the space

$$X_0 := \{x \in X |\ \varphi(x) \leq \varphi(x_0) - d(x, x_0)\}$$

with the metric induced by d. Obviously (X_0, d) is complete. Applying (c) on this space, we find $\bar{x} \in X_0$ such that

$$\varphi(x) > \varphi(\bar{x}) - d(x, \bar{x}) \text{ for all } x \in X_0 \setminus \{\bar{x}\} .$$

Let us show that the point $\bar{x}$ satisfies all the conditions (a)–(c) in (i) with $\varepsilon = \lambda = 1$. Indeed, (a) and (b) follow directly from $\bar{x} \in X_0$, i.e., from $\varphi(\bar{x}) + d(\bar{x}, x_0) \leq \varphi(x_0)$ and $\varphi(x_0) \leq \inf_X \varphi + 1$. It remains to prove (c) for $x \in X \setminus X_0$. Taking $x \notin X_0$, one has by the above construction that

$$\varphi(x) > \varphi(x_0) - d(x, x_0) \geq \varphi(\bar{x}) + d(\bar{x}, x_0) - d(x, x_0)$$

$$\geq \varphi(\bar{x}) - d(\bar{x}, x) ,$$

which ends the proof of (i).

To prove the converse statement (ii), let us consider an arbitrary Cauchy sequence $\{x_k\}$ in X and define the function

$$\varphi(x) := \lim_{k \to \infty} d(x_k, x) \text{ for all } x \in X ,$$

where the limit exists due to

$$|d(x_k, x) - d(x_n, x)| \leq d(x_k, x_n) \to 0 \text{ as } k, n \to \infty$$

by the triangle inequality. This also gives

$$|d(x_k, x) - d(x_k, u)| \leq d(x, u) \text{ for all } x, u \in X, \quad k \in \mathbb{N} ,$$

which implies the Lipschitz continuity of φ on X. Since $\{x_k\}$ is a Cauchy sequence, for every $\varepsilon > 0$ we find $k(\varepsilon) \in \mathbb{N}$ such that $d(x_k, x_n) \leq \varepsilon$ whenever $k, n \geq k(\varepsilon)$. Thus

$$\varphi(x_n) = \lim_{k \to \infty} d(x_k, x_n) \leq \varepsilon \text{ if } n \geq k(\varepsilon) ,$$

and hence φ is bounded from below with $\inf_X \varphi = 0$. To prove the completeness of X, we need to find $\bar{x} \in X$ such that $\varphi(\bar{x}) = 0$.

Choose $\varepsilon \in (0,1)$ and take $\bar{x} \in X$ satisfying (a′) and (c) with $\lambda = 1$. Then $\varphi(\bar{x}) \le \varepsilon$ due to (a′) and $\inf_X \varphi = 0$. Now pick an arbitrary small $\gamma > 0$ and put $x = x_n$ in (c) with $n \in \mathbb{N}$. From the definition of φ and the fact that $\{x_k\}$ is a Cauchy sequence, we get $d(x_n, \bar{x}) \le \varepsilon + \gamma$ when n is sufficiently large. This gives $\varphi(\bar{x}) \le \varepsilon^2$ by passing to the limit in (c) with $x = x_n$ as $n \to \infty$ and $\gamma \downarrow 0$. Repeating this procedure m times, one has $\varphi(\bar{x}) \le \varepsilon^m$ for any $m \in N$. Thus $\varphi(\bar{x}) = 0$, which justifies the completeness of X. △

Condition (c) in Theorem 2.26 means that the perturbed function $\varphi(x) + (\varepsilon/\lambda)d(x,\bar{x})$ achieves at $\bar{x}$ its *strict global minimum* over X. It has many important consequences. Let us present one, which is of special interest for subsequent discussions.

Corollary 2.27 (ε-stationary condition). *Let $\varphi\colon X \to \mathbb{R}$ be a proper l.s.c. function bounded from below on a Banach space X. Given $\varepsilon, \lambda > 0$ and $x_0 \in X$ with $\varphi(x_0) \le \inf_X \varphi + \varepsilon$, we assume that φ is Fréchet differentiable on a neighborhood U of x_0 containing $B_\lambda(x_0)$. Then there is $\bar{x} \in X$ with $\|\bar{x} - x_0\| \le \lambda$ such that $\varphi(\bar{x}) \le \varphi(x_0)$ and $\|\nabla\varphi(\bar{x})\| \le \varepsilon/\lambda$.*

Proof. Since $\bar{x}$ is a minimum point of the sum $\varphi(x) + \psi(x)$ with $\psi(x) := (\varepsilon/\lambda)\|x - \bar{x}\|$, we have $0 \in \widehat{\partial}(\varphi + \psi)(\bar{x})$ by Proposition 1.114. Now applying Proposition 1.107(i) and taking into account that $\widehat{\partial}\big(\|\cdot - \bar{x}\|\big)(\bar{x}) = \mathbb{B}^*$ for the norm function in Banach spaces, we get all the conclusions of the corollary from Theorem 2.26(i). △

Note that, since the initial ε-optimal point x_0 always exists, Corollary 2.27 ensures that every Fréchet differentiable and bounded from below function φ on a Banach space X admits an ε-optimal point $\bar{x}$ satisfying the *ε-stationary condition* $\|\nabla\varphi(\bar{x})\| \le \varepsilon$ for an arbitrary small $\varepsilon > 0$. As shown in the original paper of Ekeland [397], this result holds also for Gâteaux differentiable functions, which is a direct consequence of his variational principle.

What happens when φ is nonsmooth? This is considered next.

2.3.2 Subdifferential Variational Principles

In this subsection we first obtain a *lower* subdifferential counterpart of the ε-stationary result of Corollary 2.27 to the case of arbitrary l.s.c. functions bounded from below. We'll see that such an extension derived by using the extremal principle turns out to be a characterization of Asplund spaces. It actually plays a role of a (local) variational principle in Asplund spaces and has many important consequences, including density results for Fréchet subgradients as well as conventional forms of smooth variational principles under appropriate smoothness assumptions on Banach spaces. Finally, we derive an *upper* version of the subdifferential variational principle that holds in general Banach spaces and involves *every* upper Fréchet subgradient (provided that they exist) instead of *some* lower subgradient as in the previous lower subdifferential counterpart.

Theorem 2.28 (lower subdifferential variational principle). *Let X be a Banach space. The following are equivalent:*

(a) *The approximate extremal principle holds in X.*

(b) *For every proper l.s.c. function $\varphi\colon X \to \overline{I\!R}$ bounded from below, every $\varepsilon > 0$, $\lambda > 0$, and $x_0 \in X$ with $\varphi(x_0) < \inf_X \varphi + \varepsilon$ there are $\bar{x} \in X$ and $x^* \in \widehat{\partial}\varphi(\bar{x})$ such that $\|\bar{x} - x_0\| < \lambda$, $\varphi(\bar{x}) < \inf_X \varphi + \varepsilon$, and $\|x^*\| < \varepsilon/\lambda$.*

(c) *X is Asplund.*

Proof. Implication (c)$\Rightarrow$(a) is established in Theorem 2.20. Let us justify the other implications. We begin with (b)$\Rightarrow$(c) and then derive (a)$\Rightarrow$(b), which is the main part of the theorem.

(b)$\Rightarrow$(c). Take an arbitrary convex continuous function $\varphi\colon X \to I\!R$. Then $\widehat{\partial}\varphi(x)$ agrees with the subdifferential of convex analysis and is nonempty at every $x \in X$. To establish the Asplund property of X, it is sufficient to show that there is a dense subset $S \subset X$ such that $\widehat{\partial}(-\varphi)(x) \neq \emptyset$ for every $x \in S$. Indeed, in this case φ is Fréchet differentiable on S due to Proposition 1.87.

Fix $x_0 \in X$ and $\varepsilon > 0$. Since $\psi(x) := -\varphi(x)$ is continuous, there is a positive number $\nu < \varepsilon$ such that $\psi(x) > \psi(x_0) - \varepsilon$ for all $x \in x_0 + \nu I\!B$. Thus we have $\phi(x_0) < \inf_X \phi + 2\varepsilon$, where the function

$$\phi(x) := \psi(x) + \delta(x; x_0 + \nu I\!B), \quad x \in X ,$$

is obviously lower semicontinuous on X. Applying (b) to the latter function, we find $\bar{x} \in X$ with $\|\bar{x} - x_0\| < \nu$ such that $\widehat{\partial}\phi(\bar{x}) \neq \emptyset$. This clearly implies that $\widehat{\partial}\psi(\bar{x}) \neq \emptyset$, i.e., the set of points $x \in X$ with $\widehat{\partial}(-\varphi)(x) \neq \emptyset$ is dense in X. Hence X must be Asplund.

(a)$\Rightarrow$(b). First let us choose $0 < \tilde{\varepsilon} < \varepsilon$ with $\varphi(x_0) < \inf_X \varphi + (\varepsilon - \tilde{\varepsilon})$ and put $\tilde{\lambda} := (2\varepsilon)^{-1}(2\varepsilon - \tilde{\varepsilon})\lambda < \lambda$. Applying Theorem 2.26(i), we find $\tilde{x} \in X$ satisfying $\|\tilde{x} - x_0\| \le \tilde{\lambda}$, $\varphi(\tilde{x}) \le \inf_X \varphi + (\varepsilon - \tilde{\varepsilon})$, and

$$\varphi(\tilde{x}) < \varphi(x) + \tilde{\lambda}^{-1}(\varepsilon - \tilde{\varepsilon})\|x - \tilde{x}\| \quad \text{for all} \quad x \in X \setminus \{\tilde{x}\} . \tag{2.37}$$

Define two closed subsets of $X \times I\!R$ by

$$\Omega_1 := \operatorname{epi} \varphi, \quad \Omega_2 := \big\{(x, \alpha) \in X \times I\!R \big|\ \alpha \le \varphi(\tilde{x}) - \tilde{\lambda}^{-1}(\varepsilon - \tilde{\varepsilon})\|x - \tilde{x}\|\big\} .$$

It is easy to conclude from (2.37) that $(\tilde{x}, \varphi(\tilde{x}))$ is a *local extremal point* of the set system $\{\Omega_1, \Omega_2\}$; so we can use the extremal principle.

Consider the norm $\|(x, \alpha)\| := \|x\| + |\alpha|$ on $X \times I\!R$ with the corresponding dual norm $\|(x^*, \xi)\| = \max\{\|x^*\|, |\xi|\}$ on $X^* \times I\!R$. Applying the approximate extremal principle to the above system, for any $\hat{\varepsilon} > 0$ we find $(x_i, \alpha_i) \in \Omega_i$ and $(x_i^*, \xi_i) \in \widehat{N}((x_i, \alpha_i); \Omega_i)$, $i = 1, 2$, satisfying

$$\begin{cases} \|x_i - \tilde{x}\| + |\alpha_i - \varphi(\tilde{x})| < \hat{\varepsilon}\ , \\ \frac{1}{2} - \hat{\varepsilon} < \max\big\{\|x_i^*\|, |\xi_i|\big\} < \frac{1}{2} + \hat{\varepsilon}\ , \\ \max\big\{\|x_1^* + x_2^*\|, |\xi_1 + \xi_2|\big\} < \hat{\varepsilon}\ . \end{cases} \tag{2.38}$$

Observe that $(x_2^*, \xi_2) \neq 0$ when $\hat{\varepsilon}$ is sufficiently small. It follows from the structure of Ω_2 that $\alpha_2 = \varphi(\tilde{x}) - \tilde{\lambda}^{-1}(\varepsilon - \tilde{\varepsilon})\|x_2 - \tilde{x}\|$, which yields $\xi_2 > 0$ and thus implies

$$x_2^*/\xi_2 \in \widehat{\partial}\Big(\tilde{\lambda}^{-1}(\varepsilon - \tilde{\varepsilon})\| \cdot - \tilde{x}\|\Big)(x_2) \quad\text{and}\quad \|x_2^*\|/\xi_2 \le \tilde{\lambda}^{-1}(\varepsilon - \tilde{\varepsilon})\ .$$

Taking (2.38) into account, the latter gives the estimate

$$\xi_2 \ge \min\left\{\frac{(1 - 2\hat{\varepsilon})\tilde{\lambda}}{2(\varepsilon - \tilde{\varepsilon})},\ \frac{1}{2} - \hat{\varepsilon}\right\}\ , \tag{2.39}$$

which ensures by (2.38) that $\xi_1 < 0$ when $\hat{\varepsilon}$ is sufficiently small. This allows us to show that $\alpha_1 = \varphi(x_1)$, since the opposite implies $\xi_1 = 0$ due to $(x_1^*, \xi_1) \in \widehat{N}((x_1, \alpha_1); \operatorname{epi}\varphi)$ and the definition of $\widehat{N}$. Consequently $-x_1^*/\xi_1 \in \widehat{\partial}\varphi(x_1)$.

It follows from (2.39) that $\hat{\varepsilon}/\xi_2 \to 0$ as $\hat{\varepsilon} \downarrow 0$. Putting all the above together, we have

$$\frac{\|x_1^*\|}{|\xi_1|} < \frac{\|x_2^*\| + \hat{\varepsilon}}{\xi_2 - \hat{\varepsilon}} = \Big(\frac{\|x_2^*\|}{\xi_2} + \frac{\hat{\varepsilon}}{\xi_2}\Big)\Big/\Big(1 - \frac{\hat{\varepsilon}}{\xi_2}\Big) < \frac{\varepsilon}{\lambda}$$

when $\hat{\varepsilon}$ is sufficiently small. On the other hand, it follows from (2.38) and the choice of $\tilde{\lambda}$ that

$$\|x_1 - x_0\| < \tilde{\lambda} + \hat{\varepsilon} \quad\text{and}\quad \varphi(x_1) = \alpha_1 < \inf_X \varphi + \varepsilon - \tilde{\varepsilon} + \hat{\varepsilon}\ .$$

Finally, letting $\bar{x} := x_1$ and $x^* := -x_1^*/\xi_1$, we arrive at all the conclusions in (b) and finish the proof of the theorem. △

One can see that the major difference between the results of Theorem 2.26(i) and Theorem 2.28(b) is that, instead of the minimization condition (c) in the first theorem, we have the "almost stationary" lower subdifferential condition in the second one with the same type of estimates. The latter subdifferential condition carries essential information for local variational analysis and applications, which allows us to treat assertion (b) of Theorem 2.28 as a proper variational principle in Asplund spaces and call it the (lower) *subdifferential variational principle*. Moreover, we'll see in the next subsection that this result implies smooth variational principles in the conventional minimization/support form under additional smoothness assumptions on Asplund spaces that are *necessary* for the fulfillment of smooth variational principles but are *not needed* in Theorem 2.28.

The subdifferential variational principle of Theorem 2.28 easily implies the dense Fréchet subdifferentiability and related properties of l.s.c. functions that also turn out to be characterizations of Asplund spaces.

Corollary 2.29 (Fréchet subdiffentiability of l.s.c. functions). *Let $\mathcal{A}$ be a class of all proper l.s.c. functions $\varphi\colon X \to \overline{I\!R}$ on a Banach space X. The following properties are equivalent:*

(a) *X is Asplund.*

(b) *For every $\varphi \in \mathcal{A}$ the set of points $\{(x, \varphi(x)) \in X \times I\!R \,|\, \widehat{\partial}\varphi(x) \neq \emptyset\}$ is dense in the graph of φ.*

(c) *For every $\varphi \in \mathcal{A}$ there is $x \in \operatorname{dom}\varphi$ with $\widehat{\partial}\varphi(x) \neq \emptyset$.*

(d) *For every $\varphi \in \mathcal{A}$ and every $\varepsilon > 0$ there is $x \in \operatorname{dom}\varphi$ with $\widehat{\partial}_{g\varepsilon}\varphi(x) \neq \emptyset$.*

(e) *For every $\varphi \in \mathcal{A}$ and every $\varepsilon > 0$ there is $x \in \operatorname{dom}\varphi$ with $\widehat{\partial}_{a\varepsilon}\varphi(x) \neq \emptyset$.*

Proof. By Theorem 2.28 the smooth variational principle holds in any Asplund space. Take arbitrary $\varphi \in \mathcal{A}$, $x_0 \in \operatorname{dom}\varphi$, and $\varepsilon > 0$. Following the proof of (b)⇒(c) in the above theorem, we find $\bar{x} \in X$ such that $\|\bar{x} - x_0\| < \varepsilon$, $|\varphi(\bar{x}) - \varphi(x_0)| < 2\varepsilon$, and $\widehat{\partial}\varphi(\bar{x}) \neq \emptyset$. This justifies (a)⇒(b) in the corollary. Implications (b)⇒(c)⇒(d) are obvious, and (d)⇒(e) easily follows from Theorem 1.86. To justify the concluding implication (e)⇒(a), it is sufficient to observe that the concave continuous function $\varphi := -|\cdot|$ from Proposition 2.18 violates (e) for every $\varepsilon < \vartheta/2$. △

It follows from the proof of Corollary 2.29 that all the equivalences therein keep holding if the class $\mathcal{A}$ is replaced by more narrow classes of l.s.c. functions. In particular, one can consider only *concave continuous* functions $\varphi\colon X \to I\!R$, or proper l.s.c. functions $\varphi\colon X \to \overline{I\!R}$ *bounded from below*. The latter follows from the fact that implication (e)⇒(a) can be verified for the function $\varphi = 1/|\cdot|$, where $|\cdot|$ is taken from Proposition 2.18. Note also that the list of equivalences in Corollary 2.29 can be supplemented by counterparts of (b) and (c) in terms of *basic subgradients*. It immediately follows from the limiting representations (1.55) in Theorem 1.89.

Finally in this subsection, we establish another version of the subdifferential variational principle whose difference from that in Theorem 2.28 consists of using *upper* Fréchet subgradients instead of lower ones as above. The new version, which holds in arbitrary Banach spaces, involves *every* upper subgradient of the function in question, while it generally doesn't guarantee the existence of such subgradients. However, this result has certain essential advantages in comparison with its lower subdifferential counterpart being useful in some applications (particularly for deriving *suboptimality* conditions in constrained minimization) for important classes of functions that admit nonempty Fréchet upper subdifferential at reference points; see Chap. 5 for various results, discussions, and references.

Theorem 2.30 (upper subdifferential variational principle). *Let X be a Banach space, and let $\varphi\colon X \to I\!R$ be a l.s.c. function bounded from below. Then for every $\varepsilon > 0$, $\lambda > 0$, and $x_0 \in X$ with $\varphi(x_0) < \inf_X \varphi + \varepsilon$ there is $\bar{x} \in X$ with $\|\bar{x} - x_0\| < \lambda$ and $\varphi(\bar{x}) < \inf_X \varphi + \varepsilon$ such that*

$$\|x^*\| < \varepsilon/\lambda \quad \text{whenever} \quad x^* \in \widehat{\partial}^+\varphi(\bar{x})\,.$$

Proof. Given arbitrary numbers $\varepsilon > 0$ and $\lambda > 0$ and applying Ekeland's variational principle to the function φ and the point x_0 under consideration, we find $\bar{x} \in X$ satisfying $\|x_0 - \bar{x}\| < \lambda$, $\varphi(\bar{x}) < \inf_X \varphi(x) + \varepsilon$, and

$$\varphi(\bar{x}) \le \varphi(x) + \frac{\varepsilon}{\lambda}\|x - \bar{x}\| \quad \text{for all } x \in X .$$

Taking now any $x^* \in \widehat{\partial}^+\varphi(\bar{x}) = -\widehat{\partial}(-\varphi)(\bar{x})$ and using the smooth variational description of Fréchet subgradients from Theorem 1.88(i) held in arbitrary Banach spaces, we find a function $s\colon X \to I\!R$ Fréchet differentiable at $\bar{x}$ and such that

$$s(\bar{x}) = \varphi(\bar{x}), \quad \nabla s(\bar{x}) = x^* \text{ and } s(x) \ge \varphi(x) \text{ whenever } x \in X .$$

Combining this with the above global minimization property for the perturbation of φ at $\bar{x}$, conclude that the function $\phi(x) := s(x) + (\varepsilon/\lambda)\|x - \bar{x}\|$ attains its *global minimum* at $\bar{x}$. Then it follows from the generalized Fermat rule of Proposition 1.114, the sum rule of Proposition 1.107(i), and subdifferentiating the norm function at zero that

$$0 \in \widehat{\partial}\phi(\bar{x}) = \nabla s(\bar{x}) + \widehat{\partial}\Big(\frac{\varepsilon}{\lambda}\|\cdot - \bar{x}\|\Big)(\bar{x}) \subset x^* + \frac{\varepsilon}{\lambda} I\!B^* .$$

This gives $\|x^*\| < \varepsilon/\lambda$ and completes the proof of the theorem. △

2.3.3 Smooth Variational Principles

The crucial condition (c) in Theorem 2.26 can be interpreted as follows: for every proper l.s.c. function $\varphi\colon X \to \overline{I\!R}$ bounded from below (i.e., such that $\inf \varphi > -\infty$) there exist a point $\bar{x} \in \operatorname{dom}\varphi$ and a function $s\colon X \to I\!R$ satisfying

$$\varphi(\bar{x}) = s(\bar{x}) \text{ and } \varphi(x) \ge s(x) \text{ for all } x \in X . \tag{2.40}$$

The latter means that $s(\cdot)$ "supports φ from below." Such a function $s(\cdot)$ is usually called a *supporting function* belonging to some class $\mathcal{S}$. In these words condition(2.40), with $s(\cdot) \in \mathcal{S}$ for every l.s.c. function φ bounded from below, postulates that the *$\mathcal{S}$-variational principle* holds in X. Thus Ekeland's theorem ensures that, for the class of

$$\mathcal{S} := \big\{ -\varepsilon\|\cdot - \bar{x}\| + c \;\big|\; \varepsilon > 0,\ c \in I\!R \big\}$$

with arbitrary small positive numbers ε, the $\mathcal{S}$-variational principle holds in any Banach space. A notable limitation on applications of this result is that the supporting functions are not smooth.

If all $s(\cdot) \in \mathcal{S}$ are required to be smooth (in some sense), we speak about a *smooth variational principle* in a Banach space X. An $\mathcal{S}$-variational principle is called *concave* if $\mathcal{S}$ consists of concave functions. The afore-mentioned result of Borwein and Preiss establishes a concave smooth variational principle provided that X admits a smooth renorm with respect to some bornology. The corresponding result of Deville, Godefroy and Zizler ensures a smooth (but not concave) variational principle when the smooth renorming assumption is weaken to the existence of a smooth Lipschitzian bump function on X.

In the following theorem we consider variational principles for the three classes of $\mathcal{S}$-smooth functions on X: Fréchet differentiable ($\mathcal{S} = \mathcal{F}$), Lipschitzian and Fréchet differentiable ($\mathcal{S} = \mathcal{LF}$), and Lipschitzian and continuously differentiable ($\mathcal{S} = \mathcal{LC}^1$). Applying the lower subdifferential variational principle of Theorem 2.28 and then the variational descriptions of Fréchet subgradients established above, we derive $\mathcal{S}$-smooth variational principles in some *enhanced forms* under the corresponding smoothness assumptions on the Banach space in question, which inevitably imply the Asplund property of this space. Moreover, we show that the smoothness assumptions on X are *not only sufficient but also necessary* for the fulfillment of these smooth (resp. concave and smooth) variational principles in Asplund spaces.

Theorem 2.31 (smooth variational principles in Asplund spaces). *Let X be a Banach space, and let $\mathcal{A}$ stand for the class of all proper l.s.c. functions $\varphi\colon X \to \overline{I\!R}$ bounded from below. Given arbitrary $\varepsilon > 0$ and $\lambda > 0$, one has the following assertions:*

(i) *If X admits a Fréchet smooth renorm, then for every $\varphi \in \mathcal{A}$ and $x_0 \in X$ with $\varphi(x_0) < \inf_X \varphi + \varepsilon$ there exist $\bar{x} \in X$ and a concave Fréchet differentiable function $s\colon X \to I\!R$ such that*

$$\|\bar{x} - x_0\| < \lambda, \quad \varphi(\bar{x}) < \inf_X \varphi + \varepsilon , \tag{2.41}$$

$\|\nabla s(\bar{x})\| < \varepsilon/\lambda$, and

$$\varphi(\bar{x}) = s(\bar{x}), \quad \varphi(x) \ge s(x) + \|x - \bar{x}\|^2 \ \text{ for all } \ x \in X .$$

(ii) *Let X admit an $\mathcal{S}$-smooth bump function, where $\mathcal{S}$ stands for either $\mathcal{F}$, $\mathcal{LF}$, or $\mathcal{LC}^1$. Then for every $\varphi \in \mathcal{A}$ and $x_0 \in X$ with $\varphi(x_0) < \inf_X \varphi + \varepsilon$ there exist $\bar{x} \in X$ satisfying (2.41), an $\mathcal{S}$-smooth bump $b\colon X \to I\!R$, and a constant $c \in I\!R$ such that $\|\nabla b(\bar{x})\| < \varepsilon/\lambda$ and*

$$\varphi(\bar{x}) = b(\bar{x}) + c, \quad \varphi(x) \ge b(x) + c \ \text{ for all } \ x \in X .$$

Moreover, in this case we can find $\mathcal{S}$-smooth functions $s\colon X \to I\!R$ and $\theta\colon X \to [0, \infty)$ such that $\|\nabla s(\bar{x})\| < \varepsilon/\lambda$, $\theta(x) = 0$ only for $x = 0$, $\theta(x) \le \|x\|^2$ if $x \in I\!B$, and

$$\varphi(\bar{x}) = s(\bar{x}), \quad \varphi(x) \ge s(x) + \theta(x - \bar{x}) \ \text{ for all } \ x \in X .$$

(iii) *Conversely, the concave $\mathcal{F}$-smooth variational principle holds in X only if X admits a Fréchet smooth renorm, and the $\mathcal{S}$-smooth variational principle holds in X only if X admits an $\mathcal{S}$-smooth bump function for the corresponding classes $\mathcal{S}$ listed above.*

Proof. Assertions (i) and (ii) follow directly from the lower subdifferential variational principle in Theorem 2.28(b) due to the variational descriptions of Fréchet subgradients in Theorem 1.88. Let us justify the converse statements formulated in (iii).

First we prove that the concave $\mathcal{F}$-smooth variational principle in X implies that X admits a Fréchet smooth renorm. Applying (2.40) to the function $\varphi(x) := 1/\|x\|$, we find $0 \neq v \in X$ and a concave Fréchet differentiable function $s\colon X \to I\!R$ such that

$$s(x) \leq \varphi(x) = 1/\|x\| < 1/(2\|v\|) \ \text{ if } \ \|x\| > 2\|v\| ,$$

with $s(v) = 1/\|v\|$. Putting

$$p(x) := -s(x+v) + 1/\|v\|, \ x \in X ,$$

we conclude that p is convex and Fréchet differentiable on X due to the corresponding properties of s. Thus p is C^1-smooth on X. Moreover, one has $p(0) = 0$ and

$$p(x) > -1/(2\|v\|) + 1/\|v\| = 1/(2\|v\|) \ \text{ if } \ \|x\| > 3\|v\| ,$$

since $\|x+v\| > 2\|v\|$. Now let us consider the *Minkowski gauge functional*

$$g(x) := \inf\big\{\lambda > 0 \big|\ x \in \lambda\Omega\big\}, \ x \in X ,$$

of the set $\Omega := \big\{x \in X \big|\ p(x) \leq 1/(2\|v\|)\big\}$. It is easy to see that Ω is convex, closed, and bounded with $0 \in \text{int}\,\Omega$. In this case the Minkowski gauge is a continuous sublinear functional with $g(x) > 0$ for all $x \neq 0$ and $\Omega = \{x \in X|\ g(x) \leq 1\}$. This ensures the existence of $M > 0$ such that

$$\|x\|/(3\|v\|) \leq g(x) \leq M\|x\| \ \text{ for all } \ x \in X .$$

Now considering the function

$$n(x) := g(x) + g(-x), \ x \in X ,$$

we conclude that it defines a *norm* on X equivalent to the original one $\|\cdot\|$. To complete the proof of the first statement in (iii), it remains to justify that g is Fréchet differentiable on $X \setminus \{0\}$. The crucial step for this is to show the Gâteaux differentiability of g at every nonzero point of X. Since g is convex, the latter is equivalent to the fact that its subdifferential $\partial g(x)$ is a singleton for each $x \in X \setminus \{0\}$.

To proceed, we fix an arbitrary $x \in X$ with $g(x) = 1$ and pick $x^* \in \partial g(x)$. It can be easily derived from the definitions that

$$p(x) = 1/(2\|v\|) \ \text{ and } \ \langle x^*, x\rangle = g(x) .$$

Now taking any $t > 0$ and $h \in X$ with $\langle x^*, h\rangle = 0$, one has

$$g(x+th) \geq g(x) + \langle x^*, th \rangle = 1, \quad g(\alpha(x+th)) = \alpha g(x+th) > 1 \text{ if } \alpha > 1,$$

and hence $\alpha(x+th) \notin \Omega$. Thus $p(\alpha(x+th)) > 1/(2\|v\|)$ for all $\alpha > 1$ and all $t > 0$. Passing to the limit as $\alpha \downarrow 1$, we get $p(x+th) \geq 1/(2\|v\|)$ $(= p(x))$ for all $t > 0$. Since p is Gâteaux differentiable at x with the derivative $p'(x)$, this implies that

$$\langle p'(x), h \rangle = \lim_{t \downarrow 0} \frac{p(x+th) - p(x)}{t} \geq 0 \text{ for all } h \in X \text{ with } \langle x^*, h \rangle = 0.$$

The latter gives $\langle p'(x), h \rangle = 0$ for all such h, and so $x^* = \lambda p'(x)$ for some $\lambda \in I\!R$. Therefore

$$1 = g(x) = \langle x^*, x \rangle = \lambda \langle p'(x), x \rangle,$$

which uniquely determines $x^* \in \partial g(x)$ as $x^* = p'(x)/\langle p'(x), x \rangle$. This means that g is Gâteaux differentiable at x and $g'(x) = x^*$ when $g(x) = 1$. Considering an arbitrary nonzero $x \in X$ and taking into account that g is positively homogeneous and $g(x) \neq 0$, we get the following formula for the Gâteaux derivative of g at x:

$$g'(x) = \left\langle p'\Big(\frac{x}{g(x)}\Big), \frac{x}{g(x)} \right\rangle^{-1} p'\Big(\frac{x}{g(x)}\Big).$$

Since p is $\mathcal{C}^1$-smooth, this formula implies that g' is norm-to-norm continuous. Thus g is Fréchet differentiable at every nonzero point of X, which justifies the first part of (iii).

Next we prove the second part of (iii) simultaneously for each listed $\mathcal{S}$. Again pick the function $\varphi = 1/\|\cdot\|$ and apply to it the supporting condition (2.40) with some $v = \bar{x}$ and $\mathcal{S}$-smooth function $s : X \to I\!R$. Then consider an arbitrary $\mathcal{C}^2$-smooth function $\tau : I\!R \to [0,1]$ satisfying

$$\tau(t) = 1 \text{ if } t \geq 1/\|v\| \text{ and } \tau(t) = 0 \text{ if } t \leq 1/(2\|v\|).$$

One can easily check that $b := \tau \circ s$ is an $\mathcal{S}$-smooth bump function on X, which justifies (iii). △

Note that the supporting conditions in assertions (i) and (ii) of Theorem 2.31 carry more information in comparison with the basic supporting condition (2.40) used in the proof of assertion (iii). Observe also that the proof of Theorem 2.31(iii) holds true when the Fréchet smoothness is replaced by the Gâteaux smoothness or, generally, by any *β-smoothness* with respect to an arbitrary bornology β on X; cf. Remark 2.11. This implies that any smooth (resp. concave smooth) variational principle with the supporting condition (2.40) necessarily requires the corresponding smooth renorming/bump function assumption on the underlying Banach space X.

2.4 Representations and Characterizations in Asplund Spaces

In this section we apply the above extremal and variational principles to obtain efficient representations of the generalized differential constructions of Chap. 1 in the case of Asplund spaces. Most of these representations turn out to be characterizations of Asplund spaces. We begin with a subgradient description of the approximate extremal principle, which plays an essential role in the subsequent material. Then we derive characterizations of Asplund spaces in terms of special subdifferential sum rules involving Lipschitzian functions. This leads to simplified representations of basic subgradients, normals, and coderivatives in Asplund spaces similar to those in finite dimensions. In the last subsection we derive convenient representations of singular subgradients of extended-real-valued l.s.c. functions and related results for horizontal normals to graphs of continuous functions on Asplund spaces.

2.4.1 Subgradients, Normals, and Coderivatives in Asplund Spaces

Let $\mathcal{SL}(\bar{x})$ denote the class of pairs (φ_1, φ_2) with proper functions $\varphi_i\colon X \to \overline{I\!R}$ such that φ_1 is Lipschitz continuous around $\bar{x} \in \operatorname{dom}\varphi_1 \cap \operatorname{dom}\varphi_2$ and φ_2 is l.s.c. around this point. For brevity we say that the sum $\varphi_1 + \varphi_2$ is *semi-Lipschitzian* at $\bar{x}$ if $(\varphi_1, \varphi_2) \in \mathcal{SL}(\bar{x})$. The next result provides an equivalent description of the approximate extremal principle in terms of a "fuzzy" subgradient condition for minimum points of semi-Lipschitzian sums.

Lemma 2.32 (subgradient description of the extremal principle). *Given a Banach space X, one has the following:*

(i) *Let the approximate extremal principle hold for every extremal system of two closed sets in $X \times I\!R$. Assume that $(\varphi_1, \varphi_2) \in \mathcal{SL}(\bar{x})$ with $\varphi_i\colon X \to \overline{I\!R}$ and that the sum $\varphi_1 + \varphi_2$ attains a local minimum at $\bar{x}$. Then for any $\eta > 0$ there are $x_i \in \bar{x} + \eta I\!B$ with $|\varphi_i(x_i) - \varphi_i(\bar{x})| \le \eta$, $i = 1, 2$, such that*

$$0 \in \widehat{\partial}\varphi_1(x_1) + \widehat{\partial}\varphi_2(x_2) + \eta I\!B^* . \tag{2.42}$$

(ii) *Conversely, let for any $(\varphi_1, \varphi_2) \in \mathcal{SL}(\bar{x})$ with $\varphi_i\colon X^2 \to \overline{I\!R}$ and for any $\eta > 0$ there exist $x_i \in \bar{x} + \eta I\!B$ with $|\varphi_i(x_i) - \varphi_i(\bar{x})| \le \eta$, $i = 1, 2$, such that (2.42) is fulfilled provided that $\varphi_1 + \varphi_2$ attains a local minimum at $\bar{x}$. Then the approximate extremal principle holds for every extremal system of two closed sets in X.*

Proof. To justify (i), we consider $(\varphi_1, \varphi_2) \in \mathcal{SL}(\bar{x})$ and assume without loss of generality that $\bar{x} = 0 \in X$ is a local minimizer for $\varphi_1 + \varphi_2$ with $\varphi_1(0) = \varphi_2(0) = 0$, that φ_1 is Lipschitz continuous on $\eta I\!B$ with modulus $\ell > 0$, and that φ_2 is l.s.c. on $\eta I\!B$ for the fixed $\eta > 0$. Consider the sets

$$\Omega_1 := \operatorname{epi}\varphi_1 \quad\text{and}\quad \Omega_2 := \big\{(x, \alpha) \in X \times I\!R \big|\ \varphi_2(x) \le -\alpha\big\} ,$$

which are obviously closed around $(0,0) \in X \times I\!R$. It is easy to check that $(0,0)$ is a local extremal point of the sets $\{\Omega_1, \Omega_2\}$, since $\bar{x} = 0$ is a local minimizer for $\varphi_1 + \varphi_2$. Applying the approximate extremal principle to the system $\{\Omega_1, \Omega_2, (0,0)\}$, for any $\varepsilon > 0$ we find $(x_i, \alpha_i) \in \Omega_i$ and $(x_i^*, \lambda_i) \in X^* \times I\!R$, $i = 1, 2$, such that

$$(x_1^*, -\lambda_1) \in \widehat{N}((x_1, \alpha_1); \Omega_1), \quad (-x_2^*, \lambda_2) \in \widehat{N}((x_2, \alpha_2); \Omega_2)\,, \tag{2.43}$$

$$\|(x_i, \alpha_i)\| \le \varepsilon, \quad \tfrac{1}{2} - \varepsilon \le \|(x_i^*, \lambda_i)\| \le \tfrac{1}{2} + \varepsilon, \quad i = 1, 2\,, \tag{2.44}$$

$$\|(x_1^*, -\lambda_1) + (-x_2^*, \lambda_2)\| \le \varepsilon\,. \tag{2.45}$$

It follows from (2.43) that $\lambda_i \ge 0$ for $i = 1, 2$. Our goal is to show that choosing ε to be sufficiently small, we get $\lambda_i > 0$ and can equivalently transformed (2.43) to subgradient relations with the required estimates. For these purposes it is convenient to define the corresponding norms on $X \times I\!R$ and $X^* \times I\!R$ by

$$\|(x, \alpha)\| := \max\{\|x\|, |\alpha|\} \;\text{ and }\; \|(x^*, \lambda)\| := \|x^*\| + |\lambda|\,.$$

Then choose ε in (2.43)–(2.45) satisfying

$$0 < \varepsilon < \min\Big\{\frac{1}{4(2+\ell)}, \frac{\eta}{4(1+\ell)^2}\Big\}\,.$$

Since φ_1 is Lipschitz continuous on $\eta I\!B$, we get from $(x_1^*, -\lambda_1) \in \widehat{N}((x_1, \alpha_1); \Omega_1)$ with $\max\{\|x_1\|, |\alpha_1|\} \le \varepsilon < \eta$ that $\|x_1^*\| \le \ell\lambda_1$; see Proposition 1.85(ii). It gives by (2.44) and (2.45) that

$$\lambda_1 \ge \frac{1}{2(1+\ell)} - \frac{\varepsilon}{1+\ell} > 0 \;\text{ and }\; \lambda_2 \ge \frac{1}{2(1+\ell)} - \varepsilon\Big(\frac{2+\ell}{1+\ell}\Big) > \frac{1}{4(1+\ell)}$$

by the choice of ε. This implies by (2.43) that $\alpha_1 = \varphi_1(x_1)$, $\alpha_2 = -\varphi_2(x_2)$, and hence

$$\widetilde{x}_1^* := x_1^*/\lambda_1 \in \widehat{\partial}\varphi_1(x_1), \quad \widetilde{x}_2^* := -x_2^*/\lambda_2 \in \widehat{\partial}\varphi_2(x_2)\,.$$

By (2.44) we have

$$\|x_i\| \le \varepsilon < \eta \;\text{ and }\; |\varphi_i(x_i)| = |\alpha_i| \le \varepsilon < \eta, \quad i = 1, 2\,.$$

To justify (2.42), it remains to show that $\|\widetilde{x}_1^* + \widetilde{x}_2^*\| \le \eta$. This follows from

$$\Big\|\frac{x_1^*}{\lambda_1} - \frac{x_2^*}{\lambda_2}\Big\| = \Big\|\frac{x_1^*(\lambda_2 - \lambda_1)}{\lambda_1\lambda_2} + \frac{x_1^* - x_2^*}{\lambda_2}\Big\| \le \frac{\|x_1^*\|}{\lambda_1}\Big(\frac{|\lambda_2 - \lambda_1|}{\lambda_2}\Big) + \frac{\|x_1^* - x_2^*\|}{\lambda_2}$$

$$\le \ell\frac{\varepsilon}{\lambda_2} + \frac{\varepsilon}{\lambda_2} = \frac{\varepsilon}{\lambda_2}\Big(1 + \ell\Big) < 4\varepsilon(1+\ell)^2 < \eta$$

due to the choice of ε and the estimates above.

Next let us prove the converse assertion (ii). Take an extremal system $\{\Omega_1, \Omega_2, \bar{x}\}$ in X and find a neighborhood U of $\bar{x}$ such that, given an arbitrary $\varepsilon > 0$, there is $a \in X$ with $\|a\| < \varepsilon^2/2$ and $(\Omega_1 + a) \cap \Omega_2 \cap U = \emptyset$. Put $U = X$ for simplicity and define the function $\varphi\colon X \times X \to I\!R$ by

$$\varphi(u, v) := \tfrac{1}{2}\|u - v + a\|, \quad (u, v) \in X^2 . \tag{2.46}$$

It follows from the local extremality of $\bar{x}$ that $\varphi(\bar{x}, \bar{x}) < (\varepsilon/2)^2$ and that $\varphi(u, v) > 0$ for all $u \in \Omega_1$ and $v \in \Omega_2$.

Now we apply Ekeland's variational principle in Theorem 2.26(i) to the function φ on the complete metric space $\Omega_1 \times \Omega_2$ whose metric is induced by the norm $\|(u, v)\| := \|u\| + \|v\|$ on X^2. This gives points $(\bar{u}, \bar{v}) \in \Omega_1 \times \Omega_2$ such that $\|\bar{u} - \bar{x}\| \le \varepsilon/2$, $\|\bar{v} - \bar{x}\| \le \varepsilon/2$, and

$$\varphi(\bar{u}, \bar{v}) \le \varphi(u, v) + \frac{\varepsilon}{2}\Big(\|u - \bar{u}\| + \|v - \bar{v}\|\Big) \text{ for all } (u, v) \in \Omega_1 \times \Omega_2 .$$

The latter means that the sum of the functions

$$\varphi_1(u, v) := \varphi(u, v) + \frac{\varepsilon}{2}\Big(\|u - \bar{u}\| + \|v - \bar{v}\|\Big) \text{ and } \varphi_2(u, v) := \delta((u, v); \Omega_1 \times \Omega_2)$$

attains at $(\bar{u}, \bar{v})$ its minimum over X^2. Observe that φ_1 is Lipschitz continuous and *convex* and that φ_2 is proper and l.s.c. on X^2. By the assumptions in (ii) we find $(y_1, y_2) \in X^2$ and $(x_1, x_2) \in \Omega_1 \times \Omega_2$ such that $\|x_1 - \bar{u}\| \le \varepsilon/2$, $\|x_2 - \bar{v}\| \le \varepsilon/2$, $\varphi(y_1, y_2) > 0$, and

$$0 \in \widehat{\partial}\varphi_1(y_1, y_2) + \widehat{\partial}\varphi_2(x_1, x_2) + \frac{\varepsilon}{2}\Big(I\!B^* \times I\!B^*\Big) .$$

Note that $\widehat{\partial}\varphi_2(x_1, x_2) = \widehat{N}((x_1, x_2); \Omega_1 \times \Omega_2) = \widehat{N}(x_1; \Omega_1) \times \widehat{N}(x_2; \Omega_2)$ due to Proposition 1.2. Now using the well-known subdifferential formula for the norm function (2.46) at nonzero points, we conclude that

$$\widehat{\partial}\varphi_1(y_1, y_2) = \frac{1}{2}\Big(x^*, -x^*\Big) + \frac{\varepsilon}{2}\Big(I\!B^* \times I\!B^*\Big)$$

with some $x^* \in X^*$ of the unit norm. Finally, putting $x_1^* := -x^*/2$ and $x_2^* := x^*/2$, we get $x_i^* \in \widehat{N}(x_i; \Omega_i) + \varepsilon I\!B^*$ with $x_1^* + x_2^* = 0$ and $\|x_1^*\| + \|x_2^*\| = 1$, which justifies (ii). △

Next we obtain two subdifferential sum rules in the semi-Lipschitzian case: the *fuzzy* rule for Fréchet subgradients and ε-subgradients and the *exact* one for basic subgradients. Each of these rules applied to all semi-Lipschitzian sums is proved to be a characterization of Asplund spaces.

Theorem 2.33 (semi-Lipschitzian sum rules). *Let X be a Banach space with $\bar{x} \in X$. The following properties are equivalent:*

(a) *X is Asplund.*

(b) *For any* $(\varphi_1, \varphi_2) \in \mathcal{SL}(\bar{x})$, *for any* $\varepsilon \geq 0$, *and for any* $\gamma > 0$ *one has*

$$\widehat{\partial}_\varepsilon(\varphi_1 + \varphi_2)(\bar{x}) \subset \bigcup \Big\{ \widehat{\partial}\varphi_1(x_1) + \widehat{\partial}\varphi_2(x_2) \Big|\ x_i \in \bar{x} + \gamma I\!B\ ,$$

$$|\varphi_i(x_i) - \varphi_i(\bar{x})| \leq \gamma,\ i = 1, 2 \Big\} + (\varepsilon + \gamma) I\!B^*\ .$$

(c) *For any* $(\varphi_1, \varphi_2) \in \mathcal{SL}(\bar{x})$ *one has*

$$\partial(\varphi_1 + \varphi_2)(\bar{x}) \subset \partial\varphi_1(\bar{x}) + \partial\varphi_2(\bar{x})\ .$$

Proof. First we prove (a)$\Rightarrow$(b). Observe that if X is Asplund, then $X \times I\!R$ is Asplund as well. By Theorem 2.20 the approximate extremal principle holds in $X \times I\!R$. Hence we have property (2.42) in Lemma 2.32 for any $(\varphi_1, \varphi_2) \in \mathcal{SL}(\bar{x})$. Let us derive (b) from this property and from the variational description of analytic ε-subgradients in Proposition 1.84. Fix (ε, γ) in (b) and find η satisfying the relations

$$0 < \eta < \min\{\gamma/4,\ \bar{\eta}\}, \quad \text{where } \bar{\eta}^2 + (2 + \varepsilon)\bar{\eta} - \gamma = 0\ .$$

Then pick an arbitrary $x^* \in \widehat{\partial}_\varepsilon(\varphi_1 + \varphi_2)(\bar{x})$ and conclude by Proposition 1.84(ii) that the sum

$$\Big(\varphi_1(x) - \langle x^*, x - \bar{x}\rangle + (\varepsilon + \eta)\|x - \bar{x}\|\Big) + \varphi_2(x)$$

attains a local minimum at $\bar{x}$. Applying (2.42) with the chosen η to the above sum and then using the elementary sum rule in Proposition 1.107(i), we find $x_i \in \bar{x} + \eta I\!B$ and $x_i^* \in X^*$, $i = 1, 2$, such that

$$\Big|\varphi_1(x_1) + (\varepsilon + \eta)\|x_1 - \bar{x}\| - \varphi_1(\bar{x})\Big| \leq \eta, \quad |\varphi_2(x_2) - \varphi_2(\bar{x})| \leq \eta\ ,$$

$$x_1^* \in \widehat{\partial}\Big(\varphi_1 + (\varepsilon + \eta)\|\cdot - \bar{x}\|\Big)(x_1), \quad x_2^* \in \widehat{\partial}\varphi_2(x_2)\ ,$$

and $x^* - x_1^* - x_2^* \in \eta I\!B^*$. This implies that

$$|\varphi_1(x_1) - \varphi_1(\bar{x})| \leq \eta(\varepsilon + \eta + 1)\ .$$

Now employing Proposition 1.84(ii) in the case of the Fréchet subgradient x_1^*, we conclude that the sum $\varphi_1 + \psi$ with

$$\psi(x) := (\varepsilon + \eta)\|x - \bar{x}\| - \langle x_1^*, x - x_1\rangle + \eta\|x - x_1\|$$

attains a local minimum at x_1. Observe that ψ is convex and continuous on X with $\partial\psi(x) \subset -x_1^* + (\varepsilon + 2\eta)I\!B^*$ for any $x \in X$. Applying (2.42) to $\varphi_1 + \psi$, we find $\widetilde{x}_1 \in x_1 + \eta I\!B$ such that

$$|\varphi_1(\widetilde{x}_1) - \varphi_1(x_1)| \leq \eta \quad \text{and} \quad x_1^* \in \widehat{\partial}\varphi_1(\widetilde{x}_1) + (\varepsilon + 3\eta)I\!B^*\ .$$

We finally have

$$x^* \in \widehat{\partial}\varphi_1(\widetilde{x}_1) + \widehat{\partial}\varphi_2(x_2) + (\varepsilon + 4\eta)I\!B^*$$

with $\|\widetilde{x}_1 - \bar{x}\| \le 2\eta$ and $|\varphi_1(\widetilde{x}_1) - \varphi_1(\bar{x})| \le \eta(\varepsilon + \eta + 2)$. This gives (b) by the choice of η.

Next let us prove that (b) and the Asplund property of X implies (c). Take an arbitrary $x^* \in \partial(\varphi_1 + \varphi_2)(\bar{x})$ and by representation (1.55) in Theorem 1.89 find sequences $\varepsilon_k \downarrow 0$, $x_k \to \bar{x}$ with $\varphi_1(x_k) + \varphi_2(x_k) \to \varphi_1(\bar{x}) + \varphi_2(\bar{x})$, and $x_k^* \xrightarrow{w^*} x^*$ such that $x_k^* \in \widehat{\partial}_{\varepsilon_k}(\varphi_1 + \varphi_2)(x_k)$ as $k \to \infty$. Then employing (b) with $\gamma_k = \varepsilon_k$, we get sequences $x_{ik} \to \bar{x}$ with $\varphi_i(x_{ik}) \to \varphi_i(\bar{x})$ and $x_{ik}^* \in \widehat{\partial}\varphi_i(x_{ik})$, $i = 1, 2$, such that

$$\|x_k^* - x_{1k}^* - x_{2k}^*\| \le 2\varepsilon_k \quad \text{for all } k \in I\!N\,. \tag{2.47}$$

Since $x_k^* \to x^*$, this sequence is bounded in X^* due to the uniform boundedness principle. The sequence $\{x_{1k}^*\}$ is also bounded by modulus ℓ due to the Lipschitz continuity of φ_1 around $\bar{x}$; see Proposition 1.85(ii). Hence $\{x_{2k}^*\}$ is bounded as well. Using the weak* sequential compactness of bounded sets in duals to Asplund spaces, we find $x_i^* \in X^*$ such that $x_{ik}^* \xrightarrow{w^*} x_i^*$, $i = 1, 2$, along a subsequence of $k \to \infty$. Again employing Theorem 1.89, we get $x_i^* \in \partial\varphi_i(\bar{x})$ for $i = 1, 2$. Moreover, (2.47) implies that $x^* = x_1^* + x_2^*$, which gives (c).

It remains to show that each of the properties (b) and (c) implies that X is Asplund. Indeed, according to Proposition 2.18 and Example 2.19 for any non-Asplund space X there is an equivalent norm $|\cdot|$ on X such that

$$\widehat{\partial}\varphi(x) = \partial\varphi(x) = \emptyset \quad \text{whenever } x \in X$$

for $\varphi := -|\cdot|$. Now we can see that both properties (b) and (c) are violated for the sum $\varphi_1 + \varphi_2$ with $\varphi_1 := |\cdot|$ and $\varphi_2 := -|\cdot|$. △

The next theorem contains subdifferential characterizations of Asplund spaces via a simplified limiting representation of basic subgradients (like in finite-dimensions) and a related expansion formula for the so-called *limiting* ε*-subdifferential* of $\varphi\colon X \to \overline{I\!R}$ at $\bar{x} \in X$ with $|\varphi(\bar{x})| < \infty$ defined by

$$\partial_\varepsilon\varphi(\bar{x}) := \operatorname*{Lim\,sup}_{x \xrightarrow{\varphi} \bar{x}} \widehat{\partial}_\varepsilon\varphi(x)\,. \tag{2.48}$$

Theorem 2.34 (subdifferential representations in Asplund spaces). *Let X be a Banach space, $\bar{x} \in X$, and $\mathcal{A}(\bar{x})$ be the class of proper functions $\varphi\colon X \to \overline{I\!R}$ l.s.c. around $\bar{x} \in \operatorname{dom}\varphi$. The following properties are equivalent:*

(a) *X is Asplund.*

(b) *For every $\bar{x} \in X$ and every $\varphi \in \mathcal{A}(\bar{x})$ one has*

$$\partial\varphi(\bar{x}) = \operatorname*{Lim\,sup}_{x \xrightarrow{\varphi} \bar{x}} \widehat{\partial}\varphi(x)\,.$$

(c) *For every* $\bar{x} \in X$, *every* $\varphi \in \mathcal{A}(\bar{x})$, *and every* $\varepsilon > 0$ *one has*

$$\partial_\varepsilon \varphi(\bar{x}) = \partial\varphi(\bar{x}) + \varepsilon I\!B^* .$$

Proof. To justify (a)⇒(b), we use the fuzzy sum rule in Theorem 2.33(b) with $\varphi_1 = 0$ and $\varphi_2 = \varphi$. This gives

$$\widehat{\partial}_\varepsilon \varphi(\bar{x}) \subset \bigcup\Big\{\widehat{\partial}\varphi(x)\Big|\; x \in \bar{x} + \gamma I\!B,\; |\varphi(x) - \varphi(\bar{x})| \le \gamma\Big\} + (\varepsilon + \gamma)I\!B^* \quad (2.49)$$

for any $\varepsilon \ge 0$ and $\gamma > 0$. Passing there to the limit as $\varepsilon = \gamma \downarrow 0$, we arrive at the subdifferential representation (b).

To prove (a)⇒(c), observe that the inclusion "⊃" in (c) is trivial, and we need to show that the opposite inclusion holds in Asplund spaces. Pick $x^* \in \partial_\varepsilon\varphi(\bar{x})$ and find by (2.48) sequences $x_k \xrightarrow{\varphi} \bar{x}$ and $x_k^* \xrightarrow{w^*} x^*$ such that $x_k^* \in \widehat{\partial}_\varepsilon\varphi(x_k)$ for all $k \in I\!N$. Taking any $\gamma_k \downarrow 0$ and using (2.49) with $\gamma = \gamma_k$, one gets $u_k \in x_k + \gamma_k I\!B$ satisfying $|\varphi(u_k) - \varphi(x_k)| \le \gamma_k$ and

$$x_k^* \in \widehat{\partial}\varphi(u_k) + (\varepsilon + \gamma_k)I\!B^*, \quad k \in I\!N .$$

This allows us to find $u_k^* \in \widehat{\partial}\varphi(u_k)$ and $v_k^* \in (\varepsilon + \gamma_k)I\!B^*$ such that $x_k^* = u_k^* + v_k^*$ for all $k \in I\!N$. By the weak* sequential compactness of $I\!B^*$ and the weak* lower semicontinuity of $\|\cdot\|$ on X^* we have $v^* \in X^*$ satisfying

$$v_k^* \xrightarrow{w^*} v^* \;\text{ as }\; k \to \infty \;\text{ with }\; \|v^*\| \le \liminf_{k\to\infty} \|v_k^*\| \le \varepsilon$$

along a subsequence of $\{k\}$. This implies the existence of $u^* \in \partial\varphi(\bar{x})$ such that $u_k^* \xrightarrow{w^*} u^*$ and hence $x^* = u^* + v^* \in \partial\varphi(\bar{x}) + \varepsilon I\!B^*$, which gives (c).

To justify the opposite inclusion (c)⇒(a), one has to show that for any non-Asplund space X there are $\bar{x} \in X$, $\varphi \in \mathcal{A}(\bar{x})$, and $\bar{\varepsilon} > 0$ such that the representation in (c) doesn't hold. Taking the equivalent norm $|\cdot|$ on X and the number $\vartheta > 0$ in Proposition 2.18, let us show that this representation is violated for $\varphi = -|\cdot|$, $\bar{x} = 0$, and $\bar{\varepsilon} = 1$. Indeed, it follows from Proposition 2.18 and Definition 1.83(ii) that

$$\widehat{\partial}_\varepsilon\varphi(x) = \emptyset \;\text{ for all }\; x \in X \;\text{ if }\; 0 \le \varepsilon < \min\big\{1, \vartheta/2\big\} ,$$

which gives $\partial\varphi(0) = \emptyset$. On the other hand, one can easily check that $\widehat{\partial}_1\varphi(0) \supset \{0\} \ne \emptyset$. Hence $\partial_1\varphi(0) \ne \emptyset$ by (2.48), and thus (c) doesn't hold. Note that our proof actually shows more: if X is not Asplund, then for *any given* $\varepsilon > 0$ there is a function $\varphi \in \mathcal{A}(0)$ such that the representation in (c) is violated. Indeed, consider the function $\varphi := -\varepsilon|\cdot|$ in the above arguments.

To finish the proof of the theorem, it remains to justify (b)⇒(a), i.e., to show that the representation in (b) is violated for some $\bar{x} \in X$ and some $\varphi \in \mathcal{A}(\bar{x})$ in any non-Asplund space. Assuming that X is not Asplund, we take the equivalent norm $|\cdot|$ in Proposition 2.18, $\bar{x} = 0$, and let

$$\varphi(x) := -|x|^2 + \min\{\langle u^*, x\rangle, \langle v^*, x\rangle\}, \quad x \in X, \tag{2.50}$$

where $u^*, v^* \in X^*$ with $u^* \neq v^*$. Consider a sequence $\{x_k\} \subset X$ such that $x_k \to 0$ and $\langle u^*, x_k\rangle < \langle v^*, x_k\rangle$ for all $k \in \mathbb{N}$. Denote $\psi(x) := -|x|^2$ and observe that

$$\varphi(x) = \psi(x) + \langle u^*, x\rangle \text{ whenever } x \in U_k \text{ and } k \in \mathbb{N}$$

for some neighborhood U_k of x_k. Since $|\cdot| \le \|\cdot\|$, we have

$$|\psi(u) - \psi(v)| = \big(|u| + |v|\big) \cdot \big|(|u| - |v|)\big| \le 3|x_k| \cdot |u - v|$$

for all $u, v \in x_k + \big(\|x_k\|/2\big)\mathbb{B}$. This means that the function ψ is Lipschitzian around x_k with modulus $3|x_k|$ for any fixed $k \in \mathbb{N}$. It easily follows from the definitions that

$$u^* \in \widehat{\partial}_{3|x_k|}\varphi(x_k) \text{ for all } k \in \mathbb{N},$$

where the analytic ε-subdifferential is taken with respect to the norm $|\cdot|$. Passing to the limit as $k \to \infty$ and taking into account that representation (1.55) is invariant with respect to equivalent norms on X, we get $u^* \in \partial\varphi(0)$.

Let us show that $\widehat{\partial}\varphi(x) = \emptyset$ for all x near the origin, which violates (b) in the case of φ in (2.50) and $\bar{x} = 0$. First check that $\widehat{\partial}\varphi(0) = \emptyset$. Assuming the contrary, we get $x^* \in \widehat{\partial}\varphi(0)$ satisfying

$$\liminf_{h \to 0} \frac{1}{\|h\|}\Big[-|h|^2 + \min\{\langle u^*, h\rangle, \langle v^*, h\rangle\} - \langle x^*, h\rangle\Big] \ge 0.$$

Since the norms $|\cdot|$ and $\|\cdot\|$ are equivalent on X, we conclude that $\lim_{h\to 0}|h|^2/\|h\| = 0$ and hence

$$\liminf_{h \to 0} \frac{1}{\|h\|}\langle u^* - x^*, h\rangle \ge 0, \quad \liminf_{h \to 0} \frac{1}{\|h\|}\langle v^* - x^*, h\rangle \ge 0.$$

The latter is possible only when $u^* = x^* = v^*$, which contradicts the initial assumption that $u^* \neq v^*$; thus $\widehat{\partial}\varphi(0) = \emptyset$.

Let us finally show that $\widehat{\partial}\varphi(x) = \emptyset$ for any $x \neq 0$. If it is not the case, we take $x^* \in \widehat{\partial}\varphi(x)$ and get from (2.50) that

$$\liminf_{h \to 0} \frac{1}{\|h\|}\Big[-|x+h|^2 + |x|^2 + \min\{\langle u^*, x+h\rangle, \langle v^*, x+h\rangle\}$$
$$- \min\{\langle u^*, x\rangle, \langle v^*, x\rangle\} - \langle x^*, h\rangle\Big] \ge 0.$$

Assume first that $\langle u^*, x\rangle \le \langle v^*, x\rangle$. Then

$$\liminf_{h \to 0} \frac{1}{\|h\|}\Big[-|x+h|^2 + |x|^2 + \langle u^* - x^*, h\rangle\Big] \ge 0,$$

which means that $\widehat{\partial}\big(-|\cdot|^2\big)(x) \neq \emptyset$. Since $|\cdot|^2$ is convex and continuous, one always has $\widehat{\partial}(|\cdot|^2)(x) \neq \emptyset$. By Proposition 1.87 the function $|\cdot|^2$ is Fréchet differentiable at x, which implies the Fréchet differentiability of $|\cdot|$ at $x \neq 0$. The latter contradicts Proposition 2.18. The case of $\langle u^*, x\rangle > \langle v^*, x\rangle$ can be considered similarly. Thus $\widehat{\partial}\varphi(x) = \emptyset$ for any $x \in X$, which justifies (b)⇒(a) and completes the proof of the theorem. △

The next result related to Theorem 2.34 gives an efficient representation of basic normals to closed sets via weak* sequential limits of Fréchet normals at points nearby. It also happens to be a characterization of Asplund spaces.

Theorem 2.35 (basic normals in Asplund spaces). *Let X be a Banach space. The following properties are equivalent:*

(a) *X is Asplund.*

(b) *For every closed set $\Omega \subset X$ and every $\bar{x} \in \Omega$ one has the limiting representation*

$$N(\bar{x};\Omega) = \operatorname*{Lim\,sup}_{x\to\bar{x}} \widehat{N}(x;\Omega)\,.$$

Proof. Implication (a)⇒(b) follows from (a)⇒(b) in Theorem 2.34 for the case of set indicator functions $\varphi(x) = \delta(x;\Omega)$. It remains to prove that if X is not Asplund, representation (b) of basic normals doesn't hold for some closed set $\Omega \subset X$ and $\bar{x} \in \Omega$.

Put $X = Z \times I\!R$, where Z must be non-Asplund as well. Taking two distinct elements u^* and v^* of Z^*, define a Lipschitz function $\varphi\colon Z \to I\!R$ by (2.50), where $|\cdot|$ is the equivalent norm on Z from Proposition 2.18. We proved in Theorem 2.34 that $\widehat{\partial}\varphi(z) = \emptyset$ for every $z \in Z$. Now let us consider the epigraphical set $\Omega := \operatorname{epi}\varphi \subset X$ generated by this function and show that $\widehat{N}(x;\Omega) = \{0\}$ for every $x \in \Omega$.

It suffices to prove that $\widehat{N}((z,\varphi(z));\Omega) = \{(0,0)\}$ for all $z \in Z$. Assuming the contrary and taking into account that φ is Lipschitzian, we find

$$(z^*,\lambda) \in \widehat{N}((z,\varphi(z));\Omega) \;\;\text{with}\;\; \lambda < 0$$

due to Proposition 1.85(ii) as $\varepsilon = 0$, which gives $(-z^*/\lambda) \in \widehat{\partial}\varphi(z)$. This contradicts the fact that $\widehat{\partial}\varphi(z) = \emptyset$ proved in Theorem 2.34. Therefore

$$\operatorname*{Lim\,sup}_{x\to\bar{x}} \widehat{N}(x;\Omega) = \{0\} \;\;\text{whenever}\;\; \bar{x} \in \Omega$$

for the set Ω under consideration. On the other hand, from the proof of (b)⇒(a) in Theorem 2.34 we have $z_k \in Z$ and $\varepsilon_k > 0$ such that

$$u^* \in \widehat{\partial}_{\varepsilon_k}\varphi(z_k) \;\;\text{with}\;\; \varepsilon_k \downarrow 0 \;\;\text{and}\;\; z_k \to 0 \;\;\text{as}\;\; k \to \infty\,.$$

It implies that $(u^*,-1) \in \widehat{N}_{\varepsilon_k}((z_k,\varphi(z_k));\Omega)$ due to Theorem 1.86 and hence $(u^*,-1) \in N((0,0);\Omega)$ by definition (1.3). Thus the basic normal representation in (b) is violated for the above set Ω at the point $\bar{x} = 0$. △

Note that, for any Asplund space X, the subdifferential representation in Theorem 2.34(b) follows from the normal cone representation of Theorem 2.35 applied to epigraphical sets in the Asplund space $X \times I\!R$. The latter one is implied by the formula

$$\widehat{N}_\varepsilon(\bar{x}; \Omega) \subset \bigcup \Big\{ \widehat{N}(x; \Omega) \Big| \ x \in \Omega \cap (\bar{x} + \gamma I\!B) \Big\} + (\varepsilon + \gamma) I\!B^* \tag{2.51}$$

held for every $\varepsilon \geq 0$, $\gamma > 0$, $\bar{x} \in \Omega$, and every closed subset $\Omega \subset X$ of an Asplund space. Formula (2.51) immediately follows from (2.49) with $\varphi = \delta(\cdot; \Omega)$ and, given any $x^* \in \widehat{N}_\varepsilon(\bar{x}; \Omega)$, it can also be obtained by the direct application of the approximate extremal principle to the system of two closed sets

$$\Omega_1 := \big\{ (x, \alpha) \in X \times I\!R \big| \ x \in \Omega, \ \alpha \geq 0 \big\} ,$$

$$\Omega_2 := \big\{ (x, \alpha) \in X \times I\!R \big| \ x \in X, \ \alpha \leq \langle x^*, x - \bar{x} \rangle - (\varepsilon + \gamma) \|x - \bar{x}\| \big\}$$

for which $(\bar{x}, 0)$ is a local extremal point.

As a consequence of Theorem 2.35, we have the following simplified representations (with $\varepsilon = 0$ in Definition 1.32) of both normal and mixed coderivatives for closed-graph multifunctions between Asplund spaces.

Corollary 2.36 (coderivatives of mappings between Asplund spaces). *Let $F\colon X \rightrightarrows Y$ be a multifunction between Asplund spaces whose graph is closed around $(\bar{x}, \bar{y}) \in \operatorname{gph} F$. Then*

$$D_N^* F(\bar{x}, \bar{y})(\bar{y}^*) = \mathop{\operatorname{Lim\,sup}}_{\substack{(x,y) \to (\bar{x},\bar{y}) \\ y^* \xrightarrow{w^*} \bar{y}^*}} \widehat{D}^* F(x, y)(y^*), \quad \bar{y}^* \in Y^* ,$$

$$D_M^* F(\bar{x}, \bar{y})(\bar{y}^*) = \mathop{\operatorname{Lim\,sup}}_{\substack{(x,y) \to (\bar{x},\bar{y}) \\ y^* \to \bar{y}^*}} \widehat{D}^* F(x, y)(y^*), \quad \bar{y}^* \in Y^* .$$

Proof. Since both X and Y are Asplund, its product $X \times Y$ is Asplund as well. Hence the representation for $D_N^* F(\bar{x}, \bar{y})$ follows immediately from (1.26) and the normal cone representation of Theorem 2.35 applied to $\Omega = \operatorname{gph} F \subset X \times Y$. To prove the mixed coderivative representation, we pick any $\bar{x}^* \in D_M^* F(\bar{x}, \bar{y})(\bar{y}^*)$ and find, by Definition 1.32(iii), sequences $\varepsilon_k \downarrow 0$, $(x_k, y_k, y_k^*) \to (\bar{x}, \bar{y}, \bar{y}^*)$, and $x_k^* \xrightarrow{w^*} \bar{x}^*$ with $(x_k, y_k) \in \operatorname{gph} F$ and

$$(x_k^*, -y_k^*) \in \widehat{N}_{\varepsilon_k}((x_k, y_k); \operatorname{gph} F) \quad \text{for all } k \in I\!N .$$

Now using formula (2.51) with $\varepsilon = \gamma := \varepsilon_k$ and $\Omega = \operatorname{gph} F$, we get sequences $(\tilde{x}_k, \tilde{y}_k) \in \operatorname{gph} F$ and $(\tilde{x}_k^*, -\tilde{y}_k^*) \in \widehat{N}((\tilde{x}_k, \tilde{y}_k); \operatorname{gph} F)$ such that

$$\|(\tilde{x}_k, \tilde{y}_k) - (x_k, y_k)\| \leq \varepsilon_k \quad \text{and} \quad \|(\tilde{x}_k^*, \tilde{y}_k^*) - (x_k^*, y_k^*)\| \leq 2\varepsilon_k .$$

This implies that $\tilde{x}_k^* \xrightarrow{w^*} \bar{x}^*$ and that $(\tilde{x}_k, \tilde{y}_k, \tilde{y}_k^*) \to (\bar{x}, \bar{y}, \bar{y}^*)$ in the norm topology of $X \times Y \times Y^*$, which justifies the representation for $D_M^* F(\bar{x}, \bar{y})$. △

2.4.2 Representations of Singular Subgradients and Horizontal Normals to Graphs and Epigraphs

In Subsect. 1.3.1 we defined singular subgradients of extended-real-valued functions through horizontal normals to their epigraphs. For a number of applications of singular subgradients it is important to obtain their efficient representations via some limits of Fréchet subgradients and ε-subgradients at points nearby, similar to those available for basic subgradients. This issue is related to the possibility of approximating horizontal normals by sequences of sloping (non-horizontal) normals to epigraphs. In this subsection we consider these questions (and related ones for the case of graphs of continuous functions) in the framework of Asplund spaces.

Let us start with the basic lemma ensuring a strong approximation of horizontal Fréchet normals to epigraphs of l.s.c. functions on Asplund spaces by sequences of Fréchet subgradients.

Lemma 2.37 (horizontal Fréchet normals to epigraphs). *Let X be Asplund, and let $\varphi: X \to \overline{I\!R}$ be a proper function l.s.c. around $\bar{x} \in \operatorname{dom}\varphi$. Then for every $x^* \in X^*$ with $(x^*, 0) \in \widehat{N}((\bar{x}, \varphi(\bar{x})); \operatorname{epi}\varphi)$ there are sequences $x_k \xrightarrow{\varphi} \bar{x}$, $\lambda_k \downarrow 0$, and $x_k^* \in \lambda_k \widehat{\partial}\varphi(x_k)$ such that $\|x_k^* - x^*\| \to 0$ as $k \to \infty$.*

Proof. Fix $x^* \in X^*$ satisfying $(x^*, 0) \in \widehat{N}((\bar{x}, \varphi(\bar{x})); \operatorname{epi}\varphi)$ and assume without loss of generality that $\bar{x} = 0$, $\varphi(\bar{x}) = 0$, and $\|x^*\| = 1$. Take an arbitrary $\varepsilon \in (0, 1)$ and choose $\eta = \eta(\varepsilon) \downarrow 0$ as $\varepsilon \downarrow 0$ such that

$$\varphi(x) \geq -\varepsilon \quad \text{on} \quad \eta I\!B \quad \text{and}$$

$$\langle x^*, x\rangle < \varepsilon\big(\|x\| + |\varphi(x)|\big) \quad \text{whenever} \quad x \in (\eta I\!B) \setminus \{0\}\ . \tag{2.52}$$

Form the closed convex set

$$\Omega_\varepsilon := \big\{x \in X \big|\ \langle x^*, x\rangle \geq \varepsilon\|x\|\big\}$$

and observe that

$$\varphi(x) \geq 0 \ \text{ for all } \ x \in \Omega_\varepsilon \cap \eta I\!B\ .$$

Indeed, otherwise one has $(x, 0) \in \operatorname{epi}\varphi$, and hence (2.52) implies that $\langle x^*, x\rangle < \varepsilon\|x\|$, which contradicts the fact of $x \in \Omega_\varepsilon$. Next we show that

$$\operatorname{dist}(x; \Omega_{2\varepsilon}) \geq \frac{\varepsilon}{1 + 2\varepsilon} \quad \text{for any} \quad x \in \Omega_\varepsilon\ . \tag{2.53}$$

Assuming the opposite, we find $\widetilde{x} \in \Omega_{2\varepsilon}$ satisfying

$$\|x - \widetilde{x}\| < \frac{\varepsilon}{1 + 2\varepsilon}\ .$$

The latter inequality implies that

$$\begin{aligned}\langle x^*, \widetilde{x}\rangle &= \langle x^*, \widetilde{x} - x\rangle + \langle x^*, x\rangle \le \|x^*\| \cdot \|\widetilde{x} - x\| + \langle x^*, x\rangle \\ &< \|\widetilde{x} - x\| + \varepsilon\|x\| < \frac{\varepsilon}{1+2\varepsilon} + \varepsilon\|x\| \\ &\le 2\varepsilon\Big[\|x\| - \frac{\varepsilon}{1+2\varepsilon}\|x\|\Big] \le 2\varepsilon\big[\|x\| - \|x - \widetilde{x}\|\big] \le 2\varepsilon\|\widetilde{x}\|,\end{aligned}$$

which contradicts the fact of $\widetilde{x} \in \Omega_{2\varepsilon}$. Now given an arbitrary number $k \in I\!N$, define the function

$$\psi_{k,\varepsilon}(x) = \varepsilon\varphi(x) + k\,\mathrm{dist}(x; \Omega_{2\varepsilon}) - \langle x^*, x\rangle + 2\varepsilon\|x\|$$

that is l.s.c. and bounded from below on $\eta I\!B$. Taking $u_{k,\varepsilon} \in \eta I\!B$ with

$$\psi_{k,\varepsilon}(u_{k,\varepsilon}) \le \inf_{x \in \eta I\!B} \psi_{k,\varepsilon}(x) + \tfrac{1}{k}$$

and applying the Ekeland variational principle (Theorem 2.26) to the function $\psi_{k,\varepsilon}$ on the metric space $\eta I\!B$, we find $\bar{u}_{k,\varepsilon} \in \eta I\!B$ satisfying

$$\psi_{k,\varepsilon}(\bar{u}_{k,\varepsilon}) \le \psi_{k,\varepsilon}(x) + \tfrac{1}{k}\|x - \bar{u}_{k,\varepsilon}\| \quad \text{whenever } x \in \eta I\!B .$$

Putting $x = 0$, we arrive at the useful upper estimate

$$\psi_{k,\varepsilon}(\bar{u}_{k,\varepsilon}) \le \tfrac{1}{k}\|\bar{u}_{k,\varepsilon}\| ,$$

which means, by the construction of $\psi_{k,\varepsilon}$, that

$$\varepsilon\varphi(\bar{u}_{k,\varepsilon}) + k\,\mathrm{dist}(\bar{u}_{k,\varepsilon}; \Omega_{2\varepsilon}) = \langle x^*, \bar{u}_{k,\varepsilon}\rangle + 2\varepsilon\|\bar{u}_{k,\varepsilon}\| \le \tfrac{1}{k}\|\bar{u}_{k,\varepsilon}\| .$$

The latter clearly yields $\mathrm{dist}(\bar{u}_{k,\varepsilon}; \Omega_{2\varepsilon}) \to 0$ as $k \to \infty$.

Now we show that one can always find $k = k(\varepsilon) \in I\!N$ satisfying $\bar{u}_{k,\varepsilon} \in \mathrm{int}\,(\eta I\!B)$ whenever $\varepsilon > 0$; note that $\eta = \eta(\varepsilon)$ also depends on ε but we skip this in notation for simplicity. Assume first that $\bar{u}_{k,\varepsilon} \in \Omega_\varepsilon$, i.e.,

$$\langle x^*, \bar{u}_{k,\varepsilon}\rangle \ge \varepsilon\,\|\bar{u}_{k,\varepsilon}\| .$$

Employing (2.52), we have

$$\varepsilon\varphi(\bar{u}_{k,\varepsilon}) + \varepsilon\,\|u_{k,\varepsilon}\| - \langle x^*, u_{k,\varepsilon}\rangle \ge 0$$

with $u_{k,\varepsilon}$ chosen above, and hence

$$\psi_{k,\varepsilon}(\bar{u}_{k,\varepsilon}) \ge \varepsilon\,\|\bar{u}_{k,\varepsilon}\| + k\,\mathrm{dist}(\bar{u}_{k,\varepsilon}; \Omega_{2\varepsilon}) \ge \varepsilon\,\|\bar{u}_{k,\varepsilon}\| .$$

Combining this with the preceding upper estimate for $\psi(\bar{u}_{k,\varepsilon})$, one gets

$$\varepsilon\,\|\bar{u}_{k,\varepsilon}\| \le \tfrac{1}{k}\|\bar{u}_{k,\varepsilon}\|, \quad \text{and thus} \quad \bar{u}_{k,\varepsilon} = 0$$

for all $k \in I\!N$ sufficiently large. If $\bar{u}_{k\varepsilon} \notin \Omega_\varepsilon$, then (2.53) gives

$$\frac{\varepsilon}{1+2\varepsilon}\|\bar{u}_{k,\varepsilon}\| \le \operatorname{dist}(\bar{u}_{k,\varepsilon};\Omega_{2\varepsilon}) \to 0 ,$$

i.e., $\bar{u}_{k,\varepsilon} \to 0$ as $k \to \infty$. Thus there is a sequence of $k = k_\varepsilon \to \infty$ as $\varepsilon \downarrow 0$ for which $\|\bar{u}_{k,\varepsilon}\| \le \eta = \eta(\varepsilon)$. Taking this into account and the fact that $\bar{u}_\varepsilon := \bar{u}_{k_\varepsilon,\varepsilon}$ is a *minimizer* to the function $\psi_{k,\varepsilon} + \frac{1}{k_\varepsilon}\|x - \bar{u}_\varepsilon\|$ on $\eta I\!B$, one has

$$0 \in \widehat{\partial}\big(\varepsilon\varphi + \varphi_\varepsilon\big)(\bar{u}_\varepsilon)$$

by the generalized Fermat rule, where

$$\varphi_\varepsilon(x) := k_\varepsilon \operatorname{dist}(x;\Omega_{2\varepsilon}) - \langle x^*, x\rangle + 2\varepsilon\,\|x\| + \frac{1}{k_\varepsilon}\|x - \bar{u}_\varepsilon\| . \tag{2.54}$$

Applying the subgradient description of Lemma 2.32 to the above sum, we find elements v_ε, w_ε, v_ε^*, and w_ε^* satisfying

$$\|v_\varepsilon - \bar{u}_\varepsilon\| \le \eta, \quad \|w_\varepsilon - \bar{u}_\varepsilon\| \le \eta ,$$

$$v_\varepsilon^* \in \widehat{\partial}\varphi(v_\varepsilon), \quad w_\varepsilon^* \in \widehat{\partial}\varphi_\varepsilon(w_\varepsilon) ,$$

$$\|\varepsilon v_\varepsilon^* + w_\varepsilon^*\| \le \varepsilon \ \text{ for all } \ \varepsilon > 0 .$$

It follows from the structure of the *convex continuous* function φ_ε in (2.54), by basic convex analysis, that

$$w_\varepsilon^* \in k_\varepsilon \partial \operatorname{dist}(w_\varepsilon;\Omega_{2\varepsilon}) - x^* + \Big(2\varepsilon + \frac{1}{k_\varepsilon}\Big) I\!B^* .$$

Hence there is $\bar{w}_\varepsilon^* \in \partial\operatorname{dist}(w_\varepsilon;\Omega_{2\varepsilon})$ such that

$$\|\varepsilon\, v_\varepsilon^* + k_\varepsilon \bar{w}_\varepsilon^* - x^*\| \le 2\varepsilon + \frac{1}{k_\varepsilon} . \tag{2.55}$$

To proceed, we consider the following two cases.

Case 1. Let $w_\varepsilon \in \Omega_{2\varepsilon}$. Then, as well known from convex analysis,

$$\partial\operatorname{dist}(w_\varepsilon;\Omega_{2\varepsilon}) = N(w_\varepsilon;\Omega_{2\varepsilon}) \cap I\!B^* = \operatorname{cone}\big\{-x^* + 2\varepsilon I\!B^*\big\} \cap I\!B^*$$

due to the structure of the set $\Omega_{2\varepsilon}$; cf. Corollary 1.96. Hence

$$\bar{w}_\varepsilon^* = \alpha_\varepsilon(-x^* + 2\varepsilon e_\varepsilon^*) \ \text{ with } \ \|\bar{w}_\varepsilon^*\| \le 1 \ \text{ and } \ \|e_\varepsilon^*\| \le 1 ,$$

where $\alpha_\varepsilon \ge 0$ are uniformly bounded due to $\|x^*\| = 1$. By (2.55) one has

$$\big\|\varepsilon v_\varepsilon^* + k_\varepsilon\big(\alpha_\varepsilon(-x^* + 2\varepsilon e_\varepsilon^*)\big) - x^*\big\| \le 2\varepsilon + \frac{1}{k_\varepsilon} ,$$

which implies the estimate

$$\|\varepsilon v_\varepsilon^* - (k_\varepsilon \alpha_\varepsilon + 1)x^*\| \le 2\varepsilon k_\varepsilon \alpha_\varepsilon + 2\varepsilon + \frac{1}{k_\varepsilon} .$$

Let $\widetilde{\lambda}_\varepsilon := k_\varepsilon \alpha_\varepsilon + 1$ and observe that

$$\Big\| \frac{\varepsilon}{\widetilde{\lambda}_\varepsilon} v_\varepsilon^* - x^* \Big\| \le \frac{1}{k_\varepsilon \alpha_\varepsilon + 1}\Big(2\varepsilon k_\varepsilon \alpha_\varepsilon + 2\varepsilon + \frac{1}{k_\varepsilon}\Big) \to 0 \ \text{ as } \varepsilon \downarrow 0 .$$

Finally putting $\lambda_\varepsilon := \varepsilon/\widetilde{\lambda}_\varepsilon$, we get

$$\|\lambda_\varepsilon v_\varepsilon^* - x^*\| \to 0 \ \text{ with } \ v_\varepsilon^* \in \widehat{\partial}\varphi(w_\varepsilon) \ \text{ and } \ w_\varepsilon \to 0$$

as $\varepsilon \downarrow 0$, which justifies the lemma in Case 1 considered.

Case 2. Let $w_\varepsilon \notin \Omega_{2\varepsilon}$. First note that Theorem 1.99 implies the inclusion

$$\widehat{\partial}\text{dist}(\bar{x}; \Omega) \subset \bigcap_{\nu > 0} \bigcup \Big[\widehat{N}(x; \Omega) + \nu I\!B^* \Big| \ \|x - \bar{x}\| \le \text{dist}(\bar{x}; \Omega) + \nu \Big]$$

for any set $\Omega \subset X$ in a Banach space and any out-of-set point $\bar{x} \notin \Omega$. Putting $\bar{x} := w_\varepsilon$ and $\nu := 1/k_\varepsilon$ therein, we find $\widetilde{w}_\varepsilon \in \Omega_{2\varepsilon}$ and $\widetilde{w}_\varepsilon^* \in \widehat{N}(\widetilde{w}_\varepsilon; \Omega_{2\varepsilon}) = N(\widetilde{w}_\varepsilon; \Omega_{2\varepsilon})$ such that

$$\|\widetilde{w}_\varepsilon^* - \bar{w}_\varepsilon^*\| \le \frac{1}{k_\varepsilon} \quad \text{and}$$

$$\|\widetilde{w}_\varepsilon - w_\varepsilon\| \le \text{dist}(w_\varepsilon; \Omega_{2\varepsilon}) + \frac{1}{k_\varepsilon} \le \|w_\varepsilon\| + \frac{1}{k_\varepsilon} \to 0$$

as $\varepsilon \downarrow 0$. Then we have the representation

$$\widetilde{w}_\varepsilon^* = \alpha_\varepsilon(-x^* + 2\varepsilon e_\varepsilon^*) \ \text{ with } \ e_\varepsilon^* \in I\!B^* ,$$

where α_ε are uniformly bounded. Thus

$$\|\varepsilon v_\varepsilon^* + k_\varepsilon \bar{w}_\varepsilon^* - x^*\| \le 2\varepsilon + \frac{1}{k_\varepsilon}$$

$$\Longrightarrow \|\varepsilon v_\varepsilon^* + k_\varepsilon \widetilde{w}_\varepsilon^* - x^*\| \le \frac{1}{k_\varepsilon} + 2\varepsilon + \frac{1}{k_\varepsilon} \le \frac{2}{k_\varepsilon} + 2\varepsilon$$

$$\Longrightarrow \|\varepsilon v_\varepsilon^* + k_\varepsilon(-\alpha_\varepsilon)(-x^* + 2\varepsilon e_\varepsilon^*) - x^*\| \le \frac{2}{k_\varepsilon} + 2\varepsilon$$

$$\Longrightarrow \|\varepsilon v_\varepsilon^* - (k_\varepsilon \alpha_\varepsilon + 1)x^*\| \le 2k_\varepsilon \alpha_\varepsilon \varepsilon + \frac{2}{k_\varepsilon} + 2\varepsilon$$

$$\Longrightarrow \Big\| \frac{\varepsilon}{k_\varepsilon \alpha_\varepsilon + 1} v_\varepsilon^* - x^* \Big\| \le \frac{2}{k_\varepsilon \alpha_\varepsilon + 1}\Big[k_\varepsilon \alpha_\varepsilon \varepsilon + \frac{1}{k_\varepsilon} + \varepsilon \Big] \to 0 \ \text{ as } \ \varepsilon \downarrow 0 .$$

Finally, letting

$$\lambda_\varepsilon := \frac{\varepsilon}{k_\varepsilon \alpha_\varepsilon + 1}$$

as in Case 1, we justify the required relationships in Case 2 and thus complete the proof of the lemma. △

Theorem 2.38 (singular subgradients in Asplund spaces). *Let X be an Asplund space. Assume that $\varphi\colon X \to \overline{I\!R}$ is a proper function l.s.c. around some point $\bar{x} \in \operatorname{dom}\varphi$. Then the singular subdifferential of φ admits the following limiting representations:*

$$\partial^\infty \varphi(\bar{x}) = \mathop{\mathrm{Lim\,sup}}_{\substack{x \xrightarrow{\varphi} \bar{x} \\ \lambda \downarrow 0}} \lambda \widehat{\partial} \varphi(x) = \mathop{\mathrm{Lim\,sup}}_{\substack{x \xrightarrow{\varphi} \bar{x} \\ \varepsilon, \lambda \downarrow 0}} \lambda \widehat{\partial}_\varepsilon \varphi(x) \ .$$

Proof. The equality

$$\mathop{\mathrm{Lim\,sup}}_{\substack{x \xrightarrow{\varphi} \bar{x} \\ \lambda \downarrow 0}} \lambda \widehat{\partial} \varphi(x) = \mathop{\mathrm{Lim\,sup}}_{\substack{x \xrightarrow{\varphi} \bar{x} \\ \varepsilon, \lambda \downarrow 0}} \lambda \widehat{\partial}_\varepsilon \varphi(x)$$

for any l.s.c. function on Asplund spaces follows from formula (2.49) justified above. It remains to prove the inclusion

$$\partial^\infty \varphi(\bar{x}) \subset \mathop{\mathrm{Lim\,sup}}_{\substack{x \xrightarrow{\varphi} \bar{x} \\ \lambda \downarrow 0}} \lambda \widehat{\partial} \varphi(x) \ ,$$

since the opposite one is easily implied by the definitions. To proceed, we take an arbitrary $x^* \in \partial^\infty \varphi(\bar{x})$ for which $(x^*, 0) \in N((\bar{x}, \varphi(x)); \operatorname{epi}\varphi)$ by Definition 1.77(ii). Employing Theorem 2.35, we find sequences $(x_k, \alpha_k) \to (\bar{x}, \varphi(\bar{x}))$ and $(x_k^*, \nu_k) \xrightarrow{w^*} (x^*, 0)$ such that $\alpha_k \geq \varphi(x_k)$ and $(x_k^*, -\nu_k) \in \widehat{N}((x_k, \alpha_k); \operatorname{epi}\varphi)$, $k \in I\!N$. The latter implies that $\nu_k \geq 0$ for all k. Thus one has two possibilities for the sequence $\{(x_k^*, \nu_k)\}$: either

(a) there is a subsequence of $\{\nu_k\}$ consisting of positive numbers, or

(b) $\nu_k = 0$ for all k sufficiently large.

In case (a) we assume without loss of generality that $\nu_k > 0$ for all $k \in I\!N$, which implies that $\alpha_k = \varphi(x_k)$ and $x_k^*/\nu_k \in \widehat{\partial}\varphi(x_k)$, $k \in I\!N$. Letting $\lambda_k := \nu_k$ and $\tilde{x}_k^* := x_k^*/\nu_k$, we get $\lambda_k \tilde{x}_k^* \xrightarrow{w^*} x^*$ and $\lambda_k \downarrow 0$ as $k \to \infty$.

In case (b) one has $(x_k^*, 0) \in \widehat{N}((x_k, \varphi(x_k)); \operatorname{epi}\varphi)$ if $x_k^* \neq 0$, which we may always assume. Now employing Lemma 2.37 and the standard diagonal process, we get sequences $\tilde{x}_k \xrightarrow{\varphi} \bar{x}$, $\lambda_k \downarrow 0$, and $\tilde{x}_k^* \xrightarrow{w^*} x^*$ such that $\tilde{x}_k^* \in \lambda_k \widehat{\partial}\varphi(\tilde{x}_k)$ for large k. This completes the proof. △

Note that analytic ε-subgradients in the second representation of Theorem 2.38 can be replaced with ε-geometric subgradients due to Theorem 1.86.

We'll see further in the book many applications of both Lemma 2.37 and Theorem 2.38 to various aspects of analysis and optimization in Asplund spaces. Right now let us present a consequence of Lemma 2.37 providing a convenient subdifferential description of the SNEC property for extended-real-valued functions on Asplund spaces; cf. Definition 1.116.

Corollary 2.39 (subdifferential description of sequential normal epi-compactness). *Let X be Asplund, and let $\varphi\colon X \to \overline{I\!R}$ be a proper function l.s.c. around $\bar{x} \in \operatorname{dom}\varphi$. Then φ is SNEC at $\bar{x}$ if and only if for any sequences $x_k \xrightarrow{\varphi} \bar{x}$, $\lambda_k \downarrow 0$, and $x_k^* \in \lambda_k\widehat{\partial}\varphi(x_k)$ one has*

$$\big[x_k^* \xrightarrow{w^*} 0\big] \Longrightarrow \|x_k^*\| \to 0 \quad as \;\; k \to \infty\,.$$

Proof. Assume that φ is SNEC at $\bar{x}$. Take any sequences $x_k \xrightarrow{\varphi} \bar{x}$, $\lambda_k \downarrow 0$, and $x_k^* \in \lambda_k\widehat{\partial}\varphi(x_k)$ with $x_k^* \xrightarrow{w^*} 0$ as $k \to \infty$. Then

$$(x_k^*, -\lambda_k) \in \widehat{N}\big((x_k, \varphi(x_k)); \operatorname{epi}\varphi\big) \;\; \text{for all} \;\; k \in I\!N\,,$$

and the SNEC property of φ at $\bar{x}$ implies that $\|x_k^*\| \to 0$ as $k \to \infty$.

To prove the converse application, pick arbitrary sequences

$$(x_k, \alpha_k) \in \operatorname{epi}\varphi \;\; \text{and} \;\; (x_k^*, -\lambda_k) \in \widehat{N}\big((x_k, \alpha_k); \operatorname{epi}\varphi\big)$$

with $(x_k, \alpha_k) \to (\bar{x}, \varphi(\bar{x}))$, $\lambda_k \to 0$, and $x_k^* \xrightarrow{w^*} 0$. We need to show $\|x_k^*\| \to 0$ as $k \to \infty$; in fact it is sufficient to justify the latter holds along a subsequence.

Since $\lambda_k > 0$ for all $k \in I\!N$, there are the following two cases to consider:

(a) $\lambda_k > 0$ along a subsequence of $k \in I\!N$;

(b) $\lambda_k = 0$ for all large $k \in I\!N$.

Case (a) is simple. Indeed, we easily have $\alpha_k = \varphi(x_k)$, and hence

$$\Big(\frac{x_k^*}{\lambda_k}, -1\Big) \in \widehat{N}\big((x_k, \varphi(x_k)); \operatorname{epi}\varphi\big), \;\; \text{i.e.,} \;\; x_k^* \in \lambda_k\widehat{\partial}\varphi(x_k)\,.$$

Then $\|x_k^*\| \to 0$ by the assumption made, which yields that φ is SNEC at $\bar{x}$.

Case (b) is more involved requiring the usage of Lemma 2.37. To proceed, we suppose without lost of generality that $\lambda_k = 0$ and $\alpha_k = \varphi(x_k)$ for all $k \in I\!N$. Thus $(x_k^*, 0) \in \widehat{N}\big((x_k, \varphi(x_k)); \operatorname{epi}\varphi\big)$. Applying Lemma 2.37 for each k, we select subsequences λ_{n_k}, $\widetilde{x}_{n_k}$, and $\widetilde{x}_{n_k}^*$ so that

$$0 < \lambda_{n_k} < \frac{1}{k}, \;\; \|\widetilde{x}_{n_k} - x_k\| \le \frac{1}{k}, \;\; |\varphi(\widetilde{x}_{n_k}) - \varphi(x_k)| \le \frac{1}{k}\,,$$

$$\|\widetilde{x}_{n_k}^* - x_k^*\| \le \frac{1}{k}, \;\; \text{and} \;\; \widetilde{x}_{n_k}^* \in \lambda_{n_k}\widehat{\partial}\varphi(\widetilde{x}_{n_k})\,.$$

One clearly has $\widetilde{x}^*_{n_k} \xrightarrow{w^*} 0$ due to the construction of $\widetilde{x}^*_{n_k}$ and the assumption on $x^*_k \xrightarrow{w^*} 0$. Then $\|\widetilde{x}^*_{n_k}\| \to 0$ and hence $\|x^*_{n_k}\| \to 0$, which implies the SNEC property and completes the proof of the corollary. △

The concluding result of this section gives an efficient representation of horizontal Fréchet normals to *graphs* of continuous functions in Asplund spaces and provides a refinement of coderivative-subdifferential relations considered in Theorem 1.80.

Theorem 2.40 (horizontal normals to graphs of continuous functions). *Let X be an Asplund space, and let $\varphi\colon X \to \overline{I\!R}$ be finite and continuous around some point $x \in X$. The following hold:*

(i) *If $(x^*, 0) \in \widehat{N}((\bar{x}, \varphi(\bar{x})); \operatorname{gph}\varphi)$, then there exist sequences $x_k \to \bar{x}$, $\lambda_k \downarrow 0$, and $x^*_k \to x^*$ such that*

$$x^*_k \in \widehat{\partial}\big(\lambda_k\varphi\big)(x_k) \cup \widehat{\partial}\big(-\lambda_k\varphi\big)(x_k) \quad \textit{for all} \;\; k \in I\!N\,.$$

(ii) $D^*\varphi(\bar{x})(0) = \partial^\infty\varphi(\bar{x}) \cup \partial^\infty(-\varphi)(\bar{x})$.

Proof. To justify (i), we proceed similarly to the proof of Lemma 2.37 with a certain modification in constructions and estimates due to the continuity of φ, which makes it possible to derive two-sided formulas. For brevity we skip some details using slightly different notation.

Assume that $\bar{x} = 0$, $\varphi(\bar{x}) = 0$ and pick an arbitrary $x^* \in B^* \subset X^*$ with $(x^*, 0) \in \widehat{N}((0, 0); \operatorname{gph}\varphi)$. For each $\varepsilon > 0$ we find $\eta = \eta(\varepsilon) \downarrow 0$ as $\varepsilon \downarrow 0$ such that φ is bounded on $\eta I\!B$ and

$$\langle x^*, x\rangle < \varepsilon\big(\|x\| + |\varphi(x)|\big) \quad \text{for all} \;\; x \in \eta I\!B \setminus \{0\}\,. \tag{2.56}$$

Form the set Ω_ε as in the proof of Lemma 2.37 and observe that either

(a) $\varphi(x) \geq 0$ for all $x \in \Omega_\varepsilon \cap (\eta I\!B)$, or

(b) $\varphi(x) \leq 0$ for all $x \in \Omega_\varepsilon \cap (\eta I\!B)$.

Indeed, if there are $x_1, x_2 \in \Omega_\varepsilon \cap (\eta I\!B)$ with $\varphi(x_1) > 0$ and $\varphi(x_2) < 0$, then both x_1 and x_2 are nonzero and, by the continuity of φ, there is $x := \alpha x_1 + (1-\alpha)x_2 \in \Omega_\varepsilon \cap (\eta I\!B) \setminus \{0\}$ with $\alpha \in (0, 1)$ and $\varphi(x) = 0$. This clearly contradicts (2.56).

For each $k \in I\!N$ define the function

$$\psi_{k,\varepsilon}(x) := \begin{cases} \varepsilon\varphi(x) + k\operatorname{dist}(x; \Omega_{2\varepsilon}) - \langle x^*, x\rangle + 2\varepsilon\|x\| & \text{if (a) holds}\,, \\[2ex] -\varepsilon\varphi(x) + k\operatorname{dist}(x; \Omega_{2\varepsilon}) - \langle x^*, x\rangle + 2\varepsilon\|x\| & \text{if (b) holds} \end{cases}$$

and apply the Ekeland variational principle to this function on the metric space $\eta I\!B$. In this way we find $x_{k,\varepsilon} \in \eta I\!B$ that minimizes the function $\psi_{k,\varepsilon}(x) + \frac{1}{k}\|x - x_{k,\varepsilon}\|$ on $\eta I\!B$. In particular,

$$\psi_{k,\varepsilon}(x_{k,\varepsilon}) \le \psi_{k,\varepsilon}(0) = \tfrac{1}{k}\|x_{k,\varepsilon}\| \text{ and } \operatorname{dist}(x_{k,\varepsilon};\Omega_{2\varepsilon}) \to 0 \tag{2.57}$$

as $k \to \infty$. Let us further choose $k_\varepsilon \to \infty$ as $\varepsilon \downarrow 0$ similarly to the proof of Lemma 2.37. If $x_{k,\varepsilon} \in \Omega_\varepsilon$, then it follows from (2.56) and (2.57) that $x_{k,\varepsilon} = 0$ for $k > 1/\varepsilon$. If $x_{k,\varepsilon} \notin \Omega_\varepsilon$, then $\|x_{k,\varepsilon}\| \to 0$ as $k \to \infty$ by (2.55) and (2.57). Thus for every $\varepsilon > 0$ there are $k = k_\varepsilon$ and $x_\varepsilon := x_{k_\varepsilon,\varepsilon}$ such that $k_\varepsilon \to \infty$ as $\varepsilon \downarrow 0$, that $\|x_\varepsilon\| < \eta/2$, and that

$$0 \in \widehat{\partial}\Big(\psi_\varepsilon + \frac{1}{k}\|\cdot - x_\varepsilon\|\Big)(x_\varepsilon)\,,$$

where $\psi_\varepsilon(x) := \psi_{k_\varepsilon,\varepsilon}(x)$. Applying Lemma 2.32 and taking into account the structure of ψ_ε, we find $u_\varepsilon \in \eta I\!B$, $v_\varepsilon \in \eta I\!B$, $u_\varepsilon^* \in \widehat{\partial}\varphi(u_\varepsilon) \cup \widehat{\partial}(-\varphi)(u_\varepsilon)$, and $v_\varepsilon^* \in \partial\operatorname{dist}(v_\varepsilon;\Omega_{2\varepsilon})$ with

$$\|v_\varepsilon^*\| \le 1 \text{ and } \|\varepsilon u_\varepsilon^* + k v_\varepsilon^* - x^*\| \le 2(\varepsilon + 1/k)\,. \tag{2.58}$$

Consider again the two possible cases: $v_\varepsilon \in \Omega_{2\varepsilon}$ and $v_\varepsilon \notin \Omega_{2\varepsilon}$. In the first case we employ the representation of $\partial\operatorname{dist}(v_\varepsilon;\Omega_{2\varepsilon})$ from convex analysis and get $\alpha_\varepsilon > 0$ and $e^* \in I\!B^*$ such that $v_\varepsilon^* + \alpha_\varepsilon x^* = 2\varepsilon\alpha_\varepsilon e^*$. This implies that the sequence $\{\alpha_\varepsilon\}$ is bounded as $\varepsilon \downarrow 0$. From (2.58) one has the estimates

$$\begin{aligned}\|\varepsilon u_\varepsilon^* - (k\alpha_\varepsilon + 1)x^*\| &\le \|\varepsilon u_\varepsilon^* + k v_\varepsilon^* - x^*\| + k\|v_\varepsilon^* + \alpha_\varepsilon x^*\|\\ &\le 2(\varepsilon + 1/k) + 2k\alpha_\varepsilon\varepsilon\,.\end{aligned}$$

Dividing this by $k\alpha_\varepsilon + 1$ and denoting $\lambda_\varepsilon := \varepsilon/(k\alpha_\varepsilon + 1)$, $x_\varepsilon^* := \lambda_\varepsilon u_\varepsilon^*$, we obtain $x_\varepsilon^* \in \widehat{\partial}(\lambda_\varepsilon\varphi)(u_\varepsilon) \cup \widehat{\partial}(-\lambda_\varepsilon\varphi)(u_\varepsilon)$ with $\|x_\varepsilon^* - x^*\| \to 0$ and $\lambda_\varepsilon \downarrow 0$ as $\varepsilon \downarrow 0$. In the case of $v_\varepsilon \notin \Omega_{2\varepsilon}$ we proceed similarly to the proof of Lemma 2.37 based on the upper estimate of $\widehat{\partial}\operatorname{dist}(\bar{x};\Omega)$ with $\bar{x} \notin \Omega$ from Theorem 1.99. This completes the proof of assertion (i) in the theorem.

To justify the inclusion "$\subset$" in (ii), we argue as in the proof of Theorem 2.38. The opposite inclusion follows from Theorem 1.80. △

2.5 Versions of Extremal Principle in Banach Spaces

We have shown in the previous section that the above versions of the extremal principle and most of the related results are not only valid in Asplund spaces but happen to provide *characterizations* for this general class of Banach spaces. To cover other classes of Banach spaces, one therefore needs to employ different constructions of generalized normals involving in formulations of the extremal principle. In this section we detect those properties of axiomatically defined normal and subgradient structures that allow us to derive approximate and exact versions of the abstract extremal principle valid in appropriate classes of Banach spaces.

2.5.1 Axiomatic Normal and Subdifferential Structures

First we define an abstract *prenormal* structure on a Banach space that supports an *approximate* version of the extremal principle.

Definition 2.41 (prenormal structures). *Let X be a Banach space. We say that $\widehat{\mathcal{N}}$ defines a* PRENORMAL STRUCTURE *on X if it associates, with every nonempty set $\Omega \subset X$, a set-valued mapping $\widehat{\mathcal{N}}(\cdot;\Omega): X \rightrightarrows X^*$ such that $\widehat{\mathcal{N}}(x;\Omega) = \emptyset$ for $x \notin \Omega$, $\widehat{\mathcal{N}}(x;\Omega) = \widehat{\mathcal{N}}(x;\widetilde{\Omega})$ when Ω and $\widetilde{\Omega}$ are the same near $x \in \Omega$, and the following property holds:*

(H) *Given any small $\varepsilon > 0$, $a \in X$ with $\|a\| \le \varepsilon$, and closed sets $\Omega_1, \Omega_2 \subset X$, assume that $(\bar{x}_1, \bar{x}_2) \in \Omega_1 \times \Omega_2$ is a local minimizer for the function*

$$\psi(x_1, x_2) := \|x_1 - x_2 + a\| + \varepsilon\big(\|x_1 - \bar{x}_1\| + \|x_2 - \bar{x}_2\|\big) \tag{2.59}$$

relative to the set $\Omega_1 \times \Omega_2$ with $\bar{x}_1 - \bar{x}_2 + a \ne 0$. Then there are $\tilde{x}_i \in \bar{x}_i + \varepsilon I\!B$, $i = 1, 2$, and $x^* \in X^*$ with $\|x^*\| = 1$ such that

$$(-x^*, x^*) \in \widehat{\mathcal{N}}(\tilde{x}_1;\Omega_1) \times \widehat{\mathcal{N}}(\tilde{x}_2;\Omega_2) + \gamma\left(I\!B^* \times I\!B^*\right) \ \text{ for all } \ \gamma > \varepsilon\,. \tag{2.60}$$

We can easily check by the results above that property (H) holds for the prenormal (Fréchet normal) cone $\widehat{N}$ in Asplund spaces; cf. the proof of Lemma 2.32(ii). In general this property postulates the ability of the prenormal structure $\widehat{\mathcal{N}}$ to describe *first-order necessary optimality conditions* for minimizing functions of the norm type (2.59) over arbitrary sets. Note that (2.60) provides a "fuzzy" optimality condition, since it involves points $(\tilde{x}_1, \tilde{x}_2)$ close to the given minimizer with $\gamma > \varepsilon$ in (2.60).

Let us show that property (H) always holds for subdifferentially generated *prenormal cones* under a minimal amount of natural requirements in the corresponding Banach spaces. Given a Banach space X, we say that $\widehat{\mathcal{D}}$ defines an (abstract) *presubdifferential* on $X \times X$ if it associates, with every proper function $\varphi: X \times X \to \overline{I\!R}$, a set-valued mapping $\widehat{\mathcal{D}}\varphi: X \times X \rightrightarrows X^* \times X^*$ such that $\widehat{\mathcal{D}}\varphi(z) = \emptyset$ for $z \notin \operatorname{dom}\varphi$, $\widehat{\mathcal{D}}\varphi(z) = \widehat{\mathcal{D}}\phi(z)$ if φ and ϕ coincide around z, and one has the following:

(S1) Suppose that $\bar{z}$ provides a local minimum for the sum $\varphi_1 + \varphi_2$ of two functions finite at $\bar{z}$, where φ_1 is a convex continuous function of type (2.59) and where φ_2 is a l.s.c. function of the set indicator type. Then for any $\eta > 0$ there are $u, v \in \bar{z} + \eta I\!B$ such that $\varphi_2(v) \le \varphi_2(\bar{z}) + \eta$ and

$$0 \in \widehat{\mathcal{D}}\varphi_1(u) + \widehat{\mathcal{D}}\varphi_2(v) + \eta\big(I\!B^* \times I\!B^*\big)\,.$$

(S2) $\widehat{\mathcal{D}}\varphi(z)$ is contained in the subdifferential of convex analysis for convex continuous function of type (2.59).

(S3) If $\varphi(x_1, x_2) = \varphi_1(x_1) + \varphi_2(x_2)$, then $\widehat{\mathcal{D}}\varphi(\bar{x}_1, \bar{x}_2) \subset \widehat{\mathcal{D}}\varphi_1(\bar{x}_1) \times \widehat{\mathcal{D}}\varphi_2(\bar{x}_2)$ for any $\bar{x}_i \in \operatorname{dom}\varphi_i$, $i = 1, 2$.

Proposition 2.42 (prenormal cones from presubdifferentials). *Given a Banach space X, let $\widehat{\mathcal{D}}$ be an arbitrary presubdifferential on $X \times X$. Then $\widehat{\mathcal{N}}(x;\Omega) := \widehat{\mathcal{D}}\delta(x;\Omega)$ is a cone for any closed set $\Omega \subset X \times X$ and any $x \in \Omega$, and $\widehat{\mathcal{N}}$ defines a prenormal structure on X.*

Proof. The set $\widehat{\mathcal{N}}(x;\Omega)$ is a cone, since $\alpha\delta(x;\Omega) = \delta(x;\Omega)$ for every $\alpha > 0$. Obviously $\widehat{\mathcal{N}}(x;\Omega) = \emptyset$ if $x \notin \Omega$. We need to show that $\widehat{\mathcal{N}}$ satisfies property (H) in Definition 2.41. To proceed, take $\bar{z} = (\bar{x}_1, \bar{x}_2) \in \Omega_1 \times \Omega_2$ that provides a local minimum for ψ in (2.59) relative to $\Omega_1 \times \Omega_2$ with given $\varepsilon > 0$ and $\bar{x}_1 - \bar{x}_2 + a \neq 0$. Observe that $\bar{z}$ is a local minimizer for the function

$$\varphi(x_1, x_2) := \psi(x_1, x_2) + \delta((x_1, x_2); \Omega_1 \times \Omega_2), \quad (x_1, x_2) \in X \times X\,,$$

with no additional constraints. Pick any $\gamma > \varepsilon$ and put

$$\eta := \gamma - \varepsilon \;\text{ with }\; \eta \le \min\big\{\varepsilon, \nu/2\big\}, \quad \nu := \|\bar{x}_1 - \bar{x}_2 + a\|. \tag{2.61}$$

Applying (S1) with $\varphi_1 = \psi$ and $\varphi_2 = \delta(\cdot;\Omega_1 \times \Omega_2)$ and using the construction of $\widehat{\mathcal{N}}$, we find $u = (x_1', x_2') \in X^2$ and $v = (\tilde{x}_1, \tilde{x}_2) \in \Omega_1 \times \Omega_2$ such that

$$\max\big\{\|x_1' - \bar{x}_1\|, \|x_2' - \bar{x}_2\|, \|\tilde{x}_1 - \bar{x}_1\|, \|\tilde{x}_2 - \bar{x}_2\|\big\} \le \eta \le \varepsilon\,, \tag{2.62}$$

$$0 \in \widehat{\mathcal{D}}\psi(x_1', x_2') + \widehat{\mathcal{N}}((\tilde{x}_1, \tilde{x}_2); \Omega_1 \times \Omega_2) + \eta\big(I\!B^* \times I\!B^*\big)\,.$$

Due to (2.61) and (2.62) we get

$$\|x_1' - x_2'\| \ge \|\bar{x}_1 - \bar{x}_2 + a\| - \big(\|x_1' - \bar{x}_1\| + \|x_2' - \bar{x}_2\|\big) = \nu - 2\eta > 0\,.$$

Observe also that (S3) yields

$$\widehat{\mathcal{N}}((\bar{x}_1, \bar{x}_2); \Omega_1 \times \Omega_2) \subset \widehat{\mathcal{N}}(\bar{x}_1; \Omega_1) \times \widehat{\mathcal{N}}(\bar{x}_2; \Omega_2)\,.$$

By (S2) and the subdifferential formulas of convex analysis for function (2.59) one has the inclusion

$$\widehat{\mathcal{D}}\psi(x_1', x_2') \subset \big(x^*, -x^*\big) + \varepsilon\big(I\!B^* \times I\!B^*\big) \;\text{ with }\; \|x^*\| = 1\,. \tag{2.63}$$

Putting the above together and taking into account that $\gamma = \varepsilon + \eta$, we arrive at (2.60) and finish the proof. △

The result obtained describes an important class of prenormal structures given by subdifferentially generated conic sets. Observe that condition (2.60) with $\|x^*\| = 1$ doesn't necessarily require that $\widehat{\mathcal{N}}(x;\Omega)$ are cones or even unbounded sets. Note also that a *prenormal structure $\widehat{\mathcal{N}}$ doesn't need to be subdifferentially generated.*

Let us describe another class of prenormal structures on X involving *bounded* sets $\widehat{\mathcal{N}}(x;\Omega)$ associated with presubdifferentials of distance functions under minimal requirements. Fix an arbitrary number $\ell > 0$ and consider the class of Lipschitz continuous functions $\varphi: X \times X \to I\!R$ with modulus ℓ. We say that $\widehat{\mathcal{D}}\varphi(\cdot)$ defines an *ℓ-presubdifferential* on this class of functions if it satisfies the above presubdifferential assumptions, where (S1) and (S3) are required to hold, respectively, for functions φ_2 and φ_i, $i = 1, 2$, of this class. Then we define $\widehat{\mathcal{N}}$ on X by

$$\widehat{\mathcal{N}}(x; \Omega) := \begin{cases} \widehat{\mathcal{D}}\big(\ell \operatorname{dist}(x; \Omega)\big) & \text{if } x \in \Omega\ , \\ \emptyset & \text{otherwise} \end{cases} \tag{2.64}$$

for every closed set $\Omega \subset X$, where $\widehat{\mathcal{D}}\big(\ell \operatorname{dist}(x; \Omega)\big) := \widehat{\mathcal{D}}\big(\ell \operatorname{dist}(\cdot; \Omega)\big)(x)$.

Proposition 2.43 (prenormal structures from ℓ-presubdifferentials). *Let $\widehat{\mathcal{D}}$ be an ℓ-presubdifferential with some $\ell > 1$. Then* (2.64) *defines a prenormal structure on a Banach space X.*

Proof. Let us prove that property (H) holds for (2.64) if $\varepsilon > 0$ is sufficiently small. Fix $\ell > 1$ and take $0 < \varepsilon \le (\ell - 1)/2$. Since $(\bar{x}_1, \bar{x}_2)$ is a local minimizer of the function ψ in (2.59) over the set $\Omega_1 \times \Omega_2$, we find neighborhoods U_1 of $\bar{x}_1$ and U_2 of $\bar{x}_2$ such that ψ attains its global minimum over $(\Omega_1 \cap U_1) \times (\Omega_2 \cap U_2)$ at $(\bar{x}_1, \bar{x}_2)$. One can easily see that ψ is Lipschitz continuous on X^2 with modulus $1 + 2\varepsilon \le \ell$. It is well known that the function

$$\varphi(x_1, x_2) := \psi(x_1, x_2) + \ell \operatorname{dist}\big((x_1, x_2); (\Omega_1 \cap U_1) \times (\Omega_2 \cap U_2)\big) \tag{2.65}$$

attains its minimum over the whole space X^2 at $(\bar{x}_1, \bar{x}_2)$; see Proposition 2.4.3 from Clarke [255]. Observe that

$$\operatorname{dist}\big((x_1, x_2); (\Omega_1 \cap U_1) \times (\Omega_2 \cap U_2)\big) = \operatorname{dist}(x_1; \Omega_1 \cap U_1) + \operatorname{dist}(x_2; \Omega_2 \cap U_2)$$

due to $\|(x_1, x_2)\| = \|x_1\| + \|x_2\|$. Similarly to the proof of Proposition 2.42 we pick $\gamma > 0$ and take positive numbers η and ν satisfying (2.61). By the above property (S1) for the ℓ-presubdifferential $\widehat{\mathcal{D}}$ of the sum in (2.65) we find points $u = (x_1', x_2') \in X^2$ and $v = (\tilde{x}_1, \tilde{x}_2) \in X^2$ satisfying (2.62) so that

$$0 \in \widehat{\mathcal{D}}\psi(x_1', x_2') + \widehat{\mathcal{D}}\big(\ell \operatorname{dist}(\tilde{x}_1; \Omega_1 \cap U_1) + \ell \operatorname{dist}(\tilde{x}_2; \Omega_2 \cap U_2)\big) + \eta\big(I\!B^* \times I\!B^*\big).$$

If ε is sufficiently small, one has

$$\operatorname{dist}(x; \Omega_i \cap U_i) = \operatorname{dist}(x; \Omega_i), \quad i = 1, 2\ ,$$

for all x in some neighborhoods of $\tilde{x}_1$ and $\tilde{x}_2$, respectively. Thus

$$0 \in \widehat{\mathcal{D}}\psi(x_1', x_2') + \widehat{\mathcal{N}}(\tilde{x}_1; \Omega_1) \times \widehat{N}(\tilde{x}_2; \Omega_2) + (\gamma - \varepsilon)\big(I\!B^* \times I\!B^*\big)$$

by (2.64) and (S3). Using (S2) and (2.63), we arrive at (2.60). △

As we mentioned above, the basic property (H) of prenormal structures reflects the ability of $\widehat{\mathcal{N}}$ to describe "fuzzy" necessary optimality conditions in constrained optimization. To get "exact" conditions corresponding to $\tilde{x}_i = \bar{x}_i$, $i = 1, 2$, and $\gamma = \varepsilon$ in (2.60), one needs to employ more *robust* normal constructions. The latter can be obtained by using *limiting procedures* based on prenormals. Let us consider two kinds of such limiting constructions involving the sequential Painlevé-Kuratowski upper limit described in (1.1) and its topological closure.

Definition 2.44 (sequential and topological normal structures). *Let $\widehat{\mathcal{N}}$ be an arbitrary prenormal structure on a Banach space X. We say that $\mathcal{N}$ defines a* SEQUENTIAL NORMAL STRUCTURE *on X generated by $\widehat{\mathcal{N}}$ if*

$$\mathcal{N}(\bar{x};\Omega) = \operatorname*{Lim\,sup}_{x \to \bar{x}} \widehat{\mathcal{N}}(x;\Omega) \tag{2.66}$$

for any nonempty set $\Omega \subset X$ and any $\bar{x} \in X$. If (2.66) *is replaced with*

$$\overline{\mathcal{N}}(\bar{x};\Omega) = \mathrm{cl}^*\Big\{ \operatorname*{Lim\,sup}_{x \to \bar{x}} \widehat{\mathcal{N}}(x;\Omega) \Big\}, \tag{2.67}$$

then $\overline{\mathcal{N}}$ defines the corresponding TOPOLOGICAL NORMAL STRUCTURE *on X.*

It immediately follows from the definitions that $\mathcal{N}(\bar{x};\Omega) = \overline{\mathcal{N}}(\bar{x};\Omega) = \emptyset$ for $\bar{x} \notin \Omega$ and, moreover, one may consider only $x \in \Omega$ in (2.66) and (2.67). Obviously $\mathcal{N}(\bar{x};\Omega) \subset \overline{\mathcal{N}}(\bar{x};\Omega)$. However, sequential normal structures are mostly useful in Banach spaces X whose unit dual balls $I\!B^* \subset X^*$ are weak* sequentially compact, while topological normal structures don't need such an assumption; see, e.g., Subsect. 2.5.3.

Similarly we can define sequential and topological *subdifferential* constructions generated by presubdifferentials. It follows from Proposition 1.31 that our basic normal cone (1.3) is smaller than any other sequential (and hence topological) normal structure in Banach spaces under natural requirements. The next proposition gives a counterpart of this minimality result for the basic subdifferential in Definition 1.77(i).

Proposition 2.45 (minimality of the basic subdifferential). *Let X be a Banach space, and let $\widehat{\mathcal{D}}\varphi\colon X \rightrightarrows X^*$ satisfy the following properties on the class of proper l.s.c. functions $\varphi\colon X \to \overline{I\!R}$:*

(M1) *$\widehat{\mathcal{D}}\phi(u) = \widehat{\mathcal{D}}\varphi(x+u)$ for $\phi(u) := \varphi(x+u)$ and $x, u \in X$.*

(M2) *$\widehat{\mathcal{D}}\varphi(x)$ is contained in the subdifferential of convex analysis for convex continuous functions in the form*

$$\varphi(x) := \langle x^*, x\rangle + \varepsilon\|x\|, \quad x^* \in X^*,\ \varepsilon > 0\,. \tag{2.68}$$

(M3) *For any $\eta > 0$ and any functions φ_i, $i = 1, 2$, such that φ_1 is convex of type* (2.68) *and the sum $\varphi_1 + \varphi_2$ attains a local minimum at $x = 0$ there are $x_1, x_2 \in \eta I\!B$ with $|\varphi_2(x_2) - \varphi_2(0)| \le \eta$ and*

$$0 \in \widehat{\mathcal{D}}\varphi_1(x_1) + \widehat{\mathcal{D}}\varphi_2(x_2) + \eta I\!B^* .$$

Then for every $\bar{x} \in \operatorname{dom}\varphi$ one has the inclusion

$$\partial\varphi(\bar{x}) \subset \mathop{\mathrm{Lim\,sup}}_{x \xrightarrow{\varphi} \bar{x}} \widehat{\mathcal{D}}\varphi(x) .$$

Proof. Take $x^* \in \partial\varphi(\bar{x})$ and by Theorem 1.89 find $\varepsilon_k \downarrow 0$, $x_k \xrightarrow{\varphi} \bar{x}$, and $x_k^* \xrightarrow{w^*} x^*$ satisfying $x_k^* \in \widehat{\partial}_{\varepsilon_k}\varphi(x_k)$ for all $k \in I\!N$. Thus there are neighborhoods U_k of x_k such that

$$\varphi(x) - \varphi(x_k) - \langle x_k^*, x - x_k\rangle \ge -2\varepsilon_k\|x - x_k\| \ \text{ for all } \ x \in U_k, \quad k \in I\!N .$$

The latter means that for any fixed k the function

$$\psi_k(x) := \varphi(x_k + x) - \langle x_k^*, x\rangle + 2\varepsilon_k\|x\|$$

attains a local minimum at $x = 0$. Denoting $\varphi_1(x) := \varphi(x_k + x)$ and $\varphi_2(x) := -\langle x_k^*, x\rangle + 2\varepsilon_k\|x\|$, we represent ψ_k as the sum of two functions satisfying the assumptions in (M3). Employ (M3) with $\eta = \varepsilon_k$ and then (M1) and (M2). This gives $u_k \in X$ such that $\|u_k\| \le \varepsilon_k$, $|\varphi(x_k + u_k) - \varphi(x_k)| \le \varepsilon_k$, and

$$x_k^* \in \widehat{\mathcal{D}}\varphi(x_k + u_k) + 3I\!B^*, \quad k \in I\!N .$$

Passing to the limit as $k \to \infty$, we arrive at the desired conclusion. △

It follows from the above proof that $\widehat{\mathcal{D}}$ may be an ℓ-presubdifferential on the class of Lipschitz continuous function $\varphi\colon X \to I\!R$ with modulus $\ell > 0$ if property (M3) is required to hold only for such functions. When $\varphi = \delta(\cdot;\Omega)$, the minimality property in Proposition 2.45 corresponds to the result of Proposition 1.31 for the case of subdifferentially generated normal structures, while the latter result ensures the minimality of the basic normal cone without such an assumption.

2.5.2 Specific Normal and Subdifferential Structures

As proved in Subsect. 2.4.1, our basic normal cone and subdifferential provide a constructively defined class of *sequential* normal and subdifferential structures generated by Fréchet normals and subgradients in arbitrary Asplund spaces. Let us discuss some other remarkable classes of generalized normals and subgradients that satisfy the above requirements to abstract (pre)normal and (pre)subdifferential structures on appropriate Banach space.

A. Convex-Valued Constructions by Clarke. We start with Clarke's constructions of generalized normals to sets and subgradients of extended-real-valued functions that produce *topological* normal and subdifferential structures on arbitrary *Banach spaces* by the following *four-step* procedure; see Clarke [255] for more details and proofs. First let φ be Lipschitz continuous around $\bar{x} \in X$ with modulus ℓ. The *generalized directional derivative* of φ at $\bar{x}$ in the direction h is

$$\varphi^\circ(\bar{x}; v) := \limsup_{\substack{x \to \bar{x} \\ t \downarrow 0}} \frac{\varphi(x + tv) - \varphi(x)}{t} . \tag{2.69}$$

The function $\varphi^\circ(\bar{x}; \cdot) \colon X \to I\!R$ happens to be convex for any Lipschitzian φ; moreover, (2.69) is upper semicontinuous in both variables with $\varphi^\circ(\bar{x}; -v) = (-\varphi)^\circ(\bar{x}; v)$ and $|\varphi^\circ(\bar{x}; v)| \le \ell \|v\|$ for all $v \in X$. Then the *generalized gradient* of a locally Lipschitzian function is defined by

$$\partial_C \varphi(\bar{x}) := \big\{ x^* \in X^* \big|\ \langle x^*, v \rangle \le \varphi^\circ(\bar{x}; v) \ \text{ for any } \ v \in X \big\} . \tag{2.70}$$

It follows from (2.70) and the properties of φ° that $\partial_C \varphi(\bar{x})$ is a nonempty, weak* compact, convex subset of X^* with $\|x^*\| \le \ell$ for all $x^* \in \partial_C \varphi(\bar{x})$ and the classical plus-minus symmetry

$$\partial_C(-\varphi)(\bar{x}) = -\partial_C \varphi(\bar{x}) \ \text{ for Lipschitzian } \ \varphi . \tag{2.71}$$

The next step is to define the *Clarke normal cone* to $\Omega \subset X$ by

$$N_C(\bar{x}; \Omega) := \text{cl}^* \Big\{ \bigcup_{\lambda > 0} \lambda \partial_C \text{dist}(\bar{x}; \Omega) \Big\}, \quad \bar{x} \in \Omega , \tag{2.72}$$

through the generalized gradient of the Lipschitzian distance function, with $N_C(\bar{x}; \Omega) := \emptyset$ for $\bar{x} \notin \Omega$. Finally, the *Clarke subdifferential* of a function $\varphi \colon X \to \overline{I\!R}$ is defined by

$$\partial_C \varphi(\bar{x}) := \big\{ x^* \in X^* \big|\ (x^*, -1) \in N_C((\bar{x}, \varphi(\bar{x})); \text{epi}\, \varphi) \big\} \tag{2.73}$$

if $|\varphi(\bar{x})| < \infty$ and $\partial_C \varphi(\bar{x}) := \emptyset$ if $|\varphi(\bar{x})| = \infty$. Clearly the sets (2.72) and (2.73) are convex and weak* closed in X^*. The two basic facts ensuring that (2.72) defines a *topological* normal structure on X generated by $\bigcup_{\lambda > 0} \lambda \partial_C \text{dist}(\bar{x}; \Omega)$ are the following: the sum rule

$$\partial_C \big(\varphi_1 + \varphi_2 \big)(\bar{x}) \subset \partial_C \varphi_1(\bar{x}) + \partial_C \varphi_2(\bar{x}) \tag{2.74}$$

if φ_1 is locally Lipschitzian and φ_2 is l.s.c. around $\bar{x}$, and that the graph of $\partial_C \varphi(\cdot)$ is closed in the norm$\times$weak* topology of $X \times X^*$ if φ is Lipschitz continuous. Moreover, these facts imply by Proposition 2.43 that for any fixed $\lambda > 0$ the sets $\lambda \partial_C \text{dist}(\bar{x}; \Omega)$ define a topological normal structure on X. Note however that there are generally *strict* inclusions

$$N_C(\bar{x}; \Omega) \subset \mathop{\text{Lim}\,\text{sup}}_{x \to \bar{x}} N_C(x; \Omega) \subset \text{cl}^* \Big\{ \mathop{\text{Lim}\,\text{sup}}_{x \to \bar{x}} N_C(x; \Omega) \Big\} ,$$

where the first one may be strict even in finite dimensions unless Ω is epi-Lipschitzian at $\bar{x}$; see Rockafellar [1146]. Note also that the Clarke normal cone may be too large, especially for graphs of Lipschitzian functions when it is actually a *linear subspace*; see the proof of Theorem 1.46 and its infinite-dimensional generalizations in Subsect. 3.2.4. In particular, for $\Omega = \operatorname{gph}|x| \subset \mathbb{R}^2$ one has

$$N_C(0;\Omega) = \mathbb{R}^2, \quad \text{while } N(0;\Omega) = \big\{(v_1,v_2)\big|\, v_2 \le -|v_1|\big\} \cup \big\{(v_1,v_2)\big|\, v_2 = v_1\big\}$$

for the basic normal cone N. It follows from Proposition 2.45 that

$$\partial\varphi(\bar{x}) \subset \partial_C\varphi(\bar{x}) \;\text{ and }\; N(\bar{x};\Omega) \subset N_C(\bar{x};\Omega)$$

in general Banach spaces. More precise relationships between these objects will be obtained in Subsect. 3.2.3 in the Asplund space setting.

B. Approximate Normals and Subgradients. Another type of *topological* normal and subdifferential structures was developed by Ioffe, under the name of "approximate normals and subgradients," as an extension of Mordukhovich's construction to arbitrary *Banach spaces*; see remarks and references in Subsect. 1.4.7 and the corresponding results of Subsect. 3.2.3 on close connections with our basic constructions in the Asplund space setting. It doesn't seem that the adjective "approximate" reflects the essence of these constructions, while its usage in this context clearly contradicts the regular use of this word in the book; see Subsect. 1.4.7 and also remarks in Rockafellar and Wets [1165, p. 347] for motivations of the word "approximate" appearing in this setting. On the other hand, it has been widely spread in nonsmooth analysis. In what follows we put quotation marks when referring to "approximate" normals and subdifferentials in this context.

Let us describe the multistep procedure for these constructions from the paper of Ioffe [599], where the reader can find proofs, more discussions, and references. Given $\varphi\colon X \to \overline{\mathbb{R}}$ finite at $\bar{x}$, the constructions

$$d^-\varphi(\bar{x};v) := \liminf_{\substack{z\to v\\ t\downarrow 0}} \frac{\varphi(\bar{x}+tz)-\varphi(\bar{x})}{t},$$

$$\partial_\varepsilon^-\varphi(\bar{x}) := \big\{x^* \in X^*\big|\, \langle x^*,v\rangle \le d^-\varphi(\bar{x};v) + \varepsilon\|v\|\big\}$$

are called the *lower Dini* (or Dini-Hadamard) *directional derivative* and the *Dini* ε*-subdifferential* of φ at $\bar{x}$, respectively. As usual, we put $\partial^-\varphi(\bar{x}) := \emptyset$ if $|\varphi(\bar{x})| = \infty$. Note that the sets $\partial_\varepsilon^-\varphi(\bar{x})$ are always convex, while the function $d^-\varphi(\bar{x};\cdot)$ is not. One can check that $\partial_\varepsilon^-\varphi(\bar{x})$ reduces to the analytic ε-subdifferential from Definition 1.83(ii) if $\dim X < \infty$. In general, the A*-subdifferential* of φ at $\bar{x}$ is defined via topological limits involving finite-dimensional reductions of ε-subgradients as

$$\partial_A\varphi(\bar{x}) := \bigcap_{\substack{L\in\mathcal{L}\\ \varepsilon>0}} \overline{\operatorname*{Lim\,sup}_{x\xrightarrow{\varphi}\bar{x}} \partial_\varepsilon^-\big(\varphi+\delta(\cdot;L)\big)(x)} \tag{2.75}$$

where $\mathcal{L}$ is the collection of all finite-dimensional subspaces of X and where $\overline{\text{Lim sup}}$ stands for the *topological* counterpart of the Painlevé-Kuratowski upper limit (1.1) with sequences replaced by nets. Further, the *G-normal cone* N_G and its *nucleus* $\widetilde{N}_G$ to Ω at $\bar{x} \in \Omega$ are defined by

$$N_G(\bar{x};\Omega) := \text{cl}^*\widetilde{N}_G(\bar{x};\Omega) \text{ and } \widetilde{N}_G(\bar{x};\Omega) := \bigcup_{\lambda>0} \lambda\partial_A \text{dist}(\bar{x};\Omega)\,, \tag{2.76}$$

respectively, with $N_G(\bar{x};\Omega) = \widetilde{N}_G(\bar{x};\Omega) = \emptyset$ if $\bar{x} \notin \Omega$. Finally, the *G-subdifferential* of φ at $\bar{x}$ is defined geometrically as

$$\partial_G\varphi(\bar{x}) := \big\{x^* \in X^* \big|\ (x^*,-1) \in N_G((\bar{x},\varphi(\bar{x}));\text{epi}\,\varphi)\big\}\,, \tag{2.77}$$

while its *G-nucleus* $\widetilde{\partial}_G\varphi(\bar{x})$ corresponds to (2.77) with N_G replaced by $\widetilde{N}_G$. One always has

$$\widetilde{\partial}_G\varphi(\bar{x}) \subset \partial_G\varphi(\bar{x}) \subset \partial_A\varphi(\bar{x})\,,$$

where equalities hold if φ is locally Lipschitzian around $\bar{x}$. For closed sets Ω the graph of $N_G(\cdot;\Omega)$ is closed in the norm×weak* topology of $X \times X^*$. Moreover, both $\partial_G\varphi$ and $\widetilde{\partial}_G\varphi$ satisfy the sum rule in form (2.74) if φ_1 is locally Lipschitzian and φ_2 is l.s.c. around $\bar{x}$. Hence $N_G(\cdot;\Omega)$ and $\lambda\partial_A\text{dist}(\cdot;\Omega)$ provide topological normal structures on X and

$$\partial\varphi(\bar{x}) \subset \widetilde{\partial}_G\varphi(\bar{x}), \quad N(\bar{x};\Omega) \subset \widetilde{N}_G(\bar{x};\Omega)$$

by Proposition 2.45. Note that the latter inclusions may be *strict*, even in the case of Lipschitz continuous functions on spaces with Fréchet smooth renorms; see Example 3.61. In Subsect. 3.2.3 we obtain more precise relationships between these constructions in the general case of Asplund spaces.

C. Viscosity Subdifferentials. Next we consider normal and subgradient constructions related to the so-called *viscosity subdifferentials* that generally make sense in *smooth* Banach spaces admitting smooth renorms (or bump functions) with respect to some bornology; see Remark 2.11. The following description is based on the paper by Borwein, Mordukhovich and Shao [151], where one can find more details and references on the genesis and applications of such constructions; see also the book by Borwein and Zhu [164].

Given a bornology β on a Banach space X, we denote by X^*_β the dual space X^* endowed with the topology of uniform convergence on β-sets. The latter convergence agrees with the norm convergence in X^* when β is the (strongest) Fréchet bornology, and with the weak* convergence in X^* when β is the (weakest) Gâteaux bornology. A function $\theta\colon X \to \overline{I\!R}$ is *β-differentiable* at $\bar{x}$ with β-derivative $\nabla_\beta\theta(\bar{x}) \in X^*$ provided that

$$t^{-1}\Big(\theta(\bar{x}+tv) - \theta(\bar{x}) - t\langle\nabla_\beta\theta(\bar{x}),v\rangle\Big) \to 0$$

as $t \to 0$ uniformly in $v \in V$ for every $V \in \beta$. This function is said to be *β-smooth* around $\bar{x}$ if it is β-differentiable at each point of a neighborhood U

of $\bar{x}$ and $\nabla_\beta\theta\colon X \to X^*_\beta$ is continuous on U. The latter requirement is essential; in the case of $\beta = \mathcal{F}$, the Fréchet bornology on X, it means that $\nabla\theta\colon X \to X^*$ is norm-to-norm continuous around $\bar{x}$. Note that in the Fréchet case the β-smoothness of θ implies its Lipschitz continuity around $\bar{x}$, which may not happen for weaker bornologies $\beta < \mathcal{F}$.

Now, given $\varphi\colon X \to \overline{I\!R}$ finite at $\bar{x}$, its *viscosity β-subdifferential of rank* $\lambda > 0$ at $\bar{x}$ is the set $\partial^\lambda_\beta\varphi(\bar{x})$ of all $x^* \in X^*$ with the following properties: there are a neighborhood U of $\bar{x}$ and a β-smooth function $\theta\colon U \to I\!R$ such that θ is Lipschitz continuous on U with modulus λ, $\nabla_\beta\theta(\bar{x}) = x^*$, and $\varphi - \theta$ attains a local minimum at $\bar{x}$. The corresponding set of *β-normals of rank* λ to $\Omega \subset X$ at $\bar{x} \in \Omega$ is defined by $N^\lambda_\beta(\bar{x};\Omega) := \partial^\lambda_\beta\delta(\bar{x};\Omega)$. The unions

$$\partial_\beta\varphi(\bar{x}) := \bigcup_{\lambda>0}\partial^\lambda_\beta\varphi(\bar{x}), \quad N_\beta(\bar{x};\Omega) := \bigcup_{\lambda>0}N^\lambda_\beta(\bar{x};\Omega) \tag{2.78}$$

are called the *viscosity β-subdifferential* of φ at $\bar{x}$ and the *viscosity β-normal cone* of Ω at $\bar{x}$, respectively. Note that $\theta(\cdot)$ in the above definition can be equivalently chosen to be *concave* if X admits a β-smooth renorm.

Employing the variational descriptions of Fréchet normals and subgradients in Theorems 1.30 and 1.88, we conclude that

$$\partial_{\mathcal{F}}\varphi(\bar{x}) = \widehat{\partial}\varphi(\bar{x}) \quad \text{and} \quad N_{\mathcal{F}}(\bar{x}) = \widehat{N}(\bar{x};\Omega)$$

if X admits an *$\mathcal{F}$-smooth bump function.* These constructions may be different in more general settings of Banach and Asplund spaces. Note that, in contrast to $\widehat{\partial}\varphi(\cdot)$ and $\widehat{N}(\cdot;\Omega)$, the viscosity constructions (2.78) don't reveal useful properties without smoothness assumptions on the space in question.

It follows from the results of the afore-mentioned paper [151] that $\partial^\lambda_\beta\varphi(\cdot)$ defines a *presubdifferential* structure on a β-smooth space X for any $\lambda > 1$. Hence $N^\lambda_\beta(\cdot;\Omega)$ defines the corresponding prenormal structure under these conditions. By Proposition 2.45 we have

$$\partial\varphi(\bar{x}) \subset \mathop{\mathrm{Lim\,sup}}_{x \xrightarrow{\varphi} \bar{x}}\partial_\beta\varphi(x), \quad N(\bar{x};\Omega) \subset \mathop{\mathrm{Lim\,sup}}_{x \xrightarrow{\Omega} \bar{x}} N_\beta(x;\Omega)) \tag{2.79}$$

in β-smooth spaces. It doesn't seem to be true that viscosity subdifferentials (2.78) and their *sequential* limits in (2.79) enjoy the semi-Lipschitzian sum rules of the corresponding types (b) and (c) in Proposition 2.33 on β-smooth spaces with $\beta < \mathcal{F}$. On the other hand,

$$\widetilde{\partial}_G\varphi(\bar{x}) = \bigcup_{\lambda>0}\mathrm{cl}^*\Big\{\mathop{\mathrm{lim\,sup}}_{x \xrightarrow{\varphi} \bar{x}}\partial^\lambda_\beta\varphi\Big\}, \quad \partial_A\varphi(\bar{x}) = \overline{\mathop{\mathrm{Lim\,sup}}_{x \xrightarrow{\varphi} \bar{x}}}\,\partial_\beta\varphi(x)$$

for the *nucleus* of the G-subdifferential (2.77) and for the A-subdifferential (2.75) of any l.s.c. function on an arbitrary β-smooth Banach space; cf. Borwein and Ioffe [147, Theorem 2] and Mordukhovich, Shao and Zhu [954, Theorem 6.1], respectively.

D. Proximal Constructions. Let us consider the *Hilbert space* setting that is the closest to finite dimensions and allows one to construct prenormal and presubdifferential structures defined through the *Euclidean metric.* Given a closed subset $\Omega \subset X$ of a Hilbert space and the Euclidean projector $\Pi(\cdot;\Omega)$, the conic set

$$N_P(\bar{x};\Omega) := \text{cone}\left[\Pi^{-1}(\bar{x};\Omega) - \bar{x}\right] \tag{2.80}$$

is the *proximal normal cone* to Ω at $\bar{x} \in \Omega$. It follows from the Euclidean norm properties (cf. the proof of Theorem 1.6 above) that $x^* \in N_P(\bar{x};\Omega)$ if and only if there is $\alpha > 0$ such that

$$\langle x^*, x - \bar{x}\rangle \le \alpha\|x - \bar{x}\|^2 \quad \text{for all} \quad x \in \Omega\,.$$

This obviously implies that $N_P(\bar{x};\Omega)$ is a convex subcone of $\widehat{N}(\bar{x};\Omega)$. In contrast to the latter one, $N_P(\bar{x};\Omega)$ may not be closed even in finite dimensions; moreover, its closure may be different from $\widehat{N}(\bar{x};\Omega)$. A simple example is provided by the epigraph of a smooth function:

$$\Omega = \text{epi}\,\varphi \subset I\!R^2 \quad \text{with} \quad \varphi(x) = -|x|^{3/2} \quad \text{at} \quad \bar{x} = (0,0)\,,$$

where $N_P(\bar{x};\Omega) = \{(0,0)\}$ and $\widehat{N}(\bar{x};\Omega) = \{(v_1,v_2)|\ v_1 = 0,\ v_2 \le 0\}$.

A functional counterpart of the proximal normal cone (2.70) is the *proximal subdifferential* of a proper l.s.c. function $\varphi\colon X \to \overline{I\!R}$ at $\bar{x} \in \text{dom}\,\varphi$ defined as

$$\partial_P\varphi(\bar{x}) := \left\{x^* \in X^*\Big|\ \liminf_{x\to\bar{x}} \frac{\varphi(x) - \varphi(\bar{x}) - \langle x^*, x - \bar{x}\rangle}{\|x - \bar{x}\|^2} > -\infty\right\}, \tag{2.81}$$

which is a convex subset of the Fréchet subdifferential $\widehat{\partial}\varphi(\bar{x})$ and can be equivalently described by $(x^*, -1) \in N_P((\bar{x},\varphi(\bar{x}));\text{epi}\,\varphi)$. Note that the proximal subdifferential may be *empty* even for *smooth* functions as in the above example, where $\partial_P\varphi(0) = \emptyset$ while $\widehat{\partial}\varphi(0) = \{0\}$. Nevertheless, for every proper l.s.c. function φ finite at $\bar{x}$ the following holds: given any $x^* \in \widehat{\partial}\varphi(\bar{x})$, there are sequences $x_k \xrightarrow{\varphi} \bar{x}$ and $x_k^* \in \partial_P\varphi(x_k)$ such that $\|x_k^* - x^*\| \to 0$ as $k \to \infty$; see Loewen [802, Theorem 5.5]. Therefore

$$\partial\varphi(\bar{x}) = \mathop{\text{Lim sup}}_{x\xrightarrow{\varphi}\bar{x}} \partial_P\varphi(x) \quad \text{and} \quad N(\bar{x};\Omega) = \mathop{\text{Lim sup}}_{x\xrightarrow{\Omega}\bar{x}} N_P(x;\Omega)\,.$$

A crucial fact ensuring that (2.81) defines a *presubdifferential* structure on a Hilbert space X (hence $N_P(\cdot;\Omega)$ defines the corresponding prenormal structure) follows from the fuzzy sum rule for $\partial_P\varphi(\cdot)$ proved in Ioffe and Rockafellar [616, Theorem 2] and in Clarke et al. [265, Theorem 1.8.3].

E. Derivate Sets. In conclusion of this subsection we compare our subdifferential constructions with generalized derivatives based on the idea of *uniformly approximating* nonsmooth functions by smooth (finitely differentiable) functions. Recall that a mapping $f\colon X \to Y$ between Banach spaces is *finitely* *differentiable* at $\bar{x}$ with the derivative $\nabla f(\bar{x})$ if for every finite-dimensional subspace $X \subset X$ the mapping$z \to f(x+z)\colon Z \to Y$ is differentiable at the origin and its derivative agrees with the restriction of $\nabla f(\bar{x})$ to Z.

Given $\varphi\colon X \to \overline{I\!R}$ on a Banach space X and a point $\bar{x} \in X$ with $|\varphi(\bar{x})| < \infty$, we denote by $\mathcal{A}\varphi(\bar{x})$ a subset of X^* with the following properties: for any $\varepsilon, \alpha > 0$ there are $\gamma \in (0, \alpha]$ and a continuously finitely differentiable function $\psi\colon X \to I\!R$ such that

$$|\varphi(x) - \psi(x)| \le \varepsilon\gamma \quad \text{and} \quad \nabla\psi(x) \in \mathcal{A}\varphi(\bar{x}) \quad \text{for all} \quad x \in \bar{x} + \gamma I\!B\,.$$

The *derivate set* $\mathcal{A}\varphi(\bar{x})$ is a derivative-like object, which is not uniquely defined. If φ is continuous around $\bar{x}$ and can be represented as the uniform limit of a sequence of continuously finitely differentiable functions φ_i, $i \in I\!N$, then for any $\gamma > 0$ and $j \in I\!N$ one can take

$$\mathcal{A}\varphi(\bar{x}) = \bigcup_{\substack{\|x-\bar{x}\| \le \gamma \\ i \ge j}} \{\nabla\varphi_i(x)\}\,.$$

The following result shows that for every function φ the Fréchet subdifferential of φ at $\bar{x}$ is contained in the *norm* closure of *any* derivate set $\mathcal{A}\varphi(\bar{x})$ obtained via a uniform approximation by finitely smooth functions.

Theorem 2.46 (derivate sets and Fréchet subgradients). *Let X be a Banach space, and let $\mathcal{A}\varphi(\bar{x})$ be a derivate set of $\varphi\colon X \to \overline{I\!R}$ finite at $\bar{x}$. Then*

$$\widehat{\partial}\varphi(\bar{x}) \subset \operatorname{cl}\mathcal{A}\varphi(\bar{x}) \quad \textit{if} \quad \mathcal{A}\varphi(\bar{x}) \ne \emptyset\,.$$

Proof. Let $\bar{x}^* \notin \operatorname{cl}\mathcal{A}\varphi(\bar{x})$. Then there is $\eta > 0$ such that

$$\|\bar{x}^* - x^*\| > \eta \quad \text{for all} \quad x^* \in \mathcal{A}\varphi(\bar{x})\,. \tag{2.82}$$

Put $\bar{\varepsilon} := \eta/4$ and for each $k \in I\!N$ select a number γ_k and a function ψ_k according to the definition of the derivate set $\mathcal{A}\varphi(\bar{x})$ with $\varepsilon = \bar{\varepsilon}/4$ and $\alpha = 1/k$.

Next we define, for some positive integer N_k, a finite set of points $x_i \in X$, $i = 0, 1, \ldots, N_k$, from the following conditions:

(a) $x_0 = \bar{x}$, $\;x_{i+1} = x_i + hz_i$, $i = 0, 1, \ldots, N_k - 1$;

(b) $\|z_i\| = 1$, $i = 0, 1, \ldots, N_k - 1$;

(c) $h = \gamma_k/(2N_k)$;

(d) $\langle \bar{x}^* - \nabla\psi_k(x_i), z_i\rangle > \eta$, $i = 0, 1, \ldots, N_k - 1$.

Note that it is possible to find z_i satisfying (d) because ψ is finitely differentiable at x_i with $\nabla = \psi(x_i) \in \mathcal{A}\varphi(\bar{x})$, (2.82) holds, and

$$\|x_i - \bar{x}\| \le N_k h = \gamma_k/2 \quad \text{for} \quad i = 1, \ldots, N_k \tag{2.83}$$

due to (a), (b), and (c). When N_k is sufficiently large, one has

$$\psi_k(x_{N_k}) - \psi_k(\bar{x}) - \langle \bar{x}^*, x_{N_K} - \bar{x} \rangle$$

$$= \sum_{i=0}^{N_k-1} \Big(\int_0^h \langle \nabla \psi_k(x_i + tz_i), z_i \rangle dt - h \langle \bar{x}^*, z_i \rangle \Big)$$

$$\le h \sum_{i=0}^{N_k} \langle \psi_k(x_i) - \bar{x}^*, z_i \rangle + \frac{\eta \gamma_k}{4} .$$

This implies, by (d) and (c), that

$$\psi_k(x_{N_k}) - \psi_k(\bar{x}) - \langle \bar{x}^*, x_{N_k} - \bar{x} \rangle < -\eta\gamma_k/2 = \bar{\varepsilon}\gamma_k . \tag{2.84}$$

Now recall that ψ_k approximates the original function φ by

$$|\varphi(x) - \psi_k(x)| \le \bar{\varepsilon}\gamma_k/4 \quad \text{whenever} \quad x \in \bar{x} + \gamma_k I\!B .$$

Combining this with (2.83) and (2.84), we finally get

$$\varphi(x_{N_k} - \varphi(\bar{x}) - \langle \bar{x}^*, x_{N_k} - \bar{x} \rangle \le \bar{\varepsilon}\gamma_k/2 \le -\bar{\varepsilon}\|x_{N_k} - \bar{x}\| .$$

Since $x_{N_k} \to \bar{x}$ as $k \to \infty$, the latter means that $\bar{x}^* \notin \widehat{\partial}\varphi(\bar{x})$, which ends the proof of the theorem. △

Theorem 2.46 concerns relationships between Fréchet subgradients and derivate sets of real-valued functions that can be approximated by smooth functions near the point under consideration. It easily implies corresponding results for mappings $f\colon X \to Y$ involving their scalarization. In particular, we deduce from Theorem 2.46 the following relationship between Fréchet subgradients and screens introduced by Halkin [544] for mappings between finite-dimensional spaces.

Recall that, given $f\colon U \to I\!R^m$ defined on an open subset $U \subset I\!R^n$, a nonempty set $\mathcal{U} \subset I\!R^{mn}$ is called a *screen* of f at $\bar{x} \in U$ if for every $\varepsilon, \alpha > 0$ there exist $\gamma > 0$ and a $\mathcal{C}^1$ mapping $g\colon B^n_\gamma(\bar{x}) \to I\!R^m$ such that $B^n_\gamma(\bar{x}) \subset U$,

$$\|f(x) - g(x)\| \le \varepsilon\gamma, \quad \text{and} \quad \nabla g(x) \in \mathcal{U} + \varepsilon I\!B^{mn} \quad \text{for all} \quad x \in B^n_\gamma(\bar{x}) ,$$

where $B^n_\gamma(\bar{x}) := \bar{x} + \gamma\, I\!B_{I\!R^n}$ and $I\!B^{mn}$ stands for the closed unit ball in $I\!R^{mn}$.

Corollary 2.47 (relationship between Fréchet subgradients and screens). *Let $\mathcal{U} \subset I\!R^{mn}$ be a screen of a mapping $f\colon U \to I\!R^m$ at $\bar{x} \in U \subset I\!R^n$. Then*

$$\widehat{\partial}\langle y^*, f\rangle(\bar{x}) \subset \mathrm{cl}\,\big\{A^*y^* \big|\ A \in \mathcal{U}\big\} \quad \textit{for all} \quad y^* \in I\!R^m .$$

Proof. Given $y^* \in I\!R^m$ and a screen $\mathcal{U}$ of f at $\bar{x}$, it is not hard to check that the set $\{A^*y^*|\ A \in \mathcal{U}\}$ satisfies all the above properties of the derivate set $\mathcal{A}\varphi(\bar{x})$ for the scalarized function $\varphi(x) := \langle y^*, f\rangle(x)$ at $\bar{x}$. △

A screen of a mapping is not uniquely defined. Particular examples of screens are given by *derivate containers* of Warga [1316], which include Clarke's *generalized Jacobian* for locally Lipschitzian mappings between finite-dimensional spaces. Warga [1319] also introduced the concept of *directional derivate containers* for mappings between infinite-dimensional spaces. Theorem 2.46 allows us to obtain the following relationships between the latter construction for mappings (see the afore-mentioned papers by Warga for the exact definition) and Fréchet subgradients of their scalarizations.

Corollary 2.48 (relationship between Fréchet subgradients and derivate containers). *Consider a directional derivate container* $\{\Lambda^\varepsilon f(\bar{x})|\ \varepsilon > 0\}$ *of a mapping* $f\colon \Omega \to Y$ *at* $\bar{x} \in \operatorname{int}\Omega$*, where* $\Omega \subset X$ *is a convex compact set, and where the spaces* X *and* Y *are Banach. Then for any* $y^* \in Y^*$, $\varepsilon > 0$, *and* $\eta > 0$ *there is* $\gamma > 0$ *such that*

$$\widehat{\partial}\langle y^*, f\rangle(x) \subset \big\{A^*y^* \big|\ A \in \Lambda^\varepsilon f(\bar{x})\big\} + \eta I\!B^* \quad \textit{whenever} \quad x \in \bar{x} + \gamma I\!B\ .$$

Note that the assumption $\bar{x} \in \operatorname{int}\Omega$ is essential for the validity of the latter result. Indeed, for the function $f\colon [0,1] \to I\!R$ with $f \equiv 0$ extended by ∞ outside of $[0,1]$, we clearly have $\widehat{\partial} f(1) = [0,\infty)$, while the singleton $\{0\}$ is a directional derivate container of f at $\bar{x} = 1$.

Observe that the derivative-like constructions in Theorem 2.46 and Corollaries 2.47 and 2.48 are generally related to *presubdifferential* structures, which lead to *robust subdifferentials* and corresponding generalized derivatives of mappings via some *regularization* procedure. To this end let us recall the definition of the *minimal derivate container* by Warga

$$\begin{aligned}\Lambda^0 f(\bar{x}) :&= \operatorname*{Lim\,sup}_{\substack{x\to\bar{x}\\ k\to\infty}} \big\{\nabla f_k(x)\big\}\\ &= \bigcap_{j=1}^{\infty}\bigcap_{\gamma>0} \operatorname{cl} \bigcup_{\substack{\|x-\bar{x}\|\le\gamma\\ i\ge j}} \{\nabla f_i(x)\}\end{aligned}$$

for a continuous mapping $f\colon X \to Y$ between finite-dimensional spaces that admits a uniform approximation by a sequence of $\mathcal{C}^1$ mappings f_k. It follows from the results obtained that

$$\partial\langle y^*, f\rangle(\bar{x}) \subset \big\{A^*y^* \big|\ A \in \Lambda^0 f(\bar{x})\big\} \quad \text{for all} \quad y^* \in Y^*\ ,$$

which gives the inclusion

$$\partial^0\varphi(\bar{x}) := \partial\varphi(\bar{x}) \cup \partial^+\varphi(\bar{x}) \subset \Lambda^0\varphi(\bar{x}) \tag{2.85}$$

for the two-sided/symmetric generalized differential (1.46) of a real-valued function φ continuous around $\bar{x}$. The following example illustrates (2.85) and other relationships between various subgradients studied above.

Example 2.49 (computing subgradients of Lipschitzian functions). Consider the function

$$\varphi(x) := \big|\, |x_1| + x_2 \big|, \quad x = (x_1, x_2) \in I\!R^2 \ ,$$

which is Lipschitz continuous on $I\!R^2$. Based on representation (1.51), we compute Fréchet subgradients of φ at every point $x \in I\!R^2$ as follows:

$$\widehat{\partial}\varphi(x) = \begin{cases} (1,1) & \text{if } x_1 > 0,\ x_1 + x_2 > 0 \ , \\ (-1,-1) & \text{if } x_1 < 0,\ x_1 + x_2 < 0 \ , \\ (-1,1) & \text{if } x_1 < 0,\ x_1 - x_2 < 0 \ , \\ (1,-1) & \text{if } x_1 < 0,\ x_1 - x_2 > 0 \ , \\ \{(v,1)|\ -1 \le v \le 1\} & \text{if } x_1 = 0,\ x_2 > 0 \ , \\ \{(v_1,v_2)|\ -1 \le v \le 1\} & \text{if } x_1 > 0,\ x_1 + x_2 = 0 \ , \\ \{(v_1,-v_2)|\ -1 \le v \le 1\} & \text{if } x_1, 0,\ x_1 - x_2 = 0 \ , \\ \{(v_1,v_2)|\ |v_1| \le v_2 \le 1\} & \text{if } x_1 = 0,\ x_2 = 0 \ , \\ \emptyset & \text{if } x_1 = 0,\ x_2 < 0 \ . \end{cases}$$

Similarly, based on representation (1.52), we compute Fréchet *upper* subgradients of the above function by

$$\widehat{\partial}^+\varphi(x) = \begin{cases} (1,1) & \text{if } x_1 > 0,\ x_1 + x_2 > 0 \ , \\ (-1,-1) & \text{if } x_1 < 0,\ x_1 + x_2 < 0 \ , \\ (-1,1) & \text{if } x_1 < 0,\ x_1 - x_2 < 0 \ , \\ (1,-1) & \text{if } x_1 < 0,\ x_1 - x_2 > 0 \ , \\ \{(v,-1)|\ -1 \le v \le 1\} & \text{if } x_1 = 0,\ x_1 - x_2 < 0 \ , \\ \emptyset & \text{otherwise} \ . \end{cases}$$

Now using the limiting representation (1.56) of the basic subdifferential in Theorem 1.89 and the symmetric representation of upper subgradients, we arrive at the subgradient sets

$$\partial\varphi(0) = \{(v_1, v_2)|\ |v_1| \le v_2 \le 1\} \cup \{(v_1, v_2)|\ v_2 = -|v_1|,\ -1 \le v_1 \le 1\}\ ,$$

$$\partial^+\varphi(0) = \{(v, -1)|\ -1 \le v \le 1\} \cup \{(1, -1),\ (1, 1)\}\ ,$$

$$\partial^0\varphi(0) = \partial\varphi(0) \cup \{(v, -1)|\ -1 \le v \le 1\}\ .$$

Warga's minimal derivate container for this function is the nonconvex set

$$\Lambda^0\varphi(0) = \{\alpha(v, 1)|\ \alpha, v \in [-1, 1]\}\ ,$$

which is the union of two triangles with vertices at (0,0), (1,1), $(-1, 1)$ and (0,0), $(1, -1)$, $(-1, 1)$, respectively. Clarke's generalized gradient is the whole unit squire $[-1, 1] \times [-1, 1]$.

2.5.3 Abstract Versions of Extremal Principle

In the conclusion of this section we establish approximate and exact versions of the extremal principle valid, respectively, for abstract prenormal and normal structures considered in Subsect. 2.5.1. They hold, in particular, for the specific classes of generalized normals in appropriate Banach spaces described in Subsect. 2.5.2.

We'll see that an approximate version of the extremal principle doesn't impose any requirements on abstract prenormal structures in addition to those formulated in Definition 2.41. In contrast to Theorem 2.22, we obtain the exact extremal principle in Banach spaces in two limiting forms–sequential and topological–involving sequential and topological normal structures, respectively. Note that both limiting forms hold under the following *sequential* normal compactness condition formulated in terms of the corresponding prenormal structure similarly to Definition 1.20.

Definition 2.50 (abstract sequential normal compactness). *Let $\widehat{\mathcal{N}}$ define a prenormal structure on a Banach space X. We say that $\Omega \subset X$ is $\widehat{\mathcal{N}}$-SEQUENTIALLY NORMALLY COMPACT at $\bar{x} \in \Omega$ if for any sequence $(x_k, x_k^*) \in X \times X^*$ satisfying*

$$x_k^* \in \widehat{\mathcal{N}}(x_k; \Omega), \quad x_k \to \bar{x}, \quad x_k^* \xrightarrow{w^*} 0$$

one has $\|x_k^*\| \to 0$ as $k \to \infty$.

This property obviously holds in finite-dimensional spaces for any prenormal structure $\widehat{\mathcal{N}}$. When $\widehat{\mathcal{N}} = \widehat{N}$, the prenormal cone of Definition 1.1(i), we studied the SNC property and its modification in Subsect. 1.1.3 for arbitrary Banach spaces. In particular, we established the relationships with the compactly epi-Lipschitzian (CEL) property of sets. In addition to Remark 1.27, let us mention that, for any closed set Ω in a Banach space X, the CEL property is *equivalent* to the *topological* counterpart of the SNC property in Definition 2.50, where sequences (x_k, x_k^*) are replaced with *bounded nets* and the prenormal structure $\widehat{\mathcal{N}}$ is given by the nucleus of the G-normal cone in (2.76). It is proved by Ioffe [607, Theorem 3] and holds also for prenormal structures defined by the viscosity β-normal cones (2.78) on Banach spaces admitting a Lipschitzian β-smooth bump function. Let us call the net counterpart of the SNC property in Definition 2.50 by the *topological normal compactness* (TNC) of Ω at $\bar{x}$ with respect to $\widehat{\mathcal{N}}$ and observe that CEL$\not\Rightarrow$TNC for the case of Clarke's normal cone (2.72), as follows from Example 4.1 in Borwein [138] for $X = \ell^\infty$.

Obviously TNC$\Rightarrow$SNC for any $\widehat{\mathcal{N}}$. It is proved by Fabian and Mordukhovich [422] that these properties coincide on Banach spaces X that are *weakly compactly generated* (WCG), i.e., $X = \mathrm{cl}\,(\mathrm{span}\,K)$ for some weakly compact set $K \subset X$. This class includes all reflexive spaces as well as all separable Banach spaces. On the other hand, the SNC property may be *strictly weaker* than its TNC counterpart in general Banach (and Asplund) space settings, even for the case of convex sets; see examples in [422].

Theorem 2.51 (abstract versions of the extremal principle). *Let $\{\Omega_1, \Omega_2, \bar{x}\}$ be an extremal system of closed sets in a Banach space X, and let $\widehat{\mathcal{N}}$ define a prenormal structure on X. The following hold:*

(i) *For every $\varepsilon > 0$ there are $x_i \in \Omega_i \cap (\bar{x} + \varepsilon I\!B)$, $i = 1, 2$, and $x^* \in X^*$ with $\|x^*\| = 1$ such that*

$$x^* \in \big(\widehat{\mathcal{N}}(x_1; \Omega_1) + \varepsilon I\!B^*\big) \cap \big(-\widehat{\mathcal{N}}(x_2; \Omega_2) + \varepsilon I\!B^*\big) . \tag{2.86}$$

(ii) *Assume that one of the sets Ω_i, $i = 1, 2$, is $\widehat{\mathcal{N}}$-sequentially normally compact at $\bar{x}$. Then there is $x^* \in I\!B^* \setminus \{0\}$ such that*

$$x^* \in \overline{\mathcal{N}}(\bar{x}; \Omega_1) \cap \big(-\overline{\mathcal{N}}(\bar{x}; \Omega_2)\big) , \tag{2.87}$$

where $\overline{\mathcal{N}}$ stands for the topological normal structure (2.67) *generated by $\widehat{\mathcal{N}}$. If in addition the dual ball $I\!B^* \subset X^*$ is weak* sequentially compact, then*

$$x^* \in \mathcal{N}(\bar{x}; \Omega_1) \cap \big(-\mathcal{N}(\bar{x}; \Omega_2)\big) \tag{2.88}$$

for some $x^* \in I\!B^* \setminus \{0\}$, where $\mathcal{N}$ stands the sequential normal structure (2.66) *generated by $\widehat{\mathcal{N}}$.*

Proof. First justify (i) following basically the procedure in the proof of Lemma 2.32(ii). Fix an arbitrary $\varepsilon > 0$. Given a local extremal point $\bar{x}$ of the set system $\{\Omega_1, \Omega_2\}$, we find a neighborhood U of $\bar{x}$ and $a \in X$ such that $\|a\| \le \epsilon := \varepsilon/2$ and $(\Omega_1 + a) \cap \Omega_2 \cap U = \emptyset$. One can always assume that $\bar{x} + \epsilon I\!B \subset U$. Form the function

$$\varphi(x_1, x_2) := \|x_1 - x_2 + a\| \;\text{ for }\; (x_1, x_2) \in X^2$$

and observe that $\varphi(\bar{x},\bar{x}) = \|a\| \le \epsilon$ and

$$\varphi(x_1,x_2) > 0 \ \text{ if } \ (x_1,x_2) \in Z := \big[\Omega_1 \cap (\bar{x}+\epsilon I\!B)\big] \times \big[\Omega_2 \cap (\bar{x}+\epsilon I\!B)\big] \ .$$

We see that Z is a complete metric space with the metric induced by the sum norm on X^2, and that φ is continuous on Z. Applying Ekeland's variational principle in Theorem 2.26(i) to φ on Z, we find $(\bar{x}_1,\bar{x}_2) \in Z$ such that

$$\varphi(\bar{x}_1,\bar{x}_2) \le \varphi(x_1,x_2) + \epsilon\big(\|x_1-\bar{x}_1\| + \|x_2-\bar{x}_2\|\big) \ \text{ for all } \ (x_1,x_2) \in Z \ .$$

The latter implies that $(\bar{x}_1,\bar{x}_2) \in \Omega_1 \times \Omega_2$ is a local minimizer of the function

$$\psi(x_1,x_2) := \|x_1 - x_2 + a\| + \epsilon\big(\|x_1-\bar{x}_1\| + \|x_2-\bar{x}_2\|\big)$$

relative to the set $\Omega_1 \times \Omega_2$ with $\bar{x}_1 - \bar{x}_2 + a \ne 0$. Now applying property (H) of the prenormal structure $\widehat{\mathcal{N}}$ in Definition 2.41 with $\gamma := \varepsilon > \epsilon$, we find $\tilde{x}_i \in \bar{x} + \epsilon I\!B$, $i = 1,2$, and $x^* \in X^*$ with $\|x^*\| = 1$ such that

$$(-x^*,x^*) \in \widehat{\mathcal{N}}(\tilde{x}_1;\Omega_1) \times \widehat{\mathcal{N}}(\tilde{x}_2;\Omega_2) + \varepsilon\big(I\!B^* \times I\!B^*\big) \ .$$

It follows from the constructions above that $(\tilde{x}_1,\tilde{x}_2) \in \Omega_1 \times \Omega_2$ and $\tilde{x}_i \in \bar{x} + \varepsilon I\!B$, $i = 1,2$. Thus we get all the relationships of the approximate extremal principle in (i).

To prove (ii), we need to pass to the limit in (i) as $\varepsilon \downarrow 0$. Let us first justify the sequential version of the exact extremal principle in (ii) assuming that the dual ball $I\!B^* \subset X^*$ is weak* sequentially compact. Take a sequence $\varepsilon_k \downarrow 0$ and consider the corresponding sequences (x_{1k}, x_{2k}, x_k^*) satisfying the conclusions of (i). We have $x_{1k} \to \bar{x}$ and $x_{2k} \to \bar{x}$ as $k \to \infty$. Since $I\!B^*$ is weak* sequentially compact, we select a subsequence of $\{x_k^*\}$ (without relabeling) that converges weak* to some $x^* \in I\!B^*$. By (2.86) there are $x_{ik}^* \in \widehat{\mathcal{N}}(x_{ik};\Omega_i)$ and $b_{ik}^* \in I\!B^*$, $i = 1,2$, such that

$$x_k^* = x_{1k}^* + \varepsilon_k b_{1k}^*, \quad x_k^* = -x_{2k}^* + \varepsilon_k b_{2k}^* \quad \text{for all } \ k \in I\!N \ . \tag{2.89}$$

This implies that $x_{ik}^* \xrightarrow{w^*} x^*$ and $x_{2k}^* \xrightarrow{w^*} -x^*$ as $k \to \infty$. The latter gives, due to definition (2.66), that x^* satisfies (2.88).

To justify (ii) in the sequential case, it remains to show that $x^* \ne 0$ under the SNC assumption imposed. On the contrary, assume that $x^* = 0$, which gives $x_{ik}^* \xrightarrow{w^*} 0$ for the sequences $x_{ik}^* \in \widehat{\mathcal{N}}(x_{ik};\Omega_i)$, $i = 1,2$. Since one of the sets Ω_i (say Ω_1) is $\widehat{\mathcal{N}}$-sequentially normally compact at $\bar{x}$, we get $\|x_{1k}^*\| \to 0$. This clearly implies that $\|x_k^*\| \to 0$, which contradicts the condition $\|x_k^*\| = 1$ for all $k \in I\!N$ and ends the proof of (ii) in the sequential case.

Let us finally consider the case of general Banach spaces and justify the topological version (2.87) of the exact extremal principle under the *sequential* normal compactness condition imposed. We follow the procedure in the sequential case but now don't assume anymore that $I\!B^*$ is weak* sequentially compact, using instead the well-known fact that $I\!B^*$ is (topologically) weak* compact in arbitrary Banach spaces. This allows us to conclude that the above sequence $\{x_k^*\}$ has a weak* cluster point $x^* \in \mathrm{cl}^*\{x_k^* \mid k \in I\!N\} \cap I\!B^*$. It follows from representation (2.89) with $x_{ik}^* \in \widehat{\mathcal{N}}(x_{ik}; \Omega_i)$, $i = 1, 2$, and from definition (2.67) that x^* satisfies (2.87), where $\overline{\mathcal{N}}$ is the topological normal structure generated by $\widehat{\mathcal{N}}$. This holds for any cluster point $x^* \in \mathrm{cl}^*\{x_k^* \mid k \in I\!N\}$.

It remains to show that $x^* \neq 0$ for *some* $x^* \in \mathrm{cl}^*\{x_k^* \mid k \in I\!N\}$ if one of the sets Ω_i, $i = 1, 2$, is $\widehat{\mathcal{N}}$-sequentially normally compact at $\bar{x}$. Indeed, the opposite means the $x^* = 0$ is the only weak* cluster point of $\{x_k^*\}$. The latter yields that the whole sequence $\{x_k^*\}$ converges weak* to zero. Then it follows from (2.89) that $x_{ik}^* \xrightarrow{w^*} 0$, $i = 1, 2$, as $k \to \infty$. Hence $\|x_{ik}^*\| \to 0$ for either $i = 1$ or $i = 2$, which is impossible due to $\|x_k^*\| = 1$. This contradiction completes the proof of the theorem. △

As an immediate corollary of Theorem 2.51 we derive the following generalized versions of the Bishop-Phelps and supporting hyperplane theorems in terms of abstract prenormal and normal structures on Banach spaces.

Corollary 2.52 (prenormal and normal structures at boundary points). *Let Ω be a proper closed subset of a Banach space X, and let $\bar{x}$ be a boundary point of Ω. Consider an arbitrary prenormal structure $\widehat{\mathcal{N}}$ on X and the corresponding sequential normal structure $\mathcal{N}$ and topological normal structure $\overline{\mathcal{N}}$ generated by $\widehat{\mathcal{N}}$. Then one has:*

(i) *Given any $\varepsilon > 0$, there is $x \in \Omega \cap (\bar{x} + \varepsilon I\!B)$ such that $\widehat{\mathcal{N}}(x; \Omega) \neq \{0\}$.*

(ii) *Assume that the set Ω is $\widehat{\mathcal{N}}$-sequentially normally compact at $\bar{x}$. Then $\overline{\mathcal{N}}(\bar{x}; \Omega) \neq \{0\}$. If in addition the dual ball $I\!B^*$ is weak* sequentially compact, then $\mathcal{N}(\bar{x}; \Omega) \neq \{0\}$.*

Proof. Follows from Theorem 2.51 with $\Omega_1 := \Omega$ and $\Omega_2 := \{\bar{x}\}$. △

By the results of Subsect. 2.5.1 the abstract versions of the extremal principle in Theorem 2.51 and their corollaries hold for subdifferentially generated prenormal and normal structures under the mild requirements (S1)–(S3) on the corresponding presubdifferentials. These requirements are used in the proof of Lemma 2.32(ii) for the case of Fréchet normals and subgradients. As follows from the proof of the other statement (i) in Lemma 2.32, it holds for *any* presubdifferential $\widehat{\mathcal{D}}\varphi(\cdot)$ on the class of proper l.s.c. functions $\varphi\colon X \to \overline{I\!R}$ *generated by a prenormal cone* $\widehat{\mathcal{N}}$ on $X \times I\!R$ as

$$\widehat{\mathcal{D}}\varphi(x) := \big\{x^* \in X^* \big| \ (x^*, -1) \in \widehat{\mathcal{N}}((x, \varphi(x)); \mathrm{epi}\,\varphi)\big\}, \quad x \in \mathrm{dom}\,\varphi\ ,$$

provided that $\widehat{\mathcal{N}}(z; \Omega) \subset \{0\}$ if $z \in \mathrm{int}\,\Omega$ and that $\|x^*\| \leq \ell$ for all $x^* \in \widehat{\mathcal{D}}\varphi(x)$ if φ is locally Lipschitzian around x with modulus ℓ. Thus both statements in Lemma 2.32 are valid for general classes of normals and subgradients. It is not the case for Theorem 2.33 and most of the other material in this chapter, where the specific structure of Fréchet-like subdifferential constructions and geometric properties of Asplund spaces are essentially exploited. Note also that the structural properties of our basic constructions are utilized in Chap. 1 to build the generalized differential theory in Banach spaces.

In the subsequent chapters of this book we apply basic principles and results of the first two chapters to develop a comprehensive generalized differential calculus in Asplund spaces and give its applications to important problems in nonlinear analysis, optimization, and economics. Most of the results are formulated in terms of Fréchet-like normals/subgradients/coderivatives and their sequential limits, which is essential in the statements and proofs. As follows from the proofs (and will be explicitly mentioned in some cases), a part of the results obtained holds also for other normal and subgradient structures by the above discussions.

2.6 Commentary to Chap. 2

2.6.1. The Origin of the Extremal Principle. The chapter collects the fundamental material that is crucial for the subsequent parts of the book, in both aspects of basic theory and applications of variational analysis. Roughly speaking, all the essentials of variational analysis developed in this book largely revolve around the *extremal principle* comprehensively studied in Chap. 2. The extremal principle can be viewed as a local *variational* counterpart of the classical separation in the case of nonconvex sets; it actually plays the same role in variational analysis as separation theorems do in the presence of convexity, i.e., in the framework of convex analysis and its applications.

The term "extremal principle" was coined by Mordukhovich [910], while its first versions (in both approximate/fuzzy and exact/limiting forms of Definition 2.5) were established by Kruger and Mordukhovich [718] under the name of "generalized Euler equations" for local extremal points of finitely many sets in Fréchet smooth spaces. The essence of the exact extremal principle can be traced to the early paper by Mordukhovich [887], where the key method of *metric approximations* has been initiated in the framework of optimal control.

The properties of extremal systems and their connection with separation properties of convex and nonconvex sets presented in Subsect. 2.1.1 can be found in Kruger and Mordukhovich [719] and Mordukhovich [901]. The relationships between extremality and supporting properties from Subsect. 2.1.2 were fully investigated by Fabian and Mordukhovich [421]. To this end we mention a remarkable study of *boundary points* for sums of sets undertaken by Borwein and Jofré [148]. The latter boundary property of a set sum is actually equivalent to the local extremality of another set system; see also the recent paper by Kruger [715] for more details.

In Subsect. 2.1.3 we give a self-contained proof of the exact extremal principle in *finite-dimensional* spaces based on the *method of metric approximations*. As mentioned, this method was originated by Mordukhovich [887] and then developed in [889, 892, 719, 901, 907] in several finite-dimensional settings; see also the comments below for its infinite-dimensional counterparts with significantly more involved variational arguments. Note that the method of metric approximations contains a *constructive procedure* to study local extremal points of set systems (in particular, local solutions to various problems of constrained optimization and equilibria) based on their *symmetric approximation* by sequences of *smooth* problems of *unconstrained minimization*. The realization of this procedure as in the proof of Theorem 2.8 has actually led us to constructing the *basic/limiting normal cone* in order to describe the (exact) *generalized Euler equation*. Observe that the latter appeared in the process of passing to the limit after applying the classical *Fermat stationary rule* in the sequence of approximating problems; cf. [887]. All this indicates close relationships between *classical and modern* tools and concepts of variational analysis: the novelty comes from applying appropriate *approximation/perturbation* techniques.

2.6.2. The Extremal Principle in Fréchet Smooth Spaces and Separable Reduction. Although there are no crucial differences between finite-dimensional and infinite-dimensional settings from conceptional viewpoints, *infinite-dimensional* extensions of the above approach to the extremal principle are technically much more involved requiring the usage of refined variational arguments and delicate geometric properties of Banach spaces. There are the following three most crucial features of *finite dimensionality* significantly exploited in the construction and realization of the *metric approximation* method employed to prove the exact extremal principle in Subsect. 2.1.3:

(a) intrinsic *variational properties* of the *Euclidean norm*;

(b) the *equivalence* of *any norm* in finite dimensions to the Euclidean norm, which is *smooth* away from the origin;

(c) *compactness* of the closed unit ball (as well as the unit sphere), which is a *characterization* of finite-dimensional spaces.

Appropriate counterparts of these properties in infinite dimensions, which have nothing to do with the Euclidean norm, are among the key ingredients in deriving both approximate and exact versions of the extremal principle in the general framework of *Asplund spaces* presented in Sect. 2.2. To establish the *approximate* extremal principle in Asplund spaces, we develop a *two-step* procedure therein: first giving a direct proof of the extremal principle in Banach spaces admitting an equivalent *Fréchet smooth* norm (away from the origin), and then "rising up" the result from Fréchet smooth spaces to the general Asplund space setting by using the *method of separable reduction*.

The *variational arguments* employed in Subsect. 2.2.1 to justify the approximate extremal principle in Banach spaces with *smooth Fréchet renorms* were first developed, to the best of our knowledge, by Li and Shi [785] (preprint of

1990) in their proof of variational principles of the Ekeland and Borwein-Preiss types and then used, e.g., in [159, 265, 266, 688, 809] in parallel variational settings. We combine these arguments with the device in Mordukhovich and Shao [948] and with the subsequent induction. As mentioned in Remark 2.11, a similar device can be employed to establish the approximate extremal principle in Banach spaces admitting *smooth renorms of any kind*, with respect to natural bornologies. We refer the reader to the survey paper by Averbukh and Smolyanov [68] and to the book by Phelps [1073] for more information about bornologies. Appropriate versions of the approximate extremal principle in other (non-Fréchet) bornologically smooth spaces can be found in the paper by Borwein, Mordukhovich and Shao [151].

The method of *separable reduction* developed in Subsect. 2.2.2 in order to apply it to deriving the approximate extremal principle is probably the most difficult device given in this book. It is taken from the paper by Fabian and Mordukhovich [421], while its origin goes back to Preiss [1103] in the theory of Fréchet differentiability. Then versions of separable reduction were used by Fabian and Zhivkov [423], Fabian [413, 415], and Fabian and Mordukhovich [420, 421] in applications to various aspects of nonlinear analysis and generalized differentiability. It seems that the *Fréchet-type* differentiability and subdifferentiability is very essential in the theory and applications of this method.

2.6.3. Asplund spaces. The *Asplund property* of Banach spaces formulated in Subsect. 2.2.3 plays a crucial role in the theory and applications of variational analysis developed in this book. Although a number of important results and applications presented in the book hold in arbitrary Banach spaces, the most *comprehensive theory* of generalized differentiation, at the *same level of perfection* as in finite dimensions, is given in the Asplund space setting.

The remarkable class of Banach spaces, now called Asplund spaces, was introduced by Asplund in his 1968 paper [43] as "strong differentiability spaces." The name "Asplund spaces" was coined by Namioka and Phelps [992] soon after Asplund's death (1974). The original Asplund definition was the same one presented in Subsect. 2.2.3 with the only difference that the dense set of Fréchet differentiability points was postulated to be G_δ. The latter requirement can be *equivalently omitted* due to the fact that Fréchet differentiability points always form a G_δ set; see, e.g., Phelps [1073]. It is worth mentioning that, although the main contents of the original Asplund's paper [43] concerned the geometric theory of Banach spaces, there were nice *variational* applications therein establishing *generic* existence and unique theorems for optimal solutions to some linearly *perturbed* variational problems particularly related to Moreau's *proximal mappings* in Hilbert spaces [982].

Asplund spaces, which include all *reflexive* and many other remarkable Banach spaces, have been comprehensively investigated in the geometric theory of Banach spaces and its applications, with discovering a great number of impressive characterizations and properties; the reader may find a partial list of them in the beginning of Subsect. 2.2.3 and in the references therein. Although the Asplund property is generally related to *Fréchet differentiability*, there are Asplund spaces that fail to have *even a Gâteaux smooth renorm*; see striking examples in Haydon [553] and in Deville, Godefroy and Zizler [331]. Note that, in contrast to the class of Asplund spaces that is one of the most beautiful objects in analysis and probably in all mathematics, *weak Asplund* spaces similarly defined in [43] with the replacement of Fréchet differentiability by Gâteaux differentiability are too far from being beautiful admitting only a modest number of satisfactory results; see the book by Fabian [416]. There is an intermediate class of *Asplund generated spaces*, known also in the literature as Grothendieck-Šmulian generated spaces, which particularly include all *weakly compactly generated* (hence all separable) spaces, strongly studied geometrically in the afore-mentioned Fabian's book. An on-going research project by Fabian, Loewen and Mordukhovich [418] is devoted to certain aspects of generalized differentiation and variational analysis in the framework of Asplund generated spaces; see Remark 3.103 for some results and discussions.

2.6.4. The Extremal Principle in Asplund Spaces. The extremal *characterizations* of Asplund spaces in Theorem 2.20 via the two (equivalent) versions of the *approximate extremal principle* were established by Mordukhovich and Shao [948], while the presented proof is taken from the later papers by Fabian and Mordukhovich: from [421] for the *sufficiency* of the Asplund property to ensure the extremal principle via separable reduction and from [420], via Example 2.19 reproduced in Subsect. 2.2.3, for the *necessity* of this property to have the extremal principle. Yet another proof (actually the first one) of the validity of the approximate extremal principle in general Asplund spaces can be found in Mordukhovich and Shao [949] via a *coderivative criterion* for the covering property established in their previous paper [946].

The *boundary characterizations* of Asplund spaces from Corollary 2.21 were obtained by Fabian and Mordukhovich [420] via separable reduction, with no appeal to the extremal principle. On the other hand, assertion (c) of this corollary, which is a far-going *nonconvex extension* of the celebrated *Bishop-Phelps theorem* [116] in the framework of Asplund spaces, was first deduced by Mordukhovich and Shao [948] from the extremal principle; cf. also Borwein and Strójwas [156, 157] for other counterparts of the Bishop-Phelps theorem in nonconvex settings with other proofs. In the paper by Mordukhovich and B. Wang [960] the reader can find more *variational characterizations* of Asplund spaces via Fréchet normals and ε-normals, as well as different proofs of those mentioned above. Various subdifferential characterizations of Asplund spaces will be discussed below in the commentary to this chapter. We also refer the reader to the recent paper by Wang [1304] who derived some analogs of the afore-mentioned results and characterizations of the *reflexivity* of locally uniformly convex Banach spaces with Fréchet differentiable renorms via the *approximate extremal principle* involving *proximal* normals and subgradients.

The validity of the *exact extremal principle* in Asplund spaces under the *sequential normal compactness* conditions of Theorem 2.22 was established by Mordukhovich and Shao [949] extending the result of Kruger and Mordukhovich [718] obtained under the epi-Lipschitzian assumptions in Fréchet smooth spaces; see also the subsequent publications [707, 901]. The *converse* assertion of Theorem 2.22 was proved by Fabian and Mordukhovich [419]. Example 2.23 on the failure of the exact extremal principle in the absence of normal compactness is taken from Borwein and Zhu [162]. The nontriviality results on basic normals and subgradients from Corollaries 2.24 and 2.25, which immediately follow from the exact extremal principle, were first observed by Mordukhovich and Shao [949].

2.6.5. The Ekeland Variational Principle. According to the conventional terminology of modern nonlinear analysis, the expression "variational principle" stands for an assertion ensuring that, given a *lower semicontinuous* and *bounded from below* function φ and its arbitrary ε-minimal point x_0, there is a *small perturbation* of φ such that the perturbed function attains its *exact minimum* at some point close to x_0. The first variational principle in this sense was discovered by Ekeland in 1972 (see [396, 397, 399]) in general complete metric spaces. The exact statement of Ekeland's variational principle is presented in Theorem 2.26(i). Note that the original Ekeland's proof [396, 397] was rather complicated involving transfinite induction arguments via Zorn's lemma. It was largely similar to the proof of the Bishop-Phelps theorem [116] mentioned above, which was called by Ekeland [399] "the grandfather of it all." The much simplified proof presented in Theorem 2.26 follows the lines of Crandall's arguments reproduced in Ekeland [399] as a personal communication. The *converse* statement of Theorem 2.26(ii) ensuring that the Ekeland principle is actually a characterization of the completeness property of metric spaces is due to Sullivan [1232]. There are so many applications of Ekeland's variational principle to various areas in mathematics and related disciplines that it doesn't seem to be possible of even mentioning a great part of them in this book. The reader can find a partial list of the most important early applications with their detailed analysis in the excellent survey by Ekeland [399] of 1979.

It is worth emphasizing that among the *main motivations* for the Ekeland original study was the result of Corollary 2.27, which ensures the fulfillment of the "almost stationary" condition for "almost optimal" (*suboptimal* in our terminology) solutions to a *smooth* unconstrained minimization problem. Results of this kind are especially important for optimization problems in infinite dimensions, where optimal solutions may often not exist. Thus the principal issue of both theoretical and practical importance is to derive necessary conditions for *suboptimal solutions*, of about the same type as for optimal solutions, that eventually lead to numerical algorithms for solving optimization problems. From this viewpoint, necessary suboptimality conditions applied to solutions that always exist are not worse than those for exact optimality, which may not be reachable. We pay a strong attention to this topic throughout the book; see particularly Chaps. 5 and 6.

2.6.6. Subdifferential Variational Principles. The main result of Subsect. 2.3.2 called the *lower subdifferential variational principle* (Theorem 2.28) is a far-going development of Ekeland's ε-stationary condition in Corollary 2.27 from smooth functions to extended-real-valued l.s.c. functions; it can be applied therefore to problems of constrained optimization. This result established by Mordukhovich and B. Wang [962] is different from conventional variational principles in *only one* aspect: *instead of a perturbed minimization* condition, it contains a (lower) *subdifferential* condition of the ε-*stationary type*, which is actually a necessary condition for suboptimal solutions. The first result of this type for nonsmooth functions was obtained by Rockafellar [1147] via Clarke subgradients in Banach spaces, while for *convex* functions it actually goes back to the early work by Brøndsted and Rockafellar [179] that preceded Ekeland's variational principle; cf. also [154, 186, 501, 1165] for related results and discussions. As proved in the afore-mentioned paper [962], the subdifferential variational principle of Theorem 2.28 occurred to be an *equivalent analytic counterpart* of the *approximate extremal principle* giving hence yet another variational characterization of Asplund spaces.

The variational results of Theorem 2.28 easily imply the *subdifferential characterizations* of Asplund spaces listed in Corollary 2.29. These characterizations were first established via different devices by: Fabian [415] for (b), Fabian and Mordukhovich [419] for (c), and Fabian and Zhivkov [423] for (e); characterizations (d) follows from (e) due to Theorem 1.86. Note also that implication (e)⇒(a) was proved earlier by Ioffe [593], while the related fact that the density of the set $x \in \operatorname{dom} \varphi$ with $\widehat{\partial}_{a\varepsilon}\varphi(x) \neq \emptyset$ for any l.s.c. function $\varphi\colon X \to \overline{I\!R}$ yields the Asplund property of X goes back to Ekeland and Lebourg [400].

The *upper subdifferential variational principle* of Theorem 2.30 taken from the paper by Mordukhovich, Nam and Yen [938] is substantially different from the lower one being generally less powerful, since it applies only to special classes of functions that admit *upper* Fréchet subgradients at the points in question. However, for such classes of functions (which have been well recognized and investigated in nonsmooth analysis; see Chap. 5) the upper version involving *every* upper subgradient, has certain significant advantages in comparison with its lower counterpart from Theorem 2.28. It is particularly useful in developing necessary *suboptimality* conditions for various classes of constrained minimization problems; see Subsect. 5.1.4 for some results in this direction.

2.6.7. Smooth Variational Principles. Concerning the *conventional line* in developing variational principles, observe that the minimization condition in Ekeland's variational principle of Theorem 2.26 can be interpreted as follows: for every l.s.c. function $\varphi\colon X \to \overline{I\!R}$ with $\inf \varphi > -\infty$ there exists a function $s: X \to \overline{I\!R}$ that *supports* φ *from below* at some point $\bar{x} \in \mathrm{dom}\,\varphi$, i.e.,

$$\varphi(\bar{x}) = s(\bar{x}) \quad \text{and} \quad \varphi(x) \geq s(x) \quad \text{whenever} \quad x \in X .$$

Then Ekeland's principle ensures, in the framework of arbitrary Banach spaces, that the support $s(\cdot)$ can be chosen as a small perturbation by functions of the *norm* type. A clear disadvantage of this results is the intrinsic *nonsmoothness* of such perturbations, and so a natural question arises about conditions ensuring *smooth* perturbations, i.e., about *smooth variational principles.*

The first result of this type was obtained by Stegall in his 1978 paper [1224] who showed that, for any l.s.c. function satisfying some growth condition as $\|x\| \to \infty$ on a Banach space with the *Radon-Nikodým property* (in particular, on a reflexive space), a supporting function $s(\cdot)$ could be chosen as a *linear* functional with an arbitrarily small norm.

A more powerful smooth variational principle, in essentially more general settings, was established in the 1987 paper by Borwein and Preiss [154] who proved, assuming the existence of a *bornologically smooth renorm* on the Banach space in question, that supporting functions could be chosen as *concave and smooth* with respect to the same bornology. The Borwein-Preiss smooth variational principle was extended in some directions by Deville, Godefroy and Zizler [330, 331] who showed, in particular, that supporting functions could be chosen as bornologically *smooth* (but not concave anymore) under the more general assumption on the existence of a *smooth Lipschitzian bump* function with respect to some bornology. We refer the reader to [45, 70, 164, 265, 417, 419, 530, 531, 547, 619, 620, 785, 790, 809, 1243, 1356] among other publications for additional information about variational principles, their recent developments, and applications.

The results of Subsect. 2.3.3 are taken from the paper by Fabian and Mordukhovich [419]. Assertions (i) and (ii) of Theorem 2.31 establish *enhanced versions* of the Borwein-Preiss and Deville-Godefroy-Zizler smooth variational principles, respectively, with more *information* about supporting functions in comparison with the original versions in [154, 330]. Observe that the proof given in Theorem 2.31(i,ii) is essentially different from those of [154, 330]; it is based on the *lower subdifferential variational principle* from Theorem 2.28 and *smooth variational descriptions* of Fréchet subgradients from Theorem 1.88.

The *converse* assertion (iii) is *indeed remarkable*: it shows that the *smooth norm* and *smooth bump* assumptions in smooth variational principles of the Borwein-Preiss and Deville-Godefroy-Zizler types, respectively, are not only sufficient but also *necessary* for the validity of such results. As discussed at the end of Subsect. 2.3.3, the *Fréchet* smoothness is *not* essential for these conclusions, which hold true for *any bornology.* Observe again in this respect that *no smoothness* assumption is necessary for the fulfillment of the extremal principle and of the lower subdifferential variational principle. Furthermore, as proved in Borwein, Mordukhovich and Shao [151] (resp. in Mordukhovich [919]), the approximate extremal principle is *equivalent* to certain localized versions of the Borwein-Preiss and Deville-Godefroy-Zizler variational principles *provided* that the Banach space in question admits a Fréchet smooth renorm (resp. a Fréchet smooth and Lipschitzian bump function).

2.6.8. Limiting Normal and Subgradient Representations in Asplund Spaces. It has been mentioned above that the main results of variational analysis and its applications developed in this book are derived from the extremal principle. Section 2.4 contains the first set of results in this direction showing, in particular, that the usage of the *approximate extremal principle* and its subgradient descriptions in Asplund spaces allows us to justify simplified and *convenient representations* of basic normals, subgradients, and coderivatives in the general Asplund setting similar to those established in finite dimensions on the base of specific properties of the Euclidean norm. The power of the extremal principal and its equivalents make it possible to replace the previous arguments without any appeal to either finite dimensionality, or to the Euclidean norm, or even to smooth renorming. Moreover, the Asplund space setting happens to be also *necessary* for such representations provided that they are required for *all* sets, functions, and set-valued mappings belonging to reasonably broad families.

The subdifferential description of the approximate extremal principle given in Lemma 2.32 plays a crucial role in establishing the main results of Sect. 4. This lemma was established by Mordukhovich and Shao [948], while the essence of assertion (i) can be traced to Ioffe [600]; cf. the proof of Step 2 in Lemma 2 therein.

Results of form (2.42) known as *fuzzy sum rules* (or "zero fuzzy sum rules," or "fuzzy principles") were initiated by Ioffe [593, 594] for ε-subdifferentials ($\varepsilon > 0$) of both Fréchet and Dini types. For the case of Fréchet subgradients ($\varepsilon = 0$) on Asplund spaces, the semi-Lipschitzian result (2.42) was first established by Fabian [415] based on the Borwein-Preiss smooth variational principle and on separable reduction; cf. Ioffe [599] for Fréchet smooth spaces. There are several modifications of such fuzzy rules; all of them happens to be equivalent. The latter was first proved by Zhu [1371] for the so-called β-subdifferentials that are valuable on bornologically smooth spaces and then by Ioffe [606] and Lassonde [747] in more general settings; see also the recent book by Borwein and Zhu [164].

The *full* (not "zero") semi-Lipschitzian fuzzy sum rule of Theorem 2.33(b) was derived by Fabian first in [413] for $\varepsilon > 0$ and then in [415] for $\varepsilon = 0$ in the general Asplund space setting. Note that the structure of *Fréchet subgradients* seems to be very *essential* for this full fuzzy rule, in contrast to its zero counterpart (2.42). Some *topological* modifications of the full fuzzy sum rule (with a weak* neighborhood of the origin in X^* instead of a small dual ball) were earlier considered by Ioffe [593] who introduced Banach spaces with such properties as "trustworthy spaces" and proved that any space admitting a Fréchet smooth bump function fell into the trustworthy category. Implication (b)$\Rightarrow$(a) in Theorem 2.33 can be also deduced from [593]. We refer the reader to the afore-mentioned publications and also to [147, 151, 158, 160, 163, 164, 257, 265, 329, 413, 414, 607, 614, 616, 622, 802, 952] for more results, equivalent statements, and discussions in this direction.

The *exact/limiting* semi-Lipschitzian sum rule of Theorem 2.33(c) as well as the representations of basic subgradients and normals from Theorems 2.34 and 2.35 in Asplund spaces were established by Mordukhovich and Shao [949], while the *converse* assertions therein are due to Fabian and Mordukhovich [419]. Extended sum rules based on the extremal principle are presented in Chap. 3, where the reader can find comprehensive calculus results with more discussions.

The *limiting* ε*-subdifferential* $\partial_\varepsilon\varphi(\bar{x})$ in (2.48) for $\varepsilon > 0$ was defined by Jofré, Luc and Théra [634] (preprint of 1995) motivated by applications to ε-monotonicity and related issues. As observed by Mordukhovich and Shao [949, Proposition 2.11], this construction happened to be an ε*-enlargement* of our *basic subdifferential* (see Theorem 2.34) for any l.s.c. function on Asplund spaces; moreover, such an enlargement representation of $\partial_\varepsilon\varphi(\bar{x})$ *characterizes* the class of Asplund spaces as proved by Fabian and Mordukhovich [419].

The *singular subdifferential* limiting representation

$$\partial^\infty\varphi(\bar{x}) = \operatorname*{Lim\,sup}_{\substack{x \xrightarrow{\varphi} \bar{x} \\ \lambda \downarrow 0}} \lambda\widehat{\partial}\varphi(x) \tag{2.90}$$

from Theorem 2.38 was first obtained by Rockafellar [1150] in finite dimensions, with the *proximal subdifferential* $\partial_P\varphi(x)$ of (2.81) replacing $\widehat{\partial}\varphi(x)$ in (2.90). The latter representation was actually accepted in [1150] as the definition of $\partial^\infty\varphi(\bar{x})$. Representation (2.90) was proved by Ioffe [600] for Fréchet smooth Banach spaces, and then the full statement of Theorem 2.38 in Asplund spaces was given by Mordukhovich and Shao [949] following the approach of [600]. The proof of the preceding Lemma 2.37 presented in the book is a clarification of Ioffe's proof in [600, Theorem 4] being different from it in several significant aspects.

Assertion (i) of Theorem 2.40 on *horizontal normals* to graphs and the inclusion

$$D^*\varphi(x)(0) \subset \partial^\infty\varphi(\bar{x}) \cup \partial^\infty(-\varphi)(\bar{x})$$

for continuous functions on Asplund spaces was established by Ngai and Théra [1008]. The *opposite* inclusion to the latter one and hence the equality in the *coderivative representation* of Theorem 2.40(ii) follow from Theorem 1.80. We refer the reader to the recent papers by Zhu [1373] and Ivanov [622] (see also the book by Borwein and Zhu [164]) for other proofs of the above results and their counterparts involving β-subdifferentials in *bornologically smooth* Banach spaces.

2.6.9. Other Subdifferential Structures and Abstract Versions of the Extremal Principle. Abstract normal and subdifferential structures of

Subsect. 2.5.1 were defined and studied by Mordukhovich [920] motivated by recognizing *minimal* normal and subdifferential properties needed for deriving the extremal principle in general Banach spaces. Various axiomatic constructions of this type, with generally different properties and applications, were considered by Aussel, Corvellec and Lassonde [61], Correa, Jofré and Thibault [292], Ioffe [599, 606, 607], Ioffe and Penot [614], Lassonde [747], Mordukhovich [901], Mordukhovich and Shao [949], Thibault and Zagrodny [1254], etc. The minimality result for the basic subdifferential from Proposition 2.45 was observed by Mordukhovich and Shao [949], while the essence of such theorems (under less general assumptions) should be traced to the early work by Ioffe [596, 599] and Mordukhovich [894, 901]; see more discussions in [949, Sect. 9]. Note that Ioffe's minimality result [599] *doesn't imply*, as mistakenly stated in [599, Proposition 8.2], that the nucleus $\widetilde{\partial}_G\varphi(\bar{x})$ of his G-subdifferential belongs to our basic subdifferential $\partial\varphi(\bar{x})$ for l.s.c. functions on Fréchet smooth spaces. The point is that the mapping $\partial\varphi(\cdot)$ may *not* be of *closed-graph* for Lipschitz continuous functions as claimed in [599]. In fact, the *opposite inclusion*

$$\partial\varphi(\bar{x}) \subset \widetilde{\partial}_G\varphi(\bar{x}) \tag{2.91}$$

is fulfilled for any l.s.c. function defined on an Asplund space, where equality holds for locally Lipschitzian functions provided that the space X is weakly compactly generated (and hence automatically Fréchet smooth); see Subsect. 3.2.3 below and comments to it in Subsect. 3.4.7. Moreover, it follows from examples by Borwein and Fitzpatrick [141] that the inclusion in (2.91) may be *strict* even for concave Lipschitz continuous functions defined on some special spaces admitting $\mathcal{C}^\infty$-smooth renorms but not being weakly compactly generated; cf. Example 3.61 below.

Subsection 2.5.2 presents an overview of some remarkable normal and subdifferential structures important in the theory and applications of variational analysis via generalized differentiation. The main attention is paid to generalized normals and subgradients related to the basic constructions adopted in this book. The descriptions in Subsect. 2.5.2 are self-contained with the corresponding references to publications, where the reader can find more details and discussions; see also Commentary to Chap. 1. We just make some comments to (the last) part E of this subsection regarding the concepts and results formulated and proved therein.

The generalized differential construction $\mathcal{A}\varphi(\bar{x})$ labeled here as the "derivate set" of φ at $\bar{x}$ is inspired by Warga's *derivate containers* introduced in [1316] and then developed in many publications; see, e.g., [1317, 1318, 1319, 1320, 1321, 1370] and the more recent papers by Ermoliev, Norkin and Wets [408] and by Sussmann [1236, 1237, 1238] with the references and discussions therein. Theorem 2.46 in the form presented in this book was established by Kruger [713], while its essence and proof go back to the early work by Kruger and Mordukhovich [719] showing that the Fréchet subdifferential (and hence both lower and upper basic subdifferentials) is *smaller* than *any* Warga's derivate container for continuous functions on finite-dimensional spaces; see also [99, 304, 596, 646, 705, 901] for modifications, extensions, and applications of the latter result and its variants.

Subsection 2.5.3 is based on the paper by Mordukhovich [920], where the approximate and exact versions of the *abstract extremal principle* were derived. Previous results on the fulfillment of the *approximate* extremal principle in non-Asplund (but mostly in bornologically smooth) spaces and on its equivalence to some other basic rules of generalized differentiation were obtained by Borwein, Mordukhovich and Shao [151], Borwein, Treiman and Zhu [159], Ioffe [606], and Zhu [1371]; see also Borwein and Zhu [163, 164] for more discussions.

Regarding the *exact* version of the abstract extremal principle, observe that both its *sequential* and *topological* modifications were established in [920] under an abstract version of the *sequential* normal compactness condition. A similar observation that just a sequential compactness property is sufficient to deal with a limiting topological structure was made by Ioffe [607] in the context of metric regularity.

3

Full Calculus in Asplund Spaces

This chapter is devoted to developing a comprehensive calculus for our basic generalized differential constructions: normals to sets, coderivatives of set-valued and single-valued mappings, and subgradients of extended-real-valued functions. A useful part of the generalized differential calculus has been presented in Chap. 1 in the setting of arbitrary Banach spaces. However, a number of important results therein impose differentiability assumptions on some mappings involved in compositions. In this chapter we don't require any smoothness and/or convexity of sets and mappings under consideration developing a *full calculus* in the framework of Asplund spaces at the same level of perfection as in finite dimensions. The main impact to this development comes from the results of Chap. 2 on the extremal principle and variational properties of Fréchet-like constructions in Asplund spaces. In this way we obtain general calculus rules for our basic objects using a *geometric approach*, i.e., starting with calculus rules for normal cones and then deriving from them sum and chain rules as well as other results for coderivatives and subdifferentials. It happens that the calculus rules obtained involve *sequential normal compactness* (SNC) assumptions on sets and mappings that are automatic in finite dimensions and reveal one of the most principal differences between finite-dimensional and infinite-dimensional variational theories. For the completeness and efficient applications of variational analysis in infinite dimensions one needs to develop an *SNC calculus* ensuring that the SNC properties are preserved under various operations with sets and mappings. We conclude this chapter with such a calculus in a fairly general setting. Throughout this chapter, all the spaces are *Asplund* unless otherwise stated.

3.1 Calculus Rules for Normals and Coderivatives

In this section we obtain general calculus rules for normal cones to nonconvex sets and coderivatives of nonsmooth set-valued and single-valued mappings under natural and verifiable assumptions. We begin with calculus of normal cones and first prove a "fuzzy rule" for Fréchet normals to set intersections by using the extremal principle. Then we establish a key calculus result on representing basic normals to set intersections under appropriate qualification and sequential normal compactness conditions. Employing the normal cone calculus, we derive sum and chain rules for normal and mixed coderivatives as well as other related formulas. In the last subsection we establish relationships between normal coderivatives of Lipschitzian single-valued mappings and subgradients of the corresponding scalarized functions important for subdifferential calculus and various applications.

3.1.1 Calculus of Normal Cones

The following lemma gives a *fuzzy* relationship between Fréchet normals to sets and their intersections in Asplund spaces without *any* assumptions on the sets in question besides their local closedness. It is implied by the approximate extremal principle and plays a major technical role in further developments.

Lemma 3.1 (a fuzzy intersection rule from the extremal principle). *Let $\Omega_1, \Omega_2 \subset X$ be arbitrary sets locally closed around $\bar{x} \in \Omega_1 \cap \Omega_2$, and let $x^* \in \widehat{N}(\bar{x}; \Omega_1 \cap \Omega_2)$. Then for any $\varepsilon > 0$ there are $\lambda \geq 0$, $x_i \in \Omega_i \cap (\bar{x} + \varepsilon I\!B)$, and $x_i^* \in \widehat{N}(x_i; \Omega_i) + \varepsilon I\!B^*$, $i = 1, 2$, such that*

$$\lambda x^* = x_1^* + x_2^*, \quad \max\{\lambda, \|x_1^*\|\} = 1\,. \tag{3.1}$$

Proof. Due to Definition 1.1(i) of Fréchet normals, for any given $x^* \in \widehat{N}(\bar{x}; \Omega_1 \cap \Omega_2)$ and $\varepsilon > 0$ we find a neighborhood U of $\bar{x}$ such that

$$\langle x^*, x - \bar{x}\rangle - \varepsilon\|x - \bar{x}\| \leq 0 \;\text{ whenever }\; x \in \Omega_1 \cap \Omega_2 \cap U\,. \tag{3.2}$$

Define subsets of $X \times I\!R$ by

$$\Lambda_1 := \big\{(x, \alpha) \in X \times I\!R \big|\; x \in \Omega_1,\; \alpha \geq 0\big\} \;\text{ and}$$

$$\Lambda_2 := \big\{(x, \alpha) \in X \times I\!R \big|\; x \in \Omega_2,\; \alpha \leq \langle x^*, x - \bar{x}\rangle - \varepsilon\|x - \bar{x}\|\big\}\,.$$

Observe that $(\bar{x}, 0) \in \Lambda_1 \cap \Lambda_2$ and that the sets Λ_i are locally closed around $(\bar{x}, 0)$. Moreover, one can easily check that

$$\Lambda_1 \cap \big(\Lambda_2 - (0, \nu)\big) \cap \big(U \times I\!R\big) = \emptyset \;\text{ for all }\; \nu > 0$$

due to (3.2) and the structure of Λ_i. Thus $(\bar{x}, 0)$ is a *local extremal point* of the set system $\{\Lambda_1, \Lambda_2\}$. Applying to this system the approximate extremal principle from Theorem 2.20 in the Asplund space $X \times I\!R$ with the norm $\|(x, \alpha)\| := \|x\| + |\alpha|$, we find $(x_i, \alpha_i) \in \Lambda_i$ and $(x_i^*, \lambda_i) \in \widehat{N}((x_i, \alpha_i); \Lambda_i)$, $i = 1, 2$, such that

$$\begin{cases} \max\left\{\|x_1^*+x_2^*\|, |\lambda_1+\lambda_2|\right\} < \varepsilon \ , \\ \frac{1}{2}-\varepsilon < \max\left\{\|x_i^*\|, |\lambda_i|\right\} < \frac{1}{2}+\varepsilon \ , \\ \|x_i-\bar{x}\| + |\alpha_i| < \varepsilon \end{cases} \tag{3.3}$$

for both $i = 1, 2$. One easily has $\lambda_1 \le 0$, $x_1^* \in \widehat{N}(x_1; \Omega_1)$, and

$$\limsup_{(x,\alpha)\xrightarrow{\Lambda_2}(x_2,\alpha_2)} \frac{\langle x_2^*, x-x_2\rangle + \lambda_2(\alpha-\alpha_2)}{\|x-x_2\| + |\alpha-\alpha_2|} \le 0 \tag{3.4}$$

by the definition of Fréchet normals. It follows from the structure of Λ_2 that $\lambda_2 \ge 0$ and

$$\alpha_2 \le \langle x^*, x_2-\bar{x}\rangle - \varepsilon\|x_2-\bar{x}\| \ . \tag{3.5}$$

If inequality (3.5) is *strict*, then (3.4) yields $\lambda_2 = 0$ and $x_2^* \in \widehat{N}(x_2; \Omega_2)$. In this case we get (3.1) with $\lambda = 0$ by using (3.3).

It remains to consider the case of *equality* in (3.5). Then we take vectors $(x, \alpha) \in \Lambda_2$ with

$$\alpha = \langle x^*, x-\bar{x}\rangle - \varepsilon\|x-\bar{x}\|, \ x \in \Omega_2 \setminus \{x_2\}$$

and substitute them into (3.4). This implies that there is a neighborhood V of x_2 such that

$$\langle x_2^*, x-x_2\rangle + \lambda_2(\alpha-\alpha_2) \le \varepsilon\big(\|x-x_2\| + |\alpha-\alpha_2|\big) \tag{3.6}$$

for all $x \in \Omega_2 \cap V$ and the corresponding α satisfying

$$\alpha - \alpha_2 = \langle x^*, x-x_2\rangle + \varepsilon\big(\|x_2-\bar{x}\| - \|x-\bar{x}\|\big) \ .$$

By the triangle inequality one has

$$|\alpha-\alpha_2| \le \big(\|x^*\| + \varepsilon\big)\|x-x_2\| \ .$$

Observe that the left-hand side ϑ in (3.6) can be represented as follows:

$$\vartheta = \langle x_2^* + \lambda_2 x^*, x-x_2\rangle + \varepsilon\lambda_2\big(\|x_2-\bar{x}\| - \|x-\bar{x}\|\big) \ .$$

Thus (3.6) implies the estimate

$$\langle x_2^* + \lambda_2 x^*, x-x_2\rangle \le \varepsilon\big(1 + \|x^*\| + \lambda_2 + \varepsilon\big)\|x-x_2\|$$

for all $x \in \Omega_2 \cap V$. This gives, due to Definition 1.1(i) of ε-normals, that

$$x_2^* + \lambda_2 x^* \in \widehat{N}_{c\varepsilon}(x_2; \Omega_2) \ \text{ with } \ c := 1 + \|x^*\| + \lambda_2 + \varepsilon \ . \tag{3.7}$$

Note that $1+\|x^*\| < c < 2+\|x^*\|$ for all ε sufficiently small, i.e., the constant c in (3.7) is always positive and may be chosen depending only on the given

x^*. Now using representation (2.51) of ε-normals in Asplund spaces, we find $v \in \Omega_2 \cap (x_2 + \varepsilon I\!B)$ such that

$$x_2^* + \lambda_2 x^* \in \widehat{N}(v; \Omega_2) + 2c\varepsilon I\!B^* .$$

Denoting $\eta := \max\{\lambda_2, \|x_2^*\|\}$, we get $\frac{1}{2} - \varepsilon < \eta < \frac{1}{2} + \varepsilon$ by (3.3), with $\frac{1}{4} < \eta < \frac{3}{4}$ when ε is small. Put

$$\lambda := \lambda_2/\eta, \quad u^* := -x_2^*/\eta, \quad v^* := (x_2^* + \lambda_2 x^*)/\eta .$$

One clearly has $\lambda \geq 0$, $\max\{\lambda, \|u^*\|\} = 1$, and $\lambda x^* = u^* + v^*$. Moreover, $v^* \in \widehat{N}(v; \Omega_2) + 8c\varepsilon I\!B^*$ and

$$u^* = x_1^*/\eta - (x_1^* + x_2^*)/\eta \in \widehat{N}(x_1; \Omega_1) + 4\varepsilon I\!B^*$$

due to (3.3). Since $c > 0$ depends only on the given x^* and since ε was chosen arbitrarily, this justifies the conclusions of the lemma. △

From the proof of Lemma 3.1 we can get conditions ensuring that $\lambda \neq 0$ in (3.1) and hence

$$\widehat{N}(\bar{x}; \Omega_1 \cap \Omega_2) \subset \widehat{N}(x_1; \Omega_1) + \widehat{N}(x_2; \Omega_2) + \varepsilon I\!B^* \tag{3.8}$$

with some $x_i \in \Omega_i \cap (\bar{x} + \varepsilon I\!B)$, $i = 1, 2$, for all small $\varepsilon > 0$. It happens, in particular, when the sets Ω_i satisfy the so-called *fuzzy qualification condition*: there is $\gamma > 0$ such that

$$\big(\widehat{N}(x_1; \Omega_1) + \gamma I\!B^*\big) \cap \big(-\widehat{N}(x_2; \Omega_2) + \gamma I\!B^*\big) \cap I\!B^* \subset \tfrac{1}{2} I\!B^* \tag{3.9}$$

for all $x_i \in \Omega_i \cap (\bar{x} + \gamma I\!B)$, $i = 1, 2$. Note that under condition (3.9) we get more information in comparison with the intersection rule (3.8). Namely, (3.9) ensures in addition to (3.8) the following *uniform boundedness estimate* on x_i^*: for any given $x^* \in \widehat{N}(\bar{x}; \Omega_1 \cap \Omega_2)$, $\varepsilon > 0$, and γ from (3.9) there are $x_i \in \Omega_i \cap (\bar{x} + \varepsilon I\!B)$ and $\eta = \eta(x^*, \varepsilon, \gamma) > 0$ such that

$$\|x^* - (x_1^* + x_2^*)\| \leq \varepsilon \ \text{ for some } \ x_i^* \in \widehat{N}(x_i; \Omega_i) \cap (\eta I\!B^*), \ i = 1, 2 .$$

Our primary goal in this subsection is to obtain an intersection rule for basic normals in Asplund spaces under appropriate conditions formulated *at* a reference point of the set intersection. To achieve this goal, we are going to employ two kinds of "pointbased" conditions unified under the names of:

(a) *qualification conditions* and

(b) *sequential normal compactness conditions.*

Let us start with qualification conditions for sets that are basic for subsequent developments and applications in this book.

Definition 3.2 (basic qualification conditions for sets). *Given two subsets Ω_1, Ω_2 of a Banach space X and a point $\bar{x} \in \Omega_1 \cap \Omega_2$, we say that:*

(i) *The set system $\{\Omega_1, \Omega_2\}$ satisfies the* NORMAL QUALIFICATION CONDITION *at $\bar{x}$ if*

$$N(\bar{x}; \Omega_1) \cap \big(-N(\bar{x}; \Omega_2)\big) = \{0\} . \tag{3.10}$$

(ii) *$\{\Omega_1, \Omega_2\}$ satisfies the* LIMITING QUALIFICATION CONDITION *at $\bar{x}$ if for any sequences $\varepsilon_k \downarrow 0$, $x_{ik} \xrightarrow{\Omega_i} \bar{x}$, and $x_{ik}^* \xrightarrow{w^*} x_i^*$ with $x_{ik}^* \in \widehat{N}_{\varepsilon_k}(x_{ik}; \Omega_i)$, $i = 1, 2$, and $k \to \infty$ one has*

$$\|x_{1k}^* + x_{2k}^*\| \to 0 \Longrightarrow x_1^* = x_2^* = 0 .$$

The normal qualification condition (3.10) is formulated in terms of basic normals to both sets Ω_i at the given point $\bar{x}$ and, as we'll see below, is a proper counterpart in the general set setting of the classical constraint qualification conditions in problems of constrained optimization. By (2.51) one can equivalently put $\varepsilon_k = 0$ in Definition 3.2(ii) if X is Asplund and both sets Ω_1, Ω_2 are closed around $\bar{x}$. Taking into account the representation of basic normals in Asplund spaces from Theorem 2.35, we observe that (3.10) is equivalent to say, for locally closed sets, that for any sequences $x_{ik} \xrightarrow{\Omega_i} \bar{x}$ and $x_{ik}^* \xrightarrow{w^*} x_i^*$ with $x_{ik}^* \in \widehat{N}(x_{ik}; \Omega_i)$, $i = 1, 2$, and $k \to \infty$ one has

$$x_{1k}^* + x_{2k}^* \xrightarrow{w^*} 0 \Longrightarrow x_1^* = x_2^* = 0 .$$

This immediately implies that conditions (i) and (ii) in Definition 3.2 are equivalent in finite dimensions, but the latter condition may be substantially weaker in infinite-dimensional spaces. In particular, for the case of sets generated by *graphs* of mappings, condition (ii) can be expressed in terms of *mixed* coderivatives at reference points while (i) corresponds to normal coderivatives; see the next subsection.

In contrast to the qualification conditions in Definition 3.2, the sequential normal compactness conditions we are going to discuss next are infinite-dimensional in nature and develop the line of the SNC and PSNC properties introduced, respectively, in Subsects. 1.1.3 and 1.2.5 for sets and mappings in Banach spaces. Here we explore the *product structure* of spaces and sets under consideration. The latter makes it possible to use *partial* SNC conditions in the general intersection rule for basic normals and then to apply them to coderivative and subdifferential calculi. To establish the general intersection rule in product spaces, we need to introduce one more type of PSNC properties called "*strong* partial sequential normal compactness".

Definition 3.3 (PSNC properties in product spaces). *Let Ω belong to the product $\prod_{j=1}^m X_j$ of Banach spaces, let $\bar{x} \in \Omega$, and let $J \subset \{1, \ldots, m\}$. We say that:*

(i) *Ω is* PARTIALLY SEQUENTIALLY NORMALLY COMPACT (PSNC) *at $\bar{x}$ with respect to $\{X_j |\, j \in J\}$ (i.e., with respect to $\prod_{j \in J} X_j$, or just to J) if for* *any sequences* $\varepsilon_k \downarrow 0$, $x_k \xrightarrow{\Omega} \bar{x}$, *and* $x_k^* = (x_{1k}^*, \ldots, x_{mk}^*) \in \widehat{N}_{\varepsilon_k}(x_k; \Omega)$ *one has*

$$\Big[x_{jk}^* \xrightarrow{w^*} 0,\ j \in J \quad \& \quad \|x_{jk}^*\| \to 0,\ j \in \{1, \ldots, m\} \setminus J\Big] \Longrightarrow \|x_{jk}^*\| \to 0,\ j \in J\ .$$

(ii) Ω *is* STRONGLY PSNC *at* $\bar{x}$ *with respect to* $\{X_j |\ j \in J\}$ *if for any sequences* $\varepsilon_k \downarrow 0$, $x_k \xrightarrow{\Omega} \bar{x}$, *and* $(x_{1k}^*, \ldots, x_{mk}^*) \in \widehat{N}_{\varepsilon_k}(x_k; \Omega)$ *one has*

$$\Big[x_{jk}^* \xrightarrow{w^*} 0,\ j = 1, \ldots, m\Big] \Longrightarrow \|x_{jk}^*\| \to 0,\ j \in J\ .$$

Let us mention the two extreme cases: (a) $J = \emptyset$ when any set Ω satisfies both properties in (i) and (ii), and (b) $J = \{1, \ldots, m\}$ when both properties (i) and (ii) don't depend on the product structure and reduce to the SNC property of Definition 1.20. Note also that the PSNC property of a mapping $F\colon X \rightrightarrows Y$ in Definition 1.67 is equivalent to the above PSNC property of $\operatorname{gph} F \subset X \times Y$ with respect to X. One can equivalently put $\varepsilon_k = 0$ in Definition 3.3 if all X_j are Asplund and Ω is locally closed around $\bar{x}$.

As seen in Subsects. 1.1.3 and 1.2.5, the SNC property of sets and the PSNC property of mappings automatically hold under certain *Lipschitz-type* assumptions. Observe that Theorem 1.75 asserts, in the terminology of Definition 3.3, that if a mapping $F\colon X \rightrightarrows Y$ between Banach spaces is *partially* CEL around $(\bar{x}, \bar{y}) \in \operatorname{gph} F$, then its graph is *strongly* PSNC at this point with respect to X. Let us emphasize a crucial fact in the theory and applications of the SNC properties under consideration: they enjoy a rich *calculus*, in the sense of their preservation under natural operations with sets and mappings; see Sect. 3.3 for developments in Asplund spaces in addition to those in arbitrary Banach spaces presented in Subsects. 1.1.3 and 1.2.5.

Now we are ready to establish the main intersection rule for basic normals to arbitrary sets in products of Asplund spaces.

Theorem 3.4 (basic normals to set intersections in product spaces). *Let the sets* $\Omega_1, \Omega_2 \subset \prod_{j=1}^m X_j$ *be locally closed around* $\bar{x} \in \Omega_1 \cap \Omega_2$, *and let* $J_1, J_2 \subset \{1, \ldots, m\}$ *be such that* $J_1 \cup J_2 = \{1, \ldots, m\}$. *Assume that* Ω_1 *is PSNC at* $\bar{x}$ *with respect to* J_1, *that* Ω_2 *is strongly PSNC at* $\bar{x}$ *with respect to* J_2, *and that the system* $\{\Omega_1, \Omega_2\}$ *satisfies the limiting qualification condition at* $\bar{x}$. *Then one has the inclusion*

$$N(\bar{x}; \Omega_1 \cap \Omega_2) \subset N(\bar{x}; \Omega_1) + N(\bar{x}; \Omega_2)\ . \tag{3.11}$$

If in addition both Ω_1 *and* Ω_2 *are normally regular at* $\bar{x}$, *then* $\Omega_1 \cap \Omega_2$ *is also normally regular at this point and* (3.11) *holds as equality.*

Proof. To justify (3.11), we pick $x^* \in N(\bar{x}; \Omega_1 \cap \Omega_2)$ and by Theorem 2.35 find sequences $x_k \to \bar{x}$ and $x_k^* \xrightarrow{w^*} x^*$ such that

$$x_k \in \Omega_1 \cap \Omega_2 \quad \text{and} \quad x_k^* \in \widehat{N}(x_k; \Omega_1 \cap \Omega_2), \qquad k \in I\!N\ . \tag{3.12}$$

Take a sequence $\varepsilon_k \downarrow 0$ as $k \to \infty$ and employ Lemma 3.1 in (3.12) along this sequence for any fixed $k \in I\!N$. This gives us

$$(u_k, v_k) \in \Omega_1 \times \Omega_2, \quad \lambda_k \geq 0, \quad u_k^* \in \widehat{N}(u_k; \Omega_1), \quad v_k^* \in \widehat{N}(v_k; \Omega_2)$$

such that $\|u_k - x_k\| \leq \varepsilon_k$, $\|v_k - x_k\| \leq \varepsilon_k$, and

$$\|u_k^* + v_k^* - \lambda_k x_k^*\| \leq 2\varepsilon_k, \quad 1 - \varepsilon_k \leq \max\big\{\lambda_k, \|u_k^*\|\big\} \leq 1 + \varepsilon_k . \tag{3.13}$$

Since the sequence $\{x_k^*\}$ weak* converges, it is bounded in X^* by the uniform boundedness principle, and so are $\{u_k^*\}$ and $\{v_k^*\}$ due to (3.13). Invoking the weak* sequential compactness of bounded sets in duals to Asplund spaces, one has $u^*, v^* \in X^*$ and $\lambda \geq 0$ such that $u_k^* \xrightarrow{w^*} u^*$, $v_k^* \xrightarrow{w^*} v^*$, and $\lambda_k \to \lambda$ along a subsequence of $k \in I\!N$. Passing to the limit in (3.13) as $k \to \infty$, we conclude that $u^* \in N(\bar{x}; \Omega_1)$, $v^* \in N(\bar{x}; \Omega_2)$, and $\lambda x^* = u^* + v^*$.

To justify (3.11), it remains to show that $\lambda \neq 0$ under the assumptions made. If it is not the case, we get $\|u_k^* + v_k^*\| \to 0$ from (3.13) and hence $u^* = v^* = 0$ due to the limiting qualification condition. This implies

$$u_k^* = (u_{1k}^*, \ldots, u_{mk}^*) \xrightarrow{w^*} 0, \quad v_k^* = (v_{1k}^*, \ldots, v_{mk}^*) \xrightarrow{w^*} 0 \;\text{ as }\; k \to \infty . \tag{3.14}$$

Taking into account that Ω_2 is strongly PSNC at $\bar{x}$ with respect to J_2, we get from (3.14) that $\|v_{jk}^*\| \to 0$ for $j \in J_2$. This gives, due to (3.13) and $J_1 \cup J_2 = \{1, \ldots, m\}$, that

$$\|u_{jk}^*\| \to 0 \;\text{ for }\; j \in \{1, \ldots, m\} \setminus J_1 \;\text{ as }\; k \to 0 .$$

Using (3.14) and the PSNC property of Ω_1 with respect to J_1, we conclude that $\|u_{jk}^*\| \to 0$ for $j \in J_1$. Thus $\|u_k^*\| \to 0$ as $k \to \infty$, which contradicts the second relation in (3.13) and justifies the required inclusion (3.11).

Finally, let us prove the regularity/equality assertion of the theorem. It follows directly from the definition of Fréchet normals that they always satisfy the inclusion

$$\widehat{N}(\bar{x}; \Omega_1 \cap \Omega_2) \supset \widehat{N}(\bar{x}; \Omega_1) + \widehat{N}(\bar{x}; \Omega_2)$$

opposite to (3.11). Combining this with (3.11) and assuming the normal regularity of Ω_1 and Ω_2 at $\bar{x}$, we get

$$N(\bar{x}; \Omega_1 \cap \Omega_2) \subset \widehat{N}(\bar{x}; \Omega_1 \cap \Omega_2) ,$$

which implies the equality in (3.11) and the normal regularity of the intersection $\Omega_1 \cap \Omega_2$ at $\bar{x}$. $\triangle$

In what follows we obtain a number of important consequences of Theorem 3.4 that take into account the product structure of the space in question allowing us to use the PSNC properties and refined qualification conditions. Now let us present an immediate corollary of the theorem in spaces with *no product structure* imposed. In this case we may use just the (full) SNC property, which is required for *only one* among two sets. We don't include the equality/regularity statement in this corollary, which is not different from the one given in the theorem.

Corollary 3.5 (intersection rule under the SNC condition). *Assume that $\Omega_1, \Omega_2 \subset X$ are locally closed around $\bar{x} \in \Omega_1 \cap \Omega_2$ and that either Ω_1 or Ω_2 is SNC at this point. Then the intersection rule* (3.11) *holds provided that $\{\Omega_1, \Omega_2\}$ satisfies the limiting qualification condition at $\bar{x}$, in particular, when one has* (3.10).

Proof. This is a special case of the theorem with $m = 1$ and $J_1 = \{1\}$. △

Observe that the *SNC assumption* in Corollary 3.5 is *essential* for the fulfillment of the intersection rule (3.11) even for *convex* and norm-compact sets in infinite-dimensional spaces. Indeed, in the framework of Example 2.23 we consider the set $\Omega_1 \subset X$ defined therein and the set Ω_2 given by

$$\Omega_2 := \{ta \mid t \in [-1, 1]\} \text{ with } a := \sum_{n=1}^{\infty} \frac{e_n}{n^2} \in X .$$

One can easily check that $\Omega_1 \cap \Omega_2 = \{0\}$, $a \in \operatorname{cl}\operatorname{span}\Omega_1$,

$$N(0; \Omega_1) \cap (-N(0, \Omega_2)) = (\operatorname{span}\Omega_1)^{\perp} \cap (\operatorname{span}\Omega_2)^{\perp} = \{0\}, \quad \text{and}$$

$$X^* = N(0; \Omega_1 \cap \Omega_2) \not\subset N(0; \Omega_1) + N(0; \Omega_2) = (\operatorname{span}\Omega_1)^{\perp} .$$

Thus all but SNC assumptions of Corollary 3.5 are fulfilled, while the intersection rule (3.11) is violated.

On the other hand, the following example shows that the replacement of the SNC assumption by the *CEL* one in Corollary 3.5 may be *too restrictive* for the intersection rule to hold, even in the case of closed *convex cones* in spaces with $\mathcal{C}^{\infty}$-smooth renorms.

Example 3.6 (intersection rule with no CEL assumption). *There are a nonseparable space X with a $\mathcal{C}^{\infty}$-smooth renorm and two closed convex subcones Ω_1 and Ω_2 of X such that both Ω_i are SNC at $\bar{x}$ but not CEL around this point and that the pair $\{\Omega_1, \Omega_2\}$ satisfies the limiting qualification condition* (3.10), *and hence the intersection rule* (3.11) *holds as equality.*

Proof. Consider the space $X = \mathcal{C}_0[0, \omega_1]$ of all functions $\varphi\colon [0, \omega_1] \to I\!R$ continuous on the "long" interval $[0, \omega_1]$ with $\varphi(\omega_1) = 0$, where ω_1 means the first uncountable ordinal. The norm $\|\cdot\|$ on X is the supremum norm. It is well known that X is an Asplund space; moreover, it admits an equivalent $\mathcal{C}^{\infty}$-smooth norm; see [331, Chap. VII] for proofs and discussions. It is easy to check that for every $\varphi \in X$ there is $\alpha < \omega_1$ such that $\varphi(\beta) = 0$ whenever

$\alpha \le \beta \le \omega_1$. We further clarify what is the dual space $\mathcal{C}_0[0,\omega_1]^*$ to X. Given a set $S \subset [0,\omega_1)$, by

$$\chi_S(s) := \begin{cases} 1 & \text{if } s \in S\,, \\ 0 & \text{otherwise} \end{cases}$$

we denote the *indicatrix* (characteristic function) of S. Define the mapping $\xi \in X^* \mapsto (a_\alpha)_{\alpha<\omega_1}$ by

$$a_\alpha := \begin{cases} \langle \xi, \chi_{\{\alpha\}}\rangle & \text{if } \alpha < \omega_1 \text{ is a nonlimit ordinal}\,, \\ \lim_{\beta\uparrow\alpha}\langle \xi, \chi_{[\beta,\alpha]}\rangle & \text{if } \alpha < \omega_1 \text{ is a limit ordinal}\,. \end{cases}$$

One can check that this assignment maps X^* isometrically *onto* the space $\ell_1([0,\omega_1))$ and that

$$\langle \xi, \varphi\rangle = \sum_{\alpha<\omega_1} \varphi(\alpha) a_\alpha \quad \text{for every} \quad \varphi \in X\,.$$

Consider the closed convex subcone of X defined by

$$\Omega := \Big\{\varphi \in \mathcal{C}_0[0,\omega_1]\Big|\ \varphi \le 0\Big\}$$

and show that it is *SNC* at $\bar{x} = 0$ but *not CEL* around this point. First we justify the following description of the normal cone to Ω.

Claim. *For any $\bar{x} \in \Omega$ and any $x^* = (a_\alpha)_{\alpha<\omega_1} \in N(\bar{x};\Omega)$ one has $a_\alpha \ge 0$ whenever $\alpha \in [0,\omega_1)$.*

Indeed, take any $\bar{x} \in \Omega$ and any $0 \le \beta \le \alpha < \omega_1$. Then $x := \bar{x} - t\chi_{[\beta,\alpha]} \in \Omega$ for all $t > 0$, and hence

$$0 \ge \langle x^*, x - \bar{x}\rangle = \langle x^*, -\chi_{[\beta,\alpha]}\rangle = -\sum_{\beta\le\gamma\le\alpha} a_\gamma \ \ (\ge -\|x^*\| > -\infty)\,.$$

From these relationships and the representation

$$N(\bar{x};\Omega) = \widehat{N}(\bar{x};\Omega) = \big\{x^* \in X^*\big|\ \langle x^*, x - \bar{x}\rangle \le 0 \ \text{ for all } \ x \in \Omega\big\}$$

we subsequently get that $a_\alpha \ge 0$ whenever $\alpha < \omega_1$.

Now we are ready to show that the set Ω is SNC at $\bar{x} = 0$. Take $x_k \in \Omega$ and $x_k^* \in N(x_k;\Omega)$, $k \in I\!N$, such that $\|x_k\| \to 0$ and $x_k^* \xrightarrow{w^*} 0$ as $k \to \infty$. Let us prove that $\|x_k^*\| \to 0$. Using the isometry between X^* and $\ell_1([0,\omega_1))$, write $x_k^* = (a_\alpha^k)_{\alpha<\omega_1}$, $k \in I\!N$. The above claim says that $a_\alpha^k \ge 0$ for every $\alpha < \omega_1$ and every $k \in I\!N$. Find $\beta < \omega_1$ so large that $a_\alpha^k = 0$ whenever $\beta < \alpha < \omega_1$ and $k \in I\!N$; this can be done as we work in $\ell_1([0,\omega_1))$. Again, using the claim, we get $\|x_k^*\| = \sum_{\alpha\le\beta} a_\alpha^k = x_k^*\big(\chi_{[0,\beta]}\big) \to 0$ as $k \to \infty$, which justifies the SNC property of Ω at $\bar{x} = 0$.

Let us check that Ω is not CEL around $\bar{x} = 0$. To proceed, we use the net description of the CEL property in Asplund spaces discussed in Remark 1.27(ii). Note that whenever $(s_\alpha)_{\alpha<\omega_1}$ is a net of real numbers converging to 0 as $\alpha \uparrow \omega_1$, one necessarily has $s_\alpha = 0$ for all $\alpha < \omega_1$ sufficiently large. Taking this into account, we put $x_\alpha := 0$ and $x_\alpha^* := \delta_\alpha$ for every $\alpha < \omega_1$, where δ_α is the *Dirac measure* at α, i.e., the point mass measure at α. Since $\delta_\alpha \xrightarrow{w^*} 0$ as $\alpha \uparrow \omega_1$, the net $((x_\alpha, x_\alpha^*))_{\alpha<\omega_1}$ in $X \times X^*$ satisfies the bounded net counterpart of Definition 1.20. Yet $\|x_\alpha^*\| = 1$ for all $\alpha < \omega_1$, which proves that Ω is not CEL around the point $\bar{x} = 0$.

Note that we can also conclude that Ω is not CEL directly from the characterization of the CEL property for closed convex sets discussed in Remark 1.27(i). Observe first that the span of Ω is the whole space $\mathcal{C}_0[0,\omega_1]$. Indeed, for any $\varphi \in \mathcal{C}_0[0,\omega_1]$ there is $\alpha < \omega_1$ such that the support of φ belongs to $[0,\alpha]$. Then $\varphi = (\varphi - \|\varphi\|\chi_{[0,\alpha]}) + \|\varphi\|\chi_{[0,\alpha]}$. In order to check that $\operatorname{int}\Omega = \emptyset$, we take any $\varphi \in \Omega$ and find α for which $\varphi(\alpha) = 0$. Then $\psi_k := \varphi + \frac{1}{k}\chi_{\{\alpha\}} \notin \Omega$ and $\|\psi_k - \varphi\| = \frac{1}{k} \to 0$.

Finally, put $\Omega_1 = \Omega_2 := \Omega$ and check that the system $\{\Omega_1, \Omega_2\}$ satisfies the limiting qualification condition (3.10), which reduces in this case to

$$N(0;\Omega) \cap \big(-N(0;\Omega)\big) = \{0\}\,.$$

The latter immediately follows from the claim proved above. △

In this chapter we derive many calculus results for normal cones, coderivatives, and subdifferentials that are based on the above intersection rules and hence on the extremal principle. The first consequence gives useful rules for representing Fréchet and basic normals to *sums of sets.* It is interesting to observe that in both fuzzy and exact sum rules below don't involve *any qualification and/or SNC conditions*, which in fact hold automatically. Recall that the notions of inner semicontinuity and inner semicompactness of set-valued mappings are formulated in Definition 1.63.

Theorem 3.7 (sum rules for generalized normals). *Let Ω_1, Ω_2 be closed subsets of X, and let $\bar{x} \in \Omega_1 + \Omega_2$. Define a mapping $S\colon X \rightrightarrows X^2$ by*

$$S(x) := \big\{(x_1,x_2) \in X \times X \big|\; x_1 + x_2 = x,\ x_1 \in \Omega_1,\ x_2 \in \Omega_2\big\}\,.$$

The following assertions hold:

(i) *Given $\varepsilon > 0$, one has the inclusion*

$$\widehat{N}(\bar{x};\Omega_1+\Omega_2) \subset \bigcup_{(x_1,x_2)\in S(\bar{x})+\varepsilon I\!B} \big(\widehat{N}(x_1;\Omega_1) + \varepsilon I\!B^*\big) \cap \big(\widehat{N}(x_2;\Omega_2) + \varepsilon I\!B^*\big)\,.$$

(ii) *Assume that S is inner semicompact at $\bar{x}$. Then*

$$N(\bar{x};\Omega_1+\Omega_2) \subset \bigcup_{(x_1,x_2)\in S(\bar{x})} N(x_1;\Omega_1) \cap N(x_2;\Omega_2)\,.$$

Furthermore, if for some $(\bar{x}_1,\bar{x}_2)\in S(\bar{x})$ the mapping S is inner semicontinuous at $(\bar{x},\bar{x}_1,\bar{x}_2)$, then

$$N(\bar{x};\Omega_1+\Omega_2)\subset N(\bar{x}_1;\Omega_1)\cap N(\bar{x}_2;\Omega_2)\ .$$

Proof. To prove (i), let us take $x^*\in\widehat{N}(\bar{x};\Omega_1+\Omega_2)$ and observe that

$$(x^*,x^*)\in\widehat{N}\big((\bar{x}_1,\bar{x}_2);\widetilde{\Omega}_1\times\widetilde{\Omega}_2\big)\quad\text{whenever}\quad(\bar{x}_1,\bar{x}_2)\in S(\bar{x})\ ,$$

where $\widetilde{\Omega}_1:=\Omega_1\times X$ and $\widetilde{\Omega}_2:=X\times\Omega_2$. Now we apply the fuzzy intersection rule from Lemma 3.1 to the closed sets $\widetilde{\Omega}_1$ and $\widetilde{\Omega}_2$ noting that it holds in the "normal" form (3.8), i.e., with $\lambda=1$ in (3.1), since the fuzzy qualification condition (3.9) is obviously fulfilled. Taking into account the specific structure of the above sets $\widetilde{\Omega}_1$ and $\widetilde{\Omega}_2$, we find $x_i\in\Omega_i$ and $x_i^*\in\widehat{N}(x_i;\Omega_i)$ such that $\|x_i-\bar{x}_i\|\le\varepsilon$ and $\|x_i^*-x^*\|\le\varepsilon$ for $i=1,2$. This proves assertion (i).

To justify assertion (ii), we proceed only with the first formula; the second one can be proved similarly. Taking $x^*\in N(\bar{x};\Omega_1+\Omega_2)$ and using the definition of basic normals, we find sequences $\varepsilon_k\downarrow 0$, $x_k\to\bar{x}$ with $x_k\in\Omega_1+\Omega_2$, and $x_k^*\xrightarrow{w^*}x^*$ with $x_k^*\in\widehat{N}_{\varepsilon_k}(x_k;\Omega_1+\Omega_2)$. Note that, although X is Asplund, one cannot put $\varepsilon_k=0$ above, since the sum $\Omega_1+\Omega_2$ may not be closed under the assumptions made. By the inner semicompactness of S there is a sequence of (x_{1k},x_{2k}) that contains a subsequence converging to some $(\bar{x}_1,\bar{x}_2)$. Since Ω_1 and Ω_2 are closed, we have $(\bar{x}_1,\bar{x}_2)\in S(\bar{x})$. Defining the sets $\widetilde{\Omega}_1$ and $\widetilde{\Omega}_2$ as above, it is easy to see that

$$(x_k^*,x_k^*)\in\widehat{N}_{\varepsilon_k}\big((x_{1k},x_{2k});\widetilde{\Omega}_1\cap\widetilde{\Omega}_2\big)\quad\text{for all}\quad k\in I\!N\ ,$$

and hence $(x^*,x^*)\in N\big((\bar{x}_1,\bar{x}_2);\widetilde{\Omega}_1\cap\widetilde{\Omega}_2\big)$. To employ the intersection rule of Theorem 3.4, note that the qualification and SNC assumptions therein hold for the underlying sets $\widetilde{\Omega}_1$ and $\widetilde{\Omega}_2$. Thus there exist x_1^* and x_2^* from X^* satisfying the relations

$$\begin{aligned}&(x_1^*,0)\in N\big((\bar{x}_1,\bar{x}_2);\widetilde{\Omega}_1\big),\ (0,x_2^*)\in N\big((\bar{x}_1,\bar{x}_2);\widetilde{\Omega}_2\big),\\&(x^*,x^*)=(x_1^*,0)+(0,x_2^*)\ .\end{aligned}$$

The latter gives $x_1^*=x_2^*=x^*$. Observing that $x_i^*\in N(\bar{x}_i;\Omega_i)$ for $i=1,2$, we get $x^*\in N(\bar{x}_1;\Omega_1)\cap N(\bar{x}_2;\Omega_2)$ and complete the proof of the theorem. △

Next let us consider subsets $\Omega\subset X$ given in the form of *inverse images*

$$F^{-1}(\Theta):=\big\{x\in X\big|\ F(x)\cap\Theta\ne\emptyset\big\}$$

of some sets $\Theta\subset Y$ under set-valued mappings $F\colon X\rightrightarrows Y$ between Asplund spaces. Our goal is to represent basic normals to $F^{-1}(\Theta)$ in terms of F and Θ. We have dealt with this topic in Subsect. 1.1.2 in the case of single-valued and strictly differentiable mappings $F=f\colon X\to Y$ between Banach spaces. Now we are going to study the case of general set-valued mappings F and obtain an efficient representation formula for basic normals to $F^{-1}(\Theta)$ employing Theorem 3.4. In the following result we use the normal coderivative D_N^*F from (1.24) for the representation formula and the "reversed mixed coderivative" $\widetilde{D}_M^*F$ from (1.40) for the point qualification condition imposed on the initial system $\{F, \Theta\}$.

Theorem 3.8 (basic normals to inverse images). *Let $\bar{x} \in F^{-1}(\Theta)$, where $F: X \rightrightarrows Y$ is a closed-graph mapping and where $\Theta \subset Y$ is a closed set. Assume that the set-valued mapping $x \to F(x) \cap \Theta$ is inner semicompact at $\bar{x}$ and that for every $\bar{y} \in F(\bar{x}) \cap \Theta$ the following hold:*

(a) *Either F^{-1} is PSNC at $(\bar{y}, \bar{x})$ or Θ is SNC at $\bar{y}$.*

(b) *$\{F, \Theta\}$ satisfies the qualification condition*

$$N(\bar{y}; \Theta) \cap \ker \widetilde{D}_M^* F(\bar{x}, \bar{y}) = \{0\} .$$

Then one has

$$N(\bar{x}; F^{-1}(\Theta)) \subset \bigcup \Big[D_N^* F(\bar{x}, \bar{y})(y^*) \Big| \, y^* \in N(\bar{y}; \Theta), \ \bar{y} \in F(\bar{x}) \cap \Theta \Big] . \quad (3.15)$$

Proof. Fix $x^* \in N(\bar{x}; F^{-1}(\Theta))$ and take sequences $\varepsilon_k \downarrow 0$, $x_k \to \bar{x}$ with $x_k \in F^{-1}(\Theta)$, and $x_k^* \xrightarrow{w^*} x^*$ with $x_k^* \in \widehat{N}_{\varepsilon_k}(x_k; F^{-1}(\Theta))$ for all $k \in I\!N$; note that $F^{-1}(\Theta)$ may not be closed. Using the inner semicompactness of $F(\cdot) \cap \Theta$ at $\bar{x}$, one select a subsequence of $y_k \in F(x_k) \cap \Theta$ converging to some $\bar{y}$. The closedness assumptions on gph F and Θ ensure that $\bar{y} \in F(\bar{x}) \cap \Theta$. Construct the closed subsets

$$\Omega_1 := \operatorname{gph} F, \quad \Omega_2 := X \times \Theta$$

of the Asplund space $X \times Y$ and observe that $(x_k, y_k) \in \Omega_1 \cap \Omega_2$ for all $k \in I\!N$. It is easy to verify that

$$(x_k^*, 0) \in \widehat{N}_{\varepsilon_k}((x_k, y_k); \Omega_1 \cap \Omega_2), \quad k \in I\!N ,$$

and therefore $(x^*, 0) \in N((\bar{x}, \bar{y}); \Omega_1 \cap \Omega_2)$. To apply the intersection rule of Theorem 3.4 to the sets Ω_1, Ω_2, we need to check that its assumptions hold under the imposed conditions (a) and (b).

The set $\Omega_2 = X \times \Theta$ is obviously SNC at $(\bar{x}, \bar{y})$ if Θ is SNC at $\bar{y}$. It is also clear that the PSNC property of the mapping $F^{-1}: Y \rightrightarrows X$ at $(\bar{y}, \bar{x})$ in the sense of Definition 1.67(ii) is the same as the PSNC property of the set $\Omega_1 = \operatorname{gph} F \subset X \times Y$ at $(\bar{x}, \bar{y})$ with respect to Y. It remains to show that the qualification condition (b) implies that the constructed set system $\{\Omega_1, \Omega_2\}$ satisfies the limiting qualification condition at $(\bar{x}, \bar{y})$ in the sense of Definition 3.2(ii). Indeed, by (1.40) and Theorem 2.35, condition (b) gives that for $(x_k^*, y_{1k}^*) \in \widehat{N}((x_k, y_{1k}); \operatorname{gph} F)$ and $y_{2k}^* \in \widehat{N}(y_{2k}; \Theta)$ with $x_k \to \bar{x}$, $y_{ik} \to \bar{y}$, $i = 1, 2$, and by $y_{2k}^* \xrightarrow{w^*} y^*$ one has

$$\Big[\|x_k^*\| \to 0, \quad y_{1k}^* + y_{2k}^* \xrightarrow{w^*} 0\Big] \Longrightarrow y^* = 0\,.$$

On the other hand, the limiting qualification condition in this situation requires only that

$$\Big[\|x_k^*\| \to 0, \quad \|y_{1k}^* + y_{2k}^*\| \to 0\Big] \Longrightarrow y^* = 0\,, \tag{3.16}$$

i.e., it is definitely implied by (b) but not vice versa. Thus one can use Theorem 3.4, which ensures the existence of $(x_1^*, y_1^*) \in N((\bar{x}, \bar{y}); \operatorname{gph} F)$ and $y_2^* \in N(\bar{y}; \Theta)$ such that

$$(x^*, 0) = (x_1^*, y_1^*) + (0, y_2^*) \Longleftrightarrow x^* = x_1^*, \quad y_2^* = -y_1^*\,.$$

Taking into account description (1.26) of the normal coderivative, we get $x_1^* \in D_N^* F(\bar{x}, \bar{y})(y_2^*)$ and arrive at (3.15). △

It follows from the proof of Theorem 3.8 that condition (b) can be replaced with the weaker limiting qualification condition in (3.16). However, (b) is more convenient for applications, since it involves only the given points $(\bar{x}, \bar{y})$ and allows us to use an efficient calculus available for basic normals and coderivatives. Note that the usage of the normal qualification condition (3.10) in the proof of Theorem 3.8 leads us to the point qualification condition in terms of the normal coderivative

$$N(\bar{y}; \Theta) \cap \ker D_N^* F(\bar{x}, \bar{y}) = \{0\}\,,$$

which is more restrictive than (b).

The principal advantage of using mixed vs. normal coderivatives in Theorem 3.8 and subsequent results is as follows: in this way we can justify the validity of the main assumptions in calculus rules for important classes of multifunctions with *Lipschitzian* and/or *metric regularity* properties. This is due to coderivative results of Sect. 1.2 ensuring that the corresponding qualification and PSNC conditions *automatically hold* for such multifunctions. In what follows we mostly use local metric regularity and Lipschitz-like properties around points of graphs omitting the word "local" with no confusion.

Corollary 3.9 (inverse images under metrically regular mappings). *Let $\bar{x} \in F^{-1}(\Theta)$, where $\Theta \subset Y$ and $\operatorname{gph} F \subset X \times Y$ are closed and where $F(\cdot) \cap \Theta$ is inner semicompact at $\bar{x}$. Assume that F is metrically regular around $(\bar{x}, \bar{y})$ for every $\bar{y} \in F(\bar{x}) \cap \Theta$. Then* (3.15) *holds.*

Proof. If F is metrically regular around $(\bar{x}, \bar{y})$, then F^{-1} is Lipschitz-like around $(\bar{y}, \bar{x})$ due to Theorem 1.49(i), and hence F^{-1} is PSNC at this point by Proposition 1.68. Moreover, $\ker \widetilde{D}_M^* F(\bar{x}, \bar{y}) = \{0\}$ by Theorem 1.54(ii), i.e., (b) holds. Thus we have (3.15). △

The result obtained in Corollary 3.9 can be compared with that in Theorem 1.17 justifying the *equality*

$$N\big(\bar{x}; f^{-1}(\Theta)\big) = \nabla f(\bar{x})^* N(\bar{y}; \Theta) \ \text{ with } \ \bar{y} = f(\bar{x}) \tag{3.17}$$

in the case of single-valued mappings $f\colon X \to Y$ between Banach spaces, provided that f is strictly differentiable at $\bar{x}$ and that the operator $\nabla f(\bar{x})$ is surjective. The latter ensures that f is metrically regular around $\bar{x}$ due to the Lyusternik-Graves theorem; see Theorem 1.57. Since

$$D_N^* f(\bar{x})(y^*) = \{\nabla f(\bar{x})^* y^*\} \ \text{ whenever } \ y^* \in Y^*$$

by Theorem 1.38, the result of Corollary 3.9 corresponds to the key inclusion "$\subset$" in (3.17) proved for closed sets Θ and Asplund spaces X, Y. Note, however, that the proof of Theorem 1.17 is heavily based on the strict differentiability of f, while Theorem 3.8 and Corollary 3.9 concern general nonsmooth and set-valued mappings.

3.1.2 Calculus of Coderivatives

In this section we develop the basic calculus for normal and mixed coderivatives of set-valued mappings between Asplund spaces. The main attention is paid to sum and chain rules for coderivatives that are fundamental for the theory and applications. Let us start with *sum rules* representing coderivatives of the sum $F_1 + F_2$ in terms of the corresponding coderivatives of F_1 and F_2. Given $F_i\colon X \rightrightarrows Y$, $i = 1, 2$, we define a multifunction $S\colon X \times Y \rightrightarrows Y^2$ by

$$S(x, y) := \big\{(y_1, y_2) \in Y^2 \big|\ y_1 \in F_1(x),\ y_2 \in F_2(x),\ y_1 + y_2 = y\big\}\,. \tag{3.18}$$

The following two versions of the sum rule for coderivatives depend on the inner semicontinuity and inner semicompactness assumptions imposed on this multifunction; see Definition 1.63.

Theorem 3.10 (sum rules for coderivatives). *Let $F_i\colon X \rightrightarrows Y$, $i = 1, 2$, with $(\bar{x}, \bar{y}) \in \operatorname{gph}(F_1 + F_2)$, and let D^* stand either for the normal coderivative* (1.24) *or for the mixed coderivative* (1.25). *The following assertions hold:*

(i) *Fix $(\bar{y}_1, \bar{y}_2) \in S(\bar{x}, \bar{y})$ in* (3.18) *and let S be inner semicontinuous at $(\bar{x}, \bar{y}, \bar{y}_1, \bar{y}_2)$. Assume that the graphs of F_1 and F_2 are locally closed around $(\bar{x}, \bar{y}_1)$ and $(\bar{x}, \bar{y}_2)$, respectively, that either F_1 is PSNC at $(\bar{x}, \bar{y}_1)$ or F_2 is PSNC at $(\bar{x}, \bar{y}_2)$, and that $\{F_1, F_2\}$ satisfies the qualification condition*

$$D_M^* F_1(\bar{x}, \bar{y}_1)(0) \cap \big(-D_M^* F_2(\bar{x}, \bar{y}_2)(0)\big) = \{0\} \tag{3.19}$$

in terms of the mixed coderivative. Then for all $y^* \in Y^*$ one has

$$D^*(F_1 + F_2)(\bar{x}, \bar{y})(y^*) \subset D^* F_1(\bar{x}, \bar{y}_1)(y^*) + D^* F_2(\bar{x}, \bar{y}_2)(y^*)\,. \tag{3.20}$$

(ii) *Assume that S is inner semicompact at $(\bar{x},\bar{y})$, that F_1 and F_2 are closed-graph whenever x is near $\bar{x}$, and that (3.19) holds for every $(\bar{y}_1,\bar{y}_2)\in S(\bar{x},\bar{y})$. Then for all $y^*\in Y^*$ one has*

$$D^*(F_1+F_2)(\bar{x},\bar{y})(y^*)\subset\bigcup_{(\bar{y}_1,\bar{y}_2)\in S(\bar{x},\bar{y})}\Big[D^*F_1(\bar{x},\bar{y}_1)(y^*)+D^*F_2(\bar{x},\bar{y}_2)(y^*)\Big]$$

provided that either F_1 is PSNC at $(\bar{x},\bar{y}_1)$ or F_2 is PSNC at $(\bar{x},\bar{y}_2)$ for every $(\bar{y}_1,\bar{y}_2)\in S(\bar{x},\bar{y})$.

Proof. First we prove assertion (i). Take any (x^*,y^*) with $x^*\in D^*(F_1+F_2)(\bar{x},\bar{y})(y^*)$ and find sequences $\varepsilon_k\downarrow 0$, $(x_k,y_k)\in\operatorname{gph}(F_1+F_2)$, and $(x_k^*,-y_k^*)\in\widehat{N}_{\varepsilon_k}((x_k,y_k);\operatorname{gph}(F_1+F_2))$ such that $(x_k,y_k)\to(\bar{x},\bar{y})$, $x_k^*\xrightarrow{w^*}x^*$, and either $y_k^*\xrightarrow{w^*}y^*$ if $D^*=D_N^*$, or $y_k^*\to y^*$ if $D^*=D_M^*$. Due to the inner semicontinuity of S at $(\bar{x},\bar{y},\bar{y}_1,\bar{y}_2)$ there is a sequence $(y_{1k},y_{2k})\to(\bar{y}_1,\bar{y}_2)$ with $(y_{1k},y_{2k})\in S(x_k,y_k)$ for all $k\in\mathbb{N}$. Define the sets

$$\Omega_i:=\big\{(x,y_1,y_2)\in X\times Y\times Y\big|\;(x,y_i)\in\operatorname{gph}F_i\big\}\;\text{ for }\;i=1,2\;,$$

which are locally closed around $(\bar{x},\bar{y}_1,\bar{y}_2)$, since the graphs of F_i are assumed to be locally closed around $(\bar{x},\bar{y}_i)$, $i=1,2$. We have $(x_k,y_{1k},y_{2k})\in\Omega_1\cap\Omega_2$ and can easily check that

$$(x_k^*,-y_k^*,-y_k^*)\in\widehat{N}_{\varepsilon_k}((x_k,y_{1k},y_{2k});\Omega_1\cap\Omega_2)\;\text{ for all }\;k\in\mathbb{N}\;. \tag{3.21}$$

This gives, by passing to the limit as $k\to\infty$, that

$$(x^*,-y^*,-y^*)\in N((\bar{x},\bar{y}_1,\bar{y}_2);\Omega_1\cap\Omega_2)\;. \tag{3.22}$$

Now we apply Theorem 3.4 to the set intersection in (3.22). Observe similarly to the proof of Theorem 3.8 that (3.19) implies that the above set system $\{\Omega_1,\Omega_2\}$ satisfies the limiting qualification condition at $(\bar{x},\bar{y}_1,\bar{y}_2)$. Then assuming for definiteness that F_1 is PSNC at $(\bar{x},\bar{y}_1)$, we get that $\Omega_1\subset X\times Y\times Y$ is PSNC at $(\bar{x},\bar{y}_1,\bar{y}_2)$ with respect to $X\times Y$, where Y is the third space in the product $X\times Y\times Y$, and that Ω_2 is obviously *strongly* PSNC at this point with respect to the remaining space Y in this product. Thus there are

$$(x_1^*,-y_1^*)\in N((\bar{x},\bar{y}_1);\operatorname{gph}F_1)\;\text{ and }\;(x_2^*,-y_2^*)\in N((\bar{x},\bar{y}_2);\operatorname{gph}F_2)$$

such that $(x^*,-y^*,-y^*)=(x_1^*,-y_1^*,0)+(x_2^*,0,-y_2^*)$ by Theorem 3.4 and the structure of the sets Ω_i. This gives $x^*=x_1^*+x_2^*$ with $x_i^*\in D_N^*F_i(\bar{x},\bar{y}_i)(y^*)$, $i=1,2$, and justifies (3.20) in the case of $D^*=D_N^*$.

To prove (3.20) in the case of $D^*=D_M^*$, we apply the fuzzy rule of Lemma 3.1 to the set intersection in (3.21) along some sequence $\varepsilon_k\downarrow 0$ as $k\to\infty$. This gives $\lambda_k\ge 0$, $(\tilde{x}_{ik},\tilde{y}_{ik})\in\operatorname{gph}F_i$, and $(x_{ik}^*,-y_{ik}^*)\in\widehat{N}((\tilde{x}_{ik},\tilde{y}_{ik});\operatorname{gph}F_i)$ such that $\|(\tilde{x}_{ik},\tilde{y}_{ik})-(x_k,y_{ik})\|\le\varepsilon_k$, $i=1,2$, and

$$\|(x_{1k}^* + x_{2k}^*, -y_{1k}^*, -y_{2k}^*) - \lambda_k(x_k^*, -y_k^*, -y_k^*)\| \le 2\varepsilon_k \tag{3.23}$$

with $1 - \varepsilon_k \le \max\{\lambda_k, \|(x_{1k}^*, y_{1k}^*)\|\} \le 1 + \varepsilon_k$. Similarly to the proof of Theorem 3.4 we show that $\lambda_k \ge \lambda_0 > 0$ for large $k \in I\!N$ under the qualification and PSNC assumptions imposed, and hence one may put $\lambda_k = 1$ without loss of generality. Taking into account that $x_k^* \xrightarrow{w^*} x^*$ and $\|y_k^* - y^*\| \to 0$, we get from (3.23) that $\|y_{ik}^* - y^*\| \to 0$ and $x_{ik}^* \xrightarrow{w^*} x_i^* \in D_M^* F_i(\bar{x}, \bar{y}_i)(y^*)$, $i = 1, 2$, for some x_i^* with $x_1^* + x_2^* = x^*$. This justifies (3.20) for $D^* = D_M^*$.

To establish (ii), we proceed as in the proof of (i) observing that if $(y_{1k}, y_{2k}) \in S(x_k, y_k)$ converges to some $(\bar{y}_1, \bar{y}_2)$, then $(\bar{y}_1, \bar{y}_2)$ must belong to $S(\bar{x}, \bar{y})$ due to the closedness and lower semicompactness assumptions made in (ii). This completes the proof of the theorem. △

Observe, as in the proof of Theorem 3.8, that condition (3.19) of the above theorem can be replaced by the following more general but less convenient qualification condition: for any $(x_{ik}, y_{ik}) \in \operatorname{gph} F_i$ and $(x_{ik}^*, y_{ik}^*) \in \widehat{N}((x_{ik}, y_{ik}); \operatorname{gph} F_i)$ with $(x_{ik}, y_{ik}) \to (\bar{x}, \bar{y}_i)$, $x_{ik}^* \xrightarrow{w^*} x_i^*$, and $\|y_{ik}^*\| \to 0$ ($i = 1, 2$ as $k \to \infty$) one has

$$\|x_{1k}^* + x_{2k}^*\| \to 0 \Longrightarrow x_1^* = x_2^* = 0\,.$$

Note that the usage of the normal qualification condition (3.10) in the proof of Theorem 3.10 leads us to the replacement of (3.19) by the more restrictive qualification condition

$$D_N^* F_1(\bar{x}, \bar{y}_1)(0) \cap \big(-D_N^* F_2(\bar{x}, \bar{y}_2)(0)\big) = \{0\}$$

in terms of the *normal* coderivative, which does *not* generally imply the following important corollary ensured by (3.19). For simplicity we formulated this corollary only for the case of assertion (i).

Corollary 3.11 (coderivative sum rule for Lipschitz-like multifunctions). *Fix $(\bar{x}, \bar{y}) \in \operatorname{gph}(F_1 + F_2)$ and $(\bar{y}_1, \bar{y}_2) \in S(\bar{x}, \bar{y})$ in (3.18) and suppose that the graphs of F_i are locally closed around $(\bar{x}, \bar{y}_i)$ for $i = 1, 2$. Assume that either F_1 is Lipschitz-like around $(\bar{x}, \bar{y}_1)$ or F_2 is Lipschitz-like around $(\bar{x}, \bar{y}_2)$ and that S is inner semicontinuous at $(\bar{x}, \bar{y}, \bar{y}_1, \bar{y}_2)$. Then one has the sum rule* (3.20) *for both normal and mixed coderivatives.*

Proof. Assuming for definiteness that F_1 is Lipschitz-like around $(\bar{x}, \bar{y}_1)$, we conclude that $D_M^* F_1(\bar{x}, \bar{y}_1)(0) = \{0\}$ by Theorem 1.44 and that F_1 is PSNC at $(\bar{x}, \bar{y}_1)$ by Proposition 1.68. Thus we meet all the requirements of assertion (i) in the theorem. △

Next we compute coderivatives of *special sums* of multifunctions between Asplund spaces given in the form

$$\Phi(x) := F(x) + \Delta(x; \Omega), \quad x \in X\,, \tag{3.24}$$

where $F\colon X \rightrightarrows Y$ and where the indicator mapping $\Delta(\cdot;\Omega)$ of $\Omega \subset X$ relative to Y is defined by $\Delta(x;\Omega) := 0 \in Y$ if $x \in \Omega$ and $\Delta(x;\Omega) := \emptyset$ otherwise. Multifunctions of form (3.24) play an important role in the proof of chain rules for coderivatives of compositions considered below. To proceed, we need the following version of coderivative sum rules for mappings (3.24) that contains both inclusion and equality assertions.

Proposition 3.12 (coderivatives of special sums). *Let $\Omega \subset X$ and the graph of $F\colon X \rightrightarrows Y$ be closed around $\bar{x} \in \Omega$ and $(\bar{x},\bar{y}) \in \operatorname{gph} F$, respectively. Assume that for any sequences $x_{1k}^* \in \widehat{D}^*F(x_{1k},y_k)(y_k^*)$ and $x_{2k}^* \in \widehat{N}(x_{2k};\Omega)$ such that $(x_{1k},y_k) \to (\bar{x},\bar{y})$, $x_{2k} \to \bar{x}$, and $\{x_{1k}^*, x_{2k}^*\}$ are bounded one has*

$$\Big[\|y_k^*\| \to 0,\ \|x_{1k}^* + x_{2k}^*\| \to 0\Big] \Longrightarrow \|x_{1k}^*\| + \|x_{2k}^*\| \to 0 \quad \text{as} \quad k \to \infty\,. \tag{3.25}$$

Then the inclusion

$$D^*\big(F + \Delta(\cdot;\Omega)\big)(\bar{x},\bar{y})(y^*) \subset D^*F(\bar{x},\bar{y})(y^*) + N(\bar{x};\Omega), \qquad y^* \in Y^*\,, \tag{3.26}$$

holds for both coderivatives $D^* = D_N^*$ and $D^* = D_M^*$. Moreover, (3.26) holds as equality and $F + \Delta(\cdot;\Omega)$ is N-regular (resp. M-regular) at $(\bar{x},\bar{y})$ if F has the corresponding regularity property at $(\bar{x},\bar{y})$ and if Ω is normally regular at $\bar{x}$.

Proof. To justify (3.26), we follow the proof of Theorem 3.10 with $F_1 := F$ and $F_2 := \Delta(\cdot;\Omega)$ observing that condition (3.25) ensures in this setting that the fuzzy intersection rule holds in (3.21) with $\lambda_k \ge \lambda_0 > 0$ for large $k \in \mathbb{N}$. This implies (3.26) as in the proof above.

To justify the equality and regularity statement, we first observe that one always has

$$\widehat{D}^*\big(F + \Delta(\cdot;\Omega)\big)(\bar{x},\bar{y})(y^*) \supset \widehat{D}^*F(\bar{x},\bar{y})(y^*) + \widehat{N}(\bar{x};\Omega), \qquad y^* \in Y^*\,,$$

which follows directly from the definitions and elementary calculations of the Fréchet-like constructions under consideration. Therefore

$$D^*F(\bar{x},\bar{y})(y^*) + N(\bar{x};\Omega) = \widehat{D}^*F(\bar{x},\bar{y})(y^*) + \widehat{N}(\bar{x};\Omega) \subset D^*\big(F + \Delta(\cdot;\Omega)\big)(y^*)$$

for both cases $D^* = D_N^*$ and $D^* = D_M^*$ under the corresponding regularity assumptions of the proposition. △

Note that condition (3.25) certainly holds if

$$D_M^*F(\bar{x},\bar{y})(0) \cap \big(-N(\bar{x};\Omega)\big) = \{0\}$$

and either F is PSNC at $(\bar{x},\bar{y})$ or Ω is SNC at $\bar{x}$. In this case the inclusion part of Proposition 3.12 follows directly from Theorem 3.10(i). However, we need the full statement of Proposition 3.12 under the more precise assumption (3.25) to get the general chain rules for coderivatives considered next.

Now we are going to express normal and mixed coderivatives of *compositions* $F \circ G$ of set-valued mappings between Asplund spaces via the corresponding coderivatives of F and G, i.e., to derive *chain rules* for coderivatives. The following theorem is based on Proposition 3.12 and composition results obtained in Subsect. 1.2.4.

Theorem 3.13 (chain rules for coderivatives). *Let* $G\colon X \rightrightarrows Y$, $F\colon Y \rightrightarrows Z$, $\bar{z} \in (F \circ G)(\bar{x})$, *and*

$$S(x,z) := G(x) \cap F^{-1}(z) = \big\{ y \in G(x) \big|\ z \in F(y) \big\} .$$

The following assertions hold for both coderivatives $D^* = D^*_N$ *and* $D^* = D^*_M$ *for all* $z^* \in Z^*$*:*

(i) *Given* $\bar{y} \in S(\bar{x},\bar{z})$, *assume that* S *is inner semicontinuous at* $(\bar{x},\bar{z},\bar{y})$, *that the graphs of* F *and* G *are locally closed around the points* $(\bar{y},\bar{z})$ *and* $(\bar{x},\bar{y})$, *respectively, that either* F *is PSNC at* $(\bar{y},\bar{z})$ *or* G^{-1} *is PSNC at* $(\bar{y},\bar{x})$, *and that the mixed qualification condition*

$$D^*_M F(\bar{y},\bar{z})(0) \cap \big(-D^*_M G^{-1}(\bar{y},\bar{x})(0) \big) = \{0\} \tag{3.27}$$

is fulfilled. Then one has

$$D^*(F \circ G)(\bar{x},\bar{z})(z^*) \subset D^*_N G(\bar{x},\bar{y}) \circ D^* F(\bar{y},\bar{z})(z^*) . \tag{3.28}$$

(ii) *Assume that* S *is inner semicompact at* $(\bar{x},\bar{z})$, *that* G *and* F^{-1} *are closed-graph whenever* x *is near* $\bar{x}$ *and* z *is near* $\bar{z}$, *respectively, and that* (3.27) *holds for every* $\bar{y} \in S(\bar{x},\bar{z})$. *Then*

$$D^*(F \circ G)(\bar{x},\bar{z})(z^*) \subset \bigcup_{\bar{y} \in S(\bar{x},\bar{z})} \Big[D^*_N G(\bar{x},\bar{y}) \circ D^* F(\bar{y},\bar{z})(z^*) \Big]$$

provided that either F *is PSNC at* $(\bar{y},\bar{z})$ *or* G^{-1} *is PSNC at* $(\bar{y},\bar{x})$ *for every point* $\bar{y} \in S(\bar{x},\bar{z})$.

(iii) *Let* $G = g$ *be single-valued and Lipschitz continuous around* $\bar{x}$, *which automatically implies that* S *is inner semicompact at* $(\bar{x},\bar{z})$. *In addition to* (ii) *assume that* F *is* N*-regular (resp.* M*-regular) at* $(\bar{y},\bar{z})$ *with* $\bar{y} = g(\bar{x})$ *and that either* g *is* N*-regular at* $\bar{x}$ *while* $\dim Y < \infty$, *or* g *is strictly differentiable at* $\bar{x}$. *Then* $F \circ g$ *is* N*-regular (resp.* M*-regular) at* $(\bar{x},\bar{z})$, *and one has*

$$D^*(F \circ g)(\bar{x},\bar{z})(z^*) = D^*_N g(\bar{x}) \circ D^* F(\bar{y},\bar{z})(z^*) .$$

Proof. Let us justify assertion (i); the proof of assertion (ii) is similar. Considering the multifunction

$$\Phi(x,y) := F(y) + \Delta((x,y);\operatorname{gph} G)$$

of type (3.24), we have

$$D^*(F \circ G)(\bar{x}, \bar{z})(z^*) \subset \{x^* \in X^* |\, (x^*, 0) \in D^*\Phi(\bar{x}, \bar{y}, \bar{z})(z^*)\} \tag{3.29}$$

by Theorem 1.64(i). Then we use inclusion (3.26) of Proposition 3.12 for Φ in (3.29) observing that the qualification (3.27) and PSNC conditions of the theorem ensure the fulfillment of assumption (3.25) of the proposition. Thus (3.29) and (3.26) imply (3.28). To prove (iii), we combine the equality/regularity statements in Theorem 1.64(iii) and Proposition 3.12. △

Note that the inclusion chain rules in Theorem 3.13 may be derived directly by applying the results on basic normals to the set intersection for

$$\Omega_1 := \operatorname{gph} G \times Z \quad \text{and} \quad \Omega_2 := X \times \operatorname{gph} F$$

in the way of proving Theorem 3.10. However, in this way we cannot obtain the equality and regularity assertions in (iii). Another case of the equality chain rule for coderivatives is contained in Theorem 1.66 in the framework of arbitrary Banach spaces. Note also that, due to Corollary 3.69 established below in Subsect. 3.2.4, the N-regularity of $g\colon X \to I\!R^m$ at $\bar{x}$ in Theorem 3.13(iii) is *equivalent* to its simultaneous Fréchet differentiability and strict Hadamard differentiability at $\bar{x}$, but not to the strict Fréchet differentiability of g at this point alternatively assumed in the above theorem in infinite dimensions.

It is worth observing that we use the *mixed* coderivative qualification condition (3.27) in the chain rules for both normal and mixed coderivatives. On the other hand, the *normal* coderivative of G is involved in the chain rule (3.28) and its counterpart in assertion (ii) of Theorem 3.13 in both cases of normal and mixed coderivatives.

The next result shows that if one concerns *only with* $y^* = 0$ in the chain rule (3.28) for $D^* = D^*_M$ and its counterparts in (ii) and if F is *Lipschitz-like* around $(\bar{x}, \bar{y})$, then the *mixed coderivative* of G can be employed in such a special *zero chain rule* for mixed coderivatives, which has particularly useful applications to results of Chap. 4 ensuring the preservation of Lipschitzian and metric regularity properties under compositions of set-valued mappings.

Theorem 3.14 (zero chain rule for mixed coderivatives). *Let G, F, and S be as in Theorem 3.13, and let $\bar{z} \in (F \circ G)(\bar{x})$. The following hold:*

(i) *Given $\bar{y} \in S(\bar{x}, \bar{z})$, assume that S is inner semicontinuous at $(\bar{x}, \bar{z}, \bar{y})$, that the graphs of F and G are locally closed around the points $(\bar{y}, \bar{z})$ and $(\bar{x}, \bar{y})$, respectively, and that F is Lipschitz-like around $(\bar{y}, \bar{z})$. Then*

$$D^*_M(F \circ G)(\bar{x}, \bar{z})(0) \subset \{x^* \in X^* |\, x^* \in D^*_M G(\bar{x}, \bar{y})(0)\} .$$

(ii) *Assume that S is inner semicompact at $(\bar{x}, \bar{z})$, that G and F^{-1} are closed-graph whenever x is near $\bar{x}$ and z is near $\bar{z}$, respectively, and that F is Lipschitz-like around $(\bar{y}, \bar{z})$ for every $\bar{y} \in S(\bar{x}, \bar{z})$. Then*

$$D^*_M(F \circ G)(\bar{x}, \bar{z})(0) \subset \bigcup_{\bar{y} \in S(\bar{x}, \bar{z})} \{x^* \in X^* |\, x^* \in D^*_M G(\bar{x}, \bar{y})(0)\} .$$

Proof. Prove only (i), since the proof of (ii) is similar as above. Taking arbitrary $x^* \in D_M^*(F \circ G)(\bar{x}, \bar{z})(0)$, find by definition sequences $\varepsilon_k \downarrow 0$, $(x_k, z_k) \stackrel{\mathrm{gph}(F\circ G)}{\longrightarrow} (\bar{x}, \bar{z})$, $x_k^* \stackrel{w^*}{\to} x^*$, and $z_k^* \to 0$ (by norm) satisfying

$$(x_k^*, -z_k^*) \in \widehat{N}_{\varepsilon_k}\big((x_k, z_k); \mathrm{gph}\,(F \circ G)\big) \quad \text{for all} \quad k \in I\!N \,.$$

Since S is inner semicontinuous at $(\bar{x}, \bar{z}, \bar{y})$, there are $y_k \in S(x_k, z_k)$ such that $y_k \to \bar{y}$ along a subsequence, with no relabeling. It is easy to see that

$$(x_k^*, 0) \in \widehat{D}^*_{\varepsilon_k}\big[F + \Delta(\cdot; \mathrm{gph}\,G)\big](x_k, y_k, z_k)(z_k^*), \quad k \in I\!N \,.$$

Now we apply to the above sum the following coderivative "fuzzy sum rule" ensuring that, given closed-graph mappings $F_i\colon X \rightrightarrows Y$ between Asplund spaces and given $x^* \in \widehat{D}^*_\varepsilon(F_1 + F_2)(\bar{x}, \bar{y})(y^*)$ with $\bar{y} \in (F_1 + F_2)(\bar{x})$, for any $\eta > 0$ there are $(x_i, y_i) \in \mathrm{gph}\,F_i \cap [(\bar{x}, \bar{y}_i) + \eta I\!B]$ and $x_i^* \in \widehat{D}^* F_i(x_i, y_i)(y_i^*)$ as $i = 1, 2$ such that the *norm estimates*

$$\|y_i^* - y^*\| \le \varepsilon + \eta \;\text{ for }\; i = 1, 2 \quad \text{and} \quad \|x^* - x_1^* - x_2^*\| \le \varepsilon + \eta$$

hold provided that at least *one* of the mappings F_i is *Lipschitz-like* around the point $(\bar{x}, \bar{y}_i)$, respectively. This results follows from the fuzzy intersection rule of Lemma 3.1 being actually equivalent to the latter. Applying this result to the above sum $F + \Delta(\cdot)$ at the given points as $k \to \infty$, we take $\eta_k \downarrow 0$ and find sequences $(x_{1k}, y_{1k}) \in \mathrm{gph}\,F$, $(x_{2k}, y_{2k}) \in \mathrm{gph}\,G$,

$$y_{1k}^* \in \widehat{D}^* F(y_{1k}, z_{1k})(z_{1k}^*), \quad \text{and} \quad (x_{2k}^*, y_{2k}^*) \in \widehat{N}\big((x_{2k}, y_{2k}); \mathrm{gph}\,G\big)$$

satisfying the norm estimates:

$$\|(y_{1k}, z_{1k}) - (y_k, z_k)\| \le \eta_k, \quad \|(x_{2k}, y_{2k}) - (x_k, y_k)\| \le \eta_k \,,$$

$$\|(x_k^*, 0) - (0, y_{1k}^*) - (x_{2k}^*, y_{2k}^*)\| \le \varepsilon_k + \eta_k, \quad \text{and} \quad \|z_{1k}^* - z_k^*\| \le \varepsilon_k + \eta_k \,.$$

Since $\|z_k^*\| \to 0$ and $\|z_{1k}^* - z_k^*\| \le \varepsilon_k + \eta_k$, one has $\|z_{1k}^*\| \to 0$ as $k \to \infty$. The assumed Lipschitz-like property of F ensures that F is PSNC at $(\bar{y}, \bar{z})$, which implies that $\|y_{1k}^*\| \to 0$. Combining this with

$$\|x_k^* - x_{2k}^*\| \le \varepsilon_k + \eta_k, \quad \|y_{1k}^* + y_{2k}^*\| \le \varepsilon_k + \eta_k, \quad \text{and} \quad x_k^* \stackrel{w^*}{\to} 0 \,,$$

we conclude that $\|x_{2k}^*\| \to 0$ and $\|y_{2k}^*\| \to 0$ as $k \to \infty$. Thus one has $x^* \in D_M^* G(\bar{x}, \bar{y})(0)$, which completes the proof of the theorem. △

Note that if $D_M^* G$ is replaced by $D_N^* G$ in Theorem 3.14, then the results obtained therein are special cases of Theorem 3.13 as $z^* = 0$, since the qualification and PSNC conditions are automatic while $D_M^* F(\bar{y}, \bar{z})(0) = \{0\}$ due to the Lipschitz-like property of F. The following corollary of Theorem 3.13 explores the latter observation providing effective conditions for the fulfillment of the general coderivative chain rules in that theorem. For simplicity we present this corollary only for assertion (i).

Corollary 3.15 (coderivative chain rules for Lipschitz-like and metrically regular mappings). *Fix $\bar{z} \in (F \circ G)(\bar{x})$ and $\bar{y} \in G(\bar{x}) \cap F^{-1}(\bar{z})$ and suppose that the graphs of F and G are locally closed around $(\bar{y}, \bar{z})$ and $(\bar{x}, \bar{y})$, respectively, and that the mapping $(x, z) \to G(x) \cap F^{-1}(z)$ is inner semicontinuous at $(\bar{x}, \bar{z}, \bar{y})$. Then the chain rule (3.28) holds for both normal and mixed coderivatives if either F is Lipschitz-like around $(\bar{y}, \bar{z})$ or G is metrically regular around $(\bar{x}, \bar{y})$.*

Proof. It follows from Theorem 1.44, Proposition 1.68, and Theorem 1.49(i) that the qualification (3.27) and PSNC assumptions of Theorem 3.13(i) automatically hold for either Lipschitz-like mappings F or metrically regular mappings G. Thus we have (3.28). △

The next corollary of Theorem 3.13 concerns the case of strictly differentiable inner mappings with no surjectivity assumption on their derivatives as in Theorem 1.66.

Corollary 3.16 (coderivative chain rules with strictly differentiable inner mappings). *Let $g: X \to Y$ be strictly differentiable at $\bar{x}$, and let $\bar{z} \in (F \circ g)(\bar{x})$, where $F: Y \rightrightarrows Z$ is closed-graph around $(\bar{y}, \bar{z})$ with $\bar{y} = g(\bar{x})$. Assume that F is PSNC at $(\bar{y}, \bar{z})$ and that*

$$D_M^* F(\bar{y}, \bar{z})(0) \cap \big(\ker \nabla g(\bar{x})^*\big) = \{0\} \; ;$$

the latter two conditions automatically hold if F is Lipschitz-like around $(\bar{y}, \bar{z})$. Then one has the inclusion

$$D^*(F \circ g)(\bar{x}, \bar{z})(z^*) \subset \nabla g(\bar{x})^* D^* F(\bar{y}, \bar{z})(z^*), \quad z^* \in Z^* ,$$

for both coderivatives $D^* = D_N^*, D_M^*$. If in addition F is N-regular (resp. M-regular) at $(\bar{y}, \bar{z})$, then one has equality

$$D^*(F \circ g)(\bar{x}, \bar{z})(z^*) = \nabla g(\bar{x})^* D^* F(\bar{y}, \bar{z})(z^*), \quad z^* \in Z^* ,$$

and $F \circ g$ enjoys the corresponding regularity property at $(\bar{x}, \bar{z})$.

Proof. This follows directly from Theorem 3.13 and Corollary 3.15 due to the coderivative representations for strictly differentiable functions. △

The chain rules obtained in Corollary 3.16 allow us to establish relationships between full and partial coderivatives for set-valued mappings of two (and many) variables. Considering a multifunction $F: X \times Y \rightrightarrows Z$ of two variables $(x, y) \in X \times Y$, we denote by $D_x^* F(\bar{x}, \bar{y}, \bar{z})$ its *partial coderivative* (either normal or mixed) with respect to x at the point $(\bar{x}, \bar{y}, \bar{z}) \in \operatorname{gph} F$ that is the corresponding coderivative of the "partial" multifunction $F(\cdot, \bar{y})$ at $(\bar{x}, \bar{z})$. Let $\operatorname{proj}_x D^* F(\bar{x}, \bar{y}, \bar{z})(z^*)$ denote the *projection* of the set $D^* F(\bar{x}, \bar{y}, \bar{z})(z^*) \subset X^* \times Y^*$ on the space X^*. The following result gives a relationship between the full coderivative $D^* F$ and its partial counterpart D_x^* with respect to x; the same is valid of course for the second variable y.

Corollary 3.17 (partial coderivatives). *Let $F: X \times Y \rightrightarrows Z$, and let the graph of F be closed around $(\bar{x}, \bar{y}, \bar{z}) \in \operatorname{gph} F$. Assume that F is PSNC at $(\bar{x}, \bar{y}, \bar{z})$ and that*

$$(0, y^*) \in D_M^* F(\bar{x}, \bar{y}, \bar{z})(0) \Longrightarrow y^* = 0\,;$$

these conditions automatically hold when F is Lipschitz-like around $(\bar{x}, \bar{y}, \bar{z})$. Then one has the inclusion

$$D_x^* F(\bar{x}, \bar{y}, \bar{z})(z^*) \subset \operatorname{proj}_x D^* F(\bar{x}, \bar{y}, \bar{z})(z^*), \quad z^* \in Z^*,$$

for both normal and mixed coderivatives $D^* = D_N^*, D_M^*$, where the equality holds if F is N-regular (resp. M-regular) at $(\bar{x}, \bar{y}, \bar{z})$. Moreover, in the latter case the partial multifunction $F(\cdot, \bar{y})$ enjoys the corresponding regularity property at $(\bar{x}, \bar{z})$.

Proof. This follows from Corollary 3.16 applied to the composition $F(\cdot, \bar{y}) = F \circ g$ with $g: X \to X \times Y$ defined by $g(x) := (x, \bar{y})$. △

Next let us consider the so-called *h-composition*

$$(F_1 \overset{h}{\diamond} F_2)(x) := \bigcup \big\{h(y_1, y_2) \big|\ y_1 \in F_1(x),\ y_2 \in F_2(x)\big\}$$

of arbitrary multifunctions $F_i: X \rightrightarrows Y_i$, $i = 1, 2$, where the single-valued mapping $h: Y_1 \times Y_2 \to Z$ represents various operations on multifunctions (in particular, different kinds of product, quotient, maximum, minimum, etc.). Based on the sum and chain rules of Theorems 3.10 and 3.13, we derive general formulas for representing coderivatives of h-compositions in the case of mappings between Asplund spaces, which imply other calculus results involving special choices of the operation h. The following result is formulated and proved only in the case when the corresponding mapping S is inner semicontinuous at the given point; the case of its inner semicompactness is similar to that in Theorems 3.10 and 3.13.

Theorem 3.18 (coderivatives of *h*-compositions). *Let $F_i: X \rightrightarrows Y_i$ with $i = 1, 2$, let $h: X \times Z \to Y_1 \times Y_2$, and let $\bar{z} \in (F_1 \overset{h}{\diamond} F_2)(\bar{x})$. Define the multifunction $S: Y_1 \times Y_2 \rightrightarrows Z$ by*

$$S(x, z) := \big\{(y_1, y_2) \in Y_1 \times Y_2 \big|\ y_i \in F_i(x),\ z = h(y_1, y_2)\big\}$$

and suppose that it is inner semicontinuous at $(\bar{x}, \bar{z}, \bar{y}) \in \operatorname{gph} S$ for a given $\bar{y} = (\bar{y}_1, \bar{y}_2)$ and that the graph of F_i is locally closed around $(\bar{x}, \bar{y}_i)$ for $i = 1, 2$. Assume also that either F_1 is PSNC at $(\bar{x}, \bar{y}_1)$ or F_2 is PSNC at $(\bar{x}, \bar{y}_2)$ and that the qualification condition (3.19) *is fulfilled. The following assertions hold for all $z^* \in Z^*$:*

(i) *Let h be locally Lipschitzian around $\bar{y}$. Then*

$$D^*(F_1 \overset{h}{\diamond} F_2)(\bar{x},\bar{z})(z^*) \subset \bigcup_{y^* \in D^*h(\bar{y})(z^*)} \Big[D_N^* F_1(\bar{x},\bar{y}_1)(y_1^*) + D_N^* F_2(\bar{x},\bar{y}_2)(y_2^*)\Big] ,$$

where $y^* = (y_1^*, y_2^*)$ *and where* D^* *stands either for the normal coderivative of* $F_1 \overset{h}{\diamond} F_2$ *and* h *or for the mixed coderivative of these mappings.*

(ii) *Let* h *be strictly differentiable at* $\bar{y}$. *Then*

$$D_M^*(F_1 \overset{h}{\diamond} F_2)(\bar{x},\bar{z})(z^*) \subset D_M^* F_1(\bar{x},\bar{y}_1)(y_1^*) + D_M^* F_2(\bar{x},\bar{y}_2)(y_2^*) ,$$

where $y_i^* = \nabla_i h(\bar{y})^* z^*$, $i = 1, 2$, *in terms of the partial derivatives of* $h(y_1, y_2)$ *in the first and second variable, respectively.*

Proof. Define $F\colon X \rightrightarrows Y_1 \times Y_2$ by $F(x) := \big(F_1(x), F_2(x)\big)$ and observe that

$$D^* F(\bar{x},\bar{y})(y^*) \subset D^* F_1(\bar{x},\bar{y}_1)(y_1^*) + D^* F_2(\bar{x},\bar{y}_2)(y_2^*) \tag{3.30}$$

for both coderivatives $D^* = D_N^*$ and $D^* = D_M^*$ under the assumptions made in (i). To justify (3.30), we apply Theorem 3.10 to the sum $F = \widetilde{F}_1 + \widetilde{F}_2$, where $\widetilde{F}_1(x) := (F_1(x), 0)$ and $\widetilde{F}_2(x) := (0, F_2(x))$. Since obviously

$$(F_1 \overset{h}{\diamond} F_2)(x) = (h \circ F)(x) \tag{3.31}$$

and h is locally Lipschitzian around $\bar{y}$, we can apply the chain rule in Corollary 3.15 to the composition $h \circ F$. Taking (3.30) into account, we arrive at the conclusion in (i).

Let us prove assertion (ii). Note that its normal coderivative counterpart follows directly from (i) by Theorem 1.38, while (i) gives a bigger upper estimate of $D_M^*(F_1 \overset{h}{\circ} F_2)(\bar{x},\bar{z})(z^*)$ in comparison with (ii). This is due to using the chain rule (3.28) for $h \circ F$, which inevitably involves the normal coderivative of inner mappings. We justify the better estimate in (ii) by using the fuzzy intersection rule of Lemma 3.1 as in the proof of Theorem 3.10 for $D^* = D_M^*$.

Fix $x^* \in D_M^*(F_1 \overset{h}{\diamond} F_2)(\bar{x},\bar{z})(z^*)$ and, by Corollary 2.36, find sequences $(x_k, z_k) \in \operatorname{gph}(F_1 \overset{h}{\diamond} F_2)$ and $x_k^* \in \widehat{D}^*(F_1 \overset{h}{\diamond} F_2)(x_k, z_k)(z_k^*)$ satisfying $(x_k, z_k) \to (\bar{x},\bar{z})$, $x_k^* \overset{w^*}{\to} x^*$, and $z_k^* \to z^*$ as $k \to \infty$. Taking the usual composition form (3.31) with h strictly differentiable at $\bar{y}$ and employing our standard arguments based on the strict differentiability of h (as in the proof of Theorem 1.72) and then on representation (2.51) in Asplund spaces, we get subsequences $(\tilde{x}_k, \tilde{y}_k, \tilde{z}_k) \to (\bar{x},\bar{y},\bar{z})$, $\tilde{x}_k^* \overset{w^*}{\to} x^*$, and $\tilde{z}_k^* \to z^*$ such that $\tilde{y}_k \in F(\tilde{x}_k) \cap h^{-1}(\tilde{z}_k)$ and

$$\tilde{x}_k^* \in \widehat{D}^* F(\tilde{x}_k, \tilde{y}_k)(\tilde{y}_k^*) \ \text{ with } \ \tilde{y}_k^* := \big(\nabla h(\bar{y})\big)^* \tilde{z}_k^* \,. \tag{3.32}$$

Now taking into account that $F(x) = (F_1(x), 0) + (0, F_2(x))$ in (3.32) and following the proof of Theorem 3.10 in the case of $D^* = D_M^*$, we select subsequences $(x_{ik}, y_{ik}) \to (\bar{x}, \bar{y}_i)$, $x_{ik}^* \overset{w^*}{\to} x_i^*$, and $y_{ik}^* \to \big(\nabla_i h(\bar{y})\big)^* z^*$ with

$$x_{ik}^* \in \widehat{D}^* F_i(x_{ik}, y_{ik})(y_{ik}^*),\ i = 1, 2, \ \text{ and } x_{1k}^* + x_{2k}^* \xrightarrow{w^*} x^* \text{ as } k \to \infty .$$

Thus $x^* \in D_M^* F_1(\bar{x}, \bar{y}_1)(y_1^*) + D_M^* F_2(\bar{x}, \bar{y}_2)(y_2^*)$, where $(y_1^*, y_2^*) = \big(\nabla h(\bar{y})\big)^* z^*$. This justifies (ii) and completes the proof of the theorem. △

Note that we may always put $D_M^* h(\bar{y})(z^*) = \partial\langle z^*, h\rangle(\bar{y})$ in the framework of Theorem 3.18(i) due to the *scalarization* formula for the mixed coderivative obtained in Theorem 1.90.

To illustrate the application of Theorem 3.18, we consider the *inner product*

$$\langle F_1, F_2\rangle(x) := \big\{\langle y_1, y_2\rangle \big|\ y_i \in F_i(x),\ i = 1, 2\big\}$$

of multifunctions $F_i\colon X \rightrightarrows Y$ with the values in a Hilbert space Y. Since $\langle F_1, F_2\rangle\colon X \rightrightarrows I\!R$, there is no difference between the normal and mixed coderivatives of this mapping denoted by $D^*\langle F_1, F_2\rangle$. The next result gives an upper estimate of the latter coderivative in terms of $D_M^* F_i$, $i = 1, 2$.

Corollary 3.19 (inner product rule for coderivatives). *Given $\bar{a} \in \langle F_1, F_2\rangle(\bar{x})$ and $\bar{y}_i \in F_i(\bar{x})$ with $\bar{a} = \langle \bar{y}_1, \bar{y}_2\rangle$, suppose that the graph of F_i is locally closed around $(\bar{x}, \bar{y}_i)$ for $i = 1, 2$ and that the multifunction*

$$(x, \alpha) \to \big\{(y_1, y_2) \in Y^2 \big|\ y_i \in F_i(x),\ \alpha = \langle y_1, y_2\rangle\big\}$$

is inner semicontinuous at $(\bar{x}, \bar{a}, \bar{y}_1, \bar{y}_2)$. Assume also that either F_1 is PSNC at $(\bar{x}, \bar{y}_1)$ or F_2 is PSNC at $(\bar{x}, \bar{y}_2)$ and that the qualification condition (3.19) *holds. Then for all $\lambda \in I\!R$ one has*

$$D^*\langle F_1, F_2\rangle(\bar{x}, \bar{a})(\lambda) \subset D_M^* F_1(\bar{x}, \bar{y}_1)(\lambda\bar{y}_2) + D_M^* F_2(\bar{x}, \bar{y}_2)(\lambda\bar{y}_1) .$$

Proof. Follows from Theorem 3.18(ii) for $h(y_1, y_2) = \langle y_1, y_2\rangle$. △

Note that Theorem 3.18 allows us to derive general product and quotient rules with respect to multiplication and division defined in a *Banach algebra*; cf. Mordukhovich and Shao [950]. It also covers coderivative calculus rules for *maximum* and *minimum* of multifunctions obtained via nonsmooth h-compositions as in Mordukhovich [910].

The last result of this subsection gives a useful representation of the normal coderivative for *intersections*

$$(F_1 \cap F_2)(x) := F_1(x) \cap F_2(x)$$

of set-valued mappings that follows directly from the intersection rule for basic normals in Theorem 3.4. For simplicity we use the normal qualification condition (3.10) in the latter theorem, which is important for applications to the subdifferentiation of *maximum functions* in Subsect. 3.2.1.

Proposition 3.20 (coderivative intersection rule). *Let $F_i\colon X \rightrightarrows Y$, $i = 1, 2$, be locally closed around $(\bar{x}, \bar{y})$. Assume that*

$$N((\bar{x}, \bar{y}); \text{gph}\, F_1) \cap \big(- N((\bar{x}, \bar{y}); \text{gph}\, F_2)\big) = \{0\}$$

and that one of F_i is SNC at $(\bar{x}, \bar{y})$. Then

$$D^*(F_1 \cap F_2)(\bar{x}, \bar{y})(y^*) \subset \bigcup_{y_1^* + y_2^* = y^*} \Big[D^* F_1(\bar{x}, \bar{y})(y_1^*) + D^* F_2(\bar{x}, \bar{y})(y_2^*)\Big] \quad (3.33)$$

for all $y^* \in Y^*$, where D^* stands for the normal coderivative. Moreover, (3.33) *holds as equality and $F_1 \cap F_2$ is N-regular at $(\bar{x}, \bar{y})$ if both F_i are N-regular at this point.*

Proof. Apply Corollary 3.5 to $\Omega_i = \text{gph}\, F_i$, $i = 1, 2$, with the qualification condition (3.10). The equality/regularity assertion follows from the last part of Theorem 3.4. $\triangle$

We conclude this subsection with several remarks on other results related to coderivative calculus for set-valued mappings.

Remark 3.21 (fuzzy coderivative calculus). Based on the fuzzy intersection rule for Fréchet normals in Lemma 3.1 (i.e., actually on the *extremal principle*), one can develop a rich fuzzy calculus of ε-coderivatives $\widehat{D}_\varepsilon^*$ from (1.23) for set-valued mappings between Asplund spaces, where the crucial case is that of $\varepsilon = 0$. It can be done in the way of proving the exact calculus results for D_N^* and D_M^* in this subsection without passing to the limit. Note that we don't need any SNC conditions and can relax qualification conditions to get fuzzy calculus rules. However, results of this type are not pointbased and may be considered as a preliminary tool for the exact calculus of the limiting constructions that are the main objects in this book. More details on the fuzzy calculus for $\widehat{D}_\varepsilon^*$ and related subgradients can be found in Mordukhovich and Shao [952], where the extremal principle is directly used to derive the so-called "quantitative fuzzy sum rule" (with efficient estimates) on which other calculus results are based. Note that the fuzzy intersection rule of Lemma 3.1 is in fact *equivalent* to the Asplund property of X, which has been recently observed by Bingwu Wang (personal communication).

Remark 3.22 (calculus rules for the reversed mixed coderivative). Besides the normal and mixed coderivatives, we actively use in this book the construction $\widetilde{D}_M^*$ defined in (1.40) and called there the *reversed mixed coderivative*, since it can be obtained by reversing the convergence order in comparison with our basic mixed coderivative; cf. Penot [1071]. Although $\widetilde{D}_M^*$ is directly related to the mixed coderivative of the inverse mapping, it doesn't enjoy a comprehensive calculus similar to D_M^* and D_N^* due to the fact the many important operations and properties for mappings are *not invariant/stable* with respect to taking their inverses. As a striking example, mention *summation rules* that *cannot* be satisfactorily established for the reversed mixed coderivative even in its subdifferential specification for real-valued functions $\varphi\colon X \to \overline{I\!R}$, since the unit ball $I\!B^*$ doesn't have any compactness properties with respect to the *norm topology* of X^* in infinite dimensions. Nevertheless, some useful calculus results can be established for $\widetilde{D}^*_M$ in Asplund spaces as shown in Mordukhovich and B. Wang [963]. In particular, it follows from Theorem 3.13 and elementary transformations involving inverse mappings and their coderivatives that the *chain rule*

$$\widetilde{D}^*_M(F\circ G)(\bar{x},\bar{z})\subset\bigcup_{\bar{y}\in G(\bar{x})\cap F^{-1}(\bar{z})}\widetilde{D}^*_M G(\bar{x},\bar{y})\circ D^*_N F(\bar{y},\bar{z})$$

holds for reversed mixed coderivatives of general compositions at every point $(\bar{x},\bar{z})\in\text{gph}\,(F\circ G)$ under *exactly the same assumptions* as in Theorem 3.13(ii). Note that the qualification condition (3.27) can be equivalently written as

$$\big(\ker\widetilde{D}^*_M G(\bar{x},\bar{y})\big)\cap\big(-D^*_M F(\bar{y},\bar{z})(0)\big)=\{0\}\,.$$

The latter easily implies the inclusion

$$\ker\widetilde{D}^*_M(F\circ G)(\bar{z},\bar{x})\subset\bigcup_{\bar{y}\in G(\bar{x})\cap F^{-1}(\bar{z})}\ker D^*_N F(\bar{y},\bar{z})$$

provided that G is *metrically regular* around $(\bar{x},\bar{y})$ for every $\bar{y}\in G(\bar{x})\cap F^{-1}(\bar{z})$. Moreover, applying in this setting the *zero chain rule* of Theorem 3.14 to the inverse mappings, we arrive at the refined inclusion

$$\ker\widetilde{D}^*_M(F\circ G)((\bar{x},\bar{z})\subset\bigcup_{\bar{y}\in G(\bar{x})\cap F^{-1}(\bar{x})}\ker\widetilde{D}^*_M F(\bar{y},\bar{z})$$

involving the kernels of only the reversed mixed coderivatives; see Mordukhovich and Nam [934] for more details.

Remark 3.23 (limiting normals and coderivatives with respect to general topologies). Some of the calculus results above can be unified and generalized by considering limiting constructions with respect to an arbitrary topology τ on X^* that is compatible with the linear structure and satisfies $w^*\le\tau\le\tau_{\|\cdot\|}$, i.e., it is equal to or weaker than the norm topology on X^* and is equal to or stronger than the weak* topology on X^*. Besides $\tau=w^*$ and $\tau=\tau_{\|\cdot\|}$, valuable choices of such a topology on X^* are the weak topology, the topology generated by the convergence of bounded nets in X^*, polar topologies generated by various bornological structures in X, etc.; see the books by Holmes [580] and Phelps [1073] with their references.

Given a topology τ on X^*, we define the τ*-limiting normal cone* to $\Omega\subset X$ at $\bar{x}\in\Omega$ by

$$N_\tau(\bar{x};\Omega) := \left\{x^* \in X^* \,\middle|\, \exists \varepsilon_k \downarrow 0,\ x_k \xrightarrow{\Omega} \bar{x},\ x_k^* \xrightarrow{\tau^*} x^* \text{ with } x_k^* \in \widehat{N}_{\varepsilon_k}(x_k;\Omega)\right\},$$

where ε_k may be omitted if Ω is locally closed around $\bar{x}$ and X is Asplund. It is clear that the stronger τ is, the smaller $N_\tau(\bar{x};\Omega)$ is, and that $N_\tau(\bar{x};\Omega)$ reduces to the basic normal cone (1.3) for $\tau = w^*$. We put $\tau = \tau_{X^*} \times \tau_{Y^*}$ for the product space $X \times Y$, where τ_{X^*} and τ_{Y^*} are generally of different types, and define the *τ-limiting coderivative* of $F\colon X \rightrightarrows Y$ at $(\bar{x},\bar{y}) \in \operatorname{gph} F$ by

$$D_\tau^* F(\bar{x},\bar{y})(y^*) := \left\{x^* \in X^* \,\middle|\, (x^*,-y^*) \in N_\tau((\bar{x},\bar{y});\operatorname{gph} F)\right\},$$

which agrees with the normal coderivative (1.24) for $\tau = w^* \times w^*$, with the mixed coderivative (1.25) for $\tau = w^* \times \tau_{\|\cdot\|}$, and with the reversed mixed coderivative (1.40) for $\tau = \tau_{\|\cdot\|} \times w^*$.

Following the above geometric approach, we can develop the exact calculus of τ-limiting coderivatives based on the intersection rule for the normal cone N_τ generalizing that of Theorem 3.4. In particular, this way leads to the *symmetric coderivative chain rule*

$$D^*_{\tau_{X^*}\times\tau_{Z^*}}(F\circ G)(\bar{x},\bar{z}) \subset D^*_{\tau_{X^*}\times\tau_{Y^*}}G(\bar{x},\bar{y}) \circ D^*_{\tau_{Y^*}\times\tau_{Z^*}}F(\bar{y},\bar{z})$$

for compositions of $G\colon X \rightrightarrows Y$ and $F\colon Y \rightrightarrows Z$ under certain conditions developed by Mordukhovich and B. Wang [963], where the reader can find more results and discussions in this direction.

Remark 3.24 (coderivative calculus in bornologically smooth spaces). Another line of developing the coderivative calculus presented above is to consider appropriate coderivative constructions in Banach spaces admitting Lipschitzian bump functions that are *smooth* with respect to a given *bornology* β; see Remark 2.11. Some results in this direction, based on smooth variational principles, are obtained by Mordukhovich, Shao and Zhu [954] for *viscosity β-coderivatives* generated by the corresponding normal cone (2.78) and their *topological* limits. An essential difference between the Fréchet bornology $\beta = \mathcal{F}$ and *all the other bornologies* on X is that the corresponding topology on X^* generated by β agrees with the *norm* topology of X^* for $\beta = \mathcal{F}$. This allows us to establish in this case *exact calculus* results for *sequential* limiting constructions, in contrast to topological ones in other bornological cases.

3.1.3 Strictly Lipschitzian Behavior and Coderivative Scalarization

In Theorem 1.90 we established the scalarization formula

$$D_M^* f(\bar{x})(y^*) = \partial\langle y^*, f\rangle(\bar{x}), \quad y^* \in Y^*,$$

for the mixed coderivative of locally Lipschitzian mappings $f\colon X \to Y$ between arbitrary Banach spaces. As Example 1.35 shows, an analog of this formula doesn't hold for the normal coderivative of arbitrary locally Lipschitzian mappings without additional assumptions. In this subsection we develop conditions that ensure the normal coderivative scalarization, which is important for various applications including those to subdifferential chain rules and to necessary optimality conditions of the Lagrangian type; see below. First we define subclasses of locally Lipschitzian mappings used for these purposes and establish relationships between them.

Definition 3.25 (strictly Lipschitzian mappings). *Let $f\colon X \to Y$ be a single-valued mapping between Banach spaces. Assume that f is Lipschitz continuous around $\bar{x}$. Then:*

(i) *f is* STRICTLY LIPSCHITZIAN *at $\bar{x}$ if there is a neighborhood V of the origin in X such that the sequence*

$$y_k := \frac{f(x_k + t_k v) - f(x_k)}{t_k}, \quad k \in I\!N\,,$$

contains a norm convergent subsequence whenever $v \in V$, $x_k \to \bar{x}$, and $t_k \downarrow 0$.

(ii) *f is w^*-*STRICTLY LIPSCHITZIAN *at $\bar{x}$ if there is a neighborhood V of the origin in X such that for any $v \in X$ and any sequences $x_k \to \bar{x}$, $t_k \downarrow 0$, and $y_k^* \xrightarrow{w^*} 0$ one has $\langle y_k^*, y_k\rangle \to 0$ as $k \to \infty$, where y_k are defined in* (i).

If Y is finite-dimensional, the properties in (i) and (ii) obviously hold, so both classes in Definition 3.25 reduce to the class of locally Lipschitzian mappings $f\colon X \to I\!R^n$. It is not the case for $\dim Y = \infty$, as the mapping from Example 1.35 illustrates. One can check that both classes in Definition 3.25 are closed with respect to compositions and form linear spaces. Every mapping strictly differentiable at $\bar{x}$ is strictly Lipschitzian at this point. Moreover, the latter class includes Fredholm integral operators with Lipschitzian kernels, which are particularly important in applications to optimal control.

Proposition 3.26 (relations for strictly Lipschitzian mappings). *Every $f\colon X \to Y$ strictly Lipschitzian at $\bar{x}$ is w^*-strictly Lipschitzian at this point. The opposite holds if $I\!B_{Y^*}$ is weak* sequentially compact.*

Proof. Property (i) in Definition 3.25 obviously implies (ii) for any Banach spaces. It remains to show that (ii)$\Longrightarrow$(i) when $I\!B_{Y^*}$ is sequentially compact in the weak* topology on Y^*. Let us prove that under this assumption the convergence property in (i) follows from the one in (ii).

First we observe that the convergence property in (ii) implies the boundedness of $\{y_k\}$. On the contrary, suppose that $\|y_k\| \to \infty$ along some subsequence of $k \to \infty$ (suppose that for all $k \in I\!N$) and find, by the Hahn-Banach theorem, such $y_k^* \in Y^*$ that $\langle y_k^*, y_k\rangle = \sqrt{\|y_k\|}$ and $\|y_k^*\| = \|y_k\|^{-1/2}$, $k \in I\!N$. Then $\|y_k^*\| \to 0$ but $\langle y_k^*, y_k\rangle \not\to 0$ as $k \to \infty$, which contradicts (ii). Using this, let us show that $\{y_k\}$ is actually *totally bounded*, i.e., for every $\varepsilon > 0$ this set can be covered by a finite number of balls with radii less than ε. It is all we need to prove, since the total boundedness of a subset in a metric space is known to be equivalent to its sequential compactness; see, e.g., Dunford and Schwartz [371, p. 22].

On the contrary, assume that $\{y_k\}$ is not totally bounded. Using its boundedness, it is easy to show that there is $\alpha > 0$ such that $\{y_k\} \not\subset Z + \alpha I\!B_Y$ for any finite-dimensional subspace $Z \subset Y$. This allows us to construct a subsequence $\{z_n\}$ of $\{y_k\}$ with $z_{n+1} \notin \text{span}\{z_1, \ldots, z_n\} + \alpha I\!B_Y$ for all $n \in I\!N$. Then we can choose $y_n^* \in I\!B_{Y^*}$ such that

$$\text{span}\{z_1, \ldots, z_n\} \subset \ker y_n^* \quad \text{and} \quad \langle y_n^*, z_{n+1}\rangle \geq \alpha, \quad n \in I\!N .$$

By the assumption of the proposition, $\{y_n^*\}$ contains a subsequence $\{y_{n_m}^*\}$ that converges weak* to some $y^* \in Y^*$. We have $\langle y^*, z_n\rangle = 0$ for all $n \in I\!N$ by the construction. Hence

$$\langle y_{n_m}^* - y^*, z_{n_m+1}\rangle = \langle y_{n_m}^*, z_{n_m+1}\rangle \geq \alpha > 0, \quad m \in I\!N ,$$

which contradicts (ii) and finishes the proof. △

In the next lemma we derive an important property of w^*-strictly Lipschitzian mappings in terms of their Fréchet coderivatives, which is crucial for the proof of the scalarization formula given below. Moreover, this property completely characterizes such mappings under additional assumptions on the Banach spaces in question.

Lemma 3.27 (coderivative characterization of strictly Lipschitzian mappings). *Let $f: X \to Y$ be a mapping between Banach spaces that is locally Lipschitzian around $\bar{x}$. The following assertions hold:*

(i) *If f is w^*-strictly Lipschitzian at $\bar{x}$, then for any sequences $\varepsilon_k \downarrow 0$, $x_k \to \bar{x}$, and $(x_k^*, y_k^*) \in X^* \times Y^*$ with $x_k^* \in \widehat{D}_{\varepsilon_k}^* f(x_k)(y_k^*)$, $k \in I\!N$, one has*

$$y_k^* \xrightarrow{w^*} 0 \Longrightarrow x_k^* \xrightarrow{w^*} 0 \quad \text{as} \quad k \to \infty .$$

(ii) *If X is Asplund and Y is reflexive, then the coderivative property in* (i) *implies that f is strictly Lipschitzian at $\bar{x}$.*

Proof. To prove (i), we take sequences $x_k^* \in \widehat{D}_{\varepsilon_k}^* f(x_k)(y_k^*)$ and observe from the definitions that for any $\gamma_k \downarrow 0$ there are neighborhoods U_k of x_k with

$$\langle x_k^*, x - x_k\rangle - \langle y_k^*, f(x) - f(x_k)\rangle \leq (\gamma_k + \varepsilon_k)(\|x - x_k\| + \|f(x) - f(x_k)\|)$$

whenever $x \in U_k$ and $k \in I\!N$. By the Lipschitz continuity of f with modulus ℓ around $\bar{x}$ we get

$$\langle x_k^*, x - x_k\rangle - \langle y_k^*, f(x) - f(x_k)\rangle \leq (\gamma_k + \varepsilon_k)(1 + \ell)\|x - x_k\| \tag{3.34}$$

for all $x \in U_k$ and $k \in I\!N$. Now pick any v from the neighborhood V of the origin in Definition 3.25(ii) and choose a sequence of $t_k \downarrow 0$ such that $x_k + t_k v \in U_k$ for all $k \in I\!N$. Then (3.34) implies that

$$\langle x_k^*, v\rangle - \left\langle y_k^*, \frac{f(x_k + t_k v) - f(x_k)}{t_k}\right\rangle \le (\gamma_k + \varepsilon_k)(1 + \ell)\|v\| \,. \tag{3.35}$$

Since f is locally Lipschitzian around $\bar{x}$ and $\{y_k^*\}$ is bounded, $\{x_k^*\}$ is bounded as well due to Theorem 1.43. Hence the latter sequence is (topologically) weak* compact in X^*. Taking any $x^* \in \mathrm{cl}^*\{x_k^*\}$, we get from (3.35) and the w^*-strict Lipschitzian property of f that $\langle x^*, v\rangle \le 0$ for each $v \in V$. Thus $x^* = 0$ for every weak* cluster point of $\{x_k^*\}$, which implies that $x_k^* \xrightarrow{w^*} 0$ as $k \to \infty$ and justifies (i).

Let us prove the converse statement assuming that X is Asplund and Y is reflexive. Note that in this case the strictly Lipschitzian and w^*-strictly Lipschitzian properties of f at $\bar{x}$ are equivalent due to Proposition 3.26. Moreover, one can equivalently put $\varepsilon_k = 0$ in (i). Take $\{y_k\}$ from Definition 3.25 and show that it has a norm convergent subsequence. Since $\{y_k\}$ is bounded and Y is reflexive, we may assume that it weakly converges to some point $\bar{y} \in Y$ as $k \to \infty$. The Hahn-Banach theorem ensures the existence of $y_k^* \in Y^*$ satisfying the relations

$$\langle y_k^*, y_k - \bar{y}\rangle = \|y_k - \bar{y}\|, \quad \|y_k^*\| = 1 \text{ for all } k \in I\!N \,.$$

Suppose without loss of generality that $y_k^* \xrightarrow{w^*} \bar{y}^*$ as $k \to \infty$ for some $\bar{y}^* \in Y^*$. Now our goal is to estimate $\langle y_k^* - \bar{y}^*, y_k\rangle$. To proceed, we use the mean value inequality (3.52) from Theorem 3.49. This gives us $v_k \to \bar{x}$ and $v_k^* \in \widehat{\partial}\langle y_k^* - \bar{y}^*, f\rangle(v_k)$ satisfying

$$\langle y_k^* - \bar{y}^*, y_k\rangle \le \langle v_k^*, v\rangle + k^{-1} \text{ for all } k \in I\!N \,, \tag{3.36}$$

where y_k and v are related via Definition 3.25. One can easily check that

$$\widehat{\partial}\langle y^*, f\rangle(x) = \widehat{D}^* f(x)(y^*) \text{ for all } y^* \in Y^* \tag{3.37}$$

if f is locally Lipschitzian around x. Hence $v_k^* \in \widehat{D}^* f(v_k)(y_k^* - \bar{y}^*)$ and $v_k^* \xrightarrow{w^*} 0$ as $k \to \infty$ due to the assumption made in (ii). By (3.36) this gives $\limsup_{k\to\infty}\langle y_k^* - \bar{y}^*, y_k\rangle \le 0$. To finish the proof, we observe that

$$\|y_k - \bar{y}\| = \langle y_k^*, y_k - \bar{y}\rangle = \langle y_k^* - \bar{y}^*, y_k\rangle - \langle y_k^* - \bar{y}^*, \bar{y}\rangle + \langle \bar{y}^*, y_k - \bar{y}\rangle \,,$$

which implies the norm convergence of y_k along the chosen subsequence. △

Now we are ready to establish the required representation of the normal coderivative in terms of the basic subdifferential of the scalarized function.

Theorem 3.28 (scalarization of the normal coderivative). *Consider a mapping $f: X \to Y$ between an Asplund space X and a Banach space Y. Assume that f is w^*-strictly Lipschitzian at $\bar{x}$. Then one has*

$$D_N^* f(\bar{x})(y^*) = \partial\langle y^*, f\rangle(\bar{x}) \ne \emptyset \text{ for all } y^* \in Y^* \,.$$

Moreover, $D_M^* f(\bar{x}) = D_N^* f(\bar{x})$ under the assumptions made.

Proof. We need to show that $D_N^* f(\bar{x})(y^*) \subset \partial\langle y^*, f\rangle(\bar{x})$. The other conclusions of the theorem easily follow from Corollary 2.25 and Theorem 1.90. Pick $x^* \in D_N^* f(\bar{x})(y^*)$ and find, by definitions of the normal coderivative and ε-normals, sequences $\varepsilon_k \downarrow 0$, $x_k \to \bar{x}$, and $(x_k^*, y_k^*) \xrightarrow{w^*} (x^*, y^*)$ satisfying

$$(x_k^*, -y_k^*) \in \widehat{N}_{\varepsilon_k}\big((x_k, f(x_k)); \operatorname{gph} f\big) \quad \text{for all} \;\; k \in I\!N\,.$$

From the proof of Lemma 3.27 we get estimate (3.34) along an arbitrary sequence $\gamma_k \downarrow 0$. This gives

$$x_k^* \in \widehat{\partial}_{\tilde{\varepsilon}_k}\langle y_k^*, f\rangle(x_k) = \widehat{\partial}_{\tilde{\varepsilon}_k}\big[\langle y^*, f\rangle + \langle y_k^* - y^*, f\rangle\big](x_k)$$

with $\tilde{\varepsilon}_k := (\gamma_k + \varepsilon_k)(1 + \ell) \downarrow 0$ as $k \to \infty$. Applying the fuzzy sum rule from Theorem 2.33(b), we find sequences $u_k \to \bar{x}$, $v_k \to \bar{x}$,

$$u_k^* \in \widehat{\partial}\langle y^*, f\rangle(u_k), \quad \text{and} \quad v_k^* \in \widehat{\partial}\langle y_k^* - y^*, f\rangle(v_k)$$

such that $\|x_k^* - u_k^* - v_k^*\| \le 2\tilde{\varepsilon}_k$ for all k. It follows from (3.37) and Lemma 3.27(i) that $v_k^* \xrightarrow{w^*} 0$ as $k \to \infty$. Hence $u_k^* \xrightarrow{w^*} x^* \in \partial\langle y^*, f\rangle(\bar{x})$, which completes the proof of the theorem. △

Let us present two useful consequences of Lemma 3.27 and Theorem 3.28. The first corollary gives a convenient representation of the normal second-order subdifferential for an important subclass of $\mathcal{C}^{1,1}$ functions, while the second one proves a characterization of the SNC property for strictly Lipschitzian mappings.

Corollary 3.29 (normal second-order subdifferentials of $\mathcal{C}^{1,1}$ functions). *Let X be Asplund, and let $\varphi\colon X \to \overline{I\!R}$ be $\mathcal{C}^1$ around $\bar{x}$ with the derivative $\nabla\varphi$ that is w^*-strictly Lipschitzian at this point. Then*

$$\partial_N^2\varphi(\bar{x})(u) = \partial\langle u, \nabla\varphi\rangle(\bar{x}) \ne \emptyset \;\; \textit{for all} \;\; u \in X^{**}$$

and $\partial_M^2\varphi(\bar{x}) = \partial_N^2\varphi(\bar{x})$.

Proof. This follows directly from Theorem 3.28 with $f := \nabla\varphi\colon X \to X^*$. △

Corollary 3.30 (characterization of the SNC property for strictly Lipschitzian mappings). *Let $f\colon X \to Y$ be a mapping between Banach spaces. Assume that f is w^*-strictly Lipschitzian at $\bar{x}$ and that X is Asplund. Then f is SNC at $\bar{x}$ if and only if* $\dim Y < \infty$.

Proof. The "if" part follows from Corollary 1.69. To prove the "only if" part in the case of Asplund spaces X, we need to show that for every w^*-strictly Lipschitzian mapping $f\colon X \to Y$ at $\bar{x}$ and for every infinite-dimensional Banach space Y there are sequences $x_k \to \bar{x}$ and $(x_k^*, y_k^*) \xrightarrow{w^*} (0,0)$ satisfying

$$x_k^* \in \widehat{D}^* f(x_k)(y_k^*) \;\; \text{with} \;\; \|(x_k^*, y_k^*)\| \not\to 0 \;\; \text{as} \;\; k \to \infty\,.$$

Indeed, given a Banach space Y with $\dim Y = \infty$ and applying the fundamental Josefson-Nissenzweig theorem (cf. the proof of Theorem 1.21), we find a sequence of $y_k^* \in Y^*$ with $\|y_k^*\| = 1$ and $y_k^* \xrightarrow{w^*} 0$. By scalarization (3.37) for Lipschitzian mappings and by the density of Fréchet subgradients in Asplund spaces due to Corollary 2.29, there are sequences $(x_k, x_k^*) \in X \times X^*$ with $x_k \to \bar{x}$ and $x_k^* \in \widehat{D}^* f(x_k)(y_k^*)$ for all $k \in I\!N$. Due to Lemma 3.27(i) one has $x_k^* \xrightarrow{w^*} 0$ as $k \to \infty$. Thus f doesn't have the SNC property at $\bar{x}$. △

Note that the strict Lipschitz continuity of f is not necessary for the equivalence in Corollary 3.30. In particular, Y must be finite-dimensional for every mapping $f\colon X \to Y$ between Banach spaces that is SNC at $(\bar{x}, f(\bar{x}))$ and Fréchet differentiable at $\bar{x}$; it may not be either strictly differentiable at $\bar{x}$ or even Lipschitzian around this point. On the other hand, the above proof shows that, due to Lemma 3.27(ii), the strict Lipschitzian requirement on f is not avoidable in Corollary 3.30 if Y is assumed to be reflexive while

$$y_k^* \xrightarrow{w^*} 0 \Longrightarrow x_k^* \xrightarrow{w^*} 0 \;\text{ whenever }\; x_k^* \in \widehat{D}^* f(x_k)(y_k^*) \;\text{ and }\; x_k \to \bar{x}\,.$$

Remark 3.31 (scalarization results with respect to general topologies). One can observe from the proofs of Theorems 1.90 and 3.28 that the scalarization formulas obtained there for the mixed and normal coderivatives admit extensions to the limiting constructions with respect to general topologies described in Remark 3.23. The corresponding *τ-limiting subdifferential* of $\varphi\colon X \to \overline{I\!R}$ at $\bar{x}$ with $|\varphi(\bar{x})| < \infty$ is defined, equivalently, by

$$\partial_\tau \varphi(\bar{x}) := D_\tau^* E_\varphi(\bar{x}, \varphi(\bar{x}))(1) = \operatorname*{Lim\,sup}_{\substack{x \xrightarrow{\varphi} \bar{x} \\ \varepsilon \downarrow 0}} \widehat{\partial}_\varepsilon \varphi(x)\,,$$

where one may put $\varepsilon = 0$ provided that φ is proper and l.s.c. around $\bar{x}$ and that X and Asplund. Given a mapping $f\colon X \to Y$ between Banach spaces and an arbitrary linear topology $\tau = \tau_{X^*} \times \tau_{Y^*}$ on $X^* \times Y^*$, we get from the proof of Theorem 1.90 that

$$\partial_{\tau_{X^*}} \langle y^*, f\rangle(\bar{x}) \subset D_\tau^* f(\bar{x})(y^*), \quad y^* \in Y^*\,,$$

if f is continuous around $\bar{x}$, and that

$$D^*_{\tau_{X^*} \times \tau_{\|\cdot\|}} f(\bar{x})(y^*) = \partial_{\tau_{X^*}} \langle y^*, f\rangle(\bar{x}), \quad y^* \in Y^*\,,$$

if f is Lipschitz continuous around $\bar{x}$. This covers the case of the mixed coderivative in Theorem 1.90 when $\tau_{X^*} = w^*$. Then we observe from the proof of Theorem 3.28 that

$$D^*_{w^* \times \tau_{Y^*}} f(\bar{x})(y^*) = \partial\langle y^*, f\rangle(\bar{x}), \quad y^* \in Y^*\,,$$

if X is Asplund and f is *τ_{Y^*}-strictly Lipschitzian* at $\bar{x}$, which means that f is Lipschitz continuous around this point and satisfies the convergence condition from Definition 3.25(ii) with w^* replaced by τ_{Y^*}.

In conclusion of this section we consider a remarkable subclass of strictly Lipschitzian mappings that is related to the PSNC property of multifunctions in the sense of Definition 1.67.

Definition 3.32 (compactly strictly Lipschitzian mappings). *A single-valued mapping $f: X \to Y$ between Banach spaces is* COMPACTLY STRICTLY LIPSCHITZIAN *at $\bar{x}$ if for each sequences $x_k \to \bar{x}$ and $h_k \to 0 \in X$ with $h_k \neq 0$ the sequence*

$$\frac{f(x_k + h_k) - f(x_k)}{\|h_k\|}, \quad k \in I\!N\,,$$

has the norm convergent subsequence.

It is obvious that a compactly strictly Lipschitzian mapping is strictly Lipschitzian in the sense of Definition 3.25(i), and hence it is locally Lipschitzian around $\bar{x}$. Moreover, for $\dim Y < \infty$ the above strictly Lipschitzian notions agree and reduce to the standard local Lipschitz continuity. It is not the case when Y is infinite-dimensional, particularly Asplund. Indeed, the mapping $f: c_0 \to c_0$ given by

$$f(x) := \big\{ \sin x_k \big\} \ \text{ for } \ x := \big\{ x_k \big\}$$

is strictly Lipschitzian but not compactly strictly Lipschitzian at the origin. It is easy to check that f is compactly strictly Lipschitzian at $\bar{x}$ if it is strictly Fréchet differentiable at x with the compact derivative operator, or more generally: if f is a composition $f = g \circ f_0$, where g is strictly differentiable with the compact derivative while f_0 is locally Lipschitzian. Furthermore, the class of compactly strictly Lipschitzian mappings contains those $f: X \to Y$ that are *uniformly directionally compact* around $\bar{x}$, in the sense that there is a norm compact set $Q \subset Y$ for which

$$f(x + th) \in f(x) + t\|h\| Q + t\eta\big(\|x - \bar{x}\|, t\big) I\!B$$

whenever $h \in X$ with $\|h\| \le 1$ and x close to $\bar{x}$, where $\eta(\varepsilon, t) \to 0$ as $\varepsilon \downarrow 0$ and $t \downarrow 0$. Note that the class of compactly strictly Lipschitzian mappings forms a linear space being also closed with respect to compositions involving local Lipschitzian mappings.

It is interesting to observe that compactly strictly Lipschitzian mappings admit a coderivative characterization similar to Lemma 3.27 for strictly Lipschitzian mappings but different in one aspect, which is crucial in what follows.

Lemma 3.33 (coderivative characterization of compactly strictly Lipschitzian mappings). *Let $f: X \to Y$ be a mapping between Banach spaces that is locally Lipschitzian around $\bar{x}$. The following assertions hold:*

(i) *If f is compactly strictly Lipschitzian at $\bar{x}$, then for any sequences $\varepsilon_k \downarrow 0$, $x_k \to \bar{x}$, and $(x_k^*, y_k^*) \in X^* \times Y^*$ with $x_k^* \in \widehat{D}^*_{\varepsilon_k} f(x_k)(y_k^*)$ one has*

$$y_k^* \xrightarrow{w^*} 0 \Longrightarrow \|x_k^*\| \to 0 \quad as \quad k \to \infty\,.$$

(ii) *If X is Asplund and Y is reflexive, then the coderivative property in* (i) *implies that f is compactly strictly Lipschitzian at $\bar{x}$.*

Proof. To prove (i), we take $x_k^* \in \widehat{D}^*_{\varepsilon_k} f(x_k)(y_k^*)$ with $y_k^* \xrightarrow{w^*} 0$ and, by definition of the ε_k-coderivative, for any $\gamma_k \downarrow 0$ find $\nu_k \downarrow 0$ such that

$$\langle x_k^*, x - x_k\rangle - \langle y_k^*, f(x) - f(x_k)\rangle \le (\gamma_k + \varepsilon_k)\big(\|x - x_k\| + \|f(x) - f(x_k)\|\big)$$

whenever $x = x_k + \nu_k h_k$. Dividing this by $\nu_k > 0$, one has

$$\langle x_k^*, h_k\rangle - \left\langle y_k^*, \frac{f(x_k + \nu_k h_k) - f(x_k)}{\nu_k}\right\rangle \le \eta_k\left(1 + \left\|\frac{f(x_k + \nu_k h_k) - f(x_k)}{\nu_k}\right\|\right)$$

with $\eta_k := \gamma_k + \varepsilon_k$. Since f is compactly strictly Lipschitzian at $\bar{x}$, we may assume that the sequence $\big\{\big(f(x_k + \nu_k h_k) - f(x_k)\big)/\nu_k\big\}$, $k \in I\!N$, is norm convergent. Now passing to the limit as $k \to \infty$ and taking into account that $y_k^* \xrightarrow{w^*} 0$, we get $\langle x_k^*, h_k\rangle \to 0$, which implies that $\|x_k^*\| \to 0$ and completes the proof of assertion (i).

To justify the converse assertion (ii) of the theorem when X is Asplund and Y is reflexive, we proceed similarly to the proof of Lemma 3.27(ii) with $\varepsilon_k = 0$ in the convergence property of (i). Define

$$y_k := \frac{f(x_k + h_k) - f(x_k)}{\|h_k\|}, \quad k \in N\,,$$

and assume that $y_k \xrightarrow{w} \bar{y}$ to some $\bar{y} \in Y$ without loss of generality due to the Lipschitz continuity of f. Invoking the Hahn-Banach theorem, we find $y_k^* \in Y^*$ such that

$$\langle y_k^*, y_k - \bar{y}\rangle = \|y_k - \bar{y}\|^2, \quad \|y_k^*\| = \|y_k - \bar{y}\|, \quad \text{and} \quad y_k^* \xrightarrow{w^*} \bar{y}^*$$

for some $\bar{y}^* \in Y^*$. Then using the mean value inequality (3.52) from Theorem 3.49 and taking into account the scalarization formula (3.37) for the Fréchet coderivative, one has $v_k \to \bar{x}$ and

$$v_k^* \in \widehat{\partial}\langle y_k^* - \bar{y}^*, f\rangle(v_k) = \widehat{D}^* f(v_k)(y_k^* - \bar{y}^*)$$

satisfying the estimate

$$\langle y_k^* - \bar{y}^*\rangle \le \frac{1}{k} + \left\langle v^* + k, \frac{h_k}{\|h_k\|}\right\rangle.$$

Since $\|v_k^*\| \to 0$ by the requirement in (ii), we get $\limsup_{k\to\infty}\langle y_k^* - \bar{y}^*, y_k\rangle \le 0$. This yields $y_k \to \bar{y}$ as in Lemma 3.27(ii) and completes the proof. △

Finally, let us use the coderivative characterization of Lemma 3.33 to establish the PSNC property of the following class of mappings important in various applications.

Definition 3.34 (generalized Fredholm mappings). *A single-valued mapping $f: X \to Y$ between Banach spaces is* GENERALIZED FREDHOLM *at $\bar{x}$ if there is a mapping $g: X \to Y$, which is compactly strictly Lipschitzian at $\bar{x}$ and such that the difference $f - g$ is a linear bounded operator whose image is a closed subspace of finite codimension in Y.*

This definition extends various notions of *Fredholm-like* behavior of mappings that naturally arise in applications to optimization problems with *operator constraints* in infinite dimensions and particularly to problems of *optimal control* for dynamic systems governed by nonsmooth differential equations and inclusions; see more discussions and details in Ioffe [595, 604] and in Ginsburg and Ioffe [506] as well as in Subsects. 5.1.2 and 6.1.4 below. The principal property of generalized Fredholm mappings crucial for their applications is given in the next theorem.

Theorem 3.35 (PSNC property of generalized Fredholm mappings). *Let $f: X \to Y$ be a mapping between Banach spaces, let $\Omega \subset X$, and let*

$$f_\Omega(x) := \begin{cases} f(x) & \text{if } x \in \Omega\,, \\ \emptyset & \text{if } x \notin \Omega \end{cases}$$

be the restriction of f to Ω. Assume that f is generalized Fredholm at $\bar{x} \in \Omega$ and that:

(a) *either $\Omega = X$, or*

(b) *X and Y are Asplund, Ω is SNC at $\bar{x}$ and closed around this point.*

Then the inverse mapping f_Ω^{-1} is PSNC at $(f(\bar{x}), \bar{x})$.

Proof. Take sequences $\varepsilon_k \downarrow 0$, $x_k \xrightarrow{\Omega} \bar{x}$, $x_k^* \to 0$, and $y_k^* \xrightarrow{w^*} 0$ such that

$$x_k^* \in \widehat{D}^*_{\varepsilon_k}\big(f + \Delta(\cdot;\Omega)\big)(x_k)(y_k^*) \quad \text{for all } \; k \in I\!N\,,$$

where $\Delta(\cdot;\Omega)$ is the indicator mapping of the set Ω. To justify the PSNC property of f_Ω^{-1} at $(f(\bar{x}), \bar{x})$, we need to show, according to Definition 1.67, that $\|y_k^*\| \to 0$ as $k \to \infty$.

Consider first the case of $\Omega = X$ in the general Banach space setting and denote by $A := f - g$ the linear bounded operator from X to Y whose image/range $Y_0 := AX$ is a closed subspace of finite codimension. Thus there is a closed subspace $Y_1 \subset Y$ with $Y = Y_0 \bigoplus Y_1$ and $\operatorname{codim} Y_1 < \infty$. Due to the elementary adaptation of the sum rule from Theorem 1.62(i) to the case of ε-coderivatives (cf. the proof of Theorem 1.38), our aim is to show that $\|y_k^*\| \to$ as $k \to \infty$ whenever $y_k^* \xrightarrow{w^*} 0$, $\varepsilon_k \downarrow 0$, and $\|x_k^*\| \to 0$ provided that

$$x_k^* - A^* y_k^* \in \widehat{D}^*_{\varepsilon_k} g(x_k)(y_k^*), \quad k \in I\!N\,.$$

The latter inclusion implies by Lemma 3.33(i) that $\|x_k^* - A^* y_k^*\| \to 0$ and hence $\|A^* y_k^*\| \to 0$. On the other hand, each y_k^* is represented as $y_k^* = y_{0k}^* + y_{1k}^*$ with

$y_{ik}^* \in Y_i^*$, $i = 1, 2$, and $A^* y_k^* = A^* y_{0k}^*$. Since Y_1^* is finite-dimensional and since A maps X onto Y_0, we get $\|y_{1k}^*\| \to 0$ and also $\|A^* y_{0k}^*\| \geq \mu \|y_{0k}^*\|$ with some $\mu > 0$ by the open mapping theorem (cf. Lemma 1.18). Thus $\|y_{0k}^*\| \to 0$, which completes the proof in case (a).

Consider now case (b) with $\Omega \neq X$. Then we have

$$x_k^* \in \widehat{D}^*\big(A + g + \Delta(\cdot; \Omega)\big)(x_k)(y_k^*) .$$

Proceeding as in the proof of Theorem 3.10 in Asplund spaces, we find $\widehat{x}_k \to \bar{x}$, $u_k \to \bar{x}$, $\widetilde{x}_k^* \to 0$, $\widetilde{y}_k^* \xrightarrow{w^*} 0$, $\widehat{y}_k^* \xrightarrow{w^*} 0$, and $\widehat{x}_k^* \in \widehat{D}^* g(\widehat{x}_k)(\widehat{y}_k^*)$ such that

$$\widetilde{x}_k^* - A^* \widetilde{y}_k^* - \widehat{x}_k^* \in \widehat{N}(u_k; \Omega) \;\text{ and }\; \|\widetilde{y}_k^* - y_k^*\| \to 0 .$$

It follows from Lemma 3.33(i) that $\|\widehat{x}_k^*\| \to 0$. Furthermore, one has

$$\|\widetilde{x}_k^* - A^* \widetilde{y}_k^* - \widehat{x}_k^*\| \to 0 \;\text{ as }\; k \to \infty$$

due to the assumed SNC property of Ω at $\bar{x}$. Thus $\|A^* \widetilde{y}_k^*\| \to 0$. By the above arguments in case (a) we conclude that $\|\widetilde{y}_k^*\| \to 0$ and hence $\|y_k^*\| \to 0$, which completes the proof of the theorem. △

3.2 Subdifferential Calculus and Related Topics

This section is devoted to subdifferential calculus for extended-real-valued functions and some of its direct applications. First we develop calculus rules for basic and singular subgradients that mainly follow from the corresponding results for normal cones and coderivatives. Then we present an Asplund space version of the approximate mean value theorem that has many important applications, some of which are given in this section. Calculus results allow us to establish close relationships between graphical regularity and differentiability of Lipschitzian mappings. In the final subsection we derive an extended calculus for second-order subdifferentials in the framework of Asplund spaces.

3.2.1 Calculus Rules for Basic and Singular Subgradients

Unless otherwise stated, extended-real-valued functions under consideration are assumed to be proper and finite at references points. In this subsection we present principal calculus rules for basic and singular subgradients in fairly general settings. The results obtained include calculus for lower/epigraphical regularity of functions in the sense of Definition 1.91.

We start with a fundamental result of the first-order subdifferential calculus containing general *sum rules* for basic and singular subgradients of extended-real-valued functions.

Theorem 3.36 (sum rules for basic and singular subgradients). *Let $\varphi_i: X \to \overline{I\!R}$, $i = 1, \ldots, n \geq 2$, be l.s.c. around $\bar{x}$, and let all but one of these functions be sequentially normally epi-compact (SNEC) at $\bar{x}$. Assume that*

$$\Big[x_1^* + \ldots + x_n^* = 0,\ x_i^* \in \partial^\infty \varphi_i(\bar{x})\Big] \Longrightarrow x_i^* = 0, \quad i = 1, \ldots, n\,. \tag{3.38}$$

Then one has the inclusions

$$\partial(\varphi_1 + \ldots + \varphi_n)(\bar{x}) \subset \partial\varphi_1(\bar{x}) + \ldots + \partial\varphi_n(\bar{x})\,, \tag{3.39}$$

$$\partial^\infty(\varphi_1 + \ldots + \varphi_n)(\bar{x}) \subset \partial^\infty\varphi_1(\bar{x}) + \ldots + \partial^\infty\varphi_n(\bar{x})\,. \tag{3.40}$$

If in addition each φ_i is lower regular at $\bar{x}$, then the sum $\varphi_1 + \ldots + \varphi_n$ is lower regular at this point and (3.39) holds as equality. The equality also holds in (3.40) and $\varphi_1 + \ldots + \varphi_n$ is epigraphically regular at $\bar{x}$ if each φ_i is epigraphically regular at this point.

Proof. First consider the case of $n = 2$. In this case the qualification condition (3.38) reduces to

$$\partial^\infty\varphi_1(\bar{x}) \cap \big(-\partial^\infty\varphi_2(\bar{x})\big) = \{0\}\,,$$

and inclusions (3.39) and (3.40) follow directly from the coderivative sum rule of Theorem 3.10 applied to the epigraphical multifunctions E_{φ_i} with $E_{\varphi_1+\varphi_2} = E_{\varphi_1} + E_{\varphi_2}$. To prove the equality/regularity statements in the theorem, we observe that

$$\widehat{\partial}(\varphi_1 + \varphi_2)(\bar{x}) \supset \widehat{\partial}\varphi_1(\bar{x}) + \widehat{\partial}\varphi_2(\bar{x})\,. \tag{3.41}$$

due to representation (1.51) of Fréchet subgradients. This implies the equality in (3.39) and the lower regularity of $\varphi_1 + \varphi_2$ at $\bar{x}$ when both φ_i are lower regular at this point. By Proposition 1.92(ii) the epigraphical regularity of any $\varphi: X \to \overline{I\!R}$ requires, in addition to its lower regularity, that

$$\widehat{\partial}^\infty\varphi(\bar{x}) := \big\{x^* \in X^* \big|\ (x^*, 0) \in \widehat{N}(\bar{x}, \varphi(\bar{x})); \operatorname{epi}\varphi)\big\} = \partial^\infty\varphi(\bar{x})\,.$$

This allows us to derive the last conclusion of the theorem for the case of two functions from the inclusion

$$\widehat{\partial}^\infty(\varphi_1 + \varphi_2)(\bar{x}) \supset \widehat{\partial}^\infty\varphi_1(\bar{x}) + \widehat{\partial}^\infty\varphi_2(\bar{x})\,,$$

which follows from (3.41) and Lemma 2.37. For $n > 2$ we prove the theorem by induction, where the qualification condition (3.38) at the current step is justified by using (3.40) at the previous step. △

When all but one of φ_i are locally *Lipschitzian* around $\bar{x}$, the qualification and SNEC assumptions of the theorem are automatically satisfied due to Theorem 1.26 and Corollary 1.81. Hence we always have (3.39) in this case, which also follows from Theorem 2.33. Another special case of Theorem 3.36 concerns intersections of finitely many closed sets.

Corollary 3.37 (basic normals to finite set intersections). *Let $\Omega_1,\ldots,\Omega_n$ be subsets of X locally closed around their common point $\bar{x}$. Assume that all but one of Ω_i are SNC at $\bar{x}$ and that the qualification condition*

$$\Big[x_1^*+\ldots+x_n^*=0,\ x_i^*\in N(\bar{x};\Omega_i)\Big]\Longrightarrow x_i^*=0,\quad i=1,\ldots,n\,,$$

is satisfied. Then one has the inclusion

$$N(\bar{x};\Omega_1\cap\ldots\cap\Omega_n)\subset N(\bar{x};\Omega_1)+\ldots+N(\bar{x};\Omega_n)\,,$$

where the equality holds and $\Omega_1\cap\ldots\cap\Omega_n$ is normally regular at $\bar{x}$ if each Ω_i is normally regular at this point.

Proof. Follows from Theorem 3.36 with $\varphi_i=\delta(\cdot;\Omega_i)$ due to Proposition 1.79. It can also be derived by induction from Corollary 3.5 under the normal qualification condition (3.10). △

Our next topic is subdifferentiation of the *marginal functions*

$$\mu(x):=\inf\big\{\varphi(x,y)\big|\ y\in G(x)\big\}\ \text{ with }\ \varphi\colon X\times Y\to\overline{I\!R},\quad G\colon X\rightrightarrows Y$$

studied in Subsect. 1.3.4 in the framework of general Banach spaces. Here, considering the case of Asplund spaces, we obtain refined formulas for estimating $\partial\mu$ and $\partial^\infty\mu$ in terms of related constructions for φ and G under general assumptions on these mappings. In this way we derive efficient chain rules for basic and singular subgradients of compositions $\varphi\circ g$ involving nonsmooth mappings. The next theorem provides general results in this direction. As in Subsect. 1.3.4, we consider independent cases in (i,ii) corresponding to inner semicontinuity and inner semicompactness of the argminimum mapping $M(\cdot)$. Besides this, assertions (i,ii) are essentially different from those in (iii) and (iv) in both assumptions and conclusions. In particular, (iii) requires milder PSNC and qualification conditions in comparison with (i,ii) but for $\varphi=\varphi(y)$, while (iv) gives more precise inclusions (involving the *mixed* coderivative of G) for singular subgradients of the marginal function when φ is locally Lipschitzian.

Theorem 3.38 (basic and singular subgradients of marginal functions). *Let*

$$M(x):=\big\{y\in G(x)\big|\ \varphi(x,y)=\mu(x)\big\}$$

define the argminimum mapping for the marginal function μ generated by φ and G. The following hold:

(i) *Given $\bar{y}\in M(\bar{x})$, assume that M is inner semicontinuous at $(\bar{x},\bar{y})$, that φ is l.s.c. around $(\bar{x},\bar{y})$, and that the graph of G is closed around this point. Suppose also that either φ is SNEC at $(\bar{x},\bar{y})$ or G is SNC at $(\bar{y},\bar{x})$ and that the qualification condition*

$$(x^*,y^*)\in\partial^\infty\varphi(\bar{x},\bar{y})\cap\big(-N((\bar{x},\bar{y});\operatorname{gph}G)\big)=\{0\}$$

is satisfied. Then one has the inclusions

$$\partial\mu(\bar{x}) \subset \bigcup_{(x^*,y^*)\in\partial\varphi(\bar{x},\bar{y})} \Big[x^* + D_N^*G(\bar{x},\bar{y})(y^*)\Big], \tag{3.42}$$

$$\partial^\infty\mu(\bar{x}) \subset \bigcup_{(x^*,y^*)\in\partial^\infty\varphi(\bar{x},\bar{y})} \Big[x^* + D_N^*G(\bar{x},\bar{y})(y^*)\Big]. \tag{3.43}$$

(ii) *Assume that M is inner semicompact at $\bar{x}$, that G is closed-graph and φ is l.s.c. on* gph G *whenever x is near $\bar{x}$, and that the other assumptions of* (i) *are satisfied for every $\bar{y} \in M(\bar{x})$. Then one has analogs of inclusions* (3.42) *and* (3.43), *where the sets on the right-hand sides are replaced by their unions over $\bar{y} \in M(\bar{x})$.*

(iii) *Let $\varphi = \varphi(y)$. Assume that G^{-1} is PSNC at $(\bar{y},\bar{x})$ and that the qualification condition*

$$\partial^\infty\varphi(\bar{y}) \cap D_M^*G^{-1}(\bar{y},\bar{x})(0) = \{0\}$$

is satisfied, instead of the SNC condition on G and the qualification condition in (i) *and* (ii). *Then one has the inclusions*

$$\partial\mu(\bar{x}) \subset \bigcup_{y^*\in\partial\varphi(\bar{y})} D_N^*G(\bar{x},\bar{y})(y^*), \quad \partial^\infty\mu(\bar{x}) \subset \bigcup_{y^*\in\partial^\infty\varphi(\bar{y})} D_N^*G(\bar{x},\bar{y})(y^*);$$

$$\partial\mu(\bar{x}) \subset \bigcup_{\substack{y^*\in\partial\varphi(\bar{y})\\ \bar{y}\in M(\bar{x})}} D_N^*G(\bar{x},\bar{y})(y^*), \quad \partial^\infty\mu(\bar{x}) \subset \bigcup_{\substack{y^*\in\partial^\infty\varphi(\bar{y})\\ \bar{y}\in M(\bar{x})}} D_N^*G(\bar{x},\bar{y})(y^*)$$

under the remaining assumptions in (i) *and* (ii), *respectively.*

(iv) *Given $\bar{y} \in M(\bar{x})$ assume that $\varphi = \varphi(x,y)$ is locally Lipschitzian around $(\bar{x},\bar{y})$ and that M is inner semicontinuous around this point. Then*

$$\partial^\infty\mu(\bar{x}) \subset D_M^*G(\bar{x},\bar{y})(0).$$

If M is assumed to be inner semicompact around $\bar{x}$ while φ is locally Lipschitzian around $(\bar{x},\bar{y})$ for every $\bar{y} \in M(\bar{x})$, then one has

$$\partial^\infty\mu(\bar{x}) \subset \bigcup_{\bar{y}\in M(\bar{x})} D_M^*G(\bar{x},\bar{y})(0).$$

Proof. To justify (i) and (ii), apply first Theorem 1.108(i,ii) from Chap. 1 to get the inclusion

$$\partial\mu(\bar{x}) \subset \big\{x^* \in X^* \big|\ (x^*,0) \in \partial\big[\varphi + \delta(\cdot;\operatorname{gph} G)\big](\bar{x},\bar{y})\big\}$$

and its counterpart for $\partial^\infty\mu(\bar{x})$ with no qualification and SNC conditions in general Banach spaces. Then applying the subdifferential sum rule from

Theorem 3.36 to the sum $\varphi(x,y)+\delta\big((x,y);\operatorname{gph}G\big)$, we arrive at (3.42) and (3.43) under the assumptions made in (i) and (ii).

To justify (iii), we again use the Banach space results of Theorem 1.108 but then argue similarly to the proof of Proposition 3.12 and Theorem 3.13 replacing coderivatives by subdifferentials.

It remains to prove (iv). We justify only the first inclusion therein under the inner semicontinuity assumption on the argminimum mapping M; the proof of the second one is similar under the inner semicompactness assumption imposed on M. Observe that the marginal function μ is l.s.c. around $\bar{x}$ under the assumptions made.

To proceed, fix $x^*\in\partial^\infty\mu(\bar{x})$ and find, by Theorem 2.38 in Asplund spaces, sequences $x_k\xrightarrow{\mu}\bar{x}$, $\lambda_k\downarrow 0$, and $x_k^*\in\widehat{\partial}\mu(x_k)$ satisfying

$$\lambda_k x_k^*\xrightarrow{w^*}x^*\quad\text{as }\ k\to\infty\,.$$

By the inner semicontinuity of M at $(\bar{x},\bar{y})$, there is a sequence of $y_k\in M(x_k)$ converging to $\bar{y}$; note that it is sufficient to impose such a requirement only along of $x_k\to\bar{x}$ with $\widehat{\partial}\mu(x_k)\neq\emptyset$. Fix $k\in I\!N$ and rewrite the condition $x^*\in\widehat{\partial}\mu(x_k)$ as follows: for every $\varepsilon>0$ there is $\eta>0$ such that

$$\langle x_k^*,x-x_k\rangle\le\mu(x)-\mu(x_k)+\varepsilon\|x-x_k\|\quad\text{whenever}\ \ x\in x_k+\eta I\!B\,.$$

Invoking the function

$$\vartheta(x,y):=\varphi(x,y)+\delta\big((x,y);\operatorname{gph}G\big)\,,$$

we easily have the inequality

$$\big\langle(x_k^*,0),(x-x_k,y-y_k)\big\rangle\le\vartheta(x,y)-\vartheta(x_k,y_k)+\varepsilon\big(\|x-x_k\|+\|y-y_k\|\big)$$

whenever $(x,y)\in(x_k,y_k)+\eta I\!B$. This gives $(x_k^*,0)\in\widehat{\partial}\vartheta(x_k,y_k)$. Now taking into account the Lipschitz continuity of φ and applying the semi-Lipschitzian fuzzy sum rule from Theorem 2.33(b) to the function ϑ along some sequence $\varepsilon_k\downarrow 0$, we find $(x_{1k},y_{1k})\to(\bar{x},\bar{y})$, $(x_{2k},y_{2k})\xrightarrow{\operatorname{gph}G}(\bar{x},\bar{y})$, $(x_{1k}^*,y_{1k}^*)\in\widehat{\partial}\varphi(x_{1k},y_{1k})$, and $(x_{2k}^*,y_{2k}^*)\in\widehat{N}\big((x_{2k},y_{2k});\operatorname{gph}G\big)$ such that

$$\|x_k^*-x_{1k}^*-x_{2k}^*\|\le\varepsilon_k\ \text{ and }\ \|y_{1k}^*+y_{2k}^*\|\le\varepsilon_k\ \text{ for all }\ k\in I\!N\,.$$

Invoking again the Lipschitz continuity of φ around $(\bar{x},\bar{y})$ with some modulus ℓ, we get $\|(x_{1k}^*,y_{1k}^*)\|\le\ell$, and hence

$$\lambda_k\big\|(x_{1k}^*,y_{1k}^*)\big\|\to 0\ \text{ as }\ k\to\infty\,.$$

This implies, by the above estimates, that

$$\lambda_k x_{2k}^*\xrightarrow{w^*}x^*\ \text{ and }\ \lambda_k\|y_{2k}^*\|\to 0\ \text{ as }\ k\to\infty\,.$$

Taking into account that $\lambda_k(x_{2k}^*,y_{2k}^*)\in\widehat{N}\big((x_{2k},y_{2k});\operatorname{gph}G\big)$, we finally get $x^*\in D_M^*G(\bar{x},\bar{y})(0)$ by the construction of the mixed coderivative. This completes the proof of (iv) and of the whole theorem. △

Remark 3.39 (singular subgradients of extended marginal and distance functions). The results obtained in Theorem 3.38 can be easily extended to marginal functions of two variables defined by

$$\mu(x,y) := \inf\big\{\varphi(y,v)\big|\ v \in G(x)\big\}\ .$$

Indeed, such functions are directly reduced to the standard form considered above with respect to the new variable $z := (x,y)$. Thus all the results of Theorem 3.38 can be reformulated for $\mu(x,y)$. In particular, the counterpart of the second inclusion in (iv) is written as

$$\partial^\infty\mu(\bar{x},\bar{y}) \subset \bigcup_{\bar{v}\in M(\bar{x},\bar{y})} \big\{(x^*,0)\big|\ x^* \in D_M^*G(\bar{x},\bar{v})(0)\big\}$$

provided that the argminimum mapping

$$M(x,y) := \big\{v \in G(x)\big|\ \varphi(y,v) = \mu(x,y)\big\}$$

is inner semicompact at $(\bar{x},\bar{y})$ and that φ is locally Lipschitzian around $(\bar{y},\bar{v})$ for all $\bar{v} \in M(\bar{x},\bar{y})$. For the *distance function*

$$\rho(x,y) := \operatorname{dist}\big(y;G(x)\big)$$

to moving sets, which is a special case of the above marginal function with $\varphi(y,v) := \|y-v\|$, this gives the inclusion

$$\partial^\infty\rho(\bar{x},\bar{y}) \subset \big\{(x^*,0)\big|\ x^* \in D_M^*G(\bar{x},\bar{y})(0)\big\}$$

whenever $\bar{y} \in G(\bar{x})$. Moreover, the latter inclusion holds as *equality* if ρ is continuous around $(\bar{x},\bar{y})$. We refer the reader to the papers by Mordukhovich and Nam [935, 936] for more results, proofs, and discussions.

Let us now present efficient conditions under which the main assumptions of Theorem 3.38 automatically hold due to their characteristics in Chap. 1. For simplicity we formulate this corollary only for assertion (i).

Corollary 3.40 (marginal functions with Lipschitzian or metrically regular data). *Given $\bar{y} \in M(\bar{x})$, we assume that M is inner semicontinuous at $(\bar{x},\bar{y})$. Then inclusions* (3.42) *and* (3.43) *and their counterparts in* (iii) *hold if one of the following conditions is satisfied:*

(a) *either φ is Lipschitz continuous and the graph of G is closed around $(\bar{x},\bar{y})$, or*

(b) *$\varphi = \varphi(y)$ is l.s.c. around $\bar{y}$ and G is metrically regular around $(\bar{x},\bar{y})$.*

Proof. If φ is locally Lipschitzian around $\bar{x}$, then the SNEC and qualification conditions of the theorem hold due to Theorem 1.26 and Corollary 1.81. Note that inclusion (3.43) reduces in this case to $\partial^\infty\mu(\bar{x}) \subset D_N^*G(\bar{x},\bar{y})(0)$. Assuming (b), we immediately have $x^* = 0$ in the qualification condition of the theorem, and then $y^* = 0$ due to the condition $D^*_M G^{-1}(\bar{y}, \bar{x})(0) = \{0\}$ for the metric regularity in Theorem 1.54. Moreover, the metric regularity of G around $(\bar{x}, \bar{y})$ implies the PSNC property of G^{-1} at this point due to Proposition 1.68 and Theorem 1.49. △

When $G = g\colon X \to Y$ is single-valued, the above marginal function reduces to the composition $\varphi(x, g(x)) := (\varphi \circ g)(x)$. In this case we have the following sharpening of Theorem 3.38 that contains *subdifferential chain rules* with additional regularity and equality statements.

Theorem 3.41 (subdifferentiation of general compositions). *Let $g\colon X \to Y$ be Lipschitz continuous around $\bar{x}$, and let $\varphi\colon X \times Y \to \overline{I\!R}$ be l.s.c. around $(\bar{x}, \bar{y})$ with $\bar{y} := g(\bar{x})$. Then one has the following assertions:*

(i) *Assume that either φ is SNEC at $(\bar{x}, \bar{y})$ or g is SNC at $(\bar{y}, \bar{x})$ and that the qualification condition of Theorem* 3.38(i) *holds with $G = g$. Then the basic and singular subdifferentials of the composition $\mu = \varphi \circ g$ satisfy inclusions* (3.42) *and* (3.43), *which reduce to*

$$\partial(\varphi \circ g)(\bar{x}) \subset \bigcup_{(x^*,y^*)\in\partial\varphi(\bar{x},\bar{y})} \Big[x^* + \partial\langle y^*, g\rangle(\bar{x})\Big], \tag{3.44}$$

$$\partial^\infty(\varphi \circ g)(\bar{x}) \subset \bigcup_{(x^*,y^*)\in\partial^\infty\varphi(\bar{x},\bar{y})} \Big[x^* + \partial\langle y^*, g\rangle(\bar{x})\Big] \tag{3.45}$$

if g is strictly Lipschitzian around $\bar{x}$.

(ii) *Assume in addition to* (i) *that φ is lower regular at $(\bar{x}, \bar{y})$ and that either g is strictly differentiable at $\bar{x}$ or it is N-regular at this point with $\dim Y < \infty$. Then the equality holds in* (3.44) *and $\varphi \circ g$ is lower regular at $\bar{x}$. If in addition φ is epigraphically regular at $\bar{x}$, then the equality holds also in* (3.45) *and $\varphi \circ g$ is epigraphically regular at $\bar{x}$.*

(iii) *Let $\varphi = \varphi(y)$. Assume that either φ is SNEC at $\bar{y}$ or g^{-1} is PSNC at $(\bar{y}, \bar{x})$ and that the qualification condition of Theorem* 3.38(iii) *holds with $G = g$. Then one has the inclusions*

$$\partial(\varphi \circ g)(\bar{x}) \subset \bigcup_{y^*\in\partial\varphi(\bar{y})} D^*_N g(\bar{x})(y^*)\,,$$

$$\partial^\infty(\varphi \circ g)(\bar{x}) \subset \bigcup_{y^*\in\partial^\infty\varphi(\bar{y})} D^*_N g(\bar{x})(y^*)\,,$$

where the equalities hold under the additional assumptions of (ii).

Proof. Assertion (i) follows directly from Theorem 3.38(i) and the scalarization formula in Theorem 3.28. Note that since Y is Asplund, the strict and

w^*-strict Lipschitzian conditions for $g\colon X \to Y$ are the same due to Proposition 3.26. To prove assertion(ii), we combine the equality and regularity statements in Theorems 1.110(i) and 3.36 taking into account that g is strictly Lipschitzian around $\bar{x}$ under the assumptions made in (ii). The proof of (iii) is similar based on Theorem 3.38(iii). △

Observe that the qualification condition of Theorem 3.41(iii) reduces to

$$\partial^\infty \varphi(\bar{y}) \cap \ker \widetilde{D}^*_M g(\bar{x}) = \{0\}\,,$$

where the "reversed mixed coderivative" $\widetilde{D}^*_M$ is defined in (1.40). Since one always has $\widetilde{D}^*_M g(\bar{x})(y^*) \subset D^*_N g(\bar{x})(y^*)$, the latter qualification condition is implied by

$$\partial^\infty \varphi(\bar{y}) \cap \ker D^*_N g(\bar{x}) = \{0\}\,. \tag{3.46}$$

As a corollary of Theorem 3.41, we obtain nonsmooth extensions, in the framework of Asplund spaces, of the equality formula in Theorem 1.17 for representing basic normals to inverse images.

Corollary 3.42 (inverse images under Lipschitzian mappings). *Let $g\colon X \to Y$ be Lipschitz continuous around $\bar{x}$, and let $\Theta \subset Y$ be closed around $\bar{y} = g(\bar{x})$. Assume that either Θ is SNC at $\bar{y}$ or g^{-1} is PSNC at $(\bar{y}, \bar{x})$ and that the qualification condition*

$$N(\bar{y}; \Theta) \cap \ker \widetilde{D}^*_M g(\bar{x}) = \{0\}\,.$$

is satisfied; these hold when g is metrically regular around $\bar{x}$. Then

$$N(\bar{x}; g^{-1}(\Theta)) \subset \bigcup \Big[D^*_N g(\bar{x})(y^*) \Big|\; y^* \in N(\bar{y}; \Theta) \Big]\,,$$

where the equality is valid and $g^{-1}(\Theta)$ is normally regular at $\bar{x}$ if either g is strictly differentiable at $\bar{x}$ or it is N-regular at this point with $\dim Y < \infty$.

Proof. Putting $\varphi = \varphi(y) := \delta(y; \Theta)$, we immediately get these results from Theorem 3.41 due to the relationships of Proposition 1.79. The inclusion formula follows also from Theorem 3.4. △

The next corollary of Theorem 3.41 gives efficient chain rules for basic and singular subgradients involving only subdifferential (but not coderivative) constructions. Equality and regularity conditions are not formulated below, since they are not different from those in Theorem 3.41.

Corollary 3.43 (chain rules for basic and singular subgradients). *Let $g\colon X \to Y$ be strictly Lipschitzian at $\bar{x}$, let $\varphi\colon Y \to \overline{I\!R}$ be l.s.c. around $\bar{y} = g(\bar{x})$ and SNEC at this point, and let the qualification condition*

$$\partial^\infty \varphi(\bar{y}) \cap \ker \partial\langle \cdot, g\rangle(\bar{x}) = \{0\}$$

be satisfied. Then one has

$$\partial(\varphi \circ g)(\bar{x}) \subset \bigcup_{y^* \in \partial\varphi(\bar{y})} \partial\langle y^*, g\rangle(\bar{x}) \,,$$

$$\partial^\infty(\varphi \circ g)(\bar{x}) \subset \bigcup_{y^* \in \partial^\infty\varphi(\bar{y})} \partial\langle y^*, g\rangle(\bar{x}) \,.$$

Proof. It follows from Theorem 3.41(iii) and the scalarization formula of Theorem 3.28 for representing the qualification condition (3.46) in the given subdifferential form. It can be also derived directly from the coderivative chain rule in Theorem 3.13 with the use of scalarization. △

The chain rules obtained easily imply relationships between "full" and "partial" subgradients for functions of many variables. Given $\varphi\colon X \times Y \to \overline{I\!R}$, we denote by $\partial_x\varphi(\bar{x},\bar{y})$ and $\partial_x^\infty\varphi(\bar{x},\bar{y})$, respectively, its *basic partial subdifferential* and *singular partial subdifferential* in x at this point, i.e., the corresponding subdifferentials of the function $\varphi(\cdot,\bar{y})$ at $\bar{x}$.

Corollary 3.44 (partial subgradients). *Let $\varphi\colon X\times Y \to \overline{I\!R}$ be l.s.c. around $(\bar{x},\bar{y})$ and SNEC at this point, and let the qualification condition*

$$\big[(0,y^*) \in \partial^\infty\varphi(\bar{x},\bar{y})\big] \Longrightarrow y^* = 0$$

holds. Then one has the inclusions

$$\partial_x\varphi(\bar{x},\bar{y}) \subset \big\{x^* \in X^* \big|\ \exists y^* \in Y^* \ \ with \ \ (x^*,y^*) \in \partial\varphi(\bar{x},\bar{y})\big\} \,, \tag{3.47}$$

$$\partial_x^\infty\varphi(\bar{x},\bar{y}) \subset \big\{x^* \in X^* \big|\ \exists y^* \in Y^* \ \ with \ \ (x^*,y^*) \in \partial^\infty\varphi(\bar{x},\bar{y})\big\} \,. \tag{3.48}$$

Moreover, $\varphi(\cdot,\bar{y})$ is lower regular at $\bar{x}$ and the equality holds in (3.47) if φ is lower regular at $(\bar{x},\bar{y})$. If in addition φ is epigraphically regular at $(\bar{x},\bar{y})$, then the equality holds also in (3.48) and $\varphi(\cdot,\bar{y})$ is epigraphically regular at $\bar{x}$.

Proof. We obviously have $\varphi(x,\bar{y}) = (\varphi \circ g)(x)$, where $g\colon X \to X \times Y$ is a smooth mapping given by $g(x) := (x,\bar{y})$. Then all the results follow directly from Theorem 3.41. △

In Subsect. 1.3.4 we obtained product and quotient rules for subgradients of locally Lipschitzian functions on Banach spaces as corollaries of a chain rule.

Proposition 3.45 (refined product and quotient rules for basic subgradients). *Let $\varphi_i\colon X \to \overline{I\!R}$, $i = 1,2$, be Lipschitz continuous around $\bar{x}$. The following hold:*

(i) *One always has*

$$\partial(\varphi_1 \cdot \varphi_2)(\bar{x}) \subset \partial\big(\varphi_2(\bar{x})\varphi_1\big)(\bar{x}) + \partial\big(\varphi_1(\bar{x})\varphi_2\big)(\bar{x}) \ ,$$

where the equality holds and $\varphi_1 \cdot \varphi_2$ is lower regular at $\bar{x}$ if both functions $\varphi_2(\bar{x})\varphi_1$ and $\varphi_1(\bar{x})\varphi_2$ are lower regular at this point.

(ii) *Assume that $\varphi_2(\bar{x}) \neq 0$. Then*

$$\partial(\varphi_1/\varphi_2)(\bar{x}) \subset \frac{\partial\big(\varphi_2(\bar{x})\varphi_1\big)(\bar{x}) - \partial\big(\varphi_1(\bar{x})\varphi_2\big)(\bar{x})}{[\varphi_2(\bar{x})]^2} \ ,$$

where the equality holds and φ_1/φ_2 is lower regular at $\bar{x}$ if both functions $\varphi_2(\bar{x})\varphi_1$ and $-\varphi_1(\bar{x})\varphi_2$ are lower regular at this point.

Proof. To prove (i), we apply the Lipschitzian sum rule from Theorem 3.36 to the equality

$$\partial(\varphi_1 \cdot \varphi_2)(\bar{x}) = \partial\big(\varphi_2(\bar{x})\varphi_1 + \varphi_1(\bar{x})\varphi_2\big)(\bar{x})$$

obtained in Corollary 1.111(i). The proof of (ii) is similar involving the quotient rule of Corollary 1.111(ii). △

Next we consider *maximum functions* of the form

$$\big(\max \varphi_i\big)(x) := \max\big\{\varphi_i(x)\big|\ i = 1, \ldots, n\big\} \ ,$$

where $\varphi_i \colon X \to \overline{I\!R}$. Functions of this class are nonsmooth, and their subdifferential properties are essentially different from those for functions of the minimum type considered in Subsect. 1.3.4. In Proposition 1.113 we obtained a formula for basic subgradients of the minimum of finitely many functions in general Banach spaces. Its singular counterpart

$$\partial^\infty\big(\min \varphi_i\big)(\bar{x}) \subset \bigcup \big\{\partial^\infty \varphi_i(\bar{x})\big|\ i \in M(\bar{x})\big\}$$

is valid if X is Asplund; the proof is similar to the one in Proposition 1.113 with the use of Lemma 2.37.

The following theorem contains results for computing basic and singular subgradients of maximum functions in Asplund spaces. One can see the difference between them and the corresponding results for minimum functions. Given $\bar{x} \in X$, we define the sets

$$I(\bar{x}) := \big\{i \in \{1, \ldots, n\}\big|\ \varphi_i(\bar{x}) = \big(\max \varphi_i\big)(\bar{x})\big\} \ ,$$

$$\Lambda(\bar{x}) := \Big\{(\lambda_1, \ldots, \lambda_n)\Big|\ \lambda_i \geq 0,\ \sum_{i=1}^n \lambda_i = 1,\ \lambda_i\big(\varphi_i(\bar{x}) - \big(\max \varphi_i\big)(\bar{x})\big) = 0\Big\} \ .$$

Theorem 3.46 (subdifferentiation of maximum functions). *Let φ_i be l.s.c. around $\bar{x}$ for $i \in I(\bar{x})$ and be upper semicontinuous at $\bar{x}$ for $i \notin I(\bar{x})$. The following hold:*

(i) *Assume that the functions φ_i are SNEC at $\bar{x}$ for all but one $i \in I(\bar{x})$ and that the qualification condition* (3.38) *considered for $i \in I(\bar{x})$ is satisfied. Then one has*

$$\partial\big(\max \varphi_i\big)(\bar{x}) \subset \bigcup \Big\{ \sum_{i \in I(\bar{x})} \lambda_i \circ \partial\varphi_i(\bar{x}) \Big|\ (\lambda_1, \ldots, \lambda_n) \in \Lambda(\bar{x}) \Big\} ,$$

$$\partial^\infty\big(\max \varphi_i\big)(\bar{x}) \subset \sum_{i \in I(\bar{x})} \partial^\infty \varphi_i(\bar{x}) ,$$

where $\lambda \circ \partial\varphi(\bar{x})$ is defined as $\lambda\partial\varphi(\bar{x})$ when $\lambda > 0$ and as $\partial^\infty\varphi(\bar{x})$ when $\lambda = 0$. Moreover, the maximum function is epigraphically regular at $\bar{x}$ and both inclusions above hold as equalities if each φ_i, $i \in I(\bar{x})$, is epigraphically regular at this point.

(ii) *Assume that each φ_i is Lipschitz continuous around $\bar{x}$. Then*

$$\partial\big(\max \varphi_i\big)(\bar{x}) \subset \bigcup \Big\{ \partial\Big(\sum_{i \in I(\bar{x})} \lambda_i \varphi_i \Big)(\bar{x}) \Big|\ (\lambda_1, \ldots, \lambda_n) \in \Lambda(\bar{x}) \Big\} ,$$

where the equality holds and the maximum functions is lower regular at $\bar{x}$ if each φ_i is lower regular at this point.

Proof. Denote $\bar{\alpha} := \big(\max \varphi_i\big)(\bar{x})$ and observe that $(\bar{x}, \bar{\alpha})$ is an interior point of the set $\operatorname{epi} \varphi_i$ for any $i \notin I(\bar{x})$ due to the upper semicontinuity assumption. Then for $n = 2$ assertion (i) follows from Proposition 3.20 applied to the epigraphical multifunctions $F_i := E_{\varphi_i}$, $i = 1, 2$, and for $n > 2$ is proved by induction. It can also be derived directly from Corollary 3.37.

To prove (ii), we observe that the maximum function is represented as the composition $\varphi \circ g$ with

$$\varphi(y_1, \ldots, y_n) := \max\big\{y_1, \ldots, y_n\big\}, \quad g(x) := \big(\varphi_1(x), \ldots, \varphi_n(x)\big) .$$

Applying Corollary 3.43 to this composition and taking into account the well-known formula for subdifferentiation of the convex function g, which immediately follows from the equality in (i), we arrive at the refined inclusion in (ii). Note that

$$\sum_{i \in I(\bar{x})} \partial\big(\lambda_i \varphi_i\big)(\bar{x}) \subset \sum_{i \in I(\bar{x})} \lambda_i \partial\varphi_i(\bar{x})$$

due to Theorem 3.36 in the Lipschitz case. Since the lower regularity of a locally Lipschitzian function agrees with its epigraphical regularity, the equality/regularity statement in (ii) now follows from the one in (i). △

In conclusion of this subsection we obtain a proper extension of the classical *mean value theorem* in a general nonsmooth setting. For its formulation we involve the *two-sided* symmetric subdifferential constructions defined in (1.46). Given vectors $a, b \in X$, let us define

$$(b-a)^{\perp} := \big\{x^* \in X^* \big|\ \langle x^*, b-a\rangle = 0\big\}$$

and recall that $[a,b] := \big\{a + t(b-a) \big|\ 0 \le t \le 1\big\}$ with $[a,b]$, $[a,b)$, and $(a,b]$ defined accordingly.

Theorem 3.47 (mean values, extended). *Let $\varphi\colon X \to \overline{I\!R}$ be continuous on an open set containing $[a,b]$. Assume that for every $x \in (a,b)$ both φ and $-\varphi$ are SNEC at x (in particular, φ is SNC at this point) and that*

$$\partial^{\infty,0}\varphi(x) \cap (b-a)^{\perp} = \{0\}\ .$$

Then one has the mean value inclusion

$$\varphi(b) - \varphi(a) \in \langle \partial^0\varphi(c), b-a\rangle \ \ \textit{for some} \ \ c \in (a,b)\ . \tag{3.49}$$

Proof. It is proved in Proposition 1.115 that, for any function φ continuous on $[a,b]$, one has

$$\varphi(b) - \varphi(a) \in \partial^0_t\varphi(a + \theta(b-a)) \ \text{ with some } \ \theta \in (0,1)\ ,$$

where the set on the right-hand side stands for the symmetric subdifferential of the real function $t \to \varphi(a + t(b-a))$ at $t = \theta$. The latter function is represented as the composition

$$\varphi(a + t(b-a)) = (\varphi \circ g)(t), \quad 0 \le t \le 1\ ,$$

with a smooth mapping $g\colon [0,1] \to X$ defined by $g(t) := a + t(b-a)$. It is easy to check that the SNEC and qualification conditions imposed in the theorem ensure that all the assumptions of Corollary 3.43 are satisfied for both φ and $-\varphi$ in the composition. Applying the subdifferential chain rule from this corollary and its upper subdifferential counterpart, we arrive at the mean value inclusion (3.49) with $c := a + \theta(b-a)$. △

Finally, let us present a consequence of the above generalized mean value theorem for the case of Lipschitzian functions. In this case all the assumptions of the theorem are satisfied; moreover, we strengthen the mean value inclusion for the class of *lower regular* functions.

Corollary 3.48 (mean value theorem for Lipschitzian functions). *Let φ be Lipschitz continuous on an open set containing $[a,b]$. Then (3.49) holds. If in addition φ is lower regular on (a,b), then*

$$\varphi(b) - \varphi(a) \in \langle \partial\varphi(c), b-a\rangle \ \ \textit{for some} \ \ c \in (a,b)\ . \tag{3.50}$$

Proof. As mentioned before, the SNEC and qualification conditions automatically hold for Lipschitz continuous functions due to the results of Sect. 1.3. It remains to justify the refined mean value inclusion (3.50) under the lower regularity assumption. First we note that, by Theorem 3.41(ii), the lower regularity of φ at $c = a + \theta(b - a)$ implies the lower regularity of $t \to \varphi(a+t(b-a)) = (\varphi \circ g)(t)$ at θ. Since $\partial(\varphi \circ g)(\theta) \neq \emptyset$ due to the Lipschitz continuity of this function, its lower regularity gives $\widehat{\partial}(\varphi \circ h)(\theta) \neq \emptyset$. Hence $\widehat{\partial}^+(\varphi \circ h)(\theta) \subset \widehat{\partial}(\varphi \circ h)(\theta)$ by Proposition 1.87. In this case it follows from the proof of Proposition 1.115 that

$$\varphi(b) - \varphi(a) \in \widehat{\partial}(\varphi \circ g)(\theta) \subset \partial(\varphi \circ h)(\theta) ,$$

which implies (3.50) by Corollary 3.43. △

Note that (3.49) cannot be generally superseded by (3.50). A simple *counterexample* is provided by $\varphi(x) = -|x|$ on $[a, b] = [-1, 1]$ with $\partial\varphi(0) = \{-1, 1\}$ and $\partial^0\varphi(0) = [-1, 1]$.

3.2.2 Approximate Mean Value Theorem with Some Applications

This subsection is concerned with mean value results of a new type that are grouped around the so-called *approximate mean value theorem* for lower semicontinuous functions, which doesn't have direct analogs in the classical calculus. Based on variational arguments, we obtain an Asplund space version of the approximate mean value theorem in terms of Fréchet subgradients and derive its corollaries important for various applications, some of which are presented in this subsection. They include: characterizations of Lipschitzian behavior of l.s.c. functions in terms of Fréchet subgradients and basic subgradients, characterizations of strict Hadamard differentiability via these subgradients, subdifferential characterizations of monotonicity and constancy properties for l.s.c. functions, and relationships between the convexity of a given l.s.c. function and the monotonicity of its subdifferential mappings.

The main version of the approximate mean value theorem in Asplund spaces is as follows.

Theorem 3.49 (approximate mean values for l.s.c. functions). *Let $\varphi: X \to \overline{I\!R}$ be a proper l.s.c. function finite at two given points $a \neq b$. Consider any point $c \in [a, b)$ at which the function*

$$\psi(x) := \varphi(x) - \frac{\varphi(b) - \varphi(a)}{\|b - a\|}\|x - a\|$$

attains its minimum on $[a, b]$; such a point always exists. Then there are sequences $x_k \xrightarrow{\varphi} c$ and $x_k^* \in \widehat{\partial}\varphi(x_k)$ satisfying

$$\liminf_{k\to\infty}\langle x_k^*, b - x_k\rangle \geq \frac{\varphi(b) - \varphi(a)}{\|b - a\|}\|b - c\| , \tag{3.51}$$

$$\liminf_{k\to\infty}\langle x_k^*, b-a\rangle \ge \varphi(b)-\varphi(a)\ . \tag{3.52}$$

Moreover, when $c \neq a$ one has

$$\lim_{k\to\infty}\langle x_k^*, b-a\rangle = \varphi(b)-\varphi(a)\ .$$

Proof. The function ψ defined in the theorem is l.s.c., and hence ψ attains its minimum over $[a,b]$ at some point c. Since $\psi(a)=\psi(b)$, one can always take $c\in[a,b)$. Without loss of generality we suppose that $\varphi(a)=\varphi(b)$, i.e., $\psi(x)=\varphi(x)$ for all $x\in[a,b]$. It is easy to check that the lower semicontinuity of φ implies the existence of $r>0$ such that φ is bounded from below over the set $\Theta := [a,b]+r\mathbb{B}$ by some $\gamma\in\mathbb{R}$. Using the indicator function $\delta(\cdot;\Theta)$, we define $\vartheta(x):=\varphi(x)+\delta(x;\Theta)$, which is obviously l.s.c. on X. Then for each $k\in\mathbb{N}$ we take a real number $r_k\in(0,r)$ such that

$$\varphi(x)\ge\varphi(c)-k^{-2}\ \text{ for all }\ x\in[a,b]+r_k\mathbb{B}$$

and choose $t_k\ge k$ satisfying $\gamma+t_kr_k\ge\varphi(c)-k^{-2}$. Thus one has

$$\varphi(c)\le\inf_X\vartheta_k+k^{-2},\quad\text{where}\quad \vartheta_k(x):=\vartheta(x)+t_k\text{dist}(x;[a,b])$$

is obviously l.s.c. on X. Applying the Ekeland variational principle from Theorem 2.26(i) to this function, with the parameters $\varepsilon=k^{-2}$ and $\lambda=k^{-1}$, we find $x_k\in X$ such that

$$\|x_k-c\|\le k^{-1},\quad \vartheta_k(x_k)\le\vartheta_k(c)=\varphi(c),\quad\text{and}$$

$$\vartheta_k(x_k)\le\vartheta_k(x)+k^{-1}\|x-x_k\|\ \text{ for all }\ x\in X\ .$$

The latter means that the function $\vartheta_k(x)+k^{-1}\|x-x_k\|$ attains its minimum at $x=x_k$. Applying now Lemma 2.32(i) to this function with $\eta=\eta_k\downarrow 0$ and taking into account that $x_k\in\text{int}\,\Theta$ for large k, we find sequences $u_k\xrightarrow{\varphi}c$, $v_k\to c$, $u_k^*\in\widehat{\partial}\varphi(u_k)$, $v_k^*\in\partial\text{dist}(v_k;[a,b])$, and $e_k^*\in\mathbb{B}^*$ such that

$$\|u_k^*+t_kv_k^*+k^{-1}e_k^*\|\le\eta_k,\quad k\in\mathbb{N}\ . \tag{3.53}$$

Note that $\|v_k^*\|\le 1$ and that

$$\langle v_k^*, b-v_k\rangle\le\text{dist}(b;[a,b])-\text{dist}(v_k;[a,b])\le 0,\quad k\in\mathbb{N}\ .$$

Now we need to choose $w_k\in[a,b]$ having the same properties as v_k. Picking a projection $w_k\in\Pi(v_k;[a,b])$, we get

$$\langle v_k^*, b-w_k\rangle=\langle v_k^*, b-v_k\rangle+\langle v_k^*, v_k-w_k\rangle\le\text{dist}(b;[a,b])-\text{dist}(v_k;[a,b])$$

$$+\|v_k^*\|\cdot\|v_k-w_k\|\le-\text{dist}(v_k;[a,b])+\text{dist}(v_k;[a,b])=0\ .$$

The latter yields $\langle v_k^*, b - a\rangle \le 0$ for large $k \in N$, since $w_k \to c \neq b$ and $(x - b)\|y - b\| = (y - b)\|x - b\|$ whenever $x, y \in [a, b]$. Now using (3.53), we arrive at

$$\liminf_{k\to\infty}\langle u_k^*, b - u_k\rangle \ge 0, \quad \liminf_{k\to\infty}\langle u_k^*, b - a\rangle \ge 0\,,$$

which gives (3.51) and (3.52). Finally, let us assume that $c \neq a$. Then $v_k \neq a$ for large $k \in I\!N$, and hence $\langle v_k^*, b - c\rangle = 0$. This implies $\langle u_k^*, b - a\rangle \to 0$ by the above arguments and completes the proof of the theorem. △

It is worth mentioning that the mean value inequality (3.52) holds even in the case of $\varphi(b) = \infty$. This directly implies a useful estimate of the increment of a given function in terms of its Fréchet subgradients.

Corollary 3.50 (mean value inequality for l.s.c. functions). *Let $\varphi\colon X \to \overline{I\!R}$ be a proper l.s.c. function finite at some point $a \in X$. Then the following assertions hold:*

(i) *For any $b \in X$ there are $c \in [a, b]$ and a pair of sequences $x_k \to c$ and $x_k^* \in \widehat{\partial}\varphi(x_k)$ satisfying the mean value inequality* (3.52).

(ii) *For any $b \in X$ and $\varepsilon > 0$ one has the estimate*

$$|\varphi(b) - \varphi(a)| \le \|b - a\| \sup\Big\{\|x^*\| \;\Big|\; x^* \in \widehat{\partial}\varphi(c),\ c \in [a, b] + \varepsilon I\!B\Big\}\,.$$

Proof. To get (i), it remains to prove (3.52) when $\varphi(b) = \infty$. This follows from Theorem 3.49 applied for each $n \in I\!N$ to the sequence of functions

$$\phi_n(x) := \begin{cases} \varphi(x) & \text{if } x \neq b\,, \\ \varphi(a) + n & \text{if } x = b\,. \end{cases}$$

The estimate in (ii) follows directly from (i). △

When φ is Lipschitz continuous, we can pass to the limit in (3.52) and obtain the mean value inequality in terms of basic subgradients.

Corollary 3.51 (mean value inequality for Lipschitzian functions). *Let φ be Lipschitz continuous on an open set containing $[a, b]$. Then one has*

$$\langle x^*, b - a\rangle \ge \varphi(b) - \varphi(a) \;\; \textit{for some} \;\; x^* \in \partial\varphi(c), \quad c \in [a, b)\,.$$

Proof. By Theorem 3.49 we have a point $c \in [a, b)$ and sequences $x_k \to c$, $x_k^* \in \widehat{\partial}\varphi(x_k)$ satisfying (3.52). Since f is locally Lipschitzian, the sequence $\{x_k^*\}$ is bounded due to Proposition 1.85(ii). Remembering that X is Asplund, we select a subsequence of $\{x_k^*\}$ that converges weak* to some $x^* \in \partial\varphi(c)$. Then the result follows by passing to the limit in (3.52). △

Let us present some important applications of the approximate mean value theorem. The first application gives characterizations of the local Lipschitzian property of a l.s.c. function on Asplund spaces in terms of its Fréchet subgradients and basic subgradients.

Theorem 3.52 (subdifferential characterizations of Lipschitzian functions). *Let $\varphi\colon X \to \overline{I\!R}$ be a proper l.s.c. function finite at some point $\bar{x}$. Then the properties* (a)–(c) *involving a constant $\ell \geq 0$ are equivalent:*

(a) *There is $\gamma > 0$ such that*

$$\widehat{\partial}\varphi(x) \subset \ell I\!B^* \quad \text{whenever} \quad \|x - \bar{x}\| < \gamma, \quad |\varphi(x) - \varphi(\bar{x})| < \gamma \ .$$

(b) *There is a neighborhood U of $\bar{x}$ such that $\widehat{\partial}\varphi(x) \subset \ell I\!B^*$ for all $x \in U$.*

(c) *φ is Lipschitz continuous around $\bar{x}$ with modulus ℓ.*

Moreover, the local Lipschitz continuity of φ around $\bar{x}$ with some modulus $\ell \geq 0$ is equivalent to the following:

(d) *φ is SNEC at $\bar{x}$ with $\partial^\infty\varphi(\bar{x}) = \{0\}$.*

Proof. Without loss of generality we assume for simplicity that $\bar{x} = 0$ and $\varphi(0) = 0$. First prove that (a)$\Rightarrow$(b). To establish (b) with $U := \eta(\text{int}\, I\!B)$, it is suffices to show that there is $\eta > 0$ such that $|\varphi(x)| < \gamma$ whenever $\|x\| < \eta$. It immediately follows from the lower semicontinuity of φ at $\bar{x} = 0$ that there is $\nu > 0$ so small that $\varphi(x) > -\gamma$ if $\|x\| < \nu$. To justify (b) with $\eta := \min\{\nu, \gamma, \gamma/\ell\}$, we need to prove that $\varphi(x) < \gamma$ whenever $\|x\| < \min\{\gamma, \gamma/\ell\}$.

Suppose that the latter is not true, i.e., there is $b \in X$ satisfying $\|b\| < \min\{\gamma, \gamma/\ell\}$ and $\varphi(b) \geq \gamma$. Consider the l.s.c. function $\phi\colon X \to \overline{I\!R}$ defined by

$$\phi(x) := \min\{\varphi(x), \gamma\} \quad \text{with} \quad \phi(0) = 0, \quad \phi(b) = \gamma \ .$$

Applying to this function the mean value inequality (3.52) from Theorem 3.49 on the interval $[0, b]$, we find a point $c \in [0, b)$ and a pair of sequences $x_k \xrightarrow{\phi} c$, $x_k^* \in \widehat{\partial}\phi(x_k)$ satisfying

$$\liminf_{k\to\infty} \langle x_k^*, b\rangle \geq \phi(b) - \phi(0) = \gamma, \quad \text{hence} \quad \liminf_{k\to\infty} \|x_k^*\| \geq \gamma/\|b\| > \ell \ .$$

Recall that the chosen point c in Theorem 3.49 minimizes the function

$$\psi(x) := \phi(x) - \|b\|^{-1}\|x\|\big(\phi(b) - \phi(0)\big) \quad \text{over} \quad [0, b] \ ,$$

which implies that $\phi(c) \leq \gamma\,\|b\|^{-1}\|c\| < \gamma$. Thus $\phi(x_k) < \gamma$ along the sequence $x_k \xrightarrow{\phi} c$, and one has $\phi(x_k) = \varphi(x_k)$ for all k sufficiently large. It easily follows from the definitions that

$$\widehat{\partial}\phi(x_k) \subset \widehat{\partial}\varphi(x_k) \quad \text{due to} \quad \phi(x) \leq \varphi(x), \quad x \in X \ .$$

and hence $x_k^* \in \widehat{\partial}\varphi(x_k)$ for large k. Since $\|x_k^*\| > \ell$, this contradicts (a) and thus proves (a)$\Rightarrow$(b).

Implication (b)⇒(c) follows from the estimate in Corollary 3.50(ii), implication (c)⇒(b) is established in Proposition 1.85(ii), and implication (b)⇒(a) is trivial. It remains to prove that the local Lipschitz continuity of φ around $\bar{x}$ is equivalent to (d). In fact, we know from Chap. 1 that the local Lipschitzian property of φ implies both conditions in (d) in any Banach spaces; see Theorem 1.26 and Corollary 1.81. Now let us prove the converse implication in the Asplund space setting.

Let (d) hold. Due to the equivalence (a)⇔(c), it suffices to show that (a) is satisfied with some positive numbers ℓ and γ. Assuming the contrary, we find sequences $x_k \xrightarrow{\varphi} \bar{x}$ and $x_k^* \in \widehat{\partial}\varphi(x_k)$ with $\|x_k^*\| \to \infty$ as $k \to \infty$. Then

$$\Big(\frac{x_k^*}{\|x_k^*\|}, -\frac{1}{\|x_k^*\|}\Big) \in \widehat{N}((x_k, \varphi(x_k)); \operatorname{epi}\varphi), \quad k \in I\!N \ .$$

Putting $\widetilde{x}_k^* := x_k^*/\|x_k^*\|$ and taking into account that X is Asplund, we select a subsequence of $\{\widetilde{x}_k^*\}$ that converges weak* to some x^* with $(x^*, 0) \in N((\bar{x}, \varphi(\bar{x})); \operatorname{epi}\varphi)$. Thus $x^* \in \partial^\infty\varphi(\bar{x})$, and one gets $x^* = 0$ due to the second property in (d). Now the SNEC property of φ at $\bar{x}$ implies that $\|\widetilde{x}_k^*\| \to 0$, a contradiction. This shows that φ must be locally Lipschitzian around $\bar{x}$ with some modulus ℓ, which completes the proof of the theorem. △

The result obtained easily implies the following generalization of the fundamental fact in classical analysis ensuring that a function whose derivative is always zero must be constant. Recall that this fact is a direct corollary of the classical mean value theorem and bridges the *gap between differentiation and integration.*

Corollary 3.53 (subgradient characterization of constancy for l.s.c. functions). *Let $\varphi\colon X \to \overline{I\!R}$ be a proper l.s.c. function, and let $U \subset X$ be open. Then φ is locally constant on U if and only if*

$$x^* \in \widehat{\partial}\varphi(x) \Longrightarrow x^* = 0 \ \textit{ for all } \ x \in U \ .$$

The latter is equivalent to φ being constant on U if U is connected.

Proof. This follows from Theorem 3.52 for $\ell = 0$. △

As the next application of the approximate mean value theorem, we characterize the notion of strict differentiability in the sense of Hadamard for real-valued functions on Asplund spaces. The following characterizations involve Fréchet and basic subgradients showing, in particular, that the class of functions strictly Hadamard differentiable at a given point corresponds to the class of locally Lipschitzian functions whose basic subdifferential is a singleton.

Recall that a function $\varphi\colon X \to \overline{I\!R}$ is *strictly Hadamard differentiable* at $\bar{x}$, with the strict Hadamard derivative x^* denoted by $\nabla\varphi(\bar{x})$ if there is no confusion, provided that

$$\lim_{\substack{x \to \bar{x} \\ t \downarrow 0}} \left[\sup_{v \in C} \left| \frac{\varphi(x+tv)-\varphi(x)}{t} - \langle x^*, v \rangle \right| \right] = 0 \tag{3.54}$$

for any compact subset $C \subset X$. Clearly, every function strictly differentiable at $\bar{x}$ in the Fréchet sense (i.e., in the sense of Definition 1.13) is strictly Hadamard differentiable at $\bar{x}$, but not vice versa. In finite dimensions these notions obviously coincide.

Theorem 3.54 (subgradient characterizations of strict Hadamard differentiability). *Let $\varphi\colon X \to \overline{I\!R}$ be finite at $\bar{x}$. The following properties involving a functional $\xi \in X^*$ are equivalent:*

(a) *φ is Lipschitz continuous around $\bar{x}$, and for every sequences $x_k \to \bar{x}$ and $x_k^* \in \widehat{\partial}\varphi(x_k)$ one has $x_k^* \xrightarrow{w^*} \xi$.*

(b) *φ is Lipschitz continuous around $\bar{x}$ with $\partial\varphi(\bar{x}) = \{\xi\}$.*

(c) *φ is strictly Hadamard differentiable at $\bar{x}$ with $\nabla\varphi(\bar{x}) = \xi$.*

Proof. Without loss of generality we consider the case of $\bar{x} = 0$, $\varphi(0) = 0$, and $\xi = 0$ in the theorem. To prove (a)$\Rightarrow$(b), we pick an arbitrary $x^* \in \partial\varphi(0)$ and by Theorem 2.34 find sequences $x_k \to 0$ and $x_k^* \in \widehat{\partial}\varphi(x_k)$ with $x_k^* \xrightarrow{w^*} x^*$ as $k \to \infty$. By (a) one has $x^* = 0$, i.e., $\partial\varphi(0) = \{0\}$ and (b) holds.

Let us prove (b)$\Rightarrow$(c) arguing by contradiction. Assume that there is a compact subset $C \subset X$ for which the limit in (3.54) either doesn't exist or is different from zero. In both cases we can select subsequences (without relabeling) of $x_k \to 0$, $t_k \downarrow 0$, and $v_k \in C$ for which

$$\lim_{k \to \infty} \frac{\varphi(x_k + t_k v_k) - \varphi(x_k)}{t_k} := \alpha > 0 \, ;$$

this takes into account that the above ratio is bounded due to the Lipschitz continuity of φ. Now using Corollary 3.50(i), we find sequences $c_k \in X$ and $x_k^* \in \widehat{\partial}\varphi(c_k)$ satisfying

$$\text{dist}(c_k; [x_k, x_k + t_k v_k]) \le k^{-1}, \quad \langle x_k^*, t_k v_k \rangle \ge \varphi(x_k + t_k v_k) - \varphi(x_k) - t_k k^{-1} \, .$$

The first of the above relations implies that $c_k \to 0$. Since C is compact, there is a subsequence of $\{v_k\}$ converging to some $v \in C$. Also we have a subsequence of $\{x_k^*\}$ that converges weak* to some $x^* \in \partial\varphi(0)$; this is due to boundedness of $x_k^* \in \widehat{\partial}\varphi(c_k)$ and the Asplund property of X. Passing to the limit along these subsequences in the above relations, one has

$$\begin{aligned} \|x^*\| \cdot \|v\| \ge \langle x^*, v \rangle &= \lim_{k \to \infty} \langle x_k^*, v_k \rangle \\ &\ge \lim_{k \to \infty} \frac{\varphi(x_k + t_k v_k) - \varphi(x_k)}{t_k} := \alpha > 0 \, , \end{aligned}$$

which yields $x^* \neq 0$ and contradicts (b).

It remains to show that (c)$\Rightarrow$(a). Let $U \subset X^*$ be an arbitrary weak* neighborhood of $\xi = 0$. By shrinking U if necessary we may assume that it has the form $U = \{x^* \in X^* |\ \langle x^*, v_j \rangle < 1,\ j = 1, \ldots, n\}$ for some finite subset $v_1, \ldots, v_n$ of X with $r := \max\{\|v_1\|, \ldots, \|v_n\|\}$. Using property (c), we find $\eta > 0$ so small that

$$\big[\varphi(x + tv_j) - \varphi(x)\big]/t < 1/2 \ \text{ for all } \ j = 1, \ldots, n$$

whenever $x \in \eta I\!B$ and $0 < t < \eta$. Now picking any $x^* \in \widehat{\partial}\varphi(x)$ with some $x \in \eta I\!B$, we get from (1.51) that

$$\langle x^*, u - x \rangle \le \varphi(u) - \varphi(x) + \|u - x\|/(2r) \ \text{ for all } \ u \ \text{ near } \ x\,.$$

Putting there $u = x + tv_j$, $j = 1, \ldots, n$, one has

$$\langle x^*, v_j \rangle \le \frac{\varphi(x + tv_j) - \varphi(x) + t\|v_j\|/(2r)}{t} < \frac{1}{2} + \frac{r}{2r} = 1$$

for all $t > 0$ sufficiently small. Thus $x^* \in U$ and $\widehat{\partial}\varphi(x) \subset U$ for all x sufficiently close to the origin. This implies, by Theorem 3.52, the Lipschitz continuity of φ around $\bar{x} = 0$ and also the sequential condition in (a). $\triangle$

Next we consider an application of the approximate mean value theorem to a subgradient generalization of the classical fact that a function whose derivative is nonpositive must itself be nonincreasing.

Theorem 3.55 (subgradient characterization of monotonicity for l.s.c. functions). *Let $U \subset X$ be an open convex set on which a proper l.s.c. function φ is defined, and let $K \subset X$ be a cone with the dual/polar cone $K^* := \{x^* \in X^* |\ \langle x^*, x \rangle \le 0\}$. The following properties are equivalent:*

(a) *The function φ is K-nonincreasing, i.e.,*

$$x, u \in U,\ u - x \in K \Longrightarrow \varphi(u) \le \varphi(x)\,.$$

(b) *For every $x \in U$ one has $\widehat{\partial}\varphi(x) \subset K^*$.*

Proof. To prove (a)$\Rightarrow$(b), we take any $x \in U$ and any $x^* \in \widehat{\partial}\varphi(x)$. Then for any $\gamma >$ we find $\eta > 0$ such that

$$\langle x^*, u - x \rangle \le \varphi(u) - \varphi(x) + \gamma\, \|u - x\| \ \text{ whenever } \ u \in x + \eta I\!B\,.$$

Fix $v \in K$ and put $u = x + tv$ with $t > 0$ in this inequality. The monotonicity property in (a) implies that

$$\langle x^*, v \rangle \le \frac{\varphi(x + tv) - \varphi(x)}{t} + \gamma - \|v\| \le 0\,,$$

which therefore justifies (b).

To prove the opposite implication (b)$\Rightarrow$(a), we suppose the contrary and thus find two points $x, u \in U$ satisfying $u - x \in K$ with $\varphi(u) > \varphi(x)$. Applying Corollary 3.50(i), one gets a point $c \in [x, u]$ and a pair of sequences $x_k \to c$ and $x_k^* \in \widehat{\partial}\varphi(x_k)$ satisfying

$$\liminf_{k\to\infty}\langle x_k^*, u - x\rangle \geq \varphi(u) - \varphi(x) > 0\ .$$

Thus for large k we have $\langle x_k^*, u - x\rangle > 0$, which contradicts (b). $\triangle$

Taking $K = X$ in Theorem 3.55, we arrive at the subgradient characterization of constancy obtained above in Corollary 3.53.

Our last application in this subsection establishes the equivalence between the convexity of a l.s.c. function on an Asplund space and the monotonicity of its subdifferential mappings generated by both Fréchet and basic subgradients. Recall that a set-valued mapping $F\colon X \rightrightarrows X^*$ between a Banach space and its dual in *monotone* if

$$\langle x^* - u^*, x - u\rangle \geq 0 \ \text{ for any } \ x, u \in X \ \text{ and } \ x^* \in F(x), u^* \in F(u)\ .$$

Theorem 3.56 (subdifferential monotonicity and convexity of l.s.c. functions). *Let $\varphi\colon X \to \overline{I\!R}$ be proper and l.s.c. on X. Then each of the subdifferential mappings $\widehat{\partial}\varphi\colon X \rightrightarrows X^*$ and $\partial\varphi\colon X \rightrightarrows X^*$ is monotone if and only if φ is convex.*

Proof. If φ is convex, then both subdifferential mappings $\widehat{\partial}\varphi$ and $\partial\varphi$ reduce to the subdifferential mapping of convex analysis, which is well known to be monotone. Also, it follows from the representation of $\partial\varphi$ in Theorem 2.34 that the monotonicity of $\widehat{\partial}\varphi$ in Asplund spaces implies the monotonicity of $\partial\varphi$. Thus it remains to prove that if $\widehat{\partial}\varphi$ is monotone, then φ must be convex.

First let us show that

$$\widehat{\partial}\varphi(x) = \big\{x^* \in X^* \big|\ \langle x^*, u - x\rangle \leq \varphi(u) - \varphi(x) \ \text{ for all } \ u \in X\big\} \tag{3.55}$$

if $\widehat{\partial}\varphi$ is monotone and $x, u \in \operatorname{dom}\varphi$. The inclusion "$\supset$" in (3.55) is obvious. To prove the opposite inclusion, we consider $x, u \in \operatorname{dom}\varphi$, $x^* \in \widehat{\partial}\varphi(x)$ and use inequality (3.51) from Theorem 3.49. It gives sequences $x_k \to c \in [u, x)$ and $x_k^* \in \widehat{\partial}\varphi(x_k)$ such that

$$\varphi(x) - \varphi(u) \leq \frac{\|x - u\|}{\|x - c\|}\liminf_{k\to\infty}\langle x_k^*, x - x_k\rangle\ .$$

Then the monotonicity of the subdifferential mapping $\widehat{\partial}\varphi$ and the equality $\|x - u\|(x - c) = (x - u)\|x - c\|$ imply that

$$\varphi(x) - \varphi(u) \leq \frac{\|x - u\|}{\|x - c\|}\liminf_{k\to\infty}\langle x^*, x - x_k\rangle = \langle x^*, x - u\rangle\ ,$$

which justifies the inclusion "$\subset$" in (3.55) and hence the equality therein.

Now using (3.55), we prove that φ is convex. Take arbitrary $u, x \in \operatorname{dom}\varphi$ and consider its convex combination $v := \lambda u + (1-\lambda)x$ with $0 < \lambda < 1$. By Theorem 2.29 the domain of $\widehat{\partial}\varphi$ is dense in the graph of φ. Hence there is a sequence $u_k \xrightarrow{\varphi} u$ with $\widehat{\partial}\varphi(u_k) \neq \emptyset$. Without loss of generality we suppose that $0 \in \widehat{\partial}\varphi(u_k)$. Put $v_k := \lambda u_k + (1-\lambda)x$ and show that $v_k \in \operatorname{dom}\varphi$ for any fixed k. Assuming the contrary, we take $\alpha > \varphi(x)$ and define the function

$$\psi(z) := \begin{cases} \varphi(z) & \text{if } z \neq v_k\,, \\ \alpha & \text{if } z = v_k\,. \end{cases}$$

Applying Theorem 3.49 to this function, we get $c \in [x, v_k)$ and a pair of sequences $z_n \to c$ and $z_n^* \in \widehat{\partial}\psi(z_n)$ such that

$$\liminf_{n\to\infty}\langle z_n^*, v_k - z_n\rangle \geq \frac{\|v_k - c\|}{\|v_k - x\|}\big(\alpha - \varphi(x)\big) > 0,$$

$$\liminf_{n\to\infty}\langle z_n^*, v_k - x\rangle \geq \alpha - \varphi(x)\,.$$

It follows from the monotonicity of $\widehat{\partial}\varphi$ and the choice of $0 \in \widehat{\partial}\varphi(u_k)$ that

$$\begin{aligned} 0 &\geq \liminf_{n\to\infty}\langle z_n^*, u_k - z_n\rangle \geq \liminf_{n\to\infty}\langle z_n^*, v_k - z_n\rangle + \liminf_{n\to\infty}\langle z_n^*, u_k - v_k\rangle \\ &= \liminf_{n\to\infty}\langle z_n^*, v_k - z_n\rangle + \lambda^{-1}(1-\lambda)\liminf_{n\to\infty}\langle z_n^*, v_k - x\rangle \\ &\geq \lambda^{-1}(1-\lambda)\big(\alpha - \varphi(x)\big)\,, \end{aligned}$$

which contradicts $\alpha > \varphi(x)$. Thus $v_k \in \operatorname{dom}\varphi$ for all $k \in I\!N$. To justify the convexity of φ, we consider the following two cases:

(i) Assume that v_k is not a local minimizer for φ. Then choose $\tilde{v}_k$ so that $\|\tilde{v}_k - v_k\| < k^{-1}$ and $\varphi(\tilde{v}_k) < \varphi(v_k)$. Fix k and apply Theorem 3.49 to the function φ on the interval $[\tilde{v}_k, v_k]$. In this way we find $c_k \in [\tilde{v}_k, v_k)$ and a pair of sequences $z_n \to c_k$ as $n \to \infty$ and $z_n^* \in \widehat{\partial}\varphi(z_n)$ satisfying

$$\liminf_{n\to\infty}\langle z_n^*, v_k - z_n\rangle \geq \frac{\|v_k - c_k\|}{\|v_k - \tilde{v}_k\|}\big(\varphi(v_k) - \varphi(\tilde{v}_k)\big) > 0, \quad n \in I\!N\,.$$

This implies by (3.55) that

$$\varphi(x) - \varphi(z_n) \geq \langle z_n^*, x - z_n\rangle, \quad \varphi(u_k) - \varphi(z_n) \geq \langle z_n^*, u_k - z_n\rangle\,.$$

Involving the lower semicontinuity of φ, we therefore have

$$\lambda\varphi(u_k) + (1-\lambda)\varphi(x) \geq \liminf_{n\to\infty}\big[\varphi(z_n) + \langle z_n^*, v_k - z_n\rangle\big] \geq \varphi(c_k)$$

for all $k \in I\!N$. Passing to the limit as $k \to \infty$, one has

$$\lambda\varphi(u) + (1-\lambda)\varphi(x) \ge \varphi(v) = \varphi(\lambda u + (1-\lambda)x) . \tag{3.56}$$

(ii) Let now v_k be a local minimizer for φ. Then $0 \in \widehat{\partial}\varphi(v_k)$, and by (3.55) we get $\varphi(x) \ge \varphi(v_k)$ and $\varphi(u_k) \ge \varphi(v_k)$, which implies $\lambda\varphi(u_k) + (1-\lambda)\varphi(x) \ge \varphi(v_k)$. Passing to the limit as $k \to \infty$ in this case, we again arrive at (3.56) and complete the proof of the theorem. △

3.2.3 Connections with Other Subdifferentials

In Subsect. 2.5.2A we described the constructions of Clarke's generalized gradient/subdifferential and normal cone as well as various modifications of Ioffe's "approximate" normals and subgradients in arbitrary Banach spaces. Now we establish precise relationships between them and our basic normal and subgradient constructions in the framework of Asplund spaces. Let us start with the Clarke normal cone $N_C(\bar{x};\Omega)$ and subdifferential $\partial_C\varphi(\bar{x})$ defined in (2.72) and (2.73), respectively. Recall that the space X in question is supposed to be Asplund unless otherwise stated, and that cl^* stands for the weak* topological closure of a set in X^*.

Theorem 3.57 (relationships with Clarke normals and subgradients). *The following assertions hold:*

(i) *Let $\Omega \subset X$ be locally closed around $\bar{x} \in \Omega$. Then*

$$N_C(\bar{x};\Omega) = \mathrm{cl}^*\mathrm{co}\, N(\bar{x};\Omega) .$$

(ii) *Let $\varphi\colon X \to \overline{I\!R}$ be proper and l.s.c. around $\bar{x} \in \mathrm{dom}\,\varphi$. Then*

$$\partial_C\varphi(\bar{x}) = \mathrm{cl}^*\left[\mathrm{co}\,\partial\varphi(\bar{x}) + \mathrm{co}\,\partial^\infty\varphi(\bar{x})\right] = \mathrm{cl}^*\mathrm{co}\left[\partial\varphi(\bar{x}) + \partial^\infty\varphi(\bar{x})\right] . \tag{3.57}$$

If, in particular, φ is Lipschitz continuous around $\bar{x}$, then

$$\partial_C\varphi(\bar{x}) = \mathrm{cl}^*\mathrm{co}\,\partial\varphi(\bar{x}) . \tag{3.58}$$

Proof. According to the four-step procedure in the definition of Clarke's constructions described in Subsect. 2.5.2A, we begin with proving (3.58) and first establish the representations

$$\begin{aligned}\varphi^\circ(\bar{x};h) &= \max\left\{\langle x^*,h\rangle \,\middle|\, x^* \in \mathrm{cl}^*\partial\varphi(\bar{x})\right\}\\ &= \sup\left\{\langle x^*,h\rangle \,\middle|\, x^* \in \partial\varphi(\bar{x})\right\}\end{aligned} \tag{3.59}$$

for the generalized directional derivative (2.69) of a locally Lipschitzian function. Indeed, by definition of $\varphi^\circ(\bar{x};h)$ for each $h \in X$ one has sequences $x_k \to \bar{x}$ and $t_k \downarrow 0$ such that

$$\frac{\varphi(x_k + t_k h) - \varphi(x_k)}{t_k} \to \varphi^\circ(\bar{x};h) \quad \text{as} \quad k \to \infty .$$

Applying Theorem 3.49 to φ on the interval $[x_k, x_k + t_k h]$ for each k, we find $v_n \to c_k \in [x_k, x_k + t_k h)$ as $n \to \infty$ and $v_n^* \in \widehat{\partial}\varphi(v_n)$ with

$$\varphi(x_k + t_k h) - \varphi(x_k) \le t_k \liminf_{n\to\infty} \langle v_n^*, h\rangle, \quad k \in \mathbb{N} .$$

Passing to the limit first as $n \to \infty$ and then as $k \to \infty$, we get (3.59), which implies (3.58) due to definition (2.70) of Clarke's generalized gradient for locally Lipschitzian functions. Next we apply (3.58) to the distance function $\operatorname{dist}(\cdot;\Omega)$ for a closed set $\Omega \subset X$ and obtain

$$\bigcup_{\lambda>0} \lambda \partial_C \operatorname{dist}(\bar{x};\Omega) = \bigcup_{\lambda>0} \lambda \Big[\operatorname{cl}^* \operatorname{co} \partial \operatorname{dist}(\bar{x};\Omega)\Big] \subset \operatorname{cl}^* \operatorname{co} \Big[\bigcup_{\lambda>0} \lambda \partial \operatorname{dist}(\bar{x};\Omega)\Big] .$$

This gives $N_C(\bar{x};\Omega) \subset \operatorname{cl}^* \operatorname{co} N(\bar{x};\Omega)$ due to definition (2.72) of the Clarke normal cone and Theorem 1.97 on calculating basic normals via basic subgradients of the distance function. The opposite inclusion in (i) follows from $N(\bar{x};\Omega) \subset N_C(\bar{x};\Omega)$ and the fact that Clarke's normal cone is convex and closed in the weak* topology of X^*; see Subsect. 2.5.2A.

It remains to prove representation (3.57) for l.s.c. functions. Since $\partial^\infty\varphi(\bar{x})$ is a cone, one always has

$$\begin{aligned} &\operatorname{co}\big[\partial\varphi(\bar{x}) + \partial^\infty\varphi(\bar{x})\big] \\ &\quad = \operatorname{co}\partial\varphi(\bar{x}) + \operatorname{co}\partial^\infty\varphi(\bar{x}) ; \end{aligned}$$

thus it sufficient to justify the first equality in (3.57). Picking an arbitrary subgradient $x^* \in \partial_C\varphi(\bar{x})$ and using its definition (2.73) together with the above representation (i) of the Clarke normal cone, we find a net $x_\nu^* \xrightarrow{w^*} x^*$ satisfying $(x_\nu^*, -1) \in \operatorname{co} N((\bar{x},\varphi(\bar{x})); \operatorname{epi}\varphi)$ for all ν. Fix ν and find $p(\nu) \in \mathbb{N}$, $\alpha_{j\nu} \ge 0$, $x_{j\nu}^* \in X^*$, and $\lambda_{j\nu} \in \mathbb{R}$, $j = 1,\ldots,p(\nu)$, such that

$$(x_\nu^*, -1) = \sum_{j=1}^{p(\nu)} \alpha_{j\nu}(x_{j\nu}^*, -\lambda_{j\nu}),$$

$$(x_{j\nu}^*, -\lambda_{j\nu}) \in N((\bar{x},\varphi(\bar{x})); \operatorname{epi}\varphi), \quad \sum_{j=1}^{p(\nu)} \alpha_{j\nu} = 1 .$$

By Proposition 1.76 one has $\lambda_{j\nu} \ge 0$; so

$$x_{j\nu}^* \in \begin{cases} \lambda_{j\nu}\partial\varphi(\bar{x}) & \text{if } \lambda_{j\nu} > 0 , \\ \partial^\infty\varphi(\bar{x}) & \text{if } \lambda_{j\nu} = 0 . \end{cases}$$

This provides the representation $x_{j\nu}^* = \lambda_{j\nu} v_{j\nu}^* + u_{j\nu}^*$ with $v_{j\nu}^* \in \partial\varphi(\bar{x})$ and $u_{j\nu}^* \in \partial^\infty\varphi(\bar{x})$, where $u_{j\nu}^* = 0$ if $\lambda_{j\nu} > 0$. Observing that $\sum_{j=1}^{p(\nu)} \alpha_{j\nu}\lambda_{j\nu} = 1$ for each ν, we get

$$x_\nu^* = \sum_{j=1}^{p(\nu)} \alpha_{j\nu}\big(\lambda_{j\nu} v_{j\nu}^* + u_{j\nu}^*\big) \subset \operatorname{co}\partial\varphi(\bar{x}) + \operatorname{co}\partial^\infty\varphi(\bar{x}) ,$$

which proves the inclusion "$\subset$" in (3.57) by passing to the limit with respect to ν. To prove the opposite inclusion, take any $x^* \in \operatorname{cl}^*\big[\operatorname{co}\partial\varphi(\bar{x}) + \operatorname{co}\partial^\infty\varphi(\bar{x})\big]$ and find a bet $x_\nu^* \xrightarrow{w^*} x^*$ satisfying

$$x_\nu^* = \sum_{j=1}^{p(\nu)} \alpha_{j\nu} v_{j\nu}^* + \sum_{j=1}^{q(\nu)} \beta_{j\nu} u_{j\nu}^* \ \text{ with } \ \sum_{j=1}^{p(\nu)} \alpha_{j\nu} = 1, \quad \sum_{j=1}^{q(\nu)} \beta_{j\nu} = 1 ,$$

$p(\nu), q(\nu) \in \mathbb{N}$, $\alpha_{j\nu} \ge 0$, $\beta_{j\nu} \ge 0$, $v_{j\nu}^* \in \partial\varphi(\bar{x})$, and $u_{j\nu}^* \in \partial^\infty\varphi(\bar{x})$ for all ν. Due to the convexity of N_C we have

$$(x_\nu^*, -1) = \sum_{j=1}^{p(\nu)} \alpha_{j\nu}(v_{j\nu}^*, -1) + \sum_{j=1}^{q(\nu)} \beta_{j\nu}(u_{j\nu}^*, 0) \in N_C((\bar{x}, \varphi(\bar{x})); \operatorname{epi}\varphi) .$$

By (2.73) this yields $x^* \in \partial_C\varphi(\bar{x})$, since N_C is weak* closed. △

Next let us establish relationships between our basic normals and subgradients and the corresponding "approximate" constructions described in Subsect. 2.5.2B. First observe that due to the fuzzy sum rule from Theorem 2.33 every Asplund space is a "weakly trustworthy" space in the sense of Ioffe [593]. Hence the A-subdifferential (2.75) of any l.s.c. function on an Asplund space admits the simplified representation

$$\partial_A\varphi(\bar{x}) = \overline{\mathop{\operatorname{Lim\,sup}}_{\substack{x \xrightarrow{\varphi} \bar{x} \\ \varepsilon \downarrow 0}}} \partial_\varepsilon^- \varphi(x) \tag{3.60}$$

in terms of the *topological* Painlevé-Kuratowski upper limit of ε-Dini subgradients defined in Subsect. 2.5.2B. Along with (3.60) and the associated G-normal cone N_G, the G-subdifferential ∂_G, and their nuclei $\widetilde{N}$ and $\widetilde{\partial}_G$ described in (2.76) and (2.77), we consider the corresponding *sequential* constructions defined by

$$\partial_A^\sigma\varphi(\bar{x}) := \mathop{\operatorname{Lim\,sup}}_{\substack{x \xrightarrow{\varphi} \bar{x} \\ \varepsilon \downarrow 0}} \partial_\varepsilon^- \varphi(x), \qquad \widetilde{N}_G^\sigma(\bar{x}; \Omega) := \bigcup_{\lambda > 0} \lambda \partial_A^\sigma \operatorname{dist}(\bar{x}; \Omega) ,$$

$$\widetilde{\partial}_G^\sigma\varphi(\bar{x}) := \big\{x^* \in X^* \big|\ (x^*, -1) \in \widetilde{N}_G^\sigma((\bar{x}, \varphi(\bar{x})); \operatorname{epi}\varphi)\big\} .$$

In what follows we establish relationships between all these constructions and our basic (sequential) normal cone N and subdifferential ∂ in Asplund spaces.

Recall that a Banach space X is *weakly compactly generated* (WCG) if there is a weakly compact set $K \subset X$ such that $X = \operatorname{cl}(\operatorname{span} K)$. Canonical examples of WCG spaces are reflexive spaces that are weakly compactly generated by their balls. Every separable Banach space is also WCG, even norm compactly generated: take $K := \{k^{-1}x_k,\ k \in I\!N\} \cup \{0\}$, where $\{x_k\}$ is a dense sequence in the unit sphere of X. On the other hand, there are many Banach and Asplund spaces that are not WCG. We refer the reader to the books by Diestel [332] and Fabian [416] for various results, examples, and discussions on WCG spaces. Let us mention the following fundamental characterization of WCG spaces known in the literature as an *interpolation theorem* (see, e.g., [416, Theorem 1.2.3] with a nice and relatively simple proof): a Banach space X is WCG if and only if there is a *reflexive* space Y and an injective continuous linear operator $A\colon Y \to X$ with the dense range. Note that subspaces of WCG Banach spaces may not be themselves WCG, which is not however the case for WCG Asplund spaces. Moreover, the WCG property substantially narrows the class of Asplund spaces; it implies, in particular, the *existence of a Fréchet differentiable renorm.*

The next lemma describes connections between weak* topological and sequential limits that are important for establishing relationships between the normal cones and subdifferentials under consideration.

Lemma 3.58 (weak* topological and sequential limits). *Let X be a Banach space, and let $\{S_k\}$ be a sequence of bounded subset of X^* with $S_{k+1} \subset S_k$ for each $k \in I\!N$. The following assertions hold:*

(i) *If the closed unit ball of X^* is weak* sequentially compact, then*

$$\bigcap_{k=1}^{\infty} \mathrm{cl}^* S_k = \mathrm{cl}^* \Big\{ \lim_{k\to\infty} x_k^* \Big|\ x_k^* \in S_k \ \textit{ for all }\ k \in I\!N \Big\} .$$

(ii) *If X is a subspace of a WCG Banach space, then*

$$\bigcap_{k=1}^{\infty} \mathrm{cl}^* S_k = \Big\{ \lim_{k\to\infty} x_k^* \Big|\ x_k^* \in S_k \ \textit{ for all }\ k \in I\!N \Big\} .$$

Proof. To justify (i), we prove the inclusion "$\subset$" therein; the opposite one is obvious. Let x^* belong to the left-hand set in (i), and let W be the weak* closure of a weak* neighborhood of x^*. Then one can find $x_k^* \in W \cap S_k$ for each $k \in I\!N$. Since $I\!B_{X^*}$ is weak* sequentially compact and the sets S_k are uniformly bounded, there is a subsequence $x_{k_j}^*$, $j \in I\!N$, that converges weak* to some $z^* \in W$. Let $z_k^* := x_{k_j}^*$ for $k_{j-1} < k \le k_j$. Then $z_k^* \in S_k$ for all $k \in I\!N$, and the sequence $\{z_k^*\}$ converges weak* to z^*. Thus z^* belongs to the right-hand set in (i), which proves this assertion.

The proof of (ii) is more involved. First recall a deep and well-known fact that $I\!B_{X^*}$ is weak* sequentially compact if X is a subset of a WCG space; see, e.g., the afore-mentioned books [332, 416]. Hence the WCG assumption of (ii) ensures the equality in (i), and it remains to prove furthermore that "cl*" can be omitted on the right-hand side. To furnish this, we invoke the following two fundamental results of functional analysis:

(a) the mentioned interpolation theorem that allows us to reduce, in a sense, WCG spaces to reflexive ones, and

(b) the so-called *Whitney's construction* ensuring that every point from the weak closure of a bounded subset S of a normed space can be realized as the weak limit of a *sequence* from S; see Holmes [580, pp. 147–149], where this construction is used in the proof of the classical Eberlein-Šmulian theorem on the equivalence between weak compactness and weak sequential compactness in Banach spaces.

Let X be a subspace of a WCG Banach space Z. By the above interpolation theorem there is a reflexive space Y and an injective linear continuous operator $A: Y \to Z$ whose range is dense in Z. Let R denote the restriction mapping from Z^* onto X^* constructed via the Hahn-Banach theorem. Without loss of generality we suppose that $S_1 \subset I\!B_{X^*}$ and put

$$H_k := R^{-1}(S_k) \cap I\!B_{Z^*}, \quad K := \bigcap_{k=1}^{\infty} \operatorname{cl}^w A^* H_k ,$$

where cl^w stands for the weak closurein the reflexive space Y^*. Since the set K is bounded, it is weakly compact in Y^*. Picking an arbitrary x^* from the left-hand side set in (ii), we observe that the sets $V_k := R^{-1}x^* \cap \operatorname{cl}^* H_k$, $k \in I\!N$, are nonempty, weak* compact, and nested in Z^*. Thus there is $z^* \in \cap_{k=1}^{\infty} V_k$.

By Whitney's construction discussed in (b) we choose a sequence $z^*_{k,j} \in H_k$ such that $A^* z^*_{k,j}$ converges weakly to $A^* z^*$ as $j \to \infty$ for each $k \in I\!N$. Since the set $\{(A^* z^*, A^* z^*_{k,j}) |\ j, k \in I\!N\}$ is weakly compact and separable, it is weakly metrizable. Hence there are $j_k \in I\!N$ such that the sequence $A^* z^*_{k,j_k}$ converges weakly to $A^* z^*$ as $k \to \infty$. Taking into account that A^* is weak*-to-weak homeomorphism on $I\!B_{Z^*}$, one has that z^*_{k,j_k} converges weak* to z^*, and so $R z^*_{k,j_k}$ converges weak* ro $R z^* = x^*$. Since $R z^*_{k,j_k} \in S_k$ for all k, it follows that x^* belongs to the left-hand set in (ii). $\triangle$

The following theorem establishes relationships between our basic constructions and the various modifications of Ioffe's "approximate" normals and subgradients in Asplund spaces. It consists of three assertions involving relationships with A-subgradients, G-normals, and G-subgradients, respectively, in the sequence of their definition.

Theorem 3.59 (relationships with "approximate" normals and subgradients). *The following assertions hold:*

(i) *Let $\varphi: X \to \overline{I\!R}$ be l.s.c. around $\bar{x} \in \operatorname{dom} \varphi$. Then*

$$\partial \varphi(\bar{x}) \subset \partial^{\sigma}_{A} \varphi(\bar{x}) \subset \partial_A \varphi(\bar{x}) .$$

If in addition φ is Lipschitz continuous around $\bar{x}$, then

$$\text{cl}^*\partial\varphi(\bar{x}) = \text{cl}^*\partial_A^\sigma\varphi(\bar{x}) = \partial_A\varphi(\bar{x}) \; . \tag{3.61}$$

If in the latter case X is WCG, then the sets $\partial\varphi(\bar{x})$ and $\partial_A^\sigma\varphi(\bar{x})$ are weak closed, and one has*

$$\partial\varphi(\bar{x}) = \partial_A^\sigma\varphi(\bar{x}) = \partial_A\varphi(\bar{x}) \; . \tag{3.62}$$

(ii) *Let $\Omega \subset X$ be closed around $\bar{x} \in \Omega$. Then*

$$N(\bar{x};\Omega) \subset \widetilde{N}_G^\sigma(\bar{x};\Omega) \subset \widetilde{N}_G(\bar{x};\Omega) \subset N_G(\bar{x};\Omega) = \text{cl}^* N(\bar{x};\Omega) \; .$$

If in addition X is a WCG space, then

$$N(\bar{x};\Omega) = \widetilde{N}_G^\sigma(\bar{x};\Omega) = \widetilde{N}_G(\bar{x};\Omega) \; .$$

(iii) *If φ be l.s.c. around $\bar{x}$, then*

$$\partial\varphi(\bar{x}) \subset \widetilde{\partial}_G^\sigma\varphi(\bar{x}) \subset \widetilde{\partial}_G\varphi(\bar{x}) \subset \partial_G\varphi(\bar{x}) = \text{cl}^*\partial\varphi(\bar{x}) \; .$$

If in addition φ is Lipschitz continuous around $\bar{x}$ and X is WCG, then

$$\partial\varphi(\bar{x}) = \widetilde{\partial}_G^\sigma\varphi(\bar{x}) = \widetilde{\partial}_G\varphi(\bar{x}) = \partial_G\varphi(\bar{x}) \; . \tag{3.63}$$

Proof. It is easy to check that $\widehat{\partial}\varphi(x) \subset \partial_\varepsilon^-\varphi(x)$ for every $x \in \text{dom}\,\varphi$ and every $\varepsilon \ge 0$. Hence the inclusions in (i) follow from Theorem 2.34 and the definitions. To prove (3.61) when φ is Lipschitz continuous around $\bar{x}$, we observe based on the definitions that

$$\partial_A\varphi(\bar{x}) = \bigcap_{k=1}^\infty \text{cl}^* S_k, \quad \partial_A^\sigma\varphi(\bar{x}) = \bigcap_{k=1}^\infty \big\{ \lim_{k\to\infty} x_k^* \in S_k \;\; \text{for all} \;\; k \in I\!\!N \big\} \; ,$$

where $S_k := \bigcup\{\partial_{1/k}^-\varphi(x) |\; \|x - \bar{x}\| \le 1/k\}$. Obviously $S_{k+1} \subset S_k$ for each $k \in I\!\!N$. Moreover, all the sets S_k are bounded in X^* due to the Lipschitz continuity of φ around $\bar{x}$. Hence $\partial_A\varphi(\bar{x}) = \text{cl}^*\partial_A^\sigma\varphi(\bar{x})$, and it remains to justify $\partial_A^\sigma\varphi(\bar{x}) \subset \text{cl}^*\partial\varphi(\bar{x})$ in (3.61), which means that

$$\partial_A^\sigma\varphi(\bar{x}) \subset \partial\varphi(\bar{x}) + V$$

for any weak* neighborhood V of the origin in X^*. To verify the latter inclusion, we observe that for every neighborhood V under consideration there are a finite-dimensional subspace $L \subset X$ and a number $r > 0$ such that $L^\perp + 3r I\!\!B^* \subset V$ with the annihilator $L^\perp$ of L. $x^* \in \partial_A^\sigma\varphi(\bar{x})$ and find sequences $\varepsilon_k \downarrow 0$, $x_k \to \bar{x}$, and $x_k^* \stackrel{w^*}{\to} x^*$ with $x_k^* \in \partial_{\varepsilon_k}^-\varphi(x_k)$. Let k to be so large that $0 \le \varepsilon_k \le r$ and $1/k \le r$. Using the definition of Dini ε-subgradients from Subsect. 2.5.2B, one can easily conclude that for every $k \in I\!\!N$, $r > 0$, and finite-dimensional subspace $L \subset X$ the function

$$\psi_k(x) := \varphi(x) - \langle x_k^*, x - x_k\rangle + 2r\|x - x_k\| + \delta(x - x_k; L)$$

attains a local minimum at x_k; thus $0 \in \widehat{\partial}\psi(x_k)$. Theorem 2.33 implies due to the structure of ψ_k that

$$x_k^* \in \widehat{\partial}\varphi(u_k) + 3r I\!B^* + L^\perp \subset \widehat{\partial}\varphi(u_k) + V \quad \text{with some} \quad u_k \in x_k + \tfrac{1}{k} I\!B \, .$$

Passing there to the limit as $k \to \infty$ and taking into account that all the sets $\widehat{\partial}\varphi(u_k)$ belong to a weak* sequential compact ball in X^*, we complete the proof of (3.61). If in addition X is WCG, the same procedure gives (3.62) due to Lemma 3.58(ii).

The normal cone relationships in (ii) follow from the corresponding relationships in (i) due to the definitions of the G-normal constructions under consideration and Theorem 1.97.

To establish (iii), we only need checking that $\partial_G\varphi(\bar{x}) = \mathrm{cl}^*\partial\varphi(\bar{x})$ if φ is l.s.c. around $\bar{x}$; the other statements immediately follow from (i), (ii), and the definitions. Observe that

$$L \cap \mathrm{cl}^* N((\bar{x}, \varphi(\bar{x})); \mathrm{epi}\,\varphi) = \mathrm{cl}^*\big(L \cap N((\bar{x}, \varphi(\bar{x})); \mathrm{epi}\,\varphi)\big)$$

with $L := X^* \times \{-1\}$. This implies the mentioned equality in (iii) due to $N_G(\bar{x}; \Omega) = \mathrm{cl}^* N(\bar{x}; \Omega)$ in (ii) and completes the proof of the theorem. △

It follows from Example 1.7 and Theorem 3.59(ii) that there is a closed subset Ω of the Hilbert space ℓ^2 for which the basic normal cone $N(0; \Omega)$ is *strictly smaller* than the G-normal cone $N_G(0; \Omega)$. Indeed, in that example $N(0; \Omega)$ is not norm closed (and hence not weak closed) in ℓ^2, so $N(0; \Omega) \neq N_G(0; \Omega) = \mathrm{cl}^w N(0, \Omega)$. On the other hand, the basic subdifferential $\partial\varphi(\bar{x})$ is *weak* closed* for every locally Lipschitzian function on an arbitrary *WCG Banach space*. This follows directly from assertion (iii) of Theorem 3.59 when X is additionally assumed to be Asplund. To establish this fact in the general case of Banach spaces, one needs to use representation (1.55) of the basic subdifferential and proceed similarly to the proof of the corresponding part of Theorem 3.59(i).

We actually have the following more general fact on *robustness/graph-closeness* of the basic normal cone and subdifferential under SNC/CEL assumptions. We present this fact in the Asplund space setting; see the discussion after the proof on its counterpart in the case of Banach spaces.

Theorem 3.60 (robustness of basic normals). *Let X be a WCG Asplund space, and let $\Omega \subset X$ be its closed subset that is SNC at $\bar{x}$. Then the graph of $N(\cdot; \Omega)$ is closed near $\bar{x}$, i.e., there is $\gamma > 0$ such that the set*

$$\big(\mathrm{gph}\, N(\cdot; \Omega)\big) \cap \big((\bar{x} + \gamma I\!B) \times X^*\big)$$

is closed in the norm×weak topology of $X \times X^*$.*

Proof. The first step is to show that, for any given $\eta > 0$ and a compact set $C \subset X$, the cone

$$K(\eta;C) := \{x^* \in X^* |\ \eta\,\|x^*\| \le \max_{c\in C}\langle x^*, c\rangle\}$$

is both weak* closed and weak* locally bounded in X^*. The latter means that every point of $K(\eta;C)$ lies in a weak* open set $U \subset X^*$ such $U \cap K(\eta;C)$ is norm bounded in X^*.

The following observation will be used twice: if $\nu \in (0,\eta)$ is given, then there is a finite collection $c_1,\ldots,c_n$ in C such that

$$K(\eta;C) \subset K(\nu;c_1,\ldots,c_n)\ .$$

To prove this, consider an open covering given by $\{c + (\eta-\nu)I\!B|\ c \in C\}$. Extracting a finite subcover by the compactness of C, we find points $c_1,\ldots,c_n$ in C that ensure the inclusion

$$C \subset \bigcup_{i=1}^{n}\big(c_k + (\eta-\nu)I\!B\big)\ .$$

One therefore has

$$\eta\,\|x^*\| \le \max_{c\in C}\langle x^*, x\rangle \le \max_{i=1,\ldots,n}\langle x^*, c_k\rangle + (\eta-\nu)\|x^*\|$$

whenever $x^* \in K(\eta;C)$. Thus we arrive at the required inequality

$$\eta\,\|x^*\| \le \max_{i=1,\ldots,n}\langle x^*, c_k\rangle \quad\text{for all}\quad x^* \in K(\eta;C)\ .$$

Let us prove that the cone $K(\eta;C)$ is weak* closed. When $C = \{c\}$ is a singleton, it follows directly from the lower semicontinuity of the norm function $\|\cdot\|$ and the continuity of the linear function $\langle\cdot,c\rangle$ in the weak* topology of X^*. Thus $K(\eta;C)$ is weak* closed whenever $C = \{c_1,\ldots,c_n\}$ is a finite set, since in this case $K(\eta;C)$ is just a finite union of weak* closed sets. To prove the weak* closedness of $K(\eta;C)$ in the general case of a compact set C, suppose that $x^* \notin K(\eta;C)$ and then show that $x^* \notin \text{cl}^*K(\eta;C)$. Assume without loss of generality that $\|x^*\| = 1$ and denote $\rho := \max_{c\in C}\langle x^*, c\rangle$; this gives $\rho < \eta$ by assumption. Choose a number $\sigma \in (0,\eta)$ so small that $\rho+\sigma < \eta$. Applying the above observation, we find a finite collection of points $c_1,\ldots,c_n$ in C such that

$$K(\eta;C) \subset K(\eta-\sigma;c_1,\ldots,c_n)\ .$$

Since $K(\eta-\sigma;c_1,\ldots,c_n)$ is proved to be weak* closed, it must contain $\text{cl}^*K(\eta;C)$. On the other hand,

$$\max_{i\in 1,\ldots,n}\langle x^*, c_i\rangle \le \max_{c\in C}\langle x^*, c\rangle = \rho < \eta-\sigma = (\eta-\sigma)\|x^*\|\ ,$$

and so $x^* \notin K(\eta-\sigma;c_1,\ldots,c_n)$. Thus $x^* \notin \text{cl}^*K(\eta;C)$, which justifies the weak* closedness of $K(\eta;C)$.

Let us next show that $K(\eta; C)$ is weak* locally bounded. Fix $\widehat{x}^* \in K(\eta; C)$ and select a finite number of points in C such that

$$K(\eta; C) \subset K(\eta/2; c_1, \ldots, c_n) .$$

The given point $\widehat{x}^*$ certainly belongs to the set

$$U := \big\{x^* \in X^* \big|\ \langle x^*, c_i\rangle < 1 + \langle \widehat{x}^*, c_i\rangle,\ i = 1, \ldots, n\big\} ,$$

which is weak* open in X^*. Furthermore, every point

$$x^* \in U \cap K(\eta; C) \subset U \cap K(\eta/2; c_1, \ldots, c_n)$$

satisfies the inequalities

$$(\eta/2)\|x^*\| \le \max_{i\in 1,\ldots,n} \langle x^*, c_i\rangle < 1 + \max_{i\in 1,\ldots,n} \langle \widehat{x}^*, c_i\rangle .$$

This obviously yields the weak* local boundedness of $K(\eta; C)$.

It is proved in Theorem 1.26, assuming that C is CEL around $\bar{x}$, that there exist a compact set $C \subset X$ and positive constants η, ν such that

$$\widehat{N}(x; \Omega) \subset K(\eta; C) \quad \text{whenever} \quad x \in \Omega \cap (\bar{x} + \nu I\!B) ;$$

see (1.20) with $\varepsilon = 0$. As discussed in Remark 1.27(ii), the SNC and CEL properties are *equivalent* in the framework of WCG Asplund spaces. To complete the proof of the theorem, it therefore remains to establish the following statement with $(M, d) = \big(\Omega \cap (\bar{x} + \gamma I\!B), \|\cdot\|_X\big)$ and $F(\cdot) = \widehat{N}(\cdot; \Omega)$ in the notation above.

Claim. *Let $F\colon M \rightrightarrows X^*$ be a set-valued mapping between a metric space (M, d) and the topological dual space to a WCG Banach space X. Equip $M \times X^*$ with the $d\times$weak* topology and assume that there is a weak* closed and weak* locally bounded set $K \subset X^*$ such that*

$$F(x) \subset K \quad \textit{for all} \quad x \in M .$$

Then $(\bar{x}, x^*) \in \operatorname{cl}\operatorname{gph} F$ if and only if $x^* = \lim_{k\to\infty} x_k^*$ for some sequence $x_k^* \in F(x_k)$ with $x_k \to \bar{x}$ as $k \to \infty$.

To justify this claim, we consider a net $\{(x_\alpha, x_\alpha^*)\}_{\alpha\in\mathcal{A}} \subset M \times X^*$ such that $x_\alpha \to \bar{x}$ and $x_\alpha^* \xrightarrow{w^*} x^*$ with $x_\alpha^* \in F(x_\alpha)$ for all $\alpha \in \mathcal{A}$. The weak* closedness of K and the assumption $F(x) \subset K$ ensures that $x^* \in K$. Now taking into account the weak* boundedness of K, we find a natural number m and a subnet $\{(x_\beta, x_\beta^*)\}_{\beta\in\mathcal{B}}$, $\mathcal{B} \subset \mathcal{A}$, of $\{(x_\alpha, x_\alpha^*)\}$ such that $\|x_\alpha^*\| \le m$ for all $\beta \in \mathcal{B}$. It is easy to deduce from Lemma 3.58(ii) by the boundedness of weak* convergent sequences that for any sequence of subsets $S_k \subset X^*$ with $S_{k+1} \subset S_k$ in the dual space to a WCG Banach space X one has

$$\bigcup_{m=1}^{\infty}\bigcap_{k=1}^{\infty}\mathrm{cl}^*\big(S_k\cap m\mathbb{B}^*\big)=\Big\{\lim_{k\to\infty}x_k^*\,\Big|\;x^*\in S_k\;\text{ for all }\;k\in\mathbb{N}\Big\}\,,$$

where $\lim x_k^*$ is taken in the weak* topology of X^*. Now considering the sequence of sets

$$S_k:=\bigcup\big\{F(x)\big|\;d(x,\bar{x})\le 1/k\big\},\quad k\in\mathbb{N}\,,$$

observe that x^* belongs to the left-hand side of the latter equality. Thus we conclude that x^* lies in the set on the right-hand side therein. This completes the proof of the claim and of the whole theorem. △

It follows from the proof of Theorem 3.60 that the robustness property of the basic normal cone $N(\cdot;\Omega)$ holds true for locally closed sets Ω in any *WCG Banach* space provided that Ω is *CEL* around $\bar{x}$. To see this, we appeal to the definition of basic normals as sequential limits of ε-counterparts and to formula (1.20) for ε-normals to CEL sets valid in arbitrary Banach spaces. Note that one cannot generally replace the CEL property by the weaker SNC property of closed sets in the case of non-Asplund WCG spaces.

Combining the results in Theorems 3.59 and 3.60, we have the equalities

$$N(\bar{x};\Omega)=N_G(\bar{x};\Omega)\quad\big(=\widetilde{N}_G^{\sigma}(\bar{x};\Omega)=\widetilde{N}_G(\bar{x};\Omega)\big)$$

for SNC sets if X is a *WCG Asplund space*. Note that the CEL and SNC properties of Ω are *not necessary* for the local closedness of $\mathrm{gph}\,N(\cdot;\Omega)$. This graph-closedness holds, in particular, when $\Omega\subset X$ is a *singleton*, which is never SNC unless X is finite-dimensional; see Theorem 1.21.

Observe further that the mentioned graph-closedness of $N(\cdot;\Omega)$ near $\bar{x}$ automatically implies the local *graph-closedness of the basic subdifferential* $\partial\varphi$ in the norm×weak* topology of $X\times X^*$ provided that φ is continuous around $\bar{x}$ (or, more generally, *subdifferentially continuous* in the sense of Rockafellar and Wets [1165, Definition 13.28]). However, the graph-closedness of $\partial\varphi$ in this topology may be violated even for proper *lower semicontinuous convex* functions on separable Hilbert spaces as demonstrated in Borwein, Fitzpatrick and Girgensohn [144].

The next example shows that the *WCG requirement* imposed in Theorem 3.59 is essential for the weak* closedness of $\partial\varphi(\bar{x})$ and the validity of

$$\partial\varphi(\bar{x})=\partial_G\varphi(\bar{x})\quad\big(=\widetilde{\partial}_G\varphi(\bar{x})=\partial_A\varphi(\bar{x})\big)$$

even in the case of locally Lipschitzian functions on Asplund spaces admitting an equivalent C^∞-smooth norm.

Example 3.61 (nonclosedness of the basic subdifferential for Lipschitz continuous functions). *There are an Asplund space X admitting a C^∞-smooth renorm, a concave continuous function $\varphi\colon X\to\mathbb{R}$, and a point $\bar{x}\in X$ such that $\partial\varphi(\bar{x})$ is not weak* closed in X^*, and one has*

$$\partial\varphi(\bar{x}) \neq \partial_G\varphi(\bar{x}) = \widetilde{\partial}_G\varphi(\bar{x}) = \partial_A\varphi(\bar{x}) .$$

Proof. Consider the space $X := \mathcal{C}[0,\omega_1]$ of all functions φ continuous on the "long" interval $[0,\omega_1]$, where ω_1 is the first uncountable ordinal. The norm $\|\cdot\|$ on X is the supremum/maximum norm. It is well known that X is an Asplund space admitting an equivalent $\mathcal{C}^\infty$-smooth norm; see [331, Chap. VII] for more details and references. Define $\varphi(x) := -\|x\|$ for $x \in \mathcal{C}[0,\omega_1]$ and observe that this function is concave and continuous (hence Lipschitzian) on X. Involving Theorem 2.34 and Proposition 1.87, we conclude that

$$\partial\varphi(\bar{x}) = \operatorname*{Lim\,sup}_{x\to\bar{x}} \{\nabla\varphi(x)\}, \quad \partial_G\varphi(\bar{x}) = \widetilde{\partial}_G\varphi(\bar{x}) = \partial_A\varphi(\bar{x}) = \overline{\operatorname*{Lim\,sup}_{x\to\bar{x}}}\{\nabla\varphi(x)\}$$

in terms of Fréchet derivatives. According to Example I.1.6(b) of the mentioned book of Deville et al., the norm $\|\cdot\|$ is Fréchet differentiable at $x \in \mathcal{C}[0,\omega_1]$ if and only if there is an isolated point $\omega \in [0,\omega_1]$ (i.e., not a limit ordinal) such that $|x(\omega)| > |x(t)|$ whenever $t \neq \omega$. In this case the derivative of $\|\cdot\|$ at x is μ_ω, the point mass (Dirac measure) at ω. Take $\bar{x} \equiv 1$ and consider the perturbed functions

$$x_\nu^\omega(t) := \begin{cases} 1+\nu & \text{if } t = \omega , \\ 1 & \text{otherwise} , \end{cases}$$

where $\nu \to 0$ and where ω is any nonlimit ordinal. One clearly has $x_\nu^\omega \in \mathcal{C}[0,\omega_1]$ and $\|x_\nu^\omega - \bar{x}\| \to 0$ as $\nu \to 0$. Therefore

$$\partial\varphi(\bar{x}) = \big\{ -\mu_\omega \big|\ \omega < \omega_1 \big\} \neq \partial_G\varphi(\bar{x}) = \big\{ -\mu_\omega \big|\ \omega \in [0,\omega_1] \big\} ,$$

because ω_1 is not the limit of a sequence of countable ordinals while other $\omega \in [0,\omega_1]$ are limits of sequences of nonlimit ordinals. △

Let us emphasize that our sequential variational analysis and its applications in this book do *not generally require robustness/closedness* properties of the basic normal cone and subdifferential.

3.2.4 Graphical Regularity of Lipschitzian Mappings

This subsection contains applications of some results on subdifferential calculus and coderivative scalarization to the study of normal vectors to graphical sets and graphical regularity of Lipschitzian mappings. We prove, in particular, the subspace property of Clarke's normal cone to Lipschitzian graphs in infinite dimensions and establish relationships between graphical regularity and special kinds of differentiability for Lipschitzian mappings. The new notions of "weak differentiability" and "strict-weak differentiability" defined below may be weaker than even the classical Gâteaux differentiability for mappings into infinite-dimensional spaces.

Let us start with the subspace property of the convexified normal cone. Given $\Omega \subset X$ in a Banach space, we consider the basic normal cone $N(\bar{x};\Omega)$ to Ω at $\bar{x}$ and define its *w^*-closed convexification* by

$$\overline{N}(\bar{x};\Omega) := \mathrm{cl}^*\mathrm{co}\, N(\bar{x};\Omega), \quad \bar{x} \in \Omega\ . \tag{3.64}$$

By Theorem 3.57 the convexified normal cone (3.64) reduces to the Clarke normal cone (2.72) if Ω is locally closed around $\bar{x}$ and X is Asplund. The next theorem establishes the equivalence between the subspace property of $\overline{N}(\cdot;\Omega)$ to graphs of strictly Lipschitzian mappings $f\colon X \to Y$ and the Asplund property of the domain space X.

Theorem 3.62 (subspace property of the convexified normal cone). *Let X and Y be Banach spaces. The following properties are equivalent:*

(a) *The convexified normal cone $\overline{N}((\bar{x}, f(\bar{x}));\mathrm{gph}\, f)$ is a linear subspace of $X^* \times Y^*$ for every mapping $f\colon X \to Y$ that is w^*-strictly Lipschitzian at some point $\bar{x} \in X$.*

(b) *The space X is Asplund.*

Proof. Let us first justify (b)⇒(a) using the scalarization formula of Theorem 3.28, relationship (3.58) between basic and Clarke subgradients of locally Lipschitzian functions, and the symmetric property (2.71) of the latter construction. In this way we take any $(x^*, -y^*) \in N((\bar{x}, f(\bar{x}));\mathrm{gph}\, f)$ and get

$$x^* \in D^*_N f(\bar{x})(y^*) \subset \partial\langle y^*, f\rangle(\bar{x}) \subset \partial_C\langle y^*, f\rangle(\bar{x}) = -\partial_C\langle -y^*, f\rangle(\bar{x})$$

$$= -\mathrm{cl}^*\mathrm{co}\, \partial\langle -y^*, f\rangle(\bar{x}) \subset -\mathrm{cl}^*\mathrm{co}\, D^*_N f(\bar{x})(y^*)\ .$$

This therefore gives

$$-N((\bar{x}, f(\bar{x}));\mathrm{gph}\, f) \subset \mathrm{cl}^*\mathrm{co}\, N((\bar{x}, f(\bar{x}));\mathrm{gph}\, f)$$

and shows that the convexified cone $\overline{N}((\bar{x}, f(\bar{x}));\mathrm{gph}\, f)$ is actually a linear subspace of $X^* \times Y^*$.

To prove (a)⇒(b), let us consider an arbitrary convex function ψ on X continuous around $\bar{x} \in X$. Given Y, we represent it as $Y = I\!R \times Y_1$, where Y_1 is a subspace of Y, and define a Lipschitzian mapping $f\colon X \to Y$ by $f(x) := (\psi(x), 0)$. Then f is obviously strictly Lipschitzian at $\bar{x}$, and hence $\overline{N}((\bar{x}, f(\bar{x}));\mathrm{gph}\, f)$ is a linear subspace of $X^* \times Y^*$. Since

$$\mathrm{gph}\, f = \mathrm{gph}\, \psi \times \{0\} \quad \text{and} \quad N((\bar{x}, f(\bar{x}));\mathrm{gph}\, f) = N((\bar{x}, \psi(\bar{x}));\mathrm{gph}\, \psi) \times Y_1^*\ ,$$

it follows that $\overline{N}((\bar{x}, \psi(\bar{x}));\mathrm{gph}\, \psi)$ is a subspace of $X^* \times I\!R$. Due to the convexity and continuity of ψ we have $\partial\psi(\bar{x}) \neq \emptyset$ and

$$N((\bar{x}, \psi(\bar{x}));\mathrm{gph}\, \psi) = \big\{(x^*, -\lambda)\big|\ x^* \in \partial(\lambda\psi)(\bar{x}),\ \lambda \in I\!R\big\}$$

(the latter holds for any locally Lipschitzian function). Thus $\partial(-\psi)(\bar{x}) \neq \emptyset$; otherwise we get a contradiction with the subspace property of $\overline{N}((\bar{x}, \psi(\bar{x}));\operatorname{gph}\psi)$. Since ψ was chosen arbitrary, one has $\partial\varphi(\bar{x}) \neq \emptyset$ for any *concave* continuous function φ at every $\bar{x}$. Due to the limiting representation (1.55) of the basic subdifferential this ensures that the set $\{x \in X|\ \widehat{\partial}_\varepsilon\varphi(x) \neq \emptyset\}$ is dense in X, which implies the Asplund property of X by Proposition 2.18. △

Next we are going to establish relationships between graphical regularity and differentiability of Lipschitzian mappings acting in Banach spaces. Aside from finite dimensions, this requires new notions of differentiability that may be different from the classical differentiability and strict differentiability of mappings relative to some bornology. To proceed, we first define these notions with respect to an arbitrary bornology β discussed in Remark 2.11; actually the three main bornologies are used in what follows: Fréchet ($\beta = \mathcal{F}$), Hadamard ($\beta = \mathcal{H}$), and Gâteaux ($\beta = \mathcal{G}$).

Given a bornology β on X, we recall that a mapping $f\colon X \to Y$ is *strictly β-differentiable* at $\bar{x}$ if there is a bounded linear operator $A\colon X \to Y$ such that

$$\lim_{\substack{x \to \bar{x}\\ t \downarrow 0}} \left\| \frac{f(x+tv) - f(x)}{t} - Av \right\| = 0 \ \text{ for all } \ v \in X , \tag{3.65}$$

where the convergence is uniform relatively to v in each set belonging to β. When $x = \bar{x}$ in (3.65), f is said to be *β-differentiable* at $\bar{x}$. Prior in this book we mostly consider differentiability and strict differentiability in the sense of Fréchet; see nevertheless Theorem 3.54 involving strict differentiability in the sense of Hadamard. To simplify notation, we use the same symbol $\nabla f(\bar{x}) := A$ for all the derivatives under consideration if no confusion arises.

Definition 3.63 (weak and strict-weak differentiability). *Let $f\colon X \to Y$ be a mapping between Banach spaces, and let β be a bornology on X. Then:*

(i) *f is* STRICTLY-WEAKLY β-DIFFERENTIABLE *(abbr. $sw\beta$-differentiable) at $\bar{x}$ if the scalarized function $\langle y^*, f\rangle$ is strictly β-differentiable at $\bar{x}$ for all $y^* \in Y^*$. We say that f admits an $sw\beta$*-DERIVATIVE *at $\bar{x}$ if there is a bounded linear operator $A\colon X \to Y$ such that*

$$\lim_{\substack{x \to \bar{x}\\ t \downarrow 0}} \left\langle y^*, \frac{f(x+tv) - f(x)}{t} - Av \right\rangle = 0 \ \textit{ for all } \ v \in X,\ y^* \in Y^* , \tag{3.66}$$

where the convergence is uniform relatively to v in each set belonging to β.

(ii) *f is* WEAKLY β-DIFFERENTIABLE *at $\bar{x}$ (abbr. $w\beta$-differentiable) at $\bar{x}$ if $\langle y^*, f\rangle$ is β-differentiable at $\bar{x}$ for all $y^* \in Y^*$. If (3.66) holds with $x = \bar{x}$, the operator A is called the $w\beta$*-DERIVATIVE *of f at $\bar{x}$.*

The terminology comes from the fact that the *weak* convergence on Y is used in (3.66) instead of the norm convergence in (3.65). Observe that $w\beta$-derivatives and $sw\beta$-derivatives are unique when exist, but that the $w\beta$-differentiability and $sw\beta$-differentiability of f at $\bar{x}$ don't automatically imply the existence of the corresponding derivatives. One can check directly from the definitions that there is surely *no gap* between the above differentiability and the existence of derivatives in the following two cases:

(a) Y is reflexive and f is Lipschitz continuous at $\bar{x}$.

(b) f is *weakly* directionally differentiable at $\bar{x}$, i.e., the limit

$$\lim_{t\downarrow 0}\Big\langle y^*, \frac{f(\bar{x}+tv)-f(\bar{x})}{t}\Big\rangle$$

exists for all $y^*\in Y^*$, $v\in X$; in particular, f is Gâteaux differentiable at $\bar{x}$.

The corresponding differentiability notions in (3.65) and Definition 3.63 obviously agree if $\dim Y<\infty$. The following example shows that it is *no longer the case in infinite dimensions*: a Lipschitzian mapping may be strictly-weakly differentiable with respect to the strongest Fréchet bornology but not even Gâteaux differentiable!

Example 3.64 (weak Fréchet differentiability versus Gâteaux differentiability). *There is a Lipschitz continuous mapping $f\colon I\!R\to\ell^2$ that is strictly weakly Fréchet differentiable at $\bar{x}=0$ but doesn't admit the classical Gâteaux derivative at this point.*

Proof. Let $\varphi\colon I\!R\to I\!R$ be a $\mathcal{C}^\infty$-smooth function such that $\varphi\neq const$, $\operatorname{supp}\varphi\subset(0,1)$, and both φ and $\nabla\varphi$ are bounded by some $\alpha>0$. Consider a complete orthonormal basis $\{e_1,e_2,\ldots\}$ in the Hilbert space ℓ^2 and define the function

$$f(x):=\sum_{k=1}^\infty\varphi_k(x)e_k\ \text{ with }\ \varphi_k(x):=\frac{\varphi(2^kx-1)}{2^k},\quad x\in I\!R\,.$$

For each $k,j\in I\!N$ with $k\neq j$ one has $(\operatorname{supp}\varphi_k)\cap(\operatorname{supp}\varphi_j)=\emptyset$. Thus for every $x\in I\!R$ we get $\varphi_k(x)\neq 0$ for at most one $k\in I\!N$. This implies the Lipschitz continuity of f on $I\!R$. Define now

$$\psi(x):=\langle y^*,f\rangle(x)=\sum_{k=1}^\infty y_k\varphi_k(x),\quad y^*\in\ell^2\,,$$

where $y_k\in I\!R$ are uniquely determined by the representation $y^*=\sum y_ke_k$. Then one has the relations

$$|\psi(x_1)-\psi(x_2)|=|y_{k_1}\varphi_{k_1}(x_1)-y_{k_2}\varphi_{k_2}(x_2)|\le\big(|y_{k_1}|+|y_{k_2}|\big)\alpha|x_1-x_2|\,,$$

where $k_i\ge\log_2\eta^{-1}$ if $|x_i|<\eta$, $i=1,2$. This yields $\psi(x_1)-\psi(x_2)=o(|x_1-x_2|)$ as $x_1,x_2\to 0$, which proves the strict weak Fréchet differentiability of f at $\bar{x}=0$. If we assume that f isGâteaux differentiable at this point, then clearly $\nabla f(0)=0$ for the Gâteaux derivative. Since $\varphi\neq const$, we find $x_0\in(0,1)$ with $\varphi(x_0)\neq 0$ and put $x_k:=2^{-k}x_0+2^{-k}$. Then $x_k\to 0$ as $k\to\infty$ and

$$\frac{\|f(x_k) - f(0)\|}{x_k} = \frac{\|\varphi_k(x_k)e_k\|}{x_k} = \frac{|\varphi(x_0)|}{x_0 + 1} \quad \text{for all } k \in I\!N ,$$

which contradicts the Gâteaux differentiability of f at $\bar{x} = 0$. △

Although the differentiability properties from Definition 3.63 may be weaker than the classical notions in (3.65), they still imply a linear rate of continuity (Lipschitzian behavior) of mappings in the case of Hadamard and stronger bornologies.

Proposition 3.65 (Lipschitzian properties of weakly differentiable mappings). *The following hold for $\beta \geq \mathcal{H}$:*

(i) *If f is $w\beta$-differentiable at $\bar{x}$, then there are a neighborhood U of $\bar{x}$ and a constant $\ell > 0$ such that $\|f(x) - f(\bar{x})\| \leq \ell\|x - \bar{x}\|$ for all $x \in U$.*

(ii) *If f is strictly $w\beta$-differentiable at $\bar{x}$, then it is Lipschitz continuous around $\bar{x}$.*

Proof. It is sufficient to justify (i) for $\beta = \mathcal{H}$; the proof of (ii) is similar. Assume that the conclusion of (i) doesn't hold. Then there are x_k such that

$$\|x_k - \bar{x}\| \leq k^{-1} \quad \text{and} \quad \|f(x_k) - f(\bar{x})\| > k\|x_k - \bar{x}\| \quad \text{for all } k \in I\!N .$$

Putting $t_k := \sqrt{k}\|x_k - \bar{x}\|$ and $v_k := (x_k - \bar{x})/t_k$, one has $\|v_k\| = 1/\sqrt{k}$, $x_k = \bar{x} + t_k v_k$, and $t_k \downarrow 0$ as $k \to \infty$. Now consider a compact set $V := \{v_k |\ k \in I\!N\} \cup \{0\}$ and employ the $w\mathcal{H}$-differentiability property of f at $\bar{x}$. For every $y^* \in Y^*$, $\varepsilon > 0$, and $k \in I\!N$ sufficiently large we have

$$\left| \left\langle y^*, \frac{f(\bar{x} + t_k v) - f(\bar{x})}{t_k} \right\rangle - \left[\nabla\langle y^*, f\rangle(\bar{x})\right] v \right| \leq \varepsilon \quad \text{for all } v \in V ,$$

where $\nabla\langle y^*, f\rangle$ stands for the Hadamard derivative. This implies

$$\left| \left\langle y^*, \frac{f(\bar{x} + t_k v_k) - f(\bar{x})}{t_k} \right\rangle \right| \leq \left\|\nabla\langle y^*, f\rangle(\bar{x})\right\| \cdot \|v_k\| + \varepsilon .$$

Therefore the sequence $\{(f(\bar{x} + t_k v_k) - f(\bar{x}))/t_k\}$ weakly converges to 0 and hence bounded by the uniform bounded principle. On the other hand, $\left\|(f(\bar{x} + t_k v_k) - f(\bar{x}))/t_k\right\| \geq \sqrt{k} \to \infty$ as $k \to \infty$, a contradiction. △

Next we establish close relationships between the *single-valuedness* of the mixed and normal coderivatives for Lipschitzian mappings on Asplund spaces and their strict $w\mathcal{H}$-differentiability.

Theorem 3.66 (coderivative single-valuedness and strict-weak differentiability). *Let $f\colon X \to Y$, where X is Asplund and Y is Banach. The following hold:*

(i) *If f is strictly $w\mathcal{H}$-differentiable at $\bar{x}$, then $D^*_M f(\bar{x})$ is a single-valued bounded linear operator satisfying*

$$D_M^*f(\bar{x})(y^*) = \{\nabla\langle y^*, f\rangle(\bar{x})\}, \quad y^* \in Y^* , \tag{3.67}$$

where ∇ *stands for the strict Hadamard derivative. If in addition* f *obeys the sequential convergence condition from Definition* 3.25(ii), *then* $D_N^*f(\bar{x})$ *is also a single-valued bounded linear operator satisfying* (3.67).

(ii) *Conversely, if* f *is Lipschitz continuous around* $\bar{x}$ *and* $D_M^*f(\bar{x})$ *is single-valued, then* f *is strictly* $w\mathcal{H}$*-differentiable at* $\bar{x}$ *and* (3.67) *holds. The same is true for the case of* $D_N^*f(\bar{x})$.

Proof. Let us prove (i) for the case of $D_M^*f(\bar{x})$. First observe that f is Lipschitz continuous around $\bar{x}$ due to Proposition 3.65(ii). Hence $D_M^*f(\bar{x})(y^*) = \partial\langle y^*, f\rangle(\bar{x})$ for all $y^* \in Y^*$ by Theorem 1.90. Employing Theorem 3.54, we conclude that $\partial\langle y^*, f\rangle(\bar{x}) = \{\nabla\langle y^*, f\rangle(\bar{x})\}$ if $\langle y^*, f\rangle$ is strictly Hadamard differentiable and X is Asplund. This implies (3.67). It is easy to see that the operator in the right-hand side of (3.67) is linear and bounded due to the Lipschitz continuity of f. Thus (i) holds for the case of $D_M^*f(\bar{x})$. If in addition f satisfies the mentioned sequential convergence condition, then f is w^*-strictly Lipschitzian in the sense of Definition 3.25(ii). Thus $D_N^*f(\bar{x}) = D_M^*f(\bar{x})$ by Theorem 3.28, which completes the proof of (i).

To prove (ii) for the case of $D_M^*f(\bar{x})$, we observe that $\partial\langle y^*, f\rangle(\bar{x})$ is a singleton under the assumptions made due to the scalarization formula for the mixed coderivative; see Theorem 1.90. Involving again Theorem 3.54 (in the other direction), we conclude that $\langle y^*, f\rangle$ is strictly Hadamard differentiable at $\bar{x}$. Hence f is strictly $w\mathcal{H}$-differentiable at this point, and (3.67) follows from the above.

Finally, assume that $D_N^*f(\bar{x})$ is single-valued. Then

$$D_N^*f(\bar{x})(y^*) = D_M^*f(\bar{x})(y^*) \neq \emptyset \;\text{ for all }\; y^* \in Y^* ,$$

since X is Asplund. Thus we get back to the case of $D_M^*f(\bar{x})$ and complete the proof of the theorem. $\triangle$

Note that the sequential convergence condition in Theorem 3.66(i) holds automatically if f is strictly Gâteaux differentiable at $\bar{x}$. However, in general the strict $w\mathcal{H}$-differentiability (and even strict $w\mathcal{F}$-differentiability) of f at $\bar{x}$ doesn't imply this convergence condition, and hence it doesn't imply the w^*-strict Lipschitzian property of f around $\bar{x}$. For illustration let us consider the function $f\colon I\!R \to \ell^2$ from Example 3.64. Taking $t_k := 2^{-k}$ and $v := x_0 + 1$ with $\varphi(x_0) \neq 0$, we have $y_k := \big[f(0 + t_k v) - f(0)\big]/t_k = \varphi_k(x_0)e_k$. Hence $\langle e_k, y_k\rangle = \varphi(x_0) \not\to 0$ while $e_k \xrightarrow{w} 0$ as $k \to \infty$.

Corollary 3.67 (subspace property and strict Hadamard differentiability). *Let* X *be Asplund, and let* $f\colon X \to I\!R^m$ *be Lipschitz continuous around* $\bar{x}$. *The following properties are equivalent:*

(a) *Clarke's normal cone to* $\operatorname{gph} f$ *at* $(\bar{x}, f(\bar{x}))$ *is a linear subspace of dimension* m.

(b) *The basic normal cone* $N((\bar{x}, f(\bar{x})); \operatorname{gph} f)$ *is a linear subspace of dimension* m.

(c) f *is strictly Hadamard differentiable at* $\bar{x}$.

Proof. Equivalence (b)⇔(c) follows from Theorem 3.66 due to the fact that the graph of any bounded linear operator is isomorphic to the domain space. Equivalence (a)⇔(b) follows from Theorem 3.57. △

Now we are ready to establish relationships between the graphical regularity of Lipschitzian mappings from Definition 1.36 and the weak differentiability properties introduced above.

Theorem 3.68 (relationships between graphical regularity and weak differentiability). *Let* $f: X \to Y$, *where* X *is Asplund and* Y *is Banach. The following hold:*

(i) *Assume that* f *is both* $w\mathcal{F}$*-differentiable and strictly* $w\mathcal{H}$*-differentiable at* $\bar{x}$. *Then* f *is* M*-regular at this point. If in addition* f *obeys the sequential convergence condition from Definition* 3.25(ii), *then* f *is also* N*-regular at* $\bar{x}$.

(ii) *Conversely, the* M*-regularity (and hence* N*-regularity) of* f *at* $\bar{x}$ *implies its* $w\mathcal{F}$*-differentiability and strict* $w\mathcal{H}$*-differentiability at this point provided that* f *is Lipschitz continuous around* $\bar{x}$.

Proof. To justify (i), it is sufficient to do it for M-regularity. This implies the case of N-regularity, since $D_N^* f(\bar{x}) = D_M^* f(\bar{x})$ under the additional assumption made; see the proof of Theorem 3.66. If f is strictly $w\mathcal{H}$-differentiable at $\bar{x}$, then it is Lipschitz continuous around $\bar{x}$ and (3.67) holds by Theorem 3.66(i), where ∇ stands for the strict Hadamard derivative of $\langle y^*, f\rangle$ at $\bar{x}$. It agrees with the Fréchet derivative of $\langle y^*, f\rangle$ at $\bar{x}$ under the $w\mathcal{F}$-differentiability assumption of the theorem. On the other hand, $\widehat{\partial}\langle y^*, f\rangle(\bar{x}) = \{\nabla\langle y^*, f\rangle(\bar{x})\}$ when f is $w\mathcal{F}$-differentiable at $\bar{x}$. Involving the scalarization formula for the mixed coderivative from Theorem 1.90 and the easy one (3.37) for the Fréchet coderivative, we get

$$D_M^* f(\bar{x})(y^*) = \partial\langle y^*, f\rangle(\bar{x}) = \widehat{\partial}\langle y^*, f\rangle(\bar{x}) = \widehat{D}^* f(\bar{x})(y^*) \quad \text{for all} \quad y^* \in Y^* ,$$

which justifies the M-regularity of f at $\bar{x}$.

To prove (ii), we first observe that $\partial\langle y^*, f\rangle(\bar{x}) \neq \emptyset$ for all $y^* \in Y^*$, since f is locally Lipschitzian and X is Asplund; see Corollary 2.25. Let $x^* \in \partial\langle y^*, f\rangle(\bar{x})$. Then $x^* \in D_M^* f(\bar{x})(y^*)$ and hence $(x^*, -y^*) \in \widehat{N}((\bar{x}, f(\bar{x})); \operatorname{gph} f)$ due to the assumed M-regularity. Involving the above scalarization, we have

$$\widehat{\partial}\langle y^*, f\rangle(\bar{x}) = \partial\langle y^*, f\rangle(\bar{x}) \neq \emptyset \quad \text{for all} \quad y^* \in Y^* ,$$

which implies the Fréchet differentiability of $\langle y^*, f\rangle$ at $\bar{x}$ by Proposition 1.87. Thus $\partial\langle y^*, f\rangle(\bar{x})$ is a singleton and $\langle y^*, f\rangle$ is strictly Hadamard differentiable at $\bar{x}$ by Theorem 3.54. This justifies the $w\mathcal{H}$-differentiability of f at $\bar{x}$ and completes the proof of the theorem. △

Corollary 3.69 (graphical regularity of Lipschitzian mappings into finite-dimensional spaces). *Let X be Asplund, and let $f\colon X \to I\!R^m$ be Lipschitz continuous around $\bar{x}$. Then the following are equivalent:*

(a) *f is graphically regular at $\bar{x}$.*

(b) *f is simultaneously Fréchet differentiable and strictly Hadamard differentiable at $\bar{x}$.*

Proof. When $Y = I\!R^m$, we have only one notion of graphical regularity in Definition 1.36, and the weak differentiability notions under consideration reduce to the standard ones. Hence the desired equivalence (a)⇔(b) in this case follows directly from Theorem 3.68. △

If X is finite-dimensional, there is no difference between Fréchet differentiability and Hadamard differentiability. In this case Corollary 3.69 goes back to the claim used in the proof of Theorem 1.46.

Remark 3.70 (subspace and graphical regularity properties with respect to general topologies). One can see that the scalarization formulas for the mixed and normal coderivatives play a crucial role in the proofs of Theorems 3.62, 3.66, and 3.68. These theorems can be extended to the case of an arbitrary topology $w^* \le \tau \le \tau_{\|\cdot\|}$ based on the generalized scalarization results described in Remark 3.31. The corresponding extensions of the properties in Theorems 3.62(a), 3.66(i), and Theorem 3.68(i) for mappings $f\colon X \to Y$ require the τ_{Y^*}-counterpart of the sequential convergence condition from Definition 3.25(ii) with w^* replaced by τ_{Y^*}. This τ_{Y^*}-convergence condition is automatic for $\tau_{Y^*} = \tau_{\|\cdot\|}$ while reduces to the sequential convergence condition used in the above theorems for $\tau_{Y^*} = w^*$; see Mordukhovich and B. Wang [965] for more details.

Although the results of this subsection concern single-valued mappings, they can be used for the study of sets and set-valued mappings generated by *graphs* of single-valued Lipschitzian mappings via smooth transformations. Some definitions, discussions, and results in this direction were presented at the end of Subsect. 1.2.2 with the proofs based on finite-dimensional considerations. Now we derive infinite-dimensional analogs of these results in the case of hemi-Lipschitzian sets, which are applied to graphs of set-valued mappings as in Definition 1.45.

Definition 3.71 (hemi-Lipschitzian and hemismooth sets). *Let Ω be a subset of a Banach space Z, and let $\mathcal{B}$ stand for some differentiability concept (e.g., $\mathcal{B} = \beta, w\beta, sw\beta$). Then:*

(i) *Ω is* HEMI-LIPSCHITZIAN *around $\bar{z} \in \Omega$ if there are single-valued mappings $f\colon X \to Y$ and $g\colon Z \to X \times Y$ between Banach spaces such that $g(\bar{z}) = (\bar{x}, f(\bar{x}))$, that g is strictly Fréchet differentiable at $\bar{z}$ with the surjective derivative, that f is Lipschitz continuous around $\bar{x}$, and that*

$$\Omega \cap U = g^{-1}(V \cap \operatorname{gph} f)$$

for some neighborhoods U *of* $\bar{z}$ *and* V *of* $g(\bar{z})$*. We say that* Ω *is* STRICTLY HEMI-LIPSCHITZIAN *at* $\bar{z}$ *if* f *is additionally assumed to be* w^**-strictly Lipschitzian at* $\bar{x}$*.*

(ii) Ω *is* $\mathcal{B}$-HEMISMOOTH *at* $\bar{z}$ *if it is hemi-Lipschitzian around this point and* f *can be chosen as* $\mathcal{B}$*-differentiable at* $\bar{x}$*.*

When $\nabla g(\bar{z})$ is invertible in Definition 3.71(i), then Ω is *Lipschitzian* around $\bar{x}$. This corresponds to the notion of "Lipschitzian manifolds" in the sense of Rockafellar [1153], where g is assumed to be locally C^1 with the nonsingular Jacobian matrix in finite dimensions. The notion of $\mathcal{B}$*-smooth* sets is defined in a similar way provided that $\nabla g(\bar{z})$ is invertible.

Theorem 3.72 (properties of hemi-Lipschitzian sets). *Let* $\Omega \subset Z$ *be strictly hemi-Lipschitzian at* $\bar{z}$*, where the space* X *in Definition* 3.71(i) *can be chosen as Asplund. Then the following hold:*

(i) *The convexified normal cone* (3.64) *to* Ω *at* $\bar{z}$ *(in particular, the Clarke normal cone when* Ω *is locally closed around* $\bar{z}$ *and* Z *is Asplund) is a linear subspace of the dual space* Z^**.*

(ii) Ω *is normally regular at* $\bar{z}$ *if and only if it is simultaneously* $w\mathcal{F}$*-smooth and strictly* $w\mathcal{H}$*-smooth at* $\bar{z}$*, i.e.,* f *in Definition* 3.71(ii) *has both of these properties at* $\bar{x}$*.*

Proof. By Theorem 1.17 we have

$$N(\bar{z};\Omega) = \nabla g(\bar{z})^* N((\bar{x}, f(\bar{x})); \operatorname{gph} f)$$

provided that g is strictly Fréchet differentiable at $\bar{z}$ with the surjective derivative. This justifies (i) due to Theorem 3.62. To prove (ii), we observe that the normal regularity of Ω at $\bar{z}$ is equivalent to the N-normal regularity of f at $\bar{x}$ by Theorem 1.19. Then (ii) follows from Theorem 3.68. △

In the case of finite dimensions the simultaneous $w\mathcal{F}$-differentiability and strict $w\mathcal{H}$-differentiability of f at $\bar{x}$ reduces to the strict Fréchet differentiability of f at this point. Hence Theorem 3.71(ii) provides an infinite-dimensional extension of the set counterpart of Theorem 1.46(i) whose proof is different from the one given above (including the proof of Theorem 3.68). Similarly we can obtain infinite-dimensional extensions of Theorem 1.46(ii) involving relationships between normal regularity and $\mathcal{B}$-smoothness of Lipschitzian sets and graphically Lipschitzian mappings.

3.2.5 Second-Order Subdifferential Calculus

In this subsection we continue developing the second-order subdifferential calculus started in Subsect. 1.3.5 in the framework of general Banach spaces. Here we follow the same scheme that leads us to second-order subdifferential sum and chain rules by using coderivative calculus applied to *equality-type* sum and chain rules for first-order subgradients. In contrast to the previous consideration, we assume in this subsection that some of the spaces in question are *Asplund*. This allows us to employ extended first-order calculus rules obtained above in the framework of Asplund spaces. Note that the *norm-closedness* of $\operatorname{gph}\partial\varphi$ for some functions $\varphi\colon X \to \overline{I\!R}$ considered below is required in the *norm×norm* topology of $X \times X^*$. This is an essentially weaker assumption than the graph-closedness of $\partial\varphi$ in the norm×weak* topology of $X \times X^*$ presented in Subsect. 3.2.3; see Theorem 3.60 and the discussion after its proof. It is easy to see that the norm×norm graph-closedness of $\partial\varphi$ is similar to the one in finite dimensions and, besides continuous functions, always holds for proper convex l.s.c. functions φ and their compositions $\varphi \circ f$ with smooth mappings $f\colon X \to Y$, in particular, for the important class of amenable functions; see below. Note also that smoothness and strict differentiability in what follows are understood in the sense of Fréchet.

Most results of this subsection require the *Asplund* property of *both* the space in question and its dual. The major source of such spaces are *reflexive* Banach spaces. On the other hand, there are interesting examples of even *separable* spaces X, which are nonreflexive but Asplund together with X^*. Let us mention the famous *long James* space whose natural embedding in the second dual is of *codimension one* but which is nevertheless isometrically isomorphic to its second dual. Other examples, discussions, and references can be found, e.g., in the book by Bourgin [169].

We start as usual with sum rules and obtain the following three versions for extended-real-valued functions defined on spaces that are Asplund together with their duals. Recall that all the functions under consideration are assumed to be proper and finite at reference points.

Theorem 3.73 (second-order subdifferential sum rules). *Let $\varphi_i\colon X \to \overline{I\!R}$, $i = 1, 2$, with $\bar{y} \in \partial(\varphi_1 + \varphi_2)(\bar{x})$, and let X and X^* be Asplund. The following assertions hold for both normal ($\partial^2 = \partial^2_N$) and mixed ($\partial^2 = \partial^2_M$) second-order subdifferentials:*

(i) *Assume that $\varphi_1 \in C^1$ with $\bar{y}_1 := \nabla\varphi_1(\bar{x})$ and that the graph of $\partial\varphi_2$ is norm-closed around $(\bar{x}, \bar{y}_2)$ with $\bar{y}_2 := \bar{y} - \bar{y}_1$. Suppose also that either $\varphi_1 \in C^{1,1}$ around $\bar{x}$, or $\partial\varphi_2$ is PSNC at $(\bar{x}, \bar{y}_2)$ and*

$$\partial^2_M\varphi_1(\bar{x}, \bar{y}_1)(0) \cap \big(-\partial^2_M\varphi_2(\bar{x}, \bar{y}_2)(0)\big) = \{0\}\,. \tag{3.68}$$

Then for all $u \in X^{**}$ one has

$$\partial^2(\varphi_1 + \varphi_2)(\bar{x}, \bar{y})(u) \subset \partial^2\varphi_1(\bar{x}, \bar{y}_1)(u) + \partial^2\varphi_2(\bar{x}, \bar{y}_2)(u)\,. \tag{3.69}$$

(ii) *Let both φ_i be l.s.c. around $\bar{x}$, and let $S\colon X \times X^* \rightrightarrows X^* \times X^*$ with*

$$S(x, y) := \Big\{(y_1, y_2) \in X^* \times X^* \Big|\; y_1 \in \partial\varphi_1(x),\ y_2 \in \partial\varphi_2(x),\ y_1 + y_2 = y\Big\}$$

be inner semicontinuous at $(\bar{x}, \bar{y}, \bar{y}_1, \bar{y}_2)$ for a given $(\bar{y}_1, \bar{y}_2) \in S(\bar{x}, \bar{y})$. Suppose that the graph of each $\partial\varphi_i$ is norm-closed around $(\bar{x}, \bar{y}_i)$, that one of $\partial\varphi_i$ is

PSNC at $(\bar{x}, \bar{y}_i)$*, and that the qualification condition* (3.68) *is fulfilled. Assume also that there is a neighborhood* U *of* $\bar{x}$ *such that*

$$\partial^\infty \varphi_1(x) \cap \big(-\partial^\infty \varphi_2(x)\big) = \{0\} \quad \text{for all } \ x \in U\ ,$$

that one of φ_i *is SNEC at every* $x \in U$ *(both assumptions are fulfilled when one of* φ_i *is Lipschitz continuous around* $\bar{x}$*), and that each* φ_i *is lower regular at every* $x \in U$*. Then the sum rule* (3.69) *holds for all* $u \in X^{**}$*.*

(iii) *Assume that the above set-valued mapping* S *be inner semicompact at* $(\bar{x}, \bar{y})$*, that the graph of* $\partial\varphi_i$ *is norm-closed whenever* x *is near* $\bar{x}$*, and that the other assumptions in* (ii) *are fulfilled for any* $(\bar{y}_1, \bar{y}_2) \in S(\bar{x}, \bar{y})$*. Then for all* $u \in X^{**}$ *one has*

$$\partial^2(\varphi_1 + \varphi_2)(\bar{x}, \bar{y})(u) \subset \bigcup_{(\bar{y}_1, \bar{y}_2) \in S(\bar{x}, \bar{y})} \Big[\partial^2 \varphi_1(\bar{x}, \bar{y}_1)(u) + \partial^2 \varphi_2(\bar{x}, \bar{y}_2)(u)\Big]\ .$$

Proof. To prove (i), we use the first-order equality

$$\partial(\varphi_1 + \varphi_2)(x) = \nabla\varphi_1(x) + \partial\varphi_2(x) \ \text{ for all } \ x \in U$$

valid in some neighborhood U of $\bar{x}$ due to Proposition 1.107(ii). Since both X and X^* are Asplund, we apply to this equality the coderivative sum rule from Theorem 3.10(i) with $F_1 := \nabla\varphi_1$ and $F_2 := \partial\varphi_2$. This yields the second-order sum rule in (i). In the same way we justify the second-order sum rules in (ii) and (iii) applying Theorem 3.10(i,ii) to the first-order subdifferential equality

$$\partial(\varphi_1 + \varphi_2)(x) = \partial\varphi_1(x) + \partial\varphi_2(x), \quad x \in U\ ,$$

valid due to Theorem 3.36 under the assumptions made. △

Next we derive second-order subdifferential chain rules for compositions $(\varphi \circ g)(x) = \varphi(g(x))$ in the Asplund space framework. In contrast to Theorem 1.127, the following theorem doesn't require the surjectivity of $\nabla g(\bar{x})$ while imposing more assumptions on the outer function φ under first-order and second-order qualification conditions.

Theorem 3.74 (second-order chain rules with smooth inner mappings). *Consider the composition* $\varphi \circ g$ *of a function* $\varphi\colon Z \to \overline{I\!R}$ *and a mapping* $g\colon X \to Z$*, where the spaces* Z*,* Z^**, and* X *are Asplund. Assume that* $g \in C^1$ *around some* $\bar{x}$ *with the derivative* ∇g *strictly differentiable at this point, that* φ *is l.s.c. and lower regular around* $\bar{z} := g(\bar{x})$*, and that the inverse mapping* g^{-1} *is PSNC at* $(\bar{z}, \bar{x})$*. Suppose also that* φ *is SNEC around* $\bar{z}$ *and that the first-order qualification condition*

$$\partial^\infty \varphi(g(x)) \cap \ker \nabla g(x)^* = \{0\} \tag{3.70}$$

is satisfied around $\bar{x}$ *(the last two conditions are automatic when* φ *is locally Lipschitzian around* $\bar{x}$*). Then the following assertions hold for both second-order subdifferentials* $\partial^2 = \partial^2_N$ *and* $\partial^2 = \partial^2_M$*:*

(i) *Given $\bar{y} \in \partial(\varphi \circ g)(\bar{x})$, we assume that the mapping $S\colon X \times X^* \rightrightarrows Z^*$ with the values*

$$S(x, y) := \big\{v \in Z^* \big|\ v \in \partial\varphi(g(x)),\ \nabla g(x)^* v = y\big\}$$

is inner semicontinuous at $(\bar{x}, \bar{y}, \bar{v})$ for some fixed $\bar{v} \in S(\bar{x}, \bar{y})$, that the graph of the subdifferential mapping $\partial\varphi$ is norm-closed around $(\bar{z}, \bar{v})$, and that the mixed second-order qualification condition

$$\partial^2_M \varphi(\bar{z}, \bar{v})(0) \cap \ker \nabla g(\bar{x})^* = \{0\}$$

is satisfied. Then for all $u \in X^{**}$ one has

$$\partial^2(\varphi \circ g)(\bar{x}, \bar{y})(u) \subset \nabla^2\langle\bar{v}, g\rangle(\bar{x})^* u + \nabla g(\bar{x})^* \partial^2_N \varphi(\bar{z}, \bar{v})(\nabla g(\bar{x})^{**} u)\ .$$

(ii) *Given $\bar{y} \in \partial(\varphi \circ g)(\bar{x})$, we suppose that the above mapping S is inner semicompact at $(\bar{x}, \bar{y})$, that the graph of $\partial\varphi$ is norm-closed whenever z is near $\bar{z}$, and that the mixed second-order qualification condition in* (i) *is satisfied for every $\bar{v} \in S(\bar{x}, \bar{y})$. Then for all $u \in X^{**}$ one has*

$$\partial^2(\varphi \circ g)(\bar{x}, \bar{y})(u) \subset \bigcup_{\bar{v} \in S(\bar{x}, \bar{y})} \Big[\nabla^2\langle\bar{v}, g\rangle(\bar{x})^* u + \nabla g(\bar{x})^* \partial^2_N \varphi(\bar{z}, \bar{v})(\nabla g(\bar{x})^{**} u)\Big]\ .$$

Proof. It suffices to justify (i) for $\partial^2 = \partial^2_N$, which implies the other statements of the theorem due to the definitions. It follows from the first-order subdifferential chain rule in Theorem 3.41(ii) that the assumptions made ensure the existence of a neighborhood U of $\bar{x}$ on which $\partial(\varphi \circ g)$ admits the composite representation

$$\partial(\varphi \circ g)(x) = (f \circ G)(x), \quad x \in U\ ,$$

where $f(x, v) = \nabla g(x)^* v$ and $G(x) = \big(x, \partial\varphi(g(x))\big)$. Since f is smooth and one always has

$$D^*_N G(\bar{x}, \bar{x}, \bar{v})(x^*, v^*) = x^* + D^*_N(\partial\varphi \circ g)(\bar{x}, \bar{v})(v^*), \quad x^* \in X^*,\ v^* \in Z^{**}\ ,$$

we conclude by Theorem 1.66(i) that

$$\partial^2_N(\varphi \circ g)(\bar{x}, \bar{y})(u) \subset \nabla^2\langle\bar{v}, g\rangle(\bar{x})^*(u) + D^*_N(\partial\varphi \circ g)(\bar{x}, \bar{v})(\nabla g(\bar{x})^{**} u)$$

for all $u \in X^{**}$. It remains to compute the normal coderivative of the composition $\partial\varphi \circ g$. To furnish this, we use Theorem 3.13(i) that provides the coderivative chain rule

$$D^*_N(\partial\varphi \circ g)(\bar{x}, \bar{v})(v^*) \subset \nabla g(\bar{x})^* \circ (D^*_N \partial\varphi)(\bar{z}, \bar{v})(v^*), \quad v^* \in Z^{**}\ ,$$

under the PSNC assumption on g^{-1} and the mixed qualification condition

$$(D^*_M \partial\varphi)(\bar{z}, \bar{v})(0) \cap \ker \nabla g(\bar{x})^* = \{0\}\ ,$$

which reduces to the second-order qualification condition of the theorem. Combining these representations, we arrive at the desired second-order subdifferential chain rule in (i). △

When Z is finite-dimensional (X may be not), some of the assumptions of Theorem 3.74 either are satisfied automatically or can be essentially simplified. In this way we get the following result, where $\partial^2\varphi$ stands for the common second-order subdifferential of $\varphi\colon I\!R^m \to \overline{I\!R}$ while $\partial^2(\varphi\circ g)$ is the same as in the above theorem.

Corollary 3.75 (second-order chain rule for compositions with finite-dimensional intermediate spaces). *Let $\bar y \in \partial(\varphi\circ g)(\bar x)$, where $\varphi\colon I\!R^m \to \overline{I\!R}$ and $g\colon X \to I\!R^m$ with an Asplund space X. Assume that $g \in \mathcal{C}^1$ around $\bar x$ with the derivative strictly differentiable at $\bar x$ and that φ is l.s.c. and lower regular around $\bar z = g(\bar x)$ with closed graphs of $\partial\varphi$ and $\partial^\infty\varphi$ near $\bar z$. Suppose also that the first-order qualification condition* (3.70) *is satisfied at the point $x = \bar x$ and that one has the second-order qualification condition in the form*

$$\partial^2\varphi(\bar z,\bar v)(0)\cap\ker\nabla g(\bar x)^* = \{0\} \quad \textit{if}\ \ \bar v \in \partial\varphi(\bar z)\ \textit{with}\ \nabla g(\bar x)^*\bar v = \bar y\,. \tag{3.71}$$

Then the second-order chain rule of Theorem 3.74(ii) *holds for all $u \in X^{**}$.*

Proof. The SNEC property of φ and the PSNC property of g^{-1} are automatic when $\dim Z < \infty$. Further, one can easily check that if (3.70) holds at $\bar x$ while Z is finite-dimensional, it also holds in a neighborhood of $\bar x$. Indeed, assuming the contrary and taking into account that $\partial^\infty\varphi(\cdot)$ is a cone, we get sequences of $x_k \to \bar x$ and $z_k^* \in \partial^\infty\varphi(g(x_k))$ with $\nabla g(x_k)^* z_k^* = 0$ and $\|z_k^*\| = 1$ for all $k \in I\!N$. Then $z^* \in \partial^\infty\varphi(\bar z)$ with $\nabla g(\bar x)^* z^* = 0$ and $\|z^*\| = 1$ for a cluster point z^* of $\{z_k^*\}$ due to the graph-closedness of $\partial^\infty\varphi$ near $\bar z$; this contradicts (3.70) at $\bar x$. Similarly we check that the mapping $S\colon X\times X^* \rightrightarrows I\!R^m$ in Theorem 3.74 is always inner semicompact at $(\bar x,\bar y)$ when the qualification condition (3.70) is satisfied at $\bar x$. Thus we get the second-order chain rule from assertion (ii) of Theorem 3.74. △

The next corollary justifies the second-order chain for an important class of functions that automatically satisfy all the first-order assumptions in Corollary 3.75. Recall that a function $\psi\colon X \to \overline{I\!R}$ is *amenable* at $\bar x$ if there is a neighborhood U of $\bar x$ on which ψ can be represented in the composition form $\psi = \varphi\circ g$ with a $\mathcal{C}^1$ mapping $g\colon U \to I\!R^m$ and a proper l.s.c. convex function $\varphi\colon I\!R^m \to \overline{I\!R}$ such that the qualification condition (3.70) holds at $\bar x$. This function ψ is *strongly amenable* at $\bar x$ if such a representation exists with g not just $\mathcal{C}^1$ but $\mathcal{C}^2$. Amenable functions play a major role in the second-order variational theory in finite dimensions; see the book by Rockafellar and Wets [1165] and the references therein.

Corollary 3.76 (second-order chain rule for amenable functions). *Let $\psi\colon X \to \overline{I\!R}$ be strongly amenable at $\bar x$, and let $\varphi\colon I\!R^m \to \overline{I\!R}$ and $g\colon X \to I\!R^m$*

be mappings from its composite representation. Assume that X is Asplund and that the second-order qualification condition (3.71) *holds. Then for each $\bar{y} \in \partial\psi(\bar{x})$ and all $u \in X^{**}$ one has the inclusion*

$$\partial^2\psi(\bar{x},\bar{y})(u) \subset \bigcup_{\bar{v}\in S(\bar{x},\bar{y})} \Big[\nabla^2\langle\bar{v},g\rangle(\bar{x})^*u + \nabla g(\bar{x})^*\partial^2\varphi(\bar{z},\bar{v})(\nabla g(\bar{x})^{**}u)\Big],$$

where $\partial^2\psi$ stands for either $\partial^2_N\psi$ or $\partial^2_M\psi$ and where the point $\bar{z}$ and the mapping S are defined in Theorem 3.74.

Proof. Since φ is convex, it is lower regular on its domain, and the graphs of $\partial\varphi$ and $\partial^\infty\varphi$ are closed. Hence the result follows from Corollary 3.75. △

Finally, let us consider a second-order chain rule for compositions $\varphi\circ g$ involving $\mathcal{C}^{1,1}$ functions φ and Lipschitzian mappings g. In the next theorem we use the second-order coderivatives (normal and mixed) of Lipschitzian mappings defined in (1.63).

Theorem 3.77 (second-order chain rule with Lipschitzian inner mappings). *Let $\bar{y} \in \partial(\varphi\circ g)(\bar{x})$, where $g\colon X \to Z$ is Lipschitz continuous around $\bar{x}$, where $\varphi\colon Z \to \overline{I\!R}$ is $\mathcal{C}^{1,1}$ around $\bar{z} := g(\bar{x})$ with $\bar{v} := \nabla\varphi(\bar{z})$, and where the spaces X, X^*, Z, and Z^* are Asplund. Assume that the graph of the set-valued mapping $(x,v) \to \partial\langle v,h\rangle(x)$ is norm-closed in $X \times Z^* \times X^*$ whenever (x,v) are near $(\bar{x},\bar{v})$. Then one has the second-order chain rule*

$$\partial^2(\varphi\circ g)(\bar{x},\bar{y})(u) \subset \bigcup_{(x^*,v^*)\in D^2g(\bar{x},\bar{v},\bar{y})(u)} \Big[x^* + D^*_N g(\bar{x})\circ\partial^2_N\varphi(\bar{z})(v^*)\Big]$$

for all $u \in X^{**}$, where ∂^2 and D^2 stand for the corresponding normal and mixed second-order constructions. Moreover, this second-order inclusion holds for an arbitrary Banach space Z if $\nabla\varphi$ is strictly differentiable at $\bar{z}$.

Proof. Following the proof of Theorem 1.128, we have the representation

$$\partial(\varphi\circ g)(x) = (F\circ h)(x) \quad \text{for all } x \in U,$$

in some neighborhood U of $\bar{x}$, where the mappings $F\colon X\times Z^* \rightrightarrows X^*$ and $h\colon X \to X\times Z^*$ are defined by

$$F(x,v) := \partial\langle v,g\rangle(x), \quad h(x) := \Big(x, \nabla\varphi(g(x))\Big), \quad x \in U.$$

Let us apply to this composition the coderivative chain rule from Theorem 3.13. This gives

$$D^*(F\circ h)(\bar{x},\bar{y})(u) \subset D^*_N h(\bar{x})\circ D^*F(\bar{x},\bar{v},\bar{y})(u), \quad u \in X^{**},$$

for both normal and mixed coderivatives under the assumptions made, except that Z may be an arbitrary Banach space. If in addition Z is Asplund, one has the inclusion

$$D_N^*(\nabla\varphi \circ g)(\bar{x})(v^*) \subset D_N^* g(\bar{x}) \circ \partial_N^2 \varphi(\bar{z})(v^*) \tag{3.72}$$

from the same Theorem 3.13. Combining these two inclusions, we arrive at the second-order chain rule in the theorem when all the spaces are Asplund.

Finally, let $\nabla\varphi$ be strictly differentiable at $\bar{z}$. Then (3.72) holds in any Banach spaces, which follows from Theorem 1.65. This justifies the last statement of the theorem and completes the proof. △

3.3 SNC Calculus for Sets and Mappings

In this section we continue studying the sequential normal compactness properties of sets and mappings started in Chap. 1. These properties are crucial for the generalized differential calculus and its applications involving limiting normals to sets, coderivatives of set-valued mappings, and subgradients of extended-real-valued functions in infinite dimensions; see the results above and also in the subsequent chapters. It is important therefore to investigate how these properties behave under various operations performed on sets, functions, and set-valued mappings. This means that we need to develop an *SNC calculus* that provides efficient conditions ensuring the preservation of these properties under basic operations. We have addressed such questions in Subsects. 1.1.3 and 1.2.5, where some results have been obtained for sets and mappings in arbitrary Banach spaces. In this section we present a more developed SNC calculus in the framework of Asplund spaces, which is our standing assumption for this chapter.

As usual in this book, our approach is geometric dealing first with sets and then with functions and multifunctions. Based on the extremal principle, we obtain in Subsect. 3.3.1 efficient conditions ensuring the preservation of the SNC (and related PSNC and strong PSNC) properties for sets intersections and inverse images under nonsmooth and set-valued mappings. Subsect. 3.3.2 contains results in this direction for sums and intersections of set-valued mappings that imply the corresponding results for sums and maxima/minima of extended-real-valued functions. The final Subsect. 3.3.3 concerns general compositions of set-valued mappings and some of their specific realizations including product and quotient operations.

3.3.1 Sequential Normal Compactness of Set Intersections and Inverse Images

The basic result of this section deals with intersections of sets in *products* of Asplund spaces (that are also Asplund) and provides conditions ensuring the PSNC property in the sense of Definition 3.3. The product structure in this result is essential for subsequent applications to set-valued mappings. Of course, the initial SNC property of sets from Definition 1.20 is a special case of the PSNC property studied in Theorem 3.79. To formulate this result, we first introduce the following *mixed* qualification condition for set systems in products of arbitrary Banach spaces. It is clearly sufficient to consider the product of two spaces.

Definition 3.78 (mixed qualification condition for set systems). *Let Ω_1 and Ω_2 be subsets of the product $X \times Y$ of two Banach spaces, and let $(\bar{x}, \bar{y}) \in \Omega_1 \cap \Omega_2$. We say that the system $\{\Omega_1, \Omega_2\}$ satisfies the* MIXED QUALIFICATION CONDITION *at $(\bar{x}, \bar{y})$ with respect to Y if for any sequences $\varepsilon_k \downarrow 0$, $(x_{ik}, y_{ik}) \xrightarrow{\Omega_i} (\bar{x}, \bar{y})$, and $(x_{ik}^*, y_{ik}^*) \xrightarrow{w^*} (x_i^*, y_i^*)$ with $(x_{ik}^*, y_{ik}^*) \in \widehat{N}_{\varepsilon_k}((x_{ik}, y_{ik}); \Omega_i)$, $i = 1, 2$, and $k \to \infty$ one has*

$$\Big[x_{1k}^* + x_{2k}^* \xrightarrow{w^*} 0, \quad \|y_{1k}^* + y_{2k}^*\| \to 0\Big] \Longrightarrow (x_1^*, y_1^*) = (x_2^*, y_2^*) = 0\,.$$

As usual, we may omit ε_k in the above definition if both X and Y are Asplund and Ω_i are locally closed around $(\bar{x}, \bar{y})$. The mixed qualification condition clearly holds under the *normal qualification condition*

$$N((\bar{x}, \bar{y}); \Omega_1) \cap \big(-N((\bar{x}, \bar{y}); \Omega_2)\big) = \{(0, 0)\}\,, \tag{3.73}$$

which reduces to (3.10) from Definition 3.2(i) if there is no Y. Note that the limiting qualification condition for $\{\Omega_1, \Omega_2\}$ in the space $X \times Y$ from Definition 3.2(ii) is less restrictive than the mixed one, however, it is not sufficient for the SNC calculus.

The following principal result of the SNC calculus makes use of both PSNC and *strong* PSNC properties from Definition 3.3. The case of $m = 3$ (but not of $m = 2$) is of the main interest for applications to set-valued mappings; see the next two subsections.

Theorem 3.79 (PSNC property of set intersections). *Let the subsets $\Omega_1, \Omega_2 \subset \prod_{j=1}^m X_j$ be locally closed around $\bar{x} \in \Omega_1 \cap \Omega_2$, and let the index sets $J_1, J_2 \subset \{1, \ldots, m\}$ be such that $J_1 \cup J_2 = \{1, \ldots, m\}$. Assume that the following hold:*

(a) *For each $i = 1, 2$ the set Ω_i is PSNC at $\bar{x}$ with respect to $\{X_j |\; j \in J_i\}$.*

(b) *Either Ω_1 is strongly PSNC at $\bar{x}$ with respect to $\{X_j |\; j \in J_1 \setminus J_2\}$ or Ω_2 is strongly PSNC at $\bar{x}$ with respect to $\{X_j |\; j \in J_2 \setminus J_1\}$.*

(c) *$\{\Omega_1, \Omega_2\}$ satisfies the mixed qualification condition at $\bar{x}$ with respect to $\{X_j |\; j \in (J_1 \setminus J_2) \cup (J_2 \setminus J_1)\}$.*

Then $\Omega_1 \cap \Omega_2$ is PSNC at $\bar{x}$ with respect to $\{X_j |\; j \in J_1 \cap J_2\}$.

Proof. First observe that it is sufficient to prove the theorem in the case of $m = 3$ with $J_1 = \{1, 2\}$ and $J_2 = \{1, 3\}$. Indeed, the general case can be reduced to this one by reordering X_j and letting

$$X := \prod_{j \in J_1 \cap J_2} X_j, \quad Y := \prod_{j \in J_1 \setminus J_2} X_j, \quad Z := \prod_{j \in J_2 \setminus J_1} X_j\,.$$

In what follows we use the notation X, Y, Z for X_j, $j \in \{1, 2, 3\}$, and (x, y, z) for the corresponding points. To justify the PSNC property in the conclusion of the theorem, one needs to show that for any sequences

$$(x_k, y_k, z_k) \in \Omega_1 \cap \Omega_2, \quad (x_k^*, y_k^*, z_k^*) \in \widehat{N}((x_k, y_k, z_k); \Omega_1 \cap \Omega_2), \quad k \in I\!N ,$$

the convergence

$$(x_k, y_k, z_k) \to (\bar{x}, \bar{y}, \bar{z}), \quad x_k^* \xrightarrow{w^*} 0, \quad \|y_k^*\| \to 0, \quad \|z_k^*\| \to 0$$

implies that $\|x_k^*\| \to 0$ as $k \to \infty$. Since we are dealing with arbitrary sequences satisfying the above convergence properties, it is sufficient to show that $\|x_k^*\| \to 0$ along a subsequence. By (b), assume without loss of generality that Ω_1 is strongly PSNC at $(\bar{x}, \bar{y}, \bar{z})$ with respect to Y.

Given $(x_k^*, y_k^*, z_k^*) \in \widehat{N}((x_k, y_k, z_k); \Omega_1 \cap \Omega_2)$, we fix a sequence $\varepsilon_k \downarrow 0$ and apply Lemma 3.1 for each $k \in I\!N$. In this way we find sequences

$$(x_{ik}, y_{ik}, z_{ik}) \in \Omega_i, \quad (x_{ik}^*, y_{ik}^*, z_{ik}^*) \in \widehat{N}((x_{ik}, y_{ik}, z_{ik}); \Omega_i), \quad i = 1, 2 ,$$

and $\lambda_k \geq 0$ such that $\|(x_{ik}, y_{ik}, z_{ik}) - (x_k, y_k, z_k)\| \leq \varepsilon_k$ for $i = 1, 2$,

$$\|(x_{1k}^*, y_{1k}^*, z_{1k}^*) + (x_{2k}^*, y_{2k}^*, z_{2k}^*) - \lambda_k(x_k^*, y_k^*, z_k^*)\| \leq 2\varepsilon_k , \tag{3.74}$$

and $1 - \varepsilon_k \leq \max\{\lambda_k, \|x_{1k}^*\|, \|y_{1k}^*\|, \|z_{1k}^*\|\} \leq 1 + \varepsilon_k$. Since the sequence (x_k^*, y_k^*, z_k^*) weak* converges, it is bounded, and hence the sequences x_{ik}^*, y_{ik}^*, z_{ik}^*, $i = 1, 2$, and λ_k are bounded as well. Taking into account that the spaces X, Y, and Z are Asplund, we may suppose that $(x_{ik}^*, y_{ik}^*, z_{ik}^*)$ weak* converge to some (x_i^*, y_i^*, z_i^*) for $i = 1, 2$, and that $\lambda_k \to \lambda \geq 0$ as $k \to \infty$. This implies, by (3.74) and by the choice of (x_k^*, y_k^*, z_k^*), that

$$x_{1k}^* + x_{2k}^* \xrightarrow{w^*} 0, \quad \|y_{1k}^* + y_{2k}^*\| \to 0, \quad \text{and} \quad \|z_{1k}^* + z_{2k}^*\| \to 0 .$$

Therefore $x_i^* = y_i^* = z_i^* = 0$ for $i = 1, 2$ due to assumption (c) of the theorem. On the other hand, since Ω_1 is strongly PSNC at $(\bar{x}, \bar{y}, \bar{z})$ with respect to Y, it follows that $\|y_{1k}^*\| \to 0$, and hence $\|y_{2k}^*\| \to 0$ as $k \to \infty$. By (a) the set Ω_2 is PSNC at $(\bar{x}, \bar{y}, \bar{z})$ with respect to $\{X, Z\}$, which gives $\|x_{2k}^*\| \to 0$ and $\|z_{2k}^*\| \to 0$. This yields $\|z_{1k}^*\| \to 0$ by (3.74). Using the PSNC property of Ω_1 at $(\bar{x}, \bar{y}, \bar{z})$ with respect to $\{X, Y\}$, we similarly obtain $\|x_{1k}^*\| \to 0$. Thus $\lambda \neq 0$ by the relations above. Combining this with (3.74), we conclude that $\|x_k^*\| \to 0$, which completes the proof of the theorem. △

It is easy to see that assumptions (a) and (c) of Theorem 3.79 are essential for its conclusion. Let us show that the assumptions $J_1 \cup J_2 = \{1, \ldots, m\}$ and (b) cannot be dropped as well. To demonstrate this for the first one, we take an arbitrary Asplund space X and consider the two closed subsets

$$\Omega_1 := X \times \{0\}, \quad \Omega_2 := \{(x, x) \mid x \in X\}$$

of the product $X_1 \times X_2$ with $X_1 = X_2 = X$. Then both Ω_i are clearly PSNC at $(0,0)$ with respect to X_1, and assumptions (a)–(c) of Theorem 3.79 hold. However, the set $\Omega_1 \cap \Omega_2 = \{(0,0)\}$ is not PSNC at $(0,0)$ with respect to X_1 unless X is finite-dimensional.

In the case of (b) we take $X_1 = X_2 = X_3 := X$ for an Asplund space X and consider the sets

$$\Omega_1 := \big\{(x_1, x_2, x_3) \in X^3 \big|\ x_2 + x_3 = 0\big\}\ ,$$

$$\Omega_2 := \big\{(x_1, x_2, x_3) \in X^3 \big|\ x_1 + x_2 + x_3 = 0\big\}\ .$$

It is easy to check that Ω_1 and Ω_2 are PSNC at $(0,0,0))$ with respect to $\{X_1, X_2\}$ and $\{X_1, X_3\}$, respectively. Moreover, all the other assumptions but (b) of Theorem 3.79 hold. Nevertheless

$$\Omega_1 \cap \Omega_2 = \big\{(0, x_2, x_3)\big|\ x_2 + x_3 = 0\big\}$$

is not PSNC at $(0,0,0)$ with respect to X_1 in infinite dimensions.

Now we present two important corollaries of Theorem 3.79. The first one concerns subsets in products of two Asplund spaces.

Corollary 3.80 (PSNC sets in product of two spaces). *Let Ω_1 and Ω_2 be subsets of $X \times Y$ that are locally closed around $(\bar{x}, \bar{y}) \in \Omega_1 \cap \Omega_2$. Assume that one of the sets Ω_i is SNC at $(\bar{x}, \bar{y})$, that the other one is PSNC at this point with respect to X, and that $\{\Omega_1, \Omega_2\}$ satisfies the mixed qualification condition at $(\bar{x}, \bar{y})$ with respect to Y. Then $\Omega_1 \cap \Omega_2$ is PSNC at $(\bar{x}, \bar{y})$ with respect to X.*

Proof. Suppose that Ω_1 is SNC at $(\bar{x}, \bar{y})$. Then letting $X_1 := X$, $X_2 := Y$, $J_1 := \{1, 2\}$, and $J_2 := \{1\}$, we apply Theorem 3.79. $\triangle$

The next corollary doesn't assume any product structure on a given Asplund space X and thus provides an intersection rule for the SNC property, which is presented in the case of a finitely many sets under the normal qualification condition. Note that, in contrast to the assumptions of Corollary 3.37 ensuring the intersection formula for basic normals, the SNC property is now required for *all sets* involved in the intersection.

Corollary 3.81 (SNC property of set intersections). *Let $\Omega_1, \ldots, \Omega_n \subset X$, $n \geq 2$, be locally closed around their common point $\bar{x}$. Assume that each Ω_i is SNC at $\bar{x}$ and that*

$$\Big[x_1^* + \ldots + x_n^* = 0,\ x_i^* \in N(\bar{x}; \Omega_i)\Big] \Longrightarrow x_i^* = 0, \quad i = 1, \ldots, n\ .$$

Then the intersection $\Omega_1 \cap \ldots \cap \Omega_n$ is SNC at $\bar{x}$.

Proof. For $n = 2$ this follows from Corollary 3.80 by putting $Y = \{0\}$. In the general case we derive the result by induction. △

Intersection rules for the strong PSNC property in product spaces can be obtained similarly to the above. In particular, let us present a result for products of two Asplund spaces.

Theorem 3.82 (strong PSNC property of set intersections). *Let Ω_1 and Ω_2 be subsets of $X \times Y$ that are locally closed around $(\bar{x}, \bar{y}) \in \Omega_1 \cap \Omega_2$. Assume that Ω_1 is SNC at $(\bar{x}, \bar{y})$, that Ω_2 is strongly PSNC at this point with respect to X, and that the normal qualification condition* (3.73) *holds. Then the intersection $\Omega_1 \cap \Omega_2$ is strongly PSNC at $(\bar{x}, \bar{y})$ with respect to X.*

Proof. It is similar to the proofs of Theorem 3.79 and Corollary 3.80. △

Many applications deal with *sum of sets*, and hence it is important to clarify conditions ensuring the preservation of SNC properties under sum additions. Such conditions follow in fact from those for set intersections. The following theorem concerns the basic SNC property for sums of two sets in Asplund spaces; the corresponding results for the PSNC and strong PSNC properties can be derived similarly. Note that to derive efficient conditions for the *SNC* property of sums, we apply the ones for the *PSNC* property of intersections.

Theorem 3.83 (SNC property under set additions). *Let $\Omega_1, \Omega_2 \subset X$ be closed sets, let $\bar{x} \in \Omega_1 + \Omega_2$, and let*

$$S(x) := \big\{(x_1, x_2) \in X \times X \big|\; x_1 + x_2 = x,\; x_1 \in \Omega_1,\; x_2 \in \Omega_2\big\}\,.$$

Then the set $\Omega_1 + \Omega_2$ is SNC at $\bar{x}$ if either

(a) *S is inner semicompact at $\bar{x}$, and for each $(x_1, x_2) \in S(\bar{x})$ one of the sets Ω_1, Ω_2 is SNC at x_1 and x_2, respectively; or*

(b) *S is inner semicontinuous at $(\bar{x}_1, \bar{x}_2, \bar{x})$ with some $(\bar{x}_1, \bar{x}_2) \in S(\bar{x})$, and one of the sets Ω_1, Ω_2 is SNC at $\bar{x}_1$ and $\bar{x}_2$, respectively.*

Proof. Take a sequence of $(\varepsilon_k, x_k, x_k^*) \in I\!R_+ \times X \times X^*$ with

$$\varepsilon_k \downarrow 0,\quad x_k \to \bar{x},\quad x_k^* \in \widehat{N}_{\varepsilon_k}(x_k; \Omega_1 + \Omega_2),\quad \text{and}\quad x_k^* \xrightarrow{w^*} 0\,.$$

Considering case (a) with the inner semicompactness (the proof in case (b) is similar), we find $(u_k, v_k) \in S(x_k)$ that contains a subsequence converging to some $(\bar{x}_1, \bar{x}_2)$, which belongs to $S(\bar{x})$ to the closedness of Ω_1 and Ω_2. Define the product sets

$$\widetilde{\Omega}_1 := \Omega_1 \times X \quad\text{and}\quad \widetilde{\Omega}_2 := X \times \Omega_2\,,$$

which are closed subsets of the Asplund space X^2. It is easy to see that

$$(x_k^*, x_k^*) \in \widehat{N}_{\varepsilon_k}\big((u_k, v_k); \widetilde{\Omega}_1 \cap \widetilde{\Omega}_2\big) \quad \text{for all } k \in \mathbb{N}\,.$$

Suppose for definiteness that Ω is SNC at $\bar{x}_1$. Then $\widetilde{\Omega}_1$ is SNC at $(\bar{x}_1, \bar{x}_2)$ and $\widetilde{\Omega}_2$ is PSNC at this point with respect to the second component. Note that the mixed qualification condition from Definition 3.78 is obviously fulfilled for $\{\widetilde{\Omega}_1, \widetilde{\Omega}_2\}$. Applying Corollary 3.80, we conclude that $\widetilde{\Omega}_1 \cap \widetilde{\Omega}_2$ is PSNC at $(\bar{x}_1, \bar{x}_2)$ with respect to the first component. Thus $\|x_k^*\| \to 0$ as $k \to \infty$, which completes the proof of the theorem. △

Next let us obtain conditions ensuring the SNC property of *inverse images*

$$F^{-1}(\Theta) = \big\{x \in X \big|\ F(x) \cap \Theta \neq \emptyset\big\}$$

of sets under set-valued mappings between Asplund spaces.

Theorem 3.84 (SNC property of inverse images). *Let $\bar{x} \in F^{-1}(\Theta)$, where $F\colon X \rightrightarrows Y$ is a closed-graph mapping (near $\bar{x}$) and where Θ is a closed subset of Y. Assume that the set-valued mapping $F(\cdot)\cap\Theta$ is inner semicompact at $\bar{x}$ and that for every $\bar{y} \in F(\bar{x}) \cap \Theta$ the following hold:*

(a) *Either F is PSNC at $(\bar{x}, \bar{y})$ and Θ is SNC at $\bar{y}$, or F is SNC at $(\bar{x}, \bar{y})$.*

(b) *$\{F, \Theta\}$ satisfies the qualification condition*

$$N(\bar{y}; \Theta) \cap \ker D_N^* F(\bar{x}, \bar{y}) = \{0\}\,.$$

Then the inverse image $F^{-1}(\Theta)$ is SNC at $\bar{x}$.

Proof. Take $\{\varepsilon_k, x_k, x_k^*\}$ with

$$\varepsilon_k \downarrow 0,\ \ x_k \to \bar{x},\ \ x_k^* \in \widehat{N}_{\varepsilon_k}(x_k; F^{-1}(\Theta)),\ \text{ and }\ x_k^* \overset{w^*}{\to} 0\,.$$

Using the inner semicompactness and closedness assumptions made, we select a subsequence of $y_k \in F(x_k) \cap \Theta$ that converges (without relabeling) to some $\bar{y} \in F(\bar{x}) \cap \Theta$. One can easily check that

$$(x_k^*, 0) \in \widehat{N}_{\varepsilon_k}\big((x_k, y_k); \Omega_1 \cap \Omega_2\big) \ \text{ with } \Omega_1 := \operatorname{gph} F, \quad \Omega_2 := X \times \Theta\,. \tag{3.75}$$

Let us apply Corollary 3.80 to the set intersection in (3.75). Observe that Ω_2 is always PSNC at $(\bar{x}, \bar{y})$ with respect to X, and it is SNC at this point if and only if Θ is SNC at $\bar{y}$. Hence the assumptions in (a) ensure the fulfillment of the corresponding assumptions in Corollary 3.80. Further, due to the special structure of the sets Ω_1 and Ω_2 in (3.75), the mixed qualification condition in Corollary 3.80 is clearly equivalent in the Asplund space setting to the following: for any $(x_k, y_{1k}, y_{2k}, x_k^*, y_{1k}^*, y_{2k}^*)$ with

$$(x_k, y_{ik}) \to (\bar{x}, \bar{y}),\ (x_k, y_{1k}) \in \operatorname{gph} F,\ y_{2k} \in \Theta\,,$$

$$x_k^* \in \widehat{D}^* F(x_k, y_{1k})(y_{1k}^*),\ \text{ and }\ y_{2k}^* \in \widehat{N}(y_{2k}; \Theta)$$

one has the relation

$$\Big[x_k^* \xrightarrow{w^*} 0,\ y_{2k}^* \xrightarrow{w^*} y^*,\ \|y_{2k}^* - y_{1k}^*\| \to 0\Big] \Longrightarrow y^* = 0\ ,$$

which is implied by the qualification condition (b) of the theorem. Thus the set $\Omega_1 \cap \Omega_2$ is PSNC at $(\bar{x}, \bar{y})$ with respect to X by Corollary 3.80. It now follows from (3.75) that $\|x_k^*\| \to 0$, i.e., the set $F^{-1}(\Theta)$ is SNC at $\bar{x}$. △

Theorem 3.84 implies efficient subdifferential conditions ensuring the SNC property of *level sets* for l.s.c. functions and *solution sets* for equations given by real-valued continuous functions.

Corollary 3.85 (SNC property for level and solution sets). *Let the function $\varphi\colon X \to \overline{I\!R}$ be proper with $\varphi(\bar{x}) = 0$ for some $\bar{x}$. The following assertions hold:*

(i) *Assume that φ is l.s.c. around $\bar{x}$ and that it is SNEC at this point. Then the level set*

$$\Omega := \big\{x \in X \big|\ \varphi(x) \le 0\big\}$$

is SNC at $\bar{x}$ provided that $0 \notin \partial\varphi(\bar{x})$.

(ii) *Assume that φ is continuous around $\bar{x}$ and SNC at this point. Then the solution set*

$$\Omega := \big\{x \in X \big|\ \varphi(x) = 0\big\}$$

is SNC at $\bar{x}$ provided that $0 \notin \partial\varphi(\bar{x}) \cup \partial(-\varphi)(\bar{x})$.

Proof. Assertion (i) follows from Theorem 3.84 applied to $F := E_\varphi$ and $\Theta := (-\infty, 0]$. Assertion (ii) follows from Theorem 3.84 with $\Theta := \{0\}$ via the coderivative-subdifferential relation of Theorem 1.80. △

Note that the SNEC and SNC properties of φ in Corollary 3.85 automatically hold for locally Lipschitzian functions. Another proof of these results in the Lipschitz case is given by Mordukhovich and B. Wang [962] based on the direct application of the *extremal principle*.

It is easy to see that the subdifferential conditions are essential for the SNC properties in both assertions of Corollary 3.85, even for smooth functions φ. A simple example is provided by $\varphi(x) = \|x\|^2$ at $\bar{x} = 0$ in any infinite-dimensional space. Note also that the condition $0 \notin \partial\varphi(0)$, in contrast to its Clarke's counterpart $0 \notin \partial_C\varphi(\bar{x})$, *doesn't* ensure the *epi-Lipschitzian* property of the level set $\{x \in X|\ \varphi(x) \le 0\}$ for Lipschitzian functions. A *counterexample* is given by the function $\varphi\colon I\!R^2 \to I\!R$ defined by (1.57), whose basic subdifferential is computed in Subsect. 1.3.2. For this function we have $(0, 0) \notin \partial\varphi(0, 0)$, while the level set

$$\big\{x \in I\!R^2 \big|\ \varphi(x) \le 0\big\} = \big\{(x_1, x_2) \in I\!R^2 \big|\ |x_1| \le |x_2|\big\}$$

is obviously not epi-Lipschitzian at $(0,0)$.

The next result provides subdifferential conditions ensuring the SNC property for the class of *constraint sets* important in applications to optimization problems; see, e.g., Chap. 5.

Theorem 3.86 (SNC property of constraint sets). *Let $\varphi_i\colon X \to \overline{I\!R}$ with $\varphi_i(\bar{x}) = 0$ for $i = 1, \ldots, m + r$. Assume that φ_i are l.s.c. around $\bar{x}$ and SNEC at this point for $i = 1, \ldots, m$, and that φ_i are continuous around $\bar{x}$ and SNC at this point for $i = m + 1, \ldots, m + r$. Suppose also that the following constraint qualification conditions hold:*

(a) *$0 \notin \partial\varphi_i(\bar{x})$ for $i = 1, \ldots, m$, and $0 \notin \partial\varphi_i(\bar{x}) \cup \partial(-\varphi_i)(\bar{x})$ for $i = m + 1, \ldots, m + r$.*

(b) *one has*

$$\Big[x_1^* + \ldots + x_{m+r}^* = 0\Big] \Longrightarrow x_i^* = 0, \quad i = 1, \ldots, m + r\;,$$

for every $x_i^* \in I\!R^+\partial\varphi_i(\bar{x}) \cup \partial^\infty\varphi_i(\bar{x})$, $i = 1, \ldots, m$, and every

$$x_i^* \in I\!R^+\big[\partial\varphi_i(\bar{x}) \cup \partial(-\varphi_i)(\bar{x})\big] \cup \partial^\infty\varphi_i(\bar{x}) \cup \partial^\infty(-\varphi_i)(\bar{x}),\\ i = m + 1, \ldots, m + r\;,$$

where $I\!R^+V := \{\lambda v|\ \lambda \geq 0,\ v \in V\}$. Consider the sets

$$\Omega_i := \big\{x \in X\big|\ \varphi_i(x) \leq 0\big\}, \quad i = 1, \ldots, m\;,$$

$$\Omega_i := \big\{x \in X\big|\ \varphi_i(x) = 0\big\}, \quad i = m + 1, \ldots, m + r\;.$$

Then their intersection $\Omega_1 \cap \ldots \cap \Omega_{m+r}$ is SNC at $\bar{x}$.

Proof. Let us show that under the assumptions in (a) one has the inclusions

$$N(\bar{x}; \Omega_i) \subset I\!R^+\partial\varphi_i(\bar{x}) \cup \partial^\infty\varphi_i(\bar{x}) \;\text{ for }\; i = 1, \ldots, m\;; \tag{3.76}$$

$$N(\bar{x}; \Omega_i) \subset I\!R^+\big[\partial\varphi_i(\bar{x}) \cup \partial(-\varphi_i)(\bar{x})\big] \cup \partial^\infty\varphi_i(\bar{x}) \cup \partial^\infty(-\varphi_i)(\bar{x}) \tag{3.77}$$

for $i = m + 1, \ldots, m + r$. To establish (3.76), we observe that

$$\big\{x \in X\big|\ \varphi(x) \leq 0\big\} \times \{0\} = (\operatorname{epi}\varphi) \cap S$$

with $S := \{(x, \alpha) \in X \times I\!R|\ \alpha = 0\}$. The assumption $0 \notin \partial\varphi(\bar{x})$ ensures that the pair $\{\operatorname{epi}\varphi, S\}$ satisfies the normal qualification condition (3.10). Applying Corollary 3.5 to this intersection, we obtain inclusion (3.76) for each $i = 1, \ldots, m$. To justify (3.77) for each $i = m + 1, \ldots, m + r$, we apply the same procedure to the intersection

$$\big\{x \in X\big|\ \varphi(x) = 0\big\} \times \{0\} = (\operatorname{gph}\varphi) \cap S$$

while taking into account Theorem 2.40. Note that all the sets Ω_i, $i = 1, \ldots, m+r$ are SNC at $\bar{x}$ by Corollary 3.85. To complete the proof of the theorem, it remains to apply to the intersection $\Omega_1 \cap \ldots \cap \Omega_{m+p}$ the result of Corollary 3.81 whose qualification condition is fulfilled under the above assumption (b) due to (3.76) and (3.77). △

Note that for Lipschitzian functions φ_i the SNC and SNEC assumptions of Theorem 3.86 are fulfilled, and the qualification condition (b) is simplified by $\partial^\infty \varphi_i(\bar{x}) = \partial^\infty(-\varphi_i)(\bar{x}) = \{0\}$. If each φ_i is *strictly differentiable* at $\bar{x}$, then the qualification conditions of the theorem reduce to the classical *Mangasarian-Fromovitz constraint qualification.*

Corollary 3.87 (SNC property under the Mangasarian-Fromovitz constraint qualification). *Let $\bar{x} \in \Omega_1 \cap \ldots \cap \Omega_{m+r}$, where Ω_i are given in Theorem* 3.86 *with the functions φ_i strictly differentiable at $\bar{x}$. Put*

$$I(\bar{x}) := \big\{ i = 1, \ldots, m+r \big|\ \varphi_i(\bar{x}) = 0 \big\}$$

and assume that:

(a) $\nabla\varphi_{m+1}(\bar{x}), \ldots, \nabla\varphi_{m+r}(\bar{x})$ *are linearly independent;*

(b) *there is $u \in X$ satisfying*

$$\langle \nabla\varphi_i(\bar{x}), u \rangle < 0, \quad i \in \{1, \ldots, m\} \cap I(\bar{x}) \,,$$

$$\langle \nabla\varphi_i(\bar{x}), u \rangle = 0, \quad i = m+1, \ldots, m+r \,.$$

Then the set $\bigcap_{i \in I(\bar{x})} \Omega_i$ is SNC at $\bar{x}$.

Proof. Assume without loss of generality that $I(\bar{x}) = \{1, \ldots, m+r\}$. Then the result follows directly from Theorem 3.86 due to $\partial\varphi(\bar{x}) = \{\nabla\varphi(\bar{x})\}$ for strictly differentiable functions. △

3.3.2 Sequential Normal Compactness for Sums and Related Operations with Maps

The main results of this subsection concern the preservation of the PSNC and SNC properties under summations of set-valued mappings between Asplund spaces. The *sum operation* has certain *specific features* that distinguish it from other compositions and allow us to obtain more delicate results in this case than those in Subsect. 3.3.3. We also present here some consequences for summations, maxima, and minima of extended-real-valued functions. All the proofs are based on the SNC calculus for set intersections developed in Subsect. 3.3.1.

The first theorem ensures the preservation of the PSNC property for sums of multifunctions under the *mixed* coderivative qualification condition. Its assumptions are parallel to those in Theorem 3.10 on the coderivative sum rules, with the only difference that now the PSNC property is required for *both* mappings involved in summation.

Theorem 3.88 (PSNC property for sums of set-valued mappings). *Let $(\bar{x}, \bar{y}) \in \operatorname{gph}(F_1 + F_2)$, where both F_i are closed-graph whenever x is near $\bar{x}$. Suppose that the mapping*

$$S(x, y) := \big\{(y_1, y_2) \in Y^2 \big|\ y_1 \in F_1(x),\ y_2 \in F_2(x),\ y_1 + y_2 = y\big\}$$

is inner semicompact at $(\bar{x}, \bar{y})$ and that for every $(\bar{y}_1, \bar{y}_2) \in S(\bar{x}, \bar{y})$ the following assumptions hold:

(a) *Each F_i is PSNC at $(\bar{x}, \bar{y}_i)$, respectively.*

(b) *$\{F_1, F_2\}$ satisfies the mixed coderivative qualification condition*

$$D^*_M F_1(\bar{x}, \bar{y}_1)(0) \cap \big(- D^*_M F_2(\bar{x}, \bar{y}_2)(0)\big) = \{0\}\ .$$

Then $F_1 + F_2$ is PSNC at $(\bar{x}, \bar{y})$.

Proof. Take arbitrary sequences $\varepsilon_k \downarrow 0$, $(x_k, y_k) \in \operatorname{gph}(F_1 + F_2)$, and

$$(x_k^*, y_k^*) \in \widehat{N}_{\varepsilon_k}((x_k, y_k); \operatorname{gph}(F_1 + F_2)), \quad k \in I\!N\ , \tag{3.78}$$

satisfying $(x_k, y_k) \to (\bar{x}, \bar{y})$, $x_k^* \xrightarrow{w^*} 0$, and $\|y_k^*\| \to 0$ as $k \to \infty$. To justify the PSNC property of $F_1 + F_2$ at $(\bar{x}, \bar{y})$, it suffices to show that $\|x_k^*\| \to 0$ along a subsequence of $k \in I\!N$. Using the inner semicompactness of S and the closed-graph assumptions of the theorem, we select a subsequence of $(y_{1k}, y_{2k}) \in S(x_k, y_k)$ that converges (without relabeling) to some $(\bar{y}_1, \bar{y}_2) \in S(\bar{x}, \bar{y})$. Consider the two sets

$$\Omega_i := \big\{(x, y_1, y_2) \in X \times Y \times Y \big|\ (x, y_i) \in \operatorname{gph} F_i\big\}, \quad i = 1, 2\ ,$$

which are locally closed around $(\bar{x}, \bar{y}_1, \bar{y}_2)$. By (a) we observe that the set Ω_1 is PSNC at $(\bar{x}, \bar{y}_1, \bar{y}_2)$ with respect to the first and third components, while Ω_2 is PSNC at $(\bar{x}, \bar{y}_1, \bar{y}_2)$ with respect to the first two components and strongly PSNC at this point with respect to the second component. Using the special structure of Ω_i, one can directly check that (b) implies the mixed qualification condition for $\{\Omega_1, \Omega_2\}$ at $(\bar{x}, \bar{y}_1, \bar{y}_2)$ with respect to $Y \times Y$. Now the main Theorem 3.79 ensures, for $m = 3$, that $\Omega_1 \cap \Omega_2$ is PSNC at $(\bar{x}, \bar{y}_1, \bar{y}_2)$ with respect to X. Since

$$(x_k^*, y_k^*, y_k^*) \in \widehat{N}_{\varepsilon_k}((x_k, y_{1k}, y_{2k}); \Omega_1 \cap \Omega_2)\ ,$$

by (3.78), we conclude from here that $\|x_k^*\| \to 0$, which completes the proof of the theorem. △

Note that both assumptions (a) and (b) of Theorem 3.88 automatically hold if, for every $(\bar{y}_1, \bar{y}_2) \in S(\bar{x}, \bar{y})$, one of F_i is *Lipschitz-like* around $(\bar{x}, \bar{y}_i)$ and the other is PSNC at $(\bar{x}, \bar{y}_i)$, respectively. Also, it easily follows from the proof of Theorem 3.88 that assumptions (a) and (b) therein can be imposed only at a given point $(\bar{y}_1, \bar{y}_2) \in S(\bar{x}, \bar{y})$ if S is assumed to be inner semicontinuous at $(\bar{x}, \bar{y}, \bar{y}_1, \bar{y}_2)$.

The following corollary provides efficient conditions ensuring the preservation of the sequential normal epi-compact (SNEC) property for sums of extended-real-valued functions.

Corollary 3.89 (SNEC property for sums of l.s.c. functions). *Let $\varphi_i\colon X \to \overline{I\!R}$, $i = 1, 2$, be proper and l.s.c. around some point $\bar{x} \in (\operatorname{dom} \varphi_1) \cap (\operatorname{dom} \varphi_2)$. Assume that each φ_i is SNEC at $\bar{x}$ and that*

$$\partial^\infty \varphi_1(\bar{x}) \cap \big(-\partial^\infty \varphi_2(\bar{x})\big) = \{0\} \, . \tag{3.79}$$

Then $\varphi_1 + \varphi_2$ is SNEC at $\bar{x}$.

Proof. It follows from Theorem 3.88 applied to the epigraphical multifunctions $F_i := E_{\varphi_i}\colon X \to I\!R$ for which $F_1 + F_2 = E_{\varphi_1+\varphi_2}$. Indeed, it is clear that F_i is PSNC at $(\bar{x}, \varphi_i(\bar{x}))$ if and only if φ_i is SNEC at $\bar{x}$ for each $i = 1, 2$. Moreover, the qualification condition (b) of Theorem 3.88 obviously reduces to (3.79). Based on the lower semicontinuity of φ_i, one can directly check that the corresponding mapping S from Theorem 3.88 is inner semicompact at $(\bar{x}, \varphi_1(\bar{x}) + \varphi_2(\bar{x}))$. Hence $E_{\varphi_1} + E_{\varphi_2}$ is PSNC (i.e., SNC in this case) at the point $(\bar{x}, \varphi_1(\bar{x}) + \varphi_2(\bar{x}))$, which means that $\varphi_1 + \varphi_2$ is SNEC at $\bar{x}$. △

Next we obtain results on the preservation of the *full* SNC (not partial SNC) property for sums of set-valued mappings and real-valued functions. These results are similar to the case of PSNC with imposing *more restrictive* qualification conditions.

Theorem 3.90 (SNC property for sums of set-valued mappings). *Let $(\bar{x}, \bar{y}) \in \operatorname{gph}(F_1 + F_2)$, where both F_i are closed-graph whenever x is near $\bar{x}$. Assume that the mapping S from Theorem 3.88 is inner semicompact at $(\bar{x}, \bar{y})$ and that for every $(\bar{y}_1, \bar{y}_2) \in S(\bar{x}, \bar{y})$ the following hold:*

(a) *Each F_i is SNC at $(\bar{x}, \bar{y}_i)$, respectively.*

(b) *$\{F_1, F_2\}$ satisfies the normal coderivative qualification condition*

$$D_N^* F_1(\bar{x}, \bar{y}_1)(0) \cap \big(-D_N^* F_2(\bar{x}, \bar{y}_2)(0)\big) = \{0\} \, .$$

Then $F_1 + F_2$ is SNC at $(\bar{x}, \bar{y})$.

Proof. One can get this following the line in the proof of Theorem 3.88 with the use of Corollary 3.81 instead of Theorem 3.79. △

As a consequence of the latter result, we have a singular subdifferential condition ensuring the preservation of the SNC property for linear combinations of real-valued continuous functions.

Corollary 3.91 (SNC property for linear combinations of continuous functions). *Let $\varphi_i\colon X \to I\!R$, $i = 1, 2$, be continuous around $\bar{x}$ and SNC at this point. Assume the qualification condition*

$$\big[\partial^\infty\varphi_1(\bar{x}) \cup \partial^\infty(-\varphi_1)(\bar{x})\big] \cap \big[-\big(\partial^\infty\varphi_2(\bar{x}) \cup \partial^\infty(-\varphi_2)(\bar{x})\big)\big] = \{0\}\,. \quad (3.80)$$

Then $\alpha_1\varphi_1 + \alpha_2\varphi_2$ is SNC at $\bar{x}$ for any $\alpha_1, \alpha_2 \in I\!R$.

Proof. It follows from the above theorem due to Theorem 2.40(ii). △

Our next goal is to study the SNEC and SNC properties of *maximum functions* in the form

$$\max\{\varphi_1, \varphi_2\}(x) := \max\{\varphi_1(x), \varphi_2(x)\}$$

with $\varphi_i\colon X \to \overline{I\!R}$, $i = 1, 2$. It happens that the SNEC property of such functions is closely related to the SNC property for intersections of sets and set-valued mappings. The equivalence result below provides, in particular, a singular subdifferential condition ensuring the preservation of the SNEC property under the maximum operation over l.s.c. functions in Asplund spaces.

Proposition 3.92 (SNEC property of maximum functions). *Let $\mathcal{X}$ be a collection of Banach spaces that is closed under finite products and contains finite-dimensional spaces. Then the following assertions are equivalent:*

(i) *Take arbitrary $X \in \mathcal{X}$ and proper functions $\varphi_i\colon X \to \overline{I\!R}$, $i = 1, 2$, which are l.s.c. around some $\bar{x} \in (\operatorname{dom}\varphi_1) \cap (\operatorname{dom}\varphi_2)$ satisfying $\varphi_1(\bar{x}) = \varphi_2(\bar{x})$ and the qualification condition* (3.79). *Then $\max\{\varphi_1, \varphi_2\}$ is SNEC at $\bar{x}$ if each φ_i is SNEC at this point.*

(ii) *Take arbitrary $X, Y \in \mathcal{X}$ and mappings $(\bar{x}, \bar{y}) \in (\operatorname{gph} F_1) \cap (\operatorname{gph} F_2)$ and satisfy the qualification condition*

$$N((\bar{x}, \bar{y}); \operatorname{gph} F_1) \cap \big(-N((\bar{x}, \bar{y}); \operatorname{gph} F_2)\big) = \{(0, 0)\}$$

Then $F_1 \cap F_2$ is SNC at $(\bar{x}, \bar{y})$ if each F_i is SNC at this point.

(iii) *Take arbitrary $X \in \mathcal{X}$ and sets Ω_i, $i = 1, 2$, which are closed around some $\bar{x} \in \Omega_1 \cap \Omega_2$ and satisfy the qualification condition*

$$N(\bar{x}; \Omega_1) \cap \big(-N(\bar{x}; \Omega_2)\big) = \{0\}\,.$$

Then $\Omega_1 \cap \Omega_2$ is SNC at $\bar{x}$ if each Ω_i is SNC at this point.

In particular, the above assertions hold if $\mathcal{X}$ is the collection of Asplund spaces.

Proof. Let us show that (i)⇒(iii)⇒(ii)⇒(i). In fact, (iii)⇒(ii) is obvious. To justify (ii)⇒(i), we use (ii) for $F_i := E_{\varphi_i}$, $i = 1, 2$, at $(\bar{x}, \bar{y})$ with $\bar{y} := \varphi_1(\bar{x}) = \varphi_2(\bar{x})$. Observe that each E_{φ_i} is SNC at $(\bar{x}, \bar{y})$ and that the qualification condition in (ii) reduces to (3.79). Hence $E_{\varphi_1} \cap E_{\varphi_2}$ is SNC at $(\bar{x}, \bar{y})$. Taking into account that

$$\operatorname{gph}(E_{\varphi_1} \cap E_{\varphi_2}) = \operatorname{epi}\big(\max\{\varphi_1, \varphi_2\}\big) ,$$

we derive (i) from (ii).

To prove (i)⇒(iii), we apply (i) to the indicator functions $\varphi_i(x) = \delta(x; \Omega_i)$, $i = 1, 2$. Then each $\delta(\cdot; \Omega_i)$ is obviously SNEC at $\bar{x}$, and (3.79) reduces to the qualification condition in (iii). Since

$$\max\big\{\delta(x; \Omega_1), \delta(x; \Omega_2)\big\} = \delta(x; \Omega_1 \cap \Omega_2) ,$$

the function $\delta(\cdot; \Omega_1 \cap \Omega_2)$ is SNEC at $\bar{x}$, which is equivalent to the SNC property of $\Omega_1 \cap \Omega_2$ at this point. The last conclusion of the proposition follows from Corollary 3.81. △

The result obtained allows us to derive subgradient conditions ensuring the preservation of the SNC for continuous maximum (and minimum) functions due the following observation that holds in Asplund spaces.

Proposition 3.93 (relationship between SNEC and SNC properties of real-valued continuous functions). *Let $\varphi: X \to \overline{I\!R}$ be continuous around $\bar{x}$. Then φ is SNC at $\bar{x}$ if and only if both functions φ and $-\varphi$ are SNEC at this point.*

Proof. This easily follows from Theorem 2.40(i) held in Asplund spaces and the proof of Theorem 1.80 that gives relationships between Fréchet normals to graphs and epigraphs of continuous functions. △

Corollary 3.94 (SNC property of maximum and minimum functions). *Let $\varphi_i: X \to \overline{I\!R}$, $i = 1, 2$, be continuous around $\bar{x}$, and let $\varphi_1(\bar{x}) = \varphi_2(\bar{x})$. Assume that each φ_i is SNC at $\bar{x}$. Then:*

(i) $\max\{\varphi_1, \varphi_2\}$ *is SNC at $\bar{x}$ provided that the qualification condition* (3.79) *holds.*

(ii) $\min\{\varphi_1, \varphi_2\}$ *is SNC at $\bar{x}$ provided that*

$$\partial^\infty(-\varphi_1)(\bar{x}) \cap \big(-\partial^\infty(-\varphi_2)(\bar{x})\big) = \{0\} .$$

Proof. It follows from Proposition 3.92 that $\max\{\varphi_1, \varphi_2\}$ is SNEC at $\bar{x}$. By Proposition 3.93 it remains to show that $-\max\{\varphi_1, \varphi_2\}$ is SNEC at this point. Observe that

$$\operatorname{epi}\big(-\max\{\varphi_1, \varphi_2\}\big) = \operatorname{epi}(-\varphi_1) \cup \operatorname{epi}(-\varphi_2) .$$

Using Proposition 3.93 again, we conclude that the sets $\operatorname{epi}(-\varphi_1)$ and $\operatorname{epi}(-\varphi_2)$ are SNC at the point $(\bar{x}, \varphi_1(\bar{x})) = (\bar{x}, \varphi_2(\bar{x}))$. It easily follows from the definition of SNC sets and the decreasing property (1.5) of the sets of ε-normals that $\operatorname{epi}(-\varphi_1) \cup \operatorname{epi}(-\varphi_2)$ is also SNC at this point, which implies the SNEC property of $-\max\{\varphi_1, \varphi_2\}$. Assertion (ii) follows from (i) due to

$$\min\{\varphi_1(x), \varphi_2(x)\} = -\max\{-\varphi_1(x), -\varphi_2(x)\},$$

which completes the proof. △

Note that, in contrast to the sum operation in Corollary 3.91, the SNC property of maximum functions is ensured by the same qualification condition (3.79) as the SNEC property of such functions. Note also that the qualification conditions (3.79) and (3.80) automatically hold if one of φ_i is Lipschitz continuous around $\bar{x}$.

3.3.3 Sequential Normal Compactness for Compositions of Maps

In the final subsection of this section (and of the whole chapter) we study the PSNC and SNC properties for compositions $F \circ G$ of set-valued mappings between Asplund spaces and consider some special cases of such compositions. Based on geometric results of Subsect. 3.3.1, we obtain efficient qualification conditions for the preservation of these and related properties under various compositions. Similarly to Subsect. 3.3.2 such conditions are expressed in terms of the mixed and normal coderivatives of set-valued mappings and the singular subdifferentials of extended-real-valued functions.

The first theorem provides conditions for the preservation of the PSNC property of set-valued mappings under their general composition. Note that the qualification condition in this theorem, involving a combination of the mixed and normal coderivatives of the components, is *more restrictive* than the corresponding qualification condition sufficient for the *coderivative chain rules* derived in Theorem 3.13.

Theorem 3.95 (PSNC property of compositions). *Consider the composition* $F \circ G$ *with* $G: X \rightrightarrows Y$ *and* $F: Y \rightrightarrows Z$, *and let* $\bar{z} \in (F \circ G)(\bar{x})$. *Assume that* G *and* F^{-1} *are closed-graph near* $\bar{x}$ *and* $\bar{z}$, *respectively, and that the set-valued mapping*

$$S(x, z) := G(x) \cap F^{-1}(z) = \{y \in G(x) \mid z \in F(y)\}$$

is inner semicompact at $(\bar{x}, \bar{z})$. *Assume also that for every* $\bar{y} \in S(\bar{x}, \bar{z})$ *the following hold:*

(a) *Either* G *is PSNC at* $(\bar{x}, \bar{y})$ *and* F *is PSNC at* $(\bar{y}, \bar{z})$, *or* G *satisfies the SNC property at* $(\bar{x}, \bar{y})$.

(b) $\{F, G\}$ *satisfies the qualification condition*

$$D_M^* F(\bar{y}, \bar{z})(0) \cap \ker D_N^* G(\bar{x}, \bar{y}) = \{0\}.$$

Then the composition $F \circ G$ *is PSNC at* $(\bar{x}, \bar{z})$.

Proof. Take sequences $\varepsilon_k \downarrow 0$, $(x_k, z_k) \to (\bar{x}, \bar{z})$, $x_k^* \xrightarrow{w^*} 0$, and $\|z_k^*\| \to 0$ as $k \to \infty$ satisfying

$$z_k \in (F \circ G)(x_k) \quad \text{and} \quad x_k^* \in \widehat{D}_{\varepsilon_k}^*(F \circ G)(x_k, z_k)(z_k^*), \quad k \in I\!N\,. \tag{3.81}$$

To justify the PSNC property of $F \circ G$ at $(\bar{x}, \bar{z})$, we need to show by Definition 1.67 that $\|x_k^*\| \to 0$ along some subsequence. From the first inclusion in (3.81) one has $y_k \in S(x_k, z_k)$ for all $k \in I\!N$. Using the inner semicompactness of S and the closed-graph assumptions made, we select a subsequence of y_k that converges (without relabeling) to some $\bar{y} \in G(\bar{x}) \cap F^{-1}(\bar{z})$. Consider subsets $\Omega_1, \Omega_2 \subset X \times Y \times Z$ defined by

$$\Omega_1 := \operatorname{gph} G \times Z, \quad \Omega_2 := X \times \operatorname{gph} F\,,$$

which are locally closed around $(\bar{x}, \bar{y}, \bar{z}) \in \Omega_1 \cap \Omega_2$. It easily follows from the second inclusion in (3.81) that

$$(x_k^*, 0, -z_k^*) \in \widehat{N}_{\varepsilon_k}((x_k, y_k, z_k); \Omega_1 \cap \Omega_2), \quad k \in I\!N\,. \tag{3.82}$$

One can check that all the assumptions of Theorem 3.79 hold for the above sets Ω_1 and Ω_2 with $m = 3$ and with either $J_1 = \{1, 3\}$ and $J_2 = \{1, 2\}$, or with $J_1 = \{1, 2, 3\}$ and $J_2 = \{1\}$ depending on the alternative in (a). Applying Theorem 3.79, we conclude that the set $\Omega_1 \cap \Omega_2$ is PSNC at $(\bar{x}, \bar{y}, \bar{z})$ with respect to X. This gives by (3.82) that $\|x_k^*\| \to 0$, which completes the proof of the theorem. △

Observe that Theorem 3.84 can be derived from Theorem 3.95 with $F(y) = \delta(y; \Theta)$; this is *not* the case however for Theorem 3.88. Note also that assumptions (a) and (b) of Theorem 3.95 may be imposed only at a given point $(\bar{x}, \bar{y}, \bar{z})$ if the mapping S therein is assumed to be inner semicontinuous at this point.

Corollary 3.96 (PSNC property for compositions with Lipschitzian outer mappings). *Let $\bar{z} \in (F \circ G)(\bar{x})$, where $G\colon X \rightrightarrows Y$ and $F^{-1}\colon Z \rightrightarrows Y$ are closed-graph near $\bar{x}$ and $\bar{z}$, respectively. Assume that the mapping $G \cap F^{-1}$ is inner semicompact at $(\bar{x}, \bar{z})$ and, for every $\bar{y} \in G(\bar{x}) \cap F^{-1}(\bar{z})$, G is PSNC at $(\bar{x}, \bar{y})$ and F is Lipschitz-like around $(\bar{x}, \bar{y})$ (in particular, F is locally Lipschitzian around $\bar{x}$). Then $F \circ G$ is PSNC at $(\bar{x}, \bar{z})$.*

Proof. By Theorem 1.44 and Proposition 1.68 the main assumptions (a) and (b) of Theorem 3.95 automatically hold for Lipschitz-like mappings. △

Note that, in contrast to Corollary 3.15, the metric regularity of G at $(\bar{x}, \bar{y})$ doesn't ensure the fulfillment of assumptions (a) and (b) of Theorem 3.95 (even for $\dim Y < \infty$ when (b) automatically holds), since G may not be PSNC at $(\bar{x}, \bar{y})$ in this case.

Theorem 3.95 implies the following result on the SNEC property of compositions involving extended-real-valued outer functions and single-valued inner mappings between Asplund spaces.

Corollary 3.97 (SNEC property of compositions). *Let $g: X \to Y$ be continuous around $\bar{x}$, and let $\varphi: Y \to \overline{I\!R}$ be proper and l.s.c. around $\bar{y} := g(\bar{x})$. Assume that either g is PSNC at $\bar{x}$ and φ is SNEC at $\bar{y}$, or g is SNC at $\bar{x}$. Then $\varphi \circ g$ is SNEC at $\bar{x}$ provided that*

$$\partial^\infty \varphi(\bar{y}) \cap \ker D_N^* g(\bar{x}) = \{0\} .$$

In particular, $\varphi \circ g$ is SNEC at $\bar{x}$ if φ is locally Lipschitzian around $\bar{y}$, and if g is continuous around $\bar{x}$ and PSNC at this point.

Proof. Follows from Theorem 3.95 and Corollary 3.96 by simply putting $F := E_\varphi$ and $G := g$. △

Next we obtain conditions ensuring the preservation of the SNC property under compositions of set-valued mappings between Asplund spaces.

Theorem 3.98 (SNC property of compositions). *Let $\bar{z} \in (F \circ G)(\bar{x})$, where $G: X \rightrightarrows Y$ and $F^{-1}: Z \rightrightarrows Y$ are closed-graph near $\bar{x}$ and $\bar{z}$, respectively. Assume that $G \cap F^{-1}$ is inner semicompact at $(\bar{x}, \bar{z})$ and that for every $\bar{y} \in G(\bar{x}) \cap F^{-1}(\bar{z})$ the following hold:*

(a) *Either G is PSNC at $(\bar{x}, \bar{y})$ and F is SNC at $(\bar{y}, \bar{z})$, or G is SNC at $(\bar{x}, \bar{y})$ and F^{-1} is PSNC at $(\bar{z}, \bar{y})$; this happens, in particular, when both G and F are SNC at the corresponding points.*

(b) *$\{F, G\}$ satisfies the qualification condition*

$$D_N^* F(\bar{y}, \bar{z})(0) \cap \ker D_N^* G(\bar{x}, \bar{y}) = \{0\} .$$

Then the composition $F \circ G$ is SNC at $(\bar{x}, \bar{z})$.

Proof. To justify the SNC property of $F \circ G$ at $(\bar{x}, \bar{z})$, we need to show that for any sequences $\varepsilon_k \downarrow 0$, $(x_k, z_k) \to (\bar{x}, \bar{z})$ with $(x_k, z_k) \in \operatorname{gph}(F \circ G)$, and

$$(x_k^*, z_k^*) \in \widehat{N}_{\varepsilon_k}((x_k, z_k); \operatorname{gph}(F \circ G)) \quad \text{with} \quad (x_k^*, z_k^*) \xrightarrow{w^*} (0, 0)$$

one has $\|(x_k^*, z_k^*)\| \to 0$ along some subsequence. Following the proof of Theorem 3.95, we consider the sets Ω_1 and Ω_2 defined there and observe that

$$(x_k^*, 0, z_k^*) \in \widehat{N}_{\varepsilon_k}((x_k, y_k, z_k); \Omega_1 \cap \Omega_2), \quad k \in I\!N ,$$

with $y_k \to \bar{y} \in G(\bar{x}) \cap F^{-1}(\bar{z})$ selected by the inner semicompactness property of $G \cap F^{-1}$. Using the structure of the sets Ω_1 and Ω_2, one can check that all the assumptions of Theorem 3.79 hold with either $J_1 = \{1, 3\}$ and $J_2 = \{1, 2, 3\}$, or with $J_1 = \{1, 2, 3\}$ and $J_2 = \{1, 3\}$ depending on the alternative in (a). Hence Theorem 3.79 ensures that $\Omega_1 \cap \Omega_2$ is PSNC at $(\bar{x}, \bar{y}, \bar{z})$ with respect to $\{X, Z\}$, which implies that $\|(x_k^*, z_k^*)\| \to 0$ and completes the proof of the theorem. △

Combining Theorems 3.88, 3.90, 3.95, 3.98 and their corollaries, one can obtain results on PSNC and SNC properties of various compositions including, in particular, h-compositions considered in Subsect. 3.1.2. For example, we present below some results concerning binary operations over real-valued continuous functions that include, in particular, their products and quotients. To proceed, we first establish the following relationship between the SNC property for continuous functions $\varphi_i: X \to I\!R$ and their *aggregate mapping* $(\varphi_1, \varphi_2): X \to I\!R^2$ in Asplund spaces.

Proposition 3.99 (SNC property of aggregate mappings). *Let $\varphi_i: X \to I\!R$, $i = 1, 2$, be continuous around $\bar{x}$ and satisfy the qualification condition* (3.80). *Then both φ_i are SNC at $\bar{x}$ if and only if the aggregate mapping $(\varphi_1, \varphi_2): X \to I\!R^2$ is SNC at this point.*

Proof. Let φ_1 and φ_2 be SNC at $\bar{x}$. Then the mappings $f_i: X \to I\!R^2$ with $f_i(x) = (\varphi_i(x), 0)$, $i = 1, 2$, are clearly SNC at this point. It follows from Theorem 2.40 that

$$D^* f_i(\bar{x})(0) \subset \partial^\infty \varphi_i(\bar{x}) \cup \partial^\infty(-\varphi_i)(\bar{x}), \quad i = 1, 2 .$$

Since $(\varphi_1, \varphi_2) = f_1 + f_2$, we conclude that the mapping (φ_1, φ_2) is SNC at $\bar{x}$ due to Theorem 3.90.

Conversely, assume that (φ_1, φ_2) is SNC at $\bar{x}$. Then we derive the SNC property of each φ_i by applying Theorem 3.98 to $F_i \circ G$ with, respectively, $G(x) := (\varphi_1(x), \varphi_2(x))$ and $F_i(y_1, y_2) := y_i$, $i = 1, 2$. △

Now combining Proposition 3.99 with the above results on the SNEC and SNC properties of compositions, we derive conditions ensuring these properties for an abstract binary operation defined by some function $\upsilon: I\!R^2 \to \overline{I\!R}$.

Corollary 3.100 (SNEC and SNC properties for binary operations). *Let $\varphi_i: X \to I\!R$, $i = 1, 2$, be continuous around $\bar{x}$, and let $\upsilon: I\!R^2 \to \overline{I\!R}$ be l.s.c. around $\bar{y} := (\varphi_1(\bar{x}), \varphi_2(\bar{x}))$. Assume that each φ_i is SNC at $\bar{x}$ and that $\{\varphi_1, \varphi_2\}$ satisfies the qualification condition* (3.80). *Then the following hold:*

(i) *$\upsilon(\varphi_1, \varphi_2)$ is SNEC at $\bar{x}$ provided that $\partial^\infty \upsilon(\bar{y}) = \{0\}$.*

(ii) *$\upsilon(\varphi_1, \varphi_2)$ is SNC at $\bar{x}$ provided that υ is continuous around $\bar{y}$ and that*

$$\partial^\infty \upsilon(\bar{y}) \cup \partial^\infty(-\upsilon)(\bar{y}) = \{0\} .$$

Proof. Assertion (i) follows from Proposition 3.99 and Corollary 3.97 applied to the composition $\upsilon \circ f$ with $f(x) := (\varphi_1(x), \varphi_2(x))$. Assertion (ii) follows from Proposition 3.99 and Theorem 3.98 applied to the composition $\upsilon \circ f$, where the qualification condition (b) holds due to $D^*\upsilon(\bar{y})(0) = \partial^\infty \upsilon(\bar{y}) \cup \partial^\infty(-\upsilon)(\bar{y})$ by Theorem 2.40(ii). △

Note that Corollary 3.100 implies Corollary 3.91 but not Corollaries 3.89 and 3.94, where the qualification conditions are less restrictive due to specific features of the *unilateral* operations under consideration. Let us finally present direct consequences of Corollary 3.100 in the cases of product and quotient operations.

Corollary 3.101 (SNC property of products and quotients). *Let φ_i, $i = 1, 2$, be continuous around $\bar{x}$ and SNC at this point. Assume that the qualification condition* (3.80) *holds. Then the product $\varphi_1 \cdot \varphi_2$ is SNC at $\bar{x}$. If in addition $\varphi_2(\bar{x}) \neq 0$, then the quotient φ_1/φ_2 is also SNC at this point.*

Proof. The product and quotient results follow from Corollary 3.100(ii) with $\upsilon(y_1, y_2) := y_1 \cdot y_2$ and $\upsilon(y_1, y_2) := y_1/y_2$, respectively. △

Remark 3.102 (calculus for CEL property of sets and mappings). As mentioned in Remark 1.27(ii), the compactly epi-Lipschitzian (CEL) property of closed sets in Asplund spaces admits a complete characterization in the form similar to the SNC property with the only difference that the weak* convergence of *sequences* of Fréchet normals is replaced by the same convergence of *bounded nets.* Involving now the results from Fabian and Mordukhovich [422], we conclude that the SNC and CEL property *agree* in *weakly compactly generated* Asplund spaces (in particular, in either reflexive Banach spaces or separable Asplund spaces), while they may be different in the nonseparable setting. Thus the above results concerning the SNC property of sets and mappings provide the corresponding CEL calculus in WCG Asplund spaces.

Furthermore, it is proved by Ioffe [607] that such a weak* *topological* (bounded net) description of closed CEL sets holds true in arbitrary *Banach spaces* if the Fréchet normal cone is replaced by the *nucleus* of the *G-normal cone* defined in (2.76). Using this description and the procedure developed above, we can get results on the preservation of the CEL property under various operations on sets and mappings in Banach spaces similar to those obtained for the SNC property in Asplund spaces. The principal difference between these results is that in arbitrary Banach spaces we need to use (instead of our basic normals, subgradients, and normal coderivatives) nuclei of the G-normal cone and the associated coderivative and subdifferential constructions for mappings and functions in formulations of the corresponding *normal* qualification conditions. The latter relates to the fact that the G-normal cone provides a topological normal structure in general Banach spaces; see Sect. 2.5. In this way we get, in particular, analogs of Corollary 3.81, Theorem 3.84, Theorem 3.86 (for inequality and Lipschitzian equality constraints), Proposition 3.92, and Theorems 3.90 and 3.98 (with net counterparts of inner semicompactness) ensuring the preservation of the CEL property under general operations in arbitrary Banach spaces. Similar results in this direction related to Corollary 3.81 and to a special case of Theorem 3.98 can be found in Jourani [648] with a different proof.

Finally, note that one doesn't need any SNC calculus in finite dimensions, since every set there is automatically SNC. Hence the qualification conditions obtained in this section for SNC calculus exclusively relate to variational analysis in infinite-dimensional spaces. However, in finite dimensions they reduce to qualification conditions that are needed for *calculus rules* involving basic normals, subgradients, and coderivatives crucial for any applications of generalized differentiation. Thus the development of the SNC calculus, which is one of the most fundamental ingredients of infinite-dimensional variational analysis, leads us to a *unified theory* efficient in applications to various problems in both finite-dimensional and infinite-dimensional settings; see the subsequent chapters of this book.

Remark 3.103 (subdifferential calculus and related topics in Asplund generated spaces). Most of the results presented in this chapter involving *Fréchet-like* generalized differential constructions and their *sequential* limits require the *Asplund structure* of the Banach space in question. Our approach is mainly based on the *extremal principle* of variational analysis and its equivalent descriptions, for the validity of which the Asplund property is *necessary* as long as one deals with Fréchet-like differentiability and subdifferentiability. The Fréchet-like constructions involved and their sequential regularizations seem to be strong and natural from the viewpoints of both classical and generalized differentiation, and many crucial results and techniques developed in this book essentially employ these structures. There are other generalized differential constructions successfully used in nonsmooth analysis along with those studied in this book being, however, either essentially larger, or more complicated (involving particularly *topological/net* weak* limits), or restrictive to narrow classes of Banach spaces; see the results and discussions in Sect. 2.5 and Subsect. 3.2.3 with related comments and references.

It is interesting to clarify the possibility of extending the approach based on Fréchet-like constructions and their sequential limits to a larger class of Banach spaces that includes *all separable* spaces, which are probably the most important for applications. This work has been started by Fabian, Loewen and Mordukhovich [418] in the so-called Asplund generated spaces (AGS) that form a common roof for Asplund spaces and for weakly compactly generated spaces containing, in particular, all separable Banach spaces. A Banach space $(X, \|\cdot\|_X)$ is *Asplund generated* if there exist an Asplund space $(Y, \|\cdot\|_Y)$ and a linear bounded operator $A\colon Y \to X$ such that its range AY is dense in X; see Fabian's book [416]. Besides Asplund spaces themselves, the class of Asplund generated spaces include the following:

1. The *Lebesgue space* $X = L^1(\Omega, \Sigma, \mu, Z)$ is Asplund generated provided that (Ω, Σ, μ) is a fine measure space and Z is AGS. In this case one has $Y = L^2(\Omega, \Sigma, \mu, Z)$ and $\|\cdot\|_Y = \|\cdot\|_{L^2}$.

2. The space $\mathcal{C}(K)$ of *continuous functions* defined on a compact space K is Asplund generated *if and only if* K is homeomorphic to a weak* compact subset of Z^* for some Asplund space Z. Here the construction of Y is much more involved in comparison with the preceding example; see Theorem 1.2.4 in the afore-mentioned book by Fabian [416].

3. Every *separable Banach space* X is Asplund generated. Indeed, every such X contains the dense linear image of the Hilbert space ℓ^2. To see this, fix some countable set $\{x_k|\ k \in \mathbb{N}\}$ dense in the unit ball of X and define the mapping $A\colon \ell^2 \to X$ by

$$A(z) := \sum_{k=1}^{\infty} 2^{-k} z_k x_k \quad \text{whenever} \quad z = (z_1, z_2, \ldots) \in \ell^2 .$$

Clearly A is a linear bounded operator of dense range.

4. Every *weakly compactly generated* (WCG) Banach space X is Asplund generated. Since every separable space is WCG, this class of AGS is a generalization of the one in Item 3. However, the choice of Y in this case is much more difficult although the proof is constructive: in fact, Y may be chosen as a reflexive space as shown [416, Theorem 1.2.3]. Note to this end that, as proved in Theorem 1.2.4 of the latter book, $\mathcal{C}(K)$ is WCG *if and only if* K is an Eberlein compact; cf. Item 2.

If X is an AGS with $Y \subset X$ and with $A = \mathcal{I}\colon Y \to X$ being the injective/inclusion operator, the quadruple $(X, \|\cdot\|_X, Y, \|\cdot\|_Y)$ is called an *Asplund embedding scheme*. Note that every Asplund generated spaces can be realized as an Asplund generated scheme, and vice versa. It is more convenient to deal with Asplund generated scheme defining *normals* and *subgradients* in what follows. Given $\Omega \subset X$ and $\bar{x} \in \Omega \cap Y$ in such a scheme, we let

$$N_Y(\bar{x}; \Omega) := \mathcal{I}^{*-1}\big(N(\bar{x}; \Omega \cap Y\big) ,$$

where the basic normal cone on the right is calculated in the Asplund space Y. Similarly, given a proper function $\varphi\colon X \to \overline{\mathbb{R}}$ with $\bar{x} \in \operatorname{dom}\varphi \cap Y$, define

$$\partial_Y \varphi(\bar{x}) := \mathcal{I}^{*-1}\big(\partial(\varphi|_Y)(\bar{x})\big) \quad \text{and} \quad \partial_Y^\infty \varphi(\bar{x}) := \mathcal{I}^{*-1}\big(\partial^\infty(\varphi|_Y)(\bar{x})\big) .$$

The idea behind these definitions is to carry out the appropriate normal and subgradient computations in the Asplund space Y, thereby obtaining subsets of Y^*, and then to truncate those subsets to the space X^* by considering their inverse images under $\mathcal{I}^*$. It is shown in the afore-mentioned paper by Fabian, Loewen and Mordukhovich that for locally Lipschitzian functions φ one has

$$\mathcal{I}^*\big(\partial_Y \varphi(\bar{x})\big) = \partial(\varphi|_Y)(\bar{x}) \neq \emptyset \quad \text{and} \quad \mathcal{I}^*\big(\partial_Y^\infty \varphi(\bar{x})\big) = \partial^\infty(\varphi|_Y)(\bar{x}) = \{0\} .$$

Furthermore, there are *calculus rules*

$$N_Y(\bar{x}; \Omega_1 \cap \Omega_2) \subset N_Y(\bar{x}; \Omega_1) + N_Y(\bar{x}; \Omega_2) ,$$

$$\partial_Y(\varphi_1 + \varphi_2)(\bar{x}) \subset \partial_Y \varphi_1(\bar{x}) + \partial_Y \varphi_2(\bar{x}) ,$$

$$\partial_Y^\infty(\varphi_1 + \varphi_2)(\bar{x}) \subset \partial_Y^\infty \varphi_1(\bar{x}) + \partial_Y^\infty \varphi_2(\bar{x})$$

for normals to closed sets and subgradients of l.s.c. functions, respectively, provided the *qualification conditions*

$$N_Y(\bar{x};\Omega_1)\cap\big(-N_Y(\bar{x};\Omega_2)\big)=\{0\},\quad \partial_Y^\infty\varphi_1(\bar{x})\cap\big(-\partial_Y^\infty\varphi_2(\bar{x})\big)=\{0\}\;,$$

the *Y-SNC conditions* on one of the sets/functions naturally defined by restriction to the Asplund space Y, and the following *properness conditions*

$$\mathcal{I}^*\big(N_Y(\bar{x});\Omega_i)\big)=N(\bar{x};\Omega_i\cap Y)\quad\text{for some}\quad i\in\{1,2\}\;,$$

$$\mathcal{I}^*\big(\partial_Y\varphi_i(\bar{x})\big)=\partial(\varphi_i|_Y)(\bar{x}),\quad \mathcal{I}^*\big(\partial_Y^\infty\varphi_i(\bar{x})\big)=\partial^\infty(\varphi_i|_Y)(\bar{x})$$

for some $i\in\{1,2\}$. Note that the qualification and properness conditions are *automatic* when, respectively, one of the functions φ_i is locally *Lipschitzian* and one of the sets Ω_i is *epi-Lipschitzian* around the reference points. The presented calculus results provide the ground for deriving other calculus rules of generalized differentiation in Asplund generated spaces similarly to those developed in this chapter in the Asplund space setting.

3.4 Commentary to Chap. 3

3.4.1. The Key Role of Calculus Rules. Results of this chapter make a *bridge* between *generalized differentiation* and the majority of its *applications* to variational problems, particularly those considered in the book. Indeed, any constructions and properties introduced are of a potential use only if they enjoy satisfactory *calculus rules*, i.e., can be computed, efficiently estimated, and/or preserved under various operations. The great success of the classical differential theory with its numerous applications is mainly due to the comprehensive calculus enjoyed (almost for granted) by the classical derivatives. The same can be said about subgradients of convex analysis, where calculus rules are far to be trivial though: their proofs are strongly based on *convex separation.*

As seen in Chap. 1, a number of useful calculus rules are available for our basic generalized differential constructions in arbitrary Banach spaces. However, most of them are restricted by, e.g., smoothness requirements on some of the mappings involved in compositions. In this chapter we show based mainly on the *extremal principle* developed in Chap. 2 that none of such restrictions is needed in the framework of *Asplund spaces*, where our basic normal, coderivative, and subdifferential (of first and second order) constructions indeed enjoy fairly rich/full calculi that are the key for subsequent applications.

It should be added that, in *infinite*-dimensional spaces, *SNC calculus* rules (i.e., efficient conditions ensuring the preservation of such normal compactness properties under various operations) are also of fundamental importance for both the theory and applications. This is mainly due to the fact that SNC requirements are critical for the fulfillment of calculus rules for generalized differentiation in infinite dimensions; so one cannot proceed with applications of generalized differential calculus without ensuring the preservation of SNC properties under the corresponding operations. Such a SNC calculus has been quite recently developed (see below); it is presented in this chapter and plays a fundamental role in all the subsequent applications given in the book. This calculus is also based on the *extremal principle* of variational analysis developed in Chap. 2.

3.4.2. Dual-Space Geometric Approach to Generalized Differential Calculus. The approach to calculus presented in this book is mainly *geometric* (in dual spaces), i.e., we first establish calculus rules for generalized normals to arbitrary closed sets and then successively apply them to coderivatives of set-valued mappings and subgradients of extended-real-valued functions. This approach was initiated and developed by Mordukhovich [894, 901, 910] in the finite-dimensional framework, with using the (exact) *extremal principle* as the key tool to derive an *intersection rule* for basic normals that occurs to be the *central result* of all the *nonconvex calculus.*

Subsection 3.1.1 is mostly devoted to calculus rules for *basic normals* in the framework of Asplund spaces. From this viewpoint, Lemma 3.1 on a *fuzzy* intersection rule for Fréchet normals is a preliminary result, which however plays a major technical role in what follows. It was derived by Mordukhovich and B. Wang [963] from the approximate extremal principle. Note that, although calculus issues don't have an optimization/variational nature as given, the structure of Fréchet normals allows us to form a special *extremal* system of closed sets and then to apply the extremal principle. Observe also some similarities between employing the extremal principle in such a general nonconvex setting and the usage of the classical separation theorem in the corresponding framework of convex analysis (see, e.g., the "alternative" geometric proof of Theorem 23.8 in Rockafellar [1142]); note however that there is no need to form an extremal system of sets in the convex setting.

While the assertion of Lemma 3.1 doesn't require any qualification conditions (and it doesn't actually provide a rule to estimate Fréchet normal of $\Omega_1 \cap \Omega_2$ when $\lambda = 0$), such conditions are unavoidable to derive a "real" intersection rule for basic normals. The basic *normal qualification condition* (3.10) from Definition 3.2(i) was introduced by Mordukhovich [894] to establish the intersection rule for basic normals from Theorem 3.4 in finite dimensions. Ioffe [596] independently obtained this intersection rule, by using a penalty function method, under the more restrictive *tangential* qualification condition

$$T_C(\bar{x}; \Omega_1) - T_C(\bar{x}; \Omega_2) = I\!R^n$$

involving the Clarke tangent cone. Rockafellar [1155] (independently as well) used a counterpart of the qualification condition (3.10) formulated however in terms of the Clarke normal cone to derive an analog of the intersection rule (3.11) for *Clarke normals* in finite-dimensional spaces.

The *limiting qualification condition* from Definition 3.2(ii) was introduced by Mordukhovich and B. Wang [963]. It is equivalent to the normal condition (3.10) in finite-dimensional spaces being generally weaker in *infinite dimensions* as discussed in Subsect. 3.1.1. One of the strongest *advantages* of this limiting qualification condition in comparison with the normal one (3.10) is that it leads to significantly better results in applications to *coderivative calculus* for set-valued mappings between infinite-dimensional spaces; see Subsect. 3.1.2.

3.4.3. Normal Compactness Conditions in Infinite Dimensions. It has been well recognized starting with convex analysis that, besides qualification conditions needed in both finite and infinite dimensions, conditions of *another nature* are required to ensure the fulfillment of calculus rules in infinite-dimensional spaces; for the case of (two) convex set intersections the latter conditions usually involve the nonempty interior assumption imposed on one of the sets. The *partial sequential normal compactness* properties formulated in Definition 3.3 are probably the *weakest* conditions of the latter type; even for *convex* sets they significantly *improve* the standard assumptions involving *nonempty interiors*. For the general case of sets in *product* spaces these conditions were defined in the afore-mentioned paper [963], while the PSNC property for *graphs* of mappings was studied earlier; see Subsect. 1.2.5 and the corresponding comments to Chap. 1 given in Subsect. 1.4.15. It seems that the *strong* PSNC property haven't been explicitly recognized before Mordukhovich and B. Wang [963], although for the case of mappings it follows from the *partial CEL* property by Jourani and Thibault [655]; cf. Theorem 1.75. Note that for subsets of spaces with no product structures both PSNC properties of Definition 3.3 reduce to the basic SNC property studied in Subsect. 1.1.3; see also the comments in Subsect. 1.4.11.

3.4.4. Calculus Rules for Basic Normals. The full statement of Theorem 3.4 is due to Mordukhovich and B. Wang [963]; its important Corollary 3.5 in spaces with no product structure was derived earlier by Mordukhovich and Shao [949] under the normal qualification condition (3.10). The example presented after this corollary, which shows that the SNC assumption is essential for the validity of the intersection rule even for *convex* subsets of *any* infinite-dimensional space, is taken from Borwein and Zhu [162]. The more involved Example 3.6 showing that the SNC assumption in Corollary 3.5 is *strictly weaker* than the CEL one even for *convex subcones* in *smooth spaces* is built upon the construction from Fabian and Mordukhovich [422].

In the case of Banach spaces with *Fréchet smooth* renorms the intersection rule (3.11) was established in the paper by Kruger [708], which is largely based on his dissertation [706], under the *epi-Lipschitzian* assumption on one of the sets and an significantly more restrictive, in comparison with the normal one (3.10), *tangential* qualification condition formulated in terms of Clarke's tangent cone. Similar results with the same epi-Lipschitzian and tangential qualification conditions were obtained by Ioffe [597, 599] for his analytic and geometric "approximate" normal cones in more general spaces. Note that both latter cones may be bigger than our basic normal cone even for epi-Lipschitzian subsets of Fréchet smooth spaces; see Subsect. 2.5.2B and the subsequent discussions presented in Subsect. 3.2.3. Further extensions of the afore-mentioned results to the case of CEL subsets in Banach spaces were developed by Jourani and Thibault [658].

To best of our knowledge, the *sum rule* for basic normals from Theorem 3.7(ii) in finite-dimensional spaces was first formulated in Rockafellar and Wets [1165, Exercise 6.44], although it was actually proved earlier by Rockafellar [1155, Corollary 6.2.1] with Clarke normals replacing basic normals in the right-hand side (but not in the left-hand side) of the inclusion in Theorem 3.7(ii). The full statement of the latter result is due to another paper by Mordukhovich and B. Wang [966]. It is interesting to observe that, in contrast to the intersection rule of Theorem 3.4, we *don't* need to impose for the sum rule either qualification and SNC conditions in infinite dimensions; in fact they hold *automatically* in this setting as shown in the proof of Theorem 3.7.

Computing and estimating generalized normals to *inverse image/preimage* sets are very useful in applications, especially to optimization problems; see, e.g., Borwein and Zhu [164], Mordukhovich [901], Rockafellar and Wets [1165] with the references therein, and the subsequent material of this book. Theorem 3.8 on basic normals to inverse images of sets under set-valued mappings was derived by Mordukhovich and B. Wang [963] (as an extension of the previous results obtained Mordukhovich [908] and by Mordukhovich and Shao [950]) from the main intersection rule of Theorem 3.4. Note that all the results in [963] have been established with respect to any *reliable topology* τ used in the constructions of τ-limiting normals, subgradients, and coderivatives as well as in the definitions of the corresponding τ-SNC properties; see [963] and Remark 3.23 in this book for more details and discussions. Choosing an appropriate topology, we can get better results in comparison with the standard limiting constructions that don't take into account available *product structures* of the spaces and (graphical) sets in question. Observe, in particular, a remarkable role of the *reversed mixed coderivative* $\widetilde{D}^*_M F(\bar{x},\bar{y})$ in the qualification condition (b) of Theorem 3.8, which corresponds to the *mixed* topology $\tau = \|\cdot\| \times w^*$ on the product space $X^* \times Y^*$ and allows us to ensure the fulfillment of the inverse image rule (3.15) for *metrically regular* mappings due to the respective coderivative results of Chap. 1; see Corollary 3.9 and its proof. Note also that inverse image rules can be considered as specifications of coderivative *chain rules* for set-valued mappings and their subdifferential counterparts in the case of single-valued ones; see below.

3.4.5. Full Coderivative Calculus. The *coderivative calculus* rules presented in Subsect. 3.1.2 were first established by Mordukhovich [910] for set-valued mappings between finite-dimensional spaces, while the sum rule of

Theorem 3.10(ii) appeared a bit earlier in [908] with a somewhat different proof based directly on the method of *metric approximations*. We also refer the reader to the book by Rockafellar and Wets [1165] that reproduced the major coderivative rules of [910] in finite-dimensional spaces. Observe the *pivoting role* of *summation results* in our approach to coderivative and subdifferential calculi, while the approach of [1165] started with chain rules.

The first version of Theorem 3.10 in infinite dimensions (Asplund spaces) was obtained by Mordukhovich and Shao [950] for the case of $D^* = D^*_N$ with the more demanded qualification condition in form (3.19) formulated via the *normal* coderivative. The latter condition was improved in Mordukhovich [917] and in Mordukhovich and Shao [953] to that of (3.19) formulated via the *mixed* coderivative $D^* = D^*_M$, which was found to be sufficient for ensuring the coderivative chain rules of Theorem 3.10 in *both cases* of $D^* = D^*_N$ and $D^* = D^*_M$. The proofs given in all these papers were largely similar to the one in [910], with using first the *approximate extremal principle* in infinite-dimensional settings (instead of the *exact* extremal principle as in [910] for finite dimensions) in the coderivative framework and then passing to the limit; cf. also the subsequent paper by Mordukhovich and Shao [952] for "fuzzy" coderivative versions based on this approach.

The proof presented in the book was given by Mordukhovich and B. Wang [963] applying the normal cone intersection rules from Theorem 3.4 and Lemma 3.1, which are also based on the extremal principle while following a more direct and unified geometric approach. Note that we need to use the case of $m = 3$ in the *product structure* of Theorem 3.4 and the *limiting* (not normal) qualification condition therein to arrive at the strongest coderivative sum rules established in Theorem 3.10 with all the *pointbased* assumptions, i.e., those expressed at the reference points but not in their neighborhoods. One of the most essential advantages of using the *mixed* – in contrast to normal – coderivative in the qualification condition (3.19) and the *partial* SNC property in Theorem 3.10 is the *automatic* validity of both these assumptions for *Lipschitz-like* mappings due to the *necessary* coderivative conditions for Lipschitzian behavior established in Chap. 1; see Corollary 3.11.

The *chain rules* of Theorem 3.13 were established by Mordukhovich and Shao [917, 953] in full generality; the previous versions were given in the aforementioned [910, 950, 952]. Observe again that all the assumptions of this theorem are *pointbased* and that the *mixed* qualification condition is imposed in (3.27) to ensure the chain rules for both normal and mixed coderivatives, while the *normal* coderivative of the inner (but not of the outer) mapping is present in both – normal and mixed – coderivative chain rules. Note also that the *equality* assertion (iii) of Theorem 3.13 provides various useful conditions for preserving the *normal* and *mixed regularity* of mappings under compositions.

The chain rules of the *inclusion* type from Theorem 3.13 for the *normal* coderivatives generated by our basic normal cone in Asplund spaces and also by the *nucleus* of Ioffe's *G-normal cone* from Subsect. 2.5.2B in arbitrary

Banach spaces under the *normal* qualification condition and its G-normal counterpart, respectively, were proved by Ioffe and Penot [614] and by Jourani and Thibault [659, 660] using somewhat similar methods involving Ekeland's variational principle; see these papers for more information and discussions. Sum rules for the normal coderivatives under normal qualification conditions were deduced in [614, 659, 660] from the corresponding chain rules. We also refer the reader to the paper by Mordukhovich, Shao and Zhu [954], where sum and chain rules similar to Theorems 3.10 and 3.13 were derived for *topological/net viscosity* counterparts of our normal and mixed coderivatives under *mixed* qualification conditions in Banach spaces admitting *smooth* bump functions with respect to an arbitrary given bornology.

The so-called *zero chain rule* for mixed coderivatives was established by Mordukhovich and Nam [934]. Its main differences from the general chain rules of Theorem 3.13 are as follows:

(a) it concerns *mixed* coderivatives of compositions $F \circ G$ with *Lipschitz-like* inner mappings G and applies only to the *zero* coderivative argument ($z^* = 0$);

(b) it provided an upper estimate for the mixed coderivative of $F \circ G$ via the *mixed* coderivative of G vs. its normal coderivative as in Theorem 3.13.

This modification of the general coderivative chain rules happens to be useful for many applications; see, e.g., Chap. 4.

The usage of the mixed vs. normal coderivatives in the afore-mentioned *chain rules* allows us to *automatically* ensure the validity of these crucial results of coderivative calculus for *Lipschitz-like* outer mappings and *metrically regular* inner mappings in compositions in both cases of finite-dimensional and infinite-dimensional spaces. The corresponding Corollary 3.15 was first established by Mordukhovich [910] in finite dimensions and then by Mordukhovich and Shao [952] in Asplund spaces; see also Jourani and Thibault [660] for another proof of the latter result and its (not full) analog for "approximate" G-coderivatives required the finite-dimensionality of some spaces involved. An "approximate" coderivative chain rule for compositions $f \circ g$ of single-valued and Lipschitz continuous mappings was earlier derived by Ioffe [599] in the general Banach space setting directly from the corresponding results of subdifferential calculus. The results on *h-compositions* from Theorem 3.18 were derived by Mordukhovich and B. Wang [963] in full generality; previous calculus rules in this direction were obtained in the afore-mentioned papers [910, 950, 952].

We refer the reader to Borwein and Zhu [163, 164], Ioffe and Penot [614], Mordukhovich [917], Mordukhovich and Shao [952], and Mordukhovich, Shao and Zhu [954] concerning various versions of *fuzzy calculus* rules for coderivatives that are not considered in this book; see however some discussions in Remark 3.21.

3.4.6. Strictly Lipschitzian Behavior of Mappings in Infinite Dimensions. Strictly Lipschitzian properties considered in Subsect. 3.1.3 specifically concern single-valued mappings $f: X \to Y$ with *infinite-dimensional range* spaces; these properties obviously reduce to the classical local Lipschitzian behavior of f when the dimension of Y is finite. The main *strictly Lipschitzian* property from Definition 3.25(i) was first formulated by Mordukhovich and Shao [949], while it occurred to be equivalent to the basic version of "compactly Lipschitzian" behavior introduced and investigated much earlier by Thibault [1245, 1246] in connection with subdifferential calculus for vector-valued functions; see Thibault's paper [1252] for proving this equivalence and the joint papers by Jourani and Thibault [654, 656, 657, 658] for the study and applications of its "strongly compactly Lipschitzian" variant. The latter property is related to the existence of "strict prederivatives" in the sense of Ioffe [589] with norm compact values; see Ioffe's papers [595, 604] and his joint publication by Ginsburg [506]. It follows from the afore-mentioned papers that the collection of strictly/compactly Lipschitzian mappings includes, besides strictly differentiable ones, various classes of nonsmooth operators important for many applications; in particular, the so-called Fredholm and Fredholm-like operators arising in applications to problems of optimal control.

The w^*-*strictly Lipschitzian* property of single-valued mappings from Definition 3.25(ii) appeared in Mordukhovich and B. Wang [965], where the reader could find Proposition 3.26 on the equivalence of this modification to the basic strictly Lipschitzian property from Definition 3.25(i) for mappings with values in Banach spaces whose dual unit balls are weak* sequentially compact. The same paper [965] contains assertion (i) of Lemma 3.27 and the scalarization formula of Theorem 3.28 for the normal coderivative of w^*-strictly Lipschitzian mappings, while the proofs of these results were actually given by Mordukhovich and Shao [949] for strictly Lipschitzian mappings defined on Asplund spaces. The *converse* assertion (ii) of Lemma 3.27 for mappings with values in reflexive spaces follows from the proof given by Ngai, Luc and Théra [1007].

The *scalarization formula* of Theorem 3.28 taken from [949, 965] establishes an *precise* relationship between the *normal* coderivative of w^*-strictly Lipschitzian mappings $f: X \to Y$ and the basic subdifferential of their scalarization, which plays a crucial role in many subsequent applications presented in this book. When the range space Y is finite-dimensional, it agrees with the scalarization result of Theorem 1.90 for the mixed coderivative of locally Lipschitzian mappings; see the references and discussions in Subsect. 1.4.16. A counterpart of Theorem 3.28 involving "nuclei of G-coderivatives" (see Subsect. 2.5.2B) was obtained by Ioffe [599] for Lipschitz continuous mappings between Banach spaces admitting strict prederivatives with norm compact values; cf. also the more recent paper by Ioffe [604] for further developments and modifications of the latter result under the corresponding "directional compactness" assumptions.

The notion of *compactly strictly Lipschitzian* mappings from Definition 3.32 was introduced by Ngai, Luc and Théra [1007] who established the coderivative characterization of this property presented in Lemma 3.33. We use the latter notion to formulate the *generalized Fredholm property* of Definition 3.34, which extends the "semi-Fredholm" notion by Ioffe [604] corresponding to Definition 3.34 with $g: X \to Y$ satisfying the "uniform directional compactness" property formulated after that definition. The PSNC result of Theorem 3.35 is new, while it has its "codirectional compact" counterpart established by Ioffe [604] for semi-Fredholm mappings f and compactly epi-Lipschitzian sets Ω in the general Banach space framework of case (b).

3.4.7. Full Subdifferential Calculus. Subsection 3.2.1 contains the main *calculus rules* for our *basic and singular subgradients* of extended-real-valued functions in the Asplund space setting. Some of these subdifferential calculus rules follow directly from the corresponding calculus results for basic normals and coderivatives of general sets and mappings, while the others take into account specific features of extended-real-valued functions.

The *summation rules* from Theorem 3.36 were established by Mordukhovich and Shao [949] with the SNEC assumption replaced by somewhat more restrictive "normal compactness" property of functions corresponding in fact to the CEL property of their epigraphs; the proof given in [949] holds true nevertheless under the SNEC assumption. When $\dim X < \infty$, the sum rule (3.39) for basic subgradients under the qualification condition (3.38) goes back to Mordukhovich [894], while the *singular subdifferential* result (3.40) was first observed by Rockafellar in his privately circulated notes [1158]; see also Mordukhovich [907] and Rockafellar and Wets [1165]. The Lipschitzian as well as directionally Lipschitzian cases in (3.39) correspond to the sum rules obtained by Kruger [706, 708] for basic subgradients of functions defined on Fréchet smooth spaces and by Ioffe [590, 592, 599] for "approximate" subgradients in the general Banach space setting. The latter result was extended by Jourani and Thibault [658] under the more general CEL property of l.s.c. functions.

The first upper estimates for the basic and singular subdifferentials of the *marginal functions*

$$\mu(x) = \inf\{\varphi(x,y) \mid y \in G(x)\} \tag{3.83}$$

considered in Theorem 3.38 were obtained by Rockafellar [1150] in finite dimensions with no constraints $y \in G(x)$ in (3.83). The constrained finite-dimensional case of (3.83) with $\varphi = \varphi(y)$ was fully studied by Mordukhovich [894, 901]. Some upper estimates of $\partial\mu(\bar{x})$ and $\partial^\infty\mu(\bar{x})$ in Fréchet smooth spaces were derived by Thibault [1249], while the general statements of Theorem 3.38(i,ii) in the Asplund space setting mainly correspond to Mordukhovich and Shao [949]. The subdifferential estimates in assertion (iii) of this theorem under the *mixed qualification condition* appear here for the first time; the results of Theorem 3.38(iv) estimating $\partial^\infty\mu(\bar{x})$ via the mixed coderivative of the constraint mapping G are taken from Mordukhovich and Nam [934]. We also refer the reader to the recent paper by Mordukhovich, Nam and Yen [937] for applications of Theorem 3.38 to subdifferentiation of *value functions* in various constrained optimization problems in infinite-dimensional spaces including nonlinear and nondifferentiable programs as well as mathematical programs with equilibrium constraints considered in Sect. 5.2.

Theorem 3.41(i,ii) on the general *subdifferential chain rules* and the subsequent results of Subsect. 3.2.1, which are more or less consequences of the chain rules, were mainly derived in Mordukhovich and Shao [949]. The chain rules from assertion (iii) of Theorem 3.41 under the refined qualification and PSNC conditions have never been published. Partial results and modifications of those presented in Subsect. 3.2.1 were developed by Allali and Thibault [15], Borwein and Zhu [163, 164], Clarke et al. [265], Ioffe [590, 592, 596, 599], Ioffe and Penot [614], Jourani and Thibault [651, 652, 654, 657, 658], Kruger [706, 708, 709], Loewen [801], Mordukhovich [894, 901, 910], Mordukhovich and B. Wang [963], Ngai and Théra [1008], Rockafellar [1155, 1158], Rockafellar and Wets [1165], Thibault [1249, 1252], and Vinter [1289]; see also [949] for more comments and discussions.

3.4.8. Mean Value Theorems. The fundamental Lagrange *mean value theorem* plays an exceptionally important role in the classical mathematical analysis and its applications. It provides an *exact* relationship between a function and its derivative, thus being the basis for many crucial results of differential and integral calculus, monotonicity and convexity criteria for smooth functions, etc.

The first mean value theorem for *nonsmooth Lipschitzian* functions $\varphi\colon X \to \overline{I\!R}$ was established by Lebourg [749] via Clarke's generalized gradient in the arbitrary Banach space setting. Furthermore, it has been proved in [749] that the Clarke construction is the *smallest* among any reasonable *convex-valued* subdifferentials $\mathcal{D}\varphi(\cdot)$ of Lipschitz continuous functions φ in which terms one can obtain a natural subgradient extension

$$\varphi(b) - \varphi(a) \in \big\langle \mathcal{D}\varphi(c), b - a \big\rangle \quad \text{for some} \quad c \in (a, b) \tag{3.84}$$

of the classical mean value theorem. The result of Theorem 3.47, which origin goes back to Kruger and Mordukhovich [706, 708, 894, 901], is a significant improvement of Lebourg's mean value theorem in the Asplund space setting, since the *symmetric subdifferential* $\partial^0\varphi(c)$ is usually *nonconvex* being much smaller than Clarke's generalized gradient $\partial_C\varphi(c)$ even for simple Lipschitzian functions φ defined on $X = I\!R^2$; see the exact calculations for the function $\varphi(x_1, x_2) = |x_1| - |x_2|$ in Subsect. 1.3.2 and for the function $\varphi(x_1, x_2) = |\,|x_1| + x_2|$ in Example 2.49. Due to these simple examples, it is worth mentioning an interesting result by Borwein and Fitzpatrick [142] who proved that $\partial^0\varphi(c) = \partial_C\varphi(c)$ for every Lipschitz continuous function on the *real line* $X = I\!R$. Note also that an extended mean value theorem in form

(3.84) inevitably requires a *two-sided/symmetric* generalized differential construction like Clarke's generalized gradient for Lipschitzian functions and the symmetric subdifferential $\partial^0\varphi(\cdot)$ as in Theorem 3.47; cf. the result of Corollary 3.48 for *lower regular* functions and the *counterexample* given after it.

Approximate mean value theorems of the new type considered in Subsect. 3.2.2 are substantially different from the form of (3.84) and don't have any analogs in the classical differential calculus. The first result of this new type given in Theorem 3.49 was obtain by Zagrodny [1352] in terms of Clarke subgradients for l.s.c. extended-real-valued functions defined on general Banach spaces. As observed by Thibault [1251] (see also Thibault and Zagrodny [1254]), the main ideas developed in [1352] lead to appropriate versions of the approximate mean value theorem formulated via broad classes of subgradients satisfying natural requirements on suitable Banach spaces. Theorem 3.49 and its corollaries in terms of Fréchet subgradients were derived by Loewen [802] for l.s.c. functions on Fréchet smooth spaces; the mean value inequality from Corollary 3.50 was obtained by Borwein and Preiss [154] for Lipschitzian functions. The full statements of Theorem 3.49 and its corollaries in Asplund spaces were presented in Mordukhovich and Shao [949] with the variational proof of the main assertions, which is different at some essential points from those given in [154, 802, 1352]. Mean value inequalities of another ("multidimensional") type were established by Clarke and Ledyaev [262]; see also [61, 62, 163, 164, 265, 1371].

The *neighborhood* subgradient characterizations (a) and (b) of the local *Lipschitzian property* from Theorem 3.52 were established by Loewen [802] in Fréchet smooth spaces and then by Mordukhovich and Shao [949] in the general Asplund space setting. The *pointbased* criterion (d) of Theorem 3.52 via singular subgradients goes back to Rockafellar [1150] and Mordukhovich [894, 901] in finite-dimensional spaces. The general infinite-dimensional characterization of the local Lipschitz continuity from Theorem 3.52(d), involving the SNEC property of l.s.c. functions, appears here for the first time while partial results under stronger normal compactness conditions were obtained earlier by Loewen [802] and by Mordukhovich and Shao [949]. A subdifferential characterization of *constancy* similar to Corollary 3.53 but formulated via proximal subgradients was first established by Clarke [259] in finite dimensions and then by Clarke, Stern and Wolenski [270] in Hilbert spaces.

The subdifferential characterizations of *strict Hadamard differentiability* in Theorem 3.54 and of *function monotonicity* in Theorem 3.55 were derived by Loewen [802] based on the approximate mean value theorem for l.s.c. functions on Fréchet smooth spaces. The same proofs based on Theorem 3.49 work in the Asplund space setting as observed by Mordukhovich and Shao [949]. Another proof of the equivalency (b)$\Leftrightarrow$(c) in Theorem 3.54 with $\partial_C\varphi(\cdot)$ in (b) was given by Clarke [255] in arbitrary Banach spaces. A proximal subdifferential version of Theorem 3.55 was established by Clarke, Stern and Wolenski [270] in the Hilbert space setting.

One of the most fundamental results of convex analysis is Rockafellar's theorem on *maximal monotonicity* of the subdifferential mapping $\partial\varphi(\cdot)$ associated with a proper l.s.c. *convex* function φ on a Banach space; see [1141] and also [1073, 1142, 1213] for more discussions, applications, and references. An important question on the possibility to extend the monotonicity property for subdifferential mappings associated with *nonconvex* functions was (negatively) solved by Correa, Jofré and Thibault [292] for a large class of axiomatically defined subdifferentials satisfying certain natural properties; the preceding result in this direction was obtained by Poliquin [1088] for Clarke subgradients of l.s.c. functions on finite-dimensional spaces. Although Fréchet subgradients considered in Theorem 3.56 don't satisfy some of these properties, the given proof of Theorem 3.56 follow the procedure in [292] based on the application of the approximate mean value theorem.

3.4.9. Connections with Other Normals and Subgradients. Theorem 3.57 gives the *exact representations* of Clarke's normal and subgradient constructions, defined by polarity relations from tangential/directional derivative approximations in arbitrary Banach spaces (see Subsect. 2.5.2A), via our basic ("limiting Fréchet") normals and subgradients in the *Asplund space* setting. All the assertions of this theorem were derived in full generality by Mordukhovich and Shao [949]. In finite dimensions, these results go back to Kruger and Mordukhovich [718, 719]; cf. also Ioffe [592, 596] and the references in Subsect. 1.4.8 for equivalent representations via other (non-Fréchet type) normals and subgradients. Analogs of Theorem 3.57 in terms of Fréchet-like ε-normals and ε-subgradients were established by Treiman [1262, 1263] in Fréchet smooth spaces and then by Borwein and Strójwas [156, 157] with $\varepsilon = 0$ in reflexive spaces. Assertion (iii) of this theorem was derived by Borwein and Preiss [154] in Fréchet smooth spaces, while (i) and (ii) were given by Ioffe [600] in the same setting. It is worth mentioning that Preiss [1104] established a profound refinement of formula (3.58) for locally Lipschitzian functions φ on Asplund spaces with the replacement of Fréchet subgradients of φ in (3.58) by the classical *Fréchet derivatives*, which were proved to exist on a *dense* set.

The subsequent material of Subsect. 3.2.3 revolves around relationships between *sequential* and *net/topological* weak* limits of *Fréchet-like* and *Dini-like* subgradients in topological spaces *dual* to Banach spaces. The main motivation comes from seeking relationships between our basic generalized differential constructions involving sequential weak* limits of Fréchet-like normal and subgradients and the corresponding "approximate" constructions by Ioffe related to topological weak* limits of Dini-like subgradients described in Subsect. 2.5.2B; see also the discussion and references therein regarding the terminology used.

Observe that formula (3.60) for the A-subdifferential is different from its definition in (2.75); in fact, the "topological limiting Dini" construction (3.60) was defined by Ioffe [589] under the name of "M-subdifferential." The equivalence between (2.75) and (3.60) in *Asplund spaces* follows from combining the results by Ioffe [597], who proved this equivalence in any "weakly trustworthy" space in his sense [593], and by Fabian [413] that implies the trustworthiness property of every Asplund space.

Lemma 3.58 on the relationships between weak* sequential and topological limits in dual spaces was derived by Borwein and Fitzpatrick [141], where the proof of the main assertion (ii) in weakly compactly generated spaces was based on the fundamental *Whitney's construction* presented in Holmes [580, pp. 147–149]. This lemma is used in the proof of the major Theorem 3.59 established by Mordukhovich and Shao [949], which fully describes connections between our basic normal and subdifferential constructions and various modifications of "approximate" normals and subgradients. Note that the basic normal cone $N(\bar{x};\Omega)$ may *not* be *norm-closed* (and hence not weak* closed) even in the simplest infinite-dimensional (Hilbert) spaces; see Example 1.7 constructed by Fitzpatrick for the author's request. Thus it is *strictly smaller* than the G-normal cone $N_G(\bar{x};\Omega)$. Moreover, the basic subdifferential $\partial\varphi(\bar{x})$ may be strictly smaller than the G-subdifferential $\partial_G\varphi(\bar{x})$ not only for l.s.c. functions on Hilbert spaces but even for *Lipschitz* continuous function on (rather exotic) spaces with $\mathcal{C}^\infty$-smooth renorms as in Example 3.61 given by Borwein and Fitzpatrick [141]. The equalities

$$N_G(\bar{x};\Omega) = \mathrm{cl}^*N(\bar{x};\Omega) \quad\text{and}\quad \partial_G\varphi(\bar{x}) = \mathrm{cl}^*\partial\varphi(\bar{x})$$

in Theorem 3.59 follow also from the proofs by Ioffe [600] in the case of Fréchet smooth spaces. Actually the stronger results

$$N(\bar{x};\Omega) = \widetilde{N}_G(\bar{x};\Omega) \quad\text{and}\quad \partial\varphi(\bar{x}) = \widetilde{\partial}_G\varphi(\bar{x})\ ,$$

were formulated in [600], which however happened to be *incorrect* for *non-WCG* spaces due to Example 3.61.

The *robustness property* of basic normals in Theorem 3.60 was justified by Mordukhovich and Shao [951], although the formulation (but not the proof) in [951] involved a generally more restrictive normal compactness property, which in fact happened to be *equivalent* to the SNC property in the *WCG Asplund* setting. Previously this result was established by Loewen [800] in reflexive spaces, with the essential use of reflexivity in some points of his proof. On the other hand, the proof of Theorem 3.60 given in the book strongly follows the ideas of Loewen combined with the application of Lemma 3.58.

3.4.10. Graphical Regularity and Differentiability of Lipschitzian Mappings. The material of Subsect. 3.2.4 is mostly based on the paper by Mordukhovich and B. Wang [965]. The main motivation came from seeking appropriate *dual* infinite-dimensional counterparts of the following fundamental result by Rockafellar [1153]: for every mapping $f\colon I\!R^n \to I\!R^m$ locally Lipschitzian around $\bar{x}$ the Clarke tangent cone to the *graph* of f at $(\bar{x}, f(\bar{x}))$ is a *linear subspace* of dimension $d \le n$ in $I\!R^n \times I\!R^m$, where $d = n$ *if and only if*

f is *strictly differentiable* at $\bar{x}$. This implies, in particular, the important fact observed by Mordukhovich [912]: a *nonsmooth* Lipschitzian mappings between finite-dimensional spaces *cannot* exhibit *graphical regularity*, i.e., the Clarke normal cone to its graph never agrees with the Bouligand-Severi contingent cone at reference points (this description of graphical regularity reduces to those in Definition 1.36 in finite dimensions); cf. Claim in the proof of Theorem 1.46 in Chap. 1. Note that Rockafellar's proof in [1153] is very much involved being heavily *finite-dimensional*; it doesn't seem to be extendable to an infinite-dimensional setting.

We develop a *new scheme* to study the above questions in the *dual framework* that provides not only comprehensive and fully understood *infinite-dimensional* counterparts of the afore-mentioned results but also gives a simplified proof of Rockafellar's finite-dimensional theorem that is completely different from the original one given in [1153]. Our approach is mainly based on the *normal coderivative scalarization*, which implies in a straight way the subspace property of the convexified normal cone via the *two-sided symmetry* of Clarke's generalized gradient for Lipschitzian functions and its relationship with our nonconvex limiting subdifferential; see the proof of Theorem 3.62 for more details.

The above scalarization scheme is the key ingredient to derive the afore-mentioned results in finite dimensions; more is required however in *infinite-dimensional spaces.* There are two major issues on *differentiability* that distinguish the infinite-dimensional setting from the finite-dimensional one in order to establish an *equivalence* between *graphical regularity* and some *smoothness* of Lipschitzian mappings:

(a) we need to use simultaneously *different bornologies* (namely, Fréchet and Hadamard) to characterize graphical regularity via bornological smoothness;

(b) we need to introduce new notions of differentiability of functions on infinite-dimensional spaces (called conditionally *weak differentiability* and *strict-weak differentiability*) to appropriately described the equivalence we are looking for.

It surprisingly happens that these "weak" and "strict-weak" differentiability notions, classical in nature, can be dramatically different from the conventional differentiability concepts even for simple functions with values in Hilbert spaces. In particular, Example 3.64 shows that there exist Lipschitzian functions, which are *strictly-weakly differentiable* with respect to the strongest *Fréchet* bornology while *not being differentiable* in the classical *Gâteaux* sense.

Following the pattern suggested by Rockafellar [1153] who used smooth *nonsingular* transformations (actually the change of coordinates) in finite-dimensional spaces, the above results for single-valued Lipschitzian mappings were extended to "hemi-Lipschitzian" sets and set-valued mappings in Mordukhovich and B. Wang [965]; see Definition 3.71 and Theorem 3.72. The main difference between hemi-Lipschitzian (resp. hemismooth) manifolds in [965] and their Lipschitzian (resp. smooth) analogs from [1153] consists of using smooth (actually strictly differentiable) graph transformations with *surjective* derivatives instead of invertible/nonsingular ones as in [1153]. Then the corresponding *equality-type* calculus of basic and Fréchet normals available in both finite and infinite dimensions allows us to reduce the set-valued case to the single-valued one.

3.4.11. Second-Order Subdifferential Calculus in Asplund Spaces. Subsection 3.2.5 is mainly based on the paper by Mordukhovich [923]. Considering the Asplund space framework, we derive significantly more developed second-order subdifferential calculus in comparison with the general Banach space setting of Subsect. 1.3.5. Note that the results presented in Subsect. 3.2.5 are different and *generally independent*, even in the case of finite-dimensional case, from those presented in Subsect. 1.3.5, where mostly *equality* relations were obtained under certain second-order smoothness and surjectivity requirements on some components of compositions. Now we develop an *inclusion-type* calculus with *no* second-order smoothness and surjectivity assumptions involved.

The second-order subdifferential *sum rules* of Theorem 3.73 were first obtained by Mordukhovich [910] in finite dimensions. *Amenable functions* used in the second-order *chain rule* of Corollary 3.76 were introduced in Poliquin and Rockafellar [1089] and were thoroughly studied in Rockafellar and Wets [1165]; see also the references therein. Another proof of the second-order subdifferential chain rule involving such functions in Corollary 3.76 was independently developed by Rockafellar (personal communication) by using *quadratic penalties* in the case of $\dim X < \infty$. A modification of this result for the so-called "amenable functions with compatible parametrization" was given in Levy and Mordukhovich [769]. Some special second-order chain rules for finite-dimensional compositions with Lipschitzian inner mappings, different from Theorem 3.77 and not presented here, were derived in the paper by Mordukhovich and Outrata [939], where the reader can find applications of these results to *stability issues* and *mechanical equilibria.*

3.4.12. SNC Calculus for Sets and Mappings in Asplund Spaces. Section 3.3 contains basic calculus of *sequential normal compactness* for sets, set-valued mappings, and extended-real-valued functions in the framework of Asplund spaces. As mentioned, by *SNC calculus* we understand efficient conditions ensuring the preservation of SNC/PSNC properties under various operations performed on sets and mappings. Since such properties are automatic in finite dimensions and for Lipschitzian real-valued functions, SNC calculus is not needed in these cases. However, in more general settings, SNC and related normal compactness properties are unavoidably involved in major results concerning limiting generalized differential constructions and their applications in infinite-dimensional spaces; thus it is difficult to overestimate the importance of such calculus from the viewpoint of both theory and applications. The absence of SNC calculus till the recent work by Mordukhovich and B. Wang [961, 964], on which the material of Sect. 3.3 is mainly based, has been indeed a serious obstacle for broad applications of generalized differentiation in infinite dimensions.

The *extremal principle* plays the major role in deriving results of the SNC calculus presented in Sect. 3.3. Observe the *difference* as well as *similarity* between the *qualification conditions* ensuring the rules of generalized differentiation developed above and the corresponding SNC calculus relations derived in this section. Usually conditions required for SNC calculus are *stronger* than those for rules of generalized differentiation. Let us mention a rather surprising result of Corollary 3.87 concerning the standard *smooth constraint systems* in *nonlinear programming*. It happens, as a simple consequence of significantly more general relations, that the classical *Mangasarian-Fromovitz constraint qualification*, designed for completely different reasons, ensures the fulfillment of the SNC property for the most conventional set of feasible solutions in constrained optimization! This seems indeed to be of undoubted interest even in the simplest case of linear constraints.

4

Characterizations of Well-Posedness and Sensitivity Analysis

The primary goal of this chapter is to show that the basic principles and tools of variational analysis developed above allow us to provide *complete characterizations* and efficient applications of fundamental properties in nonlinear studies related to Lipschitzian stability, metric regularity, and covering/openness at a linear rate. These properties indicate a certain *well-posedness* (i.e., "good behavior") of set-valued mappings and play a principal role in many aspects of nonlinear analysis, particularly those concerning optimization and sensitivity. We have considered these properties in Chap. 1 in the framework of arbitrary Banach spaces, where *necessary* conditions for their fulfillment were obtained via *coderivatives* of set-valued mappings. These conditions were efficiently used in Chaps. 1 and 3 for developing the generalized differential calculus and related issues. In this chapter we show, based on *variational arguments*, that the conditions obtained are not only necessary but also *sufficient* for the validity of the mentioned properties in the framework of *Asplund spaces*. Moreover, we compute the *exact bounds* of the corresponding moduli in terms of coderivatives and subdifferentials. Two kinds of dual characterizations are derived in this way: *neighborhood* criteria involving generalized differential constructions *around* reference points, and *pointbased* criteria expressed only *at* the points under consideration. Then we apply the obtained characterizations for Lipschitzian behavior of set-valued mappings and comprehensive calculus rules of generalized differentiation to *sensitivity analysis* for parametric constraint and variational systems including those described by implicit multifunctions, by the so-called generalized equations/variational conditions that arise in numerous optimization and equilibrium models, by variational and hemivariational inequalities, etc. Let us emphasize that sensitivity/stability analysis is of particular importance from both *qualitative and numerical* viewpoints. The latter involves the justification of *successful numerical solution* by treating perturbations as *errors* typically occurring in computations, and also as a tool of determining a *convergence rate* of solution algorithms; here estimations of *Lipschitzian moduli* play a crucial role.

4.1 Neighborhood Criteria and Exact Bounds

In this section we obtain neighborhood dual characterizations of covering, metric regularity, and Lipschitzian properties of closed-graph multifunctions between Asplund spaces. The conditions obtained are expressed in terms of Fréchet coderivatives of set-valued mappings considered in neighborhoods of reference points. We also derive coderivative formulas for computing the exact bounds of the corresponding covering, regularity, and Lipschitzian moduli.

The fundamental properties under consideration have been defined in Sect. 1.3, where we established relationships between them and obtained necessary coderivative conditions for their validity in arbitrary Banach spaces. Now we show the necessary conditions obtained happen to be sufficient and the one-sided estimates for the exact bounds become *equalities* in the framework of Asplund spaces.

We begin with studying the covering properties from Definition 1.51 and consider their local and semi-local versions, which are generally independent. Then we derive the corresponding results for the metric regularity and Lipschitzian properties of set-valued mappings taking into account the equivalencies established in Sect. 1.3.

4.1.1 Neighborhood Characterizations of Covering

First we consider the *local covering* property of a set-valued mapping $F\colon X \rightrightarrows Y$ around $(\bar{x}, \bar{y}) \in \operatorname{gph} F$, which means, according to Definition 1.51(ii), that there are a neighborhood U of $\bar{x}$, a neighborhood V of $\bar{y}$, and a number (modulus) $\kappa > 0$ satisfying

$$F(x) \cap V + \kappa r I\!B \subset F(x + r I\!B) \text{ whenever } x + r I\!B \subset U \text{ as } r > 0\,. \tag{4.1}$$

The supremum of all moduli $\{\kappa\}$ satisfying (4.1) with some neighborhoods U and V is called the *exact covering bound* of F around $(\bar{x}, \bar{y})$ and is denoted by $\operatorname{cov} F(\bar{x}, \bar{y})$. Let us emphasize that the modulus κ gives a rate of the *uniform linear dependence* between the F-image of the ball $x + r I\!B$ and the corresponding ball around $F(x) \cap V$ covered by $F(x + r I\!B)$.

To obtain the main neighborhood characterization of the local covering, we define the constant

$$\begin{aligned}\widehat{a}(F, \bar{x}, \bar{y}) := \sup_{\eta>0} \inf \Big\{ \|x^*\| \;\Big|\; & x^* \in \widehat{D}^* F(x, y)(y^*),\ x \in B_\eta(\bar{x}),\\ & y \in F(x) \cap B_\eta(\bar{y}),\ \ \|y^*\| = 1 \Big\}\end{aligned} \tag{4.2}$$

computed via the Fréchet coderivative of F at neighboring points to $(\bar{x}, \bar{y})$.

Theorem 4.1 (neighborhood characterization of local covering). *Let $F\colon X \rightrightarrows Y$ be a set-valued mapping between Asplund spaces. Assume that F is closed-graph around $(\bar{x}, \bar{y}) \in \operatorname{gph} F$. Then the following are equivalent:*

(a) *F enjoys the local covering property around $(\bar{x}, \bar{y})$.*
(b) *One has $\widehat{a}(F, \bar{x}, \bar{y}) > 0$ for the constant $\widehat{a}(F, \bar{x}, \bar{y})$ defined in (4.2).*

Moreover, the exact covering bound of F around $(\bar{x}, \bar{y})$ is computed by

$$\operatorname{cov} F(\bar{x}, \bar{y}) = \widehat{a}(F, \bar{x}, \bar{y}) \,.$$

Proof. If F enjoys the local covering property around $(\bar{x}, \bar{y})$, then one has

$$\widehat{a}(F, \bar{x}, \bar{y}) \geq \operatorname{cov} F(\bar{x}, \bar{y}) > 0$$

due to Theorem 1.54(i) valid in Banach spaces. It remains to show that $\widehat{a}(F, \bar{x}, \bar{y}) \leq \operatorname{cov} F(\bar{x}, \bar{y})$ if both X and Y are Asplund and if F is closed-graph around $(\bar{x}, \bar{y})$. The latter surely implies that (b)$\Rightarrow$(a).

To proceed, we pick any number $0 < \kappa < \widehat{a}(F, \bar{x}, \bar{y})$ and show that it is a covering modulus for F around $(\bar{x}, \bar{y})$. Suppose that it is not true for some fixed positive number $\kappa < \widehat{a}(F, \bar{x}, \bar{y})$. Then using (4.1), we find sequences $x_k \to \bar{x}$, $y_k \to \bar{y}$, $r_k \downarrow 0$, and $z_k \in Y$ such that

$$y_k \in F(x_k), \quad \|z_k - y_k\| \leq \kappa r_k, \quad \text{and} \quad z_k \notin F(x) \quad \text{for all} \quad x \in B_{r_k}(x_k) \,. \tag{4.3}$$

Fix an arbitrary number $\nu > \kappa$, choose some $\alpha \in (\kappa/\nu, 1)$, and pick a sequence $\gamma_k \downarrow 0$ satisfying

$$0 < \gamma_k < \min\Big\{r_k, \ \frac{1}{2(\nu\alpha + 1)}, \ \frac{\nu(1-\alpha)}{1 + \nu(\nu\alpha + 1)}\Big\}, \quad k \in I\!N \,. \tag{4.4}$$

For any fixed $k \in I\!N$ we define the norm

$$\|(x, y)\|_{\gamma_k} := \|x\| + \gamma_k \|y\|$$

on the product space $X \times Y$, which is clearly equivalent to the standard sum norm $\|x\| + \|y\|$. Since both X and Y are Asplund, their product endowing with the norm $\|(\cdot, \cdot)\|_{\gamma_k}$ is Asplund as well. Note that Fréchet normals on $X \times Y$ used below don't depend on the choice of equivalent norms.

Consider the closed subset $E_k \subset X \times Y$ defined by

$$E_k := (\operatorname{gph} F) \cap \big((x_k, y_k) + \gamma_k I\!B_{X \times Y}\big)$$

and view it as a complete metric space with the metric induced by $\|(\cdot, \cdot)\|_{\gamma_k}$ for every fixed $k \in I\!N$. Let

$$\varphi_k(x, y) := \|y - z_k\| \quad \text{for} \quad (x, y) \in E_k, \quad k \in I\!N \,.$$

Since $\varphi_k \colon E_k \to I\!R$ is a nonnegative l.s.c. function on a complete metric space, we apply to it the Ekeland variational principle (Theorem 2.26) at the point (x_k, y_k) with $\varepsilon_k := \kappa r_k$ and $\lambda_k := \kappa r_k / \nu\alpha$ for each k. Noting that $\varphi_k(x_k, y_k) \leq \varepsilon_k$ due to (4.3), we find a point $(\tilde{x}_k, \tilde{y}_k) \in E_k$ satisfying

$$0 < \rho_k := \|\tilde{y}_k - z_k\| \le \|y_k - z_k\| \le \kappa r_k, \quad \|(\tilde{x}_k, \tilde{y}_k) - (x_k, y_k)\|_{\gamma_k} \le \lambda_k < r_k \ ,$$

$$\|\tilde{y}_k - z_k\| \le \|y - z_k\| + \nu\alpha\|(x, y) - (\tilde{x}_k, \tilde{y}_k)\|_{\gamma_k} \text{ for all } (x, y) \in E_k \ .$$

The latter implies that the sum $\psi_k(x, y) + \delta((x, y); \operatorname{gph} F)$ with

$$\psi_k(x, y) := \|y - z_k\| + \nu\alpha\|(x, y) - (\tilde{x}_k, \tilde{y}_k)\|_{\gamma_k}$$

attains its unconditional local minimum on $X \times Y$ at the point $(\tilde{x}_k, \tilde{y}_k)$. Note that ψ_k is a convex continuous function whose Fréchet subdifferential agrees with the subdifferential ∂ of convex analysis. Since the space $X \times Y$ is Asplund, we apply the subgradient description of the extremal principle from Lemma 2.32 to the semi-Lipschitzian sum $\psi_k + \delta(\cdot; \operatorname{gph} F)$ taking there $\eta = \min\{\gamma_k, \rho_k\gamma_k/2\}$. This gives points $(x_{1k}, y_{1k}) \in X \times Y$ and $(x_{2k}, y_{2k}) \in \operatorname{gph} F$ such that

$$\|(x_{ik}, y_{ik}) - (\tilde{x}_k, \tilde{y}_k)\| \le \rho_k\gamma_k/2 \text{ with } y_{ik} \ne z_k \text{ for } i = 1, 2, \quad \text{and}$$

$$0 \in \partial\big[\|\cdot - z_k\| + \nu\alpha\|(\cdot,\cdot) - (\tilde{x}_k, \tilde{y}_k)\|_{\gamma_k}\big](x_{1k}, y_{1k})$$

$$+ \widehat{N}((x_{2k}, y_{2k}); \operatorname{gph} F) + \gamma_k(I\!B_{X^*} \times I\!B_{Y^*}) \ .$$

Now using standard convex analysis and taking into account that $y_{ik} \ne z_k$, we get elements $u_k^* \in X^*$, $v_k^* \in Y^*$, $w_k^* \in Y^*$, $z_k^* \in X^*$, $p_k^* \in Y^*$, and $(x_k^*, -y_k^*) \in \widehat{N}((x_{2k}, y_{2k}); \operatorname{gph} F)$ such that

$$\|u_k^*\| \le \gamma_k, \quad \|v_k^*\| \le \gamma_k, \quad \|w_k^*\| = 1, \quad \|z_k^*\| \le 1, \quad \|p_k^*\| = 1, \quad \text{and}$$

$$(u_k^*, v_k^*) = (0, w_k^*) + \nu\alpha(z_k^*, 0) + \nu\alpha\gamma_k(0, p_k^*) + (x_k^*, -y_k^*) \ .$$

Therefore one has

$$\|x_k^*\| \le \nu\alpha + \gamma_k \quad \text{and} \quad \|w_k^* - y_k^*\| \le \gamma_k(\nu\alpha + 1) \ ,$$

which implies, due to the choice of γ_k in (4.4), that

$$\|y_k^*\| \ge \|w_k^*\| - \gamma_k(\nu\alpha + 1) = 1 - \gamma_k(\nu\alpha + 1) > 1/2 \ .$$

Denoting $\tilde{x}_k^* := x_k^*/\|y_k^*\|$, $\tilde{y}_k^* := y_k^*/\|y_k^*\|$ and using (4.4) again, we get

$$\tilde{x}_k^* \in \widehat{D}^* F(x_{2k}, y_{2k})(\tilde{y}_k^*), \quad \|\tilde{y}_k^*\| = 1, \quad \text{and} \quad \|\tilde{x}_k^*\| \le \frac{\nu\alpha + \gamma_k}{1 - \gamma_k(\nu\alpha + 1)} < \nu \ .$$

Now passing to the limit as $k \to \infty$ and taking into account definition (4.2) of the constant $\widehat{a}(F, \bar{x}, \bar{y})$, one has $\widehat{a}(F, \bar{x}, \bar{y}) \le \nu$. Since $\nu > \kappa$ was chosen arbitrary, we finally obtain $\widehat{a}(F, \bar{x}, \bar{y}) \le \kappa$. This contradiction completes the proof of the theorem. △

If the graph of F is *convex*, we have an explicit formula for computing the Fréchet coderivative that implies the following corollary.

Corollary 4.2 (neighborhood characterization of local covering for convex-graph multifunctions). *Suppose that F is convex-graph under the assumptions of Theorem 4.1. Then the conclusions of this theorem hold with the covering constant $\widehat{a}(F,\bar{x},\bar{y})$ computed by*

$$\widehat{a}(F,\bar{x},\bar{y}) := \sup_{\eta>0}\inf\Big\{\|x^*\| \;\Big|\; \langle x^*,x\rangle - \langle y^*,y\rangle = \sup_{(u,v)\in\operatorname{gph}F}\big[\langle x^*,u\rangle - \langle y^*,v\rangle\big]\,,$$

$$x\in B_\eta(\bar{x}),\;\; y\in F(x)\cap B_\eta(\bar{y}),\;\; \|y^*\|=1\Big\}\,.$$

Proof. It follows from Theorem 4.1 due to Proposition 1.37. △

In the case of single-valued and locally Lipschitzian mappings the covering constant (4.2) is expressed in terms of Fréchet subgradients.

Corollary 4.3 (neighborhood covering criterion for single-valued mappings). *Let $f\colon X \to Y$ be a single-valued mapping between Asplund spaces. Assume that f is Lipschitz continuous around some point $\bar{x}$. Then the conclusions of Theorem 4.1 hold with the covering constant $\widehat{a}(f,\bar{x})$ computed by*

$$\widehat{a}(f,\bar{x}) = \sup_{\eta>0}\inf\big\{\|x^*\| \;\big|\; x^*\in\widehat{\partial}\langle y^*,f\rangle(x),\;\; x\in B_\eta(\bar{x}),\;\; \|y^*\|=1\big\}\,.$$

Proof. Since f is Lipschitz continuous on $B_\eta(\bar{x})$ for small $\eta>0$, one has the scalarization formula

$$\widehat{D}^*f(x)(y^*) = \widehat{\partial}\langle y^*,f\rangle(x)\;\text{ for all } x\in B_\eta(\bar{x}) \;\text{ and }\; y^*\in Y^*\,,$$

which easily follows from the definitions. Thus (4.2) reduces to the form presented in the corollary. △

Next let us consider the *semi-local covering* property of $F\colon X\rightrightarrows Y$ around $\bar{x}\in\operatorname{dom}F$ in the sense of Definition 1.51(iii), which corresponds to (4.1) with $V=Y$. The exact covering bound is denoted by $\operatorname{cov}F(\bar{x})$ in this case. If F is closed-graph and locally compact around $\bar{x}$, then Theorem 4.1 immediately implies the corresponding characterization of the semi-local covering property due to the relationships of Corollary 1.53. The following theorem justifies this characterization with *no local compactness* assumption.

Theorem 4.4 (neighborhood characterization of semi-local covering). *Let $F\colon X\rightrightarrows Y$ be a set-valued mapping between Asplund spaces. Assume that F is closed-graph near $\bar{x}\in\operatorname{dom}F$. Then the following are equivalent:*

(a) *F enjoys the semi-local covering property around $\bar{x}$.*

(b) *One has $\widehat{a}(F,\bar{x})>0$ for the constant $\widehat{a}(F,\bar{x})$ defined by*

$$\widehat{a}(F,\bar{x}) := \sup_{\eta>0} \inf \Big\{ \|x^*\| \ \Big| \ x^* \in \widehat{D}^* F(x,y)(y^*),\ x \in B_\eta(\bar{x}),$$

$$y \in F(x),\ \ \|y^*\| = 1 \Big\}.$$

Moreover, $\widehat{a}(F,\bar{x})$ *is the exact covering bound* $\operatorname{cov} F(\bar{x})$ *of* F *around* $\bar{x}$.

Proof. If F has the semi-local covering property around $\bar{x}$, then

$$\widehat{a}(F,\bar{x}) \ge \operatorname{cov} F(\bar{x}) > 0$$

due to Corollary 1.55 valid in any Banach spaces. To prove the opposite estimate for closed-graph mappings between Asplund spaces, we suppose on the contrary that there is a positive number $\kappa < \widehat{a}(F,\bar{x})$, which is not a modulus of semi-local covering. Involving the definition of this property, we find sequences $x_k \to \bar{x}$, $r_k \downarrow 0$, and $(y_k, z_k) \in Y \times Y$ such that relations (4.3) hold. In contrast to the local covering property in the proof of Theorem 4.1, we don't specify the convergence of y_k, which is actually not needed to establish the required estimate due to the definition of the semi-local covering constant $\widehat{a}(F,\bar{x})$. Now proceeding similarly to the proof of Theorem 4.1, we arrive at the contradiction $\widehat{a}(F,\bar{x}) \le \kappa$. △

4.1.2 Neighborhood Characterizations of Metric Regularity and Lipschitzian Behavior

The above characterizations of covering properties and relationships of Sect. 1.3 allow us to derive neighborhood criteria and exact bound formulas for metric regularity and Lipschitzian properties of set-valued mappings between Asplund spaces.

We start with the *metric regularity* properties of $F\colon X \rightrightarrows Y$ and consider first its *local* version from Definition 1.47(ii), where $\operatorname{reg} F(\bar{x},\bar{y})$ denotes the *exact regularity bound* of F around $(\bar{x},\bar{y})$.

Theorem 4.5 (neighborhood characterization of local metric regularity). *Let* $F\colon X \rightrightarrows Y$ *be a set-valued mapping between Asplund spaces. Assume that* F *is closed-graph around* $(\bar{x},\bar{y}) \in \operatorname{gph} F$. *Then the following assertions are equivalent:*

(a) F *is locally metrically regular around* $(\bar{x},\bar{y})$.

(b) *One has* $\widehat{b}(F,\bar{x},\bar{y}) < \infty$, *where*

$$\widehat{b}(F,\bar{x},\bar{y}) := \inf_{\eta>0} \inf \Big\{ \mu > 0 \ \Big| \ \|y^*\| \le \mu \|x^*\|,\quad x^* \in \widehat{D}^* F(x,y)(y^*),$$

$$x \in B_\eta(\bar{x}),\quad y \in F(x) \cap B_\eta(\bar{y}) \Big\}.$$

Moreover, the exact regularity bound of F *around* $(\bar{x},\bar{y})$ *is computed by*

$$\operatorname{reg} F(\bar{x}, \bar{y}) = \widehat{b}(F, \bar{x}, \bar{y})$$

$$= \inf_{\eta>0} \sup \left\{ \|\widehat{D}^* F(x, y)^{-1}\| \;\Big|\; x \in B_\eta(\bar{x}),\; y \in F(x) \cap B_\eta(\bar{y}) \right\} .$$

Proof. If F is locally metrically regular around $(\bar{x}, \bar{y})$, then

$$\widehat{b}(F, \bar{x}, \bar{y}) \le \operatorname{reg} F(\bar{x}, \bar{y}) < \infty ,$$

which follows directly from estimate (1.41) in Theorem 1.54. To justify the opposite inequality $\widehat{b}(F, \bar{x}, \bar{y}) \ge \operatorname{reg} F(\bar{x}, \bar{y})$ under the assumptions made, we observe that

$$\mu > \widehat{b}(F, \bar{x}, \bar{y}) \Longrightarrow \mu^{-1} < \widehat{a}(F, \bar{x}, \bar{y}) ,$$

which easily follows from the definitions of these constants and the fact that $\widehat{D}^* F(\bar{x}, \bar{y})(\cdot)$ is positively homogeneous. Thus assuming $\widehat{b}(F, \bar{x}, \bar{y}) < \operatorname{reg} F(\bar{x}, \bar{y})$, we find $0 < \mu < \operatorname{reg} F(\bar{x}, \bar{y})$ such that $\mu^{-1} < \widehat{a}(F, \bar{x}, \bar{y})$. Theorem 4.1 allows us to conclude that μ^{-1} is a covering modulus for F around $(\bar{x}, \bar{y})$. Then Theorem 1.52(ii) ensures that μ is a modulus of local metric regularity for F around this point, which is impossible due to $\mu < \operatorname{reg}(F, \bar{x}, \bar{y})$. We therefore arrive at a contradiction that justifies the equality

$$\operatorname{reg} F(\bar{x}, \bar{y}) = \widehat{b}(F, \bar{x}, \bar{y}) .$$

To establish the second representation for $\operatorname{reg} F(\bar{x}, \bar{y})$, observe that the inequality "$\ge$" is proved in Theorem 1.54(i). The opposite one follows directly from the comparison of $\widehat{b}(F, \bar{x}, \bar{y})$ and last constant of the theorem. △

Involving Proposition 1.50 about relationships between local and *semi-local* metric regularity, Theorem 4.5 immediately implies criteria and exact bound formulas for both semi-local metric regularity properties of $F\colon X \rightrightarrows Y$ with respect to domain and range spaces from Definition 1.47(iii) assuming the *local compactness* of F around $\bar{x}$ and of F^{-1} around $\bar{y}$, respectively. The next result provides a complete characterization of the semi-local metric regularity of F around $\bar{x} \in \operatorname{dom} F$ with *no* local compactness assumption.

Theorem 4.6 (neighborhood characterization of semi-local metric regularity). *Let $F\colon X \rightrightarrows Y$ be a set-valued mapping between Asplund spaces. Assume that F is closed-graph near $\bar{x} \in \operatorname{dom} F$. Then the following assertions are equivalent:*

(a) *F is semi-locally metrically regular around $\bar{x} \in \operatorname{dom} F$.*

(b) *One has $\widehat{b}(F, \bar{x}) < \infty$, where*

$$\widehat{b}(F, \bar{x}) := \inf_{\eta>0} \inf \Big\{ \mu > 0 \;\Big|\; \|y^*\| \le \mu \|x^*\|, \quad x^* \in \widehat{D}^* F(x, y)(y^*) ,$$

$$x \in B_\eta(\bar{x}), \quad y \in F(x) \Big\} .$$

Moreover, the exact regularity bound of F around $\bar{x}$ is computed by

$$\operatorname{reg} F(\bar{x}) = \widehat{b}(F,\bar{x}) = \inf_{\eta>0} \sup \left\{ \|\widehat{D}^* F(x,y)^{-1}\| \;\Big|\; x \in B_\eta(\bar{x}),\; y \in F(x) \right\}.$$

Proof. It is similar to the proof of Theorem 4.5 with the use of relationships between the semi-local covering and metric regularity properties from Theorem 1.52(i) and the characterization of semi-local covering in Theorem 4.4. △

In conclusion of this subsection let us obtain neighborhood characterizations of Lipschitzian properties of set-valued mappings from Definition 1.40. We present results for the (local) *Lipschitz-like* property of F around $(\bar{x},\bar{y}) \in \operatorname{gph} F$, which are the most useful for subsequent applications. Due to relationships of Theorem 1.42, the results obtained below immediately imply the corresponding characterizations of the classical local Lipschitzian property of F around $\bar{x}$ for locally compact multifunctions.

Theorem 4.7 (neighborhood characterization of Lipschitz-like multifunctions). *Let $F: X \rightrightarrows Y$ be a set-valued mapping between Asplund spaces. Assume that F is closed-graph around $(\bar{x},\bar{y}) \in \operatorname{gph} F$. Then the following properties are equivalent:*

(a) *F is Lipschitz-like around $(\bar{x},\bar{y})$.*

(b) *There are positive numbers ℓ and η such that*

$$\sup \left\{ \|x^*\| \;\Big|\; x^* \in \widehat{D}^* F(x,y)(y^*) \right\} \le \ell \|y^*\|$$

whenever $x \in B_\eta(\bar{x})$, $y \in F(x) \cap B_\eta(\bar{y})$, and $y^* \in Y^*$.

Moreover, the exact Lipschitzian bound of F around $(\bar{x},\bar{y})$ is computed by

$$\operatorname{lip} F(\bar{x},\bar{y}) = \inf_{\eta>0} \sup \left\{ \|\widehat{D}^* F(x,y)\| \;\Big|\; x \in B_\eta(\bar{x}),\; y \in F(x) \cap B_\eta(\bar{y}) \right\}.$$

Proof. Property (b) of Lipschitz-like mappings and the lower estimate of the exact Lipschitzian modulus are proved in Theorem 1.43(i) for general Banach spaces. We know from Theorem 1.49(i) that the Lipschitz-like property of F around $(\bar{x},\bar{y})$ is equivalent to the local metric regularity of F^{-1} around $(\bar{y},\bar{x})$ with the same exact bounds. Taking into account the norm definition (1.22) for positively homogeneous mappings and the equality

$$\|\widehat{D}^* F^{-1}(y,x)\| = \|\widehat{D}^* F(x,y)^{-1}\| \quad \text{for any} \quad (x,y) \in \operatorname{gph} F,$$

we deduce this theorem from Theorem 4.5. △

4.2 Pointbased Characterizations

It is more convenient for applications to get *pointbased* criteria for covering, metric regularity, and Lipschitzian properties of multifunctions considered above. This means that one needs results expressed in terms of derivative-like constructions *at* the references points $(\bar{x}, \bar{y})$ alone (but not at all points of their neighborhoods). To derive such conditions, we have to impose additional assumptions on the mappings under consideration. A fundamental result of this type is given in Theorem 1.57, which shows that the classical Lyusternik-Graves surjectivity condition is necessary and sufficient for the metric regularity and covering around a given point $\bar{x}$ of strictly differentiable mappings $f: X \to Y$ between Banach spaces; moreover, the corresponding exact bounds are expressed in terms of the strict derivative of f at $\bar{x}$. Section 1.3 also contains some *necessary* pointbased conditions for the mentioned properties and one-sided modulus estimates expressed in terms of *mixed* coderivatives at references points for set-valued mappings between Banach spaces.

In this section we show that the conditions obtained are also *sufficient* for the validity of these fundamental properties for set-valued mappings $F: X \rightrightarrows Y$ between Asplund spaces, provided that *partial sequential normal compactness* assumptions on F are imposed. Moreover, the latter PSNC conditions happen to be also *necessary* for the fulfillment of the properties under consideration.

For computing the *exact bounds* of the corresponding moduli, we need to involve not only mixed coderivatives but also *normal* coderivatives at given points to furnish estimates in the opposite direction. In this way we obtain *precise formulas* to express the exact bounds for rather broad classes of set-valued mappings, where the *norms* of mixed and normal coderivatives agree at reference points. The final subsection of this section contains applications of the results obtained to computing the so-called *radius of metric regularity* that gives a measure of the extent to which a set-valued mapping can be perturbed before metric regularity is lost.

4.2.1 Lipschitzian Properties via Normal and Mixed Coderivatives

We start with pointbased characterizations of Lipschitzian properties for set-valued mappings between Asplund spaces. The main result of this section, Theorem 4.10, gives necessary and sufficient conditions for the Lipschitz-like property of F around $(\bar{x}, \bar{y})$ in terms of the mixed coderivative $D^*_M F(\bar{x}, \bar{y})$ and the PSNC property of F at $(\bar{x}, \bar{y})$, while the principal upper estimate of the exact Lipschitzian bound $\operatorname{lip} F(\bar{x}, \bar{y})$ is expressed via the normal coderivative $D^*_N F(\bar{x}, \bar{y})$. This implies the precise formula for computing the exact bound $\operatorname{lip} F(\bar{x}, \bar{y})$ for set-valued mappings satisfying the following requirement.

Definition 4.8 (coderivatively normal mappings). *Let $F: X \rightrightarrows Y$ be a set-valued mapping between Banach spaces, and let $(\bar{x}, \bar{y}) \in \operatorname{gph} F$. Then:*

(i) *F is* CODERIVATIVELY NORMAL *at $(\bar{x}, \bar{y})$ if*

$$\|D^*_M F(\bar{x}, \bar{y})\| = \|D^*_N F(\bar{x}, \bar{y})\| .$$

(ii) *F is* STRONGLY CODERIVATIVELY NORMAL *at $(\bar{x}, \bar{y})$ if*

$$D_M^* F(\bar{x}, \bar{y}) = D_N^* F(\bar{x}, \bar{y}) := D^* F(\bar{x}, \bar{y}) .$$

Example 1.35 shows that coderivative normality may not always hold even for M-regular mappings $f: X \to \ell^2$, which happen to be Lipschitz continuous around $\bar{x} = 0$ and strictly-weakly Fréchet differentiable at this point (in the sense of Definition 3.63). Indeed, for the mapping f from Example 1.35 one has $\|D_M^* f(0)\| = 0$ while $\|D_N^* f(0)\| = \infty$. The next proposition lists some important classes of mappings that are strongly coderivatively normal (and hence coderivatively normal) at reference points.

Proposition 4.9 (classes of strongly coderivatively normal mappings). *A set-valued mapping $F: X \rightrightarrows Y$ between Banach spaces is strongly coderivatively normal at $(\bar{x}, \bar{y}) \in \operatorname{gph} F$ if it satisfies one of the following conditions:*

(a) *Y is finite-dimensional.*

(b) *F is the indicator mapping of a set $\Omega \subset X$ relative to Y.*

(c) *F is N-regular at $(\bar{x}, \bar{y})$; in particular, either it is strictly differentiable at $\bar{x}$ or its graph is convex around $(\bar{x}, \bar{y})$.*

(d) *F is single-valued and w^*-strictly Lipschitzian at $\bar{x}$, and X is Asplund.*

(e) *$F = f \circ g$, where $g: X \to \mathbb{R}^n$ is Lipschitz continuous around $\bar{x}$ and $f: \mathbb{R}^n \to Y$ is strictly differentiable at $g(\bar{x})$.*

(f) *$F = f + F_1$, where $f: X \to Y$ is strictly differentiable at $\bar{x}$ and $F_1: X \rightrightarrows Y$ is strongly coderivatively normal at $(\bar{x}, \bar{y} - f(\bar{x}))$.*

(g) *$F = F_1 \circ g$, where $g: X \to Z$ is strictly differentiable at $\bar{x}$ with the surjective derivative and where $F_1: Z \rightrightarrows Y$ is strongly coderivatively normal at $(g(\bar{x}), \bar{y})$.*

(h) *$F = f \circ G$, where $f(x, \cdot)$ is a bounded linear operator from Z into Y for every x around $\bar{x}$ such that $x \to f(x, \cdot)$ is strictly differentiable at $\bar{x}$ while $f(\bar{x}, \cdot)$ is injective with the w^*-extensible range in Y, and where $G: X \rightrightarrows Z$ is strongly coderivatively normal at $(\bar{x}, \bar{z})$ with $\bar{y} = f(\bar{x}, \bar{z})$.*

(i) *$F = \partial(\varphi \circ g)$, where $\varphi: Z \to \overline{\mathbb{R}}$ and $g \in C^2$ with the surjective derivative $\nabla g(\bar{x})$ such that the range of $\nabla g(\bar{x})^*$ is w^*-extensible in X^*, and where $\partial\varphi$ is strongly coderivatively normal at $(\bar{z}, \bar{v})$ with $\bar{z} := g(\bar{x})$ and $\bar{v}$ uniquely defined by the relations*

$$\bar{y} = \nabla g(\bar{x})^* \bar{v} \quad \textit{and} \quad \bar{v} \in \partial\varphi(\bar{z}) .$$

Proof. Assertions (a) and (c) are obvious; the specifications of (c) for convex-graph and for strictly differentiable mappings follow from Propositions 1.37 and 1.38, respectively. Assertion (b) is taken from Proposition 1.33. Assertion (d) is a part of Theorem 3.28, while (e) is proved in Theorem 1.65(iii). Assertions (f)–(i) follow from the calculus rules for the normal and mixed coderivatives established in Theorems 1.62(ii), 1.66, Lemma 1.126, and Theorem 1.127, respectively. △

Note that further sufficient conditions for strong coderivative normality follows from calculus rules for N-regularity of set-valued mappings between Asplund spaces; see Subsect. 3.1.2.

Theorem 4.10 (pointbased characterizations of Lipschitz-like property). *Let $F\colon X \rightrightarrows Y$ be a set-valued mapping between Asplund spaces that is assumed to be closed-graph around $(\bar{x}, \bar{y}) \in \operatorname{gph} F$. Then the following properties are equivalent:*

(a) *F is Lipschitz-like around $(\bar{x}, \bar{y})$.*

(b) *F is PSNC at $(\bar{x}, \bar{y})$ and $\|D_M^* F(\bar{x}, \bar{y})\| < \infty$.*

(c) *F is PSNC at $(\bar{x}, \bar{y})$ and $D_M^* F(\bar{x}, \bar{y})(0) = \{0\}$.*

Moreover, in this case one has the estimates

$$\|D_M^* F(\bar{x}, \bar{y})\| \leq \operatorname{lip} F(\bar{x}, \bar{y}) \leq \|D_N^* F(\bar{x}, \bar{y})\| \tag{4.5}$$

for the exact Lipschitzian bound of F around $(\bar{x}, \bar{y})$, where the upper estimate holds if $\dim X < \infty$. *If in addition F is coderivatively normal at $(\bar{x}, \bar{y})$, then*

$$\operatorname{lip} F(\bar{x}, \bar{y}) = \|D_M^* F(\bar{x}, \bar{y})\| = \|D_N^* F(\bar{x}, \bar{y})\| \, . \tag{4.6}$$

Proof. The necessity of the PSNC and coderivative conditions in (b) and (c) for the Lipschitz-like property of F follows from Proposition 1.68 and Theorem 1.44(i), where the latter result implies also the lower bound estimate in (4.5) for any Banach spaces. Since

$$\|D_M^* F(\bar{x}, \bar{y})\| < \infty \Longrightarrow D_M^* F(\bar{x}, \bar{y})(0) = \{0\} \, ,$$

it remains to show that (c)$\Rightarrow$(a) in the Asplund space setting, and that the upper bound estimate holds in (4.5) if in addition X is finite-dimensional.

To prove (c)$\Rightarrow$(a) by contradiction, we suppose that F is not Lipschitz-like around $(\bar{x}, \bar{y})$. Then the neighborhood criterion from Theorem 4.7(b) doesn't hold. Hence there are sequences $(x_k, y_k) \in \operatorname{gph} F$ and $(x_k^*, -y_k^*) \in \widehat{N}((x_k, y_k); \operatorname{gph} F)$ with $(x_k, y_k) \to (\bar{x}, \bar{y})$ and

$$\|x_k^*\| > k\|y_k^*\| \quad \text{for all} \quad k \in I\!N \, .$$

Letting $\tilde{x}_k^* := x_k^*/\|x_k^*\|$ and $\tilde{y}_k^* := y_k^*/\|y_k^*\|$, we have

$$\tilde{x}_k^* \in \widehat{D}^* F(x_k, y_k)(\tilde{y}_k^*), \quad \|\tilde{x}_k^*\| = 1, \quad \text{and} \quad \|\tilde{y}_k^*\| \leq \tfrac{1}{k} \to 0 \quad \text{as} \quad k \to \infty \, . \tag{4.7}$$

Since X is Asplund, there is a subsequence of $\{\tilde{x}_k^*\}$ that weak* converges to some $x^* \in X^*$. Passing to the limit in (4.7) and using the definition of the mixed coderivative, we arrive at $x^* \in D_M^* F(\bar{x}, \bar{y})(0)$. Hence $x^* = 0$ due to the condition $D_M^* F(\bar{x}, \bar{y})(0) = \{0\}$ in (c). Employing further the PSNC property of F imposed in (c), we conclude that $\|\tilde{x}_k^*\| \to 0$ along a subsequence. This contradicts the condition $\|\tilde{x}_k^*\| = 1$ in (4.7) and completes the proof of the equivalencies in (a)–(c).

Let us finally justify the upper estimate in (4.5) assuming that X is finite-dimensional. To furnish this, we use the neighborhood formula for computing the exact Lipschitzian bound of F around $(\bar{x}, \bar{y})$ from Theorem 4.7. According to this formula and the norm definition (1.22) in the case of $\widehat{D}^* F(x, y)$, pick any $\nu > 0$ and find sequences $(x_k, y_k) \to (\bar{x}, \bar{y})$ and $(x_k^*, y_k^*) \in X^* \times Y^*$ such that $(x_k, y_k) \in \operatorname{gph} F$, $\{x_k^*\}$ is bounded, and

$$\operatorname{lip} F(\bar{x}, \bar{y}) < \|x_k^*\| + \nu, \quad x_k^* \in \widehat{D}^* F(x_k, y_k)(y_k^*), \quad \|y_k^*\| \le 1 \tag{4.8}$$

whenever $k \in I\!N$. Since X is finite-dimensional and Y is Asplund, there is a pair $(x^*, y^*) \in X^* \times Y^*$ for which $x_k^* \to x^*$ and $y_k^* \xrightarrow{w^*} y^*$ along a subsequence of $\{k\}$. Then $\|x_k^*\| \to \|x^*\|$ along this subsequence and

$$\|y^*\| \le \liminf_{k \to \infty} \|y_k^*\| \le 1$$

due to the continuity of the norm function in finite dimensions and its lower semicontinuity in the weak* topology of Y^*. Passing to the limit in (4.8) as $k \to \infty$ and taking into account the definition of the normal coderivative, we conclude that

$$\operatorname{lip} F(\bar{x}, \bar{y}) \le \|x^*\| + \nu \;\text{ with }\; x^* \in D_N^* F(\bar{x}, \bar{y})(y^*), \quad \|y^*\| \le 1\,.$$

Since $\nu > 0$ was chosen arbitrary, the latter implies the upper estimate in (4.5) under the assumptions made. Equalities (4.6) for the exact Lipschitzian bound immediately follow from estimates (4.5) provided that F is coderivatively normal at the reference point $(\bar{x}, \bar{y})$. △

The results obtained allow us to establish pointbased characterizations of the *classical local Lipschitzian* property of set-valued mappings formulated in Definition 1.40(iii).

Corollary 4.11 (pointbased characterizations of local Lipschitzian property). *Let $F\colon X \rightrightarrows Y$ be a set-valued mapping between Asplund spaces whose graph is closed near some point $\bar{x} \in \operatorname{dom} F$. Assume that F is locally compact around $\bar{x}$. Then the following are equivalent:*

(a) *F is locally Lipschitzian around $\bar{x}$.*

(b) *For every $\bar{y} \in F(\bar{x})$, F is PSNC at $(\bar{x}, \bar{y})$ and $\|D_M^* F(\bar{x}, \bar{y})\| < \infty$.*

(c) *For every $\bar{y} \in F(\bar{x})$, F is PSNC at $(\bar{x}, \bar{y})$ and $D_M^* F(\bar{x}, \bar{y})(0) = \{0\}$.*

Moreover, in this case one has the estimates

$$\sup_{\bar{y} \in F(\bar{x})} \|D_M^* F(\bar{x}, \bar{y})\| \le \operatorname{lip} F(\bar{x}) \le \sup_{\bar{y} \in F(\bar{x})} \|D_N^* F(\bar{x}, \bar{y})\|\,,$$

for the exact Lipschitzian bound of F around $\bar{x}$, where the upper estimate holds if $\dim X < \infty$. If in addition F is coderivatively normal at $(\bar{x}, \bar{y})$ for all $\bar{y} \in F(\bar{x})$, then

$$\operatorname{lip} F(\bar{x}) = \sup_{\bar{y} \in F(\bar{x})} \|D_M^* F(\bar{x}, \bar{y})\| = \sup_{\bar{y} \in F(\bar{x})} \|D_N^* F(\bar{x}, \bar{y})\|\,.$$

Proof. This is implied by Theorem 4.10 due to the relationships between the local Lipschitzian and Lipschitz-like properties of locally compact multifunctions established in Theorem 1.42. △

In what follows we mostly consider applications of the criteria obtained in Theorem 4.10 for the Lipschitz-like property; one has similar results for the classical Lipschitzian property of locally compact multifunctions due to Corollary 4.11. Note that the criteria obtained above are simplified when $\dim X < \infty$, since F is automatically PSNC in this case. If both X and Y are finite-dimensional, then we have the unified characterization

$$D^*F(\bar{x},\bar{y})(0)=\{0\} \ \text{ with } \ \operatorname{lip} F(\bar{x},\bar{y})=\|D^*F(\bar{x},\bar{y})\| \tag{4.9}$$

of the Lipschitz-like property for set-valued mappings in terms of the common coderivative $D^*F(\bar{x},\bar{y})$.

Another important situation when the conditions of Theorem 4.10 can be essentially simplified and efficiently specified concerns set-valued mappings with closed and *convex graphs*. In contrast to (4.9), the next result is not a straightforward corollary of Theorem 4.10, although its proof is primarily based on the above coderivative criteria specified for convex-graph mappings.

Theorem 4.12 (Lipschitz-like property of convex-graph multifunctions). *Let $F\colon X \rightrightarrows Y$ be a convex-graph multifunction between Asplund spaces. Given $\bar{x}\in\operatorname{dom} F$, assume that the graph of F is closed near $\bar{x}$. Then the following are equivalent:*

(a) *There is $\bar{y}\in F(\bar{x})$ such that F is Lipschitz-like around $(\bar{x},\bar{y})$.*

(b) *The range of F^{-1} is SNC at $\bar{x}$ and $N(\bar{x};\operatorname{rge} F^{-1})=\{0\}$.*

(c) *$\bar{x}$ is an interior point of the range of F^{-1}.*

(d) *F is Lipschitz-like at $(\bar{x},\bar{y})$ for every $\bar{y}\in F(\bar{x})$.*

If in addition $\dim X<\infty$, then one has

$$\operatorname{lip} F(\bar{x},\bar{y})=\sup_{\|y^*\|\le 1}\Big\{\|x^*\|\,\Big|\,\langle x^*,x-\bar{x}\rangle\le\langle y^*,y-\bar{y}\rangle \ \text{ for all } \ (x,y)\in\operatorname{gph} F\Big\}$$

whenever $\bar{y}\in F(\bar{x})$.

Proof. Implication (a)⇒(b) (actually the equivalence between these properties) follows from (a)⟹(c) in Theorem 4.10 due to the coderivative representation for convex-graph mappings in Proposition 1.37. Note that in this setting $\operatorname{int}(\operatorname{rge} F^{-1})\neq\emptyset$, which can be easily observed from the local covering property of F^{-1} that is equivalent to the Lipschitz-like property of F. Thus $\operatorname{int}\Omega=\operatorname{int}(\operatorname{cl}\Omega)$ for the convex set $\Omega=\operatorname{rge} F^{-1}$, which is well known from convex analysis. By this we may assume without loss of generality that the range of F^{-1} is locally closed around $\bar{x}$. Then implication (b)⇒(c) follows directly from the normal characterization of boundary points for closed SNC sets in Corollary 2.24. To prove (c)⇒(d), we first observe that, due to the convexity of the sets gph F and rge F^{-1}, the SNC property of rge F^{-1} at $\bar{x}$ is equivalent to the PSNC property of F at $(\bar{x}, \bar{y})$ for every $\bar{y} \in F(\bar{x})$. Since (c) implies that rge F^{-1} is SNC at $\bar{x}$ by Proposition 1.25 and Theorem 1.26, and since one obviously has

$$(\mathrm{c}) \Longrightarrow D^*F(\bar{x}, \bar{y})(0) = \{0\} \ \text{ for every } \ \bar{y} \in F(\bar{x}),$$

then (c)⇒(d) follows from Theorem 4.10. Implication (d)⇒(a) is trivial, and the exact bound formula in the theorem is a direct consequence of (4.6), Proposition 1.37, and the norm definition (1.22). △

Note that implication (c)⇒(d) of Theorem 4.12 follows also from the inverse version of the classical Robinson-Ursescu theorem on metric regularity of closed- and convex-graph mappings between arbitrary Banach spaces; cf. Theorem 4.21 stated below.

Remark 4.13 (Lipschitzian properties via Clarke normals). Theorem 4.10 immediately implies a sufficient condition for the Lipschitz-like property of $F: X \rightrightarrows Y$ around $(\bar{x}, \bar{y})$, where $D^*_M F(\bar{x}, \bar{y})(0) = \{0\}$ in (c) is replaced by its counterpart in terms of Clarke normals:

$$\big[(x^*, 0) \in N_C((\bar{x}, \bar{y}); \mathrm{gph}\, F)\big] \Longrightarrow x^* = 0\,. \tag{4.10}$$

Recall that the latter cone agrees in Asplund spaces with the convexified cone $\mathrm{cl}^*\mathrm{co}\, N$ due to Theorem 3.57. Note that there is no difference between (4.10) and the basic condition $D^*_M F(\bar{x}, \bar{y})(0) = \{0\}$ for convex-graph mappings, while these conditions may be essentially different in *nonconvex* settings.

Indeed, it follows from Theorems 3.62 and 3.72(i), that the Clarke normal cone in (4.10) is actually a *linear subspace* if F is single-valued and w^*-strictly Lipschitzian at $\bar{x}$ or, more general, if the graph of $F: X \rightrightarrows Y$ is strictly hemi-Lipschitzian at $(\bar{x}, \bar{y})$. This means that condition (4.10) is *far removed from necessity* for such mappings F to be Lipschitz-like around $(\bar{x}, \bar{y})$, even in finite-dimensional spaces.

To demonstrate this, we consider a mapping $f: I\!R^n \to I\!R^m$ locally Lipschitzian around $\bar{x}$. Then it follows from the proof of Theorem 1.46 that condition (4.10) holds *if and only if* this mapping is strictly differentiable at $\bar{x}$; so it is *never fulfilled* for *nonsmooth* Lipschitzian mappings. In contrast, condition (4.9) via basic normals completely characterizes Lipschitz-like mappings between finite-dimensional spaces.

It is crucial for applications of Theorem 4.10 that the PSNC property and both coderivatives used in its formulation enjoy the fairly *rich calculi* developed above. Due to these calculi, the obtained characterizations can be efficiently employed in typical situations when mappings F are given in special forms arising in applications. Some of such applications will be considered in

Sect. 4.3, where we study Lipschitzian stability of parametric constrained and variational systems related to optimization and equilibrium models.

Now let us show that the dual characterizations of Theorem 4.10 and the mentioned coderivative and PSNC calculi allow us to derive efficient conditions ensuring the *preservation of Lipschitz continuity* under various operations performed on set-valued mappings. To obtain results in this direction, we essentially use the fact that Theorem 4.10 provides *necessary and sufficient* conditions for Lipschitz continuity.

The first theorem deals with general compositions of set-valued mappings between Asplund spaces. As usual, we present results for the Lipschitz-like property, which automatically imply similar conditions for the preservation of classical Lipschitz continuity of locally compact multifunctions.

Theorem 4.14 (Lipschitz-like property under compositions). *Consider $\bar{z} \in (F \circ G)(\bar{x})$, where $G\colon X \rightrightarrows Y$ and $F\colon Y \rightrightarrows Z$ are set-valued mappings between Asplund spaces such that the graphs of G and F^{-1} are locally closed near $\bar{x}$ and $\bar{z}$, respectively. Assume that:*

(a) *The set-valued mapping $(x, z) \to G(x) \cap F^{-1}(z)$ is inner semicompact around $(\bar{x}, \bar{z})$.*

(b) *For every $\bar{y} \in G(\bar{x}) \cap F^{-1}(\bar{z})$ both mappings G and F are Lipschitz-like around $(\bar{x}, \bar{y})$ and $(\bar{y}, \bar{z})$, respectively.*

Then $F \circ G$ is Lipschitz-like around $(\bar{x}, \bar{z})$. If in addition $\dim X < \infty$ *and for every $\bar{y} \in G(\bar{x}) \cap F^{-1}(\bar{z})$ both F and G are coderivatively normal at the points $(\bar{y}, \bar{z})$ and $(\bar{x}, \bar{y})$, respectively, then one has*

$$\operatorname{lip}(F \circ G)(\bar{x}, \bar{z}) \le \max_{\bar{y} \in G(\bar{x}) \cap F^{-1}(\bar{z})} \operatorname{lip} G(\bar{x}, \bar{y}) \cdot \operatorname{lip} F(\bar{y}, \bar{z})\,. \tag{4.11}$$

Proof. Due to assumption (b) of the theorem and implication (a)⇒(c) of Theorem 4.10, we have that for every $\bar{y} \in G(\bar{x}) \cap F^{-1}(\bar{z})$ the mappings G and F are PSNC at $(\bar{x}, \bar{y})$ and $(\bar{y}, \bar{z})$, respectively, with

$$D_M^* G(\bar{x}, \bar{y})(0) = \{0\} \quad \text{and} \quad D_M^* F(\bar{y}, \bar{z})(0) = \{0\}\,. \tag{4.12}$$

Then by the zero chain rule of Theorem 3.14 we have that

$$D_M^*(F \circ G)(\bar{x}, \bar{z})(0) = \{0\}\,.$$

Furthermore, Corollary 3.96 ensures the PSNC property of $F \circ G$ at $(\bar{x}, \bar{z})$. Employing now the opposite implication (c)⇒(a) of Theorem 4.10, we conclude that $F \circ G$ is Lipschitz-like around $(\bar{x}, \bar{z})$.

It remains to justify the bound inequality (4.11). By Theorem 3.13(ii) we have the chain rule

$$D^*(F \circ G)(\bar{x}, \bar{z})(z^*) \subset \bigcup_{\bar{y} \in G(\bar{x}) \cap F^{-1}(\bar{z})} D_N^* G(\bar{x}, \bar{y}) \circ D^* F(\bar{y}, \bar{z})(z^*) \tag{4.13}$$

for both coderivatives $D^* = D^*_M$ and $D^* = D^*_N$. Using (4.13) and taking into account that $\|H_1 \circ H_2\| \le \|H_1\| \cdot \|H_2\|$ for any positively homogeneous multifunctions, and also that both F and G are coderivatively normal, we get from (4.5) the following relations when $\dim X < \infty$:

$$\begin{aligned}\operatorname{lip}(F \circ G)(\bar{x}, \bar{z}) &\le \sup_{\bar{y} \in G(\bar{x}) \cap F^{-1}(\bar{z})} \|D^*_N G(\bar{x}, \bar{y})\| \cdot \|D^*_N F(\bar{y}, \bar{z})\| \\ &= \sup_{\bar{y} \in G(\bar{x}) \cap F^{-1}(\bar{z})} \|D^*_M G(\bar{x}, \bar{y})\| \cdot \|D^*_M F(\bar{y}, \bar{z})\| \\ &\le \max_{\bar{y} \in G(\bar{x}) \cap F^{-1}(\bar{z})} \operatorname{lip} G(\bar{x}, \bar{y}) \cdot \operatorname{lip} F(\bar{y}, \bar{z}) ,\end{aligned}$$

which implies (4.11). Note that we use "max" in the latter inequality and in (4.11), since the set $G(\bar{x}) \cap F^{-1}(\bar{z})$ is compact under assumption (a) and since the function $(x, y) \to \operatorname{lip} S(x, y)$ is upper semicontinuous on the graph of any Lipschitz-like multifunction S. △

Observe that if the mapping $G \cap F^{-1}$ is *inner semicontinuous* vs. inner semicompact at $(\bar{x}, \bar{z}, \bar{y})$ for some $\bar{y} \in G(\bar{x}) \cap F^{-1}(\bar{z})$ and if the graph of $F \circ G$ is locally closed around $(\bar{x}, \bar{z})$, then all the other assumptions and conclusions in Theorem 4.14 (and similarly in the subsequent results) are applied to this particular point $\bar{y}$. Let us specify the assumptions of Theorem 4.14 in the situation when the inner mapping $G = g$ is single-valued.

Corollary 4.15 (compositions with single-valued inner mappings). *Take $\bar{z} \in (F \circ g)(\bar{x})$, where $g\colon X \to Y$ and $F\colon Y \rightrightarrows Z$ are mappings between Asplund spaces. Assume that g is Lipschitz continuous around $\bar{x}$ and that F is closed-graph around $(g(\bar{x}), \bar{z})$. Then $F \circ g$ is Lipschitz-like around $(\bar{x}, \bar{z})$ provided that F is Lipschitz-like around $(g(\bar{x}), \bar{z})$. Moreover,*

$$\operatorname{lip}(F \circ g)(\bar{x}, \bar{z}) \le \operatorname{lip} g(\bar{x}) \cdot \operatorname{lip} F(g(\bar{x}), \bar{z})$$

if $\dim X < \infty$ *and if both F and g are coderivatively normal at the points $(g(\bar{x}), \bar{z})$ and $\bar{x}$, respectively.*

Proof. Under the assumptions made the mapping $(g \cap F^{-1})(x, z) = \{g(x)\}$ is obviously inner semicompact around $(\bar{x}, \bar{z})$, and so we have a direct specification of Theorem 4.14. △

The next result presents conditions ensuring the preservation of the Lipschitz-like property for *sums* of set-valued mappings, with relationships between the exact Lipschitzian bounds. It is sufficient to consider the sum of two multifunctions, which implies the general summation case by induction. For brevity we formulate this result only under the corresponding inner semicompactness assumption.

Theorem 4.16 (Lipschitz-like property under summation). *Consider two mappings $F_i\colon X \rightrightarrows Y$ between Asplund spaces whose graphs are locally closed near some point $\bar{x} \in (\operatorname{dom} F_1) \cap (\operatorname{dom} F_2)$. Take $\bar{y} \in F_1(\bar{x}) + F_2(\bar{x})$ and assume that the mapping $S\colon X \times Y \rightrightarrows Y^2$ defined by*

$$S(x,y) := \big\{(y_1,y_2) \in Y^2 \big|\ y_1 \in F_1(x),\ y_2 \in F_2(x),\ y_1 + y_2 = y\big\}$$

is inner semicompact around $(\bar{x},\bar{y})$. Then the sum $F_1 + F_2$ is Lipschitz-like around $(\bar{x},\bar{y})$ provided that for every $(\bar{y}_1,\bar{y}_2) \in S(\bar{x},\bar{y})$ each F_i is Lipschitz-like around $(\bar{x},\bar{y}_1)$ and $(\bar{x},\bar{y}_2)$, respectively. Moreover,

$$\operatorname{lip}(F_1 + F_2)(\bar{x},\bar{y}) \le \max_{(\bar{y}_1,\bar{y}_2)\in S(\bar{x},\bar{y})} \big\{\operatorname{lip} F_1(\bar{x},\bar{y}_1) + \operatorname{lip} F_2(\bar{x},\bar{y}_2)\big\}$$

if $\dim X < \infty$ *and F_i is coderivatively normal at $(\bar{x},\bar{y}_i)$ for both $i = 1,2$ and for all vectors $(\bar{y}_1,\bar{y}_2) \in S(\bar{x},\bar{y})$.*

Proof. It follows from Theorem 3.10 that the sum rule

$$D^*(F_1 + F_2)(\bar{x},\bar{y})(y^*) \subset \bigcup_{(y_1,y_2)\in S(\bar{x},\bar{y})} \big[D^*F_1(\bar{x},\bar{y}_1)(y^*) + D^*F_2(\bar{x},\bar{y}_2)(y^*)\big]$$

holds for both coderivatives $D^* = D^*_N, D^*_M$ under the assumptions made. Putting $y^* = 0$ in this coderivative sum rule for the case of $D^* = D^*_M$, we get by Theorem 1.44 that

$$D^*_M(F_1 + F_2)(\bar{x},\bar{y})(0) = \{0\}\,.$$

Furthermore, the PSNC property of the sum $F_1 + F_2$ follows from the PSNC calculus result of Theorem 3.88. Invoking the coderivative criterion for the Lipschitz-like property from Theorem 4.10, we conclude that $F_1 + F_2$ is Lipschitz-like at $(\bar{x},\bar{y})$. Finally, using the above some rule for both coderivatives $D^* = D^*_N, D^*_M$ and the obvious inequality

$$\|H_1 + H_2\| \le \|H_1\| + \|H_2\|$$

for the norms of positively homogeneous multifunctions, we arrive at the exact bound formula of the theorem similarly to the proof of Theorem 4.14. △

The next consequence of Theorems 4.14 and 4.16 concerns *h-compositions* $F_1 \overset{h}{\diamond} F_2$ of set-valued mappings that cover many various operations on multifunctions; see Subsect. 3.1.2.

Corollary 4.17 (Lipschitz-like property under h-compositions). *Take $\bar{z} \in (F_1 \overset{h}{\diamond} F_2)(\bar{x})$ with $F_i\colon X \rightrightarrows Y_i$, $i = 1,2$, and $h\colon X \times Z \to Y_1 \times Y_2$ in the Asplund space setting. Define the multifunction $S\colon Y_1 \times Y_2 \rightrightarrows Z$ by*

$$S(x,z) := \big\{(y_1,y_2) \in Y_1 \times Y_2 \big|\ y_i \in F_i(x),\ z = h(y_1,y_2)\big\}$$

and suppose that it is inner semicompact at $(\bar{x}, \bar{z})$. *Assume also that for every* $\bar{y} = (\bar{y}_1, \bar{y}_2) \in S(\bar{x}, \bar{z})$ *the mappings* F_i *are closed-graph and Lipschitz-like around* $(\bar{x}, \bar{y}_1)$ *and* $(\bar{x}, \bar{y}_2)$, *respectively, and that* h *is locally Lipschitzian around* $(\bar{x}, \bar{y})$. *Then* $F_1 \overset{h}{\diamond} F_2$ *is Lipschitz-like around* $(\bar{x}, \bar{z})$.

Proof. Define $F\colon X \rightrightarrows Y_1 \times Y_2$ by $F(x) := \big(F_1(x), F_2(x)\big)$ and observe that $F = \widetilde{F}_1 + \widetilde{F}_2$, where $\widetilde{F}_1(x) := \big(F_1(x), 0\big)$ and $\widetilde{F}_2(x) := \big(0, F_2(x)\big)$. It follows from Corollary 4.16 that F is Lipschitz-like around $(\bar{x}, \bar{y})$ for all $\bar{y} = (\bar{y}_1, \bar{y}_2) \in S(\bar{x}, \bar{z})$. Since clearly

$$(F_1 \overset{h}{\diamond} F_2)(x) = (h \circ F)(x)$$

and since h is locally Lipschitzian, we apply now Theorem 4.14 to the latter composition and thus complete the proof of the corollary. △

4.2.2 Pointbased Characterizations of Covering and Metric Regularity

In this subsection we obtain pointbased characterizations of the covering and metric regularity properties of multifunctions between Asplund spaces, with formulas for estimating and computing the corresponding exact bounds. We also present results on the preservation of the mentioned properties under general compositions. The results obtained are derived from the above characterizations of the Lipschitzian properties due to relationships between all these properties established in Subsect. 1.2.3. We start with pointbased criteria and exact bound formulas for local covering and metric regularity. For these characterizations it is convenient to use, together with the mixed and normal coderivatives, the *reversed* mixed coderivative defined by

$$\widetilde{D}^*_M F(\bar{x}, \bar{y})(y^*) := \Big\{x^* \in X^* \big|\ y^* \in -D^*_M F^{-1}(\bar{y}, \bar{x})(-x^*)\Big\}\,.$$

Theorem 4.18 (pointbased characterizations of local covering and metric regularity). *Let* $F\colon X \rightrightarrows Y$ *be a set-valued mapping between Asplund spaces that is assumed to be closed-graph around* $(\bar{x}, \bar{y}) \in \operatorname{gph} F$. *Then the following are equivalent:*

(a) F *is locally metrically regular around* $(\bar{x}, \bar{y})$.

(b) F *enjoys the local covering property around* $(\bar{x}, \bar{y})$.

(c) F^{-1} *is PSNC at* $(\bar{y}, \bar{x})$ *with the equivalent conditions*

$$D^*_M F^{-1}(\bar{y}, \bar{x})(0) = \{0\} \Longleftrightarrow \ker \widetilde{D}^*_M F(\bar{x}, \bar{y}) = \{0\}\,.$$

(d) F^{-1} *is PSNC at* $(\bar{y}, \bar{x})$ *and*

$$\|D^*_M F^{-1}(\bar{y}, \bar{x})\| = \|\widetilde{D}^*_M F(\bar{x}, \bar{y})^{-1}\| < \infty\,.$$

(e) F^{-1} *is PSNC at* $(\bar{y}, \bar{x})$ *and*

$$\inf\Big\{\|x^*\|\;\Big|\;x^*\in \widetilde{D}^*_M F(\bar{x},\bar{y})(y^*),\ \|y^*\|=1\Big\}>0\,.$$

Moreover, one has the estimates

$$\operatorname{reg} F(\bar{x},\bar{y})\le \|D^*_N F^{-1}(\bar{y},\bar{x})\| = \|D^*_N F(\bar{x},\bar{y})^{-1}\|\,, \tag{4.14}$$

$$\operatorname{cov} F(\bar{x},\bar{y})\ge \inf\Big\{\|x^*\|\;\Big|\;x^*\in D^*_N F(\bar{x},\bar{y})(y^*),\ \|y^*\|=1\Big\} \tag{4.15}$$

when $\dim Y<\infty$. *If in addition* F^{-1} *is coderivatively normal at* $(\bar{y},\bar{x})$, *then*

$$\operatorname{reg} F(\bar{x},\bar{y}) = \|D^* F^{-1}(\bar{y},\bar{x})\| = \|D^* F(\bar{x},\bar{y})^{-1}\|\,, \tag{4.16}$$

$$\operatorname{cov} F(\bar{x},\bar{y}) = \inf\Big\{\|x^*\|\;\Big|\;x^*\in D^* F(\bar{x},\bar{y})(y^*),\ \|y^*\|=1\Big\}\,, \tag{4.17}$$

where D^* *stands for either* D^*_N, *or* D^*_M, *or* $\widetilde{D}^*_M$.

Proof. Equivalence (a)⇔(b) is proved in Theorem 1.52(i) for any Banach spaces. Equivalences (a)⇔(c) and (a)⇔(d) follow from the relationships between the metric regularity and Lipschitz-like property of Theorem 1.49(i) and the characterizations of the latter property from Theorem 4.10. Equivalence (a)⇔(d) implies the one of (b)⇔(e) due to Theorem 1.52(i) by taking into account the relationship

$$1\Big/\|H^{-1}\| = \inf\big\{\|y\|\;\big|\;y\in H(x),\ \|x\|=1\big\} \tag{4.18}$$

valid for any positively homogeneous multifunction. Estimates (4.14) and (4.15) follow then from the upper estimate in (4.5) applied to the inverse mapping F^{-1} and from formula (4.18) applied to the coderivative $H=D^*_N F(\bar{x},\bar{y})$. Employing (4.14) and the opposite inequality

$$\operatorname{reg} F(\bar{x},\bar{y})\ge \|D^*_M F^{-1}(\bar{y},\bar{x})\|$$

established in Theorem 1.54(ii) in arbitrary Banach spaces, we get equality (4.16) for the case of $D^*=D^*_N$ when F^{-1} is coderivatively normal at $(\bar{y},\bar{x})$. Note that the latter is clearly equivalent to $\|\widetilde{D}^*_M F(\bar{x},\bar{y})\| = \|D^*_N F(\bar{x},\bar{y})\|$. Moreover, $D^*_M F(\bar{x},\bar{y}) = D^*_N F(\bar{x},\bar{y})$ when Y is finite-dimensional. Thus (4.16) holds also for $D^*=D^*_M$ and $D^*=\widetilde{D}^*_M$ under the assumptions made. Finally, (4.16) is equivalent to (4.17) in this case due to (4.18). △

The following example shows that the *PSNC* condition is *essential* for the point characterizations of the covering and metric regularity properties of multifunctions between infinite-dimensional spaces in Theorem 4.18 (and for the equivalent characterizations of Lipschitzian stability in Theorem 4.10).

Example 4.19 (violation of covering and metric regularity in the absence of PSNC). *For any separable Banach space X there is a convex-valued mapping $F\colon X \rightrightarrows X$ that doesn't have covering and metric regularity properties around $(\bar{x},\bar{y}) \in \operatorname{gph} F$ while* $\ker D_N^* F(\bar{x},\bar{y}) = \{0\}$.

Proof. Let X be an arbitrary separable Banach space, and let $\{e_n\}_{n=1}^{\infty}$ be unit independent vectors that densely span X. Form the convex sets

$$\Omega_1 := \operatorname{clco}\Big\{\frac{e_n}{2^n},\ -\frac{e_n}{2^n}\Big\} \quad\text{and}\quad \Omega_2 := \Big\{ta\Big|\ t \in [-1,1]\Big\} \quad\text{with}\quad a := \sum_{n=1}^{\infty}\frac{e_n}{n^2} \in X$$

that are obviously norm-compact and satisfies $\Omega_1 \cap \Omega_2 = \{0\}$. Define the set-valued mapping $F\colon X \rightrightarrows X$ by

$$F(x) := \begin{cases} x + \Omega_1 & \text{if } x \in \Omega_2\,, \\ \emptyset & \text{otherwise} \end{cases}$$

for which $(0,0) \in \operatorname{gph} F$. Since $\operatorname{span}\Omega_1$ is dense in X, we have

$$N((0,0);\operatorname{gph} F) \subset \big[\{0\} \times \Omega_1\big]^{\perp} = X^* \times \{0\}\,,$$

and hence $\ker D_N^* F(0,0) = \{0\}$. It remains to check that F doesn't have the local covering property around $(0,0)$. It is sufficient to show that for any $r > 0$ the image set

$$F(r I\!B) = \bigcup_{\alpha \in [0, r/\|a\|]} \big[\alpha a + \Omega_1\big]$$

doesn't contain an open ball around the origin. Indeed, taking $b := \sum_{n=1}^{\infty}\frac{e_n}{n^3}$ and an arbitrarily small number $\beta > 0$, we observe that $\beta b - \alpha a \in \Omega_1$ for some $\alpha \in [0, r/\|a\|]$, which can only happen if $\beta b - \alpha a = 0$. $\triangle$

Theorem 4.18 and the relationships between local and semi-local properties established in Subsect. 1.2.3 imply pointbased characterizations of semi-local covering and two kinds of metric regularity for locally compact multifunctions acting in Asplund spaces.

Corollary 4.20 (pointbased characterizations of semi-local covering and metric regularity). *Let $F\colon X \rightrightarrows Y$ be a set-valued mapping between Asplund spaces. The following assertions hold:*

(i) *Given $\bar{x} \in \operatorname{dom} F$, we assume that F is locally compact around $\bar{x}$ and that its graph is closed whenever x is near this point. Then F enjoys the semi-local covering property around $\bar{x}$ if and only if each of the equivalent conditions* (c)–(e) *of Theorem* 4.18 *is fulfilled for every vector $\bar{y} \in F(\bar{x})$. If in addition* $\dim Y < \infty$, *then*

$$\operatorname{cov} F(\bar{x}) \geq \inf \Big\{ \|x^*\| \ \Big|\ x^* \in D_N^* F(\bar{x}, \bar{y})(y^*),\ \bar{y} \in F(\bar{x}),\ \|y^*\| = 1 \Big\}\,.$$

The latter estimate holds as equality (with $D_N^* = D_M^* = \widetilde{D}_M^*$) if F^{-1} is coderivatively normal at $(\bar{y}, \bar{x})$ for all $\bar{y} \in F(\bar{x})$.

(ii) *Under the corresponding assumptions of* (i), *F is semi-locally metrically regular around $\bar{x} \in \operatorname{dom} F$ if and only if each of the equivalent conditions* (c)–(e) *of Theorem* 4.18 *is fulfilled for every vector $\bar{y} \in F(\bar{x})$. Moreover, one has the estimate*

$$\operatorname{reg} F(\bar{x}) \leq \max_{\bar{y} \in F(\bar{x})} \|D_N^* F^{-1}(\bar{y}, \bar{x})\| = \max_{\bar{y} \in F(\bar{x})} \|D_N^* F(\bar{x}, \bar{y})^{-1}\|\,,$$

which holds as equality (with $D_N^* = D_M^* = \widetilde{D}_M^*$) if F^{-1} is coderivatively normal at $(\bar{y}, \bar{x})$ for all $\bar{y} \in F(\bar{x})$.

(iii) *Given $\bar{y} \in \operatorname{rge} F$, we assume that F^{-1} is locally compact around $\bar{y}$ and its graph is closed whenever y is near this point. Then F is semi-locally metrically regular around $\bar{y}$ if and only if, for all $\bar{y} \in F^{-1}(\bar{x})$, F^{-1} is PSNC at $(\bar{y}, \bar{x})$ and each of the following equivalent conditions holds:*

$$D_M^* F^{-1}(\bar{y}, \bar{x})(0) = \{0\}, \quad \ker \widetilde{D}_M^* F(\bar{x}, \bar{y}) = \{0\}\,,$$

$$\|D_M^* F^{-1}(\bar{y}, \bar{x})\| = \|\widetilde{D}_M^* F(\bar{x}, \bar{y})^{-1}\| < \infty\,.$$

When $\dim Y < \infty$, *one has the estimate*

$$\operatorname{reg} F(\bar{y}) \leq \max_{\bar{x} \in F^{-1}(\bar{y})} \|D_N^* F^{-1}(\bar{y}, \bar{x})\| = \max_{\bar{x} \in F^{-1}(\bar{y})} \|D_N^* F(\bar{x}, \bar{y})^{-1}\|\,,$$

which holds as equality (with $D_N^* = D_M^* = \widetilde{D}_M^*$) if F^{-1} is coderivatively normal at $(\bar{y}, \bar{x})$ for all $\bar{x} \in F^{-1}(\bar{y})$.

Proof. All the conclusions follow from the corresponding results of Theorem 4.18 due to the equivalence between the local and semi-local properties established in Proposition 1.50 and Corollary 1.53. △

In the rest of this subsection we consider various results related to the local metric regularity and covering properties, which imply similar results for the semi-local counterparts due to Corollary 4.20.

Observe that for single-valued mappings $F = f\colon X \to Y$ strictly differentiable at $\bar{x}$, Theorem 4.18 goes back to the characterizations of Theorem 1.57 the sufficient part of which (Lyusternik-Graves' theorem) is proved there for general Banach spaces.

The next result can be derived from Theorem 4.18 similarly to the proof of Theorem 4.12; it is actually a direct corollary of Theorem 4.12 applied to the inverse mapping. Note that implication (c)⇒(d) below is the main contents of the *Robinson-Ursescu closed graph/metric regularity theorem* valid in arbitrary Banach spaces; see, e.g., Theorem 3.3.1 in Aubin and Ekeland [52].

Theorem 4.21 (metric regularity and covering of convex-graph mappings). *Let $F: X \rightrightarrows Y$ be a convex-graph multifunction between Asplund spaces. Given $\bar{y} \in \operatorname{rge} F$, we assume that the graph of F is closed near $\bar{y}$. Then the following are equivalent:*

(a) *There is $\bar{x} \in F^{-1}(\bar{y})$ such that F is locally metrically regular (that is, it enjoys the local covering property) around $(\bar{x}, \bar{y})$.*

(b) *The convex set $\operatorname{rge} F$ is SNC at $\bar{y}$ and $N(\bar{y}; \operatorname{rge} F) = \{0\}$.*

(c) *$\bar{y}$ is an interior point of the range of F.*

(d) *F is locally metrically regular (that is, it enjoys the local covering property) around $(\bar{x}, \bar{y})$ for every $\bar{x} \in F^{-1}(\bar{y})$.*

If in addition $\dim Y < \infty$, then one has

$$\operatorname{reg} F(\bar{x}, \bar{y}) = \sup_{\|x^*\| \leq 1} \Big\{ \|y^*\| \Big| \ \langle x^*, x - \bar{x}\rangle \leq y^*, y - \bar{y}\rangle \ \textit{ for all } \ (x, y) \in \operatorname{gph} F \Big\},$$

$$\operatorname{cov} F(\bar{x}, \bar{y}) = \inf_{\|y^*\| = 1} \Big\{ \|x^*\| \Big| \ \langle x^*, x - \bar{x}\rangle \leq y^*, y - \bar{y}\rangle \ \textit{ for all } \ (x, y) \in \operatorname{gph} F \Big\}$$

whenever $\bar{y} \in F(\bar{x})$.

Proof. It is the inverse version of Theorem 4.12 applied to F^{-1}, which is Lipschitz-like around $(\bar{y}, \bar{x})$ in this setting. The precise formulas for the regularity and covering bounds follow directly from (4.16) and (4.17) due to Proposition 1.37 for convex-graph multifunctions. △

As in the case of Lipschitz continuity in Subsect. 4.2.1, the obtained characterizations imply efficient conditions ensuring the preservation of the metric regularity and covering properties under general compositions.

Theorem 4.22 (metric regularity and covering under compositions). *Let $\bar{z} \in (F \circ G)(\bar{x})$, where $G: X \rightrightarrows Y$ and $F: Y \rightrightarrows Z$ are set-valued mappings between Asplund spaces. Assume that the graphs of G and F^{-1} are locally closed near $\bar{x}$ and $\bar{z}$, respectively, and that the following conditions hold:*

(a) *The set-valued mapping $(x, z) \to G(x) \cap F^{-1}(z)$ is inner semicompact around $(\bar{x}, \bar{z})$.*

(b) *For every $\bar{y} \in G(\bar{x}) \cap F^{-1}(\bar{z})$ both mappings G and F are locally metrically regular (have the local covering property) around $(\bar{x}, \bar{y})$ and $(\bar{y}, \bar{z})$, respectively.*

Then $F \circ G$ is locally metrically regular (has the local covering property) around $(\bar{x}, \bar{z})$. If in addition $\dim Z < \infty$ and for every $\bar{y} \in G(\bar{x}) \cap F^{-1}(\bar{z})$ both mappings F^{-1} and G^{-1} are coderivatively normal at $(\bar{z}, \bar{y})$ and $(\bar{y}, \bar{x})$, respectively, then one has

$$\operatorname{reg}\,(F \circ G)(\bar{x}, \bar{z}) \leq \max_{\bar{y} \in G(\bar{x}) \cap F^{-1}(\bar{z})} \operatorname{reg} G(\bar{x}, \bar{y}) \cdot \operatorname{reg} F(\bar{y}, \bar{z}),$$

$$\operatorname{cov}(F \circ G)(\bar{x}, \bar{z}) \geq \min_{\bar{y} \in G(\bar{x}) \cap F^{-1}(\bar{z})} \operatorname{cov} G(\bar{x}, \bar{y}) \cdot \operatorname{cov} F(\bar{y}, \bar{z}) \, .$$

Proof. We derive this result from Theorem 4.14. Indeed, it is easy to check that for any set-valued mappings one has

$$(F \circ G)^{-1} = G^{-1} \circ F^{-1} \, .$$

Therefore, Theorem 4.14 applied to the composition $G^{-1} \circ F^{-1}$ gives the listed conditions for the preservation of the metric regularity and covering properties under the composition $F \circ G$. The exact bound inequalities for metric regularity and covering follow directly from (4.11) and the relationships between the exact bounds of all three properties under consideration established in Subsect. 1.2.3. △

4.2.3 Metric Regularity under Perturbations

An important issue in numerical work concerns the study of how large a perturbation can be before good behavior of a solution map breaks down. This relates to the classical Eckart-Young theorem in numerical analysis and to the so-called *distance to infeasibility* and the *condition number* theorems in mathematical programming.

Metric regularity and equivalent Lipschitzian and openness notions are key properties of "good behavior" in variational analysis. The following constant measures the extent to which a set-valued mapping can be *perturbed* by the addition of a linear mapping *without destroying* the metric regularity.

Definition 4.23 (radius of metric regularity). *Let $F\colon X \rightrightarrows Y$ be a set-valued mapping between Banach spaces, and let $(\bar{x}, \bar{y}) \in \operatorname{gph} F$. The* RADIUS OF METRIC REGULARITY *of F around $(\bar{x}, \bar{y})$ is*

$$\operatorname{rad} F(\bar{x}, \bar{y}) := \inf_{g \in \mathcal{L}(X,Y)} \Big\{ \|g\| \; \Big| \; \textit{metric regularity fails for } \; F + g \Big\} \, ,$$

where $\mathcal{L}(X, Y)$ stands for the space of linear bounded operators from X into Y and where the metric regularity of $F + g$ is considered around $(\bar{x}, \bar{y} + g(\bar{x}))$.

The radius value in Definition 4.23 could equally well be called the *distance to irregularity*, with respect to adding a linear mapping to F. Our main goal in what follows is to relate this value to the *exact regularity* bound $\operatorname{reg} F(\bar{x}, \bar{y})$ introduced in Definition 1.47(ii).

First we obtain a generalization of the Eckart-Young theorem for *positively homogeneous* multifunctions. Recall that the norm of a positively homogeneous multifunction is defined in (1.22). It is easy to observe that the inverse mapping F^{-1} is positively homogeneous if and only if F has this property.

Theorem 4.24 (extended Eckart-Young). *Let $F\colon X \rightrightarrows Y$ be a positively homogeneous multifunction between Banach spaces. Then*

$$\inf_{g\in\mathcal{L}(X,Y)}\Big\{\|g\|\;\Big|\;F+g\;\;with\;\;\|(F+g)^{-1}\|=\infty\Big\}=1/\|F^{-1}\|\,,$$

where the infimum is the same if restricted to mappings $g \in \mathcal{L}(X,Y)$ of rank one. If moreover X is a dual space to some Banach space Z, the additional restriction can be made that g is weak*-to-norm continuous.

Proof. First note that if $\|F^{-1}\| = \infty$, then the equality in the theorem holds with 0 in both sides; so we can assume that $\|F^{-1}\| < \infty$. Furthermore, we can always assume that $\|F^{-1}\| > 0$, since the opposite corresponds to $\operatorname{dom} F = \{0\}$, which implies that $\operatorname{dom}(g+F) = \{0\}$ and hence $\|(F+g)^{-1}\| = 0$. In this case the equality in the theorem holds with ∞ in both sides.

Taking now any $g \in \mathcal{L}(X,Y)$ with $\|(F+g)^{-1}\| = \infty$, we find by definition a sequence $(x_k, y_k) \in \operatorname{gph}(F+g)$ with $\|y_k\| \le 1$ and $0 < \|x_k\| \to \infty$ as $k \to \infty$. It follows from $y_k \in (F+g)(x_k)$ that $x_k \in F^{-1}(y_k - g(x_k))$, hence $\|x_k\| \le \|F^{-1}\| \cdot \|y_k - g(x_k)\|$ and consequently

$$1/\|F^{-1}\| \le \big(\|y_k\| + \|g(x_k)\|\big)/\|x_k\| \le (1/\|x_k\|) + \|g\|\,.$$

Passing to the limit as $k \to \infty$, we get $1/\|F^{-1}\| \le \|g\|$ and hence justify the inequality "$\ge$" in the theorem. It remains to prove the opposite inequality.

Take any finite number $\gamma > 1/\|F^{-1}\|$ and find, by definition of the norm (1.22), a pair $(\widehat{x}, \widehat{y}) \in \operatorname{gph} F$ with $\|\widehat{y}\| = 1$ and $\|\widehat{x}\| > 1/\gamma$. Then there is $\widehat{x}^* \in X^*$ such that $\langle \widehat{x}, \widehat{x}^* \rangle = \|\widehat{x}\|$ and $\|\widehat{x}^*\| = 1$. Now define the rank-one mapping $\widehat{g} \in \mathcal{L}(X,Y)$ by $\widehat{g}(x) := -\|\widehat{x}\|^{-1}\langle x, \widehat{x}^*\rangle \widehat{y}$. Then

$$\widehat{g}(\widehat{x}) = -\widehat{y} \;\text{ and }\; 0 \in F(\widehat{x}) - \widehat{y} = F(\widehat{x}) + \widehat{g}(\widehat{x}) = (F+\widehat{g})(\widehat{x})\,.$$

Hence $\widehat{x} \in (F+\widehat{g})^{-1}(0)$, which implies that $\|(F+\widehat{g})^{-1}\| = \infty$. On the other hand, $\|\widehat{g}\| = \|\widehat{y}\|/\|\widehat{x}\| = 1/\|\widehat{x}\| < \gamma$. By the choice of γ we arrive at the inequality "$\le$" in the theorem. Finally, for $X = Z^*$ the latter argument can be refined by taking $\widehat{x}^* \in I\!B_Z$ with $\langle \widehat{x}, \widehat{x}^* \rangle > 1 - \delta$ for small $\delta > 0$, and the proof goes much as before. △

Note that the classical Eckart-Young theorem (that measures the extent to which a nonsingular $n \times n$ matrix can be perturbed by the addition of an $n \times n$ matrix without destroying the nonsingularity) corresponds to Theorem 4.24 with a linear operator $F\colon I\!R^n \to I\!R^n$. In this case Theorem 4.24 can be obviously reformulated in terms of quadratic matrices, where the condition $\|(F+g)^{-1}\| = \infty$ corresponds to the matrix singularity.

We are going to apply Theorem 4.24 to *coderivatives* as positively homogeneous multifunctions and, combining this with the *precise coderivative formula* (4.16) for the regularity bound $\operatorname{reg} F(\bar{x}, \bar{y})$ as well as with the coderivative calculus, to establish relationships between $\operatorname{reg} F(\bar{x}, \bar{y})$ and the radius of metric regularity from Definition 4.23. To proceed, we also need the following estimate of the exact regularity bound under the addition of single-valued Lipschitzian perturbations. The proof of this result is based on the

Lyusternik-Graves iterative procedure similar to the one used in the proof of Theorem 1.57. It is easy to see that, for single-valued mappings $g\colon X \to Y$, the *exact Lipschitzian bound* from Definition 1.40 is computed by

$$\operatorname{lip} g(\bar{x}) = \limsup_{\substack{x \to \bar{x} \\ u \to \bar{x}}} \frac{\|g(x) - g(u)\|}{\|x - u\|} .$$

Theorem 4.25 (metric regularity under Lipschitzian perturbations). *Let $F\colon X \rightrightarrows Y$ be a set-valued mapping between Banach spaces the graph of which is locally closed around $(\bar{x}, \bar{y}) \in \operatorname{gph} F$. Consider also a single-valued mapping $g\colon X \to Y$ and positive constants μ, ℓ with $\operatorname{reg} F(\bar{x}, \bar{y}) < \mu < \infty$ and $\operatorname{lip} g(\bar{x}) < \ell < \mu^{-1}$. Then*

$$\operatorname{reg}(F + g)\big(\bar{x}, \bar{y} + g(\bar{x})\big) < (\mu^{-1} - \ell)^{-1} = \frac{\mu}{1 - \ell\mu} .$$

Proof. Recall that $B_\alpha(\bar{x}) := \bar{x} + \alpha I\!B_X$ and $B_\alpha(\bar{y}) := \bar{x} + \alpha I\!B_Y$, and with this notation take $\alpha > 0$ so small that $\operatorname{gph} F$ is closed relative to $B_\alpha(\bar{x}) \times B_\alpha(\bar{y})$, g is Lipschitz continuous on $B_\alpha(\bar{x})$ with constant ℓ, and

$$\operatorname{dist}(x; F^{-1}(y)) \le \mu \operatorname{dist}(y; F(x)) \quad \text{for all} \quad (x, y) \in B_\alpha(\bar{x}) \times B_\alpha(\bar{y})$$

due to the metric regularity of F around $(\bar{x}, \bar{y})$. This implies that

$$\operatorname{dist}(\bar{x}; F^{-1}(y)) \le \mu\|y - \bar{y}\| \quad \text{whenever} \quad y \in B_\alpha(\bar{y})$$

and hence $F^{-1}(y) \ne \emptyset$ for all these y. Choose ν such that

$$0 < \nu < \tfrac{1}{4}\alpha(1 - \mu\ell) \min\big\{1, \mu\big\}$$

and take $x \in B_{\nu/4}(\bar{x})$ and $y \in B_{\nu/4\mu}(\bar{y})$. Then

$$\|y - g(x) + g(\bar{x}) - \bar{y}\| \le \ell\|x - \bar{x}\| + \|y - \bar{y}\| \le (\ell\nu/4) + (\nu/4\mu) \le \alpha .$$

Now selecting ε from

$$0 < \varepsilon < \tfrac{1}{4}\alpha(1 - \mu\ell) \min\big\{1; 1/\ell\big\} ,$$

we find $z_1 \in F^{-1}(y - g(x) + g(\bar{x}))$ satisfying

$$\|z_1 - x\| \le \operatorname{dist}\big(x; F^{-1}(y - g(x) + g(\bar{x}))\big) + \varepsilon .$$

It follows from the metric regularity of F around $(\bar{x}, \bar{y})$ that

$$\begin{aligned} \|z_1 - x\| &\le \|x - \bar{x}\| + \operatorname{dist}\big(\bar{x}; F^{-1}(y - g(x) + g(\bar{x}))\big) + \varepsilon \\ &\le \|x - \bar{x}\| + \mu \operatorname{dist}\big(y - g(x) + g(\bar{x}); F(\bar{x})\big) + \varepsilon \\ &\le \|x - \bar{x}\| + \mu\|y - \bar{y}\| + \mu\ell\|x - \bar{x}\| + \varepsilon \\ &\le (\nu/4) + (\mu\nu/4\mu) + (\mu\ell\nu/4) + \varepsilon \le (3\nu/4) + \varepsilon , \end{aligned}$$

which consequently implies that

$$\|z_1 - \bar{x}\| \le \|z_1 - x\| + \|x - \bar{x}\| \le (3\nu/4) + \varepsilon + (\nu/4) \le \nu + \varepsilon \le \alpha\,.$$

This procedure allows us to construct by induction a sequence of elements $z_j \in X$, $j = 1, 2, \ldots$, satisfying

$$z_{j+1} \in F^{-1}(y - g(z_j) + g(\bar{x})) \quad \text{and} \quad \|z_{j+1} - z_j\| \le (\mu\ell)^j \|z_1 - \bar{x}\|\,.$$

Indeed, suppose that we have generated such $z_2, \ldots, z_k$ from z_1. Then

$$\begin{aligned}\|z_j - \bar{x}\| &\le \sum_{i=1}^{j-1} \|z_{i+1} - z_i\| + \|z_1 - \bar{x}\| \le \sum_{i=0}^{j-1} (\mu\ell)^i \|z_1 - x\| + \|z_1 - \bar{x}\| \\ &\le \frac{1}{1-\mu\ell}\|z_1 - x\| + \|z_1 - \bar{x}\| \le \frac{1}{1-\mu\ell}(3\nu/4 + \varepsilon) + \nu + \varepsilon \le \alpha\end{aligned}$$

for $j = 1, \ldots, k$ due to the above choice of the constants ν and ε. Also

$$\|y - g(z_j) + g(\bar{x}) - \bar{y}\| \le \frac{\nu}{4\mu} + \frac{\ell}{1-\mu\ell}(3\nu/4 + \varepsilon) + \ell(\nu + \varepsilon) \le \alpha\,.$$

By the metric regularity of F around $\bar{x}$ we find

$$\begin{aligned}&z_{k+1} \in F^{-1}(y - g(z_k) + g(\bar{x})) \text{ with} \\ &\quad \|z_{k+1} - z_k\| \le \mu \operatorname{dist}\big(y - g(z_k) + g(\bar{x}); F(z_k)\big)\,.\end{aligned}$$

Since $z_k \in F^{-1}(y - g(z_{k-1}) + g(\bar{x}))$, the latter implies that

$$\|z_{k+1} - z_k\| \le \mu\|g(z_k) - g(z_{k-1})\| \le \mu\ell\|z_k - z_{k-1}\|$$

and completes the induction procedure.

It is easy to see that $\{z_1, z_2, \ldots\}$ is a Cauchy sequence, hence it converges to some z from the graph of F due to its local closedness. Moreover, $z \in F^{-1}(y - g(z) + g(\bar{x}))$, which means that $z \in (F+g)^{-1}(y + g(\bar{x}))$ and that

$$\begin{aligned}\operatorname{dist}(x; (F+g)^{-1}(y + g(\bar{x})) &\le \|z - x\| \le \lim_{k\to\infty} \sum_{i=1}^{k} \|z_{i+1} - z_i\| + \|z_1 - \bar{x}\| \\ &\le \lim_{k\to\infty} \sum_{i=0}^{k} (\mu\ell)^i \|z_1 - x\| \le \frac{1}{1-\mu\ell}\|z_1 - x\| \\ &\le \frac{\mu}{1-\mu\ell}\Big[\operatorname{dist}\big(y + g(\bar{x}); (F+g)^{-1}(x)\big) + \varepsilon\Big]\,.\end{aligned}$$

Since the left-hand side above doesn't depend on ε, which can be arbitrary small, the latter justifies the metric regularity of $F + g$ around $(\bar{x}; \bar{y} + g(\bar{x}))$ with modulus $\mu(1 - \ell\mu)^{-1}$. △

Corollary 4.26 (lower estimate for Lipschitzian perturbations). *Let $F: X \rightrightarrows Y$ be locally closed graph around $(\bar{x}, \bar{y})$, and let $g: X \to Y$ be Lipschitz continuous around $\bar{x}$. Then*

$$\operatorname{lip} g(\bar{x}) \geq 1/\operatorname{reg} F(\bar{x}, \bar{y})$$

for every $g(\cdot)$ such that $F + g$ is not metrically regular around $(\bar{x}, \bar{y} + g(\bar{x}))$.

Proof. If $\operatorname{lip} g(\bar{x}) < 1/\operatorname{reg} F(\bar{x}, \bar{y})$, then there are constants $\ell > \operatorname{lip} g(\bar{x})$ and $\mu > \operatorname{reg} F(\bar{x}, \bar{y})$ with $\ell < 1/\mu$. Thus $F + g$ must be metrically regular around $(\bar{x}, \bar{y} + g(\bar{x}))$ by Theorem 4.25. △

Now we are ready to establish the *main result* of this subsection that gives relationships between the radius of metric regularity and the exact regularity bound for set-valued mappings. Note that efficient conditions and calculus rules ensuring the coderivative normality property in the following theorem are listed in Proposition 4.9.

Theorem 4.27 (relationships between the radius and exact bound of metric regularity). *Let $F: X \rightrightarrows Y$ be a set-valued mapping between Banach spaces, and let $(\bar{x}, \bar{y}) \in \operatorname{gph} F$ be a point around which the graph of F is locally closed. Then one has*

$$\operatorname{rad} F(\bar{x}, \bar{y}) \geq 1/\operatorname{reg} F(\bar{x}, \bar{y})\ .$$

If in addition X is Asplund, $\dim Y < \infty$, and F^{-1} is coderivatively normal at $(\bar{y}, \bar{x})$, then the equality holds:

$$\operatorname{rad} F(\bar{x}, \bar{y}) = 1/\operatorname{reg} F(\bar{x}, \bar{y})\ .$$

Furthermore, in this case the infimum in the definition of $\operatorname{rad} F(\bar{x}, \bar{y})$ is unchanged if taken with respect to $g \in \mathcal{L}(X, Y)$ of rank one, but also is unchanged when the space of perturbations g is enlarged from linear bounded operators to locally Lipschitzian mappings:

$$\operatorname{rad} F(\bar{x}, \bar{y}) = \inf_{g: X \to Y} \Big\{ \operatorname{lip} g(\bar{x}) \Bigm| \textit{metric regularity fails for } \ F + g \Big\}\ .$$

Proof. The inequality $\operatorname{rad} F(\bar{x}, \bar{y}) \geq 1/\operatorname{reg} F(\bar{x}, \bar{y})$ follows directly from Corollary 4.26 and the definitions. Moreover, Corollary 4.26 ensures, since $\operatorname{lip} g(\bar{x}) = \|g\|$ for linear continuous mappings g, that the second equality in the theorem follows from the first one. Thus it remains to show that $\operatorname{rad} F(\bar{x}, \bar{y}) = 1/\operatorname{reg} F(\bar{x}, \bar{y})$ under the assumptions made, along with verifying that the infimum in the definition of $\operatorname{rad} F(\bar{x}, \bar{y})$ is unchanged when restricted to linear operators $g \in \mathcal{L}(X, Y)$ of rank one. We are going to prove it by using the *pointbased coderivative characterization* of metric regularity in Theorem 4.18 together with simple rules of *coderivative calculus.*

Applying Theorem 4.18 to the mapping $(F+g)\colon X \rightrightarrows Y$, we first observe that $(F+g)^{-1}\colon Y \rightrightarrows X$ is automatically PSNC at $(\bar{y}+g(\bar{x}),\bar{x})$ by $\dim Y < \infty$. Thus, by the equivalence (a)⇔(d) in Theorem 4.18, we conclude that $F+g$ is *not* metrically regular around $(\bar{x},\bar{y}+g(\bar{x}))$ *if and only if*

$$\|D_M^*(F+g)^{-1}(\bar{y}+g(\bar{x}),\bar{x})\| = \|\widetilde{D}_M^*(F+g)(\bar{x},\bar{y}+g(\bar{x}))^{-1}\| = \infty\,.$$

Let us show that

$$\widetilde{D}_M^*(F+g)(\bar{x},\bar{y}+g(\bar{x}))(y^*) = \widetilde{D}_M^*F(\bar{x},\bar{y})(y^*) + g^*(y^*), \quad g \in \mathcal{L}(X,Y)\,,$$

provided that the space Y is *finite-dimensional*; the latter actually holds for any $g\colon X \to Y$ strictly differentiable at $\bar{x}$ with the replacement of the adjoint operator to g by the one to $\nabla g(\bar{x})$.

Indeed, taking $x^* \in \widetilde{D}_M^*(F+g)(\bar{x},\bar{y}+g(\bar{x}))(y^*)$ and using the representation of $\widetilde{D}_M^*$ in Asplund spaces (see Corollary 2.36) as well as $\dim Y < \infty$, we find sequences $x_k \to \bar{x}$, $y_k \to \bar{y}$ with $y_k \in F(x_k)$, and $(x_k^*, y_k^*) \to (x^*, y^*)$ such that $x_k^* \in \widehat{D}^*(F+g)(x_k, y_k+g(x_k))(y_k^*)$ for all $k \in I\!N$. It follows from Proposition 1.62(i) that

$$\widehat{D}^*(F+g)(x_k,y_k+g(x_k))(y_k^*) = \widehat{D}^*F(x_k,y_k)(y_k^*) + g^*(y_k^*)\,,$$

which gives $x_k^* - g^*(y_k^*) \in \widehat{D}^*F(x_k,y_k)(y_k^*)$. Since $x_k^* - g^*(y_k^*) \to x^* - g^*(y^*)$, the latter ensures by passing to the limit as $k \to \infty$ that $x^* \in \widetilde{D}_M^*F(\bar{x},\bar{y})(y^*) + g^*(y^*)$. This justifies the inclusion "$\subset$" for $\widetilde{D}_M^*(F+g)(\bar{x},\bar{y}+g(\bar{x}))$ in the above formula. The opposite inclusion follows from

$$\widetilde{D}_M^*\big[(F+g)+(-g)\big](\bar{x},\bar{y})(y^*) \subset \widetilde{D}_M^*(F+g)(\bar{x},\bar{y}+g(\bar{x}))(y^*) - g^*(y^*)\,.$$

Thus $F+g$ is not metrically regular around $(\bar{x},\bar{y}+g(\bar{x}))$ *if and only if*

$$\|(\widetilde{D}_M^*F(\bar{x},\bar{y}) + g^*)^{-1}\| = \infty, \quad g \in \mathcal{L}(X,Y)\,.$$

Now we apply the *exact bound formula* (4.16) of Theorem 4.18 to the mapping F^{-1} that is assumed to be *coderivatively normal* at $(\bar{x},\bar{y})$. Taking into account that $\|g^*\| = \|g\|$ for $g \in \mathcal{L}(X,Y)$, the targeted equality $\operatorname{rad} F(\bar{x},\bar{y}) = 1/\operatorname{reg} F(\bar{x},\bar{y})$ can be identified with

$$\inf_{g\in\mathcal{L}(X,Y)} \Big\{\|g^*\| \;\Big|\; \|\widetilde{D}_M^*F(\bar{x},\bar{y}) + g^*\| = \infty\Big\} = 1/\|\widetilde{D}_M^*F(\bar{x},\bar{y})^{-1}\|\,.$$

Observe that every $h \in \mathcal{L}(Y^*,X^*)$ is represented as the adjoint operator $g^*\colon Y^* \to X^*$ for some $g \in \mathcal{L}(X,Y)$ provided that Y is reflexive (in our case $\dim Y < \infty$). Indeed, since $X \subset X^{**}$ and $Y^{**} = Y$, we construct $g \in \mathcal{L}(X,Y)$ as the restriction on X of $h^*\colon X^{**} \to Y^{**}$. Finally applying Theorem 4.24 to the positively homogeneous mapping $\widetilde{D}_M^*F(\bar{x},\bar{y})\colon Y^* \rightrightarrows X^*$, we complete the proof of the theorem. △

Theorem 4.27 also gives information on what happens to the radius of metric regularity under perturbations.

Corollary 4.28 (perturbed radius of metric regularity). *Let* $F\colon X \rightrightarrows Y$ *and* $g\colon X \to Y$. *Assume that* X *is Asplund and* $\dim Y < \infty$, *that the graph of* F *is locally closed around* $(\bar{x}, \bar{y}) \in \operatorname{gph} F$, *and that* F^{-1} *is coderivatively normal at* $(\bar{y}, \bar{x})$. *Then*

$$\operatorname{rad}(F + g)(\bar{x}, \bar{y} + g(\bar{x})) \geq \operatorname{rad} F(\bar{x}, \bar{y}) - \operatorname{lip} g(\bar{x}) \quad \textit{if} \quad \operatorname{lip} g(\bar{x}) < \operatorname{rad} F(\bar{x}, \bar{y}) .$$

Proof. Consider a mapping $h\colon X \to Y$ with $\operatorname{lip} h(\bar{x}) < \operatorname{rad} F(\bar{x}, \bar{y}) - \operatorname{lip} g(\bar{x})$. Then we conclude that $(F+g)+h$ is metrically regular around $(\bar{x}, \bar{y}+g(\bar{x})+h(\bar{x}))$. Indeed, this is the same as the metric regularity of $F + \widetilde{g}$ around $(\bar{x}, \bar{y}+\widetilde{g}(\bar{x}))$ with $\widetilde{g} := g+h$, and the latter is true due to the last equality in Theorem 4.27, since $\operatorname{lip} \widetilde{g}(\bar{x}) \leq \operatorname{lip} g(\bar{x}) + \operatorname{lip} h(\bar{x}) < \operatorname{rad} F(\bar{x}, \bar{y})$. △

Another conclusion can be drawn from Theorem 4.27. Recall that a mapping $G\colon X \rightrightarrows Y$ is said to give a *first-order approximation* to a mapping $F\colon X \rightrightarrows Y$ around $(\bar{x}, \bar{y})$ if on some neighborhood U of $\bar{x}$ there is a mapping $g\colon U \to Y$ such that

$$G = F + g, \quad g(\bar{x}) = 0, \quad \text{and} \quad \operatorname{lip} g(\bar{x}) = 0 .$$

Corollary 4.29 (radius of metric regularity under first-order approximations). *Let* $F\colon X \rightrightarrows Y$ *satisfy the assumptions of Corollary* 4.28, *and let* $G\colon X \rightrightarrows Y$ *furnish a first-order approximation to* F *around* $(\bar{x}, \bar{y})$. *Then one has the equality*

$$\operatorname{rad} F(\bar{x}, \bar{y}) = \operatorname{rad} G(\bar{x}, \bar{y}) .$$

Proof. Consider $g\colon U \to Y$ from the definition of first-order approximation and extend it in any way to a mapping from X to Y. Since $g(\bar{x}) = 0$ and $F+g$ agrees with G around $\bar{x}$, we have $\operatorname{rad} G(\bar{x}, \bar{y}) = \operatorname{rad}(F+g)(\bar{x}, \bar{y})$. On the other hand, $\operatorname{rad}(F+g)(\bar{x}, \bar{y}) \geq \operatorname{rad} F(\bar{x}, \bar{y})$ by Corollary 4.28, since $\operatorname{lip} g(\bar{x}) = 0$. Thus $\operatorname{rad} G(\bar{x}, \bar{y}) \geq \operatorname{rad} F(\bar{x}, \bar{y})$. The opposite inequality follows from the fact that F also gives a first-order approximation to G; the relationship is *symmetric* with $-g$ replacing g. △

An example of a first-order approximation to which Corollary 4.28 can be applied is seen when

$$F(x) = F_0(x) + f(x) \quad \text{with} \quad F_0\colon X \rightrightarrows Y ,$$

where $f\colon X \to Y$ is *strictly differentiable* at $\bar{x}$. In this case a first-order approximation to F is given by $G(x) = F_0(x) + g(x)$, where

$$g(x) := f(\bar{x}) + \langle \nabla f(\bar{x}), x - \bar{x} \rangle .$$

A partial parametric version of such a first-order approximation will be used below in Subsect. 4.4.3.

Remark 4.30 (computing and estimating the radius of metric regularity via coderivative calculus). The results obtained above relate the radius of metric regularity of general mappings to computing the *exact regularity bound*, which has been *characterized* or *estimated* via the corresponding *coderivatives*. In this way, given a specific constraint and/or variational system and employing coderivative and SNC calculi, we can derive efficient results for computing/estimating the regularity radius in terms of the initial data of the given system. In what follows we present such coderivative calculations for a number of constraint and variational systems typically arising in applications. These results are then used to study Lipschitzian stability of constraint and variational systems. Based on the relationships between the Lipschitzian and regularity bounds, one may utilize the results obtained to compute or estimate the radius of metric regularity in concrete settings.

4.3 Sensitivity Analysis for Constraint Systems

In this section we present efficient applications of the above pointbased characterizations and calculus rules of generalized differentiation to *local sensitivity analysis* for general classes of *constraint systems depending on parameters*. Such systems cover, in particular, parametric sets of feasible solutions for problems of mathematical programming. Our primary interest is *robust Lipschitzian stability* of multivalued solution maps with respect to parameter perturbations. The main attention is paid to results on the Lipschitz-like property of solution maps to constraint systems that easily imply the corresponding results for classical local Lipschitzian behavior. Note that both Lipschitz-like and classical local Lipschitzian properties are *robust* (stable) with respect to perturbations of initial data, which is of great significance for sensitivity analysis. Coderivative characterizations of robust Lipschitzian behavior and efficient calculus rules for the basic generalized differential constructions and the corresponding sequential normal compactness allow us to derive effective *sufficient* (as well as *necessary and sufficient*) conditions for Lipschitzian stability with evaluating the exact Lipschitzian bounds.

To conduct such a local sensitivity analysis, we first express *coderivatives* of general *parametric constraint systems* and their important specifications in terms of the initial data. This is certainly of independent interest while playing a crucial role (along with the SNC calculus in infinite dimensions) for the subsequent study of Lipschitzian stability via the pointbased coderivative criteria of the preceding section.

4.3.1 Coderivatives of Parametric Constraint Systems

Let us consider a class of multifunctions $F\colon X \rightrightarrows Y$ given in the form

$$F(x) := \big\{ y \in Y \big|\, g(x,y) \in \Theta, \quad (x,y) \in \Omega \big\}\,, \tag{4.19}$$

where $g: X \times Y \to Z$ is a single-valued mapping between Banach spaces, and where Θ and Ω are subsets of the spaces Z and $X \times Y$, respectively. Such set-valued mappings describe *constraint systems* depending on a parameter $x \in X$. One can view the parametric system (4.19) as a natural generalization of the *feasible solution sets* to perturbed problems in nonlinear programming with inequality and equality constraints given by

$$F(x) := \Big\{ y \in Y \Big|\ \varphi_i(x, y) \le 0, \quad i = 1, \ldots, m; \qquad \varphi_i(x, y) = 0, \quad i = m+1, \ldots, m+r \Big\}, \tag{4.20}$$

where φ_i are real-valued functions on $X \times Y$. Clearly (4.20) is a special case of (4.19) with $g = (\varphi_1, \ldots, \varphi_{m+r})$, $\Omega = X \times Y$, $Z = I\!R^{m+r}$, and

$$\Theta := \Big\{ (\alpha_1, \ldots, \alpha_{m+r}) \Big|\ \alpha_i \le 0 \ \text{ for } \ i = 1, \ldots, m \quad \text{and} \qquad \alpha_i = 0 \ \text{ for } \ i = m+1, \ldots, m+r \Big\}. \tag{4.21}$$

Another special case of (4.19) with $\Theta = \{0\}$ and $\Omega = X \times Y$ is addressed by the classical implicit function theorem when the mapping

$$F(x) := \big\{ y \in Y \big|\ g(x, y) = 0 \big\} \tag{4.22}$$

is single-valued and smooth. In general we have *implicit multifunctions* in (4.22) and are interested in properties of their Lipschitz continuity. More examples of parametric systems that can be reduced to form (4.19) are given in the next section.

In this subsection we express the normal and mixed coderivatives of set-valued mappings defined by (4.19), (4.20), and (4.22) in terms of the initial data $\{g, \Theta, \Omega\}$, which is an important part of the subsequent sensitivity analysis. The next theorem provides *precise formulas* (equalities) for computing these coderivatives in general Banach space and Asplund space settings. The proofs of this theorem as well as other results given below are based on the generalized differential and SNC calculi developed in Chaps. 1 and 3.

Theorem 4.31 (computing coderivatives of constraint systems). *Let $F: X \times Y \rightrightarrows Z$ be given in (4.19) with $g: X \times Y \to Z$, $\Theta \subset Z$, and $\Omega \subset X \times Y$. Take $(\bar{x}, \bar{y}) \in \operatorname{gph} F$ and put $\bar{z} := g(\bar{x}, \bar{y}) \in \Theta$. The following assertions hold:*

(i) *Assume that X, Y, Z are Banach spaces, that $\Omega = X \times Y$, and that g is strictly differentiable at $(\bar{x}, \bar{y})$ with the surjective derivative $\nabla g(\bar{x}, \bar{y})$. Then for all $y^* \in Y^*$ one has*

$$D_N^* F(\bar{x}, \bar{y})(y^*) = \big\{ x^* \in X^* \big|\ (x^*, -y^*) \in \nabla g(\bar{x}, \bar{y})^* N(\bar{z}; \Theta) \big\}. \tag{4.23}$$

If moreover $\dim Z < \infty$, then representation (4.23) holds also the mixed coderivative $D_M^* F(\bar{x}, \bar{y})$, i.e., F is strongly coderivatively normal at $(\bar{x}, \bar{y})$.

(ii) *Let X, Y, Z be Asplund, and let g be Lipschitz continuous around $(\bar{x}, \bar{y})$. Assume that*

$$\big(D_N^* g(\bar{x}, \bar{y}) \circ N(\bar{z}; \Theta)\big) \cap \big(- N((\bar{x}, \bar{y}); \Omega)\big) = \{0\}, \tag{4.24}$$

that either g is N-regular at $(\bar{x}, \bar{y})$ with $\dim Z < \infty$ or g is strictly differentiable at $(\bar{x}, \bar{y})$, and that the sets Ω and Θ are locally closed around $(\bar{x}, \bar{y})$ and $\bar{z}$ and normally regular at these points, respectively. Then one has

$$D^* F(\bar{x}, \bar{y})(y^*) = \Big\{x^* \in X^* \Big| \; (x^*, -y^*) \in D^* g(\bar{x}, \bar{y}) \circ N(\bar{z}; \Theta) + N((\bar{x}, \bar{y}); \Omega)\Big\}, \quad y^* \in Y^*, \tag{4.25}$$

for both coderivatives $D^* = D_N^*, D_M^*$ provided that

$$N(\bar{z}; \Theta) \cap \ker D_N^* g(\bar{x}, \bar{y}) = \{0\} \tag{4.26}$$

and that either Ω is SNC at $(\bar{x}, \bar{y})$ while g^{-1} is PSNC at $(\bar{z}, \bar{x}, \bar{y})$, or Θ is SNC at $\bar{z}$. Under the assumptions made F is N-regular at $(\bar{x}, \bar{y})$, and hence it is strongly coderivatively normal at this point.

Proof. To justify (i), observe that

$$\operatorname{gph} F = g^{-1}(\Theta) \quad \text{when} \quad \Omega = X \times Y$$

for the mapping F in (4.19). Thus representation (4.23) follows directly from Theorem 1.17. Let us prove that (4.23) holds true for the mixed coderivative $D_M^* F(\bar{x}, \bar{y})$ provided that the space Z is finite-dimensional. It is sufficient to observe in this case that

$$N_{w^* \times \|\cdot\|}((\bar{x}, \bar{y}); g^{-1}(\Theta)) = \nabla g(\bar{x}, \bar{y})^* N(\bar{x}; \Theta),$$

where $N_{w^* \times \|\cdot\|}(\cdot; \Omega)$ stands for the limiting normal cone to a set $\Omega \subset X \times Y$ defined in Remark 3.23 with respect to the weak* topology on X^* and the norm topology on Y^*. The latter easily follows from the proof of Theorem 1.17.

Now we show that, under the assumptions made in (ii), representation (4.25) holds for $D^* = D_N^*$ and also that F is N-regular at $(\bar{x}, \bar{y})$. Note that in general one has the representation

$$\operatorname{gph} F = g^{-1}(\Theta) \cap \Omega \tag{4.27}$$

for the mapping F in (4.19). To prove (4.25) and the N-regularity of F at $(\bar{x}, \bar{y})$, we start with the case when Ω is SNC at $(\bar{x}, \bar{y})$. Taking into account that $g^{-1}(\Theta)$ is normally regular at $(\bar{x}, \bar{y})$ due to Theorem 3.13(iii) and applying the equality/regularity statement of Theorem 3.4, we conclude that

$$N((\bar{x}, \bar{y}); \operatorname{gph} F) = N((\bar{x}, \bar{y}); g^{-1}(\Theta)) + N((\bar{x}, \bar{y}); \Omega) \tag{4.28}$$

and the graph of F is normally regular at $(\bar{x}, \bar{y})$ provided that

$$N((\bar{x}, \bar{y}); g^{-1}(\Theta)) \cap \big(- N((\bar{x}, \bar{y}); \Omega)\big) = \{0\} . \tag{4.29}$$

Using the chain rule of Theorem 3.13(iii) when the outer mapping is the indicator of Θ, one has

$$N((\bar{x}, \bar{y}); g^{-1}(\Theta)) = D_N^* g(\bar{x}, \bar{y}) \circ N(\bar{z}; \Theta)$$

provided that the qualification condition (4.26) holds and that either Θ is SNC at $\bar{z}$ or g^{-1} is PSNC at $(\bar{z}, \bar{x}, \bar{y})$. Substituting the latter equality into (4.28) and (4.29), we justify representation (4.25) for $D^* = D_N^*$ and the N-regularity of F at $(\bar{x}, \bar{y})$ under the assumptions made.

When Ω is not assumed to be SNC at $(\bar{x}, \bar{y})$, we still get equality (4.28) and the N-regularity of F at $(\bar{x}, \bar{y})$ by Theorem 3.4 under condition (4.29) if the set $g^{-1}(\Theta)$ is SNC at $(\bar{x}, \bar{y})$. Let us show that the latter holds under the assumptions imposed on g and Θ. To furnish this, we apply the SNC calculus rule of Theorem 3.98 when the outer mapping is the indicator function $\delta(\cdot; \Theta)$. Observing that the inner mapping g is PSNC at $(\bar{x}, \bar{y})$ due to Proposition 1.68 and that the SNC property of $\delta(\cdot; \Theta)$ and Θ are equivalent, we conclude that $g^{-1}(\Theta)$ is SNC at $(\bar{x}, \bar{y})$ if either g is SNC at $(\bar{x}, \bar{y})$ or Θ is SNC at $\bar{z}$ under the qualification condition (4.26). When g is strictly differentiable at $(\bar{x}, \bar{y})$, the SNC property of g implies, by Corollary 3.30 that Z is finite dimensional, i.e., Θ is automatically SNC at $\bar{z}$. Combining all the above, we complete the proof of the theorem. △

If the mapping g is assumed to be *strictly Lipschitzian* in Theorem 4.31(ii), then one has, by the scalarization results of Theorem 3.28, that

$$D^* g(\bar{x}, \bar{y})(z^*) = \partial\langle z^*, g\rangle(\bar{x}, \bar{y}), \quad z^* \in Z^* ,$$

$$D^* g(\bar{x}, \bar{y}) \circ N(\bar{z}; \Theta) = \bigcup \Big\{ \partial\langle z^*, g\rangle(\bar{x}, \bar{y})(z^*) \Big| \; z^* \in N(\bar{z}; \Theta) \Big\}$$

for both coderivatives $D^* = D_N^*, D_M^*$. Moreover, by Corollary 3.69 we conclude that the N-regularity assumption on g at $(\bar{x}, \bar{y})$ and $\dim Z < \infty$ in Theorem 4.31(ii) imply that g is strictly Hadamard differentiable at this point. Thus Theorem 3.66(i) ensures that $D^* g(\bar{x}, \bar{y})$ in (4.25) is actually a (single-valued) *bounded linear operator.*

The next theorem gives *upper estimates* for the normal and mixed coderivatives of F under less restrictive assumptions on the initial data in comparison with Theorem 4.31(ii). For simplicity we present identical upper estimates of both coderivatives; see also Remark 4.33 formulated after the theorem.

Theorem 4.32 (upper estimates for coderivatives of constraint systems). *Let $g: X \times Y \to Z$ be a mapping between Asplund spaces continuous* *around* $(\bar{x}, \bar{y}) \in \operatorname{gph} F$ *for the constraint system* F *defined in* (4.19), *where* $\Omega \subset X \times Y$ *and* $\Theta \subset Z$ *are locally closed around* $(\bar{x}, \bar{y})$ *and* $\bar{z} = g(\bar{x}, \bar{y})$, *respectively. Assume that* $\{g, \Theta, \Omega\}$ *satisfies* (4.24) *and that one of the following conditions holds:*

(a) Ω *is SNC at* $(\bar{x}, \bar{y})$, Θ *is SNC at* $\bar{z}$, *and* $\{g, \Theta\}$ *satisfies*

$$N(\bar{z}; \Theta) \cap \ker \widetilde{D}_M^* g(\bar{x}, \bar{y}) = \{0\} . \tag{4.30}$$

(b) Ω *is SNC at* $(\bar{x}, \bar{y})$, g^{-1} *is PSNC at* $(\bar{z}, \bar{x}, \bar{y})$, *and* $\{g, \Theta\}$ *satisfies the constraint qualification* (4.30).

(c) g *is SNC at* $(\bar{x}, \bar{y})$, *and* $\{g, \Theta\}$ *satisfies* (4.26).

(d) g *is PSNC at* $(\bar{x}, \bar{y})$, Θ *is SNC at* $\bar{z}$, *and* $\{g, \Theta\}$ *satisfies* (4.26).

Then for all $y^* \in Y^*$ *one has the inclusion*

$$D^*F(\bar{x}, \bar{y})(y^*) \subset \Big\{x^* \in X^* \Big| \, (x^*, -y^*) \in D_N^* g(\bar{x}, \bar{y}) \circ N(\bar{z}; \Theta) + N((\bar{x}, \bar{y}); \Omega)\Big\} \tag{4.31}$$

for both coderivatives $D^* = D_N^*, D_M^*$ *of* F *at* $(\bar{x}, \bar{y})$.

Proof. It is sufficient to justify (4.31) for $D^* = D_N^*$. Applying Corollary 3.5 to the set intersection in (4.27), we get the inclusion

$$N((\bar{x}, \bar{y}); \operatorname{gph} F) \subset N((\bar{x}, \bar{y}); g^{-1}(\Theta)) + N((\bar{x}, \bar{y}); \Omega) \tag{4.32}$$

under the qualification condition (4.29) provided that either Ω is SNC at $(\bar{x}, \bar{y})$ or $g^{-1}(\Theta)$ is SNC at $(\bar{x}, \bar{y})$. Then we have

$$N((\bar{x}, \bar{y}); g^{-1}(\Theta)) \subset D_N^* g(\bar{x}, \bar{y}) \circ N(\bar{z}; \Theta) \tag{4.33}$$

from Theorem 3.8 under the qualification condition (4.30) if either g^{-1} is PSNC at $(\bar{z}, \bar{x}, \bar{y})$ or Θ is SNC at $\bar{z}$.

Further, recall the conditions ensuring that $g^{-1}(\Theta)$ is SNC at $(\bar{x}, \bar{y})$, which are needed if Ω is not assumed to be SNC at $(\bar{x}, \bar{y})$. By Theorem 3.84 on the SNC property of inverse images one has that $g^{-1}(\Theta)$ is SNC at $(\bar{x}, \bar{y})$ when $\{g, \Theta\}$ satisfies (4.26) and either g is SNC at $(\bar{x}, \bar{y})$, or Θ is SNC at $\bar{z}$ while g is PSNC at $(\bar{x}, \bar{y})$ (in particular, when g is locally Lipschitzian around this point). Combining all these conditions and substituting (4.33) into (4.29) and (4.32), we complete the proof of the theorem. △

Remark 4.33 (refined estimates for mixed coderivatives of constraint systems). Following the proof of Theorem 4.32, we can obtain more *subtle upper estimates* of the *mixed coderivative* $D_M^* F(\bar{x}, \bar{y})$ for the constraint system (4.19) in terms of a modified coderivative construction for the mapping $g: X \times Y \to Z$. As observed in Remark 3.23, the mixed coderivative $D_M^* F(\bar{x}, \bar{y})$ admits the geometric representation

$$D_M^* F(\bar{x}, \bar{y})(y^*) = \{x^* \in X^* | \ (x^*, -y^*) \in N_\tau((\bar{x}, \bar{y}); \text{gph}\, F)\} \qquad (4.34)$$

via the τ-limiting normal cone N_τ defined and discussed in Remark 3.23. In (4.34), $\tau = w^* \times \|\cdot\|$ is the weak*×norm topology on $X^* \times Y^*$. It is proved in Mordukhovich and B. Wang [963] that τ-limiting normals and related coderivative and subgradient constructions enjoy rich calculi for general topologies τ satisfying appropriate conditions. In particular, we have the corresponding τ-analogs of the intersection and inverse image formulas (4.32) and (4.33). For $\tau = w^* \times \|\cdot\|$, the latter τ-analog is given by

$$N_\tau((\bar{x}, \bar{y}); g^{-1}(\Theta)) \subset D^*_{\tau \times w^*} g(\bar{x}, \bar{y}) \circ N(\bar{z}; \Theta)\ ,$$

where $D^*_{\tau \times w^*} g(\bar{x}, \bar{y})\colon Z^* \rightrightarrows X^* \times Y^*$ is defined similarly to the mixed coderivative (1.25) by using the $w^* \times \|\cdot\| \times w^*$-topology on $X^* \times Y^* \times Z^*$. In this way one can get refined estimates of $D_M^* F(\bar{x}, \bar{y})$ in (4.19) via $D^*_{\tau \times w^*} g(\bar{x}, \bar{y})$ with $\tau = w^* \times \|\cdot\|$. The reader may develop such estimates in more details based on the techniques from the afore-mentioned paper [963].

Next let us present a consequence of Theorems 4.31 and 4.32 concerning coderivatives of set-valued mappings given in the classical *implicit function form* (4.22) without imposing the classical assumptions.

Corollary 4.34 (coderivatives of implicit multifunctions). *Let*

$$F(x) := \{y \in Y | \ g(x, y) = 0\}\ ,$$

where $g\colon X \times Y \to Z$ *with* $g(\bar{x}, \bar{y}) = 0$. *The following assertions hold for both coderivatives* $D^* = D_N^*, D_M^*$*:*

(i) *Assume that* X, Y, Z *are Banach spaces and that* g *is strictly differentiable at* $(\bar{x}, \bar{y})$ *with the surjective derivative* $\nabla g(\bar{x}, \bar{y})$. *Then*

$$D_N^* F(\bar{x}, \bar{y})(y^*) = \{x^* \in X^* | \ (x^*, -y^*) = \nabla g(\bar{x}, \bar{y})^* z^* \ \ \textit{for some}\ \ z^* \in Z^*\}\ .$$

If moreover $\dim Z < \infty$, *the latter representation holds also the mixed coderivative* $D_M^* F(\bar{x}, \bar{y})$.

(ii) *Let* X *and* Y *be Asplund, and let* $\dim Z < \infty$. *Assume that* g *is Lipschitz continuous around* $(\bar{x}, \bar{y})$, N*-regular at this point, and satisfies the subdifferential condition*

$$\ker \partial\langle\cdot, g\rangle(\bar{x}, \bar{y}) = \{0\}\ .$$

Then F *is* N*-regular at* $(\bar{x}, \bar{y})$ *with*

$$D^* F(\bar{x}, \bar{y})(y^*) = \{x^* \in X^* | \ (x^*, -y^*) \in \partial\langle z^*, g\rangle(\bar{x}, \bar{y}) \ \ \textit{for some}\ \ z^* \in Z^*\}\ .$$

(iii) *Let* X, Y, Z *be Asplund. Assume that* g^{-1} *is PSNC at* $(\bar{z}, \bar{x}, \bar{y})$ *and*

$$\ker \widetilde{D}_M^* g(\bar{x}, \bar{y}) = \{0\}\ .$$

Then for all $y^* \in Y^*$ *one has the inclusion*

$$D^* F(\bar{x}, \bar{y})(y^*) \subset \{x^* \in X^* | \ (x^*, -y^*) \in \text{rge}\, D_N^* g(\bar{x}, \bar{y})\}\ .$$

Proof. Assertion (i) follows immediately from Theorem 4.31(i) with $\Theta = \{0\}$. Assertion (ii) is a direct consequence of Theorem 4.31(ii) and the coderivative scalarization. Note that, in this setting, the strict differentiability assumption on g reduces (ii) to (i) in Theorem 4.31, since the condition $\ker \nabla g(\bar{x}, \bar{y})^* = \{0\}$ is equivalent to the surjectivity of $\nabla g(\bar{x}, \bar{y})$.

To prove (iii), we use Theorem 4.32 and observe that conditions (b) there are the most general among (a)–(d) ensuring inclusion (4.31) in the setting under consideration when $\Omega = X \times Y$ is always SNC while $\Theta = \{0\}$ is never SNC unless Z is finite-dimensional; see Theorem 1.21. Note that in the latter case g^{-1} is always PSNC at $(\bar{z}, \bar{x}, \bar{y})$. △

Next let us consider consequences of Theorems 4.31 and 4.32 for parametric constraint systems given in form (4.20), which describe sets of *feasible solutions* to perturbed problems of mathematical programming in infinite-dimensional spaces. We present two results for such constraint systems. The first corollary concerns classical constraint systems in (smooth) *nonlinear programming* with equality and inequality constraints given by strictly differentiable functions. In this framework we obtain an *exact formula* for computing coderivatives of feasible solution maps under a parametric version of the *Mangasarian-Fromovitz constraint qualification.*

Corollary 4.35 (coderivatives of constraint systems in nonlinear programming). *Let $F\colon X \rightrightarrows Y$ be a multifunction between Asplund spaces given in form* (4.20), *where all $\varphi_i\colon X \times Y \to I\!R$, $i = 1, \ldots, m+r$, are strictly differentiable at $(\bar{x}, \bar{y}) \in \operatorname{gph} F$. Denote $\bar{z} := \big(\varphi_1(\bar{x}, \bar{y}), \ldots, \varphi_{m+r}(\bar{x}, \bar{y})\big)$,*

$$I(\bar{x}, \bar{y}) := \big\{ i \in \{1, \ldots, m+r\} \big| \ \varphi_i(\bar{x}, \bar{y}) = 0 \big\}$$

and assume that:

(a) *$\nabla\varphi_{m+1}(\bar{x}, \bar{y}), \ldots, \nabla\varphi_{m+r}(\bar{x}, \bar{y})$ are linearly independent;*

(b) *there is $u \in X \times Y$ satisfying*

$$\langle \nabla\varphi_i(\bar{x}, \bar{y}), u \rangle < 0, \quad i \in \{1, \ldots, m\} \cap I(\bar{x}, \bar{y}) \, ,$$

$$\langle \nabla\varphi_i(\bar{x}, \bar{y}), u \rangle = 0, \quad i = m+1, \ldots, m+r \, .$$

Then F is N-regular at $(\bar{x}, \bar{y})$, and one has

$$D^*F(\bar{x}, \bar{y})(y^*) = \Big\{ x^* \in X^* \Big| \ (x^*, -y^*) = \sum_{i \in I(\bar{x}, \bar{y})} \lambda_i \nabla\varphi_i(\bar{x}, \bar{y}) \, , \quad \lambda_i \ge 0 \ \text{if} \ i \in \{1, \ldots, m\} \cap I(\bar{x}, \bar{y}) \Big\} \tag{4.35}$$

with arbitrary $\lambda_i \in I\!R$ for $i = m+1, \ldots, m+r$.

Proof. Use Theorem 4.31(ii) with $g = (\varphi_1, \ldots, \varphi_{m+r})$, $\Omega = X \times Y$, and Θ given in (4.21). The set Θ is convex (thus normally regular at every point), and one has

$$N(\bar{z};\Theta) = \big\{(\lambda_1, \ldots, \lambda_{m+r}) \in I\!R^{m+r}\big|\ \lambda_i \ge 0,\ \ \lambda_i\varphi_i(\bar{x}, \bar{y}) = 0 \ \text{ if } \ i = 1, \ldots, m\big\}\ .$$

In this case the qualification condition (4.26) is equivalent to the fulfillment of (a) and (b) in the corollary, and (4.25) reduces to (4.35). △

In the nonparametric case ($\varphi_i(x, y) = \varphi_i(y)$), conditions (a) and (b) of Theorem 4.35 reduce to the classical Mangasarian-Fromovitz constraint qualification; see Corollary 3.87. Note that these conditions automatically hold if the full gradients $\nabla\varphi_i(\bar{x}, \bar{y})$ therein are replaced by the partial ones with respect to y.

The following corollary of Theorem 4.32 gives upper estimates for both coderivatives of feasible solution maps in parametric problems of *nondifferentiable programming* with equality and inequality constraints described by Lipschitz continuous functions on Asplund spaces.

Corollary 4.36 (coderivatives of constraint systems in nondifferentiable programming). *Let $F\colon X \rightrightarrows Y$ be a multifunction between Asplund spaces given in* (4.20), *let $(\bar{x}, \bar{y}) \in \operatorname{gph} F$, and let $\bar{z}$ and $I(\bar{x}, \bar{y})$ be defined in Corollary* 4.35. *Assume that all φ_i, $i = 1, \ldots, m + r$, are Lipschitz continuous around $(\bar{x}, \bar{y})$ and that*

$$\Big[\sum_{i\in I(\bar{x},\bar{y})} \lambda_i x_i^* = 0\Big] \Longrightarrow \Big[\lambda_i = 0,\ i \in I(\bar{x}, \bar{y})\Big] \tag{4.36}$$

whenever $\lambda_i \ge 0$ for $i \in I(\bar{x}, \bar{y})$, $x_i^* \in \partial\varphi_i(\bar{x}, \bar{y})$ for $i \in \{1, \ldots, m\} \cap I(\bar{x}, \bar{y})$, and $x_i^* \in \partial\varphi_i(\bar{x}, \bar{y}) \cup \partial(-\varphi_i)(\bar{x}, \bar{y})$ for $i = m + 1, \ldots, m + r$. Then one has the inclusion

$$D^*F(\bar{x}, \bar{y})(y^*) \subset \Big\{x^* \in X^*\Big|\ (x^*, -y^*) \in \sum_{i\in\{1,\ldots,m\}\cap I(\bar{x},\bar{y})} \lambda_i\partial\varphi_i(\bar{x}, \bar{y})$$

$$+ \sum_{i=m+1}^{m+r} \lambda_i\Big(\partial\varphi_i(\bar{x}, \bar{y}) \cup \partial(-\varphi_i)(\bar{x}, \bar{y})\Big),\ \ \lambda_i \ge 0 \ \ \text{as} \ \ i \in I(\bar{x}, \bar{y})\Big\}$$

for both coderivatives $D^* = D^*_N, D^*_M$.

Proof. Use Theorem 4.32 in case (d), where g is automatically PSNC at $(\bar{x}, \bar{y})$, and where $\Theta \subset I\!R^{m+r}$ is SNC at every point. Due to the scalarization formula for $D^*_N g(\bar{x}, \bar{y})$ with $g = (\varphi_1, \ldots, \varphi_{m+r})$ from Theorem 3.28 (or from Theorem 1.90 in this case) and due to the subdifferential sum rule from Theorem 2.33(c), one has

$$D_N^*g(\bar{x},\bar{y})(z^*) = \partial\Big(\sum_{i=1}^{m+r} \lambda_i\varphi_i\Big)(\bar{x},\bar{y}) \subset \sum_{i=1}^{m} \lambda_i\varphi_i(\bar{x},\bar{y}) + \partial\Big(\sum_{m+1}^{m+r} \lambda_i\varphi_i\Big)(\bar{x},\bar{y})$$

for $z^* = (\lambda_1, \ldots, \lambda_{m+r}) \in I\!R^{m+r}$ provided that $\lambda_i \ge 0$ as $i = 1, \ldots, m$. Taking into account the above expression for $N(\bar{z};\Theta)$, we derive the coderivative inclusion of the corollary under the qualification condition (4.36) from the corresponding relations (4.26) and (4.31) of Theorem 4.32. △

4.3.2 Lipschitzian Stability of Constraint Systems

Now we are ready to derive efficient conditions for robust Lipschitzian stability of constraint systems based on the coderivative characterizations of the Lipschitz-like property in Theorem 4.10 and the coderivative representations for parametric constraint systems obtained in the previous subsection. Let us first consider constraint systems under *regularity* assumptions, which allow us to obtain *necessary and sufficient* conditions for Lipschitzian stability in terms of the initial data. The proofs of the next theorem and subsequent results require applications of the SNC calculus in infinite dimensions together with the coderivative characterizations and representations mentioned above.

Theorem 4.37 (Lipschitzian stability of regular constraint systems). *Let $F\colon X \rightrightarrows Y$ be a set-valued mapping between Asplund spaces defined by the constraint system* (4.19), *let $\bar{z} := g(\bar{x},\bar{y})$ with $(\bar{x},\bar{y}) \in \operatorname{gph} F$, and let Θ be locally closed around $\bar{z}$ and SNC at this point. The following assertions hold:*

(i) *Assume that Z is Banach, that $\Omega = X \times Y$, and that g is strictly differentiable at $(\bar{x},\bar{y})$ with the surjective derivatives $\nabla g(\bar{x},\bar{y})$. Then the condition*

$$(x^*, 0) \in \nabla g(\bar{x},\bar{y})^* N(\bar{z};\Theta) \Longrightarrow x^* = 0 \qquad (4.37)$$

is sufficient for the Lipschitz-like property of F around $(\bar{x},\bar{y})$ being necessary and sufficient for this property if F is strongly coderivatively normal at $(\bar{x},\bar{y})$ (in particular, when Y is finite-dimensional). If in addition $\dim X < \infty$, *then*

$$\operatorname{lip} F(\bar{x},\bar{y}) = \sup\Big\{\|x^*\|\Big|(x^*, -y^*) \in \nabla g(\bar{x},\bar{y})^* N(\bar{z};\Theta),\ \|y^*\| \le 1\Big\}, \qquad (4.38)$$

where the maximum is attained provided that the graph of $N(\cdot;\Theta)$ is locally closed near $\bar{z}$ in the norm$\times$weak topology of $Z \times Z^*$.*

(ii) *Assume that Z is Asplund; that Θ is normally regular at $\bar{z}$; that Ω is locally closed around $(\bar{x},\bar{y})$, normally regular at $(\bar{x},\bar{y})$, and PSNC at this point with respect to X; and that g is either strictly differentiable at $(\bar{x},\bar{y})$ or N-regular at this point with* $\dim Z < \infty$. *Suppose also that both qualification conditions* (4.24) *and* (4.26) *are fulfilled. Then the implication*

$$(x^*, 0) \in D^*g(\bar{x},\bar{y}) \circ N(\bar{z};\Theta) + N((\bar{x},\bar{y});\Omega) \Longrightarrow x^* = 0 \qquad (4.39)$$

is necessary and sufficient for the Lipschitz-like property of F around $(\bar{x},\bar{y})$. If in addition $\dim X < \infty$, *then*

$$\operatorname{lip} F(\bar{x},\bar{y}) = \sup\Big\{\|x^*\| \,\Big|\, (x^*,-y^*) \in D^*g(\bar{x},\bar{y}) \circ N(\bar{z};\Theta)$$

$$+N((\bar{x},\bar{y});\Omega),\ \ \|y^*\| \le 1\Big\}.$$

Proof. We use characterization (c) and the exact bound formula (4.6) from Theorem 4.10 for the Lipschitz-like property of general closed-graph multifunctions between Asplund spaces. To justify (i), observe first that F is SNC at $(\bar{x},\bar{y})$ under the assumptions made due to $\operatorname{gph} F = g^{-1}(\Theta)$ and Theorem 1.22. Then using (4.23), we get characterization (4.37) from the condition $D_M^*F(\bar{x},\bar{y})(0)=\{0\}$ and the exact bound formula (4.38) from the one in (4.6). When the graph of $N(\cdot;\Theta)$ is locally closed near $\bar{z}$, it is possible to put "max" instead of "sup" in (4.38) due to $\|D^*F(\bar{x},\bar{y})\| < \infty$ and the surjectivity of $\nabla g(\bar{x},\bar{y})$ involving Lemma 1.18.

To prove (ii), we represent the graph of F in the intersection form (4.27) and deduce from Corollary 3.80 that F is PSNC at $(\bar{x},\bar{y})$ if the qualification condition (4.29) is fulfilled and if Ω is PSNC at $(\bar{x},\bar{y})$ with respect to X while $g^{-1}(\Theta)$ is SNC at this point. By Theorem 3.84 the latter property holds if Θ is SNC at $\bar{z}$ under the qualification condition (4.26). Moreover, these assumptions ensure that the qualification conditions (4.24) and (4.26) imply (4.29) due to the inclusion for $N((\bar{x},\bar{y});g^{-1}(\Theta))$ from Theorem 3.8. Involving the other assumptions in (ii), we get equality (4.25) for both normal and mixed coderivatives of F at $(\bar{x},\bar{y})$ by Theorem 4.31(ii). Thus the condition $D^*F(\bar{x},\bar{y})(0)=\{0\}$ is equivalent to (4.39), and the exact bound formula of the theorem reduces to (4.6) in Theorem 4.10. △

Note that the graph of $N(\cdot;\Theta)$ is indeed locally closed near $\bar{z}$ in the norm×weak* topology of $Z \times Z^*$ if Z is a weakly compactly generated Banach space while Θ is its closed subset having the CEL property at $\bar{z}$ (the latter agrees with the SNC property of Θ when Z is WCG and Asplund; see Remark 1.27 and Theorem 3.60). On the other hand, the graph of $N(\cdot;\Theta)$ is obviously closed for $\Theta=\{0\}$, which is the case of the next corollary.

Corollary 4.38 (Lipschitzian implicit multifunctions defined by regular mappings). *Let $F: X \rightrightarrows Y$ be an "implicit multifunction" defined in* (4.22) *by the mapping $g: X \times Y \to Z$ with $g(\bar{x},\bar{y})=0$. The following hold:*

(i) *Assume that* $\dim Z < \infty$ *while X and Y are Asplund and that g is strictly differentiable at $(\bar{x},\bar{y})$ with the surjective derivative $\nabla g(\bar{x},\bar{y})$. Then the condition*

$$\Big[\nabla_y g(\bar{x},\bar{y})^*z^* = 0\Big] \Longrightarrow \Big[\nabla_x g(\bar{x},\bar{y})^*z^* = 0\Big] \ \ \text{for any } z^* \in Z^*$$

is necessary and sufficient for the Lipschitz-like property of F around $(\bar{x},\bar{y})$. If in addition $\dim X < \infty$, *then one has*

$$\operatorname{lip} F(\bar{x},\bar{y}) = \max\Big\{\|\nabla_x g(\bar{x},\bar{y})^*z^*\| \,\Big|\, \|\nabla_y g(\bar{x},\bar{y})^*z^*\| \le 1\Big\}.$$

(ii) *Let X and Y be Asplund, and let Z be finite-dimensional. Assume that g is Lipschitz continuous around* $(\bar{x}, \bar{y})$ *and N-regular at this point and that* $\ker \partial\langle\cdot, g\rangle(\bar{x}, \bar{y}) = \{0\}$. *Then the condition*

$$(x^*, 0) \in \partial\langle z^*, g\rangle(\bar{x}, \bar{y}) \Longrightarrow x^* = 0 \ \text{ for any } \ z^* \in Z^* \tag{4.40}$$

is necessary and sufficient for the Lipschitz-like property of F around $(\bar{x}, \bar{y})$. *If in addition* $\dim X < \infty$, *then*

$$\operatorname{lip} F(\bar{x}, \bar{y}) = \sup\Big\{\|x^*\|\Big|\ \exists z^* \in Z^* \ \text{with} \ (x^*, -y^*) \in \partial\langle z^*, g\rangle(\bar{x}, \bar{y}), \|y^*\| \le 1\Big\}.$$

Moreover, (4.40) *holds when*

$$0 \in \partial_y\langle z^*, g\rangle(\bar{x}, \bar{y}) \Longrightarrow \partial_x\langle z^*, g\rangle(\bar{x}, \bar{y}) = \{0\} \ \text{ whenever } \ z^* \in Z^*; \tag{4.41}$$

also one has the upper bound estimate

$$\operatorname{lip} F(\bar{x}, \bar{y}) \le \sup\Big\{\|x^*\|\Big|\ \exists z^* \in Z^* \ \text{ with } \ x^* \in \partial_x\langle z^*, g\rangle(\bar{x}, \bar{y}),$$
$$-y^* \in \partial_y\langle z^*, g\rangle(\bar{y}, \bar{x}), \ \|y^*\| \le 1\Big\}$$

when X is finite-dimensional.

Proof. Assertion (i) follows from Theorem 4.37(i) with $\Theta = \{0\}$ and the strong coderivative normality of F in this case. The first part of assertion (ii), with characterization (4.40) and the equality for the exact Lipschitzian bound, follows from Theorem 4.37(ii). Now employing the relationship between full and partial coderivatives of N-regular mappings from Corollary 3.17 and the coderivative scalarization, we conclude that (4.41) implies (4.40), and that the upper bound estimate holds. △

The next corollary characterizes Lipschitzian stability of the classical feasible solution sets in parametric nonlinear programming.

Corollary 4.39 (Lipschitzian stability of constraint systems in nonlinear programming). *Let* $F\colon X \rightrightarrows Y$ *be a constraint system given in* (4.20), *where X and Y are Asplund and where* $\varphi_i\colon X \times Y \to I\!R$ *are strictly differentiable at* $(\bar{x}, \bar{y}) \in \operatorname{gph} F$ *for all* $i = 1, \ldots, m + r$. *Denote* $\bar{z}$ *and* $I(\bar{x}, \bar{y})$ *as in Corollary* 4.35 *and assume that the parametric Mangasarian-Fromovitz constraint qualification imposed therein holds. Then the condition*

$$\Big[\sum_{i \in I(\bar{x}, \bar{y})} \lambda_i \nabla_y \varphi_i(\bar{x}, \bar{y}) = 0\Big] \Longrightarrow \Big[\sum_{i \in I(\bar{x}, \bar{y})} \lambda_i \nabla_x \varphi_i(\bar{x}, \bar{y}) = 0\Big]$$

$$\text{for any } \ \lambda_i \in I\!R \ \text{ with } \ \lambda_i \ge 0 \ \text{ if } \ i \in \{1, \ldots, m\} \cap I(\bar{x}, \bar{y})$$

is necessary and sufficient for the Lipschitz-like property of F *around* $(\bar{x}, \bar{y})$. *If in addition* $\dim X < \infty$, *then*

$$\operatorname{lip} F(\bar{x}, \bar{y}) = \max\Big\{\Big\| \sum_{i\in I(\bar{x},\bar{y})} \lambda_i \nabla_x \varphi_i(\bar{x}, \bar{y})\Big\| \ \text{subject to}\ \lambda_i \in I\!R\,,$$

$$\Big\| \sum_{i\in I(\bar{x},\bar{y})} \lambda_i \nabla_y \varphi_i(\bar{x}, \bar{y})\Big\| \le 1, \ \text{and}\ \lambda_i \ge 0 \ \text{if}\ i \in \{1, \ldots, m\} \cap I(\bar{x}, \bar{y})\Big\}. \tag{4.42}$$

Proof. The necessary and sufficient condition of the corollary and the exact bound formula (4.42) with "sup" instead of "max" follow directly from Theorem 4.37(ii) with $\Omega = X \times Y$, $g = (\varphi_1, \ldots, \varphi_{m+r})$, and Θ defined in (4.21). The only thing we need to prove is that the maximum is attained in (4.42). Assuming the contrary, find sequences $\lambda_{ik} \in I\!R$, with $i \in I(\bar{x}, \bar{y})$ and $k \in I\!N$, satisfying the relations

$$\lambda_{ik} \ge 0 \ \text{ for } \ i \in \{1, \ldots, m\} \cap I(\bar{x}, \bar{y}), \quad \lambda_k := \sum_{i\in I(\bar{x},\bar{y})} \big|\lambda_{ik}\big| \to \infty \ \text{ as } \ k \to \infty\,,$$

$$\lim_{k\to\infty} \Big\| \sum_{i\in I(\bar{x},\bar{y})} \lambda_{ik} \nabla_x \varphi_i(\bar{x}, \bar{y})\Big\| = \ell, \quad \limsup_{k\to\infty} \Big\| \sum_{i\in I(\bar{x},\bar{y})} \lambda_{ik} \nabla_y \varphi_i(\bar{x}, \bar{y})\Big\| \le 1$$

with $\ell := \operatorname{lip} F(\bar{x}, \bar{y}) < \infty$. Consider the numbers

$$\widetilde{\lambda}_{ik} := \frac{\lambda_{ik}}{\lambda_k}, \ \ i \in I(\bar{x}, \bar{y}), \ k \in I\!N, \ \text{ with } \sum_{i\in I(\bar{x},\bar{y})} \big|\widetilde{\lambda}_{ik}\big| = 1$$

and find subsequences (without relabeling) such that $\widetilde{\lambda}_{ik} \to \widetilde{\lambda}_i$ as $k \to \infty$ for $i \in I(\bar{x}, \bar{y})$. Then $\widetilde{\lambda}_i$ are not equal to zero simultaneously for $i \in I(\bar{x}, \bar{y})$, and one has $\widetilde{\lambda}_i \ge 0$ for $i \in \{1, \ldots, m\} \cap I(\bar{x}, \bar{y})$,

$$\sum_{i\in I(\bar{x},\bar{y})} \widetilde{\lambda}_i \nabla_x \varphi_i(\bar{x}, \bar{y}) = 0, \quad \sum_{i\in I(\bar{x},\bar{y})} \widetilde{\lambda}_i \nabla_y \varphi_i(\bar{x}, \bar{y}) = 0\,.$$

The latter contradicts the assumed Mangasarian-Fromovitz constraint qualification and thus proves that the maximum is attained in (4.42). △

Now we obtain *sufficient* conditions for Lipschitzian stability of the general constraint systems (4.19) and their special cases with *no regularity* assumptions on the initial data.

Theorem 4.40 (Lipschitzian stability of general constraint systems). *Let* $F\colon X \rightrightarrows Y$ *be a set-valued mapping defined by the constraint system* (4.19), *and let* $(\bar{x}, \bar{y}) \in \operatorname{gph} F$. *Suppose that* $g\colon X \times Y \to Z$ *is continuous around* $(\bar{x}, \bar{y})$, *that the spaces* X, Y, Z *are Asplund, and that the sets* Ω *and* Θ *are locally closed around* $(\bar{x}, \bar{y})$ *and* $\bar{z} = g(\bar{x}, \bar{y})$, *respectively. Assume also that:*

(a) *Ω is PSNC at $(\bar{x},\bar{y})$ with respect to X.*
(b) *Either g is PSNC at $(\bar{x},\bar{y})$ and Θ is SNC at $\bar{z}$, or g is SNC at $(\bar{x},\bar{y})$.*
(c) *One has the qualification conditions* (4.24), (4.26), *and*

$$\Big[(x^*,0)\in D^*_N g(\bar{x},\bar{y})\circ N(\bar{z};\Theta)+N((\bar{x},\bar{y});\Omega)\Big]\Longrightarrow x^*=0\,. \tag{4.43}$$

Then F is Lipschitz-like around $(\bar{x},\bar{y})$. If in addition $\dim X<\infty$, *then*

$$\operatorname{lip}F(\bar{x},\bar{y})\le\sup\Big\{\|x^*\|\,\Big|\,(x^*,-y^*)\in D^*_N g(\bar{x},\bar{y})\circ N(\bar{z};\Theta)$$
$$+N((\bar{x},\bar{y});\Omega),\quad \|y^*\|\le 1\Big\}\,.$$

Proof. To establish the Lipschitz-like property of the constraint system (4.19) and the exact bound estimate, we employ the pointbased characterization (c) with the upper estimate (4.5) from Theorem 4.10 and the corresponding calculus rules of Sects. 3.1 and 3.3. Let us first check that the assumptions made ensure that F is PSNC at $(\bar{x},\bar{y})$. Following the proof of Theorem 4.37 and using the SNC calculus rules from Corollary 3.80 and Theorem 3.84 as well as the representation of $N((\bar{x},\bar{y});g^{-1}(\Theta))$ from Theorem 3.8, we conclude that F is PSNC at $(\bar{x},\bar{y})$ under assumptions (a), (b) of the theorem and the qualification conditions (4.24) and (4.26). Observe that these assumptions ensure the fulfillment of the coderivative inclusion (4.31) from Theorem 4.32. Thus $D^*_M F(\bar{x},\bar{y})(0)=\{0\}$ if the qualification condition (4.43) also holds. If in addition X is finite-dimensional, we derive the exact bound estimate in the theorem from (4.5) and (4.31) with $D^*=D^*_N$. △

The next corollary shows that all three qualification conditions in Theorem 4.40(c) can be equivalently unified into *one* provided that g is strictly Lipschitzian around $(\bar{x},\bar{y})$.

Corollary 4.41 (constraint systems generated by strictly Lipschitzian mappings). *Let $F\colon X\rightrightarrows Y$ be given in* (4.19), *where $g\colon X\times Y\to Z$ is a mapping between Asplund spaces that is assumed to be strictly Lipschitzian at $(\bar{x},\bar{y})\in\operatorname{gph}F$. Then conditions* (4.24), (4.26), *and* (4.43) *are fulfilled simultaneously if and only if*

$$\Big[(x^*,0)\in\partial\langle z^*,g\rangle(\bar{x},\bar{y})+N((\bar{x},\bar{y});\Omega),\quad z^*\in N(\bar{z};\Theta)\Big]$$
$$\Longrightarrow z^*=0\quad\text{and}\quad x^*=0\,. \tag{4.44}$$

If in this setting Ω and Θ are locally closed around $(\bar{x},\bar{y})$ and $\bar{z}=g(\bar{x},\bar{y})$, respectively, then condition (4.44) *is sufficient for the Lipschitz-like property of F around $(\bar{x},\bar{y})$ provided that Ω is PSNC at $(\bar{x},\bar{y})$ with respect to X and that Θ is SNC at $\bar{z}$. If in addition* $\dim X<\infty$, *then*

$$\text{lip}\, F(\bar{x}, \bar{y}) \le \sup\Big\{\|x^*\| \Big| \; (x^*, -y^*) \in \partial\langle z^*, g\rangle(\bar{x}, \bar{y}) + N((\bar{x}, \bar{y}; \Omega)\,,$$

$$z^* \in N(\bar{z}; \Theta), \quad \|y^*\| \le 1\Big\}\,.$$

Proof. By Theorem 3.28 we have

$$D_N^* g(\bar{x}, \bar{y})(z^*) = \partial\langle z^*, g\rangle(\bar{x}, \bar{y}) \;\; \text{for all} \;\; z^* \in Z^*$$

when $g\colon X \times Y \to Z$ is a strictly Lipschitzian mapping between Asplund spaces. Corollary 3.30 implies in this case that the SNC assumption on g in Theorem 4.40(b) is redundant in comparison with the SNC property of Θ. Hence the only thing we need to prove is that (4.44) is equivalent to the simultaneous fulfillment of (4.24), (4.26), and (4.43).

Let (4.44) hold. It obviously contains (4.43). To justify (4.24), we take any $(x^*, y^*) \in \partial\langle z^*, g\rangle(\bar{x}, \bar{y})$ satisfying the inclusions $(-x^*, -y^*) \in N((\bar{x}, \bar{y}); \Omega)$ and $z^* \in N(\bar{z}; \Theta)$. Then one has

$$(0, 0) \in \partial\langle z^*, g\rangle(\bar{x}, \bar{y}) + N((\bar{x}, \bar{y}); \Omega), \quad z^* \in N(\bar{z}; \Theta)\,,$$

and hence $z^* = 0$ due to (4.44). Thus $(x^*, y^*) = (0, 0)$, which gives (4.24). Similarly, if z^* belongs to the intersection in (4.26), then

$$(0, 0) \in \partial\langle z^*, g\rangle(\bar{x}, \bar{y}), \quad z^* \in N(\bar{z}; \Theta)\,, \tag{4.45}$$

and hence $z^* = 0$ by (4.44), i.e., (4.26) holds.

Now let us justify the opposite implication, that is, (4.44) is implied by (4.24), (4.26), and (4.43). Taking (x^*, z^*) from the set on the left-hand side of (4.44), we immediately have $x^* = 0$ by (4.43). It remains to show that $z^* = 0$ is the only solution to system (4.45). Indeed, if z^* satisfies (4.45), then there is $(x^*, y^*) \in \partial\langle z^*, g\rangle(\bar{x}, \bar{y})$ with $(-x^*, -y^*) \in N((\bar{x}, \bar{y}); \Omega)$. By (4.24) one has $(x^*, y^*) = (0, 0)$, and thus

$$z^* \in N(\bar{z}; \Theta) \cap \ker \partial\langle \cdot, g\rangle(\bar{x}, \bar{y})\,.$$

Hence $z^* = 0$ due to (4.26), which completes the proof of the corollary. △

It is easy to see from the above arguments that for $\Omega = X \times Y$ the condition

$$\Big[(x^*, 0) \in D_N^* g(\bar{x}, \bar{y})(z^*), \;\; z^* \in N(\bar{z}; \theta)\Big] \Longrightarrow z^* = 0, \;\; x^* = 0 \tag{4.46}$$

is equivalent to the simultaneous fulfillments of (4.26) and (4.43) even without the strict Lipschitzian assumption on g. If in this case g is strictly Lipschitzian at $(\bar{x}, \bar{y})$, then one can only require that $z^* = 0$ in (4.44) and (4.46), which obviously implies that $x^* = 0$.

We conclude this subsection with two corollaries of Theorem 4.40 that give efficient conditions for Lipschitzian stability of two remarkable constraint systems: implicit multifunctions defined by general/irregular mappings and feasible solution maps in problems of nondifferentiable programming.

Corollary 4.42 (Lipschitzian implicit multifunctions defined by irregular mappings). *Let $g: X \times Y \to Z$ be a mapping between Asplund spaces, and let $g(\bar{x}, \bar{y}) = 0$. Assume that g is SNC at $(\bar{x}, \bar{y})$, which is automatic if g is Lipschitz continuous around $(\bar{x}, \bar{y})$ and $\dim Z < \infty$. Then the condition*

$$(x^*, 0) \in D^*_N g(\bar{x}, \bar{y})(z^*) \Longrightarrow z^* = 0,\ x^* = 0$$

is sufficient for the Lipschitz-like property of the implicit multifunction

$$F(x) := \big\{ y \in Y \big|\ g(x, y) = 0 \big\}$$

around $(\bar{x}, \bar{y})$. If in addition $\dim X < \infty$, then

$$\operatorname{lip} F(\bar{x}, \bar{y}) \le \sup \Big\{ \|x^*\| \Big|\ (x^*, -y^*) \in \operatorname{rge} D^*_N g(\bar{x}, \bar{y}),\ \ \|y^*\| \le 1 \Big\}\,.$$

Proof. This is a special case of Theorem 4.40 with $\Theta = \{0\}$ and $\Omega = X \times Y$. Note that in this case the alternative assumption in Theorem 4.40(c) holds only when Z is finite-dimensional, and hence the PSNC property of g reduces to the SNC one. △

Corollary 4.43 (Lipschitzian stability of constraint systems in nondifferentiable programming). *Let $F: X \rightrightarrows Y$ be a multifunction between Asplund spaces given in (4.20), let $(\bar{x}, \bar{y}) \in \operatorname{gph} F$, and let $\bar{z}$ and $I(\bar{x}, \bar{y})$ be defined in Corollary 4.35. Assume that all φ_i, $i = 1, \ldots, m + r$, are Lipschitz continuous around $(\bar{x}, \bar{y})$ and that the constraint qualification (4.36) holds. Then the condition*

$$\Big[(x^*, 0) \in \sum_{i \in \{1,\ldots,m\} \cap I(\bar{x},\bar{y})} \lambda_i \partial\varphi_i(\bar{x}, \bar{y}) + \sum_{i=m+1}^{m+r} \lambda_i \Big(\partial\varphi_i(\bar{x}, \bar{y}) \cup \partial(-\varphi_i)(\bar{x}, \bar{y}) \Big)\,,$$

$$\lambda_i \ge 0 \ \ \text{for}\ \ i \in I(\bar{x}, \bar{y})\Big] \Longrightarrow x^* = 0$$

is sufficient for the Lipschitz-like property of F around $(\bar{x}, \bar{y})$. If in addition $\dim X < \infty$, then one has the upper estimate

$$\operatorname{lip} F(\bar{x}, \bar{y}) \le \sup \Big\{ \|x^*\| \Big|\ (x^*, -y^*) \in \sum_{i \in \{1,\ldots,m\} \cap I(\bar{x},\bar{y})} \lambda_i \partial\varphi_i(\bar{x}, \bar{y})$$

$$+ \sum_{i=m+1}^{m+r} \lambda_i \Big(\partial\varphi_i(\bar{x}, \bar{y}) \cup \partial(-\varphi_i)(\bar{x}, \bar{y}) \Big),\ \ \lambda_i \ge 0 \ \ \text{for}\ \ i \in I(\bar{x}, \bar{y}),\ \ \|y^*\| \le 1 \Big\}\,.$$

Proof. This follows from Theorem 4.40 with $g = (\varphi_1, \ldots, \varphi_{m+r}): X \times Y \to I\!R^{m+r}$, $\Omega = X \times Y$, and Θ defined in (4.21) due to the coderivative formula of Corollary 4.36. Note that g is automatically SNC at $(\bar{x}, \bar{y})$, since it is locally Lipschitzian and its range space is finite-dimensional. △

4.4 Sensitivity Analysis for Variational Systems

In this section we consider the so-called *generalized equations* given by

$$0 \in f(y) + Q(y) , \tag{4.47}$$

where f is a single-valued mapping while Q is a set-valued mapping between Banach spaces. For convenience we use the terms *base* and *field* referring to the single-valued and set-valued part of (4.47), respectively.

Generalized equations were introduced by Robinson [1130] as an extension of standard equations with no multivalued part. It has been well recognized that this model provide a convenient framework for the unified study of *optimal solutions* in many optimization-related areas including mathematical programming, complementarity, variational inequalities, optimal control, mathematical economics, equilibria, game theory, etc. In particular, generalized equations (4.47) reduce to the classical *variational inequalities*:

$$\text{find } \ y \in \Omega \ \text{ with } \ \big\langle f(y), v - y \big\rangle \geq 0 \ \text{ for all } \ v \in \Omega \tag{4.48}$$

when $Q(y) = N(y;\Omega)$ is the normal cone mapping generated by a *convex* set Ω. The classical *complementarity* problem corresponds to (4.48) when Ω is the nonnegative orthant in $I\!R^n$. It is well known that the latter form covers sets of optimal solutions with the corresponding Lagrange multipliers, or sets of KKT (Karush-Kuhn-Tucker) vectors, satisfying first-order necessary optimality conditions in problems of nonlinear programming.

Observe that the variational inequality (4.48) can be written in form (4.47) with the *subdifferential mapping* $Q(y) = \partial\varphi(y)$ for $\varphi(y) = \delta(y;\Omega)$. Thus the generalized equation model (4.47) covers also natural generalizations of variational inequalities when φ is not an indicator function and may even be *nonconvex*; the latter case relates to the so-called *hemivariational inequalities*.

The primary goal of this section is to conduct sensitivity analysis for generalized equations (4.47) and their specifications under perturbations of the initial data. For these purposes we consider a *parametric version* of (4.47) given in the form

$$0 \in f(x,y) + Q(x,y) \tag{4.49}$$

with a perturbation parameter x, where y is usually called the *decision* variable. Following the terminology of the previous section, we label (4.49) as *parametric variational systems*, since this model is suitable to describe sets of *optimal solutions* to parameter-dependent variational and related problems. The central question of local sensitivity analysis for (4.49) is to clarify how the following *solution map*

$$S(x) := \big\{ y \in Y \big|\ 0 \in f(x,y) + Q(x,y) \big\} \tag{4.50}$$

depends on the parameter x while (x, y) vary around the reference point $(\bar{x}, \bar{y}) \in \operatorname{gph} S$. As before, we are mostly concerned with *robust Lipschitzian stability* of solution maps paying the main attention to establishing efficient conditions for the Lipschitz-like property of multifunction (4.50) around $(\bar{x}, \bar{y})$. Based on the above *coderivative characterizations* of the Lipschitz-like property, we start sensitivity analysis for variational systems with evaluating coderivatives of the solution map (4.50) and its specifications.

4.4.1 Coderivatives of Parametric Variational Systems

First we obtain conditions that ensure *precise formulas* for computing the normal and mixed coderivatives of the solution map (4.50). These conditions require a smoothness (strict differentiability) assumption on the base f in the generalized equation (4.49). Given $f: X \times Y \to Z$ strictly differentiable at the reference point $(\bar{x}, \bar{y})$ satisfying (4.49), define the *adjoint generalized equation*

$$0 \in \nabla f(\bar{x}, \bar{y})^* z^* + D_N^* Q(\bar{x}, \bar{y}, \bar{z})(z^*) , \tag{4.51}$$

where $\bar{z} := -f(\bar{x}, \bar{y}) \in Q(\bar{x}, \bar{y})$.

Theorem 4.44 (computing coderivatives for regular variational systems). *Let $f: X \times Y \to Z$ be strictly differentiable at $(\bar{x}, \bar{y})$, let $Q: X \times Y \rightrightarrows Z$ with $\bar{z} = -f(\bar{x}, \bar{y}) \in Q(\bar{x}, \bar{y})$, and let $S: X \rightrightarrows Y$ be the solution map (4.50). The following assertions hold:*

(i) *Assume that X, Y, Z are Banach, that $\nabla_x f(\bar{x}, \bar{y})$ is surjective, and that Q doesn't depend on x. Then*

$$D_N^* S(\bar{x}, \bar{y})(y^*) = \Big\{ x^* \in X^* \Big| \; \exists z^* \in Z^* \;\; \text{with} \;\; x^* = \nabla_x f(\bar{x}, \bar{y})^* z^* ,$$
$$-y^* \in \nabla_y f(\bar{x}, \bar{y})^* z^* + D_N^* Q(\bar{y}, \bar{z})(z^*) \Big\} .$$

Moreover, S^{-1} is strongly coderivatively normal at $(\bar{y}, \bar{x})$ if Q is strongly coderivatively normal at $(\bar{y}, \bar{z})$.

(ii) *Assume that X, Y, Z are Asplund and that Q is locally closed-graph around $(\bar{x}, \bar{y}, \bar{z})$ and N-regular at this point. Suppose also that either Z is finite-dimensional or Q is SNC at $(\bar{x}, \bar{y}, \bar{z})$. Then S is N-regular at $(\bar{x}, \bar{y})$ and*

$$D^* S(\bar{x}, \bar{y})(y^*) = \Big\{ x^* \in X^* \Big| \; \exists z^* \in Z^* \;\; \text{with} \;\; \Big(x^* - \nabla_x f(\bar{x}, \bar{y})^* z^* ,$$
$$-y^* - \nabla_y f(\bar{x}, \bar{y})^* z^* \Big) \in D_N^* Q(\bar{x}, \bar{y}, \bar{z})(z^*) \Big\}$$

provided that the adjoint generalized equation (4.51) *admits only the trivial solution $z^* = 0$.*

Proof. We prove assertions (i) and (ii) in a parallel way using the corresponding assertions of Theorem 4.31. Observe that the graph of the solution map S in (4.50) is represented as

$$\operatorname{gph} S = \big\{(x, y) \in X \times Y \big|\, g(x, y) \in \Theta\big\} \text{ with } \Theta := \operatorname{gph} Q\ , \tag{4.52}$$

where g is defined by

$$g(x, y) := \big(y, -f(x, y)\big) \text{ if } Q = Q(y) \quad \text{and by} \tag{4.53}$$

$$g(x, y) := \big(x, y, -f(x, y)\big) \text{ if } Q = Q(x, y)\ . \tag{4.54}$$

In case (4.53), apply Theorem 4.31(i) and observe that $\nabla g(\bar{x}, \bar{y})$ is surjective if and only if $\nabla_x f(\bar{x}, \bar{y})$ is surjective. Then we arrive at the representation of $D_N^* S(\bar{x}, \bar{y})$ in this case by computing $\nabla g(\bar{x}, \bar{y})$ from (4.53) via representation (1.26) of the normal coderivative $D_N^* Q(\bar{y}, \bar{z})$ and elementary calculations. Furthermore, it is easy to check that

$$\widetilde{D}_M^* S(\bar{x}, \bar{y})(y^*) \supset \Big\{x^* \in X^* \Big|\, \exists z^* \in Z^* \text{ with } x^* = \nabla_x f(\bar{x}, \bar{y})^* z^*\ ,$$
$$-y^* \in \nabla_y f(\bar{x}, \bar{y})^* z^* + D_M^* Q(\bar{y}, \bar{z})(z^*)\Big\}$$

under the assumptions made in (i). To furnish this, we follow the above proofs for the case of D_N^* while using the definitions of D_M^* and $\widetilde{D}_M^*$ and taking into account that Fréchet-like normals and coderivatives enjoy required calculus rules under the imposed smoothness and surjectivity assumptions on the mappings involved; cf. Lemma 1.16 and Theorem 1.62. The latter inclusion and the above representation for $D^* S(\bar{x}, \bar{y})$ imply that

$$\widetilde{D}_M^* S(\bar{x}, \bar{y})(y^*) = D_N^* S(\bar{x}, \bar{y})(y^*) \text{ for all } y^* \in Y^*$$

provided that $D_M^* Q(\bar{y}, \bar{z}) = D_N^* Q(\bar{y}, \bar{z})$. Thus S^{-1} is strongly coderivatively normal at $(\bar{y}, \bar{x})$.

To prove (ii), we cannot use assertion (i) of Theorem 4.31, since $\nabla g(\bar{x}, \bar{y})$ is *never* surjective in case (4.54). Let us apply assertion (ii) of that theorem. First observe that there is *no* alternative assumption to the strict differentiability in Theorem 4.31(ii), since $\dim(X \times Y \times Z) < \infty$ in (4.54), and since the N-regularity of g at $(\bar{x}, \bar{y})$ implies the strict differentiability of g (and hence of f) at this point due to Theorem 1.46(ii). Then applying Theorem 4.31(ii) in this case, we check that the qualification condition (4.26) is equivalent to the fact that the adjoint generalized equation (4.51) has only the *trivial* solution. Thus S is N-regular at $(\bar{x}, \bar{y})$, and we derive the stated representation of $D^* S(\bar{x}, \bar{y})$ from the one in (4.25) provided that either Q is SNC at $(\bar{x}, \bar{y}, \bar{z})$, or g^{-1} is PSNC at $(\bar{w}, \bar{x}, \bar{y})$ with $\bar{w} := (\bar{x}, \bar{y}, \bar{z})$.

To complete the proof of the theorem, it remains to show that the latter assumption is equivalent to $\dim Z < \infty$. Indeed, due to Theorem 1.38 and the definition of strict derivative we conclude that the PSNC property of g^{-1} at $(\bar{w}, \bar{x}, \bar{y})$ is equivalent to the fact that for any sequences $(x_k, y_k) \to (\bar{x}, \bar{y})$, $(u_k^*, v_k^*, z_k^*) \xrightarrow{w^*} (0, 0, 0)$, and

$$(x_k^*, y_k^*) = (u_k^*, v_k^*) - \nabla f(\bar{x}, \bar{y})^* z_k^* \quad \text{with} \quad \|(x_k^*, y_k^*)\| \to 0 \tag{4.55}$$

one has $\|(u_k^*, v_k^*, z_k^*)\| \to 0$ as $k \to \infty$. It immediately follows from (4.55) that this property is fulfilled if Z is finite-dimensional. On the other hand, for *any* space Z of *infinite dimension* we find (by the Josefson-Nissenzweig theorem) a sequence of unit vectors $z_k^* \in Z^*$ that converges weak* to zero. Then taking an arbitrary sequence (x_k^*, y_k^*) with $\|(x_k^*, y_k^*)\| \to 0$, we define the sequence (u_k^*, v_k^*) by (4.55) and observe that $(u_k^*, v_k^*) \xrightarrow{w^*} (0, 0)$. Since $\|(u_k^*, v_k^*, z_k^*)\| \not\to 0$, this contradicts the PSNC property of g^{-1} at $(\bar{w}, \bar{x}, \bar{y})$. △

When $Q = Q(y)$ and f is strictly differentiable at $(\bar{x}, \bar{y})$, it is convenient to consider the following *partial adjoint generalized equation*

$$0 \in \nabla_y f(\bar{x}, \bar{y})^* z^* + D_N^* Q(\bar{y}, \bar{z})(z^*) \tag{4.56}$$

with $\bar{z} = -f(\bar{x}, \bar{y}) \in Q(\bar{y})$. In this setting z^* is a solution to the (full) adjoint generalized equation (4.51) if and only if it satisfies the partial one (4.56) together with $z^* \in \ker \nabla_x f(\bar{x}, \bar{y})^*$, where the latter requirement is redundant when $\nabla_x f(\bar{x}, \bar{y})$ is surjective. Thus the qualification condition of Theorem 4.44(ii) on the triviality of solutions to (4.51) reduces for $Q = Q(y)$ to the triviality of those solutions to (4.56), which belong to the kernel of $\nabla_x f(\bar{x}, \bar{y})^*$. This observation is useful in what follows.

One can get various consequences of Theorem 4.44 when the field Q of the generalized equation (4.49) is given in special forms allowing us to evaluate/estimate the normal coderivative $D_N^* Q$. We may employ for these purposes calculus rules for coderivatives as well as specific formulas obtained, e.g., in Subsect. 4.3.1. Let us present efficient results for the case of *convex-graph* multifunctions Q.

Given $Q: X \times Y \rightrightarrows Z$ and $f: X \times Y \to Z$ strictly differentiable at $(\bar{x}, \bar{y})$, we consider the *linearized* set-valued operator $L: X \times Y \rightrightarrows Z$ with

$$\begin{aligned} L(x, y) :&= f(\bar{x}, \bar{y}) + \nabla_x f(\bar{x}, \bar{y})(x - \bar{x}) \\ &+ \nabla_y f(\bar{x}, \bar{y})(y - \bar{y}) + Q(x, y) \end{aligned} \tag{4.57}$$

as well as, in the case of $Q = Q(y)$, the *partial linearized* operator $\widetilde{L}: Y \rightrightarrows Z$ defined by

$$\widetilde{L}(y) := f(\bar{x}, \bar{y}) + \nabla_y f(\bar{x}, \bar{y})(y - \bar{y}) + Q(y)\,. \tag{4.58}$$

Corollary 4.45 (coderivatives of solution maps to generalized equations with convex-graph fields). *Let $(\bar{x}, \bar{y})$ satisfy the generalized equation (4.49), where $f: X \times Y \to Z$ is strictly differentiable at $(\bar{x}, \bar{y})$ and where the graph of $Q: X \times Y \rightrightarrows Z$ is convex. The following hold for the coderivatives of the solution map (4.50):*

(i) *Assume that X, Y, Z are Banach, that $\nabla_x f(\bar{x}, \bar{y})$ is surjective, and that Q doesn't depend on x. Then S is N-regular at $(\bar{x}, \bar{y})$ and one has*

$$D^* S(\bar{x}, \bar{y})(y^*) = \Big\{ \nabla_x f(\bar{x}, \bar{y})^* z^* \Big| \; -(y^*, z^*) \in N\big((0,0); \operatorname{rge} \widetilde{M}\big) \Big\} ,$$

where $\widetilde{M}: Y \rightrightarrows Y \times Z$ is defined by

$$\widetilde{M}(y) := \big(y - \bar{y}, \widetilde{L}(y) \big) .$$

(ii) *Assume that X, Y, Z are Asplund and that Q is locally closed-graph around $(\bar{x}, \bar{y}, \bar{z})$ with $\bar{z} = -f(\bar{x}, \bar{y})$. Suppose also that either Z is finite-dimensional or Q is SNC at $(\bar{x}, \bar{y}, \bar{z})$, and that*

$$N(0; \operatorname{rge} L) = \{0\} , \tag{4.59}$$

where the mapping L is given in (4.57). Then S is N-regular at $(\bar{x}, \bar{y})$ and

$$D^* S(\bar{x}, \bar{y})(y^*) = \Big\{ x^* \in X^* \Big| \; \exists z^* \in Z^* \;\; \text{with} \quad (x^*, -y^*, -z^*) \in N\big((0,0,0); \operatorname{rge} M\big) \Big\} ,$$

where $M: X \times Y \rightrightarrows X \times Y \times Z$ is defined by

$$M(x, y) := \big(x - \bar{x}, y - \bar{y}, L(x, y) \big) .$$

Proof. We prove both assertions (i) and (ii) simultaneously based on the corresponding results of Theorem 4.44. Let us first check that the triviality of solutions to the adjoint equation (4.51) can be formulated as the qualification condition (4.59) in this case.

To proceed, employ the coderivative representation for convex-graph mappings from Proposition 1.37 and rewrite (4.51) as

$$\big\langle \nabla f(\bar{x}, \bar{y})^* z^*, (x, y) - (\bar{x}, \bar{y}) \big\rangle + \big\langle z^*, f(\bar{x}, \bar{y}) + z \big\rangle \ge 0 \;\; \text{for} \;\; (x, y, z) \in \operatorname{gph} Q .$$

This is equivalent to

$$\big\langle z^*, L(x, y) + z \big\rangle \ge 0 \;\; \text{whenever} \;\; (x, y, z) \in \operatorname{gph} Q \tag{4.60}$$

with L defined in (4.57). The latter means that $\bar{w} = 0$ is an optimal solution to the *convex minimization problem:*

$$\text{minimize} \;\; \langle z^*, w \rangle \;\; \text{subject to} \;\; w \in \Omega := \operatorname{rge} L .$$

Employing the generalized Fermat rule $0 \in \partial\varphi(\bar{w})$ as a *necessary and sufficient* condition for minimization of the convex function $\varphi(w) := \langle z^*, w\rangle + \delta(w; \Omega)$ and then using the subdifferential sum rule from Proposition 1.107, we conclude that (4.60) is equivalent to $-z^* \in N(0; \operatorname{rge} L)$. Thus the adjoint generalized equation (4.51) has only the trivial solution *if and only if* the qualification condition (4.59) holds.

To justify the coderivative representations in (i) and (ii) under the assumptions made, we involve similar arguments applied to the corresponding representations of Theorem 4.44. Since convex-graph mappings are N-regular at every point of their graph, we conclude that the solution map (4.50) is N-regular at $(\bar{x}, \bar{y})$ under the assumptions of this corollary. △

The qualification condition (4.59) obviously holds if $0 \in \operatorname{int}(\operatorname{rge} L)$, which is actually equivalent to (4.59) if the range of L is locally closed around $\bar{w} = 0$ and SNC at this point. Note that, due to convexity, the SNC property of the sets $\operatorname{rge} L$ and $\operatorname{gph} Q$ can be characterized via their finite codimensionality by Theorem 1.21. Observe also that for $Q = Q(y)$ the qualification condition (4.59) is clearly equivalent to

$$\ker \nabla_x f(\bar{x}, \bar{y})^* \cap N(0; \operatorname{rge} \widetilde{L}) = \{0\} , \tag{4.61}$$

where $\widetilde{L}$ is defined in (4.58).

Let us mention a special case of (4.49) when Q is given by

$$Q(x, y) := \begin{cases} E & \text{if } (x, y) \in \Omega , \\ \emptyset & \text{otherwise} , \end{cases} \tag{4.62}$$

where $E \subset Z$ and $\Omega \subset X \times Y$ are closed convex sets. In this case the interiority condition $0 \in \operatorname{int}(\operatorname{rge} L)$ reduces to

$$0 \in \operatorname{int}\Big\{ f(\bar{x}, \bar{y}) + \nabla f(\bar{x}, \bar{y})\big(\Omega - (\bar{x}, \bar{y})\big) + E \Big\}$$

When $Q = Q(y)$ in (4.62), the corresponding qualification (4.61) automatically holds under the *Robinson qualification condition*

$$0 \in \operatorname{int}\Big\{ f(\bar{x}, \bar{y}) + \nabla_y f(\bar{x}, \bar{y})\big(\Omega - \bar{y}\big) + E \Big\} .$$

In case (4.62) the coderivative formulas from Corollary 4.45 can be modified accordingly.

Next we obtain efficient conditions under which the equalities in Theorem 4.44 turn into *upper estimates* for coderivatives of solution maps (4.50) with *no surjectivity* and/or *normal regularity* assumptions made above. Moreover, we consider general cases of nonsmooth bases f in (4.49).

Theorem 4.46 (coderivative estimates for general variational systems). *Let* $(\bar{x}, \bar{y})$ *satisfy* (4.49), *where* X, Y, Z *are Asplund,* $f: X \times Y \to Z$ *is continuous around* $(\bar{x}, \bar{y})$, *and the graph of* Q *is closed around* $(\bar{x}, \bar{y}, \bar{z})$ *with* $\bar{z} = -f(\bar{x}, \bar{y})$. *Then*

$$D^*S(\bar{x}, \bar{y})(y^*) \subset \Big\{x^* \in X^* \Big| \ \exists z^* \in Z^* \ \text{ with } \\ (x^*, -y^*) \in D^*_N f(\bar{x}, \bar{y})(z^*) + D^*_N Q(\bar{x}, \bar{y}, \bar{z})(z^*)\Big\} \tag{4.63}$$

for both coderivatives $D^* = D^*_N, D^*_M$ *of the solution map* (4.50) *at* $(\bar{x}, \bar{y})$ *provided that either one of the following conditions holds:*

(a) Q *is SNC at* $(\bar{x}, \bar{y}, \bar{z})$, *and* $(x^*, y^*, z^*) = (0, 0, 0)$ *is the only triple satisfying the inclusion*

$$(x^*, y^*) \in D^*_N f(\bar{x}, \bar{y})(z^*) \cap \big(-D^*_N Q(\bar{x}, \bar{y}, \bar{z})(z^*)\big); \tag{4.64}$$

the latter is equivalent to

$$\Big[0 \in \partial\langle z^*, f\rangle(\bar{x}, \bar{y}) + D^*_N Q(\bar{x}, \bar{y}, \bar{z})(z^*)\Big] \Longrightarrow z^* = 0 \tag{4.65}$$

if f *is strictly Lipschitzian at* $(\bar{x}, \bar{y})$.

(b) f *is Lipschitz continuous around* $(\bar{x}, \bar{y})$, $\dim Z < \infty$, *and the triviality condition* (4.65) *is satisfied.*

Proof. First prove (4.63) under conditions (a) and (b) in a parallel way based on Theorem 4.32 and the graph representation (4.52) for the mapping S in (4.50) with g and Θ defined in (4.54). Applying Theorem 4.32, we use those assumptions therein that include the qualification condition (4.26), but not the ones with (4.30). The reason is that the latter condition involves the "reversed" coderivative $\widetilde{D}^*_M g$, which *doesn't* possess a satisfactory calculus allowing us to deal efficiently with functions g of type (4.54). Employing the normal coderivative $D^*_N g$ and taking into account that

$$g(x, y) = \big(x, y, 0\big) + \big(0, 0, -f(x, y)\big)$$

for g in (4.54) and that $D^*_N(-f)(\bar{x}, \bar{y})(z^*) = D^*_N f(\bar{x}, \bar{y})(-z^*)$, we get

$$D^*_N g(\bar{x}, \bar{y})(x^*, y^*, z^*) = (x^*, y^*) + D^*_N f(\bar{x}, \bar{y})(-z^*)$$

by Theorem 1.62(ii). Then it is easy to check that the qualification condition (4.26) for g and Θ from (4.54) is equivalent to $(x^*, y^*, z^*) = (0, 0, 0)$ for every triple satisfying (4.64). The latter reduces to (4.65) for strictly Lipschitzian mappings f due to Theorem 3.28 and Proposition 3.26. Similarly we can check that the coderivative inclusion (4.31) in Theorem 4.32 reduces to (4.63) if the above triviality condition for (4.64) holds and if either Q is SNC at $(\bar{x}, \bar{y}, \bar{z})$

or g^{-1} from (4.54) is PSNC at $(\bar{w}, \bar{x}, \bar{y})$ with $\bar{w} := (\bar{x}, \bar{y}, \bar{z})$. This justifies the conclusion of the theorem under the assumptions in (a).

To prove the theorem under the assumptions in (b), it remains to show that the PSNC property of g^{-1} holds if f is Lipschitz continuous around $(\bar{x}, \bar{y})$ while Z is finite-dimensional. By the structure of g in (4.54) and by the easy scalarization formula for the Fréchet coderivative of locally Lipschitzian mappings we conclude that the PSNC property of g^{-1} at $(\bar{w}, \bar{x}, \bar{y})$ means in this setting that for every sequences $(x_k, y_k) \to (\bar{x}, \bar{y})$, $(u_k^*, v_k^*) \xrightarrow{w^*} (0, 0)$, and

$$(x_k^*, y_k^*) - (u_k^*, v_k^*) \in \widehat{\partial}\langle -z_k^*, f\rangle(x_k, y_k) \quad \text{with} \quad \|(x_k^*, y_k^*, z_k^*)\| \to 0$$

one has $\|(u_k^*, v_k^*)\| \to 0$ as $k \to \infty$. This directly follows from the above inclusion due to the definition of Fréchet subgradients. △

Let us formulate a specification of Theorem 4.46 in the case of parametric generalized equations with smooth (strictly differentiable) bases; this case is of particular importance for applications.

Corollary 4.47 (coderivative estimates for generalized equations with smooth bases). *Let $f: X \times Y \to Z$ be a mapping between Asplund spaces that is strictly differentiable at a point $(\bar{x}, \bar{y})$ satisfying the generalized equation* (4.49)*, and let $Q: X \times Y \rightrightarrows Z$ be locally closed-graph around $(\bar{x}, \bar{y}, \bar{z})$ with $\bar{z} = -f(\bar{x}, \bar{y})$. Then*

$$D^*S(\bar{x}, \bar{y})(y^*) \subset \Big\{x^* \in X^* \Big|\ \exists z^* \in Z^* \ \text{ with } \ \Big(x^* - \nabla_x f(\bar{x}, \bar{y})^* z^*,$$

$$-y^* - \nabla_y f(\bar{x}, \bar{y})^* z^*\Big) \in D_N^* Q(\bar{x}, \bar{y}, \bar{z})(z^*)\Big\}$$

for both coderivatives $D^* = D_N^*, D_M^*$ of the solution map (4.50) *if the adjoint generalized equation* (4.51) *has only the trivial solution and if either Q is SNC at $(\bar{x}, \bar{y}, \bar{z})$ or* $\dim Z < \infty$.

Proof. This follows directly from Theorem 4.46 due to the coderivative representation for strictly differentiable mappings. △

The next corollary that concerns generalized equations with parameter-independent fields. For simplicity we formulate results only in the case when bases of generalized equations are smooth.

Corollary 4.48 (coderivatives of solution maps to HVIs with smooth bases). *Let $(\bar{x}, \bar{y})$ satisfy* (4.49)*, where X, Y, Z are Asplund, where $f: X \times Y \to Z$ is strictly differentiable at $(\bar{x}, \bar{y})$, and where $Q: Y \rightrightarrows Z$ is closed-graph around $(\bar{y}, \bar{z})$ with $\bar{z} = -f(\bar{x}, \bar{y})$. Assume that the partial adjoint generalized equation* (4.56) *has only the trivial solution on* $\ker \nabla_x f(\bar{x}, \bar{y})^*$ *and that either Q is SNC at $(\bar{y}, \bar{z})$ or* $\dim Z < \infty$. *Then one has the inclusion*

$$D^*S(\bar{x},\bar{y})(y^*) \subset \Big\{\nabla_x f(\bar{x},\bar{y})^*z^* \Big| -y^* \in \nabla_y f(\bar{x},\bar{y})^*z^* + D_N^*Q(\bar{y},\bar{z})(z^*)\Big\}$$

for both coderivatives $D^* = D_N^*, D_M^*$ *of the solution map* (4.50), *where equality holds if either* $\nabla_x f(\bar{x},\bar{y})$ *is surjective or* Q *is* N*-regular at* $(\bar{y},\bar{z})$.

Proof. The coderivative inclusion in this corollary follows directly from Corollary 4.47 when the field Q doesn't depend on x. The equality cases are contained in Theorem 4.44. △

Recall two simple and useful settings when Q is automatically SNC at every point of its graph: if either X, Y, Z are finite-dimensional or Q is convex-graph with nonempty interior. More general sufficient conditions for the SNC property of Q can be extracted from the results of Subsect. 1.2.5 and the SNC calculus developed in Sect. 3.3. Comprehensive coderivative calculus in Asplund spaces allows us to apply the above results to derive efficient coderivative estimates for fields Q and thus for solution maps (4.50) to parametric variational systems.

Many important applications of variational systems (4.49) relate to the case when $Q = \partial\varphi$ is a *subdifferential operator* generated by a l.s.c. function φ. In this case we have $D_N^*Q(\bar{x},\bar{y}) = \partial_N^2\varphi(\bar{x},\bar{y})$ by Definition 1.118(i) of the normal second-order subdifferential, and hence one can use advantages of the *second-order subdifferential calculus* developed in Subsects. 1.3.5 and 3.2.5. Borrowing mechanical terminology, we label φ as *potential.*

As mentioned in the beginning of this section, potentials φ are convex and parameter-independent in the classical settings of variational inequalities and complementarity problems. In the case of nonconvex and parameter-independent potentials the corresponding generalized equations relate to the so-called *hemivariational inequalities* (HVIs), which are conventionally considered in terms of Clarke subgradients for Lipschitz continuous functions. For convenience we use this terminology also in the case of our *basic subgradients* for l.s.c. *parameter-independent* potentials.

The main attention is paid in what follows to general classes of (4.49), where the *parameter-dependent* field $Q = Q(x,y)$ is given in two composite forms involving the basic first-order subdifferential. For convenience we call such generalized equations with subdifferential fields by *generalized variational inequalities* (GVIs).

The first class of GVIs under consideration concerns fields with *composite potentials* of the type $\varphi\circ g$, where $g\colon X\times Y \to W$ and $\varphi\colon W \to \overline{I\!R}$ are mappings between Banach spaces. On the other words, we'll study solutions maps given in the composite form

$$S(x) := \Big\{y \in Y \Big| 0 \in f(x,y) + \partial(\varphi\circ g)(x,y)\Big\}. \tag{4.66}$$

Note that the range space for f and $Q = \partial(\varphi\circ g)$ in (4.66) is either $X^*\times Y^*$ when $g = g(x,y)$, or Y^* when $g = g(y)$.

The second class of GVIs considered below involves *composite fields* of the form $Q(x,y) = \partial\varphi \circ g$ with $g\colon X \times Y \to W$ and $\varphi\colon W \to \overline{I\!R}$. Solution maps for such GVIs are given by

$$S(x) := \Big\{ y \in Y \Big|\ 0 \in f(x,y) + (\partial\varphi \circ g)(x,y) \Big\}, \tag{4.67}$$

where $f\colon X \times Y \to W^*$. By definition of the basic subdifferential we have that $\partial(\varphi \circ g)(x,y) = \emptyset$ in both (4.66) and (4.67) if $g(x,y) \notin \operatorname{dom}\varphi$.

Besides the classical variational inequalities and related systems mentioned above, models (4.66) and (4.67) cover a broad range of parametric variational systems important in applications. In particular, framework (4.66) is convenient for describing stationary point maps and stationary point-multiplier maps in problems of *composite optimization* with parameter-dependent constraints. Form (4.67) includes perturbed *implicit complementarity* problems of the type: find $y \in Y$ satisfying

$$f(x,y) \ge 0, \quad y - g(x,y) \ge 0, \quad \big\langle f(x,y), y - g(x,y)\big\rangle = 0,$$

where the inequalities are understood in the sense of some order on Y (in particular, component-wisely in finite-dimensions). Problems of this kind frequently arise in a large spectrum of mathematical models involving various types of economic and mechanical equilibria; see Commentary to this chapter for more references and discussions.

Our objective is to derive efficient coderivative representations/estimates for the solution maps (4.66) and (4.67) in terms of their initial data. We start with model (4.66) and first obtain conditions ensuring an *upper estimate* and an *exact formula* for computing coderivatives of (4.67) in general Banach spaces. These conditions apply to the case of *parameter-independent* potentials, i.e., they concern solution maps to parametric *hemivariational inequalities* (in our terminology) with potentials given in a composite form.

Theorem 4.49 (computing coderivatives of solution maps to HVIs with composite potentials). *Let X, Y, and W be Banach spaces, and let $(\bar x, \bar y) \in \operatorname{gph} S$ for S defined in (4.66) with $g\colon Y \to W$ and $\varphi\colon W \to \overline{I\!R}$. Put $\bar q := -f(\bar x,\bar y) \in \partial(\varphi \circ g)(\bar y)$ and assume the following:*

(a) *$f\colon X \times Y \to Y^*$ is strictly differentiable at $(\bar x,\bar y)$ with the surjective partial derivative $\nabla_x f(\bar x,\bar y)\colon X \to Y^*$.*

(b) *$g \in \mathcal{C}^1$ around $\bar y$ with the surjective derivative $\nabla g(\bar y)\colon Y \to W$, and the mapping $\nabla g\colon Y \to \mathcal{L}(Y,W)$ is strictly differentiable at $\bar y$.*

Let $\bar v \in W^*$ be a unique functional satisfying the relations

$$\bar q = \nabla g(\bar y)^*\bar v \quad \text{and} \quad \bar v \in \partial\varphi(\bar w) \quad \text{with} \quad \bar w := g(\bar y).$$

Then one has the inclusion

$$\begin{aligned} D_N^* S(\bar x,\bar y)(y^*) \subset \Big\{ x^* \in X^* \Big|\ &\exists u \in Y^{**} \quad \text{with} \quad x^* = \nabla_x f(\bar x,\bar y)^* u, \\ -y^* \in \nabla_y f(\bar x,\bar y)^* u + \nabla^2\langle \bar v, g\rangle(\bar y)^* u &+ \nabla g(\bar y)^* \partial_N^2 \varphi(\bar w,\bar v)\big(\nabla g(\bar y)^{**}u\big) \Big\}, \end{aligned} \tag{4.68}$$

which holds as equality if the range of $\nabla g(\bar{y})^*$ *is* w^**-extensible in* Y^**, in particular, when either this subspace is complemented in* Y^* *or the closed unit ball of* Y^{**} *is weak* sequentially compact. If in addition the subdifferential mapping* $\partial\varphi\colon W \rightrightarrows W^*$ *is strongly coderivatively normal at* $(\bar{w}, \bar{v})$*, then* S^{-1} *is strongly coderivatively normal at* $(\bar{y}, \bar{x})$*.*

Proof. Using first Theorem 4.44(i) and the definition of $\partial_N^2(\varphi \circ g)$, we get

$$D_N^* S(\bar{x}, \bar{y})(y^*) = \Big\{x^* \in X^* \Big|\ \exists u \in Y^{**} \text{ with } x^* = \nabla_x f(\bar{x}, \bar{y})^* u\ ,$$

$$-y^* \in \nabla_y f(\bar{x}, \bar{y})^* u + \partial_N^2(\varphi \circ g)(\bar{y}, \bar{q})(u)\Big\}$$

under assumption (a). Next applying the second-order subdifferential chain rule for $\partial_N^2(\varphi \circ g)$ in the inclusion form of Theorem 1.127, we arrive at (4.68) if both (a) and (b) are assumed. The equality case for in (4.68) follows from the one in Theorem 1.127. Finally, assuming that the first-order subdifferential mapping $\partial\varphi$ is strongly coderivatively normal at $(\bar{w}, \bar{v})$ and applying equality chain rules for both $\partial_N^2(\varphi \circ g)$ and $\partial_M^2(\varphi \circ g)$ in Theorem 1.127, we have

$$\partial_M^2(\varphi \circ g)(\bar{y}, \bar{q})(u) = \partial_N^2(\varphi \circ g)(\bar{y}, \bar{q})(u), \quad u \in Y^{**}\ .$$

This implies the strong coderivative normality of the inverse mapping S^{-1} at $(\bar{y}, \bar{x})$ by Theorem 4.44(i). △

Note that we didn't present an application of the equality case in Theorem 4.44(ii) to the solution map (4.66). The reason is that the N-regularity assumption on the field Q in Theorem 4.44(ii) is *not realistic* for *subdifferential mappings.* Indeed, even in the case of *convex* (as well as of more general) potentials φ in finite dimensions, $\partial\varphi$ is *graphically Lipschitzian,* and hence its *regularity is equivalent to its smoothness,* which actually excludes variational inequalities from consideration; see Definition 1.45, Theorem 1.46, and related discussions in Subsect. 1.2.2.

Next we obtain *upper coderivative estimates* for solution maps to GVIs with composite potentials (4.66) depending on the parameter x under significantly *less restrictive* assumptions on the mappings f and g in comparison with those in Theorem 4.49. To proceed, one may combine the upper coderivative estimates for general variational systems from Theorem 4.46 with the second-order chain rules for $\partial^2(\varphi \circ g)$ from Theorem 3.74. We are not going to present here the most general case in this direction, confining for simplicity our consideration to finite-dimensional spaces.

Theorem 4.50 (coderivative estimates for solution maps to GVIs with composite potentials). *Let* $(\bar{x}, \bar{y}) \in \operatorname{gph} S$ *with* S *defined in* (4.66)*, where* X, Y, W *are finite-dimensional,* $g\colon X \times Y \to W$ *is* C^2 *around* $(\bar{x}, \bar{y})$*,* $f\colon X \times Y \to X^* \times Y^*$ *is continuous around* $(\bar{x}, \bar{y})$*, and* $\varphi\colon W \to \overline{I\!R}$ *is l.s.c. around* $\bar{w} := g(\bar{x}, \bar{y})$*. Denote* $\bar{q} := -f(\bar{x}, \bar{y}) \in \partial(\varphi \circ g)(\bar{x}, \bar{y})$ *and*

$$M(\bar{x},\bar{y}) := \{\bar{v} \in W^* \big| \, \bar{v} \in \partial\varphi(\bar{w}), \quad \nabla g(\bar{x},\bar{y})^*\bar{v} = \bar{q}\}$$

and assume that:

(a) *The graphs of $\partial\varphi$ and $\partial^\infty\varphi$ are closed when w is near $\bar{w}$ (in particular, φ is either locally continuous or convex), and φ is lower regular around $\bar{w}$.*

(b) *The qualification conditions*

$$\partial^\infty\varphi(\bar{w}) \cap \ker \nabla g(\bar{x},\bar{y})^* = \{0\}\,, \tag{4.69}$$

$$\partial^2\varphi(\bar{w},\bar{v})(0) \cap \ker \nabla g(\bar{x},\bar{y})^* = \{0\} \tag{4.70}$$

are fulfilled whenever $\bar{v} \in M(\bar{x},\bar{y})$ in (4.70).

(c) *The relation*

$$(x^*,y^*) \in \bigcup_{\bar{v}\in M(\bar{x},\bar{y})} \Big[\nabla^2\langle\bar{v},g\rangle(\bar{x},\bar{y})u + \nabla g(\bar{x},\bar{y})^*\partial^2\varphi(\bar{w},\bar{v})\big(\nabla g(\bar{x},\bar{y})u\big)\Big]$$

$$\bigcap\Big[-D^*f(\bar{x},\bar{y})(u)\Big]$$

holds only for the trivial triple $(x^*,y^*,u) = (0,0,0)$ in $X^* \times Y^* \times (X \times Y)$.

Then one has the inclusion

$$D^*S(\bar{x},\bar{y})(y^*) \subset \Big\{x^* \in X^* \Big| \, \exists u \in X \times Y \ \ \textit{with} \ \ (x^*,-y^*) \in D^*f(\bar{x},\bar{y})(u)$$
$$+ \bigcup_{\bar{v}\in M(\bar{x},\bar{y})} \Big[\nabla^2\langle\bar{v},g\rangle(\bar{x},\bar{y})(u) + \nabla g(\bar{x},\bar{y})^*\partial^2\varphi(\bar{w},\bar{v})\big(\nabla g(\bar{x},\bar{y})u\big)\Big]\Big\} \tag{4.71}$$

for the basic coderivative of the solution map (4.66).

Proof. Applying Theorem 4.46 with $Q(x,y) = \partial(\varphi \circ g)$, we have the inclusion

$$D^*S(\bar{x},\bar{y})(y^*) \subset \Big\{x^* \in X^* \Big| \, \exists u \in X \times Y \ \ \text{with}$$
$$(x^*,-y^*) \in D^*f(\bar{x},\bar{y})(u) + \partial^2(\varphi \circ g)(\bar{x},\bar{y},\bar{q})(u)\Big\}$$

for S from (4.66) under the assumptions of that theorem with $D^*_N Q$ replaced by $\partial^2(\varphi \circ g)$, which is automatically SNC in finite dimensions. Now we employ the second-order chain rule from Theorem 3.74(ii) to get an upper estimate for $\partial^2(\varphi \circ g)(\bar{x},\bar{y},\bar{q})$. Using this result (actually Corollary 3.75 under the assumptions made) and taking into account the symmetry of the classical Hessian matrix in finite dimensions, we arrive at the conclusion of the theorem. △

An efficient consequence of the latter result concerns the case when the potential in (4.66) is a *strongly amenable* function; see the definition in Subsect. 3.2.5. In this case, which is especially important in applications to parametric optimization, assumption (a) of Theorem 4.50 and the first-order qualification condition (4.69) are *automatically fulfilled.*

Corollary 4.51 (coderivatives of solution maps to GVIs with amenable potentials). *Let S be the GVI solution map* (4.66) *whose potential $\psi = \varphi \circ g$ is strongly amenable at $(\bar{x}, \bar{y}) \in \operatorname{gph} S$ in finite dimensions. Assume that, in the notation of Theorem* 4.50, *f is continuous around $(\bar{x}, \bar{y})$, and that* (c) *and the second-order qualification condition* (4.70) *are fulfilled. Then the coderivative estimate* (4.71) *holds.*

Proof. It follows from Theorem 4.50 due to the definition of strongly amenable functions; it can be also derived from Theorem 4.46 and Corollary 3.76. △

The next corollary gives simplifications of the results in Theorem 4.50 and Corollary 4.51 under the *strict differentiability* assumption on the base f.

Corollary 4.52 (coderivatives of solution maps to GVIs with composite potentials and smooth bases). *Let f be strictly differentiable at $(\bar{x}, \bar{y})$ under the other assumptions of Theorem* 4.50 *and Corollary* 4.51. *Then condition* (c) *in Theorem* 4.50 *is equivalent to*

$$\Big[0 \in \nabla f(\bar{x}, \bar{y})^* u + \nabla^2 \langle \bar{v}, g\rangle(\bar{x}, \bar{y}) u + \nabla g(\bar{x}, \bar{y})^* \partial^2 \varphi(\bar{w}, \bar{v})\big(\nabla g(\bar{x}, \bar{y}) u\big)\Big] \Longrightarrow u = 0\,,$$

and one has the upper coderivative estimate

$$D^* S(\bar{x}, \bar{y})(y^*) \subset \Big\{x^* \in X^* \Big|\ \exists u \in X \times Y \quad \text{with} \quad \big(x^* - \nabla_x f(\bar{x}, \bar{y})^* u, -y^* - \nabla_y f(\bar{x}, \bar{y})^* u\big) \in \bigcup_{\bar{v} \in M(\bar{x}, \bar{y})} \big[\nabla^2 \langle \bar{v}, g\rangle(\bar{x}, \bar{y})(u) + \nabla g(\bar{x}, \bar{y})^* \partial^2 \varphi(\bar{w}, \bar{v}) \big(\nabla g(\bar{x}, \bar{y}) u\big)\big]\Big\}\,.$$

Proof. This follows directly from the coderivative representation for strictly differentiable mappings. △

Note that for $g = g(y)$ we get back to the coderivative inclusion (4.68), which is proved in Theorem 4.49 as equality in finite dimensions under the surjectivity assumptions on $\nabla_x f(\bar{x}, \bar{y})$ and $\nabla g(\bar{y})$. Here an upper estimate is proved in the same form with *no surjectivity* assumptions.

In the conclusion of this subsection we evaluate coderivatives of solution maps to GVIs with *composite fields* (4.67). First present an *exact formula* for computing the normal coderivative of (4.67) with $g = g(y)$ under surjectivity assumptions in arbitrary Banach spaces.

Proposition 4.53 (computing coderivatives of solution maps to HVIs with composite fields). *Let X, Y, W be Banach, and let $(\bar{x}, \bar{y}) \in \operatorname{gph} S$ for S defined in* (4.67) *with $g\colon Y \to W$ and $\varphi\colon W \to \overline{I\!R}$. Denote $\bar{w} := g(\bar{y})$, $\bar{q} := -f(\bar{x}, \bar{y}) \in \partial \varphi(\bar{w})$ and assume that:*

(a) *$f: X \times Y \to W^*$ is strictly differentiable at $(\bar{x}, \bar{y})$ with the surjective partial derivative $\nabla_x f(\bar{x}, \bar{y})$.*

(b) *g is strictly differentiable at $\bar{y}$ with the surjective derivative $\nabla g(\bar{y})$.*

Then one has the normal coderivative representation

$$D_N^* S(\bar{x}, \bar{y})(y^*) = \Big\{ x^* \in X^* \Big| \ \exists u \in W^{**} \ \text{ with } \ x^* = \nabla_x f(\bar{x}, \bar{y})^* u \, ,$$

$$-y^* \in \nabla_y f(\bar{x}, \bar{y})^* u + \nabla g(\bar{y})^* \partial_N^2 \varphi(\bar{w}, \bar{q})(u) \Big\}$$

for the solution map S in (4.67). Moreover, S^{-1} is strongly coderivatively normal at $(\bar{x}, \bar{y})$ if $\partial\varphi$ is strongly coderivatively normal at $(\bar{w}, \bar{q})$.

Proof. First we use Theorem 4.44(i) and get the equality

$$D_N^* S(\bar{x}, \bar{y})(y^*) = \Big\{ x^* \in X^* \Big| \ \exists u \in W^{**} \ \text{ with } \ x^* = \nabla_x f(\bar{x}, \bar{y})^* u \, ,$$

$$-y^* \in \nabla_y f(\bar{x}, \bar{y})^* u + D_N^* (\partial\varphi \circ g)(\bar{y}, \bar{q})(u) \Big\}$$

provided that $\nabla_x f(\bar{x}, \bar{y})$ is surjective. If in addition $\partial\varphi \circ g$ is strongly coderivatively normal at $(\bar{y}, \bar{q})$, than S^{-1} has this property at $(\bar{x}, \bar{y})$. Now we apply the coderivative chain rules (for $D^* = D_N^*$ and $D^* = D_M^*$) from Theorem 1.66 to the composition $\partial\varphi \circ g$ with $\nabla g(\bar{x}, \bar{y})$ surjective and arrive in this way at both conclusions of the proposition. △

Next we obtain an upper estimate for the coderivatives of (4.67) in general parameter-dependent settings of $g = g(x, y)$ employing coderivative chain rules and the SNC calculus in Asplund spaces.

Theorem 4.54 (coderivative estimates for solution maps to GVIs with composite fields). *Let X, Y, W be Asplund spaces, and let the dual space W^* be Asplund as well. Take $(\bar{x}, \bar{y}) \in \operatorname{gph} S$ for S defined in (4.67) with $\bar{q} = -f(\bar{x}, \bar{y}) \in \partial\varphi(\bar{w})$ and $\bar{w} = g(\bar{x}, \bar{y})$. Assume that $g: X \times Y \to W$ and $f: X \times Y \to W^*$ are continuous around $(\bar{x}, \bar{y})$, that the graph of $\partial\varphi: W \rightrightarrows W^*$ is norm-closed around $(\bar{w}, \bar{q})$, and that*

$$\partial_N^2 \varphi(\bar{w}, \bar{q})(0) \cap \ker D_N^* g(\bar{x}, \bar{y}) = \{0\} \, . \tag{4.72}$$

Suppose also that one of the conditions (a) and (b) is satisfied:

(a) *The implication*

$$\Big[(x^*, y^*) \in D_N^* f(\bar{x}, \bar{y})(u) \cap \big(- D_N^* g(\bar{x}, \bar{y}) \circ \partial_N^2 \varphi(\bar{w}, \bar{q})(u) \big) \Big]$$
$$\Longrightarrow (x^*, y^*, u) = (0, 0, 0) \, , \tag{4.73}$$

is fulfilled, and either g is PSNC at $(\bar{x}, \bar{y})$ and $\partial\varphi$ is SNC at $(\bar{w}, \bar{q})$, or g is SNC at $(\bar{x}, \bar{y})$ and $\partial\varphi^{-1}$ is PSNC at $(\bar{q}, \bar{w})$.

(b) *f is Lipschitz continuous around $(\bar{x}, \bar{y})$, W is finite-dimensional, and the constraint qualification* (4.73) *holds.*

Then the inclusion

$$D^*S(\bar{x}, \bar{y})(y^*) \subset \Big\{x^* \in X^* \Big| \ \exists u \in W^{**} \ \text{ with } \\ (x^*, -y^*) \in D^*_N f(\bar{x}, \bar{y})(u) + D^*_N g(\bar{x}, \bar{y}) \circ \partial^2_N \varphi(\bar{w}, \bar{q})(u)\Big\} \tag{4.74}$$

is valid for both coderivatives $D^* = D^*_N, D^*_M$ of the solution map (4.67).

Proof. Applying Theorem 4.46 to (4.67), we get the inclusion

$$D^*S(\bar{x}, \bar{y})(y^*) \subset \Big\{x^* \in X^* \Big| \ \exists u \in W^{**} \ \text{ with } \\ (x^*, -y^*) \in D^*_N f(\bar{x}, \bar{y})(u) + D^*_N (\partial\varphi \circ g)(\bar{x}, \bar{y}, \bar{q})(u)\Big\}$$

for both coderivatives $D^* = D^*_N, D^*_M$ under the assumptions of that theorem with $Q = \partial\varphi \circ g$. To pass from the latter inclusion to (4.74) and to efficiently express the assumptions of Theorem 4.46 for such Q in terms of the initial data of (4.67), we need to employ chain rules for $D^*_N(\partial\varphi \circ g)$ and SNC calculus results for this composition. An appropriate chain rule for this setting is obtained in Theorem 3.13(i), and the corresponding SNC calculus rule is given in Theorem 3.98. Applying these results to $\partial\varphi \circ g$, one can check that the assumptions made in the theorem ensure the fulfillment of those in Theorem 4.46. In this way we complete the proof. △

The final result of this subsection unifies and simplifies the assumptions of the latter theorem when W is finite-dimensional.

Corollary 4.55 (coderivatives for GVIs with composite fields of finite-dimensional range). *Let $(\bar{x}, \bar{y}) \in \operatorname{gph} S$ with S defined in* (4.67), *where X and Y are Asplund, and where $g: X \times Y \to I\!R^m$ and $f: X \times Y \to I\!R^m$ are continuous around $(\bar{x}, \bar{y})$. Using the notation of Theorem* 4.54, *we assume that the graph of $\partial\varphi: I\!R^m \rightrightarrows I\!R^m$ is closed around $(\bar{w}, \bar{q})$ (which is automatic for continuous and for amenable functions), that* (4.72) *and* (4.73) *hold, and that f is Lipschitz continuous around $(\bar{x}, \bar{y})$, and that either g is Lipschitz continuous around this point or X and Y are finite-dimensional. Then the coderivative estimate* (4.74) *is satisfied for the solution map* (4.67).

Proof. It follows from the observation that, if W is finite-dimensional and either g is locally Lipschitzian or X and Y are finite-dimensional, the SNC assumptions of g and $\partial\varphi$ are automatic in Theorem 4.54(a). △

4.4.2 Coderivative Analysis of Lipschitzian Stability

This subsection is devoted to coderivative analysis of Lipschitzian stability for parametric variational systems considered above. We mainly focus on the Lipschitz-like property of solution maps to variational systems providing *sufficient* (as well as *necessary and sufficient*) conditions for its fulfillment with *evaluating the exact Lipschitzian bound.* Our basic tool for this analysis is the pointbased criteria and exact bound formulas established in Theorem 4.10 and its corollaries. To conduct such an analysis, we need to use the coderivative representations and estimates for solution maps from the preceding subsection as well as efficient results of the SNC calculus.

Let us start with *characterizations* of Lipschitzian stability for variational systems described by generalized equations under *regularity* conditions.

Theorem 4.56 (characterizations of Lipschitzian stability for regular generalized equations). *Let S be the solution map (4.50), where $f\colon X \times Y \to Z$ is strictly differentiable at $(\bar{x}, \bar{y}) \in \operatorname{gph} S$, where$Q\colon X \times Y \rightrightarrows Z$ is locally closed-graph around $(\bar{x}, \bar{y}, \bar{z})$ with $\bar{z} := -f(\bar{x}, \bar{y})$ and SNC at this point, and where the spaces X, Y are Asplund. The following hold:*

(i) *Assume that Z is Banach, that $\nabla_x f(\bar{x}, \bar{y})$ is surjective, and that Q doesn't depend on x. Then S is Lipschitz-like around $(\bar{x}, \bar{y})$ if the partial adjoint generalized equation (4.56) has only the trivial solution. This condition is also necessary for the Lipschitz-like property of S around $(\bar{x}, \bar{y})$ if S is strongly coderivatively normal at $(\bar{x}, \bar{y})$, in particular, when Y is finite-dimensional. If in addition the space X is finite-dimensional, then*

$$\operatorname{lip} S(\bar{x}, \bar{y}) = \sup \Big\{ \|\nabla_x f(\bar{x}, \bar{y})^* z^*\| \;\Big|\; \exists y^* \in D_N^* Q(\bar{y}, \bar{z})(z^*) \;\; \textit{with}$$

$$\|\nabla_y f(\bar{x}, \bar{y})^* z^* + y^*\| \le 1 \Big\},$$

where the maximum is attained provided that the graph of the set-valued mapping $(y, z, z^*) \to D_N^* Q(y, z)(z^*)$ is locally closed near $(\bar{y}, \bar{z})$ in the norm $\times$ weak topology of $(Y \times Z) \times (Y^* \times Z^*)$.*

(ii) *Assume that Z is Asplund and that Q is N-regular at $(\bar{x}, \bar{y}, \bar{z})$. Then S is N-regular at $(\bar{x}, \bar{y})$, and the condition*

$$\Big[(x^*, 0) \in \nabla f(\bar{x}, \bar{y})^* z^* + D^* Q(\bar{x}, \bar{y}, \bar{z})(z^*)\Big] \Longrightarrow x^* = z^* = 0 \tag{4.75}$$

is sufficient for the Lipschitz-like property of S around $(\bar{x}, \bar{y})$. This condition is also necessary for the Lipschitz-like property of S provided that the adjoint generalized equation (4.51) has only the trivial solution. If in addition the space X is finite-dimensional, then

$$\operatorname{lip} S(\bar{x}, \bar{y}) = \sup \Big\{ \|x^*\| \;\Big|\; \exists z^* \in Z^* \;\; \textit{with} \;\; \Big(x^* - \nabla_x f(\bar{x}, \bar{y})^* z^*,$$

$$-y^* - \nabla_y f(\bar{x}, \bar{y})^* z^*\Big) \in D^* Q(\bar{x}, \bar{y}, \bar{z})(z^*), \;\; \|y^*\| \le 1 \Big\}.$$

In particular, for $Q = Q(y)$ *the solution map* S *is Lipschitz-like around* $(\bar{x}, \bar{y})$ *if the partial adjoint generalized equation* (4.56) *has only the trivial solution. This condition is also necessary for the Lipschitz-like property of* S *provided that* (4.56) *admits only the trivial solution on* $\ker \nabla_x f(\bar{x}, \bar{y})^*$. *If in addition* $\dim X < \infty$, *then* $\operatorname{lip} S(\bar{x}, \bar{y})$ *is computed by the formula from* (i).

Proof. As before, we are based on criteria (c) and the exact bound formula (4.6) from Theorem 4.10 characterizing the Lipschitz-like property of general closed-graph multifunctions between Asplund spaces. To proceed in this way for the solution map (4.50), we need to employ the coderivative formulas from Theorem 4.44 together with appropriate results of the SNC calculus.

Let us first prove assertion (i). By Theorem 4.44(i) we have

$$D_N^* S(\bar{x}, \bar{y})(0) = \Big\{\nabla_x f(\bar{x}, \bar{y})^* z^* \text{ with } z^* \in Z^* \text{ satisfying } (4.56)\Big\}, \quad (4.76)$$

from which and the surjectivity of $\nabla_x f(\bar{x}, \bar{y})$ we conclude that $D_N^* S(\bar{x}, \bar{y}) = \{0\}$ if and only if the partial adjoint generalized equation (4.56) has only the trivial solution. Further, the representation

$$\operatorname{gph} S = \big\{(x, y) \in X \times Y \big|\ g(x, y) \in \operatorname{gph} Q\big\} \ \text{ with } \ g(x, y) = \big(y, -f(x, y)\big)$$

and Theorem 1.22 imply, in any Banach spaces, that S is SNC at $(\bar{x}, \bar{y})$ if and only if Q is SNC at $(\bar{y}, \bar{z})$ provided that $\nabla g(\bar{x}, \bar{y})$ is surjective. Since the latter condition is equivalent to the surjectivity of $\nabla_x f(\bar{x}, \bar{y})$ and since $D_M^* S(\bar{x}, \bar{y})(y^*) \subset D_N^* S(\bar{x}, \bar{y})(y^*)$ with the equality for strongly coderivatively normal mappings, we arrive at the conclusions of (i) on the Lipschitz-like property of S with the exact bound formula. It remains to observe that the local closedness property of $D_N^* Q$ assumed in the last part of (i) is clearly equivalent to the one for $N(\cdot; \operatorname{gph} Q)$, and hence the maximum is attained in the formula for $\operatorname{lip} S(\bar{x}, \bar{y})$ in (i) due to the corresponding conclusion of Theorem 4.37(i); see also the discussion after that theorem.

To justify (ii), observe that condition (4.75) implies that the (full) adjoint generalized equation (4.51) has only the trivial solution. Then Theorem 4.44(ii) ensures that S is N-regular at $(\bar{x}, \bar{y})$ and that $D^* S(\bar{x}, \bar{y})$ is computed by the formula therein. Thus

$$D^* S(\bar{x}, \bar{y})(0) = \Big\{x^* \in X^* \Big|\ \exists z^* \in Z^* \text{ with } \Big(x^* - \nabla_x f(\bar{x}, \bar{y})^* z^*, \\ -\nabla_y f(\bar{x}, \bar{y})^* z^*\Big) \in D^* Q(\bar{x}, \bar{y}, \bar{z})(z^*\Big\}.$$

Hence condition (4.75) implies that $D^* S(\bar{x}, \bar{y})(0) = \{0\}$. Furthermore, by Theorem 3.98 and representation (4.52) with g defined in (4.54), we conclude that S is SNC at $(\bar{x}, \bar{y})$ under the assumptions made. Now Theorem 4.10 ensures the Lipschitz-like property of S around $(\bar{x}, \bar{y})$ and the exact bound formula in (ii). It follows from the above arguments that condition (4.75) is also necessary for the Lipschitz-like property of S provided that (4.51) has only the trivial solution.

It remains to justify the last conclusion of the theorem for $Q = Q(y)$. In this case the equation (4.51) has only the trivial solution if and only if

$$\ker \nabla_x f(\bar{x}, \bar{y})^* \cap \big\{z^* \in Z^* \ \text{satisfying} \ (4.56)\big\} = \{0\} .$$

Using this together with (4.76), we complete the proof of the theorem. △

Corollary 4.57 (Lipschitzian stability for generalized equations with convex-graph fields). *Let S be the solution map* (4.50) *under the common assumptions of Theorem* 4.56, *let the graph of Q be convex, and let the mappings $M, \widetilde{M}, L$ be defined in Corollary* 4.45. *The following assertions hold:*

(i) *Assume that Z is Banach, that $\nabla_x f(\bar{x}, \bar{y})$ is surjective, and that Q doesn't depend on x. Then the condition*

$$(0, z^*) \in N((0,0); \operatorname{rge} \widetilde{M}) \Longrightarrow z^* = 0$$

is necessary and sufficient for the Lipschitz-like property of S around $(\bar{x}, \bar{y})$. Moreover, in this case

$$\operatorname{lip} S(\bar{x}, \bar{y}) = \sup\Big\{\|\nabla_x f(\bar{x}, \bar{y})^* z^*\| \;\Big|\; -(y^*, z^*) \in N((0,0); \operatorname{rge} \widetilde{M}),\ \ \|y^*\| \le 1\Big\}$$

if X is finite-dimensional.

(ii) *Assume that Z is Asplund. Then the condition*

$$(x^*, 0, z^*) \in N((0,0,0); \operatorname{rge} M) \Longrightarrow x^* = z^* = 0$$

is sufficient for the Lipschitz-like property of S around $(\bar{x}, \bar{y})$ being also necessary for this property if $N(0; \operatorname{rge} L) = \{0\}$. In this case

$$\operatorname{lip} S(\bar{x}, \bar{y}) = \sup\Big\{\|x^*\| \;\Big|\; \exists (y^*, z^*) \in Y^* \times Z^* \ \ \text{with}$$
$$(x^*, -y^*, -z^*) \in N((0,0,0); \operatorname{rge} M),\ \ \|y^*\| \le 1\Big\}$$

if X is finite-dimensional.

Proof. If follows from Theorem 4.56 due to the coderivative representation for convex-graph mappings from Proposition 1.37; cf. the proof of Corollary 4.45. It can be also derived directly from Theorem 4.10 and Corollary 4.45 similarly to the proof of Theorem 4.56. △

Remark 4.58 (basic normals versus Clarke normals in Lipschitzian stability). Observe that Theorem 4.56(ii) doesn't distinguish between the usage of basic and Clarke normals to the graph of Q provided that the basic normal cone $N((\bar{x}, \bar{y}, \bar{z}); \operatorname{gph} Q)$ is weak* closed (this is the case, in particular, when either X, Y, Z are finite-dimensional or the graph of Q is convex as in

Corollary 4.57). On the contrary, Theorem 4.56(i) strikingly does. Indeed, a counterpart of Theorem 4.56(i) with $D_N^*Q(\bar{y},\bar{z})(z^*)$ replaced by the cone

$$\big\{(y^*,z^*)\in Y^*\times Z^*\big|\ (y^*,-z^*)\in N_C\big((\bar{y},\bar{z});\operatorname{gph} Q\big)\big\}$$

obviously provides a sufficient condition for the Lipschitz-like property of the solution map (4.50) at $(\bar{x},\bar{y})$. However, the latter condition is *far removed from necessity* and actually *doesn't hold at all* for a large class of set-valued mappings Q. Let us present *two examples* demonstrating this phenomenon.

First consider the parametric generalized equation

$$0\in x+[-y,y]\ \text{ with } x,y\in I\!R\ .$$

In this case $Q(y)=[-y,y]$, and one may directly check that

$$N((0,0);\operatorname{gph} Q)=\big\{(v,u)\in I\!R^2\big|\ |u|=|v|\big\}\ \text{ and }\ N_C((0,0);\operatorname{gph} Q)=I\!R^2\ .$$

Hence $D^*Q(0,0)(u)=\{-u,u\}$ and the condition $D^*Q(0,0)(0)=\{0\}$ is obviously fulfilled characterizing Lipschitzian stability of (4.50), while its Clarke counterpart

$$\Big[\big(-\nabla_y f(\bar{x},\bar{y})z^*,-z^*\big)\in N_C\big((\bar{y},\bar{z});\operatorname{gph} Q\big)\Big]\Longrightarrow z^*=0 \qquad (4.77)$$

doesn't hold although the solution map $S(x)=\big\{y\in I\!R\big|\ -x\in[-y,y]\big\}$ is clearly Lipschitz-like around $(0,0)$.

The second example concerns the classical framework of perturbed *variational inequalities/complementarity problems*:

$$\text{find }\ y\geq 0\ \text{ with }\ (ay+x)(v-y)\geq 0\ \text{ for all }\ v\geq 0\ , \qquad (4.78)$$

where $a\in I\!R$ is a given number and $x\in I\!R$ is a perturbation parameter. This example can be written in the generalized equation form (4.49) with

$$f(x,y):=ay+x\ \text{ and }\ Q(y):=\begin{cases}0 & \text{if } y>0\ ,\\ I\!R_- & \text{if } y=0\ ,\\ \emptyset & \text{if } y<0\ .\end{cases}$$

It is easy to see that $Q(y)=N(y;\Omega)=\partial\delta(y;\Omega)$ for $\Omega:=I\!R_+$, and therefore Q has the *nonconvex graph*

$$\operatorname{gph} Q=\big\{(y,z)\in I\!R^2\big|\ y\geq 0,\ \ z\leq 0,\ \ yz=0\big\}\ .$$

Invoking Theorem 1.6, we compute the basic normal cone to this graph

$$N((0,0);\operatorname{gph} Q)=\big\{(v,u)\in I\!R^2\big|\ v\leq 0,\ \ u\geq 0\big\}\ ,$$

which gives the coderivative expression

$$D^*Q(0,0)(u) = \begin{cases} 0 & \text{if } u > 0\,, \\ I\!R & \text{if } u = 0\,, \\ I\!R_- & \text{if } u < 0\,. \end{cases}$$

This allows us to conclude by Theorem 4.56(i) that the solution map to (4.78) is Lipschitz-like around (0, 0) *if and only if* $a > 0$. On the other hand, one has $N_C((0,0); \operatorname{gph} Q) = I\!R^2$ for the Clarke normal cone, and hence the sufficient condition (4.77) carries *no information* about Lipschitzian stability of the perturbed variational inequality (4.78).

It turns out that the situation in the above examples is typical for a sufficiently broad class of variational systems including the classical variational inequalities. Considering the case when $Q\colon I\!R^n \rightrightarrows I\!R^n$ is a graphically Lipschitzian mapping of dimension n around $(\bar{y}, \bar{z})$ (this includes maximal monotone relations, particularly *subdifferential mappings* $Q = \partial\varphi$ for convex and other nice functions; see the discussion after Definition 1.45), we conclude from the proof of Theorem 1.46 that $N_C((\bar{y}, \bar{z}); \operatorname{gph} Q)$ is a *subspace* of $I\!R^{2n}$ having dimension *at least* n. It is easy to check that in this setting the sufficient condition (4.77) implies that the dimension of the subspace $N_C((\bar{y}, \bar{z}); \operatorname{gph} Q)$ is *exactly* n, and hence the set gph Q is *graphically smooth* at $(\bar{y}, \bar{z})$; see Theorem 1.46(ii). Moreover, if $Q = \partial\varphi$ with a proper l.s.c. convex function on $I\!R^n$, then the latter property corresponds to some second-order differentiability of φ, which is very close to the classical contents; see Rockafellar [1153]. Hence condition (4.77) involving Clarke normals cannot actually cover standard settings of variational inequalities and complementarity problems in finite dimensions, where φ is the indicator function of a convex set. In contrast to this, we present here characterizations and efficient sufficient conditions for Lipschitzian stability of such and more general variational systems in terms of our basic normals and second-order subdifferentials.

Next we consider variational systems (4.50) with *no regularity* assumptions on the initial data in a general nonsmooth setting. The following theorem gives sufficient conditions for their Lipschitzian stability in terms of normal coderivatives of both bases and fields.

Theorem 4.59 (Lipschitzian stability for irregular generalized equations). *Let S be the solution map (4.50), where $f\colon X \times Y \to Z$ is continuous around $(\bar{x}, \bar{y}) \in \operatorname{gph} S$, where $Q\colon X \times Y \rightrightarrows Z$ is locally closed-graph around $(\bar{x}, \bar{y}, \bar{z})$ with $\bar{z} = -f(\bar{x}, \bar{y})$ and SNC at this point, and where the spaces X, Y, Z are Asplund. Assume further that f is PSNC at $(\bar{x}, \bar{y})$ and that one has the qualification conditions*

$$\Big[(x^*,0)\in D_N^*f(\bar{x},\bar{y})(z^*)+D_N^*Q(\bar{x},\bar{y},\bar{z})(z^*)\Big]\Longrightarrow x^*=0\,,$$

$$\Big[(x^*,y^*)\in D_N^*f(\bar{x},\bar{y})(z^*)\cap\big(-D_N^*Q(\bar{x},\bar{y},\bar{z})(z^*)\big)\Big]\Longrightarrow x^*=y^*=z^*=0\,.$$

Then S is Lipschitz-like around $(\bar{x},\bar{y})$. If in addition $\dim X<\infty$, *then*

$$\operatorname{lip} S(\bar{x},\bar{y})\le\sup\Big\{\|x^*\|\;\Big|\;\exists z^*\in Z^*\ \ \textit{with}$$

$$(x^*,-y^*)\in D_N^*f(\bar{x},\bar{y})(z^*)+D_N^*Q(\bar{x},\bar{y},\bar{z})(z^*),\ \ \|y^*\|\le 1\Big\}\,.$$

Proof. Observe that the assumptions made in this theorem imply the fulfillment of all the assumptions in Theorem 4.46. Hence the coderivative inclusion (4.63) holds, and thus the first qualification condition of the theorem ensures that $D_M^*S(\bar{x},\bar{y})(0)=\{0\}$. By Theorem 4.10 it remains to show that S is PSNC at $(\bar{x},\bar{y})$.

Let us prove that S is actually SNC at $(\bar{x},\bar{y})$ if f is assumed to be PSNC at this point in addition to the second qualification condition of the theorem and the SNC property of Q at $(\bar{x},\bar{y},\bar{z})$. To furnish this, we apply Theorem 3.84 to the inverse image

$$\operatorname{gph} S=g^{-1}(\operatorname{gph} Q)\ \text{ with }\ g(x,y)=\big(x,y,-f(x,y)\big)\,.$$

The only thing one needs to check is that g is PSNC at $(\bar{x},\bar{y})$ if f is PSNC at this point. Indeed, taking sequences $(x_k^*,y_k^*)\in\widehat{D}^*g(x_k,y_k)(u_k^*,v_k^*,z_k^*)$ with $(x_k^*,y_k^*)\xrightarrow{w^*}(0,0)$ and $\|(u_k^*,v_k^*,z_k^*)\|\to 0$, we get

$$(x_k^*,y_k^*)=(u_k^*,v_k^*)+(\hat{x}_k^*,\hat{y}_k^*)\ \text{ with }\ (\hat{x}_k^*,\hat{y}_k^*)\in\widehat{D}^*f(x_k,y_k)(-z_k^*)$$

due to the representation

$$g(x,y)=(x,y,0)+(0,0,-f(x,y))$$

and the sum rule of Theorem 1.62(i). This implies that $(\hat{x}_k^*,\hat{y}_k^*)\xrightarrow{w^*}(0,0)$, and hence $\|(\hat{x}_k^*,\hat{y}_k^*)\|\to 0$ by the PSNC property of f. Thus $\|(x_k^*,y_k^*)\|\to 0$ as well, i.e., g is PSNC at $(\bar{x},\bar{y})$. △

When f is strictly Lipschitzian around $(\bar{x},\bar{y})$, the two qualification conditions in Theorem 4.59 can be unified, and the above result admits the following simplified formulation.

Corollary 4.60 (stability for generalized equations with strictly Lipschitzian bases). *Let f be strictly Lipschitzian at $(\bar{x},\bar{y})$ in the framework of Theorem 4.59, and let Q be closed-graph and SNC at $(\bar{x},\bar{y},\bar{z})$. Then the solution map (4.50) is Lipschitz-like around $(\bar{x},\bar{y})$ provided that*

$$\Big[(x^*,0) \in \partial\langle z^*, f\rangle(\bar{x},\bar{y}) + D_N^* Q(\bar{x},\bar{y},\bar{z})(z^*)\Big] \Longrightarrow x^* = z^* = 0\,, \quad (4.79)$$

which reduces to (4.75) *with* $D^* = D_N^*$ *when* f *is strictly differentiable at* $(\bar{x},\bar{y})$. *If in addition* $\dim X < \infty$, *then*

$$\operatorname{lip} S(\bar{x},\bar{y}) \le \sup\Big\{\|x^*\|\ \Big|\ \exists z^* \in Z^* \ \ \textit{with}\ \ (x^*,-y^*) \in \partial\langle z^*, f\rangle(\bar{x},\bar{y})$$
$$+D_N^* Q(\bar{x},\bar{y},\bar{z})(z^*),\ \ \|y^*\| \le 1\Big\}\,.$$

Proof. If f is strictly Lipschitzian at $(\bar{x},\bar{y})$, then the second qualification condition in Theorem 4.59 is equivalent to (4.65) by the scalarization formula of Theorem 3.28. Finally, it is easy to check that the unified qualification condition (4.79) is equivalent to the simultaneous fulfillment of (4.65) and the first qualification condition in Theorem 4.59. △

The following corollary concerns Lipschitzian stability of solution maps to perturbed generalized equations with parameter-independent fields $Q = Q(y)$.

Corollary 4.61 (stability of solution maps to general HVIs). *Let* $Q = Q(y)$ *be closed-graph and SNC at* $(\bar{y},\bar{z})$ *in the framework of Theorem* 4.59. *Then the solution map* (4.50) *is Lipschitz-like around* $(\bar{x},\bar{y})$ *if*

$$\Big[(x^*,y^*) \in D_N^* f(\bar{x},\bar{y})(z^*),\ \ -y^* \in D_N^* Q(\bar{y},\bar{z})(z^*)\Big] \Longrightarrow x^* = y^* = z^* = 0\,,$$

which is equivalent to

$$\Big[\emptyset \ne \operatorname{proj}_y \partial\langle z^*, f\rangle(\bar{x},\bar{y}) \cap \big(-D_N^* Q(\bar{y},\bar{z})(z^*)\big)\Big] \Longrightarrow z^* = 0 \qquad (4.80)$$

(proj_y *stands here for the projection on* Y^*) *when* f *is strictly Lipschitzian at the reference point* $(\bar{x},\bar{y})$.

Proof. It is easy to see that for $Q = Q(y)$ the two qualification conditions of Theorem 4.59 hold simultaneously if and only if the qualification conditions of this corollary is fulfilled. This reduces to

$$\Big[(x^*,y^*) \in \partial\langle z^*, f\rangle(\bar{x},\bar{y}),\ \ -y^* \in D_N^* Q(\bar{y},\bar{z})(z^*)\Big] \Longrightarrow z^* = 0$$

by the coderivative scalarization for strictly Lipschitzian mappings. The latter is obviously equivalent to (4.80). △

Next we study the Lipschitz-like property of solution maps (4.66) to generalized variational inequalities (GVIs) with *composite potentials.* The following theorem gives sufficient conditions, as well as characterizations, for Lipschitzian stability of such systems in terms of their initial data. For simplicity we consider only Lipschitz continuous bases in assertion (ii).

Theorem 4.62 (Lipschitzian stability for GVIs with composite potentials). *Let* $(\bar{x}, \bar{y}) \in \operatorname{gph} S$ *for* S *defined in* (4.66), *where* $f\colon X \times Y \to X^* \times Y^*$ *with* $\bar{q} := -f(\bar{x}, \bar{y})$, *where* $g\colon X \times Y \to W$ *with* $\bar{w} := g(\bar{x}, \bar{y})$, *and where* $\varphi\colon W \to \overline{I\!R}$. *The following assertions hold:*

(i) *Suppose that* W *is Banach,* X *is Asplund while* $Y = I\!R^m$, *that* $g = g(y)$, *and that assumptions* (a) *and* (b) *of Theorem* 4.49 *are fulfilled with* $\bar{v}$ *defined therein. Then* S *is Lipschitz-like around* $(\bar{x}, \bar{y})$ *if and only if* $u = 0 \in I\!R^m$ *is the only vector satisfying*

$$0 \in \nabla_y f(\bar{x}, \bar{y})^* u + \nabla^2 \langle \bar{v}, g \rangle (\bar{y})^* u + \nabla g(\bar{y})^* \partial_N^2 \varphi(\bar{w}, \bar{v})\big(\nabla g(\bar{y}) u\big) .$$

In in addition X *is finite-dimensional, then one has*

$$\operatorname{lip} S(\bar{x}, \bar{y}) = \sup \Big\{ \|\nabla_x f(\bar{x}, \bar{y})^* u\| \;\; \text{with} \;\; -y^* \in \nabla_y f(\bar{x}, \bar{y})^* u$$
$$+ \nabla^2 \langle \bar{v}, g \rangle (\bar{y})^* u + \nabla g(\bar{y})^* \partial_N^2 \varphi(\bar{w}, \bar{v})\big(\nabla g(\bar{y}) u\big), \;\; \|y^*\| \le 1 \Big\} ,$$

where the maximum is attained when, in particular, W *is finite-dimensional.*

(ii) *Suppose that all three spaces* X, Y, W *are finite-dimensional, that* $g \in C^2$ *and* f *is Lipschitz continuous around* $(\bar{x}, \bar{y})$, *that* φ *is l.s.c. around* $\bar{w}$, *and that assumptions* (a) *and* (b) *of Theorem* 4.50 *are fulfilled with* $M(\bar{x}, \bar{y})$ *defined therein. Then the condition*

$$\begin{aligned} &\Big[(x^*, 0) \in \partial \langle u, f \rangle (\bar{x}, \bar{y}) + \bigcup_{\bar{v} \in M(\bar{x}, \bar{y})} \big[\nabla^2 \langle \bar{v}, g \rangle (\bar{x}, \bar{y})(u) \\ &+ \nabla g(\bar{x}, \bar{y})^* \partial^2 \varphi(\bar{w}, \bar{v})\big(\nabla g(\bar{x}, \bar{y}) u\big)\big]\Big] \Longrightarrow x^* = u = 0 \end{aligned} \tag{4.81}$$

is sufficient for the Lipschitz-like property of S, *and one has*

$$\operatorname{lip} S(\bar{x}, \bar{y}) \le \sup \Big\{ \|x^*\| \,\Big|\, \exists u \in X \times Y \;\; \text{with} \; (x^*, -y^*) \in \partial \langle u, f \rangle (\bar{x}, \bar{y})$$
$$+ \bigcup_{\bar{v} \in M(\bar{x}, \bar{y})} \big[\nabla^2 \langle \bar{v}, g \rangle (\bar{x}, \bar{y})(u) + \nabla g(\bar{x}, \bar{y})^* \partial^2 \varphi(\bar{w}, \bar{v})\big(\nabla g(\bar{x}, \bar{y}) u\big)\big], \;\; \|y^*\| \le 1 \Big\} .$$

Proof. To prove (i), we use the coderivative representation (4.68) in Theorem 4.49 for $D_N^* S(\bar{x}, \bar{y}) = D_M^* S(\bar{x}, \bar{y})$, which holds as equality due to the finite dimensionality of Y. Moreover, the graph of S is SNC at $(\bar{x}, \bar{y})$, since it is the inverse image of $\operatorname{gph} Q$ under a strictly differentiable mapping with the surjective derivative, where $Q = \partial(\varphi \circ g)\colon Y \rightrightarrows Y^*$ is automatically SNC; cf. the proof of Theorem 4.56(i). Thus the condition $D_M^* S(\bar{x}, \bar{y})(0) = \{0\}$ reduces to the one assumed in (i), which is therefore necessary and sufficient for the Lipschitz-like property of S by Theorem 4.10. This also implies the exact bound formula in (i), where the maximum is attained in finite dimensions due to Theorem 4.56(i).

To justify (ii), we employ the coderivative upper estimate (4.71) from Theorem 4.50, where the qualification condition in (c) is fulfilled due to (4.81). Moreover, the latter assumption ensures that $D^*S(\bar{x}, \bar{y})(0) = \{0\}$ by (4.71), and hence S is Lipschitz-like around $(\bar{x}, \bar{y})$ with the exact bound estimate in (ii) due to Theorem 4.10. △

Corollary 4.63 (Lipschitzian stability of GVIs with amenable potentials). *Let S be the GVIs solution map* (4.66) *whose potential $\psi = \varphi \circ g$ is strongly amenable at $(\bar{x}, \bar{y}) \in \operatorname{gph} S$ in finite dimensions. Assume that f is locally Lipschitzian around $(\bar{x}, \bar{y})$ and that conditions* (4.70) *and* (4.81) *are fulfilled. Then S is Lipschitz-like around $(\bar{x}, \bar{y})$ with the exact bound estimate from Theorem* 4.62(ii).

Proof. This follows from Theorem 4.62(ii) due to the definition and properties of strongly amenable functions discussed in Subsect. 3.2.5. △

The next corollary concerns variational systems (4.66) with *smooth potentials*, in which case

$$S(x) = \left\{ y \in Y \,\middle|\, 0 \in f(x, y) + \nabla(\varphi \circ g)(x, y) \right\}$$

describes solutions to perturbed *gradient equations.* Note that φ may be nonsmooth in the second-order, in particular, $\varphi \in \mathcal{C}^{1,1}$. In the latter case one gets the following efficient conditions ensuring Lipschitzian stability for gradient equations. For brevity we formulate this result only in finite dimensions unifying both assertions of Theorem 4.62 and omitting formulas for the exact Lipschitzian bound.

Corollary 4.64 (Lipschitzian stability for gradient equations). *Let S be the solution map* (4.66) *in finite dimensions with $\varphi \in \mathcal{C}^{1,1}$ around $\bar{w}$. Assume that $g \in \mathcal{C}^2$ and f is Lipschitz continuous around $(\bar{x}, \bar{y})$. Then condition* (4.81) *is sufficient for the Lipschitz-like property of S around $(\bar{x}, \bar{y})$. Moreover, it is also necessary for this property if $g = g(y)$ with the surjective derivative $\nabla g(\bar{y})$ and f is strictly differentiable at $(\bar{x}, \bar{y})$ with the surjective partial derivative $\nabla_x f(\bar{x}, \bar{y})$.*

Proof. One just needs to observe that all the assumptions in (a) and (b) of Theorem 4.50 automatically hold provided that $\varphi \in \mathcal{C}^{1,1}$ around $\bar{w}$. △

To illustrate the application of Corollary 4.64, we consider the parameterized gradient equation

$$0 = f(x, y) + \nabla\varphi(y) \quad \text{with} \quad \varphi(y) = \tfrac{1}{2}\operatorname{sign}(y), \qquad f(x, y) = ay + x\,,$$

where $x, y \in I\!R$ and $a > 1$. Here $\varphi \in \mathcal{C}^{1,1}$ with the nonsmooth derivative $\nabla\varphi(y) = |y|$. One easily has

$$N((0,0);\operatorname{gph}\nabla\varphi)=\big\{(v,u)\big|\,u\le|v|\ \text{ for }\ u\le 0\big\}\cap\big\{(v,u)\big|\,u=|v|\ \text{ for }\ u>0\big\}\,,$$

which implies that

$$\partial^2\varphi(0)(u)=[-u,u]\ \text{ if }\ u\ge 0\qquad\text{and}\qquad \partial^2\varphi(0(u))=\{u,-u\}\ \text{ if }\ u<0\,.$$

According to Corollary 4.64, the solution map to the gradient equation under consideration is Lipschitz-like around $(0,0)$ *if and only if* the inclusion

$$0\in\begin{cases}[a-1)u,(a+1)u] & \text{if } u\ge 0\,,\\[1ex] \{(a-1)u,(a+1)u\} & \text{if } u<0\end{cases}$$

is fulfilled only for $u=0$. This is definitely true when $a>1$. Note that the sufficient condition (4.77) in terms of Clarke normals doesn't hold in this example, since $N_C((0,0);\operatorname{gph}\nabla\varphi)=I\!R^2$.

The final results in this subsection concern Lipschitzian stability of solution maps (4.67) to GVIs with *composite fields.* For simplicity we consider only strictly Lipschitzian bases in assertion (ii).

Theorem 4.65 (Lipschitzian stability for GVIs with composite fields). *Let S be defined in* (4.67) *with $g\colon X\times Y\to W$, $\varphi\colon W\to\overline{I\!R}$, and $f\colon X\times Y\to W^*$. Given $(\bar x,\bar y)\in\operatorname{gph}S$, we denote $\bar w:=g(\bar x,\bar y)$ and $\bar q:=-f(\bar x,\bar y)$. The following assertions hold:*

(i) *Assume that X,Y are Asplund while W is Banach, that $g=g(y)$ is strictly differentiable at $\bar y$ with the surjective derivative $\nabla g(\bar y)$, and that f is strictly differentiable at $(\bar x,\bar y)$ with the surjective partial derivative $\nabla_x f(\bar x,\bar y)$. Then the condition*

$$\Big[0\in\nabla_y f(\bar x,\bar y)^*u+\nabla g(\bar y)^*\partial^2_N\varphi(\bar w,\bar q)(u)\Big]\Longrightarrow u=0\,.$$

is sufficient for the Lipschitz-like property of S around $(\bar x,\bar y)$ being also necessary for this property if S is strongly coderivatively normal at $(\bar x,\bar y)$ (in particular, when $\dim Y<\infty$). If in addition $\dim X<\infty$, then one has

$$\operatorname{lip}S(\bar x,\bar y)=\sup\Big\{\|\nabla_x f(\bar x,\bar y)^*u\|\ \Big|\ \exists z^*\in\partial^2_N\varphi(\bar w,\bar q)(u)\ \text{ with }$$
$$\|\nabla_y f(\bar x,\bar y)^*u+\nabla g(\bar y)^*z^*\|\le 1\Big\}$$

provided that $\partial\varphi$ is SNC at $(\bar w,\bar q)$.

(ii) *Assume that X,Y,W,W^* are Asplund, that g is continuous around $(\bar x,\bar y)$ while f is strictly Lipschitzian at this point, that the graph of $\partial\varphi$ is norm-closed around $(\bar w,\bar q)$, and that the qualification condition* (4.72) *is fulfilled. Then the implication*

$$\Big[(x^*,0)\in\partial\langle u,f\rangle(\bar{x},\bar{y})+D^*_N g(\bar{x},\bar{y})\circ\partial^2_N\varphi(\bar{w},\bar{q})(u)\Big]\Longrightarrow x^*=u=0$$

is sufficient for the Lipschitz-like property of S around $(\bar{x},\bar{y})$ provided that either g is PSNC at $(\bar{x},\bar{y})$ and $\partial\varphi$ is SNC at $(\bar{w},\bar{q})$, or g is SNC at $(\bar{x},\bar{y})$ and $\partial\varphi^{-1}$ is PSNC at $(\bar{q},\bar{w})$. If in addition $\dim X<\infty$, *then one has the exact bound estimate*

$$\operatorname{lip}S(\bar{x},\bar{y})\le\sup\Big\{\|x^*\|\;\Big|\;\exists u\in W^{**}\;\;\text{with}$$

$$(x^*,-y^*)\in\partial\langle u,f\rangle(\bar{x},\bar{y})+D^*_N g(\bar{x},\bar{y})\circ\partial^2_N\varphi(\bar{w},\bar{q})(u),\;\;\|y^*\|\le 1\Big\}.$$

Proof. To prove (i), we use the coderivative representation from Proposition 4.53 and then apply Theorem 4.10 observing that the SNC property of $\partial\varphi$ at $(\bar{w},\bar{q})$ yields the one for S at $(\bar{x},\bar{y})$ due to

$$\operatorname{gph}S=\big\{(x,u)\in X\times Y\big|\;g(x,y)\in\operatorname{gph}(\partial\varphi\circ g)\big\}$$

and Theorems 1.22, 3.98. This assertion can be also derived from Theorem 4.56(i) using the coderivative chain from Theorem 1.66 and the mentioned results of the SNC calculus.

To prove (ii), we apply the coderivative inclusion from Theorem 4.54(a) and the basic characterization of Theorem 4.10. Note that, when f is strictly Lipschitzian at $(\bar{x},\bar{y})$, both conditions (4.73) and $D^*_M S(\bar{x},\bar{y})(0)=\{0\}$ in (4.74) are satisfied if the implication in (ii) holds; cf. the proof of Corollary 4.60. It remains to observe that, as shown in the proof of Theorem 4.54, the composition $\partial\varphi\circ g$ is SNC at $(\bar{x},\bar{y},\bar{q})$ under the assumptions made. Hence S is SNC at $(\bar{x},\bar{y})$, which completes the proof of the theorem. △

Note that g is automatically PSNC at $(\bar{x},\bar{y})$ if it is Lipschitz continuous around this point, and that $\partial\varphi$ is SNC if W is finite-dimensional. In the latter case one has (4.72) if g is metrically regular around $(\bar{x},\bar{y})$. Let us present an efficient corollary of Theorem 4.65(ii) when both f and g are strictly differentiable at $(\bar{x},\bar{y})$. For simplicity we formulate it in the case of $W=I\!R^m$.

Corollary 4.66 (GVIs with composite fields under smoothness assumptions). *Let $(\bar{x},\bar{y})\in\operatorname{gph}S$ with S defined in (4.67), where X and Y are Asplund, and where $g\colon X\times Y\to I\!R^m$ and $f\colon X\times Y\to I\!R^m$ are strictly differentiable at $(\bar{x},\bar{y})$. Assume that the graph of $\partial\varphi$ is closed around $(\bar{w},\bar{q})$ (which holds, in particular, for continuous and for amenable functions), that*

$$\begin{aligned}&\partial^2\varphi(\bar{w},\bar{q})(0)\cap\ker\nabla g(\bar{x},\bar{y})^*=\{0\},\\&\partial^2\varphi(\bar{w},\bar{q})(0)\subset\ker\nabla_x g(\bar{x},\bar{y})^*,\end{aligned}\tag{4.82}$$

and that one has

$$\Big[0 \in \nabla_y f(\bar{x},\bar{y})^* u + \nabla_y g(\bar{x},\bar{y})^* \partial^2 \varphi(\bar{w},\bar{q})(u)\Big] \Longrightarrow u = 0 . \tag{4.83}$$

Then S is Lipschitz-like around $(\bar{x},\bar{y})$. If in addition $\dim X < \infty$*, then*

$$\operatorname{lip} S(\bar{x},\bar{y}) \le \sup\Big\{\|x^*\| \;\Big|\; \exists u \in I\!R^m,\ y^* \in \nabla_y g(\bar{x},\bar{y})^* \partial^2 \varphi(\bar{w},\bar{q})(u) \ \textit{with}$$

$$x^* - \nabla_x f(\bar{x},\bar{y})^* u \in \nabla_x g(\bar{x},\bar{y})^* \partial^2 \varphi(\bar{w},\bar{q})(u),\ \ \|\nabla_y f(\bar{x},\bar{y})^* u + y^*\| \le 1\Big\} .$$

Proof. We use Theorem 4.65(ii) with $W = I\!R^m$ and observe that (4.72) reduces under the strict differentiability assumptions to the first condition in (4.82), while (4.83) and the second condition in (4.82) clearly imply (actually are equivalent to) the qualification condition of Theorem 4.65(ii). △

We conclude this subsection with an example of applications of the results obtained to a practical problem of *continuum mechanics*. We refer the reader to the paper of Mordukhovich and Outrata [939] for more details, illustrations, and other applications.

Example 4.67 (Lipschitzian stability for a contact problem with nonmonotone friction). The underlying mechanical problem is taken from the book by Haslinger, Miettinen and Panagiotopoulos [551] and can be described as follows.

There is an elastic body Ω supported from below by a rigid obstacle and exposed to external forces that represent our perturbation vector x. Vectors y_t, y_n represent, respectively, tangential and normal displacements of the discretization nodes lying on the contact boundary Γ_c. In many situations it is possible to replace the "nonpenetrability condition" $y_n \ge 0$ with the equality $y_n = 0$. Then we put $y := y_t \in I\!R^m$ and describe the equilibrium in this mechanical problem by the following hemivariational inequality with a composite filed of type (4.67):

$$0 \in Ay + p(x) + \partial\varphi(By) , \tag{4.84}$$

where m is the number of nodes on Γ_c, n is the dimension of external forces $x \in I\!R^n$, A is an $m \times m$ positively definite "stiffness" matrix, $p : I\!R^n \to I\!R^m$ is a continuously differentiable mapping related to external forces, and B is an $m \times m$ nonsingular matrix defined by a quadrature formula that is used for the boundary integral along Γ_c. The function φ in (4.84) is given in the form

$$\varphi(z) = \sum_{i=1}^m \varphi_i(z_i) \quad \text{with} \ \ z = (z_1, \ldots, z_m) \in I\!R^m , \tag{4.85}$$

where $\varphi_i : I\!R \to I\!R$ represents the *nonmonotone friction law* described by

$$\varphi_i(z_i) := \begin{cases} (-k_1 + k_2 z_0)\, z_i + \frac{k_2}{2}(z_0)^2 & \text{if } z_i < -z_0\,, \\ -k_1 z_i - \frac{k_2}{2}(z_i)^2 & \text{if } z_i \in [-z_0, 0)\,, \\ k_1 z_i - \frac{k_2}{2}(z_i)^2 & \text{if } z_i \in [0, z_0)\,, \\ (k_1 - k_2 z_0)\, z_i + \frac{k_2}{2}(z_0)^2 & \text{if } z_i \ge z_0 \end{cases} \tag{4.86}$$

with the given parameters $z_0 > 0$, $k_1 > 0$, and $k_2 > 0$. Functions $\varphi\colon I\!R^m \to \overline{I\!R}$ of type (4.85), (4.86) belong to the class of *separable piecewise* $\mathcal{C}^2$ functions for which the second-order subdifferential $\partial^2\varphi$ is efficiently computed in Mordukhovich and Outrata [939]. Here we present calculations in the case given (4.84)–(4.86) to be able to check efficiently the conditions of Theorem 4.65 on Lipschitzian stability of the solution map $S\colon I\!R^n \to I\!R^m$ to the hemivariational inequality (4.84). Given a point $(\bar{x}, \bar{y}) \in \operatorname{gph} S$, we associate with it the following index sets:

$$I_1(\bar{x},\bar{y}) := \big\{i \in \{1,\ldots,m\}\big|\ (B\bar{y})_i < -z_0\big\}\,,$$

$$I_2(\bar{x},\bar{y}) := \big\{i \in \{1,\ldots,m\}\big|\ (B\bar{y})_i = -z_0\big\}\,,$$

$$I_3(\bar{x},\bar{y}) := \big\{i \in \{1,\ldots,m\}\big|\ (B\bar{y})_i \in (-z_0, 0)\big\}\,,$$

$$I_4(\bar{x},\bar{y}) := \big\{i \in \{1,\ldots,m\}\big|\ (B\bar{y})_i = 0, (-A\bar{y} - p(\bar{x}))_i = -k_1\big\}\,,$$

$$I_5(\bar{x},\bar{y}) := \big\{i \in \{1,\ldots,m\}\big|\ (B\bar{y})_i = 0, (-A\bar{y} - p(\bar{x}))_i \in (-k_1, k_1)\big\}\,,$$

$$I_6(\bar{x},\bar{y}) := \big\{i \in \{1,\ldots,m\}\big|\ (B\bar{y})_i = 0, (-A\bar{y} - p(\bar{x}))_i = k_1\big\}\,,$$

$$I_7(\bar{x},\bar{y}) := \big\{i \in \{1,\ldots,m\}\big|\ (B\bar{y})_i \in (0, z_0)\big\}\,,$$

$$I_8(\bar{x},\bar{y}) := \big\{i \in \{1,\ldots,m\}\big|\ (B\bar{y})_i = z_0\big\}\,,$$

$$I_9(\bar{x},\bar{y}) := \big\{i \in \{1,\ldots,m\}\big|\ (B\bar{y})_i > z_0\big\}\,.$$

These sets completely describe the position of the point $\big(B\bar{y}, -A\bar{y} - p(\bar{x})\big)$ on the graph of $\partial\varphi$; their union is exactly the whole index set $\{1,\ldots,m\}$. Now we compute the basic normal cones to the graph of $\partial\varphi_i$ at the points $\big((B\bar{y})_i, (-A\bar{y} - p(\bar{x}))_i\big)$ associated with the above index sets. To simplify the notation, we give (as a subscript) only the number of the index set to which the corresponding component of $(B\bar{y}, -A\bar{y} - p(\bar{x}))$ belongs:

$$N_1 = N_9 = \{0\} \times I\!R \,,$$

$$N_2 = N_1 \cup \Big\{(w,u) \in I\!R^2 \Big|\; u = \tfrac{1}{k_2}\, w\Big\} \cup \big\{(w,u) \in I\!R^2 \big|\; 0 \le w \le k_2 u\big\} \;,$$

$$N_3 = N_7 = \Big\{(w,u) \in I\!R^2 \Big|\; u = \tfrac{1}{k_2}\, w\Big\} \;,$$

$$N_4 = N_3 \cup \big\{(w,u) \in I\!R^2 \big|\; u = 0\big\} \cup \big\{(w,u) \in I\!R^2 \big|\; w - k_2 u \ge 0,\; u \le 0\big\} \;,$$

$$N_5 = \big\{(w,u) \in I\!R^2 \big|\; u = 0\big\} \;,$$

$$N_6 = N_3 \cup N_5 \cup \big\{(w,u) \in I\!R^2 \big|\; w - k_2 u \le 0,\; u \ge 0\big\} \;,$$

$$N_8 = N_1 \cup N_3 \cup \big\{(w,u) \in I\!R^2 \big|\; k_2 u \le w \le 0\big\} \;.$$

To formulate verifiable conditions for Lipschitzian stability of the solution set to the hemivariational inequality (4.84) with the potential φ defined in (4.85) and (4.86), we consider the following generalized equation

$$0 \in A^* u + \Xi(\bar{x}, \bar{y}, u) \tag{4.87}$$

with the field Ξ given by

$$\Xi(\bar{x}, \bar{y}, u) = \prod_{i=1}^{m} \Xi_i(\bar{x}, \bar{y}, u_i) \;\text{ for }\; \Xi_i(\bar{x}, \bar{y}, u_i) := \big\{w_i \in I\!R \big|\; (w_i, -u_i) \in N_j\big\} \;,$$

where the cones N_j are computed above and where j is a uniquely determined index from $\{1, \ldots, m\}$ for which $i \in I_j(\bar{x}, \bar{y})$. According to Theorem 4.65 and its concretization in Corollary 4.66 the *solution map to* (4.84) *is Lipschitz-like around* $(\bar{x}, \bar{y})$ *if the adjoint generalized equation* (4.87) *has only the trivial solution*. Moreover, the latter condition is also *necessary* for this stability property if the Jacobian matrix $\nabla p(\bar{x})$ has *full rank*.

Finally, we consider a two-dimensional hemivariational inequality of type (4.84) and check the above conditions for Lipschitzian stability of its solution map. Let $n = m = 2$, $p(x) = x$, $A = \begin{bmatrix} 5 & 4 \\ 4 & 5 \end{bmatrix}$, $B = \begin{bmatrix} 1 & 0 \\ 0 & 1 \end{bmatrix}$ in (4.84), and let φ is given by (4.85) and (4.86) with $k_1 = 1$, $k_2 = \frac{1}{2}$, $z_0 = 1$. Taking the reference point $(\bar{x}_1, \bar{x}_2, \bar{y}_1, \bar{y}_2) = \big(3, \frac{13}{4}, 0, -\frac{1}{2}\big)$, we get from the above formulas that $I_3(\bar{x}, \bar{y}) = \{2\}$, $I_4(\bar{x}, \bar{y}) = \{1\}$, and all the other index sets are empty. The adjoint equation in (4.83) attains the form

$$\begin{aligned} &0 \in 5u_1 + 4u_2 + \big\{w_1 \in I\!R \big|\; (w_1, -u_1) \in N_4\big\}, \\ &0 \in 4u_1 + 5u_2 + \big\{w_2 \in I\!R \big|\; u_2 = -2w_2\big\} \end{aligned} \tag{4.88}$$

with the cone N_4 computed above. In this case Theorem 4.65(i) ensures that the solution map to (4.84) with the given data is Lipschitz-like around $(\bar{x}, \bar{y})$ *if and only* the adjoint generalized equation (4.88) admits only the trivial solution. Let us check that it holds true. Indeed, from the second relation in (4.88) we have $u_2 = -\frac{8}{9}u_1$. By inserting this into the first relation in (4.88), one gets the inclusion

$$0 \in \frac{13}{9}u_1 + \big\{w_1 \in I\!R \big|\ (w_1, -u_1) \in N_4\big\}$$

that is fulfilled only when $u_1 = 0$ due to the above expression for N_4. Hence $u_2 = 0$ as well, which justifies the Lipschitz-like property of the solution map in the example under consideration.

4.4.3 Lipschitzian Stability under Canonical Perturbations

In this subsection we consider parametric variational systems obtained via the so-called *canonical perturbations* of generalized equations (4.47). Such systems are given in the form

$$\Sigma(x,q) := \big\{y \in Y \big|\ q \in f(x,y) + Q(x,y)\big\}\ , \tag{4.89}$$

where $f\colon X \times Y \to Z$ and $Q\colon X \times Y \rightrightarrows Z$ are mappings between Banach spaces. In contrast to the solution map S from (4.50), there is a pair of parameters $p := (x,q)$ in (4.89), where the *canonical parameters* q correspond to the perturbation of the left-hand side in (4.47). One clearly has $S(x) = \Sigma(x,0)$ for the solution map (4.50). On the other hand, (4.89) can be viewed as a special case of (4.50) with respect to the parameter pair $p = (x,q)$. Therefore the results of Subsect. 4.4.2 readily induce the corresponding conditions for Lipschitzian stability of variational systems under canonical perturbations.

In this subsection we explore *another approach* to the study of robust Lipschitzian stability of canonically perturbed systems (4.89) that allows us to get more subtle results for such systems by taking into account a specific parametric structure of (4.89). Unlike the one developed in the previous subsection, this approach in *not* based on the *direct* application of characterizations of Theorem 4.10 via coderivative formulas from Subsect. 4.4.1, but involves a preliminary *first-order approximation* of the original system. In such a way, establishing relationships between Lipschitzian stability of the original and approximating systems and applying the results of Subsect. 4.4.2 to the latter one, we derive *characterizations* as well as *sufficient conditions* for Lipschitzian stability of canonically perturbed variational systems. The sufficient conditions obtained for Σ clearly ensure the Lipschitz-like property of the solution map S from (4.50) being generally *independent* of those obtained in Subsect. 4.4.2 even in finite dimensions.

Let us start with an appropriate concept of first-order approximation, which is a natural generalization of the classical linearization idea.

Definition 4.68 (strong approximation). *Let $f\colon X\times Y\to Z$ be a mapping between Banach spaces. The mapping $h\colon Y\to Z$* STRONGLY APPROXIMATES *f in y at $(\bar{x},\bar{y})$ if $h(\bar{y})=f(\bar{x},\bar{y})$ and for each $\varepsilon>0$ there are neighborhoods U of $\bar{x}$ and V of $\bar{y}$ such that*

$$\big\|[f(x,y_1)-h(y_1)]-[f(x,y_2)-h(y_2)]\big\|\le\varepsilon\big\|y_1-y_2\big\|$$

whenever $x\in U$ and $y_1,y_2\in V$.

This definition actually means that, although both f and h may not be differentiable in any sense, its difference $g(x,y):=f(x,y)-h(y)$ is *strictly differentiable in* y at $(\bar{x},\bar{y})$ in the sense of

$$\lim_{\substack{y,v\to\bar{y}\\ x\to\bar{x}}}\Big[\frac{g(x,y)-g(x,v)-\nabla_y g(\bar{x},\bar{y})(y-v)}{\|y-v\|}\Big]=0 \tag{4.90}$$

with $\nabla_y g(\bar{x},\bar{y})=0$. Observe that (4.90) holds, in particular, when g is (Fréchet) differentiable in y around $(\bar{x},\bar{y})$ and $\nabla_y g$ is continuous with respect to x and y at this point.

Note that any mapping f in the *separated form*

$$f(x,y)=f_1(x)+f_2(y)$$

admits an obvious strong approximation in y given by f_2. If f itself is strictly differentiable in y at $(\bar{x},\bar{y})$ in the sense of (4.90), its efficient strong approximation can be obtained by the *linearization*

$$h(y):=f(\bar{x},\bar{y})+\nabla_y f(\bar{x},\bar{y})(y-\bar{y})\,. \tag{4.91}$$

Also one can check that the *composite mapping* $p(x,y)=f(x,s(y))$ admits a strong approximation in y at $(\bar{x},\bar{y})$ if $f(x,z)$ is strictly differentiable in z at $(\bar{x},\bar{z})$ with $\bar{z}:=s(\bar{y})$ while s is Lipschitz continuous around $\bar{y}$.

Let $h\colon Y\to Z$ strongly approximate f in y at the point $(\bar{x},\bar{y})$ in the sense of Definition 4.68. Along with (4.89) we consider the *approximating system*

$$\Xi(x,q):=\big\{y\in Y\big|\;q\in h(y)+Q(x,y)\big\}\,. \tag{4.92}$$

The following lemma shows that the Lipschitz-like property is preserved under such a first-order approximation. Recall that $f\colon X\times Y\to Z$ is *locally Lipschitzian in x uniformly in y* around $(\bar{x},\bar{y})$ if there are neighborhoods U of $\bar{x}$ and V of $\bar{y}$ and a number $\ell\ge 0$ such that

$$\|f(x_1,y)-f(x_2,y)\|\le\ell\|x_1-x_2\|$$

whenever $x_1,x_2\in U$ and $y\in V$.

Lemma 4.69 (Lipschitz-like property under strong approximation). *Let X,Y,Z be Banach, let Σ and Ξ be given in* (4.89) *and* (4.92), *and let*

$\bar{y} \in \Sigma(\bar{p})$ with $\bar{p} := (\bar{x}, \bar{q})$. Assume that both Σ and Ξ are closed-valued around $\bar{p}$, that f is locally Lipschitzian in x uniformly in y around $(\bar{x}, \bar{y})$, and that h strongly approximates f in y at this point. Then the following are equivalent:

(a) *Ξ is Lipschitz-like around $(\bar{p}, \bar{y})$.*
(b) *Σ is Lipschitz-like around $(\bar{p}, \bar{y})$.*

Proof. We show that (a)$\Rightarrow$(b) employing on the Lyusternik-Graves iterative procedure used in the proofs of Theorems 1.57 and 4.25. By the Lipschitz-like property of Ξ around $(\bar{p}, \bar{y})$ there are positive numbers μ, ν, η for which

$$\Xi(\tilde{w}) \cap B_\nu(\bar{y}) \subset \Xi(\hat{w}) + \mu\|\tilde{w} - \hat{w}\| I\!B_Y \quad \text{whenever} \quad \tilde{w}, \hat{w} \in B_\eta(\bar{p}) .$$

Let ε be taken from Definition 4.68, and let ℓ be the Lipschitzian modulus of f in y given above. Choose $\alpha > 0$ and $\beta > 0$ such that

$$\alpha < \min\{\nu, \eta/\varepsilon\}, \quad \beta \le \min\left\{\frac{\alpha(1-\varepsilon\mu)}{4\mu(1+\ell)}, \frac{\eta - \varepsilon\alpha}{1+\ell}\right\},$$

and that for all $\tilde{y}, \hat{y} \in B_\alpha(\bar{y})$ and $x \in B_\beta(\bar{x})$ one has

$$\|f(x, \tilde{y}) - h(\tilde{y}) - f(x, \hat{y}) + h(\hat{y})\| \le \varepsilon\|\tilde{y} - \hat{y}\| .$$

Fix $\tilde{p}, \hat{p} \in B_\beta(\bar{p})$ with $\tilde{p} = (\tilde{x}, \tilde{q})$ and $\hat{p} = (\hat{x}, \hat{q})$, and let $\tilde{y} \in \Sigma(\tilde{p}) \cap B_{\alpha/2}(\bar{y})$. Observe that $\tilde{y} \in \Xi(\tilde{w}) \cap B_{\alpha/2}(\bar{y})$ for $\tilde{w} := \big(\tilde{x}, \tilde{q} + h(\tilde{y}) - f(\tilde{x}, \tilde{y})\big)$. It follows from the above constructions and the choice of β that

$$\|\tilde{w} - \bar{p}\| \le \|\tilde{p} - \bar{p}\| + \varepsilon\|\tilde{y} - \bar{y}\| + \ell\|\tilde{x} - \bar{x}\| \le \beta(1+\ell) + \varepsilon\alpha/2 \le \eta ,$$

i.e., $\tilde{w} \in B_\eta(\bar{p})$. Similarly we have $\hat{w} \in B_\eta(\bar{p})$ for $\hat{w} := \big(\hat{x}, \hat{q} + h(\hat{y}) - f(\hat{x}, \hat{y})\big)$. Now denote $y_1 := \tilde{y}$ and by the Lipschitz-like property of Ξ find y_2 such that $\hat{q} + h(\tilde{y}) - f(\hat{x}, y_1) \in h(y_2) + Q(\hat{x}, y_2)$ and

$$\|y_2 - y_1\| \le \mu\|\tilde{w} - \hat{w}\| \le \mu(\ell + 1)\|\tilde{p} - \hat{p}\| .$$

Proceeding by induction, suppose that there are $y_2, \ldots, y_{n-1}$ with the following properties:

$$\hat{q} + h(y_{i-1}) - f(\hat{x}, y_{i-1}) \in h(y_i) + Q(\hat{x}, y_i) ,$$

$$\|y_i - y_{i-1}\| \le \mu(\ell + 1)\|\tilde{p} - \hat{p}\|(\mu\varepsilon)^{i-2}$$

for $i = 2, \ldots, n-1$. By the above choice of β one has

$$\begin{aligned}\|y_i - \bar{y}\| &\le \|y_1 - \bar{y}\| + \sum_{j=2}^{i}\|y_j - y_{j-1}\| \le \alpha/2 + \mu(\ell+1)\|\tilde{p} - \hat{p}\|\sum_{j=2}^{i}(\mu\varepsilon)^{j-2} \\ &\le \alpha/2 + \frac{\mu(\ell+1)}{1-\mu\varepsilon}\|\tilde{p} - \hat{p}\| \le \alpha/2 + \frac{2\mu\beta(\ell+1)}{1-\mu\varepsilon} \le \alpha .\end{aligned}$$

Similarly to the first step of induction we find, using the Lipschitz-like property of Ξ, a point $y_n \in B_\alpha(\bar{y})$ satisfying

$$\hat{q} + h(y_{n-1}) - f(\hat{x}, y_{n-1}) \in h(y_n) + Q(\hat{x}, y_n) ,$$

$$\|y_n - y_{n-1}\| \le \mu\big(\|\tilde{q} - \hat{q}\| + \ell\|\tilde{x} - \hat{x}\|\big)(\mu\varepsilon)^{n-2} .$$

Thus we get $\|y_n - y_{n-1}\| \to 0$ as $n \to \infty$ and, moreover, $\{y_n\}$ is a Cauchy sequence converging to some $\hat{y}^0 \in B_\alpha(\bar{y})$. Let us show that $\hat{y}^0 \in \Sigma(\hat{p})$ and obtain the estimate of $\|\tilde{y} - \hat{y}^0\|$ with $\hat{p}$ chosen above, which allow us to justify the Lipschitz-like property of Σ around $(\bar{p}, \bar{y})$.

Putting $\hat{w}^0 := \big(\hat{x}, \hat{q} + h(\hat{y}^0) - f(\hat{x}, \hat{y}^0)\big)$, one easily has

$$\|\hat{w}^0 - \bar{p}\| \le (1 + \ell)\beta + \varepsilon\alpha \le \eta .$$

Furthermore, by the construction of y_n and the Lipschitz-like property of Ξ around $(\bar{p}, \bar{y})$ we get

$$\operatorname{dist}(y_n; \Xi(\hat{w}^0)) \le \mu\varepsilon\|y_{n-1} - \hat{y}^0\| \to 0 \quad \text{as} \quad n \to \infty ,$$

which gives $\hat{y}^0 \in \Xi(\hat{w}^0)$ due to the closed-valuedness of Ξ. This implies that $\hat{y}^0 \in \Sigma(\hat{p})$. Moreover,

$$\|y_n - \tilde{y}\| \le \sum_{i=2}^{n} \|y_i - y_{i-1}\| \le \mu(\ell + 1)\|\tilde{p} - \hat{p}\| \sum_{i=2}^{n} (\mu\varepsilon)^{i-2} \le \frac{\mu(\ell + 1)}{1 - \mu\varepsilon}\|\tilde{p} - \hat{p}\|,$$

and hence one has the desired estimate

$$\|\hat{y}^0 - \tilde{y}\| \le \frac{\mu(\ell + 1)}{1 - \mu\varepsilon}\|\tilde{p} - \hat{p}\|$$

by passing to the limit as $n \to \infty$. This ends the proof of (a)⇒(b).

To prove the opposite implication (b)⇒(a), we denote $G(x, y) := f(x, y) + Q(x, y)$ and observe that $\Sigma(x, q) = \{y \in Y |\; q \in G(x, y)\}$ and

$$\Xi(x, q) = \big\{y \in Y \big|\; q \in h(y) - f(x, y) + G(x, y)\big\} .$$

Since $g(y) := 0$ strongly approximates $h - f$ in y, we derive the Lipschitz-like property of Ξ from the one for Σ due to (a)⇒(b) proved above. △

The established relationship between the Lipschitz-like property of Σ and Ξ allows us to derive efficient coderivative conditions for Lipschitzian stability of the solution map (4.89) from those for the (apparently more simple) approximating system (4.92) using the results of Subsect. 4.4.2. Let us first present a counterpart of Theorem 4.56 in the case of canonical perturbations with refined *necessary and sufficient* conditions obtained in this way.

Theorem 4.70 (characterizations of Lipschitzian stability for canonically perturbed systems). *Let $\bar{y} \in \Sigma(\bar{x}, \bar{q})$ for $\Sigma: X \times Z \rightrightarrows Y$ given in (4.89), where the spaces X, Y, Z are Asplund. Suppose that $f: X \times Y \to Z$ is strictly differentiable in y at $(\bar{x}, \bar{y})$ and locally Lipschitzian in x uniformly in y around this point, and that $Q: X \times Y \rightrightarrows Z$ is closed-graph and SNC at $(\bar{x}, \bar{y}, \bar{s})$ with $\bar{s} := \bar{q} - f(\bar{x}, \bar{y})$. The following hold:*

(i) *Assume that $Q = Q(y)$. Then Σ is Lipschitz-like around $(\bar{x}, \bar{q}, \bar{y})$ if the partial adjoint generalized equation (4.56) with $\bar{z} = \bar{s}$ has only the trivial solution. This condition is also necessary for the Lipschitz-like property of Σ when either* $\dim Y < \infty$ *or Q is N-regular at $(\bar{y}, \bar{s})$.*

(ii) *Assume that $Q = Q(x, y)$ is N-regular at $(\bar{x}, \bar{y}, \bar{s})$. Then the condition*

$$\left(x^*, -\nabla_y f(\bar{x}, \bar{y})^* z^*\right) \in D^* Q(\bar{x}, \bar{y}, \bar{s})(z^*) \Longrightarrow x^* = z^* = 0 \tag{4.93}$$

is necessary and sufficient for the Lipschitz-like property of Σ around $(\bar{x}, \bar{q}, \bar{y})$.

Proof. As mentioned above, if f is strictly differentiable in y at $(\bar{x}, \bar{y})$, then its linearization $h(y)$ defined in (4.91) strongly approximates f in y at $(\bar{x}, \bar{y})$. Note that $\nabla h(\bar{y}) = \nabla_y f(\bar{x}, \bar{y})$. We conclude from Lemma 4.69 that the Lipschitz-like property of Σ around $(\bar{x}, \bar{q})$ is equivalent to this property of Ξ in (4.92) with h defined by (4.91). Let us apply Theorem 4.56 to $\Xi: P \rightrightarrows Y$ with $p = (x, q) \in P := X \times Z$ written in the standard form

$$\Xi(p) = \left\{y \in Y \middle|\, 0 \in \widetilde{h}(p, y) + \widetilde{Q}(p, y)\right\}, \tag{4.94}$$

where $\widetilde{h}(p, y) := h(y) - q$ and $\widetilde{Q}(p, y) := Q(x, y)$. Clearly the strict derivative of $\widetilde{h}$ at $(\bar{p}, \bar{y})$ is surjective and

$$\nabla \widetilde{h}(\bar{p}, \bar{y})^* z^* = (0, -z^*, \nabla_y f(\bar{x}, \bar{y})^* z^*) \quad \text{for all} \quad z^* \in Z^* .$$

If $Q = Q(y)$, we apply Theorem 4.56(i) to (4.94) and conclude that the triviality of solutions to the partial adjoint generalize equation (4.56) is necessary and sufficient for the Lipschitz-like property of Ξ (and hence of Σ) around $(\bar{p}, \bar{y})$ provided that Y is finite-dimensional. The remaining part of (i) under the regularity assumption immediately follows from assertion (ii) when Q doesn't depend on x.

Let us show that (ii) holds in the general case of $Q = Q(x, y)$ by applying Theorem 4.56(ii) to the solution map (4.94). Indeed, one can easily check that the adjoint generalized equation (4.51) to (4.94) admits only the trivial solution in the case under consideration. Furthermore, criterion (4.75) applied to the above $\widetilde{h}$ and $\widetilde{Q}$ clearly reduces to (4.93). △

Note that Theorem 4.70 can be derived directly from Theorem 4.56 in the case of canonical parameters provided that f is strictly differentiable at $(\bar{x}, \bar{y})$ with respect to *both* variables x and y, while the preliminary strong approximation allows us to justify this result when f is strictly differentiable *only in* y. On the other hand, Theorem 4.56 gives criteria of Lipschitzian stability not only for canonical perturbations. Similarly to Corollary 4.57 one can get efficient specifications of Theorem 4.70 in the case of convex-graph mappings Q in (4.89).

Next we obtain sufficient conditions for Lipschitzian stability of canonically perturbed variational systems (4.89) with nonsmooth and irregular data. In the rest of this subsection D^*F stands for the *normal coderivative* of $F = F(x,y)$, while D_y^*F is its *partial* normal coderivative with respect to y.

Theorem 4.71 (Lipschitzian stability of irregular systems under canonical perturbations). *Let* $\bar{y} \in \Sigma(\bar{x},\bar{q})$ *for* Σ *given in* (4.89) *with* $\bar{s} = \bar{q} - f(\bar{x},\bar{y})$. *Assume that* X, Y, Z *are Asplund, that* f *admits a strong approximation in* y *at* $(\bar{x},\bar{y})$, *and that the following hold:*

(a) f *is continuous in* (x,y) *and locally Lipschitzian in* x *uniformly in* y *around* $(\bar{x},\bar{y})$. *Moreover,* $f(\bar{x},\cdot)$ *is PSNC at* $\bar{y}$, *which is automatic if* $f(\bar{x},\cdot)$ *is Lipschitz continuous around* $\bar{y}$.

(b) Q *is closed-graph around* $(\bar{x},\bar{y},\bar{s})$ *and SNC at this point.*

Then Σ *is Lipschitz-like around* $(\bar{x},\bar{q},\bar{y})$ *provided the qualification condition*

$$\Big[y^* \in D_y^* f(\bar{x},\bar{y})(z^*), \quad (x^*,-y^*) \in D^*Q(\bar{x},\bar{y},\bar{s})(z^*)\Big] \Longrightarrow x^* = y^* = z^* = 0\,,$$

which is equivalent to

$$\Big[y^* \in \partial_y\langle z^*, f\rangle(\bar{x},\bar{y}), \quad (x^*,-y^*) \in D^*Q(\bar{x},\bar{y},\bar{s})(z^*)\Big] \Longrightarrow x^* = z^* = 0$$

if $f(\bar{x},\cdot)$ *is strictly Lipschitzian at* $\bar{y}$.

Proof. By Lemma 4.69 it is equivalent to consider the Lipschitz-like property of the solution map Ξ defined in (4.94), where $h\colon Y \to Z$ strongly approximates f in y at $(\bar{x},\bar{y})$. Applying Theorem 4.59 to (4.94), we need to check that the assumptions made imply (actually are equivalent to) the assumptions of Theorem 4.59 for the data of (4.94). Since h strongly approximates f, the mapping $g(y) := f(\bar{x},y) - h(y)$ is strictly differentiable at $\bar{y}$ with $\nabla g(\bar{y}) = 0$. This implies, by Theorems 1.62(ii) and 1.70, that

$$D^*h(\bar{y})(z^*) = D_y^* f(\bar{x},\bar{y})(z^*) \quad \text{for all} \quad z^* \in Z^*$$

and that h is PSNC at $\bar{y}$ if and only if $f(\bar{x},\cdot)$ is PSNC at this point. Furthermore, it follows from the structure of $\widetilde{h}$ and $\widetilde{Q}$ in (4.94) that the qualification conditions of Theorem 4.59 are simultaneously fulfilled for these mappings if and only if one has the general qualification condition formulated in the theorem. When $f(\bar{x},\cdot)$ is strictly Lipschitzian at $\bar{y}$, the latter is equivalent to the second qualification condition of the theorem due to the scalarization formula of Theorem 3.28. This justifies the Lipschitz-like property of Σ around $(\bar{x},\bar{q},\bar{y})$ and completes the proof of theorem. △

Let us present some corollaries of Theorem 4.71. The first one concerns the case of parameter-independent fields in (4.89).

Corollary 4.72 (canonical perturbations with parameter-independent fields). *Let $Q = Q(y)$ under the assumptions of Theorem 4.71. Then Σ is Lipschitz-like around $(\bar{x}, \bar{q}, \bar{y})$ provided that*

$$\Big[0 \in D_y^* f(\bar{x}, \bar{y})(z^*) + D^* Q(\bar{y}, \bar{s})(z^*)\Big] \Longrightarrow z^* = 0$$

and that one has

$$D_y^* f(\bar{x}, \bar{y})(0) \cap \big(- D^* Q(\bar{y}, \bar{s})(0)\big) = \{0\} \, .$$

The latter condition is automatic when either $f(\bar{x}, \cdot)$ is strictly Lipschitzian at $\bar{y}$, or Q is Lipschitz-like around $(\bar{y}, \bar{s})$ and strongly coderivatively normal at this point.

Proof. It is easy to see that for $Q = Q(y)$ the qualification condition of Theorem 4.71 is equivalent to the simultaneous fulfillment of both conditions of the corollary. The last statement follows from the coderivative scalarization and from the necessity of $D_M^* Q(\bar{y}, \bar{s})(0) = \{0\}$ for Lipschitz-like mappings due to Theorem 1.44. △

The next corollary gives sufficient conditions for Lipschitzian stability of solutions maps to canonically perturbed generalized equations with smooth bases. They are in the same form as in Theorem 4.70(ii) without imposing the regularity assumption on Q.

Corollary 4.73 (canonical perturbations of generalized equations with smooth bases). *In addition to the common assumptions of Theorem* 4.70 *suppose that the qualification condition* (4.93) *holds. Then Σ is Lipschitz-like around $(\bar{x}, \bar{q}, \bar{y})$.*

Proof. Follows from Theorem 4.71 and the fact that the base mapping f smooth in y always admits a strong approximation of form (4.93). △

Observe that for $Q = Q(y)$ condition (4.93) reduces to the triviality of solutions to the partial adjoint generalized equation (4.56) with $\bar{z} = \bar{s}$, the sufficiency of which for the Lipschitz-like property of Σ has been established in Theorem 4.70(i). Note also that, since $S(x) = \Sigma(x, 0)$, Corollary 4.73 *unreservedly* improves the sufficient conditions for the Lipschitz-like property of the standard solution map S from (4.50) in the case of smooth mappings f assuming the strict differentiability of f only in y but not in (x, y). In general the sufficient conditions for Lipschitzian stability of S obtained in Theorems 4.59 and 4.71 are *independent*. Indeed, one can check involving Corollary 3.44 that the second qualification condition in Theorem 4.71 always implies (4.79). However, Theorem 4.59 and its corollaries don't require the strong approximation of f as in Theorem 4.71. Furthermore, Theorem 4.71 requires the Lipschitz continuity of f in x, which is not assumed in Theorem 4.59.

Let us give applications of the results obtained to Lipschitzian stability of solution maps to canonically perturbed *generalized variational inequalities* (GVIs) with *composite potentials*:

$$\Sigma(x,q) := \Big\{ y \in Y \Big|\, q \in f(x,y) + \partial(\varphi \circ g)(x,y) \Big\}\,, \tag{4.95}$$

where $g\colon X \times Y \to W$, $\varphi\colon W \to \overline{I\!R}$, and $f\colon X \times Y \to X^* \times Y^*$. Mappings (4.95) are a special case of those in (4.89) with *subdifferential* fields given by $Q = \partial(\varphi \circ g)$. Thus one can derive efficient conditions for the Lipschitz-like property of (4.95) from the corresponding conditions for (4.89) by using the second-order subdifferential calculus; cf. Subsect. 4.4.2. In the next corollary we formulate some results in this direction considering for simplicity only the case of *strongly amenable* functions in assertion (ii).

Corollary 4.74 (canonical perturbations of GVIs with composite potentials). *Let $\bar y \in \Sigma(\bar x,\bar q)$ for Σ defined in (4.95), where $Y = I\!R^m$, X and W are Asplund, and g is $\mathcal{C}^2$ around $(\bar x,\bar y)$. The following hold with $\bar s := \bar q - f(\bar x,\bar y)$ and $\bar w := g(\bar x,\bar y)$.*

(i) *Assume that $g = g(y)$ with the surjective derivative $\nabla g(\bar y)$, and that f is strictly differentiable in y at $(\bar x,\bar y)$ and locally Lipschitzian in x uniformly in y around this point. Let $\bar v \in W^*$ be a unique functional satisfying the relations*

$$\bar s = \nabla g(\bar y)^*\bar v, \quad \bar v \in \partial\varphi(\bar w)\,.$$

Then Σ is Lipschitz-like around $(\bar x,\bar q,\bar y)$ if and only if $u = 0 \in I\!R^m$ is the only vector satisfying

$$0 \in \nabla_y f(\bar x,\bar y)^* u + \nabla^2\langle \bar v, g\rangle(\bar y)u + \nabla g(\bar y)^*\partial^2\varphi(\bar w,\bar v)\big(\nabla g(\bar y)u\big)\,,$$

where $\partial^2\varphi$ stands for the normal second-order subdifferential.

(ii) *Assume that X and Y are finite-dimensional, that f is Lipschitz continuous around $(\bar x,\bar y)$ and admits a strong approximation in y at this point, and that the potential $\psi := \varphi \circ g$ is strongly amenable at $(\bar x,\bar y)$. Denoting*

$$M(\bar x,\bar y) := \big\{ \bar v \in W^* \big|\, \bar v \in \partial\varphi(\bar w), \quad \nabla g(\bar x,\bar y)^*\bar v = \bar s \big\}\,,$$

we assume also the qualification conditions

$$\partial^2\varphi(\bar w,\bar v)(0) \cap \ker \nabla g(\bar x,\bar y)^* = \{0\} \text{ for all } \bar v \in M(\bar x,\bar y)\,,$$

$$\Big[y^* \in \partial_y\langle u, f\rangle(\bar x,\bar y), \quad (x^*,-y^*) \in \bigcup_{\bar v \in M(\bar x,\bar y)} \big[\nabla^2\langle \bar v, g\rangle(\bar x,\bar y)(u)$$

$$+\nabla g(\bar x,\bar y)^*\partial^2\varphi(\bar w,\bar v)\big(\nabla g(\bar x,\bar y)u\big)\big]\Big] \Longrightarrow x^* = u = 0\,,$$

where the latter reduces to

$$\Big[0 \in \partial_y\langle u, f\rangle(\bar{x},\bar{y}) + \bigcup_{\bar{v}\in M(\bar{x},\bar{y})} \big[\nabla^2\langle\bar{v}, g\rangle(\bar{y})(u)$$

$$+\nabla g(\bar{y})^*\partial^2\varphi(\bar{w},\bar{v})\big(\nabla g(\bar{y})u\big)\big]\Big] \Longrightarrow u = 0$$

if $g = g(y)$. *Then* Σ *is Lipschitz-like around* $(\bar{x},\bar{q},\bar{y})$.

Proof. To justify (i), we use Theorem 4.70(i) and then the second-order chain rule from Theorem 1.127. Assertion (ii) follows from Theorem 4.71 and Corollary 4.72 due to the second-order chain rule in Corollary 3.76 for strongly amenable functions. △

The last corollary concerns Lipschitzian stability of solutions maps to canonically perturbed generalized variational inequalities with *composite fields*:

$$\Sigma(x,q) := \Big\{y \in Y\Big|\ 0 \in f(x,y) + (\partial\varphi\circ g)(x,y)\Big\}, \tag{4.96}$$

where $g\colon X\times Y\to W$, $\varphi\colon W\to\overline{I\!R}$, and $f\colon X\times Y\to W^*$.

Corollary 4.75 (canonical perturbations of GVIs with composite fields). *Let* $\bar{y}\in\Sigma(\bar{x},\bar{q})$ *with* $\bar{s} := \bar{q} - f(\bar{x},\bar{y})$ *and* $\bar{w} := g(\bar{x},\bar{y})$ *for* Σ *given in* (4.96, *where* X, Y, W *are Asplund and where the first-order subdifferential mapping* $\partial\varphi$ *is SNC at* $(\bar{w},\bar{s})$. *The following hold with* $\partial^2\varphi$ *standing for the normal second-order subdifferential of* φ.

(i) *Assume that* $g = g(y)$ *is strictly differentiable at* $\bar{y}$ *with the surjective derivative* $\nabla g(\bar{y})$, *and that* f *is strictly differentiable in* y *at* $(\bar{x},\bar{y})$ *and locally Lipschitzian in* x *uniformly in* y *around this point. Then the condition*

$$\Big[0 \in \nabla_y f(\bar{x},\bar{y})^*u + \nabla g(\bar{y})^*\partial^2\varphi(\bar{w},\bar{s})(u)\Big] \Longrightarrow u = 0$$

is necessary and sufficient for the Lipschitz-like property of Σ *around* $(\bar{x},\bar{z})$ *provided that* $\dim Y < \infty$.

(ii) *Assume that* W^* *is Asplund, that* g *is continuous around* $(\bar{x},\bar{y})$ *and PSNC at this point, that the graph of* $\partial\varphi$ *is norm-closed around* $(\bar{w},\bar{s})$, *and that* f *is strictly Lipschitzian around* $(\bar{x},\bar{y})$ *and admits a strong approximation in* y *at this point. Assume also the qualification conditions*

$$\partial^2\varphi(\bar{w},\bar{s})(0)\cap\ker D^*g(\bar{x},\bar{y}) = \{0\},$$

$$\Big[y^* \in \partial_y\langle u, f\rangle(\bar{x},\bar{y}),\ \ (x^*,-y^*) \in D^*g(\bar{x},\bar{y})\circ\partial^2\varphi(\bar{w},\bar{s})(u)\Big] \Longrightarrow x^* = u = 0,$$

where the latter reduces to

$$\Big[0 \in \partial_y\langle u, f\rangle(\bar{x},\bar{y}) + D^*g(\bar{x},\bar{y})\circ\partial^2\varphi(\bar{w},\bar{s})(u)\Big] \Longrightarrow u = 0$$

when $g = g(y)$. *Then* Σ *is Lipschitz-like around* $(\bar{x},\bar{q},\bar{y})$.

Proof. To prove (i), we use Theorem 4.70(i) and then employ the chain rule of Theorem 1.66 and the SNC calculus rule of Theorem 1.74 to the composition $\partial\varphi \circ g$. Assertion (ii) follows from Theorem 4.71 and Corollary 4.72 due to the coderivative chain rule of Theorem 3.13(ii) and the SNC calculus rule of Theorem 3.98 applied to the composition $\partial\varphi \circ g$. △

It is easy to see that, if f is strictly differentiable in y at $(\bar{x}, \bar{y})$ and g is strictly differentiable in both variables at this point, the last qualification condition in Corollary 4.75(ii) is *equivalent* to

$$\Big[0 \in \nabla_y f(\bar{x}, \bar{y})^* u + \nabla_y g(\bar{x}, \bar{y})^* \partial^2\varphi(\bar{w}, \bar{s})(u)\Big] \Longrightarrow u = 0$$

and $\partial^2\varphi(\bar{w}, \bar{s})(0) \subset \ker \nabla_x g(\bar{x}, \bar{y})^*$.

Remark 4.76 (Robinson strong regularity). The property of solution maps to parametric generalized equations to be *single-valued and Lipschitz continuous* around a reference point relates to *Robinson strong regularity.* Actually this property was defined by Robinson for solution maps to *linearized* generalized equations and then was shown to imply the same property for the original system; see [1131]. The results presented above allow us to obtain *sufficient* as well as *necessary and sufficient* conditions for Robinson strong regularity in the case of *monotone* fields $Q = Q(y)$ in the original generalized equation (4.47), which particularly covers subdifferential operators $Q = \partial\varphi$ with a proper convex function φ (e.g., the classical variational inequalities and complementarity problems). This relates to the well-known fact that a *monotone map has to be single-valued and continuous wherever it is lower/inner semicontinuous.* Thus the above conditions for the Lipschitz-like property of solution maps to the variational systems under consideration ensure actually their strong regularity provided monotonicity. Such a monotonicity of solution maps follows from the monotonicity of Q and the corresponding monotonicity of a strong approximation to f in the sense of Definition 4.68; cf. Mordukhovich [912, Sect. 7] for more discussions and coderivative conditions for strong regularity obtained in this way for generalized equations in finite dimensions. Note that in the case when the base f is *strictly differentiable* in y the *monotonicity* of strong approximations corresponds to the *positive semidefiniteness* of $\nabla_y f(\bar{x}, \bar{y})$.

If $Q = \delta(y; \Omega)$ is the indicator function of a *convex polyhedron* $\Omega \subset I\!R^n$ and f is smooth in y, efficient characterizations of strong regularity for canonically perturbed variational inequalities are obtained by Dontchev and Rockafellar [364] *with no positive semidefiniteness* assumption on $\nabla_y f(\bar{x}, \bar{y})$. Their main result establishes the *equivalence* between strong regularity of the original generalized equation and the Lipschitz-like property of the solution map to its linearization, for which a verifiable "critical face" condition is derived on the base of the coderivative criterion from Theorem 4.70(i). In that framework new characterizations of strong regularity are obtained for nonlinear complementarity problems and for variational inequalities associated with the Karush-Kuhn-Tucker conditions in standard problems of nonlinear programming with twice differentiable data under canonical perturbations.

Remark 4.77 (Lipschitzian stability of solution maps in parametric optimization). The above coderivative analysis is useful for studying Lipschitzian stability of solution maps in *parameterized minimization problems* given in the form:

$$\text{minimize } \varphi_0(x,y)+\varphi(x,y)\,, \tag{4.97}$$

where φ_0 is a *cost function* depending on the parameter x and the decision variable y, and where $\varphi(x,y)$ is a l.s.c. extended-real-valued *constraint function* incorporating parameter-dependent constraints in the problem under consideration. In particular, model (4.97) covers parameterized problems of nonlinear programming, where the focus is on sensitivity analysis of *stationary point multifunctions* and *stationary point-multiplier multifunctions* involving Karush-Kuhn-Tucker (KKT) vectors associated with first-order necessary optimality conditions. Such an analysis is conducted in the paper by Levy and Mordukhovich [769] in the case of $\mathcal{C}^2$ cost functions φ_0 defined on $I\!R^n\times I\!R^m$. The stationary point multifunction for the optimization problem (4.97) is given as a solution map to the *parameter-dependent generalized equation*

$$S(x):=\Big\{y\in I\!R^m\Big|\;0\in\nabla\varphi_0(x,y)+\partial_y\varphi(x,y)\Big\}\,, \tag{4.98}$$

where $\partial_y\varphi(x,y)$ stands for the set of *partial* basic subgradients of the constraint functions with respect to the decision variable. The results of Sect. 4.4 allow us to compute/estimate the coderivative of the stationary point multifunction (4.98) and to derive conditions for the Lipschitz-like property of (4.98) around the reference point $(\bar{x},\bar{y})\in\operatorname{gph}S$ via the *partial second-order subdifferential* defined by

$$\partial_y^2\varphi(\bar{x},\bar{y},\bar{z}):=D^*(\partial_y\varphi)(\bar{x},\bar{y},\bar{z})$$

of the constraint function φ at $(\bar{x},\bar{y},\bar{z})$ with $\bar{z}:=-\nabla\varphi_0(\bar{x},\bar{y})$. Further analysis taking into account specific features of (4.97) and (4.98) leads to enhanced conditions for Lipschitzian stability of (4.98) in terms of the “full” second-order subdifferential $\partial^2\varphi(\bar{x},\bar{y},\bar{z})$, which enjoys a rich calculus developed in Subsects. 1.3.5 and 3.2.5. More efficient results based on second-order chain rules are obtained for parameterized problems involving constraint functions $\varphi(x,y)$ that are *strongly amenable* in y with *compatible parameterization* in x, especially in the case of *canonical perturbations.* Similar conditions hold for stationary point-multiplier multifunctions involving KKT vectors together with stationary points associated with (4.97). We also refer the reader to the recent paper by Dutta and Dempe [377] for further developments of this approach and its applications to bilevel programming.

Remark 4.78 (coderivative analysis of metric regularity). In Sects. 4.3 and 4.4 wee paid the main attention to applications of the pointbased coderivative criteria from Sect. 4.2 to Lipschitzian stability of parametric constraint and variational systems. These results may be applied to the study of *metric regularity* of such systems due to the known equivalence between the metric regularity of a mapping and the Lipschitz-like property of its inverse.

On the other hand, in this way we can derive efficient coderivative conditions for metric regularity of constraint and variational systems by using the characterizations from Theorem 4.18 and refined formulas for computing/estimating the *reversed mixed coderivative* of the corresponding solution maps; see Geremew, Mordukhovich and Nam [503] for more details. In particular, we have the representation

$$\widetilde{D}^*_M F(\bar{x},\bar{y})(y^*) = \big\{x^* \in X^* \big|\ (x^*,-y^*) \in \nabla g(\bar{x},\bar{y})^* N(\bar{z};\Theta)\big\} \quad \text{with } \bar{z} := g(\bar{x},\bar{y})$$

for the Banach space constraint system

$$F(x) := \big\{y \in Y \big|\ g(x,y) \in \Theta\big\}\ ,$$

where g is strictly differentiable at $(\bar{x},\bar{y})$ with the surjective derivative, and where Θ is *reliable* at $\bar{z}$ in the sense that the basic normal cone at this point agrees with the collection of *norm sequential limits* of $z_k^* \in \widehat{N}_{\varepsilon_k}(z_k;\Theta)$ as $\varepsilon_k \downarrow 0$ and $z_k \to \bar{z}$. The latter reliability clearly includes any subsets of finite-dimensional spaces and any sets that are normally regular at $\bar{z}$. This immediately implies representation (4.35) for the reversed mixed coderivative of the classical constraint systems in Corollary 4.35, which in turn gives efficient (Mangasarian-Fromovitz type) conditions for their metric regularity.

Concerning solutions maps to parametric variational systems of the type

$$S(x) := \big\{y \in Y \big|\ 0 \in f(x,y) + Q(y)\big\}$$

we have, under the assumptions of Theorem 4.44(i), that

$$\widetilde{D}^*_M S(\bar{x},\bar{y})(y^*) = \Big\{x^* \in X^* \Big|\ \exists z^* \in Z^* \text{ with } x^* = \nabla_x f(\bar{x},\bar{y})^* z^*,\ -y^* \in \nabla_y f(\bar{x},\bar{y})^* z^* + D^*_M Q(\bar{y},\bar{z})(z^*)\Big\}$$

and, moreover,

$$\ker \widetilde{D}^*_M S(\bar{x},\bar{y}) = -D^*_M Q(\bar{y},\bar{z})(0)\ .$$

Due to the *equality* chain rule for the *mixed* second-order subdifferential from Theorem 1.127, the latter implies the corresponding representations of $\widetilde{D}^*_M S(\bar{x},\bar{y})$ and its kernel for *composite subdifferential systems*

$$S(x) = \Big\{y \in Y \Big|\ 0 \in f(x,y) + \partial(\varphi \circ g)(y)\Big\}$$

with the corresponding conditions for metric regularity.

4.5 Commentary to Chap. 4

4.5.1. Variational Approach to Metric Regularity and Related Properties. The notions of metric regularity, linear openness/covering, and robust Lipschitzian behavior of set-valued and single-valued mappings are among the *most fundamental* concepts in nonlinear analysis and its applications, especially to optimization and related problems. We have discussed these and close to them properties, with their history and interrelations, in Chap. 1 and the corresponding comments to it. Recall that the results of Sect. 1.2 establish *necessary coderivative conditions* for the fulfillment of these properties in general Banach spaces. Furthermore, the realizations of the coderivative conditions in the case of *strictly differentiable* single-valued mappings give the classical Lyusternik-Graves regularity assumptions, which are proved to provide *full criteria* for the equivalent properties of metric regularity and covering of mappings as well as for the Lipschitz-like property of their (generally set-valued) inverses.

The primary goal of this chapter is to show that the above coderivative conditions happen to be *necessary and sufficient* for the fulfillment of the aforementioned properties in the framework of *Asplund spaces*; moreover, they induce *precise formulas* for computing the *exact bounds* of the Lipschitzian, metric regularity, and covering moduli. Invoking the coderivative and related calculus established in Chap. 3, these results allow us to develop an efficient *sensitivity analysis* for parametric constraint and variational systems.

The approach adopted in this chapter to derive *sufficient* conditions for the fundamental properties under consideration is *significantly different* from the Lyusternik-Graves iterative procedure used in Chap. 1; see also the corresponding comments therein on further developments and modifications of the above procedure. The main difference is in using the *Ekeland variational principle* that leads to a sequence of *nonsmooth minimization* problems requiring the application of appropriate results of generalized differential calculus. This approach was initiated by Ioffe [587] in the context of studying a *one-point* (metric) regularity property of single-valued Lipschitzian mappings relative to sets; cf. Definition 5.15 of "weakened metric regularity." First sufficient conditions for the latter property were obtained in [587] via Clarke's subgradient and tangential constructions defined in a *neighborhood* of the point in question. These results were improved in Ioffe's subsequent papers [589, 594, 598], by using more advanced generalized differential constructions, to derive neighborhood sufficient conditions for the related *surjection* property close to one-point counterparts of the covering/openness at a linear rate. Furthermore, papers [589, 598] contain *lower estimates* for the "constant of surjection" related to the covering/regularity bounds considered in this book.

Note that, although Ioffe originally dealt with regularity and surjection properties *at* reference points, his approach based on Ekeland's variational principle and subdifferential calculus leads in fact to sufficient conditions for stronger properties *around* the point in question, the importance of which was first emphasized by Milyutin; see the corresponding comments in Subsect. 1.4.14 and also in Ioffe's recent papers [607, 608]. This approach to deriving *sufficient conditions* for metric regularity and related properties was later developed in many publications, e.g., [49, 52, 53, 57, 69, 70, 71, 88, 137, 164, 165, 166, 282, 339, 506, 647, 651, 652, 655, 656, 657, 661, 563, 686, 709, 727, 728, 751, 901, 909, 946, 951, 1008, 1066, 1068, 1070, 1071].

4.5.2. First Characterizations of Covering and Metric Regularity. It seems that the first *necessary and sufficient* condition for the covering/linear openness property with a *precise/equality* formula for the *exact bound* was established by Mordukhovich [894] for set-valued mappings between *finite-dimensional* spaces; a complete proof of these *coderivative* results, corresponding to the *pointbased* criterion (e) and formula (4.17) of Theorem 4.18, appeared in Mordukhovich's book [901]. Note that the *necessity* part of this criterion and the *upper estimate* for the exact bound were significantly based on the "*around*" covering property under consideration and on the usage of our basic "*limiting Fréchet*" constructions.

The results obtained in [894, 901] implied, in particular, that the assumptions of the classical *smooth* Lyusternik-Graves and *convex-graph* Robinson-Ursescu theorems happened to be not only *sufficient* but also *necessary* in the settings under consideration. Moreover, these results provided for the first time the *exact bound formulas* of the *equality* type in the smooth and convex setting, which had *never been an issue* in the classical framework. To the best of our knowledge, the previous literature didn't pay attention to the *necessity* of the classical regularity/openness conditions, except mentioning without proof in (written by Milyutin) Sect. 5 of [337] that "Lyusternik's condition is necessary and sufficient for covering in the class of smooth operators." We also refer the reader to the discussion in Subsect. 1.4.14 and to the subsequent paper by Cominetti [282] containing a full (necessary and sufficient) treatment of the classical *smooth* and *convex-graph* settings in arbitrary Banach spaces, with no however exact bound considerations.

4.5.3. Neighborhood Dual and Primal Criteria. *Neighborhood* criteria for the covering property in infinite dimensions were first obtained by Kruger [709] for set-valued mappings between *Fréchet smooth* spaces. His results of the *dual* nature were formulated in terms of neighborhood constants defined via two-parametric constructions of the ε-*coderivative* type. All the neighborhood characterizations and the exact bound formulas given in Sect. 4.1 were established by Mordukhovich and Shao [946] in the framework of *Asplund spaces*. Some partial analogs of such neighborhood criteria via other subdifferentials in appropriate *Banach spaces* were later derived by Ioffe [607] under certain *tangentially* formulated additional assumptions.

Among *primal-space* developments, we mention the results by Kummer [727, 728] (see also the book by Klatte and Kummer [686]) who obtained primal *neighborhood* criteria for metric regularity via the so-called "Ekeland points" following the approach by Aubin and Ekeland [52]. Other primal criteria for metric regularity of set-valued mappings between complete *metric spaces* were developed be Ioffe [608] by using the notion of *strong slope* introduced by De Giorgi, Marino and Tosques [312] in the theory of evolution equations and first applied to the study of metric regularity and related topics by Azé, Corvellec and Lucchetti [70]; see also their paper [69] and the references therein. Very recent results in the primal direction with various applications were established by Dontchev, Quincampoix and Zlateva [363].

4.5.4. Pointbased Coderivative Characterizations of Robust Lipschitzian Behavior. Section 4.2 is devoted to *pointbased* characterizations of Lipschitzian, metric regularity, and covering/openness properties. Pointbased conditions for these properties, which are expressed via generalized differential constructions defined at the reference point *alone*, seem to be significantly *more attractive* and *convenient for applications* in comparison with the corresponding neighborhood conditions that invoke *all* points of a neighborhood. One of the major advantages of the pointbased conditions given in Sect. 4.2 is that they characterize the above fundamental properties via *robust* generalized differential constructions enjoying *full calculi.* This makes them suitable for applications to *structured* constraint and variational systems defined in specific ways via various compositions.

The main result of Sect. 4.2 is the *complete pointbased characterizations* of the *Lipschitz-like* property in Theorem 4.10 taken from Mordukhovich [924]. For set-valued mappings F between *finite-dimensional* spaces with $(\bar{x},\bar{y}) \in \operatorname{gph} F$, these characterizations reduce to the elegant *coderivative criterion* with the precise formula for the *exact Lipschitzian bound*

$$D^*F(\bar{x},\bar{y})(0)=\{0\} \quad \text{and} \quad \operatorname{lip} F(\bar{x},\bar{y})=\|D^*F(\bar{x},\bar{y})\| \tag{4.99}$$

established earlier by Mordukhovich [907]. Actually both results in (4.99) follow from the pointbased characterizations [894, 901] for the covering property of F around $(\bar{x},\bar{y})$ (see Subsect. 4.5.2) due to the equivalence between the covering/metric regularity of set-valued mappings and the Lipschitz-like property of their inverses, but at that time the author was not familiar with this equivalence. Taking into account such an equivalence, the *sufficiency* part of the coderivative condition in (4.99) could be also derived from Ioffe's results [596] on the surjection property. Comprehensive treatments of these and related topics in finite-dimensional spaces were given in the subsequent paper by Mordukhovich [909], where the reader can find various extensions and further developments. Another proof of the coderivative criterion for the Lipschitz-like (Aubin) property and the modulus formula in (4.99) was developed by Rockafellar and Wets [1165], whose book strongly demonstrated a *fundamental role* of these results in the basic theory of finite-dimensional variational analysis and its applications to optimization-related problems.

Note that the coderivative criterion in (4.99) strikingly illustrates a *major difference* between our basic normals and Clarke normals regarding applications to Lipschitzian stability/metric regularity/covering issues. Indeed, a counterpart of the criterion $D^*F(\bar{x},\bar{y})(0)=\{0\}$ expressed in terms of Clarke normals, being of course sufficient for the Lipschitz-like property of F around $(\bar{x},\bar{y})$ as was first shown by Aubin [49] (cf. also Rockafellar [1154]), is *never fulfilled* for nonsmooth Lipschitzian single-valued mappings as well as their set-valued graphically Lipschitzian extensions; see more discussions in [909] and in Remark 4.13 of this book.

4.5.5. Pointbased Criteria in Infinite Dimensions Involving Partial Normal Compactness. In *infinite-dimensional* settings (what really matters is infinite dimensionality of the range space in the case of metric regularity/covering and infinite dimensionality of the domain space in the case of Lipschitzian properties) pointbased generalized differential conditions alone are *not sufficient* for the validity of these fundamental properties. Some additional *amount of compactness* (of a non-conventional type) is required. In the classical frameworks of the Lyusternik-Graves and Robinson-Ursescu theorems for, respectively, smooth and convex-graph mappings such a compactness is (hiddenly) ensured by the surjectivity and interiority conditions therein; cf. Theorems 1.57 and 4.21.

For *nonsmooth* single-valued mappings defined on closed sets, the first condition of this type was introduced by Ioffe [595] under the name "finite codimension condition" formulated in terms of Clarke's generalized differential constructions. The original motivation came from the development of a nonsmooth analog of the abstract "principle of Lagrange" suggested by Ioffe and Tikhomirov [618] for extremal problems with mixed smooth-convex structure involving operator constraints, where the finite codimension requirement on the derivative range for the operator constraint mapping played a crucial role. Modifications and extensions of the finite codimension condition formulated via "approximate" normals and subgradients were given by Ioffe in [598] and in his joint paper with Ginsburg [506]. It has been shown in [506, 595, 604, 618] that the versions of the finite codimension property under consideration are fulfilled for various classes of infinite-dimensional operators, in particular, for those of "Fredholm type" important for applications in optimal control.

Further progress in this direction was achieved by Ioffe [607], Jourani and Thibault [655, 661], Mordukhovich and Shao [950, 951], and Penot [1068, 1071]. The weakest property of this kind was defined in [950] under the name of *partial sequential normal compactness* – it is the same PSNC property we have strongly used in the previous chapters to establish calculus rules for generalized differentiation. Other names for this property and its *net counterparts* formulated via various normals and subgradients are "sequential codirectional compactness" in [607], "partial coderivative compactness" in [661], and "coderivative compactness" in [1071].

Theorem 4.10 that provides complete *pointbased characterizations* of Lipschitzian behavior for set-valued mappings between *Asplund spaces* was given by Mordukhovich [917], except that the notions of *coderivative normality* from Definition 4.8 with the efficient conditions of Proposition 4.9 needed for the exact bound formula (4.6) were taken from his subsequent paper [924]. A complete proof of criteria (b) and (c) of the theorem with the lower bound estimate in (4.5) appeared in paper [953] by Mordukhovich and Shao, while the *upper bound* estimate therein was derived earlier in their paper [951]; one can also find therein previous sufficient conditions for Lipschitzian stability/metric regularity/covering expressed via the normal coderivative $D_N^* F(\bar{x}, \bar{y})$ of set-valued mappings between Asplund spaces.

Theorem 4.18 is a metric regularity/covering counterpart of the Lipschitzian characterizations from Theorem 4.10. An analog of criterion (c) from this theorem via topological/net limiting constructions was derived by Penot [1071] with *no bound* estimates. *Sufficient* conditions of this type for metric regularity of set-valued mappings between appropriate "trustworthy" Banach spaces were obtained by Ioffe [607] and by Jourani and Thibault [661] via corresponding analogs of the *normal* coderivative; see also [598, 655] for previous results in this direction in terms of "approximate" subdifferentials and coderivatives in arbitrary Banach spaces. Furthermore, in [607, 661] there were some pointwise *necessary* coderivative conditions for $F\colon X \rightrightarrows Y$ to be metrically regular provided that the domain space X is *finite-dimensional.* Example 4.19 illustrating the significance of normal compactness conditions for metric regularity and related properties of mappings between infinite-dimensional spaces is taken from Borwein and Zhu [162].

Among numerous applications of subdifferential/coderivative conditions for covering and metric regularity (some of them are considered in this book) we mention the usage of such conditions for deriving various rules of *generalized differential calculus.* It seems that this approach was initiated by Kruger [709] and then was strongly developed by Jourani and Thibault [651, 652, 656] and by Ioffe [607, 608].

4.5.6. Preservation of Lipschitzian Behavior and of Metric Regularity Under Compositions. Due to a fundamental role played by Lipschitz-like, metric regularity, and covering/openness properties in nonlinear analysis and applications, it is important to find efficient conditions ensuring the *preservation* of these properties under various operations. Concerning the Lipschitz-like ("pseudo-Lipschitzian") property of set-valued mappings between finite-dimensional spaces, such issues were first studied by Rockafellar [1154] who established various results in this direction with rather involved proofs based mainly on definitions and finite-dimensional geometry. The achieved pointbased coderivative *characterizations* of Lipschitzian and related properties allow us to use *coderivative calculus* for these purposes, which leads in addition to establishing relationships between *exact bounds* under compositions.

In this way Rockafellar and Wets [1165] derived, based on characterizations (4.99), verifiable conditions ensuring the preservation of robust Lipschitzian stability with the corresponding bound estimates under various operations performed on set-valued mappings between finite-dimensional spaces. Theorems 4.14, 4.16 and their corollaries established in Subsect. 4.2.1 in infinite dimensions are taken from Mordukhovich and Nam [934]. The related preservation results for metric regularity and covering from Theorem 4.22 have never been published.

4.5.7. Good Behavior under Perturbations. The study of well-posedness (i.e., "good behavior") *under perturbations* is a major topic in *sensitivity analysis* important from both theoretical and numerical viewpoints. It seems that *quantitative* aspects of these issues on measuring the *exact bounds* of perturbations that *don't violate well-posedness* (i.e., don't cause irregular/ill-posed behavior) were first addressed in the classical Eckart-Young theorem on nonsingular matrix perturbations [388] that was motivated by applications to numerical analysis. The Eckart-Young theorem identifies the minimal norm of $n \times n$ matrices, the addition of which to a given nonsingular matrix A breaks down its *nonsingularity*, with the reciprocal of $\|A^{-1}\|$. Results of this type, called often "distance to ill-posedness theorems" and also "condition number theorems," play a prominent role in various problems of numerical analysis; see, e.g., Demmel [315] and the references therein.

In optimization theory and its applications, results of this type were first developed in the thorough study by Renegar [1122, 1123] who introduced a concept of the *distance to infeasibility* for constraint systems in conic linear programming and who related this concept to the complexity of solving associated linear and semidefinite programs. Renegar's characterizations of the distance to infeasibility can be treated as appropriate extensions of the Eckart-Young theorem to conic linear systems, although his motivation mainly came from *complexity analysis* of interior point methods developed by Nesterov and Nemirovsky [999].

Most on the subsequent research on conditioning in optimization was built on (or was strongly influenced by) Renegar's seminal work; see, e.g., [219, 361, 366, 405, 475, 776, 777, 780, 996, 1055, 1056, 1057, 1058, 1061, 1206, 1376, 1377, 1332] and the references therein. We particularly mention the work by Peña [1055, 1056, 1057, 1058, 1059] who introduced and developed the technique of *rank-one perturbations*, which is fundamental for the theory of the distance to ill-posedness, and the work by Lewis [776, 777, 780] who extended Renegar's results to set-valued *convex processes* (in the sense of Rockafellar [1142]) bringing to the area of conditioning the elegant language and constructions of convex analysis.

A crucial contribution to this subject was made by Dontchev, Lewis and Rockafellar [361] who introduced the *radius of metric regularity* from Definition 4.23 and related it to the reciprocal of the *exact bound* for regularity moduli. Furthermore, involving a homogenization procedure and the

Robinson-Ursescu theorem, they established a link between the radius of metric regularity and Renegar's distance to infeasibility for conic constraint systems (actually for convex-graph multifunctions), which thus paves the way for extensive applications of modern variational analysis and advanced tools of generalized differentiation to conditioning in optimization.

Theorem 4.24 extending the classical Eckart-Young result to positively homogeneous set-valued mappings between Banach spaces was proved in [361], while its preceding version for sublinear mappings (i.e., for closed convex processes) was established by Lewis [775]. Theorem 4.25 providing an upper estimate of the exact regularity bound under Lipschitzian perturbations goes back to Milyutin (see [337]) who proved it for single-valued mappings by using the Lyusternik-Graves iterative process; the full version presented above was given in [361]; cf. also Ioffe [598, 608]. This fact easily implies the *lower estimate*

$$\operatorname{rad} F(\bar{x}, \bar{y}) \geq \frac{1}{\operatorname{reg} F(\bar{x}, \bar{y})} \tag{4.100}$$

of the metric regularity radius in Theorem 4.27, while the main result on the *equality* in (4.100), and also the statement that the infimum in the definition of $\operatorname{rad} F(\bar{x}, \bar{y})$ is *unchanged* if taken with respect to *linear* perturbations of *rank one*, was established by Dontchev, Lewis and Rockafellar [361] in the case of general set-valued mappings between *finite-dimensional* spaces.

It is worth mentioning that the proof in [361] was heavily based on the finite-dimensional *coderivative characterization* (4.99) of Lipschitzian stability/metric regularity, where *both* coderivative criterion and the exact bound formula are important, and also on the *homogenization* of the original mapping via its coderivative. We refer the reader to the paper by Dontchev and Rockafellar [366] for more recent results concerning related properties of "strong metric regularity" and "strong metric subregularity." A generalization of the radius theorem from [361] to set-valued mappings defined on *Riemannian manifolds* (associated with finite-dimensional spaces) has been recently obtained by Dontchev and Lewis [360].

In [924], Mordukhovich extended the approach of [361] to set-valued mappings acting from *Asplund to finite-dimensional* spaces and established Theorem 4.27 in *full generality* based on the coderivative characterization of metric regularity for infinite-dimensional multifunctions from Theorem 4.18 and on some amount of coderivative calculus. Subsequently Ioffe showed [609] that the inequality in (4.100) may be *strict* for a Lipschitz continuous single-valued mapping, even with good *weak differentiability* properties, from a *Hilbert space into itself*. Furthermore, Ioffe proved in [610] that an analog of (4.100) held as *equality* for *single-valued* mappings between general infinite-dimensional spaces if the infimum in the definition of the radius of metric regularity was computed over all the *Lipschitzian perturbations* but *not* over *linear* ones as in Definition 4.23 and therefore not over those of *rank one* as justified in Theorem 4.27 under the assumptions made therein.

Finally, we mention the recent paper by Cánovas, Dontchev, López and Parra [219] who established a counterpart of the Eckart-Young theorem in the *equality* form of (4.100) for a special class of set-valued mappings acting from a *finite-dimensional space to the Banach space* of continuous functions over a compact Hausdorff space. Such mappings were defined by the so-called *linear semi-infinite systems* of equalities and inequalities (with a compact index set) and described feasible constraints in *semi-infinite programming.* Based on the Lyusternik-Graves theorem, the authors also extended their results to *nonlinear* semi-infinite constraint systems.

4.5.8. Sensitivity Analysis of Parametric Constraint Systems via Generalized Differentiation. Sections 4.3 and 4.4 are devoted to *sensitivity analysis* of *constraints* and *variational systems* by means of *set-valued differentiation.* There are a great many developments on sensitivity analysis for optimization-related problems based on various approaches; see, e.g., [45, 46, 47, 54, 56, 57, 60, 70, 133, 134, 137, 164, 255, 348, 355, 356, 367, 424, 448, 447, 469, 519, 523, 562, 563, 584, 623, 639, 640, 641, 681, 685, 686, 692, 697, 698, 727, 729, 734, 751, 763, 766, 768, 773, 797, 816, 820, 832, 834, 907, 911, 912, 929, 939, 1030, 1031, 1043, 1044, 1047, 1128, 1131, 1136, 1138, 1154, 1183, 1191, 1196, 1203, 1205, 1225, 1378] and the references therein. In this book we mainly concern *robust Lipschitzian* stability of parametric systems and develop an approach based on applying *pointbased coderivative* characterizations of this stability via available generalized differential and SNC *calculi.*

General parametric *constraint systems* of form (4.19) considered in Sect. 4.3 were introduced by Rockafellar [1154] as extensions of the standard constraint systems (4.20) in nonlinear programming. The primal motivation to study parametric systems of type (4.19) and (4.20) is to conduct a local *sensitivity analysis* of feasible solutions under perturbations. On the other hand, the general constraint form (4.19) is convenient to describe *implicit multifunctions* (4.22) providing a natural formalism to extend the classical implicit and inverse function theorems.

Rockafellar's study [1154] was devoted to *Lipschitzian stability* of constraint systems of type (4.19) in *finite dimensions* and their specifications. He obtained *sufficient* conditions for the Lipschitz-like ("pseudo-Lipschitzian") and related properties of multifunctions in terms Clarke generalized normals and subgradients. In fact, sufficient conditions for Lipschitzian properties of set-valued mappings were derived in [1154] via *scalarization* of the Lipschitz-like property (see Theorem 1.41 established in [1154]) and applying subdifferential calculus rules for constructions of Clarke. This approach was developed in Mordukhovich [907, 911], by applying our basic/nonconvex generalized differential constructions and their calculus instead of Clarke ones. In the way more delicate sufficient conditions for Lipschitzian stability of constraint systems were derived in [907, 911]; moreover, these conditions happened to be also *necessary* in some settings including the classical framework of (4.20) under the *Mangasarian-Fromovitz constraint qualification.*

Coderivatives of parametric constraint systems (4.19) and their specifications (4.20) and (4.22) in finite-dimensional spaces were computed/estimated in Mordukhovich's papers [910, 913] based on coderivative calculus. By the coderivative characterizations in (4.100), these results allowed us not only to derive efficient conditions for Lipschitzian stability but also to compute/estimate the *exact Lipschitzian bounds* for the corresponding constraint systems; see [913]. Various developments in this direction for specific classes of constraint systems arising in applications were given by Avelin [66, 67], Dontchev, Lewis and Rockafellar [361], Dontchev and Rockafellar [364, 366], Dutta and Dempe [377], Flegel, Kanzow and Outrata [457], Henrion [557, 558], Henrion and Outrata [561, 562], Henrion and Römisch [563, 564], Jourani [647], Kočvara, Kružik and Outrata [689], Kočvara and Outrata [690], Lee, Tam, and Yen [755], Levy [768], Levy and Mordukhovich [769], Levy and Poliquin [770], Lucet and Ye [816], Mordukhovich and Outrata [939], Outrata [1024, 1025, 1027, 1030], Ye [1338, 1339], Ye and Ye [1343], Ye and Zhu [1345], etc. Some partial infinite-dimensional extensions of those in [910, 911, 913] were derived by Mordukhovich and Shao [951] in the framework of Asplund spaces, while Ledyaev and Zhu [751] conducted a thorough study of Lipschitzian and related properties for *implicit multifunctions* in Fréchet smooth spaces, with computing and estimating the corresponding Fréchet and normal coderivatives.

Among many other publications on sensitivity analysis via set-valued differentiation, we mention the research by Aubin [49], Aubin and Frankowska [53, 54], Dontchev and Rockafellar [365], Frankowska [467, 469], Fusek, Klatte and Kummer [482], King and Rockafellar [681], Klatte and Kummer [686, 687], Kummer [725, 726, 728], Levy [766, 767], Levy and Rockafellar [773, 774], Rockafellar and Wets [1165], and Zhang [1360] that particularly contain exact formulas and upper estimates for various *graphical derivatives* (in primal spaces) and their applications to the study of Lipschitzian properties of various constraint systems.

Most of the results presented in Sect. 4.3 in *infinite-dimensional* (largely Asplund) spaces are taken from Mordukhovich [927]; in fact, they are extensions of the corresponding finite-dimensional results of [910, 911, 913]. In contrast to the finite-dimensional framework, the *SNC conditions* and their calculus play a *prominent role* in infinite-dimensional settings.

4.5.9. Generalized Equations and Variational Conditions. The framework of *generalized equations* (4.47) and their parameter *perturbations* (4.49) was introduced by Robinson [1130]. It seems that his primal motivation was to include *variational inequalities* (4.48) and its *complementarity* specification into the "equation" setting (4.47), which indeed reduces to the standard equations $f(x) = 0$ when the set-valued part Q disappears. Such a "generalized equation" viewpoint happened to be very convenient to develop qualitative and numerical results for variational inequalities and complementarity problems by analogy with those known for standard equations (e.g., the corresponding versions of Newton-type methods).

Although generalized equations (4.47) are well defined in infinite-dimensional spaces, they were originally introduced and studied in *finite dimensions*, since motivations and applications were related at that time to finite-dimensional optimization, particularly to nonlinear programming and complementarity problems; see the surveys in [294, 424, 550, 1134]. On the other hand, variational inequalities of type (4.48) in *infinite-dimensional* (mostly in Hilbert) spaces have been studied in connection with *nonlinear partial differential equations* and their mechanical applications starting with Stampacchia's work in the early 1960s; see, e.g., [1223, 680, 795], the recent book by Giannessi [504], and the references therein.

The variational inequality (4.48) obviously reduces to the generalized equation (4.47) for the normal cone mapping $Q(y) = N(y; \Omega)$ corresponding to a convex set Ω. Variational inequalities of the other ("second") classical type correspond to the generalized equation model (4.47) with $Q(y) = \partial\varphi(y)$, where φ is a *convex continuous* function. An extension of the latter model to $Q(y) = \partial_C\varphi(y)$, where φ is a Lipschitz continuous function and where $\partial_C\varphi$ stands for its Clarke generalized gradient, was introduced by Panagiotopoulos (see [1042] and also [551, 994]) under the name of *hemivariational inequalities.* In [911], Mordukhovich first studied more general variational systems in form (4.47) with $Q(y) = \partial\varphi(y)$, where $\partial\varphi$ stands for the basic/limiting subdifferential of an arbitrary l.s.c. function φ; such systems were called *variational conditions* by Rockafellar and Wets [1165]; see also the recent papers by Robinson [1137, 1138, 1139].

It has been well recognized starting with Robinson's seminal work [1130, 1131, 1132, 1133] that generalized equations provide a convenient model for *sensitivity analysis* of *optimal solution* and associated maps under parameter perturbations. In particular, they describe perturbed sets of *stationary* and *Karush-Kuhn-Tucker* (KKT) points in problems of nonlinear programming.

4.5.10. Robust Lipschitzian Stability of Generalized Equations and Variational Inequalities. Robinson's original efforts and many subsequent publications mostly aimed to derive efficient conditions ensuring the local *single-valuedness and Lipschitz continuity* of solution maps; this property closely relates to *Robinson strong regularity* defined via linearization; see more discussions in Remark 4.76 as well comments given in Subsect. 4.5.11 below. On the other hand, Robinson was the first to recognize the *upper Lipschitzian* property of *set-valued* mappings (now often used under the name of "calmness") and to establish its validity for *polyhedral* multifunctions and for the corresponding solution maps to generalized equations. The reader can find more information on these and related developments in the afore-mentioned publications, particularly in the recent book by Facchinei and Pang [424] with the comprehensive bibliography therein.

It seems to be a disadvantage of the upper Lipschitzian property that is *not robust* with respect to perturbations of the initial data; it is even doesn't go back to the classical local Lipschitz continuity for single-valued mappings. A new *robust* Lipschitzian property of multifunction was introduced by Aubin [49] under the "pseudo-Lipschitzian" name; we broadly use it in this book as the "Lipschitz-like" property, since it is the most natural extension – actually just the graphical localization – of the classical Lipschitz continuity. In [49], Aubin derived sufficient conditions for this robust Lipschitzian behavior of set-valued solution maps to perturbed *convex* optimization problems in terms of generalized differential constructions by Clarke.

Soon after that, Rockafellar [1154] considered the perturbed generalized equation

$$0 \in f(x, y) + Q(y) \tag{4.101}$$

with a locally Lipschitzian mapping $f\colon I\!R^n \times I\!R^m \to I\!R^d$ and derived sufficient conditions for the Lipschitz-like property of the solution map to (4.101) via Clarke's constructions of the generalized Jacobian to f and the normal cone to the graph of Q. Rockafellar clearly realized that the results obtained were *significantly restrictive*, especially in the most interesting cases of *subdifferential mappings* $Q = \partial\varphi$ generated by *convex* functions φ, and that they were far removed from Robinson's conditions obtained for some special cases of (4.101). This is due to the *subspace property* of Clarke's normal cone (never employed by Robinson), which happens to have a *fundamental drawback* for applications to this type of subdifferential variational systems; see examples and discussions in Remark 4.58. However, *no alternative* constructions of nonsmooth analysis were suggested to use in [1154].

Adequate sufficient conditions, and also *necessary and sufficient* conditions in some settings, for the Lipschitz-like property of solution maps to the perturbed generalized equation (4.101) in *finite-dimensional* spaces were established by Mordukhovich [911] via the nonconvex generalized differential constructions of this book. In the case of *smooth* (strictly differentiable) mappings f, the conditions obtained were formulated as the requirement that the *adjoint generalized equation*

$$0 \in \nabla_y f(\bar{x}, \bar{y})^* z^* + D^* Q\big(\bar{y}, -f(\bar{x}, \bar{y})\big)(z^*)$$

has *only the trivial solution* $z^* = 0$. This form, reminding us the *Fredholm alternative* in the theory of integral equations, reflects deep relationships between "good behavior" of primal systems and the triviality of solutions to their adjoint/dual ones.

Most of the material presented Subsects. 4.4.1 and 4.4.2 is based on the recent papers by Mordukhovich [924, 931, 933], where new results were established in both *finite* and *infinite* dimensions; some of their finite-dimensional analogs can be found in [911] for the case of (4.101). Note that our analysis concerns solution maps to the perturbed generalized equations

$$0 \in f(x,y) + Q(x,y) ,$$

where *both* single-valued and set-valued parts depend on the parameter x. Such systems are extensions of (4.101) and happen to be essentially more complicated. They particularly cover the so-called *quasivariational inequalities* when the set Ω in (4.48) is *moving*, i.e., also depends on parameters. Let us also mention results by Mordukhovich and Outrata [939] related to parametric variational systems of the special *composite subdifferential structure*

$$Q(x,y) = \partial\varphi\big(g(x,y)\big)$$

with a smooth mapping g between finite-dimensional spaces; see Theorem 4.65 and Corollary 4.66 for generalizations.

Another approach to sensitivity analysis of perturbed variational systems

$$0 \in f(x,y) + N\big(y; \Omega(x)\big)$$

with a moving and generally *nonconvex* set $\Omega(x)$ in finite dimensions has been recently developed by Robinson [1137, 1138, 1139]; see also the references therein. Example 4.67 on applications to Lipschitzian stability for some practical problems of *continuum mechanics* is taken from Mordukhovich and Outrata [939], where the reader can find more results and applications in this direction.

4.5.11. Strong Approximation and Canonical Perturbations. Subsection 4.4.3 concerns developments on robust Lipschitzian stability of variational systems under perturbations that are significantly different from those presented in Subsect. 4.4.2. Indeed, the approach to sensitivity analysis developed in Subsect. 4.4.2 is based on *reducing* parametric *variational* systems of the general type (4.49) to a special kind of *constraint systems* studied in Sect. 4.3. Then powerful calculus results for the basic generalized differential constructions of this book allowed us to compute *coderivatives* for solution maps to parametric variational systems and thus to derive efficient conditions for their Lipschitzian stability (with calculation/estimation of the exact Lipschitzian bounds) based on the *pointbased coderivative characterizations* established in Subsect. 4.2.1.

The approach of Subsect. 4.4.3 invokes other ideas that are in fact a continuation of a long tradition coming from the classical inverse and implicit function theorems, while they have been championed and strongly developed by Robinson in the framework of optimization and variational analysis. Roughly speaking, they revolve around the fundamental idea of *linearization* (or, more generally, of an appropriate *approximation*) of the original system in such a way that the approximating system is easily to analyze, but at the same time the required properties of the original system follow from those for the approximating one.

Implementing this procedure, Robinson introduced in [1131] the property of *strong regularity* as the local single-valuedness and Lipschitz continuity of solution maps to the linearized generalized equation of type (4.101) with a *smooth* base function f and showed that such a regular behavior is inherent in the original nonlinear system. Developing the approximating idea in the case of *nonsmooth* bases in the implicit function and generalized equation frameworks, Robinson originated the concept of *strong approximation* considered in Subsect. 4.4.3. These ideas have been significantly developed and applied to various optimization and equilibrium problems in many publications; see, e.g., [133, 348, 350, 351, 352, 355, 356, 364, 365, 424, 686, 639, 640, 641, 692, 767, 768, 797, 820, 912, 929, 1043, 1044, 1047, 1092, 1133, 1134, 1205] and the references therein.

In Mordukhovich [912], we employed Robinson's *linearization/strong approximation* ideas, combined with our advanced tools of generalized differentiation, to the study of robust Lipschitzian stability for parametric generalized equations of type (4.101) in finite-dimensional spaces. Using this procedure, in contrast to the approach developed in Subsects. 4.4.1 and 4.4.2, we applied the coderivative criterion (4.100) *not* directly to the original generalized equation but to the *approximating one*, while taking into account that the Lipschitz-like property of the approximating system implies this property for the original solution map due to Dontchev and Hager [356]. The results obtained in this way are generally *independent* of those derived in [911]; see more discussions in Subsect. 4.4.3.

It seems that the linearlization/strong approximation approach happens to be the most efficient for *canonically perturbed* systems of type (4.49). Although it has been realized for a long time that the structure of perturbations should be "rich enough" to ensure better results in sensitivity analysis, the main emphasis of "canonical" perturbations and the very terminology probably first appeared in Rockafellar's work [1160]. This structure was strongly exploited in the excellent paper by Dontchev and Rockafellar [364] devoted to canonically perturbed variational inequalities over *polyhedral convex sets* in finite dimensions. They established the *equivalence* between Robinson strong regularity and the Lipschitz-like/Aubin property of solution maps for such variational systems and, based on the *coderivative criterion* (4.100), derived a "critical face" characterization of these properties. As an application of their critical face characterization, they obtained verifiable *necessary and sufficient* conditions for strong regularity of the general *nonlinear complementarity problem* with canonical perturbations and finally applied these results to characterizing strong regularity of the *KKT systems* in nonlinear programming with twice differentiable data. In this way they solved a long-standing problem about the *necessity* of the so-called "strong second-order sufficient condition" for strong regularity; the sufficiency of the latter condition for strong regularity was established in Robinson's landmark paper [1131]. We also refer the reader to the papers by Bonnans and Sulem [134], Jongen et al. [640], and Kojima [692] for related developments that didn't employ tools of nonsmooth analysis.

Canonical perturbations play a strong role in the concept of *tilt-stable* local minimum introduced by Poliquin and Rockafellar [1092] from the viewpoint of sensitivity analysis. As established in [1092], the *positive definiteness* of the *second-order subdifferential* from [907] (see Subsect. 1.3.5) is a *characterization* of a tilt-stable minimum. Further strong developments in this direction can be found in the subsequent paper by Levy, Poliquin and Rockafellar [771].

The material of Subsect. 4.4.3 is mostly based on Mordukhovich's paper [929]. Lemma 4.69 on the equivalence between Lipschitzian stability of canonically perturbed generalized equations and their strong approximations was proved by Dontchev [350] by applying the Lyusternik-Graves iterative process.

References

1. Y. A. Abramovich and C. D. Aliprantis (2002), *An Invitation to Operator Theory*, American Mathematical Society, Providence, Rhode Island.
2. A. A. Agrachev and Y. L. Sachkov (2004), *Control Theory from the Geometric Viewpoint*, Springer, Berlin.
3. N. U. Ahmed (2005), Necessary conditions of optimality with discontinuous vector fields, *Nonlinear Funct. Anal. Appl.* **10**, 129–150.
4. N. U. Ahmed and K. L. Teo (1981), *Optimal Control of Distributed Parameter Systems*, Elsevier, New York.
5. N. U. Ahmed and X. Xiang (1997), Nonlinear uncertain systems and necessary conditions of optimality, *SIAM J. Control Optim.* **35**, 1755–1772.
6. N. U. Ahmed and X. Xiang (1997), Necessary conditions of optimality for differential inclusions in Banach spaces, *Nonlinear Anal.* **30**, 5437–5445.
7. V. M. Alekseev, V. M. Tikhomirov and S. V. Fomin (1987), *Optimal Control*, Consultants Bureau, New York.
8. A. D. Alexandrov (1939), Almost everywhere existence of the second differential of a convex function and some properties of convex surfaces connected with it, *Uchenye Zapiski Leningrad Gos. Univ., Ser. Math.* **6**, 3–35.
9. J. J. Alibert and J.-P. Raymond (1998), A Lagrange multiplier theorem for control problems with state constraints, *Numer. Funct. Anal. Optim.* **19**, 697–704.
10. C. D. Aliprantis, D. J. Brown and O. Burkinshaw (1990), *Existence and Optimality of Competitive Equilibria*, Springer, New York.
11. C. D. Aliprantis and O. Burkinshaw (1988), The fundamental theorems of welfare economics without proper preferences, *J. Math. Econ.* **17**, 41–54.
12. C. D. Aliprantis, M. Florenzano and R. Tourky (2005), Linear and non-linear price decentralization, *J. Econ. Theory*, to appear.
13. C. D. Aliprantis, P. K. Monteiro and R. Tourky (2004), Non-marketed options, non-existence of equilibria, and non-linear prices, *J. Econ. Theory* **114**, 345–357.
14. C. D. Aliprantis, R. Tourky and N. C. Yannelis (2001), A theory of value with non-linear prices: Equilibrium analysis beyond vector lattices, *J. Econ. Theory* **100**, 22–72.
15. K. Allali and L. Thibault (2005), Fréchet subdifferentials of marginal functions, *Control Cybernet.*, to appear.

16. T. Amahroq and N. Gadhi (2001), On the regularity conditions for vector programming problems, *J. Global Optim.* **21**, 435–443.
17. L. Ambrosio, O. Ascenzi and G. Buttazzo (1989), Lipschitz regularity of intergral functionals with highly discontinuous integrands, *J. Math. Anal. Appl.* **142**, 301–316.
18. R. M. Anderson (1988), The second welfare theorem with non-convex preferences, *Econometrica* **56**, 361–282.
19. T. S. Angell and A. Kirsch (1990), On the necessary conditions for optimal control of retarded systems, *Appl. Math. Optim.* **22**, 117–145.
20. M. Anitescu (2005), Global convergence of an elastic mode approach for a class of mathematical programs with complementarity constraints, *SIAM J. Control Optim.*, to appear.
21. N. Arada (2001), Minimax Dirichlet boundary control problem with state constraints, *Nonlinear Anal.* **46**, 653–673.
22. N. Arada, M. Bergounioux and J.-P. Raymond (2000), Minimax controls for uncertain distributed parabolic systems, *SIAM J. Control Optim.* **38**, 1481–1500.
23. N. Arada and J.-P. Raymond (1999), Optimality conditions for state-constrained Dirichlet boundary control problems, *J. Optim. Theory Appl.* **102**, 51–68.
24. N. Arada and J.-P. Raymond (2002), Dirichlet boundary control of semilinear parabolic equations, II: Problems with pointwise state constraints, *Appl. Math. Optim.* **45**, 145–167.
25. V. I. Arkin and V. L. Levin (1973), Convexity of the values of vector integrals, theorems on measurable selection, and variational problems, *Russian Math. Surveys* **27**, 21–85
26. K. J. Arrow (1951), An extension of the basic theorems of classical welfare economics, in *Proceedings of the Second Berkeley Symposium on Mathematical Statistics and Probability*, pp. 507–532, University of California, Berkeley, California.
27. K. J. Arrow and L. Hurwicz (1960), Decentralization and computation in resource allocation, in *Essays in Economics and Econometrics in Honor of Harold Hotelling*, edited by R. W. Pfouts, pp. 34–104, University of North Carolina Press, Chapel Hills, North Carolina.
28. Z. Artstein (1994), First order approximations for differential inclusions, *Set-Valued Anal.* **2**, 7–17.
29. Z. Artstein (1995), A calculus for set-valued maps and set-valued evolution equations, *Set-Valued Anal.* **3**, 213–261.
30. Z. Artstein and V. Gaitsgory (2000), The value function of singularly perturbed control systems, *Appl. Math. Optim.* **41**, 425–445.
31. Z. Artstein and C. C. Popa (2004), Orlicz-Young structures of Young measures, *Acta Appl. Math.* **80**, 1–33.
32. A. V. Arutyunov (2000), *Optimality Conditions: Abnormal and Degenerate Problems*, Kluwer, Dordrecht, The Netherlands.
33. A. V. Arutyunov and S. M. Aseev (1997), Investigation of the degeneracy phenomenon of the maximum principle for optimal control problems with state constraints, *SIAM J. Control Optim.* **35**, 930–952.
34. A. V. Arutyunov, S. M. Aseev and V. I. Blagodatskikh (1994), Necessary conditions of the first order in the problem of optimal control of a differential inclusion with phase constraints, *Math. Sbornik* **79**, 117–139.

35. A. V. ARUTYUNOV AND A. F. IZMAILOV (2005), Sensitivity analysis for cone-constrained optimization problems under relaxed constraint qualifications, *Math. Oper. Res.* **30**, 333–353..
36. A. V. ARUTYUNOV AND A. F. IZMAILOV (2005), Directional stability theorem and directional metric regularity, *Math. Oper. Res.*, to appear.
37. A. V. ARUTYUNOV AND F. L. PEREIRA (2005), Second-order necessary conditions of optimality for problems without a priori normality assumptions, *Math. Oper. Res.*, to appear.
38. A. V. ARUTYUNOV AND R. B. VINTER (2004), A simple 'finite approximations' proof of the Pontryagin maximum principle under reduced differentiability hypotheses, *Set-Valued Anal.* **12**, 5–24.
39. S. M. ASEEV (1991), Smooth approximations of differential inclusions and the time-optimality problem, *Proc. Steklov Inst. Math.* **200**, 27–34.
40. S. M. ASEEV (1997), A method of smooth approximations in the theory of necessary optimality conditions for differential inclusions, *Izvestia: Mathematics* **61**, 235–258.
41. S. M. ASEEV (2001), Extremal problems for differential inclusions with phase constraints, *Proc. Steklov Inst. Math.* **233**, 1–63.
42. L. T. ASHCHEPKOV AND O. V. VASILIEV (1975), On optimality of singular controls in Goursat-Darboux systems, *USSR Comput. Maths. Math. Phys.* **15**, 1157–1167.
43. E. ASPLUND (1968), Fréchet differentiability of convex functions, *Acta Math.* **121**, 31–47.
44. H. ATTOUCH (1984), *Variational Convergence for Functions and Operators*, Pitman, Boston, Massachusetts.
45. H. ATTOUCH AND H. RUANI (1993), Stability results for Ekeland's ε-variational principle and cone extremal solutions, *Math. Oper. Res.* **18**, 173–201.
46. H. ATTOUCH AND R. J-B. WETS (1991), Quantitative stability of variational systems, I: The epigraphical distance, *Trans. Amer. Math. Soc.* **338**, 695–729.
47. H. ATTOUCH AND R. J-B. WETS (1993), Quantitative stability of variational systems, III: ε-approximate solutions, *Math. Progr.* **61**, 197–214.
48. J.-P. AUBIN (1981), Contingent derivatives of set-valued maps and existence of solutions to nonlinear inclusions and differential inclusions, in *Mathematical Analysis and Applications*, edited by L. Nachbin, pp. 159–229, Academic Press, New York.
49. J.-P. AUBIN (1984), Lipschitz behavior of solutions to convex minimization problems, *Math. Oper. Res.* **9**, 87–111.
50. J.-P. AUBIN AND A. CELLINA (1984), *Differential Inclusions*, Springer, Berlin.
51. J.-P. AUBIN AND I. EKELAND (1976), Estimates of the duality gap in nonconvex programming, *Math. Oper. Res.* **1**, 225–245.
52. J.-P. AUBIN AND I. EKELAND (1984), *Applied Nonlinear Analysis*, Wiley, New York.
53. J.-P. AUBIN AND H. FRANKOWSKA (1987), On inverse function theorems for set-valued maps, *J. Math. Pures Appl.* **66**, 71–89.
54. J.-P. AUBIN AND H. FRANKOWSKA (1990), *Set-Valued Analysis*, Birkhäuser, Boston, Massachusetts.
55. R. J. AUMANN (1965), Integrals of set-valued functions, *J. Math. Anal. Appl.* **12**, 1–12.

56. A. Auslender (1978), Differential stability in nonconvex and nondifferentiable programming, *Math. Progr. Study* **10**, 29–41.
57. A. Auslender (1984), Stability in mathematical programming with nondifferentiable data, *SIAM J. Control Optim.* **22**, 239–254.
58. A. Auslender and R. Cominetti (1991), A comparable study of multifunction differentiability with applications in mathematical programming, *Math. Oper. Res.* **16**, 240–258.
59. A. Auslender and J.-P. Crouzeix (1988), Global regularity theorems, *Math. Oper. Res.* **13**, 243–253.
60. A. Auslender and M. Teboulle (2003), *Asymptotic Cones and Functions in Optimization and Variational Inequalities*, Springer, New York.
61. D. Aussel, J.-N. Corvellec and M. Lassonde (1995), Mean value property and subdifferential criteria for lower semicontinuous functions, *Trans. Amer. Math. Soc.* **347**, 4147–4161.
62. D. Aussel, J.-N. Corvellec and M. Lassonde (1995), Nonsmooth constrained optimization and multidimensional mean value inequalities, *SIAM J. Optim.* **9**, 690–706.
63. D. Aussel, A. Daniilidis and L. Thibault (2004), Subsmooth sets: Functional characterizations and related concepts, *Trans. Amer. Math. Soc.* **357**, 1275–1302.
64. E. P. Avakov (1988), Necessary conditions of minima for nonregular problems in Banach spaces: Maximum principle for abnormal problems of optimal control, *Proc. Steklov Inst. Math.* **185**, 3–29.
65. E. R. Avakov, A. A. Agrachev and A. V. Arutyunov (1992), The level set of a smooth mapping in a neighborhood of a singular point, *Math. Sbornik* **73**, 455–466.
66. J. Avelin (1997), *Differential Calculus cfor Multifunctions and Nonsmooth Functions*, Ph.D. dissertation, Department of Mathematics, Uppsala University, Sweden.
67. J. Avelin (2000), Differential calculus for complex-valued multifunctions, *J. Appl. Anal.* **6**, 47–76.
68. V. I. Averbukh and O. G. Smolyanov (1968), The various definitions of the derivative in linear topological spaces, *Russian Math. Surveys* **23**, 67–113.
69. D. Azé and J.-N. Corvellec (2004), Characterizations of error bounds for lower semicontinuous functions on metric spaces, *Control Optim. Calc. Var.*, **10**, 409-425.
70. D. Azé, J.-N. Corvellec and R. E. Lucchetti (2002), Variational pairs and applications to stability in nonsmooth analysis, *Nonlinear Anal.* **49**, 643–670.
71. D. Azé and J.-B. Hiriart-Urruty (2002), Optimal Hoffman type estimates in eigenvalue and semidefinite inequality constraints, *J. Global Optim.* **24**, 133–147.
72. R. L. Baire (1905), *Lecons sur les Fonctions Discontinues*, Gauthier-Villars, Paris.
73. A. V. Balakrishnan (1965), Optimal control problems in Banach spaces, *SIAM J. Control* **3**, 152–180.
74. A. V. Balakrishnan (1981), *Applied Functional Analysis*, 2nd edition, Springer, New York.
75. E. J. Balder (2000), Lectures on Young measure theory and its applications to economics, *Rens. Istit. Mat. Univ. Trieste* **31**, 1–69.

76. S. BANACH (1932), *Théorie des Opérations Linéaires*, Monografje Matematyczne, Warsaw.
77. B. BANK, J. GUDDAT, D. KLATTE, B. KUMMER AND K. TAMMER (1982), *Non-Linear Parametric Optimization*, Birkhaüser, Basel.
78. H. T. BANKS (1968), Necessary conditions for control problems with variable time lags, *SIAM J. Control* **6**, 9–47.
79. H. T. BANKS AND G. A. KENT (1972), Control of functional differential equations of retarded and neutral type to target sets in function spaces, *SIAM J. Control* **10**, 567–591.
80. H. T. BANKS AND K. KUNISCH (1989), *Estimation Techniques for Distributed Parameter Systems*, Birkhäuser, Boston, Massachusetts.
81. H. T. BANKS AND A. MANITIUS (1974), Applications of abstract variational theory to hereditary systems – a survey, *IEEE Trans. Autom. Control* **19**, 524–533.
82. V. BARBU (1993), *Analysis and Control of Nonlinear Infinite Dimensional Systems*, Academic Press, Boston, Massachusetts.
83. V. BARBU, I. LASIECKA AND R. TRIGGIANI (2005), Boundary stabilization of Navier-Stokes equations, *Mem. Amer. Math. Soc.*, to appear.
84. V. BARBU AND T. PRECUPANU (1986), *Convexity and Optimization in Banach Spaces*, D. Reidel, Dordrecht, The Netherlands.
85. M. BARDI AND I. CAPUZZO DOLCETTA (1997), *Optimal Control and Viscosity Solutions of Hamilton-Jacobi Equations*, Birkhäuser, Boston, Massachusetts.
86. N. E. BARRON AND R. JENSEN (1991), Optimal control and semicontinuous viscosity solutions, *Proc. Amer. Math. Soc.* **113**, 397–402.
87. T. BAŞAR AND P. BERNHARD (1995), *H_∞-Optimal Control and Related Minimax Design Problems*, 2nd edition, Birkhäuser, Boston, Massachusetts.
88. H. BAUSCHKE, J. M. BORWEIN AND W. LI (1999), Strong conical hull intersection property, bounded linear regularity, Jameson's property (G), and error bounds in convex optimization, *Math. Progr.* **86**, 135–160.
89. M. S. BAZARAA, J. J. GOODE AND M. Z. NASHED (1974), An the cone of tangents with applications to mathematical programming, *J. Optim. Theory Appl.* **13**, 389–426.
90. C. R. BECTOR, S. CHANDRA AND J. DUTTA (2004), *Principles of Optimization Theory*, Alpha Science International Publishers, UK.
91. E. M. BEDNARCZUK (2003), Order-Lipschitzian properties of multifunctions with applications to stability of efficient points, *Control Cybernet.* **32**, 491–502.
92. G. BEER (1993), *Topologies on Closed and Convex Sets*, Kluwer, Dordrecht, The Netherlands.
93. S. BELLAASSALI AND A. JOURANI (2004), Necessary optimality conditions in multiobjective dynamic optimization, *SIAM J. Control Optim.* **42**, 2043–2061.
94. R. BELLMAN AND K. L. COOKE (1963), *Differential-Difference Equations*, Academic Press, New York.
95. R. BELLMAN, I. GLICKSBERG AND O. GROSS (1956), On the "bang-bang" control problem, *Quart. Appl. Math.* **14**, 11–18.
96. H. BENABDELLAH (2000), Existence of solutions to the nonconvex sweeping process, *J. Diff. Eq.* **164**, 286-295.
97. H. BENABDELLAH, C. CASTAING, A. SALVADORI AND A SYAM (1996), Nonconvex sweeping process, *J. Appl. Anal.* **2**, 217–240.

98. M. Benaim, J. Hofbauer and S. Sorin (2005), Stochastic approximations and differential inclusions, *SIAM J. Control Optim.* **44**, 328–348.
99. J. Benoist (1994), Approximation and regularization of arbitrary sets in finite dimensions, *Set-Valued Anal.* **2**, 95–115.
100. A. Bensoussan (1988), *Perturbation Methods in Optimal Control*, Wiley, New York.
101. A. Bensoussan, G. Da Prato, M. Delfour, S. K. Mitter (1992–93), *Representation and control of infinite-dimensional systems*, published in two volumes, Birkhäuser, Boston, Massachusetts.
102. A. Ben-Tal and J. Zowe (1985), Directional derivatives in nonsmooth optimization, *J. Optim. Theory App.* **47**, 483–490.
103. M. Bergounioux and D. Tiba (1996), General optimality conditions for constrained convex control problems, *SIAM J. Control Optim.* **34**, 698–711.
104. M. Bergounioux and F. Tröltzsch (1996), Optimality conditions and generalized bang-bang principle for a state-constrained semilinear parabolic problem, *Numer. Func. Anal. Optim.* **17**, 517–536.
105. L. D. Berkovitz (1974), *Optimal Control Theory*, Springer, New York.
106. L. D. Berkovitz (1976), A penalty function proof of the maximum principle, *Appl. Math. Optim.* **2**, 291–303.
107. H. Berliocchi and J. M. Lasry (1973), Principe de Pontryagin pour des systémes régis par une equation différentielle multivoque, *C. R. Acad. Sci. Paris* **277**, 1103–1105.
108. F. Bernard and L. Thibault (2004), Prox-regularity of functions and sets in Banach spaces, *Set-Valued Anal.* **12**, 25–47.
109. F. Bernard and L. Thibault (2005), Prox-regular functions in Hilbert spaces, *J. Math. Anal. Appl.* **303**, 1–14.
110. F. Bernard and L. Thibault (2005), Uniform prox-regularity of functions and epigraphs in Hilbert spaces, *Nonlinear Anal.* **60**, 187–207.
111. D. P. Bertsekas (1999), *Nonlinear Programming*, Athena Scientific, Boston, Massachusetts.
112. D. P. Bertsekas and A. E. Ozdaglar (2002), Pseudonormality and a Lagrange multiplier theory for constraint optimization, *J. Optim. Theory Appl.* **114**, 287–343.
113. D. Bessis, Y. S. Ledyaev and R. B. Vinter (2001) Dualization of the Euler and Hamiltonian inclusions, *Nonlinear Anal.* **43**, 861–882.
114. R.-M. Bianchini and M. Kawski (2003), Needle variations that cannot be summed, *SIAM J. Control Optim.* **42**, 218–238.
115. J. Birge and L. Qi (1995), Subdifferential convergence in stochastic programming, *SIAM J. Optim.* **5**, 436–453.
116. E. Bishop and R. R. Phelps (1963), The support functionals of convex sets, in *Convexity*, edited by V. Klee, vol. VII of *Proceedings of Symposia in Pure Mathematics*, pp. 27–35, American Mathematical Society, Providence, Rhode Island.
117. V. I. Blagodatskikh (1984), The maximum principle for differential inclusions, *Proc. Steklov Inst. Math.* **166**, 23–43.
118. V. I. Blagodatskikh and A. F. Filippov (1986), Differential inclusions and optimal control, *Proc. Steklov Inst. Math.* **169**, 199–259.
119. G. A. Bliss (1946), *Lectures on the Calculus of Variations*, University of Chicago Press, Chicago, Illinois.

120. N. A. Bobylev, S. V. Emel'yanov and S. K. Korovin (2004), Convexity of images of convex sets under smooth maps, *Comput. Math. Model.* **15**, 213–222.
121. N. N. Bogolyubov (1930), Sur quelques methodes nouvelles dans le calculus des variations, *Ann. Math. Pura Appl.* **7**, 249–271.
122. J. Bolte, A. Daniilidis and A. Lewis (2004), The Lojasiewicz inequality for nonsmooth subanalytic functions with applications to subgradient dynamical systems, preprint.
123. J. Bolte, A. Daniilidis and A. Lewis (2005), A nonsmooth Morse-Sard theorem for subanalytic functions, *J. Math. Anal. Appl.*, to appear.
124. V. G. Boltyanskii (1958), The maximum principle in the theory of optimal processes, *Dokl. Akad. Nauk SSSR* **119**, 1070–1073.
125. V. G. Boltyanskii (1968), Local section method in the theory of optimal processes, *Diff. Eq.* **4**, 2166–2183.
126. V. G. Boltyanskii (1973), The maximum principle for problems of optimal steering, *Diff. Eq.* **9**, 1363–1370.
127. V. G. Boltyanskii (1973), *Optimal Control of Discrete Systems*, Nauka, Moscow.
128. V. G. Boltyanskii (1994), The maximum principle – how it came to be?, Report No. 526, Math. Institut, Tech. Univ. München, München, Germany.
129. V. G. Boltyanskii, R. V. Gamkrelidze and L. S. Pontryagin (1956), On the theory of optimal processes, *Dokl. Akad. Nauk SSSR* **110**, 7–10.
130. O. Bolza (1913), Über den anormalen fall beim Lagrangeschen und Mayerschen problem mit gemischten behingugen und variabeln endpunkten, *Mathematische Annalen* **74**, 430–446.
131. J. F. Bonnans and E. Casas (1992), A boundary Pontryagin's principle for the optimal control of state-constrained elliptic systems, *Int. Ser. Numer. Math.* **107**, 241–249.
132. J. F. Bonnans, R. Cominetti and A. Shapiro (1999), Second-order optimality conditions based on parabolic second-order tangent sets, *SIAM J. Optim.* **9**, 466–492.
133. J. F. Bonnans and A. Shapiro (2000), *Perturbation Analysis of Optimization Problems*, Springer, New York.
134. J. F. Bonnans and A. Sulem (1995), Pseudopower expansion of solutions of generalized equations and constraint optimization problems, *Math. Progr.* **70**, 123–148.
135. J.-M. Bonnisseau and B. Cornet (1988), Valuation equilibrium and Pareto optimum in non-convex economies, *J. Math. Econ.* **17**, 293–308.
136. D. Borwein, J. M. Borwein and X. Wang (1996), Approximate subgradients and coderivatives in $I\!R^n$, *Set-Valued Anal.* **4**, 375–398.
137. J. M. Borwein (1986), Stability and regular points of inequality systems, *J. Optim. Theory Appl.* **48**, 9–52.
138. J. M. Borwein (1987), Epi-Lipschitz-like sets in Banach spaces: Theorems and examples, *Nonlinear Anal.* **11**, 1207–1217.
139. J. M. Borwein, M. Fabian, I. Kortezov and P. D. Loewen (2001), The range of the gradient of a continuously differentiable bump, *J. Nonlinear Convex Anal.* **2**, 1–19.
140. J. M. Borwein, M. Fabian and P. D. Loewen (2002), The shape of the range of a C^1 smooth bump in infinite dimensions, *Israel J. Math.* **132**, 239–251.

141. J. M. Borwein and S. P. Fitzpatrick (1995), Weak* sequential compactness and bornological limit derivatives, *J. Convex Anal.* **2**, 59–68.
142. J. M. Borwein and S. P. Fitzpatrick (1995), Characterization of Clarke subgradients among one-dimensional multifunctions, in *Proc. Optimization Miniconference II*, edited by B. M. Glover and V. Jeyakumar, pp. 61–64, University of New South Wales, Sydney, Australia.
143. J. M. Borwein and S. P. Fitzpatrick (2001), Duality inequalities and sandwiched functions, *Nonlinear Anal.* **46**, 365–380.
144. J. M. Borwein, S. P. Fitzpatrick and R. Girgensohn (2003), Subdifferentials whose graphs are not norm×weak* closed, *Canad. Math. Bull.* **46**, 538–545.
145. J. M. Borwein, S. P. Fitzpatrick and J. R. Giles (1987), The differentiability of real functions on normed linear spaces using generalized subgradients, *J. Math. Anal. Appl.* **128**, 512–534.
146. J. M. Borwein and J. R. Giles (1987), The proximal normal formula in Banach spaces, *Trans. Amer. Math. Soc.* **302**, 371–381.
147. J. M. Borwein and A. D. Ioffe (1996), Proximal analysis in smooth spaces, *Set-Valued Anal.* **4**, 1–24.
148. J. M. Borwein and A. Jofré (1998), A nonconvex separation property in Banach spaces, *Math. Meth. Oper. Res.* **48**, 169–179.
149. J. M. Borwein and A. S. Lewis (2000), *Convex Analysis and Nonlinear Optimization: Theory and Examples*, Springer, New York.
150. J. M. Borwein, Y. Lucet and B. S. Mordukhovich (2000), Compactly epi-Lipschitzian convex sets and functions in normed spaces, *J. Convex Anal.* **7**, 375–393.
151. J. M. Borwein, B. S. Mordukhovich and Y. Shao (1999), On the equivalence of some basic principles of variational analysis, *J. Math. Anal. Appl.* **229**, 228–257.
152. J. M. Borwein, W. B. Moors and X. Wang (2001), Generalized subdifferentials: A Baire categorical approach, *Trans. Amer. Math. Soc.* **353**, 3875–3893.
153. J. M. Borwein and D. Noll (1994), Second order differentiability of convex functions in Banach spaces, *Trans. Amer. Math. Soc.* **342**, 43–81.
154. J. M. Borwein and D. Preiss (1987), A smooth variational principle with applications to subdifferentiability and differentiability of convex functions, *Trans. Amer. Math. Soc.* **303**, 517–527.
155. J. M. Borwein and H. M. Strójwas (1985), Tangential approximations, *Nonlinear Anal.* **9**, 1347–1366.
156. J. M. Borwein and H. M. Strójwas (1986), Proximal analysis and boundaries of closed sets in Banach spaces, I: Theory, *Canad. J. Math.* **38**, 431–452.
157. J. M. Borwein and H. M. Strójwas (1987), Proximal analysis and boundaries of closed sets in Banach spaces, II: Applications, *Canad. J. Math.* **39**, 428–472.
158. J. M. Borwein, J. S. Treiman and Q. J. Zhu (1998), Necessary conditions for constrained optimization problems with semicontinuous and continuous data, *Trans. Amer. Math. Soc.* **350**, 2409–2429.
159. J. M. Borwein, J. S. Treiman and Q. J. Zhu (1999), Partially smooth variational principles and applications, *Nonlinear Anal.* **35**, 1031–1059.

160. J. M. Borwein and Q. J. Zhu (1996), Viscosity solutions and viscosity subderivatives in smooth Banach spaces with applications to metric regularity, *SIAM J. Control Optim.* **34**, 1568–1591.
161. J. M. Borwein and Q. J. Zhu (1997), Variational analysis in nonreflexive spaces and applications to control problems with L^1 perturbations, *Nonlinear Anal.* **28**, 889–915.
162. J. M. Borwein and Q. J. Zhu (1998), Limiting convex examples for nonconvex subdifferential calculus, *J. Convex Anal.* **5**, 221–235.
163. J. M. Borwein and Q. J. Zhu (1999), A survey of subdifferential calculus with applications, *Nonlinear Anal.* **38**, 687–773.
164. J. M. Borwein and Q. J. Zhu (2005) *Techniques of Variational Analysis*, Springer, New York.
165. J. M. Borwein and D. M. Zhuang (1988), Verifiable necessary and sufficient conditions for openness and regularity of set-valued and single-valued maps, *J. Math. Anal. Appl.* **134**, 441–459.
166. P. Bosch, A. Jourani and R. Henrion (2004), Sufficient conditions for error bounds and applications, *Appl. Math. Optim.* **50**, 161–181.
167. G. Bouligand (1930), Sur les surfaces dépourvues de points hyperlimits, *Ann. Soc. Polon Math.* **9**, 32–41.
168. G. Bouligand (1932), *Introduction à la Géométrie Infinitésimale Directe*, Gauthier-Villars, Paris.
169. R. D. Bourgin (1983), *Geometric Aspects of Convex Sets with the Radon-Nicodým Property*, Springer, New York, 1993.
170. M. Bounkhel (2004), Scalarization of the normal Fréchet regularity of set-valued mappings, *New Zealand J. Math.* **33**, 1–18.
171. M. Bounkhel and A. Jofré (2004), Subdifferential stability of the distance function and its applications to nonconvex economics and equilibrium, *J. Nonlinear Convex Anal.* **5**, 331–347.
172. M. Bounkhel and L. Thibault (2002), On various notions of regularity of sets in nonsmooth analysis, *Nonlinear Anal.* **48**, 223–246.
173. M. Bounkhel and L. Thibault (2005), Further characterizations of regular sets in Hilbert spaces and their applications to nonconvex sweeping processes, *J. Nonlinear Convex Anal.*, to appear.
174. K. E. Brennan, S. L. Campbell and L. R. Pretzold (1989), *Numerical Solution of Initial Value Problems in Differential-Algebraic Equations*, North-Holland, New York.
175. A. Bressan (1990), Upper and lower semicontinuous differential inclusions: A unified approach, in *Nonlinear Controllability and Optimal Control*, edited by H. J. Sussmann, pp. 21-31, Marcel Dekker, New York.
176. O. A. Brezhneva and A. A. Tret'yakov (2003), Optimality conditions for degenerate extremum problems with equality constraints, *SIAM J. Control Optim.* **42**, 729–745.
177. H. Bréziz and W. Strauss (1973), Semilinear second-order elliptic equations in L^1, *J. Math. Soc. Japan* **25**, 565–590.
178. J. Brinkhuis and V. M. Tikhomirov (2005), *Optimization: Motivations and Applications*, Princeton University Press, Princeton, New Jersey.
179. A. Brøndsted and R. T. Rockafellar (1965), On the subdifferentiability of convex functions, *Proc. Amer. Math. Soc.* **16**, 605–611.
180. L. E. J. Brouwer (1910), Über eineindeutige, stetige transformationen von flächen in sich, *Math. Ann.* **69**, 176–180.

181. D. J. Brown (1991), Equilibrium analysis with nonconvex technologies, in *Handbook of Mathematical Economics*, Vol. 4, edited by W. Hildenbrand and H. Sonnenschein, North-Holland, Amsterdam, The Netherlands.
182. A. Bruckner (1994), *Differentiation of Real Functions*, American Mathematical Society, Providence, Rhode Island.
183. F. Bucci (1992), A Dirichlet boundary control problem for the strongly damped wave equation, *SIAM J. Control Optim.* **30**, 1092–1100.
184. B. M. Budak, E. M. Berkovich and E. N. Solovieva (1969), Difference approximations in optimal control problems, *SIAM J. Control* **7**, 18–31.
185. B. M. Budak and F. P. Vasiliev (1975), *Some Computational Aspects of Optimal Control Problems*, Moscow University Press, Moscow.
186. R. S. Burachik and A. N. Iusem (2006), *Set-Valued Mappings and Enlargements of Monotone Operators*, SIAM Publications, Philadelphia, Pennsylvania.
187. R. S. Burachik, A. N. Iusem and B. F. Svaiter (1997), Enlargement of monotone operators with applications to variational inequalities, *Set-Valued Anal.* **5**, 159–180.
188. J. V. Burke (1991), Calmness and exact penalization, *SIAM J. Control Optim.* **29**, 493–497.
189. J. V. Burke (1991), An exact penalization viewpoint of constrained optimization, *SIAM J. Control Optim.* **29**, 968–998.
190. J. V. Burke and S. Deng (2002), Weak sharp minima revisted, I: Basic theory, *Control Cybernet.* **31**, 439–469.
191. J. V. Burke and S. Deng (2005), Weak sharp minima revisted, II: Application to linear regularity and error bounds, *Math. Progr.*, to appear.
192. J. V. Burke and M. C. Ferris (1993), Weak sharp minima in mathematical programming, *SIAM J. Control Optim.* **31**, 1340–1359.
193. J. V. Burke, M. C. Ferris and M. Quian (1992), *J. Math. Anal. Appl.* **166**, 199–213.
194. J. V. Burke, A. S. Lewis and M. L. Overton (2000), Optimizing matrix stability, *Proc. Amer. Math. Soc.* **129**, 1635–1642.
195. J. V. Burke, A. S. Lewis and M. L. Overton (2001), Optimal stability and eigenvalue multiplicity, *Found. Comput. Math.* **1**, 205–225.
196. J. V. Burke, A. S. Lewis and M. L. Overton (2002), Approximating subdifferentials by random sampling of gradients, *Math. Oper. Res.* **27**, 567–584.
197. J. V. Burke, A. S. Lewis and M. L. Overton (2003), Robust stability and a criss-cross algorithm for pseudospectra, *IMA J. Numerical Anal.* **23**, 1–17.
198. J. V. Burke, A. S. Lewis and M. L. Overton (2004), Variational analysis of the abscissa mapping for polynomials via the Gauss-Lucas theorem, *J. Global Optim.* **28**, 259–268.
199. J. V. Burke, A. S. Lewis and M. L. Overton (2004), A robust gradient sampling algorithm for nonsmooth, nonconvex optimization, *SIAM J. Optim.* **15**, 751–779.
200. J. V. Burke, A. S. Lewis and M. L. Overton (2005), Variational analysis of functions of the roots of polynomials, *Math. Progr.*, to appear.
201. J. V. Burke and D. R. Luke (2003), Variational analysis applied to the problem of optical phase retrieval, *SIAM J. Control Optim.* **42**, 576–595.

202. J. V. Burke and M. L. Overton (1992), in *Nonsmooth Optimization Methods and Applications*, edited by F. Giannessi, pp. 19–29, Gordon and Breach, Philadelphia, Pennsylvania.
203. J. V. Burke and M. L. Overton (2000), Variational analysis of the abscissa mapping for polynomials, *SIAM J. Control Optim.* **39**, 1651–1676.
204. J. V. Burke and M. L. Overton (2001), Variational analysis of non-Lipschitz spectral functions, *Math. Progr.* **90**, 317–351.
205. J. V. Burke and P. Tseng (1996), A unified analysis of Hoffman's bound via Fenchel duality, *SIAM J. Optim.* **6**, 265–282.
206. J. A. Burns (2003), Nonlinear distributed parameter control systems with non-normal linearizations: Applications and approximations, *Front. Appl. Math.* **27**, 17–53.
207. M. Bustos (1994), Epsilon gradients pour functions localements Lipschitziennes et applications, *Numer. Funct. Anal. Optim.* **15**, 435–453.
208. A. G. Butkovsky (1963), Necessary and sufficient optimality conditions for impulse control systems, *Autom. Remote Control* **24**, 1056–1064.
209. A. G. Butkovsky (1969), *Distributed Control Systems*, Elsevier, New York.
210. A. G. Butkovsky, A. I. Egorov and K. A. Lurie (1968), Optimal control of distributed systems (a survey of Soviet publications), *SIAM J. Control* **6**, 437–476.
211. A. G. Butkovsky and A. J. Lerner (1960), Optimal control systems with distributed parameters, *Autom. Remote Control* **21**, 472–477.
212. G. Buttazzo and G. Dal Maso (1982), Γ-convergence and optimal control problems, *J. Optim. Theory Appl.* **38**, 385–407.
213. G. Buttazzo, M. Giaquinta and S. Hildebrandt (1998), *One-Dimensional Variational Problems: An Introduction*, Oxford University Press, New York.
214. F. Camilli and M. Falcone (1996), Approximation of optimal control problems with state constraints: Estimates and applications, in *Nonsmooth Analysis and Geometric Methos in Deterministic Optimal Control*, edited by B. S. Mordukhovich and H. J. Sussmann, pp. 23–58, Springer, New York.
215. P.-M. Cannarsa and G. Da Prato (1991), Second-order Hamilton-Jacobi equations in infinite dimensions, *SIAM J. Control Optim.* **29**, 474–492.
216. P.-M. Cannarsa and H. Frankowska (1991), Some characterizations of optimal trajectories in control theory, *SIAM J. Control Optim.* **29**, 1322–1347.
217. P.-M. Cannarsa and C. Sinestrari (2004), *Semiconvex Functions, Hamilton-Jacobi Equations, and Optimal Control*, Birkhäuser, Boston, Massuchusetts.
218. M. Cannon, C. Cullum and E. Polak (1970), *Theory of Optimal Control and Mathematical Programming*, McGraw-Hill, New York.
219. M. J. Cánovas, A. L. Dontchev, M. A. López and J. Parra (2005), Metric regularity of semi-infinite constraint systems, *Math. Progr.*, to appear.
220. I. Capuzzo Dolcetta (1983), On a discrete approximation of the Hamilton-Jacobi bequation of dynamic programming, *Appl. Math. Optim.* **10**, 367–377.
221. I. Capuzzo Dolcetta and M. Falcone (1989), Discrete dynamic programming and visocity solutions of the Bellman equation, *Ann. Inst. H. Poincaré: Analyse Non Linéaire* **6**, 161–184.
222. C. Carathéodory (1935), *Variationsrechung und Partielle Differentialgleichungen erster Ordnung*, B. G. Teubner, Leipzig and Berlin, 1935.

223. O. CÂRJĂ (1984), The time optimal problem for boundary-distributed control systems, *Bull. Unione Mat. Italiana* **3-B**, 563–581.
224. D. A. CARLSON, A. HAURIE AND A. LEIZAROWITZ (1991), *Infinite Horizon Optimal Control*, Springer, Berlin.
225. E. CASAS (1997), Pontryagin's principle for state-constrained boundary control problems of semilinear parabolic equations, *SIAM J. Control Optim.* **35**, 1297–1327.
226. E. CASAS, J.-P. RAYMOND AND H. ZIDANI (2000), Pontryagin's principle for local solutions of control problems with mixed control-state constraints, *SIAM J. Control Optim.* **39**, 1182–1203.
227. E. CASAS AND J. YONG (1995), Maximum principle for state-constrained optimal control problems governed by quasilinear elliptic equations, *Diff. Integ. Eq.* **8**, 1–18.
228. C. CASTAING, A. SALVADORI AND L. THIBAULT (2001), Functional evolution equations governed by nonconvex sweeping process, *J. Nonlinear Convex Anal.* **2**, 217–241.
229. C. CASTAING AND M. VALADIER (1977), *Convex Analysis and Measurable Multifunctions*, Springer, New York.
230. E. CAVAZZUTI, M. PAPPALARDO AND M. PASSACANTANDO (2002), Nash equilibria, variational inequalities, and dynamical systems, *J. Optim. Theory Appl.* **114**, 491–505.
231. A. CELLINA (1988), On the set of solutions to Lipschitzian differential inclusions, *Diff. Integ. Eq.* **1**, 459–500.
232. A. CELLINA (2003), The classical problem of the calculus of variations in the autonomous case: Relaxation and Lipschitzianity of solutions, *Trans. Amer. Math. Soc.* **356**, 415–426.
233. A. CERNEA (2001), Some second-order necessary conditions for non-convex hyperbolic differential inclusions problems, *J. Math. Anal. Appl.* **253**, 616–639.
234. A. CERNEA (2005), Second-order necessary conditions for differential-difference inclusion problems, *Nonlinear Anal.* **62**, 963–974.
235. L. CESARI (1983), *Optimization – Theory and Applications*, Springer, New York.
236. R. W. CHANEY (1987), Second-order directional derivatives for nonsmooth functions, *J. Math. Anal. Appl.* **128**, 495–511.
237. A. V. CHERKAEV (2000), *Variational Methods for Structural Optimization*, Springer, New York.
238. F. L. CHERNOUSKO (1994), *State Estimation for Dynamic Systems*, CRC Press, Boca Raton, Florida.
239. A. A. CHIKRII (1997), *Conflict-Controlled Processes*, Kluwer, Dordrecht, The Netherlands.
240. I. CHRYSSOCHOOS AND R. B. VINTER (2003), Optimal control on manifolds: A dynamic programming approach, *J. Math. Anal. App.* **287**, 118–140.
241. E. N. CHUKWU (1992), *Stability and Time-Optimal Control of Hereditary Systems*, Academic Press, Boston, Massachusetts.
242. M. CILIGOT-TRAVAIN AND S. TRAORE (2002), On subgradients of spectral functions, *J. Convex Anal.* **9**, 401–414.
243. F. H. CLARKE (1973), *Necessary Conditions for Nonsmooth Problems in Optimal Control and the Calculus of Variations*, Ph.D. dissertation, Department of Mathematics, University of Washington, Seattle.

244. F. H. CLARKE (1975), Generalized gradients and applications, *Trans. Amer. Math. Soc.* **205**, 247–262.
245. F. H. CLARKE (1975), The Euler-Lagrange differential inclusion, *J. Diff. Eq.* **19**, 80–90.
246. F. H. CLARKE (1975), Admissible relaxation in variational and control problems, *J. Math. Anal. Appl.* **51**, 557–576.
247. F. H. CLARKE (1976), Optimal solutions to differential inclusions, *J. Optim. Theory Appl.* **19**, 469–478.
248. F. H. CLARKE (1976), The generalized problem of Bolza, *SIAM J. Control Optim.* **14**, 682–699.
249. F. H. CLARKE (1976), A new approach to Lagrange multipliers, *Math. Oper. Res.* **2**, 165–174.
250. F. H. CLARKE (1976), The maximum principle under minimal hypotheses, *SIAM J. Control Optim.* **14**, 1078–1091.
251. F. H. CLARKE (1976), Necessary conditions for a general control problem, in *Calculus of Variations and Control Theory*, edited by D. L. Russel, pp. 257–278, Academic Press, New York.
252. F. H. CLARKE (1976), On the inverse function theorem, *Pacific J. Math.* **64**, 97–102.
253. F. H. CLARKE (1977), Extremal arcs and extended Hamiltonian systems, *Trans. Amer. Math. Soc.* **231**, 349–367.
254. F. H. CLARKE (1980), The Erdmann condition and Hamiltonian inclusions in optimal control and the calculus of variations, *Can. J. Math.* **32**, 494–509.
255. F. H. CLARKE (1983), *Optimization and Nonsmooth Analysis*, Wiley, New York.
256. F. H. CLARKE (1987), Hamiltonian analysis of the generalized problem of Bolza, *Trans. Amer. Math. Soc.* **301**, 385–400.
257. F. H. CLARKE (1989), *Methods of Dynamic and Nonsmooth Optimization*, SIAM Publications, Philadelphia, Pennsylvania.
258. F. H. CLARKE (1993), A decoupling principle in the calculus of variations, *J. Math. Anal. Appl.* **172**, 92–105.
259. F. H. CLARKE (1993), An indirect method in the calculus of variations, *Trans. Amer. Math. Soc.* **336**, 655-673.
260. F. H. CLARKE (2003), Necessary conditions in optimal control: A new approach, *Math. Progr.* **97**, 71–89.
261. F. H. CLARKE (2005), Necessary conditions in dynamic optimization, *Mem. Amer. Math. Soc.* **173**, No. 816.
262. F. H. CLARKE AND Y. S. LEDYAEV (1994), Mean value inequality in Hilbert spaces, *Trans. Amer. Math. Soc.* **344**, 307–324.
263. F. H. CLARKE, Y. S. LEDYAEV, E. D. SONTAG AND A. I. SUBBOTIN (1997), Asymptotic controllability implies feedback stabilization, *IEEE Trans. Autom. Control* **42**, 1394–1407.
264. F. H. CLARKE, Y. S. LEDYAEV, R. J. STERN AND P. R. WOLENSKI (1995), Qualitative properties of trajectories of control systems: A survey, *J. Dynam. Control Systems* **1**, 1–47.
265. F. H. CLARKE, Y. S. LEDYAEV, R. J. STERN AND P. R. WOLENSKI (1998), *Nonsmooth Analysis and Control Theory*, Springer, New York.
266. F. H. CLARKE, Y. S. LEDYAEV AND P. R. WOLENSKI (1995), Proximal analysis and minimization principles, *J. Math. Anal. Appl.* **196**, 722–735.

267. F. H. Clarke and P. D. Loewen (1986), The value function in optimal control: Sensitivity, controllability and time-optimality, *SIAM J. Control Optim.* **24**, 243–263.
268. F. H. Clarke and P. D. Loewen (1987), State constraints in optimal control: A case study in proximal normal analysis, *SIAM J. Control Optim.* **25**, 1440–1456.
269. F. H. Clarke and R. J. Stern (2003), State constrained feedback stabilization, *SIAM J. Control Optim.* **42**, 422–441.
270. F. H. Clarke, R. J. Stern and P. R. Wolenski (1993), Subgradient criteria for monotonicity and the Lipschitz condition, *Canad. J. Math.* **45**, 1167–1183.
271. F. H. Clarke, R. J. Stern and P. R. Wolenski (1995), Proximal smoothness and the lower-C^2 property, *J. Convex Anal.* **2**, 117–144.
272. F. H. Clarke and R. B. Vinter (1989), Applications of optimal multiprocesses, *SIAM J. Control Optim.* **25**, 1048–1071.
273. F. H. Clarke and R. B. Vinter (1989), Optimal multiprocesses, *SIAM J. Control Optim.* **25**, 1072–1091.
274. F. H. Clarke and G. G. Watkins (1986), Necessary conditions, controllability and the value function for differential-difference inclusions, *Nonlinear Anal.* **10**, 1155–1179.
275. F. H. Clarke and P. R. Wolenski (1991), The sensitivity of optimal control problems to time delay, *SIAM J. Control Optim.* **29**, 1176–1215.
276. F. H. Clarke and P. R. Wolenski (1996), Necessary conditions for functional differential inclusions, *Appl. Math. Optim.* **34**, 34–51.
277. G. Colombo and V. V. Goncharov (1999), The sweeping processes without convexity, *Set-Valued Anal.* **7**, 357–374.
278. G. Colombo and V. V. Goncharov (2001), Variational inequalities and regularity properties of closed sets in Hilbert spaces, *J. Convex Anal.* **8**, 197–221.
279. G. Colombo and A. Marigonda (2005), Differentiability properties for a class of nonconvex functions, *Calc. Var. Partial Diff. Eq.*, to appear.
280. G. Colombo and P. R. Wolenski (2004), Variational analysis for a class of minimal time functions in Hilbert spaces, *J. Convex Anal.* **11**, 335–361.
281. F. Colonius (1982), The maximum principle for relaxed hereditary differential systems with function space end conditions, *SIAM J. Control Optim.* **20**, 695–712.
282. R. Cominetti (1990), Metric regularity, tangent cones, and second-order optimality conditions, *Appl. Math. Optim.* **21**, 265–287.
283. R. Cominetti and R. Correa (1990), A generalized second order derivative in nonsmooth optimization, *SIAM J. Optim.* **28**, 789–809.
284. R. Conti (1968), Time-optimal solution of a linear evolution equation in Banach spaces, *J. Optim. Theory Appl.* **2**, 277–284.
285. B. Cornet (1981), Regular properties of tangent and normal cones, CEREMADE Publication 8130, Universiteé de Paris IX "Dauphine".
286. B. Cornet (1986), The second welfare theorem in nonconvex economies, CORE discussion paper No. 8630.
287. B. Cornet (1988), General equilibrium theorem and increasing returns: Presentation, *J. Math. Econ.* **17**, 103–118.

288. B. CORNET (1990), Marginal cost pricing and Pareto optimality, in *Essays in Honor of Edmond Malinvaud*, edited by P. Champsaur, Vol. 1, pp. 14–53, MIT Press, Cambridge, Massachusetts.
289. B. CORNET AND M.-O. CZARNECKI (1999), Smooth normal approximations of epi-Lipschitz subset in $\mathbb{R}^n$, *SIAM J. Control Optim.* **37**, 710–730.
290. B. CORNET AND M.-O. CZARNECKI (2001), Existence of generalized equilibria: Necessary and sufficient conditions, *Nonlinear Anal.* **44**, 555–574.
291. R. CORREA, P. GAJARDO AND L. THIBAULT (2005), Subdifferential representation formula and subdifferential criteria for the behavior of nonsmooth functions, *Nonlinear Anal.*, to appear.
292. R. CORREA, A. JOFRÉ AND L. THIBAULT (1994), Subdifferential monotonicity as characterization of convex functions, *Numer. Funct. Anal. Optim.* **15**, 1167–1183.
293. R. W. COTTLE, F. GIANNESSI AND J.-L. LIONS, ed. (1980), *Variational Inequalities and Complementarity Problems*, Wiley, New York.
294. R. W. COTTLE, J.-S. PANG AND R. E. STONE (1992), *The Linear Complementarity Problem*, Academic Press, Boston, Massachusetts.
295. M. G. CRANDALL, L. C. EVANS AND P.-L. LIONS (1984), Some properties of viscosity solutions, *Trans. Amer. Math. Soc.* **282**, 487–502.
296. M. G. CRANDALL, H. ISHII AND P.-L. LIONS (1992), User's guide to viscosity solutions of second-order partial differential equations, *Bull. Amer. Math. Soc.* **27**, 1–67.
297. M. G. CRANDALL AND P.-L. LIONS (1983), Viscosity solutions of Hamilton-Jacobi equations, *Trans. Amer. Math. Soc.* **277**, 1–42.
298. B. D. CRAVEN (1994), Convergence of discrete approximations for constrained minimization, *J. Austral. Math. Soc.* **35**, 1–12.
299. B. D. CRAVEN (1995), *Control and Optimization*, Chapman and Hall, London, UK.
300. B. D. CRAVEN AND D. V. LUU (1997), A method for establishing optimality conditions for a nonconvex vector-valued minimax problem, *J. Optim. Theory Appl.* **95**, 295–304.
301. G. P. CRESPI, D. LA TORRE AND M. ROCCA (2003), Second-order mollified derivatives and optimization, *J. Nonlinear Convex Anal.* **3**, 437-454.
302. J. CULLUM (1969), Discrete approximations to continuous optimal control problems, *SIAM J. Control* **7**, 32–49.
303. J. CULLUM (1972), Finite-dimensional approximations of state constrained continuous optimal control problems, *SIAM J. Control* **10**, 649–670.
304. M.-O. CZARNECKI AND L. RIFFORD (2005), Approximation and regularization of Lipschitz functions: Convergence of the gradients, *Trans. Amer. Math. Soc.*, to appear.
305. L. DAI (1989), *Singular Control Systems*, Springer, Berlin.
306. P. DANIELE AND A. MAUGERI (2001), On dynamical equilibrium problems and variational inequalities, in *Equilibrium Problems: Nonsmooth Optimization and Variational Inequality Models*, edited by F. Giannessi, A. Maugeri and P. Pardalos, pp. 59–69, Kluwer, Dordrecht, The Netherlands.
307. J. M. DANSKIN (1967), *The Theory of Min-Max and Its Application to Weapon Allocations Problems*, Springer, New York.
308. F. S. DE BLASI, G. PIANIGIANI AND A. A. TOLSTONOGOV (2004), A Bogolyubov type theorem with a nonconvex constraint in Banach spaces, *SIAM J. Control Optim.* **43**, 466-476.

309. G. Debreu (1951), The coefficient of resource utilization, *Econometrica* **19**, 273–292.
310. G. Debreu (1959), *Theory of Values*, Yale University Press, New Haven, Connecticut.
311. G. Debreu (1970), Economies with a finite set of equilibria, *Econometrica* **38**, 387–392.
312. E. De Giorgi, A. Marino and M. Tosques (1980), Problemi di evoluzione in spazi metric e curve di massima pendenza, *Atti. Accad. Naz. Lincei Rend. Cl. Sci. Fis. Mat. Natur.* **68**, 180–187.
313. M. Degiovanni, A. Marino and M. Tosques (1985), Evolution equations with lack of convexity, *Nonlinear Anal.* **9**, 1401–1443.
314. K. Deimling (1992), *Multivalued Differential Equations*, De Gruyter, Berlin.
315. J. Demmel (1987), The condition number and the distance to the nearest ill-posed problem, *Numerische Math.* **51**, 251–289.
316. S. Dempe (2002), *Foundations of Bilevel Programming*, Kluwer, Dordrecht, The Netherlands.
317. S. Dempe (2003), Annotated bibliography on bilevel programming and mathematical programs with equilibrium constraints, *Optimization* **52**, 333–359.
318. V. F. Demyanov (2005), *Extremality Conditions and Calculus of Variations*, Vysshaya Shkola, Moscow.
319. V. F. Demyanov and V. N. Malozemov (1974), *Introduction to Minimax*, Wiley, New York.
320. V. F. Demyanov amd A. M. Rubinov (1980), On quasidifferentiable functionals, *Dokl. Akad. Nauk SSSR* **250**, 21–25.
321. V. F. Demyanov and A. M. Rubinov (1995), *Constructive Nonsmooth Analysis*, Peter Lang, Frankfurt, Germany.
322. V. F. Demyanov and A. M. Rubinov, eds. (1995), *Quasidifferentiability and Related Topics* (2000), Kluwer, Dordrecht, The Netherlands.
323. Z. Denkowski, S. Migórski and N. S. Papageorgiou (2004), *An Introduction to Nonlinear Analysis and Its Applications*, published in two volumes, Kluwer, Boston, Massachusetts.
324. D. Dentcheva and W. Römisch (2000), Differential stability of two-stage stochastic programs, *SIAM J. Optim.* **11**, 87–112.
325. D. Dentcheva and A. Ruszczyński (2004), Optimization with stochastic dominance constraints, *SIAM J. Optim.* **14**, 548-566.
326. A. De Simone and M. G. Graziano (2004), The pure theory of public goods: The case of many commodities, *J. Math. Econ.* **40**, 847–868.
327. E. N. Devdariani and Y. S. Ledyaev (1999), Maximum principle for implicit control systems, *Appl. Math. Optim.* **41**, 79–103.
328. R. Deville (1994), Stability of subdifferentials of nonconvex functions in Banach spaces, *Set-Valued Anal.* **2**, 141–157.
329. R. Deville and E. M. El Haddad (1996), The viscosity subdifferential of the sum of two functions in Banach spaces, I: First order case, *J. Convex Anal.* **3**, 295–308.
330. R. Deville, G. Godefroy and V. Zizler (1993), A smooth variational principle with applications to Hamilton-Jacobi equations in infinite dimensions, *J. Funct. Anal.* **111**, 197–212.
331. R. Deville, G. Godefroy and V. Zizler (1993), *Smoothness and Renorming in Banach Spaces*, Wiley, New York.

332. J. DIESTEL (1975), *Geometry of Banach Spaces–Selected Topics*, Springer, Berlin.
333. J. DIESTEL (1984) *Sequences and Series in Banach Spaces*, Springer, New York.
334. J. DIESTEL AND J. J. UHL, JR. (1977), *Vector Measures*, American Mathematical Society, Providence, Rhode Island.
335. U. DINI (1878), *Fondamenti per la Teoria delle Funzioni di Variabili Reali*, Pisa, Italy.
336. A. V. DMITRUK (2002), An nonlocal Lyusternik estimate and its applications to control systems with sliding modes, in *Nonlinear Control Systems 2001*, edited by A. B. Kurzhanskii and A. L. Fradkov, Vol. 2, pp. 1061–1064, Elsevier, Exeter, UK.
337. A. V. DMITRUK, A. A. MILYUTIN AND N. P. OSMOLOVSKII (1980), Lyusternik's theorem and the theory of extrema, *Russian Math. Surveys* **35**, 11–51.
338. B. DOITCHINOV AND V. VELIOV (1993), Parameterizations of integrals of set-valued mappings and applications, *J. Math. Anal. Appl.* **179**, 483–499.
339. S. DOLECKI (1980), A general theory of necessary optimality conditions, *J. Math. Anal. Appl.* **78**, 267–308.
340. S. DOLECKI (1982), Tangency and differentiation: Some applications of convergence theory, *Ann. Mat. Pura Appl.* **130**, 223–255.
341. S. DOLECKI AND S. ROLEWICZ (1979), Exact penalties for local minima, *SIAM J. Control Optim.* **17**, 596–606.
342. T. DONCHEV (1997), Lower semicontinuous differential inclusions: One-sided Lipschitz approach, *Coll. Math.* **74**, 177–184.
343. T. DONCHEV (2005), Approximation of solution sets for optimal control problems, in *Large-Scale Scientific Computations*, edited by I. Lirkov, S. Margenov and J. Wasniewski, Lecture Notes Comp. Sci., Springer, Berlin.
344. T. DONCHEV AND E. FARKHI (1998), Stability and Euler approximation of one-sided Lipschitz differential inclusions, *SIAM J. Control Optim.* **36**, 780–796.
345. T. DONCHEV AND E. FARKHI (1999), Euler approximation of discontinuous one-sided Lipschitz convex differential inclusions, in *Calculus of Variations and Differential Equations*, edited by A. Ioffe, S. Reich and I. Shafrir, pp. 101–118, Chapman and Hall/CRC Press, Boca Raton, Florida.
346. T. DONCHEV AND B. S. MORDUKHOVICH (2005), Strong convergence of discrete approximations for one-sided Lipschitzian differential inclusions, preprint.
347. A. L. DONTCHEV (1981), Error estimates for s discrete approximation to constrained control problems, *SIAM J. Numer. Anal.* **18**, 500–514.
348. A. L. DONTCHEV (1983), *Perturbations, Approximations and Sensitivity Analysis of Optimal Control Systems*, Springer, Berlin.
349. A. L. DONTCHEV (1988), Equivalent perturbations and approximations in optimal control, in *International Series of Numerical Mathematics* **84**, pp. 43–54, Birkhäuser, Basel, Switzerland.
350. A. L. DONTCHEV (1995), Characterization of Lipschitz stability in optimization, in *Recent Developments in Well-Posed Variational Problems and Related Topics*, edited by R. Lucchetti and J. Revalski, pp. 95–116, Kluwer, Dordrecht, The Netherlands.

351. A. L. DONTCHEV (1995), Implicit function theorems for generalized equations, *Math. Progr.* **70**, 91–106.
352. A. L. DONTCHEV (1996), The Graves theorem revisited, *J. Convex Anal.* **3**, 45–54.
353. A. L. DONTCHEV (1996), Discrete approximations in optimal control, in *Nonsmooth Analysis and Geometric Methods in Deterministic Optimal Control*, edited by B. S. Mordukhovich and H. J. Sussmann, pp. 59–81, Springer, New York.
354. A. D. DONTCHEV AND E. M. FARKHI (1989), Error estimates for discretized differential inclusions, *Computing* **41**, 349–358.
355. A. L. DONTCHEV AND W. W. HAGER (1993), Lipschitzian stability in nonlinear control and optimization, *SIAM J. Control Optim.* **31**, 569–603.
356. A. L. DONTCHEV AND W. W. HAGER (1994), Implicit functions, Lipschitz maps and stability in optimization, *Math. Oper. Res.* **19**, 753–768.
357. A. L. DONTCHEV AND W. W. HAGER (1998), A new approach to Lipschitz continuity in state constrained optimal control, *Syst. Control Lett.* **35**, 137–143.
358. A. L. DONTCHEV AND W. W. HAGER (2001), The Euler approximation in state constrained optimal control, *Math. Comp.* **70**, 173–203.
359. A. L. DONTCHEV AND F. LEMPIO (1992), Difference methods for differential inclusions: A survey, *SIAM Rev.* **34**, 263–294.
360. A. L. DONTCHEV AND A. S. LEWIS (2005), Perturbations and metric regularity, *Set-Valued Anal.*, to appear.
361. A. L. DONTCHEV, A. S. LEWIS AND R. T. ROCKAFELLAR (2003), The radius of metric regularity, *Trans. Amer. Math. Soc.* **355**, 493–517.
362. A. L. DONTCHEV AND B. S. MORDUKHOVICH (1983), Relaxation and well-posedness of nonlinear optimal processes, *Syst. Control Lett.* **3**, 177–179.
363. A. L. DONTCHEV, M. QUINCAMPOIX AND N. ZLATEVA (2005), Aubin criterion for metric regularity, *J. Convex Anal.*, to appear.
364. A. L. DONTCHEV AND R. T. ROCKAFELLAR (1996), Characterizations of strong regularity for variational inequalities over polyhedral convex sets, *SIAM J. Optim.* **7**, 1087–1105.
365. A. L. DONTCHEV AND R. T. ROCKAFELLAR (2001), Ample parameterization of variational inclusions, *SIAM J. Optim.* **12**, 170–187.
366. A. L. DONTCHEV AND R. T. ROCKAFELLAR (2004), Regularity and conditioning of solution mappings in variational analysis, *Set-Valued Anal.* **12**, 79–109.
367. A. L. DONTCHEV AND T. ZOLEZZI (1993), *Well-Posed Optimization Problems*, Springer, New York.
368. A. DOUGLIS (1961), The continuous dependence of generalized solutions of non-linear partial differential equations upon initial data, *Comm. Pure Appl. Math.* **14**, 267–284.
369. A. Y. DUBOVITSKII AND A. A. MILYUTIN (1963), Extremum problems in the presence of constraints, *Dokl. Akad. Nauk SSSR* **149**, 759–762.
370. A. Y. DUBOVITSKII AND A. A. MILYUTIN (1965), Extremum problems in the presence of restrictions, *USSR Comput. Maths. Math. Phys.* **5**, 1–80.
371. N. DUNFORD AND J. T. SCHWARTZ (1958), *Linear Operators, I: General Theory*, Interscience, New York.
372. J. C. DUNN (1995), Second order optimality conditions in sets of L^∞ functions with range in a polyhedron, *SIAM J. Control Optim.* **33**, 1603–1635.

373. M. Dür, R. Horst and M. Locatelli (1998), Necessary and sufficient conditions for convex maximization revisited, *J. Math. Anal. Appl.* **217**, 637–649.
374. J. Dutta (2005), Necessary optimality conditions and saddle points for approximate optimization in Banach spaces, *TOP: Spanish J. Oper. Res.*, **13**, 127–143.
375. J. Dutta (2005), Optimality conditions for maximizing a locally Lipschitz function, *Optimization* **54**, 377–389.
376. J. Dutta (2005), Generalized derivatives and nonsmooth optimization: A finite-dimensional tour, *TOP: Spanish J. Oper. Res.*, **13**.
377. J. Dutta and S. Dempe (2005), Bilevel programming with convex lower level problems, in *Optimization with Multivalued Mappings: Theory, Applications and Algorithms*, edited by S. Dempe and V. V. Kalashnikov, Springer, Berlin.
378. J. Dutta and C. Tammer (2005), Lagrangian conditions for vector optimization on Banach spaces, *Math. Meth. Oper. Res.*, to appear.
379. A. N. Dyukalov (1983), *Problems of Applied Mathematical Economics*, Nauka, Moscow.
380. Z. Dzalilov, A. F. Ivanov and A. M. Rubinov (2001), Difference inclusions with delay of economic growth, *Dynam. Syst. Appl.* **10**, 283–293.
381. A. Eberhard (2001), Prox-regularity and subjects, in *Optimization and Related Topics*, edited by A. Rubinov and B. Glover, *Applied Optimization* **47**, pp. 237–313, Kluwer, Dordrecht, The Netherlands.
382. A. Eberhard and M. Nyblom (1998), Jets, generalized convexity, proximal normality and differences of functions, *Nonlinear Anal.* **34**, 319–360.
383. A. Eberhard, M. Nyblom and D. Ralph (1998), Applying generalized convexity notions to jets, in *Generalized Convexity, Generalized Monotonicity: Recent Results*, edited by J.-P. Crouzeix, J. E. Martinez-Legaz and M. Volle, pp. 111–157, Kluwer, Dordrecht, The Netherlands.
384. A. Eberhard and C. E. M. Pearce (2005), A comparison of two approaches to second-order subdifferentiability concepts with application to optimality conditions, in *Optimization and Control with Applications*, edited by L. Qi, K. L. Teo and X. Yang, pp. 36–100, Springer, Berlin.
385. A. Eberhard, C. E. M. Pearce and D. Ralph (1998), A comparative study of jets, graphical derivatives and coderivatives, RMIT Research Report, No. 8, Melbourne, Australia.
386. A. Eberhard, R. Sivakumaran and R. Wenczel (2005), On the variational behavior of the subhessians of the Lasry-Lions envelope, *J. Convex Anal.*, to appear.
387. A. Eberhard and R. Wenczel (2004), Optimality conditions in nonsmooth analysis, RMIT Research Report, No. 3, Melbourne, Australia.
388. C. Eckart and G. Young (1936), The approximation of one matrix by another of lower rank, *Psychometrika* **1**, 211–218.
389. J. E. Edmond and L. Thibault (2002), Inclusions and integration of subdifferentials, *J. Nonlinear Convex Anal.* **3**, 411–434.
390. J. E. Edmond and L. Thibault (2005), Relaxation of an optimal control problem involving a perturbed sweeping process, *Math. Progr.*, to appear.
391. A. I. Egorov (1967), Necessary optimality conditions for distributed parameter systems, *SIAM J. Control* **5**, 352–408.
392. A. I. Egorov (1978), *Optimal Control of Heat and Diffusion Processes*, Nauka, Moscow.

393. Y. V. Egorov (1963), Certain problems in optimal control theory, *USSR Comput. Maths. Math. Phys.* **3**, 1209–1232.
394. Y. V. Egorov (1964), Some necessary conditions for optimality in Banach spaces, *Math. Sbornik* **64**, 79–101.
395. K. J. Eisenhart (2003), *Multiobjective Optimal Control Problems with Endpoint and State Constraints*, Ph.D. dissertation, Departments of Mathematics, Western Michigan University, Kalamazoo, Michigan.
396. I. Ekeland (1972), Sur les problémes variationnels, *C. R. Acad. Sci. Paris* **275**, 1057–1059.
397. I. Ekeland (1974), On the variational principle, *J. Math. Anal. Appl.* **47**, 324–353.
398. I. Ekeland (1974), Une estimation a priori en programmation non convexe, *C. R. Acad. Sci. Paris* **277**, 149–151.
399. I. Ekeland (1979), Nonconvex minimization problems, *Bull. Amer. Math. Soc.* **1**, 432–467.
400. I. Ekeland and G. Lebourg (1976), Generic Fréchet differentiability and perturbed optimization problems in Banach spaces, *Trans. Amer. Math. Soc.* **224**, 193–216.
401. I. Ekeland and R. Temam (1976), *Convex Analysis and Variational Problems*, North-Holland, Amsterdam, The Netherlands.
402. B. El Abdouni and L. Thibault (1992), Lagrange multipliers for Pareto nonsmooth programming problems in Banach spaces, *Optimization* **26**, 277–285.
403. L. E. Élsgolts (1955), *Qualitative Methods in Mathematical Analysis* (1955). GITTL, Moscow.
404. L. E. Élsgolts and S. B. Norkin (1973), *Introduction to the Theory and Applications of Differential Equations with Deviating Arguments*, Academic Press, New York.
405. M. Epelman and R. M. Freund (2002), A new condition measure, preconditioners, and relations between different measures of conditioning for conic linear systems, *SIAM J. Optim.* **13**, 627–655.
406. I. I. Eremin (1966), The penalty method in convex programming, *Soviet Math. Dokl.* **8**, 458–462.
407. Y. M. Ermoliev, V. P. Gulenko and T. I. Tzarenko (1978), *Finite Difference Method in Optimal Control Problems*, Naukova Dumka, Kiev.
408. Y. M. Ermoliev, V. I. Norkin and R. J-B. Wets (1995), The minimization of semicontinuous functions: Mollifier subgradients, *SIAM J. Control Optim.* **33**, 149–167.
409. E. Ernst and M. Théra (2005), A converse to the Eidelgeit theorem in real Hilbert spaces, *Bull. Sci. Math.* **129**, 381–397.
410. E. Ernst and M. Théra (2005), Necessary and sufficient conditions for the existence of a global maximum for convex functions in reflexive Banach spaces, *J. Convex Anal.*, to appear.
411. L. Euler (1774), *Methodus Inveniendi Curvas Lineas Maximi Minimive Proprietate Gaudentes Sive Solution Problematis Viso Isoperimetrici Latissimo Sensu Accepti*, Lausanne; reprinted in *Opera Omnia*, Ser. 1, Vol. 24, 1952.
412. U. G. Evtushenko (1982), *Methods of Solving Extremal Problems and Their Application in Optimization Systems*, Nauka, Moscow.
413. M. Fabian (1986), Subdifferentials, local ε-supports and Asplund spaces, *J. London Math. Soc.* **34**, 568–576.

414. M. FABIAN (1988), On classes of subdifferentiability spaces of Ioffe, *Nonlinear Anal.* **12**, 568–576.
415. M. FABIAN (1989), Subdifferentiability and trustworthiness in the light of a new variational principle of Borwein and Preiss, *Acta Univ. Carolina, Ser. Math. Phys.* **30**, 51–56.
416. M. FABIAN (1997), *Gâteaux Differentiability of Convex Functions and Topology. Weak Asplund Spaces*, Wiley, New York.
417. M. FABIAN, P. HÁJEK AND J. VANDERWERFF (1996), On smooth variational principles in Banach spaces, *J. Math. Anal. Appl.* **197**, 153–172.
418. M. FABIAN, P. D. LOEWEN AND B. S. MORDUKHOVICH (2005), Subdifferential calculus in Asplund generated spaces, *J. Math. Anal. Appl.*, to appear.
419. M. FABIAN AND B. S. MORDUKHOVICH (1998), Smooth variational principles and characterizations of Asplund spaces, *Set-Valued Anal.* **6**, 381–406.
420. M. FABIAN AND B. S. MORDUKHOVICH (1999), Separable reduction and supporting properties of Fréchet-like normals in Banach spaces, *Canad. J. Math.* **51**, 26–48.
421. M. FABIAN AND B. S. MORDUKHOVICH (2002), Separable reduction and extremal principles in variational analysis, *Nonlinear Anal.* **49**, 265–292.
422. M. FABIAN AND B. S. MORDUKHOVICH (2003), Sequential normal compactness versus topological normal compactness in variational analysis, *Nonlinear Anal.* **54**, 1057–1067.
423. M. FABIAN AND N. V. ZHIVKOV (1985). A characterization of Asplund spaces with help of local ε-supports of Ekeland and Lebourg, *C. R. Acad. Bulgare Sci.* **38**, 671–674.
424. F. FACCHINEI AND J.-S. PANG (2003), *Finite-Dimensional Variational Inequalities and Complementary Problems*, published in two volumes, Springer, New York.
425. M. FALCONE (1994), Discrete time high-order schemes for viscosity solutions of the Hamilton-Jacobi equation, *Numer. Math.* **67**, 315–344.
426. L.-T. FAN AND C.-S. WANG (1964), *The Discrete Maximum Principle: A Study of Multistage Systems Optimization*, Wiley, New York.
427. H. O. FATTORINI (1964), Time-optimal control of solutions of operational differential equations, *SIAM J. Control* **2**, 54–59.
428. H. O. FATTORINI (1968), Boundary control problems, *SIAM J. Control* **6**, 349–385.
429. H. O. FATTORINI (1987), A unified theory of necessary conditions for nonlinear nonconvex control systems, *Appl. Math. Optim.* **15**, 141–185.
430. H. O. FATTORINI (1996), Optimal control with state constraints for semilinear distributed parameter systems, *J. Optim. Theory Appl.* **88**, 25–59.
431. H. O. FATTORINI (1997), Nonlinear infinite dimensional optimal control problems with state constraints and unbounded control sets, *Rend. Instit. Mat. Univ. Trieste* **28**, 127–146.
432. H. O. FATTORINI (1999), *Infinite Dimensional Optimization and Control Theory*, Cambridge University Press, Cambridge, Massachusetts.
433. H. O. FATTORINI (2001), The maximum principle for control systems described by linear parabolic equations, *J. Math. Anal. Appl.* **259**, 630–651.
434. H. O. FATTORINI AND H. FRANKOWSKA (1991), Necessary conditions for infinite dimensional control problems, *Math. Contr. Signal Syst.* **4**, 41–67.

435. H. O. FATTORINI AND T. MURPHY (1994), Optimal problems for nonlinear parabolic boundary control problems: Dirichlet boundary conditions, *Diff. Integ. Eq.* **7**, 1367–1388.
436. H. O. FATTORINI AND T. MURPHY (1994), Optimal problems for nonlinear parabolic boundary control systems, *SIAM J. Control Optim.* **32**, 1577-1596.
437. H. FEDERER (1959), Curvature measures, *Trans. Amer. Math. Soc.* **93**, 418–491.
438. R. P. FEDORENKO (1970), Maximum principle for differential inclusions, *USSR Comput. Maths. Math. Phys.* **10**, 57–68.
439. R. P. FEDORENKO (1971), Maximum principle for differential inclusions (necessity), *USSR Comput. Maths. Math. Phys.* **11**, 885–893.
440. A. A. FELDBAUM (1953), Optimal process in systems of automatic control, *Autom. Telemekh* **14**, No. 6.
441. H. FENCHEL (1951), *Convex Cones, Sets and Functions*, Lecture Notes, Princeton University, Princeton, New Jersey.
442. P. DE FERMAT (1636), Letters to M. Mersenne and J. Roberval, Toulouse, France.
443. M. M. A. FERREIRA, F. A. C. C. FONTES AND R. B. VINTER (1999), Nondegenerate necessary conditions for nonconvex optimal control problems with state constraints, *J. Math. Anal. Appl.* **233**, 116–129.
444. M. M. A. FERREIRA AND R. B. VINTER (1994), When is the maximum principle for state-constrained problems degenerate?, *J. Math. Anal. Appl.* **187**, 432–467.
445. M. C. FERRIS AND O. L. MANGASARIAN (1993), Error bounds and strong upper semicontinuity for monotone affine variational inequalities, *Ann. Oper. Res.* **47**, 293–305.
446. M. C. FERRIS AND J.-S. PANG (1997), Engineering and economic applications of complementarity problems, *SIAM Rev.* **39**, 669–713.
447. A. V. FIACCO (1983), *Introduction to Sensitivity and Stability Analysis in Nonlinear Programming*, Academic Press, New York.
448. A. V. FIACCO AND G. P. MCCORMICK (1968), *Nonlinear Programming: Sequential Unconstrained Minimization Techniques*, Wiley, New York.
449. A. F. FILIPPOV (1962), On certain questions in the theory of optimal control, *SIAM J. Control* **1**, 76–84.
450. A. F. FILIPPOV (1988), *Differential Equations with Discontinuous Right-Hand Sides*, Kluwer, Dordrecht, The Netherlands.
451. S. P. FITZPATRICK (1980), Metric projections and the differentiability of distance functions, *Bull. Austral. Math. Soc.* **22**, 291–312.
452. S. D. FLÅM (2005), Upward slopes and inf-convolutions, *Math. Oper. Res.*, to appear.
453. S. D. FLÅM AND A. JOURANI (2000), Prices and Pareto optima, Research Report.
454. M. L. FLEGEL (2005), *Constraint Qualifications and Stationarity Concepts for Mathematical Programs with Equilibrium Constraints*, Ph.D. Dissertation, Facult. Math., Univ. Würzburg, Germany.
455. M. L. FLEGEL AND C. KANZOW (2003), A Fritz John approach to first order optimality conditions for mathematical programs with equilibrium constraints, *Optimization* **52**, 277–296.

456. M. L. FLEGEL AND C. KANZOW (2005), On *M*-stationarity points for mathematical programs with equilibrium constraints, *J. Math. Anal. Appl.* **310**, 286–302.
457. M. L. FLEGEL, C. KANZOW AND J. V. OUTRATA (2005), Optimality conditions for disjunctive programs with equilibrium constraints, preprint.
458. W. H. FLEMING AND H. M. SONER (1993), *Controlled Markov Processes and Viscosity Solutions*, Springer, New York.
459. M. FLORENZANO (2003), *General Equilibrium Analysis: Existence and Optimality Properties of Equilibria*, Kluwer, Dordrecht, The Netherlands.
460. M. FLORENZANO, P. GOURDEL AND A. JOFRÉ (2003), Supporting weakly Pareto optimal allocations in infinite dimensional nonconvex economies, preprint.
461. F. FLORES-BAZÁN (1997), On minima of the difference functions, *J. Optim. Theory Appl.* **93**, 525–531.
462. F. FLORES-BAZÁN AND W. OETTLI (2001), Simplified optimality conditions for minimizing the difference of vector-valued convex functions, *J. Optim. Theory Appl.* **108**, 571–586.
463. D. FOLEY (1967), Resource allocation and the public sector, *Yale Econ. Essays* **7**, 43–98.
464. H. FRANKOWSKA (1985), Necessary conditions for the Bolza problem, *Math. Oper. Res.* **10**, 361–366.
465. H. FRANKOWSKA (1987), The maximum principle for an optimal control to a differential inclusion with endpoint constraints, *SIAM J. Control Optim.* **25**, 145–157.
466. H. FRANKOWSKA (1987), An open mapping principle for set-valued maps, *J. Math. Anal. Appl.* **127**, 172–180.
467. H. FRANKOWSKA (1989), Higher order inverse function theorems, *Ann. Inst. H. Poincaré: Analyse Non Linéaire* **6**, 283–304.
468. H. FRANKOWSKA (1989), Contingent cones to reachable sets of control systems, *SIAM J. Control Optim.* **27**, 170–198.
469. H. FRANKOWSKA (1990), Some inverse mappings theorems, *Ann. Inst. H. Poincaré: Analyse Non Linéaire* **7**, 183–234.
470. H. FRANKOWSKA (1990), A priori estimates for operational differential inclusions, *J. Diff. Eq.* **84**, 100–128.
471. H. FRANKOWSKA (1993), Lower semicontinuous solutions of Hamilton-Jacobi-Bellman equations, *SIAM J. Control Optim.* **31**, 257–272.
472. H. FRANKOWSKA (2005), Optimal synthesis via superdifferentials of value functions, *Control Cybernet.* **34** (2005), No. 3.
473. M. FRÉCHET (1911), Sur la notion de différentielle, *C. R. Acad. Sci. Paris* **152**, 845–847.
474. R. A. FREEMAN AND P. V. KOKOTOVIĆ (1996), *Robust Nonlinear Control Design: State-Space and Lyapunov Techniques*, Birkhäuser, Boston, Massachusetts.
475. R. M. FREUND AND J. R. VERA (1999), Some characterizations and properties of the 'distance to ill-posedness' and the condition measure of a conic linear system, *Math. Progr.* **86**, 225–260.
476. A. FRIEDMAN (1964), Optimal control for hereditary processes, *J. Math. Anal. Appl.* **15**, 396–414.
477. A. FRIEDMAN (1967), Optimal control in Banach spaces, *J. Math. Anal. Appl.* **19**, 35–55.

478. A. FRIEDMAN (1982), *Variational Principles and Free-Boundary Problems*, Wiley, New York.
479. A. FRIEDMAN (1987), Optimal control for parabolic variational inequalities, *SIAM J. Control Optim.* **25**, 482–497.
480. M. FUKUSHIMA AND J.-S. PANG (2005), Quasi-variational inequalities, generalized Nash equilibria, and multi-leader-follower games, *Comput. Management Sci.* **1**, 21–56.
481. A. V. FURSIKOV (2000), *Optimal Control of Distributed Systems. Theory and Applications*, American Mathematical Society, Prividence, Rhode Island.
482. P. FUSEK, D. KLATTE AND B. KUMMER (2002), Examples and couterexamples in Lipschitz analysis, *Control Cybernet.* **31**, 471–492.
483. R. GABASOV AND S. V. CHURAKOVA (1968), Necessary optimality conditions in time-lag systems, *Autom. Remote Control* **29**, 37–54.
484. R. GABASOV AND F. M. KIRILLOVA (1966), On the extension of the maximum principle by L. S. Pontryagin to discrete systems, *Autom. Remote Control* **27**, 1878–1882.
485. R. GABASOV AND F. M. KIRILLOVA (1974), *Maximum Principle in the Theory of Optimal Processes*, Nauka i Tekhnika, Minsk, Belarus.
486. R. GABASOV AND F. M. KIRILLOVA (1976), *Qualitative Theory of Optimal Processes*, Marcel Dekker, New York.
487. R. GABASOV AND F. M. KIRILLOVA (1977), Methods of optimal control, *J. Soviet Math.* **7**, 805–849.
488. R. GABASOV, F. M. KIRILLOVA AND B. S. MORDUKHOVICH (1983), The ε-maximum principle for suboptimal controls, *Soviet Math. Dokl.* **27**, 95–99.
489. N. GADHI (2005), Optimality conditions for a d. c. set-valued problem via the extremal principle, in *Optimization with Multivalued Mappings: Theory, Applications and Algorithms*, edited by S. Dempe and V. V. Kalashnikov, Springer, Berlin.
490. G. N. GALBRAITH (2001), Extended Hamilton-Jacobi characterization of value functions in optimal control, *SIAM J. Control Optim.* **39**, 281–305.
491. G. N. GALBRAITH (2002), Cosmically Lipschitz set-valued mappings, *Set-Valued Anal.* **10**, 331–360.
492. G. N. GALBRAITH (2004), Solution regularity in optimal control via subgradient analysis of the value function, *Set-Valued Anal.* **12**, 111–126.
493. G. N. GALBRAITH AND R. B. VINTER (2003), Lipschitz continuity of optimal controls for state constrained problems, *SIAM J. Control Optim.* **42**, 1727–1744.
494. R. V. GAMKRELIDZE (1957), On the theory of optimal processes in linear systems, *Dokl. Akad. Nauk SSSR* **116**, 9–11.
495. R. V. GAMKRELIDZE (1962), On sliding optimal regimes, *Soviet Math. Dokl.* **3**, 559–561.
496. R. V. GAMKRELIDZE (1965), On some extremal problem in the theory of differential equations with applications to the theory of optimal control, *SIAM J. Control* **3**, 106–128.
497. R. V. GAMKRELIDZE (1978), *Principles of Optimal Control Theory*, Plenum Press, New York.
498. R. V. GAMKRELIDZE (1999), Discovery of the maximum principle, *J. Dynam. Control Syst.* **5**, 437–451.
499. J. GAUVIN AND F. DUBEAU (1982), Differential properties of the marginal function in mathematical programming, *Math. Progr. Study* **19**, 101–119.

500. V. S. Gavrilov and M. I. Sumin (2004), Parametric optimization of nonlinear Goursat-Darboux systems with state constraints, *Comput. Maths. Math. Phys.* **44**, 949–968.
501. M. Geoffroy and M. Lassonde (2000), On a convergence of lower semicontinuous functions linked with the graph convergence of their subdifferentials, in *Constructive, Experimental and Nonlinear Analysis*, edited by M. Théra, Canad. Math. Soc. Conf. Proc. **27**, 93–109.
502. P. G. Georgiev and N. P. Zlateva (1996), Second-order subdifferential of $C^{1,1}$ functions and optimality conditions, *Set-Valued Anal.* **4**, 101–117.
503. W. Geremew, B. S. Mordukhovich and N. M. Nam (2005), Coderivative analysis of metric regularity for constraint and variational systems, preprint.
504. F. Giannessi (2005), *Constrained Optimization and Image Space Analysis*, published in two volumes Springer, Berlin.
505. J. R. Giles (1982), On the characterization of Asplund spaces, *J. Aust. Math. Soc.* **32**, 134–144.
506. B. Ginsburg and A. D. Ioffe (1996), The maximum principle in optimal control of systems governed by semilinear equations, in *Nonsmooth Analysis and Geometric Methos in Deterministic Optimal Control*, edited by B. S. Mordukhovich and H. J. Sussmann, pp. 81–110, Springer, New York.
507. I. V. Girsanov (1972), *Lectures on Mathematical Theory of Extremum Problems*, Springer, Berlin.
508. B. M. Glover and B. D. Craven (1994), A Fritz John optimality condition using the approximate subdifferential, *J. Optim. Theory Appl.* **82**, 253–265.
509. B. M. Glover, B. D. Craven and S. D. Flåm (1993), A generalized Karush-Kuhn-Tucker optimality condition without constraint qualification using the approximate subdifferential, *Numer. Func. Anal. Optim.* **14**, 333–353.
510. B. M. Glover and D. Ralph (1994), First order approximations to nonsmooth mappings with applications to metric regularity, *Numer. Func. Anal. Optim.* **15**, 599–620.
511. R. Goebel (2004), Regularity of the optimal feedback and the value function in convex problems of optimal control, *Set-Valued Anal.* **12**, 127–145.
512. B. Gollan (1984), On the marginal function in nonlinear programming, *Math. Oper. Res.* **9**, 208–221.
513. S. B. Gorelik and B. S. Mordukhovich (1985), High-order necessary optimality conditions in neutral systems with some applications, *Izv. Akad Nauk BSSR. Ser. Fiz.-Mat. Nauk*, No. 6, 111–112; Depon. VINITI # 1154-85, Moscow.
514. S. B. Gorelik and B. S. Mordukhovich (1986), Legendre-Clebsch and Kelley conditions for nonlinear systems of neutral type with applications to existence of optimal control, *Izv. Akad Nauk BSSR. Ser. Fiz.-Mat. Nauk*, No. 65, 119; Depon. VINITI # 4057-85, Moscow.
515. V. V. Gorokhovik (1986), Epsilon-quasidifferentiability of real-valued functions and optimality conditions in extremal problems, *Math. Progr. Study* **29**, 203–218.
516. V. V. Gorokhovik (1990), *Convex and Nonsmooth Problems of Vector Optimization*, Nauka i Tekhnika, Minsk, Belarus.
517. V. V. Gorokhovik and P. P. Zabreiko (1998), Fréchet differentiability of multimappings, in *Nonlinear Analysis and Applications*, edited by I. V. Gaishun, pp. 34–49, Minsk, Belarus.

518. M. G. GOVIL AND A. MEHRA (2005), Epsilon-optimality for nonsmooth programming on a Hilbert space, in *Generalized Convexity, Generalized Monotonicity and Applications*, edited by A. Eberhard et al., Nonconvex Optimization and Its Applications, pp. 287–298, Springer, New York.
519. M. S. GOWDA AND R. SZNAJDER (1996), On the Lipshitzian properties of polyhedral multifunctions, *Math. Progr.* **74**, 267–278.
520. G. GRAMMEL (2003), Towards fully discretized differential inclusions, *Set-Valued Anal.* **11**, 1–8.
521. K. A. GRASSE (1985), A higher order sufficient condition for surjectivity, *Nonlinear Anal.* **9**, 87–96.
522. L. M. GRAVES (1950), Some mapping theorems, *Duke Math. J.* **17**, 111–114.
523. J. GUDDAT, F. GUERRA VASQUEZ AND H. T. JONGEN (1990), *Parametric Optimization: Singularities, Pathfollowing and Jumps*, B. G. Tebner and John Wiley & Sons, Stuttgart and Chichester.
524. R. GUESNERIE (1975), Pareto optimality in non-convex economies, *Econometrica* **43**, 1–29.
525. S. GUILLAUME (2000), Subdifferential evolution inclusions in nonconvex analysis, *Positivity* **4**, 357–395.
526. P. GUPTA, S. SHIRAISHI AND K. YOKOYAMA (2005), Epsilon-optimality without constraint qualifications for multiobjective fractional programs, *J. Nonlinear Convex Anal.* **6**, 347–357.
527. V. I. GURMAN (1997), *The Extension Principle in Control Problems*, 2nd edition, Nauka, Moscow.
528. M. L. GUSAKOVA (1972), Necessary optimality conditions for systems that are not solved with respect to the derivative, *Diff. Eq.* **8**, 1498–1500.
529. M. L. GUSAKOVA (1974), On necessary optimality conditions for systems of neutral type, *Diff. Eq.* **810**, 1894–1897.
530. T. X. D. HA (2003), The Ekeland variational principle for set-valued maps involving coderivatives, *J. Math. Anal. Appl.* **286**, 509–523.
531. T. X. D. HA (2005), Some variants of the Ekeland variational principle for a set-valued map, *J. Optim. Theory Appl.* **124**, 187–206.
532. T. X. D. HA (2005), Lagrange multipliers for set-valued problems associated with coderivatives, *J. Math. Anal. Appl.*, to appear.
533. A. HABTE (2005), *Applications of Variational Analysis to Welfare Economics*, Ph.D. dissertation, Department of Mathematics, Wayne State University, Detroit, Michigan.
534. N. HADJISAVVAS, S. KOMLOSI AND S. SCHAIBLE, eds. (2005), *Handbook of Generalized Convexity and Generalized Monotonicity*, Springer, Berlin.
535. W. W. HAGER (1976), Rates of convergence for discrete approximations to unconstrained control problems, *SIAM J. Numer. Anal.* **13**, 1–14.
536. W. W. HAGER (1979), Lipschitz continuity for constrained processes, *SIAM J. Control Optim.* **17**, 321–332.
537. A. HALANAY (1968), Optimal control for systems with time lag, *SIAM J. Control* **6**, 215–234.
538. J. HALE (1977), *Theory of Functional Differential Equations*, Springer, New York.
539. H. HALKIN (1964), On the necessary condition for the optimal control of nonlinear systems, *J. Analyse Math.* **12**, 1–82.

540. H. HALKIN (1964), Optimal control for systems by nonlinear difference equations, in *Advances in Control Systems*, Vol. 1, pp. 173–196, Academic Press, New York.
541. H. HALKIN (1970), A satisfactory treatment of equality and operator constraints in Dubovitskii-Milyutin optimization formalism, *J. Optim. Theory Appl.* **6**, 138–149.
542. H. HALKIN (1972), Extremal properties of biconvex contingent equations, in *Ordinary Differential Equations*, edited by L. Weiss, pp. 109–119, Academic Press, New York.
543. H. HALKIN (1974), Implicit functions and optimization problems without continuous differentiability of the data, *SIAM J. Control* **12**, 229–236.
544. H. HALKIN (1976), Mathematical programming without differentiability, in *Calculus of Variations and Control Theory*, edited by D. L. Russel, pp. 279–287, Academic Press, New York.
545. H. HALKIN (1978), Necessary conditions for optimal control problems with differentiable and nondifferentiable data, in *Mathematical Control Theory*, Lecture Notes Math. **680**, pp. 77–118, Springer, Berlin.
546. A. H. HAMEL (2001), An epsilon-Lagrange multiplier rule for a mathematical programming problem on a Banach space, *Optimization* **49**, 137–150.
547. A. H. HAMEL (2003), Phelps' lemma, Daneš' drop theorem and Ekeland's variational principle in locally convex spaces, *Proc. Amer. Math. Soc.* **131**, 3025–3038.
548. W. R. HAMILTON (1834), On a general method employed in dynamics, by which the study of the motions of all free systems of attracting or repelling points is reduced to the search and differentiation of one central solution or characteristic function, *Philosophical Transactions of the Royal Society of London* **124**, 247–308.
549. W. L. HARE AND A. LEWIS (2005). Estimating tangent and normal cones, *Math. Oper. Res.*, to appear.
550. P. T. HARKER AND J.-S. PANG (1990), Finite-dimensional variational inequalities and complementarity problems: a survey of theory, algorithms and applications, *Math. Progr.* **60**, 161–220.
551. J. HASLINGER, M. MIETTINEN AND P. D. PANAGIOTOPOULOS (1999), *Finite Elements Methods for Hemivariational Inequalities*, Kluwer, Dordrecht, The Netherlands.
552. F. HAUSDORFF (1927), *Mengenlehre*, Walter de Gruyter, Berlin.
553. R. HAYDON (1990), A counterexample in several quastions about scattered compact spaces, *Bull. London Math. Soc.* **22**, 261–268.
554. Z.-X. HE (1987), State constrained control problems governed by variational inequalities, *SIAM J. Control Optim.* **25**, 1119–1144.
555. R. HENRION (1995), Topological characterization of the approximate subdifferential in the finite-dimensional cases, *Math. Methods Oper. Res.* **41**, 161–173.
556. R. HENRION (1997), Topological properties of the approximate subdifferential, *J. Math. Anal. Appl.* **207**, 345–360.
557. R. HENRION (1997), Characterization of stability for cone increasing constraints in stochastic programming, *Set-Valued Anal.* **5**, 323–349.
558. R. HENRION (2004), Perturbation analysis of chance-constrained programs under variation of all constraint data, in *Dynamic Stochastic Optimization*, Lecture Notes Econ. Math. Syst. **523**, pp. 257–274, Springer, Berlin.

559. R. Henrion and A. Jourani (2002), Subdifferential conditions for calmness of convex constraints, *SIAM J. Optim.* **13**, 520–534.
560. R. Henrion, A. Jourani and J. V. Outrata (2002), On the calmness of a class of multifunctions, *SIAM J. Optim.* **13**, 603–618.
561. R. Henrion and J. V. Outrata (2001), A subdifferential condition for calmness of multifunctions, *J. Math. Anal. Appl.* **258**, 110–130.
562. R. Henrion and J. V. Outrata (2005), Calmness of constraint systems with applications, *Math. Progr.*, to appear.
563. R. Henrion and W. Römisch (1999), Metric regularity and quantitative stability in stochastic programming with probabilistic constraints, *Math. Progr.* **84**, 55–88.
564. R. Henrion and W. Römisch (2000), Stability of solutions to chance constrained stochastic programs, in *Parametric Optimization and Related Topics V*, edited by J. Guddat et al., pp. 95–114, Peter Lang, Frankfurt, Germany.
565. M. R. Hestenes (1966), *Calculus of Variations and Optimal Control Theory*, Wiley, New York.
566. J. R. Hicks (1939), The foundations of welfare economics, *Econ. J.* **49**, 696–712.
567. D. Hilbert (1902), Mathematical problems, *Bull. Amer. Math. Soc.* **7**, 437–479.
568. W. Hildenbrand (1969), Pareto optimality for a measure space of economic agents, *Intern. Econ. Rev.* **10**, 363–372.
569. J.-B. Hiriart-Urruty (1979), New concepts in nondifferentiable programming, *Bull. Soc. Math. France* **60**, 57–85.
570. J.-B. Hiriart-Urruty (1979), Tangent cones, generalized gradients and mathematical programming in Banach spaces, *Math. Oper. Res.* **4**, 79–97.
571. J.-B. Hiriart-Urruty (1979), Refinements of necessary optimality conditions in nondifferentiable programming I, *Appl. Math. Optim.* **5**, 63–82.
572. J.-B. Hiriart-Urruty (1982), Refinements of necessary optimality conditions in nondifferentiable programming II, *Math. Progr. Study* **19**, 120–139.
573. J.-B. Hiriart-Urruty (1989), From convex optimization to nonconvex optimization: Necessary and sufficient conditions for global optimality, in *Nonconvex Optimization and Related Topics*, pp. 219–239, Plenum Press, New York.
574. J.-B. Hiriart-Urruty and Y. S. Ledyaev (1996), A note on the characterization of the global maxima of a (tangentially) convex function over a convex set, *J. Convex Anal.* **3**, 55–61.
575. J.-B. Hiriart-Urruty and C. Lemaréchal (1993), *Convex Analysis and Minimization Algorithms*, published in two volumes, Springer, Berlin.
576. J.-B. Hiriart-Urruty and P. Plazanet (1989), Moreau's decomposition theorem revisited, *Ann. Inst. H. Poincaré: Analyse Non Linéaire* **6**, 325–338.
577. J.-B. Hiriart-Urruty and A. Seeger (1989), Calculus rules on a new set-valued second-order derivative for convex functions, *Nonlinear Anal.* **13**, 721–738.
578. I. Hlaváček, J. Haslinger, J. Nečas and J. Lovišek (1988), *Solution of Variational Inequalities in Mechanics*, Springer, New York.
579. A. J. Hoffman (1952), On approximate solutions to systems of linear inequalities, *J. Nat. Bureau Stand.* **49**, 263–265.
580. R. B. Holmes (1975), *Geometric Functional Analysis and Its Applications*, Springer, New York.

581. L. HÖRMANDER (1954), Sur la function d'appui des ensdembles convexes dans un espace localement convexe, *Arkiv för Mat.* **3**, 181–186.
582. L. HÖRMANDER (1990), *The Analysis of Linear Partial Differential Operators I*, Springer, Berlin.
583. R. HORST, P. D. PARDALOS AND N. V. THOAI (2000), *Introduction to Global Optimization*, 2nd edition, Kluwer, Dordrecht, The Netherlands.
584. X. HU AND D. RALPH (2002), A note of sensitivity of value functions of mathematical programs with complementarity constraints, *Math. Progr.* **93**, 265–279.
585. X. HU, D. RALPH, E. K. RALPH, P. BARDSLEY AND M. C. FERRIS (2002), The effect of transmission capacities on competition in deregulated electricity markets, preprint.
586. D. K. HUGHES (1968), Variational and optimal control problems with delayed argument, *J. Optim. Theory Appl.* **2**, 1–14.
587. A. D. IOFFE (1979), Regular points of Lipschitz functions, *Trans. Amer. Math. Soc.* **251**, 61–69.
588. A. D. IOFFE (1979), Necessary and sufficient conditions for a local minimum, I: A reduction theorem and first order conditions, *SIAM J. Control Optim.* **17**, 245–250.
589. A. D. IOFFE (1981), Nonsmooth analysis: Differential calculus of non-differentiable mappings, *Trans. Amer. Math. Soc.* **266** (1981), 1–56.
590. A. D. IOFFE (1981), Sous-differentielles approaches de fonctions numériques, *C.R. Acad. Sci. Paris* **292**, 675–678.
591. A. D. IOFFE (1981), Calculus of Dini subdifferentials, CEREMADE Publication 8110, Universiteé de Paris IX "Dauphine".
592. A. D. IOFFE (1981), Approximate subdifferentials of nonconvex functions, CEREMADE Publication 8120, Universiteé de Paris IX "Dauphine".
593. A. D. IOFFE (1983), On subdifferentiability spaces, *Ann. New York Acad. Sci.* **410**, 107–119.
594. A. D. IOFFE (1984), Calculus of Dini subdifferentials and contingent coderivatives of set-valued maps, *Nonlinear Anal.* **8**, 517–539.
595. A. D. IOFFE (1984), Necessary conditions in nonsmooth optimization, *Math. Oper. Res.* **9**, 159–188.
596. A. D. IOFFE (1984), Approximate subdifferentials and applications, I: The finite dimensional theory, *Trans. Amer. Math. Soc.* **281**, 389–415.
597. A. D. IOFFE (1986), Approximate subdifferentials and applications, II: Functions on locally convex spaces, *Mathematika* **33**, 111–128.
598. A. D. IOFFE (1987), On the local surjection property, *Nonlinear Anal.* **11**, 565–592.
599. A. D. IOFFE (1989), Approximate subdifferentials and applications, III: The metric theory, *Mathematika* **36**, 1–38.
600. A. D. IOFFE (1990), Proximal analysis and approximate subdifferentials, *J. London Math. Soc.* **41**, 175–192.
601. A. D. IOFFE (1991), Variational analysis of composite functions: A formula for the lower second order epi-derivative, *J. Math. Anal. Appl.* **160**, 379–405.
602. A. D. IOFFE (1993), A Lagrange multiplier rule with small convex-valued subdifferentials for non-smooth problems of mathematical programming involving equality and non-functional constraints, *Math. Progr.* **72**, 137–145.

603. A. D. Ioffe (1996), Nonsmooth subdifferentials: Their calculus and applications, in *Proceedings of the First World Congress of Nonlinear Analysts*, edited by V. Lakshmikantham, pp. 2299–2310, De Gruyter, Berlin.

604. A. D. Ioffe (1997), Directional compactness and nonsmooth semi-Fredholm mappings, *Nonlinear Anal.* **29**, 201–219.

605. A. D. Ioffe (1997), Euler-Lagrange and Hamiltonian formalisms in dynamic optimization, *Trans. Amer. Amer. Soc.* **349**, 2871–2900.

606. A. D. Ioffe (1998), Fuzzy principles and characterization of trustworthiness, *Set-Valued Anal.* **6**, 265–276.

607. A. D. Ioffe (2000), Codirectional compactness, metric regularity and subdifferential calculus, in *Constructive, Experimental and Nonlinear Analysis*, edited by M. Théra, Canad. Math. Soc. Conf. Proc. **27**, 123–164.

608. A. D. Ioffe (2000), Metric regularity and subdifferential calculus, *Russian Math. Surveys* **55**, 501–558.

609. A. D. Ioffe (2003), On stability estimates for the regularity property of maps, in *Topological Methods, Variational Methods and Their Applications*, edited by H. Breziz et al., pp. 133–142, World Scientific Publishing, River Edge, New Jersey.

610. A. D. Ioffe (2003), On robustness for the regularity property of maps, *Control Cybernet.* **32**, 543–554.

611. A. D. Ioffe (2005), Optimality alternative: A non-variational approach to necessary conditions, in *Variational Analysis and Applications*, edited by F. Giannessi and A. Maugeri, pp. 531–552, Springer, Berlin.

612. A. D. Ioffe and V. L. Levin (1972), Subdifferentials of convex functions, *Trudy Moscow Mat. Ob.* **26**, 3–73.

613. A. D. Ioffe and T. Milosz (2002), On a characterization of $C^{1,1}$ functions, *Cybernetics and Systems Analysis*, No. 3, 3–13.

614. A. D. Ioffe and J.-P. Penot (1996), Subdifferentials of performance functions and calculus of coderivatives of set-valued mappings, *Serdica Math. J.* **22**, 359–384.

615. A. D. Ioffe and J.-P. Penot (1997), Limiting subhessians, limiting subjects and their calculus, *Trans. Amer. Math. Soc.* **349**, 789–807.

616. A. D. Ioffe and R. T. Rockafellar (1996), The Euler and Weierstrass conditions for nonsmooth variational problems, *Calc. Var. Partial Diff. Eq.* **4**, 59–87.

617. A. D. Ioffe and V. M. Tikhomirov (1968), Extensions of variational problems, *Trans. Moscow Math. Soc.* **18**, 207–273.

618. A. D. Ioffe and V. M. Tikhomirov (1979), *Theory of Extremal Problems*, North-Holland, Amsterdam, The Netherlands.

619. A. D. Ioffe and V. M. Tikhomirov (1997), Some remarks on variational principles, *Math. Notes* **61**, 248–253.

620. A. D. Ioffe and A. J. Zaslavski (2000), Variational principles and well-posedness in optimization and calculus of variations, *SIAM J. Control Optim.* **38**, 566–581.

621. A. N. Iusem (1998), On some properties of generalized proximal point methods for variational inequalities, *J. Optim. Theory Appl.* **96**, 337–362.

622. M. Ivanov (2004), Sequential representation formulae for G-subdifferential and Clarke subdifferential in smooth Banach spaces, *J. Convex Anal.* **11**, 179–196.

623. A. F. Izmailov (2006), *Sensitivity in Optimization*, Fizmatlit, Moscow.
624. A. F. Izmailov and M. V. Solodov (2001), Error bounds for 2-regular mappings with Lipschitzian derivatives and its applications, *Math. Progr.* **89**, 413–435.
625. C. G. J. Jacobi (1866), *Volesungen über Dynamik*, edited by A. Clebsch, Verlag Georg Reimer, Berlin.
626. M. Q. Jacobs and C. E. Langenhop (1976), Criteria for function space controllability of linear neutral systems, *SIAM J. Control Optim.* **14**, 1009–1048.
627. J. Jahn (2004), *Vector Optimization: Theory, Applications and Extensions*, Springer, Berlin.
628. J. Jahn, A. A. Khan and P. Zeilinger (2005), Second order optimality conditions in set-valued optimization, *J. Optim. Theory Appl.* **125**, 331–347.
629. R. Janin (1973), Sur une classe de fonctions sous-linéarisables, *C. R. Acad. Sci. Paris* **277**, 265–267.
630. V. Jeyakumar and D. T. Luc (1998), Approximate Jacobian matrices for nonsmooth continuous maps and C^1-optimization, *SIAM J. Control Optim.* **36**, 1815–1832.
631. V. Jeyakumar and N. D. Yen (2004), Solution stability, regularity and implicit functions for nonsmooth continuous systems, *SIAM J. Optim.* **14**, 1106–1127.
632. H. Jiang and D. Ralph (2000), Smooth SQP methods for mathematical programs with nonlinear complementarity constraints, *SIAM J. Optim.* **10**, 779–808.
633. A. Jofré (2000), A second-welfare theorem in nonconvex economics, in *Constructive, Experimental and Nonlinear Analysis*, edited by M. Théra, Canad. Math. Soc. Conf. Proc. **27**, 175–184.
634. A. Jofré, D. T. Luc and M. Théra (1998), Epsilon-sundifferentials and epsilon-monotonicity, *Nonlinear Anal.* **33**, 71–90.
635. A. Jofré and J. Rivera (2005), A nonconvex separation property and some applications, *Math. Progr.*, to appear.
636. A. Jofré, R. T. Rockafellar and R. J-B. Wets (2005), A variational inequality scheme for determining an economic equilibrium of classical or extended type, in *Variational Analysis and Applications*, edited by F. Giannessi and A. Maugeri, pp. 553–578, Springer, Berlin.
637. A. Jofré, R. T. Rockafellar and R. J-B. Wets (2005), Variational inequalities and economic equilibrium, *Math. Oper. Res.*, to appear.
638. F. John (1948), Extremum problems with inequalities as side conditions, in *Studies and Essays: Courant Anniversary Volume*, edited by K. O. Fredrichs, O. E. Neugebauer and J. J. Stoker, pp. 187–204, Wiley, New York.
639. H. T. Jongen, D. Klatte and K. Tammer (1990), Implicit functions and sensitivity of stationary points, *Math. Progr.* **49**, 123–138.
640. H. T. Jongen, T. Möbert, J.-J. Rückmann and K. Tammer (1987), Implicit functions and sensitivity analysis of stationary points, *Linear Algerbra Appl.* **95**, 97–109.
641. H. T. Jongen, J.-J. Rückmann and G.-W. Weber (1994), One-parametric semi-infinite optimization: On the stability of the feasible set, *SIAM J. Optim.* **4**, 637–648.
642. E. Jouini (1988), A remark on Clarke's normal cone and the marginal cost pricing rule, *J. Math. Econ.* **17**, 309–315.

643. A. JOURANI (1995), Compactly epi-Lipschitzian sets and A-subdifferentials in WT-spaces, *Optimization* **34**, 1–17.
644. A. JOURANI (1995), Intersection formulae and the marginal function in Banach spaces, *J. Math. Anal. Appl.* **192**, 867–891.
645. A. JOURANI (1998), Necessary conditions for extremality and separation theorems with applications to multobjective optimization, *Optimization* **44**, 327–350.
646. A. JOURANI (1999), Limit superior of subdifferentials of uniformly convergent functions, *Positivity* **3**, 33–47.
647. A. JOURANI (2000), Hoffman's error bound, local controllability and sensitivity analysis, *SIAM J. Control Optim.* **38**, 947–970.
648. A. JOURANI (2003), On a class of compactly epi-Lipschitzian sets, *Nonlinear Anal.* **54**, 471–483.
649. A. JOURANI (2005), Weak regularity of functions and sets in Asplund spaces, *Nonlinear Anal.*, to appear.
650. A. JOURANI AND M. THÉRA (1998), On the limiting Fréchet ε-sibdifferentials, in *Generalized Convexity, Generalized Monotonicity: Recent Results*, edited by J. P. Crouzeix et al., Nonconvex Optim. Appl. **27**, pp. 185–198, Kluwer, Dordrecht, The Netherlands.
651. A. JOURANI AND L. THIBAULT (1990), Approximate subdifferential and metric regularity: The finite-dimensional case, *Math. Progr.* **47**, 203–218.
652. A. JOURANI AND L. THIBAULT (1990), The use of metric graphical regularity in approximate subdifferential calculus rules in finite dimensions, *Optimization* **21**, 1–11.
653. A. JOURANI AND L. THIBAULT (1993), The approximate subdifferential of composite functions, *Bull. Austral. Math. Soc.* **47**, 443–445.
654. A. JOURANI AND L. THIBAULT (1994), A note on Fréchet and approximate subdifferentials of composite functions, *Bull. Austral. Math. Soc.* **49**, 111–116.
655. A. JOURANI AND L. THIBAULT (1995), Verifiable conditions for openness, metric regularity of multivalued mappings, *Trans. Amer. Math. Soc.* **347**, 1225–1268.
656. A. JOURANI AND L. THIBAULT (1995), Metric regularity for strongly compactly Lipschitzian mappings, *Nonlinear Anal.* **24**, 229–240.
657. A. JOURANI AND L. THIBAULT (1996), Metric regularity and subdifferential calculus in Banach spaces, *Set-Valued Anal.* **3**, 87–100.
658. A. JOURANI AND L. THIBAULT (1996), Extensions of subdifferential calculus rules in Banach spaces, *Canad. J. Math.* **48**, 834–848.
659. A. JOURANI AND L. THIBAULT (1998), Chain rules for coderivatives of multivalued mappings in Banach spaces, *Proc. Amer. Math. Soc.* **126**, 1479–1485.
660. A. JOURANI AND L. THIBAULT (1998), Qualification conditions for calculus rules of coderivatives of multivalued mappings, *J. Math. Anal. Appl.* **218**, 66–81.
661. A. JOURANI AND L. THIBAULT (1999), Coderivatives of multivalued mappings, locally compact cones and metric regularity, *Nonlinear Anal.* **35**, 925–945.
662. A. JOURANI AND J. J. YE (2005), Error bounds for eigenvalue and semidefinite matrix inequality systems, *Math. Progr.*, to appear.
663. G. A. KAMENSKII AND E. A. HVILON (1968), Optimality conditions for systems with deviating argument, in *Trudy Sem. Teor. Diff. Urav. Otklon. Argum.* **6**, 213–222.

664. L. V. Kantorovich (1940), On an efficient method for solving some classes of extremum problems, *Dokl. Akad. Nauk SSSR* **28**, 212–215.
665. W. Karush (1939), *Minima of Functions of Several Variables with Inequalities as Side Conditions*, Master Thesis, Department of Mathematics, University of Chicago.
666. B. Kaśkosz and S. Lojasiewicz, Jr. (1985), A maximum principle for generalized control systems, *Nonlinear Anal.* **9**, 109–130.
667. B. Kaśkosz and S. Lojasiewicz, Jr. (1992), Lagrange-type extremal trajectories in differential inclusions, *Syst. Control Lett.* **19**, 241–247.
668. G. A. Kent (1971), A maximum principle for optimal control problems with neutral functional differential systems, *Bull. Amer. Math. Soc.* **77**, 565–570.
669. M. A. Khan (1988) Ioffe's normal cone and the foundations of welfare economics: An example, *Econ. Lett.* **28**, 5–19.
670. M. A. Khan (1991), Ioffe's normal cone and the foundations of welfare economics: The infinite dimensional theory, *J. Math. Anal. Appl.* **161**, 284–298.
671. M. A. Khan (1999), The Mordukhovich normal cone and the foundations of welfare economics, *J. Public Econ. Theory* **1**, 309–338.
672. M. A. Khan and S. Rashid (1975), Nonconvexity and Pareto optimality in large markets, *Inter. Econ. Rev.* **16**, 222–245.
673. M. A. Khan and R. Vohra (1987), An extension of the second welfare theorem to economies with non-convexities and public goods, *Quartery J. Econ.* **102**, 223–245.
674. M. A. Khan and R. Vohra (1988), Pareto optimal allocations of nonconvex economies in locally convex spaces, *Nonlinear Anal.* **12**, 943–950.
675. M. A. Khan and R. Vohra (1988), On approximate decentralization of Pareto optimal allocations in locally convex spaces, *J. Approx. Theory* **52**, 149–161.
676. P. Q. Khanh (1986), An induction theorem and general open mapping theorem, *J. Math. Anal. Appl.* **118**, 519–536.
677. P. Q. Khanh (1989), On general open mapping theorems, *J. Math. Anal. Appl.* **144**, 305–312.
678. G. L. Kharatishvili (1961), The maximum principle in the theory of optimal processes with a delay, *Soviet Math. Dokl.* **2**, 28–32.
679. G. L. Kharatishvili and T. A. Tadumadze (1998), A nonlinear optimal control problem with variable delays, nonfixed initial moment, and piecewise-continuous prehistory, *Proc. Steklov Inst. Math.* **220**, 233–252.
680. D. Kinderlehrer and G. Stampacchia (1980), *An Introduction to Variational Inequalities and Their Applications*, Academic Press, New York.
681. A. J. King and R. T. Rockafellar (1992), Sensitivity analysis for nonsmooth generalized equations, *Math. Progr.* **55**, 193–212.
682. M. Kisielewicz (1991), *Differential Inclusions and Optimal Control*, Kluwer, Dordrecht, The Netherlands.
683. K. C. Kiwiel (1985), *Methods of Descent for Nondifferentiable Optimization*, Springer, Berlin.
684. D. Klatte (1994), On quantitative stability for non-isolated minima, *Control Cybernet.* **23**, 183–200.
685. D. Klatte and R. Henrion (1998), Regularity and stability in nonlinear semi-infinite optimization, in *Semi-Infinite Programming*, edited by R. Reemtsen and J.-J. Rückmann, pp. 69–102, Kluwer, Dordrecht, The Netherlands.

686. D. KLATTE AND B. KUMMER (2002), *Nonsmooth Equations in Optimization: Regularity, Calculus, Methods, and Applications*, Kluwer, Boston, Massachusetts,
687. D. KLATTE AND B. KUMMER (2002), Constrained minima and Lipschitzian penalties in metric spaces, *SIAM J. Optim.* **13**, 619–633.
688. M. KOCAN AND A. ŚWIECH (1996), Perturbed optimization on product spaces, *Nonlinear Anal.* **26**, 81–90.
689. M. KOČVARA, M. KRUŽIK AND J. V. OUTRATA (2005), On the control of an evolutionary equilibrium in micromagnetics, in *Optimization with Multivalued Mappings: Theory, Applications and Algorithms*, edited by S. Dempe and V. V. Kalashnikov, Springer, Berlin.
690. M. KOČVARA AND J. V. OUTRATA (2004), Optimization problems with equilibrium constraints and their numerical solutions, *Math. Progr.* **101**, 119–149.
691. M. KOČVARA AND J. V. OUTRATA (2005), On the modeling and control of delamination processes, in *Control and Boundary Analysis*, edited by J. Cagnol and J.-P. Zolésio, Lecture Notes Pure Applied Math. **240**, pp. 169–185, Marcel Dekker, New York.
692. M. KOJIMA (1980), Strongly stable stationary solutions in nonlinear programs, in *Analysis and Computation of Fixed Points*, edited by S. M. Robinson, pp. 93–138, Academic Press, New York.
693. P. V. KOKOTOVIĆ, H. K. KHALIL AND J. O'REILLY (1986), *Singular Perturbations in Control Analysis and Design*, Academic Press, New York.
694. V. B. KOLMANOVSKII AND A. D. MYSHKIS (1992), *Applied Theory of Functional Differential Equations*, Kluwer, Dordrecht, The Netherlands.
695. V. B. KOLMANOVSKII AND V. R. NOSOV (1984), Neutral-type systems with aftereffects, *Autom. Remote Control* **45**, 1–28.
696. V. B. KOLMANOVSKII AND L. E. SHAIKHET (1996), *Control of Systems with Aftereffect*, Academic Press, New York.
697. S. KOMLOSI, T. RAPCSÁK AND S. SCHAIBLE, eds. (1994), *Generalized Convexity*, Lecture Notes Econ. Math. Syst. **405**, Springer, Berlin.
698. E. KOSTINA AND O. KOSTYUKOVA (2005), Generalized implicit function theorem and its applications to parametric optimal control, *J. Math. Anal. Appl.*, to appear.
699. S. G. KRANTZ AND H. R. PARKS (2002), *The Implicit Function Theorem: History, Theory, and Applications*, Birkhäuser, Boston, Massachusetts.
700. N. N. KRASOVSKII (1964), Optimal processes in systems with time lag, in *Automatics and Remote Control Theory*, edited by L. Butterths, pp. 327–332, Oldenburg, Munish, Germany.
701. N. N. KRASOVSKII AND N. Y. LUKOYANOV (2000), Equations of Hamilton-Jacobi type in hereditary systems: Minimax solutions, *Proc. Steklov Inst. Math.: Control in Dynamic Systems, suppl. 1*, 136–153.
702. N. N. KRASOVSKII AND A. I. SUBBOTIN (1988), *Game-Theoretical Control Problems*, Springer, New York.
703. A. J. KRENER (1977), The high order maximum principle and its applications to singular extremals, *SIAM J. Control Optim.* **15**, 256–293.
704. V. F. KROTOV (1996), *Global Methods in Optimal Control*, Marcel Dekker, New York.
705. A. Y. KRUGER (1981), Epsilon-semidifferentials and epsilon-normal elements, Depon. VINITI #1331-81, Moscow.

706. A. Y. Kruger (1981), *Generalized Differentials of Nonsmooth Functions and Necessary Conditions for an Extremum*, Ph.D. dissertation, Department of Mathematics, Belarus State University, Minsk, Belarus.
707. A. Y. Kruger (1985), Generalized differentials of nonsmooth functions and necessary conditions for an extremum, *Siberian Math. J.* **26**, 370–379.
708. A. Y. Kruger (1985), Properties of generalized differentials, *Siberian Math. J.* **26**, 822–832.
709. A. Y. Kruger (1988), A covering theorem for set-valued mappings, *Optimization* **19**, 763–780.
710. A. Y. Kruger (1997), Strict ε-semidifferentials and extremality conditions, *Dokl. Nat. Akad. Belarus* **41**, 21–26.
711. A. Y. Kruger (1998), On extremality of set systems, *Dokl. Nat. Akad. Belarus* **42**, 24–28.
712. A. Y. Kruger (2002), Strict (ε, δ)-semidifferentials and extremality conditions, *Optimization* **51**, 539–554.
713. A. Y. Kruger (2003), On Fréchet subdifferentials, *J. Math. Sci.* **116**, 3325–3358.
714. A. Y. Kruger (2004), Weak stationarity: Eliminating the gap between necessary and sufficient conditions, *Optimization* **53**, 147–164.
715. A. Y. Kruger (2005), Stationarity and regularity of set systems, *Pacific J. Optim.* **1**, 101–126.
716. A. Y. Kruger (2005), About regularity of set systems, *SIAM J. Optim.*, to appear.
717. A. Y. Kruger and B. S. Mordukhovich (1978), Minimization of nonsmooth functionals in optimal control problems, *Eng. Cybernetics* **16**, 126–133.
718. A. Y. Kruger and B. S. Mordukhovich (1980), Extremal points and the Euler equation in nonsmooth optimization, *Dokl. Akad. Nauk BSSR* **24**, 684–687.
719. A. Y. Kruger and B. S. Mordukhovich (1980), Generalized normals and derivatives, and necessary optimality conditions in nondifferential programming, Parts I and II, Depon. VINITI: I# 408-80, II# 494-80, Moscow.
720. S. N. Kruzhkov (1960), The Cauchy problem in the large for certain nonlinear first order differential equations, *Soviet Math. Dokl.* **1**, 474–477.
721. A. V. Kryazhimskii and Y. S. Osipov (1995), On evolutionary-differential games, *Proc. Steklov Math. Inst.* **211**, 257–287.
722. E. I. Kugushev (1973), Maximum principle in optimal control systems with nonsmooth right-hand side, *Vest. Moskov. Univ. Ser. Mat. Mekh.*, No. 3, 107–113.
723. H. W. Kuhn (1976), Nonlinear programming: A historical survey, in *Nonlinear Programming*, edited by R. W. Cottle and C. E. Lemke, pp. 1–26, American Mathematical Society, Providence, Rhode Island.
724. H. W. Kuhn and A. W. Tucker (1951), Nonlinear programming, in *Proceedings of the Second Berkley Symposium on Mathematical Statistics and Probability*, edited by J. Neyman, pp. 481–492, University of California Press, Berkeley, California.
725. B. Kummer (1991), Lipschitzian inverse functions, directional derivatives and applications in $C^{1,1}$ optimization, *J. Optim. Theory Appl.* **70**, 561–582.
726. B. Kummer (1991), An implicit function theorem for $C^{0,1}$-equations and parametric $C^{1,1}$-optimization, *J. Math. Anal. Appl.* **158**, 35–46.

727. B. KUMMER (1999), Metric regularity: Characterizations, nonsmooth variations and successive approximation, *Optimization* **46**, 247–281.
728. B. KUMMER (2000), Inverse functions of pseudo regular mappings and regularity conditions, *Math. Progr.* **88**, 313–339.
729. L. KUNTZ AND S. SCHOLTES (1994), Structural analysis of nonsmooth mappings, inverse functions, and metric projections, *J. Math. Anal. Appl.* **188**, 346–386.
730. G. A. KURINA (1979), Application of the method of tents to an optimal control problem for a differential equation with a singular matrix multiplying the derivative, *Diff. Eq.* **15**, 417–423.
731. A. B. KURZHANSKII (1977), *Control and Observation under Conditions of Uncertainty*, Nauka, Moscow.
732. A. B. KURZHANSKII AND I. VÁLYI (1997), *Ellipsoidal Calculus for Estimation and Control*, Burlhäuser, Boston, Massachusetts.
733. A. G. KUSRAEV AND S. S. KUTATELADZE (1995), *Subdifferentials: Theory and Applications*, Kluwer, Dordrecht, The Netherlands.
734. J. KYPARISIS (1992), Parametric variational inequalities with multivalued solution sets, *Math. Oper. Res.* **17**, 341–364.
735. O. A. LADYZHENSKAYA, V. A. SOLONNIKOV AND N. N. URALZEVA (1968), *Linear and Quasilinear Equations of Parabolic Type*, American Mathematical Society, Providence, Rhode Island.
736. J. E. LAGNESE (1989), *Boundary Stabilization of Thin Plates*, SIAM Publications, Philadelphia, Pennsylvania.
737. J. L. LAGRANGE (1813), *Thérie des fonctions analytique*, Paris.
738. O. LANGE (1942), The foundations of welfare economics, *Econometrica* **10**, 215–228.
739. I. LASIECKA (1980), Unified theory of abstract parabolic boundary problems – a semigroup approach, *Appl. Math. Optim.* **6**, 287–333.
740. I. LASIECKA, J.-L. LIONS AND R. TRIGGIANI (1986), Nonhomogeneous boundary value problems for second-order hyperbolic operators, *J. Mat. Pures Appl.* **65**, 149–192.
741. I. LASIECKA AND J. SOKOLOWSKI (1988), Regularity and strong convergence of a variational approximation to a nonhomogeneous Dirichlet hyperbolic boundary problem, *SIAM J. Math. Anal.* **19**, 528–540.
742. I. LASIECKA AND R. TRIGGIANI (1983), Dirichlet boundary control problems for parabolic equations with quadratic cost: Analyticity and Riccati feedback synthesis, *SIAM J. Control Optim.* **21**, 41–67.
743. I. LASIECKA AND R. TRIGGIANI (1987), The regulator problem for parabolic equations with Dirichlet boundary control, I: Riccati's feedback synthesis and regularity of optimal solutions, *Appl. Math. Optim.* **16**, 147–168.
744. I. LASIECKA AND R. TRIGGIANI (1990), Sharp regularity theory for second order hyperbolic equations of Neumann type, I: L_2 nonhomogeneous data, *Ann. Mat. Pura Appl.* **157**, 285–367.
745. I. LASIECKA AND R. TRIGGIANI (1991), Regularity theory of hyperbolic equations with non-homogeneous Neumann boundary conditions, II: General boundary data, *J. Diff. Eq.* **94**, 112–164.
746. I. LASIECKA AND R. TRIGGIANI (2000), *Control Theory for Partial Differential Equations: Continuous and Approximation Theory*, published in two volumes, Cambridge University Press, Cambridge, UK.

747. M. LASSONDE (2001), First-order rules for nonsmooth constrained optimization, *Nonlinear Anal.* **44**, 1031–1056.
748. E. B. LEACH (1961), A note on inverse function theorem, *Proc. Amer. Math. Soc.* **12**, 694–697.
749. G. LEBOURG (1975), Valeur moyenne pour gradient généraliseé, *C. R. Acad. Sci. Paris* **281**, 795–798.
750. Y. S. LEDYAEV (2004), On generic existence and uniqueness in nonconvex optimal control, *Set-Valued Anal.* **12**, 147–162.
751. Y. S. LEDYAEV AND Q. J. ZHU (1999), Implicit multifunction theorem, *Set-Valued Anal.* **7**, 209–238.
752. Y. S. LEDYAEV AND Q. J. ZHU (2004), Techniques for nonsmooth analysis on smooth manifolds I: Techniques for local problems, in *Optimal Control, Stabilization and Nonsmooth Analysis*, edited by M. de Queiroz, M. Malisoff and P. Wolenski, Lecture Notes Cont. Inf. Sci. **301**, pp. 283–287, Springer, New York.
753. Y. S. LEDYAEV AND Q. J. ZHU (2004), Techniques for nonsmooth analysis on smooth manifolds II: Using deformations and flowes, in *Optimal Control, Stabilization and Nonsmooth Analysis*, edited by M. de Queiroz, M. Malisoff and P. Wolenski, Lecture Notes Cont. Inf. Sci. **301**, pp. 299–311, Springer, New York.
754. Y. S. LEDYAEV AND Q. J. ZHU (2005), Nonsmooth analysis on smooth manifolds, *Trans. Amer. Math. Soc.*, to appear.
755. G. M. LEE, N. N. TAM AND N. D. YEN (2005), Normal coderivative for multifunctions and implicit function theorems, preprint.
756. U. LEDZEWICZ AND H. SCHÄTTLER (1998), High-order approximations and generalized necessary conditions for optimality, *SIAM J. Control Optim.* **37**, 33–56.
757. G. W. LEIBNIZ (1696), Letter to Johann Bernoulli of July 31, 1696, Hanover, Germany.
758. G. LEITMANN (1978), On generalized Stackelberg strategies, *J. Optim. Theory Appl.* **59**, 637–643.
759. C. LEMARÉCHAL, F. OUSTRY AND C. SAGASTIZÁBAL (2000), The $\mathcal{U}$-Lagrangian of a convex function, *Trans. Amer. Math. Soc.* **352**, 711–729.
760. F. LEMPIO (1995), Euler's method revisited, *Proc. Steklov Inst. Math.* **211**, 473–494.
761. F. LEMPIO AND V. M. VELIOV (1998), Discrete approximations to differential inclusions, *Mitteilungen der GAMM* **21**, 101–135.
762. S. LENHART, V. PROTOPESCU AND S. STOJANOVIĆ (1993), A minimax problem for semilinear nonlocal competitive systems, *Appl. Math. Optim.* **28**, 113–132.
763. E. S. LEVITIN (1994), *Perturbation Theory in Mathematical Programming and Its Applications*, Wiley, New York.
764. E. S. LEVITIN, A. A. MILYUTIN AND N. P. OSMOLOVSKII (1978), Higher order conditions for a local minimum in the problems with constraints, *Russian Math. Surveys* **33**, 97–168.
765. A. B. LEVY (1993), Second-order epi-derivatives of integral functionals, *Set-Valued Anal.* **1**, 379–392.
766. A. B. LEVY (1996), Implicit multifunction theorems for the sensitivity analysis of variational conditions, *Math. Progr.* **74**, 333–350.

767. A. B. LEVY (2001), Lipschitzian multifunctions and a Lipschitzian inverse function theorem, *Math. Oper. Res.* **26**, 105–118.
768. A. B. LEVY (2001), Solution stability from general principles, *SIAM J. Control Optim.* **40**, 209–238.
769. A. B. LEVY AND B. S. MORDUKHOVICH (2004), Coderivatives in parametric optimization, *Math. Progr.* **99**, 311–327.
770. A. B. LEVY AND R. A. POLIQUIN (1997), Characterizing the single-valuedness of multifunctions, *Set-Valued Anal.* **5**, 351–364.
771. A. B. LEVY, R. A. POLIQUIN AND R. T. ROCKAFELLAR (2000), Stability of locally optimal solutions, *SIAM J. Optim.* **10**, 580–604.
772. A. B. LEVY, R. A. POLIQUIN AND L. THIBAULT (1995), A partial extension of Attouch's theorem and its applications to second-order epi-differentiation, *Trans. Amer. Math. Soc.* **347**, 1269–1294.
773. A. B. LEVY AND R. T. ROCKAFELLAR (1994), Sensitivity analysis of solutions to generalized equations, *Trans. Amer. Math. Soc.* **345**, 661–671.
774. A. B. LEVY AND R. T. ROCKAFELLAR (1996), Variational conditions and the proto-differentiation of partial subgradient mappings, *Nonlinear Anal.* **26**, 1951–1964.
775. A. S. LEWIS (1999), Nonsmooth analysis of engenvalues, *Math. Progr.* **84**, 1–24.
776. A. S. LEWIS (1999), Ill-conditioned convex processes and conic linear systems, *Math. Oper. Res.* **23**, 829–834.
777. A. S. LEWIS (2001), Ill-conditioned inclusions, *Set-Valued Anal.* **9**, 375–381.
778. A. S. LEWIS (2003), Active sets, nonsmoothness and sensitivity, *SIAM J. Optim.* **13**, 702–725.
779. A. S. LEWIS (2003), The mathematics of eigenvalue optimization, *Math. Progr.* **97**, 155–176.
780. A. S. LEWIS (2004), The structured distance to ill-posedness for conic systems, *Math. Oper. Res.* **29**, 776–785.
781. A. S. LEWIS AND M. L. OVERTON (1996), Eigenvalue optimization, *Acta Numerica* **5**, 149–190.
782. A. S. LEWIS AND H. S. SENDOV (2005), Nonsmooth analysis of singular values. Part I: Theory, *Set-Valued Anal.* **13**, 213–241.
783. A. S. LEWIS AND H. S. SENDOV (2005), Nonsmooth analysis of singular values. Part II: Applications, *Set-Valued Anal.* **13**, 243–264.
784. W. LI AND I. SINGER (1998), Global error bounds for convex multifunctions, *Math. Oper. Res.* **23**, 443–462.
785. Y. LI AND S. SHI (2000), A generalization of Ekeland's ε-variational principle and its Borwein-Preiss variant, *J. Math. Anal. Appl.* **246**, 308–319.
786. X. LI AND Y. YAO (1981), On optimal control for distributed parameter systems, in *Proc. IFAC 8th Triennial World Congress*, pp. 207–212, Kyoto, Japan.
787. X. LI AND Y. YAO (1985), Maximum principle of distributed parameter systems with time lags, in *Lecture Notes Cont. Inform. Sci.* **75**, pp. 410–427, Springer, Berlin.
788. X. LI AND J. YONG (1991), Necessary conditions of optimal control for distributed parameter systems, *SIAM J. Control Optim.* **29**, 895–908.
789. X. LI AND J. YONG (1995), *Optimal Control Theory for Infinite-Dimensional Systems*, Birkhäuser, Boston, Massachusetts.

790. L.-J. Lin and W.-S. Du (2005), Ekeland's variational principles and the existence of nonconvex equilibria and minimax inequalities in complete metric spaces, *J. Math. Anal. Appl.*, to appear.
791. J.-L. Lions (1971), *Optimal Control of Systems Governed by Partial Differential Equations*, Springer, Berlin.
792. J.-L. Lions (1983), *Contrôle des Systémes Distribués Singuliers*, Gauthier-Villars, Paris.
793. P.-L. Lions (1982), *Generalized Solutions of Hamilton-Jacobi Equations*, Pitman, Boston, Massachusetts.
794. J.-L. Lions and E. Magenes (1972), *Nonhomogeneous Boundary Value Problems and Applications*, Springer, New York.
795. J.-L. Lions and G. Stampacchia (1967), Variational inequalities, *Comm. Pure Appl. Math.* , 493–519.
796. R. Lipschitz (1877), *Lehrbuch der Analysis*, Cohen & Sohn, Bonn.
797. J. M. Liu (1995), Strong stability in variational inequalities, *SIAM J. Control Optim.* **33**, 725–749.
798. P. D. Loewen (1987), The proximal normal formula in Hilbert spaces, *Nonlinear Anal.* **11**, 979–995.
799. P. D. Loewen (1988), The proximal subgradient formula in Banach spaces, *Canad. Math. Bull.* **31**, 353–361.
800. P. D. Loewen (1992), Limits of Fréchet normals in nonsmooth analysis, in *Optimization and Nonlinear Analysis*, edited by A. Ioffe, L. Marcus and S. Reich, Pitman Research Notes Math. Ser. **244**, pp. 178–188, Longman, Harlow, Essex, UK.
801. P. D. Loewen, *Optimal Control via Nonsmooth Analysis* (1993), American Mathematical Society, Providence, Rhode Island.
802. P. D. Loewen (1994), A mean value theorem for Fréchet subgradients, *Nonlinear Anal.* **23**, 1365–1381.
803. P. D. Loewen, F. H. Clarke and R. B. Vinter (1988), Differential inclusions with free time, *Ann. Inst. H. Poincaré: Analyse Non Linéaire* **5**, 573–593.
804. P. D. Loewen and R. T. Rockafellar (1991), The adjoint arcs in nonsmooth optimization, *Trans. Amer. Math. Soc.* **325**, 39–72.
805. P. D. Loewen and R. T. Rockafellar (1994), Optimal control of unbounded differential inclusions, *SIAM J. Control Optim.* **32**, 442–470.
806. P. D. Loewen and R. T. Rockafellar (1996), New necessary conditions for the generalized problem of Bolza, *SIAM J. Control Optim.* **34**, 1496–1511.
807. P. D. Loewen and R. T. Rockafellar (1997), Bolza problem with general time constraints, *SIAM J. Control Optim.* **35**, 2050–2069.
808. P. D. Loewen and R. B. Vinter (1987), Pontryagin type necessary conditions for differential inclusions, *Syst. Control Lett.* **9**, 263–265.
809. P. D. Loewen and X. Wang (2001), A generalized variational principle, *Canad. J. Math.* **53**, 1174–1193.
810. S. Lojasiewicz, Jr. (1992), Local controllability of parameterized differential equations, preprint.
811. P. Loridan (1982), Necessary conditions for ε-optimality, *Math. Progr. Study* **19**, 140–152.
812. P. Loridan and J. Morgan (1989), New results on approximate solutions in two level optimization, *Optimization* **20**, 819–836.

813. D. T. LUC (1989), *Theory of Vector Optimization*, Springer, Berlin.
814. D. T. LUC (1996), A strong mean value theorem and applications, *Nonlinear Anal.* **26**, 915–923.
815. D. T. LUC (2005), Chain rules for approximate Jacobians of continuous functions, *Nonlinear Anal.* **61**, 97–114.
816. Y. LUCET AND J. J. YE (2001), Sensitivity analysis of the value function for optimization problems with variational inequality constraints, *SIAM J. Optim.* **40**, 699–723.
817. D. R. LUKE, J. V. BURKE AND R. G. LYON (2002), Optical wavefront reconstruction: Theory and numerical methods, *SIAM Rev.* **44**, 169–224.
818. Y. LUO AND A. EBERHARD (2005), Comparizon principles for viscosity solutions of elliptic equations via fuzzy sum rule, *J. Math. Anal. Appl.* **307**, 736–752.
819. Y. LUO AND A. EBERHARD (2005), An application of $C^{1,1}$ approximation to comparison principles for viscosity solutions of curvature equations, preprint.
820. Z. Q. LUO, J.-S. PANG AND D. RALPH (1996), *Mathematical Programs with Equilibrium Constraints*, Cambridge University Press, Cambridge, UK.
821. K. A. LURIE (1993), *Applied Optimal Control of Distributed Systems*, Plenum Press, New York.
822. A. A. LYAPUNOV (1940), Sur les founctions-vecteurs complétement additives, *Izvest. Akad. Nauk SSSR, Ser. Mat.* **3**, 465–478.
823. S. I. LYASHKO (2002), *Generalized Optimal Control of Linear Systems with Distributed Parameters*, Kluwer, Dordrecht, The Netherlands.
824. L. A. LYUSTERNIK (1934), On conditional extrema of functionals, *Math. Sbornik* **41**, 390–401.
825. U. MACKENROTH (1982), Convex parabolic boundary control problems with pointwise state constraints, *J. Math. Anal. Appl.* **87**, 256–277.
826. G. G. MAGARIL-IL'YAEV (1978), An implicit function theorem for Lipschitzian mappings, *Russian Math. Surveys* **33**, 209–210.
827. E. N. MAHMUDOV (2005), On duality in problems of optimal control described by convex differential inclusions of Goursat-Darboux type, *J. Math. Anal. Appl.* **307**, 628–640.
828. E. N. MAHMUDOV (2005), The optimality principle for discrete and first order partial differential inclusions, *J. Math. Anal. Appl.* **308**, 605–619.
829. V. L. MAKAROV, M. J. LEVIN AND A. M. RUBINOV (1995), *Mathematical Economic Theory: Pure and Mixed Types of Economic Mechanisms*, North-Holland, Amsterdam, The Netherlands.
830. K. MALANOWSKI (1969), On optimal control of the vibrating string, *SIAM J. Control* **7**, 260–271.
831. K. MALANOWSKI (1979), On convergence of finite difference approximations to control and state constrained convex optimal control problems, *Arch. Autom. Telemech.* **24**, 319–337.
832. K. MALANOWSKI (1987), *Stability of Solutions to Convex Problems of Optimization*, Springer, Berlin.
833. K. MALANOWSKI (1994), Regularity of solutions in stability analysis of optimization and optimal control problems, *Control Cybernet.* **23**, 61–86.
834. K. MALANOWSKI (2003), Two-norm approach in stability and sensitivity analysis of optimization and optimal control problems, *Advan. Math. Sci. Appl.* **2**, 397–443.

835. K. Malanowski and Y. Sokolowski (1986), Sensitivity of solutions to convex, control constrained optimal control problems for distributed parameter systems, *J. Math. Anal. Appl.* **120**, 240–263.
836. G. G. Malcolm and B. S. Mordukhovich (2001), Pareto optimality in nonconvex economies with infinite-dimensional commodity spaces, *J. Global Optim.* **20**, 323–346.
837. M. Malisoff, L. Rifford and E. Sontag (2003), Remarks on input to state stabilization, *Proc. 42rd IEEE Conf. Dec. Cont.*, pp. 1053–1058, Maui, Hawaii.
838. M. Malisoff, L. Rifford and E. Sontag (2004), Global asymptotic controllability implies input to state stabilization, *SIAM J. Control Optim.* **42**, 1121–1138.
839. O. L. Mangasarian (1985), A condition number for differentiable convex inequalities, *Math. Oper. Res.* **10**, 175–179.
840. O. L. Mangasarian (1994), *Nonlinear Programming*, SIAM Publications, Philadelphia, Pennsylvania.
841. O. L. Mangasarian and S. Fromovitz (1967), The Fritz John condition in the presence of equality and inequality constraints, *J. Math. Anal. Appl.* **17**, 34–47.
842. O. L. Mangasarian and T. H. Shiau (1986), Error bounds for monotone linear complementarity problems, *Math. Progr.* **36**, 81–89.
843. K. B. Mansimov (1998), Singular controls in systems of neutral type, *Autom. Remote Control* **59**, 653–661.
844. K. B. Mansimov (2001), On the theory of necessary conditions for optimality in a problem with distributed parameters, *Comput. Maths. Math. Phys.* **41**, 1429–1443.
845. C. Marcelli (2002), Variational problems with nonconvex, noncoercive, highly discontinuous integrands: characterization and existence of minimizers, *SIAM J. Control Optim.* **40**, 1473–1490.
846. C. Marcelli (2005), Necessary and sufficient conditions for optimality of nonconvex, noncoercive autonomous variational problems with constraints, preprint.
847. C. Marcelli, E. Outkine and M. Sytchev (2002), Remarks on necessary conditions for minimizers of one-dimensional variational problems, *Nonlinear Anal.* **48**, 979–993.
848. S. Marcellin (2004), *Intégration D'epsilon-sous-différentiels et Problémes D'évolution Non Convexes*, Ph.D. dissertation, Department of Mathematics, University of Montpellier II, Montpellier, France.
849. S. Marcellin and L. Thibault (2005), Integration of ε-Fenchel subdifferentials and maximal cyclic monotonicity, *J. Global Optim.* **32**, 83–91.
850. A. Marchaud (1934), Sur les champs de demi-cônes et les équations différentielles du premier ordre, *Bull. Sc. Math.* **62**, 12–38.
851. A. Marino and M. Tosques (1990), Some variational problems with lack of convexity and some partial differential inequalities, in *Methods of Nonconvex Analysis*, edited by A. Cellina, Lecture Notes in Mathematics **1446**, pp. 58–83, Springer, Berlin.
852. J. E. Martinez-Legaz and P. H. Sach (1999), A new subdifferential in quasiconvex analysis, *J. Convex Anal.* **6**, 1–12.
853. J. E. Martinez-Legaz and M. Théra (1996), Epsilon-subdifferentials in terms of subdifferentials, *Set-Valued Anal.* **4**, 327–332.

854. A. Mas-Colell (1985), Pareto optima and equilibria: The inifinite-dimensional case, in *Advances in Economics*, edited by C. D. Aliprantis et al., pp. 25–42, Springer, New York.
855. A. Mas-Colell (1986), Valuation equilibrium and Pareto optimum revisited, in *Contributions to Mathematical Economics*, edited by W. Hildenbrand and A. Mas-Collel, pp. 317–331, North-Holland, Amsterdam, The Netherlands.
856. A. Mas-Colell, M. D. Whinston and J. R. Green (1995), *Microeconomic Theory*, Oxford University Press, Oxford, UK.
857. H. Maurer and N. P. Osmolovskii (2003), Second order conditions, *Control Cybernet.* **32**, 555–584.
858. H. Maurer and J. Zowe (1979), First and second-order necessary and sufficient optimality conditions for infinite-dimensional programming problems, *Math. Progr.* **16**, 98–110.
859. A. Mayer (1884), Zur aufstellung der kriterien des maximums und minimums der einfachen integrale bei variabeln grenwerten, *Leipziger Berichte* **36**, 99–128.
860. E. J. McShane (1939), On multipliers for Lagrange problem, *Amer. J. Math.* **91**, 809–819.
861. E. J. McShane (1940), Generalized curves, *Duke Math. J.* **6**, 513–536.
862. E. J. McShane (1940), Necessary conditions for generalized curve problems in the calculus of variations, *Duke Math. J.* **7**, 1–27.
863. E. J. McShane (1967), Relaxed controls and vbariational problems, *SIAM J. Contr.* **5**, 438–485.
864. E. J. McShane (1973), The Lagrange multiplier rule, *Amer. Math. Monthly* **80**, 922–925.
865. E. J. McShane (1989), The calculus of variations from the beginning through optimal control theory, *SIAM J. Control Optim.* **27**, 916–939.
866. N. G. Medhin (1990), Minimizing sequences for differential inclusion problems in the presence of state constraints, *Comput. Math. Appl.* **19**, 127–134.
867. N. G. Medhin (1995), On optimal control of functional-differential systems, *J. Optim. Theory Appl.* **85**, 363–376.
868. T. K. Melikov (1996), An analogue of Pontryagin's maximum principle in systems with aftereffect of neutral type, *Comput. Maths. Math. Phys.* **26**, 1541–1546.
869. A. A. Melikyan (1998), *Generalized Characteristics of First Order PDEs*, Birkhäuser, Boston, Massachusetts.
870. P. Michel and J.-P. Penot (1984), Calcul sous-différentiel pour des fonctions lipschitziennes et non-lipschitziennes, *C. R. Acad. Sci. Paris* **298**, 684–687.
871. P. Michel and J.-P. Penot (1992), A generalized derivative for calm and stable functions, *Diff. Integ. Eq.* **5**, 189–196.
872. R. Mifflin (1977), Semismooth and semiconvex functions in constrained optimization, *SIAM J. Control Optim.* **15**, 957–972.
873. R. Mifflin and C. Sagastizábal (2003), Primal-dual gradient structured functions: Second-order results; links to epi-derivatives and partly smooth functions, *SIAM J. Optim.* **13**, 1174–1194.
874. R. Mifflin and C. Sagastizábal (2004), $\mathcal{VU}$-smoothness and proximal point results for some nonconvex functions, *Optim. Meth. Soft.* **19**, 463-478.

875. A. A. MILYUTIN (1999), Convex-valued Lipschitz differential inclusions and Pontryagin's maximum principle, in *Optimal Control*, edited by R. V. Gamkrelidge, Vol. 4, pp. 175–187, VINITI, Moscow.
876. A. A. MILYUTIN (2003), On strengthening the conditions by Clarke and Smirnov for convex-valued differential inclusions, *Math. Sbornik* **194**, 261–280.
877. A. A. MILYUTIN AND N. P. OSMOLOVSKII (1998), *Calculus of Variations and Optimal Control*, American Mathematical Society, Providence, Rhode Island.
878. L. I. MINCHENKO (1999), Necessary optimality conditions for differential-difference inclusions, *Nonlinear Anal.* **35**, 307–322.
879. L. I. MINCHENKO (2003), Multivalued analysis and differential properties of multivalued mappings and marginal functions, *J. Math. Sci.* **116**, 3266–3302.
880. L. I. MINCHENKO AND S. SIROTKO (2002), Controllability of nonsmooth discrete systems with delays, *Optimization* **51**, 161–174.
881. L. I. MINCHENKO AND A. A. VOLOSEVICH (2003), Value function and necessary optimality conditions in optimal control problems for differential-difference inclusions, *Nonlinear Anal.* **53**, 407–424.
882. H. MINKOWSKI (1911), *Theorie der Konvexen Körper, Insbesondere Begründung inhres Ober Flächenbegriffs*, Gesammelte Abhandlungen, II, B. G. Teubner, Leipzig.
883. V. J. MIZEL AND T. I. SEIDMAN (1997), An abstract bang-bang principle and time-optimal boundary control of the heat equation, *SIAM J. Control Optim.* **35**, 1204–1216.
884. N. N. MOISEEV (1971), *Numerical Methods in Optimal Systems Theory*, Nauka, Moscow.
885. N. N. MOISEEV (1975), *Elements of Optimal Systems Theory*, Nauka, Moscow.
886. M. D. P. MONTEIRO MARQUES (1993), *Differential Inclusions in Nonsmooth Mechanical Problems*, Birkhäuser, Boston, Massachusetts.
887. B. S. MORDUKHOVICH (1976), Maximum principle in problems of time optimal control with nonsmooth constraints, *J. Appl. Math. Mech.* **40**, 960–969.
888. B. S. MORDUKHOVICH (1977), Existence of optimal controls, *J. Soviet Math.* **7**, 850–886.
889. B. S. MORDUKHOVICH (1977), Approximation and maximum principle for nonsmooth problems of optimal control, *Russian Math. Surveys* **196**, 263–264.
890. B. S. MORDUKHOVICH (1978), On difference approximations of optimal control systems, *J. Appl. Math. Mech.* **42**, 452–461.
891. B. S. MORDUKHOVICH (1978), On the theory of difference approximations of continuous-time control systems, in *Third USSR Conference on Operations Research*, edited by N. N. Moiseev, pp. 330–331, Gorky, USSR.
892. B. S. MORDUKHOVICH (1980), Metric approximations and necessary optimality conditions for general classes of extremal problems, *Soviet Math. Dokl.* **22**, 526–530.
893. B. S. MORDUKHOVICH (1981), Penalty functions and necessary conditions for an extemum in nonsmooth and nonconvex optimization problems, *Russian Math. Surveys* **36**, 242–243.
894. B. S. MORDUKHOVICH (1984), Nonsmooth analysis with nonconvex generalized differentials and adjoint mappings, *Dokl. Akad. Nauk BSSR* **28**, 976–979.

895. B. S. Mordukhovich (1984), Duality theory in systems with aftereffect, *J. Appl. Math. Mech.* **48**, 440–447.
896. B. S. Mordukhovich (1984), Controllability, observability and duality in dynamical systems with aftereffect. *Autom. Remote Control* **45**, 1019–1027.
897. B. S. Mordukhovich (1985), On necessary conditions for an extremum in nonsmooth optimization, *Soviet Math. Dokl.* **32**, 215–220.
898. B. S. Mordukhovich (1986), Optimal control of the groundwater regime on engineering reclamation systems, *Water Resources* **12**, 244–253.
899. B. S. Mordukhovich (1986), On the theory of difference approximations in optimal control, *Dokl. Akad. Nauk BSSR* **30**, 1964–1067.
900. B. S. Mordukhovich (1987), Approximation methods and optimality conditions in nonconvex control problems, *Soviet Math. Dokl.* **36**, 164–168.
901. B. S. Mordukhovich (1988), *Approximation Methods in Problems of Optimization and Control*, Nauka, Moscow.
902. B. S. Mordukhovich (1988), Approximation and optimization of differential inclusions, *Cybernetics* **24**, 781–788.
903. B. S. Mordukhovich (1988), Approximate maximum principle for finite-difference control systems, *USSR Comput. Maths. Math. Phys.* **28**, 106–114.
904. B. S. Mordukhovich (1989), Necessary optimality conditions for nonsmooth control systems with free time, *Diff. Eq.* **25**, 290–299.
905. B. S. Mordukhovich (1990), Minimax design for a class of distributed parameter systems, *Autom. Remote Control* **50**, 262–283.
906. B. S. Mordukhovich (1990), Maximum principle for nonconvex finite difference systems, in *Analysis and Optimization of Systems*, edited by A. Bensoussan and J.-L. Lions, Lecture Notes Cont. Inf. Sci. **144**, pp. 539–548, Springer, Berlin.
907. B. S. Mordukhovich (1992), Sensitivity analysis in nonsmooth optimization, in *Theoretical Aspects of Industrial Design*, edited by D. A. Field and V. Komkov, SIAM Proc. Appl. Math. **58**, pp. 32–46, Philadelphia, Pennsylvania.
908. B. S. Mordukhovich (1992), On variational analysis of differential inclusions, in *Optimization and Nonlinear Analysis*, edited by A. Ioffe, L. Marcus and S. Reich, Pitman Researh Notes Math. Ser. **244**, pp. 199–213, Longman, Harlow, Essex, UK.
909. B. S. Mordukhovich (1993), Complete characterization of openness, metric regularity, and Lipschitzian properties of multifunctions, *Trans. Amer. Math. Soc.* **340**, 1–35.
910. B. S. Mordukhovich (1994), Generalized differential calculus for nonsmooth and set-valued mappings, *J. Math. Anal. Appl.* **183**, 250–288.
911. B. S. Mordukhovich (1994), Lipschitzian stability of constraint systems and generalized equations, *Nonlinear Anal.* **22**, 173–206.
912. B. S. Mordukhovich (1994), Stability theory for parametetic generalized equations and variational inequalities via nonsmooth analysis, *Trans. Amer. Math. Soc.* **343**, 609–658.
913. B. S. Mordukhovich (1994), Sensitivity analysis for constraint and variational systems by means of set-valued differentiation, *Optimization* **31**, 13–46.
914. B. S. Mordukhovich (1994), Necesary optimality and controllability conditions for nonsmooth control systems, in *Proc. 33rd IEEE Conf. Dec. Cont.*, pp. 3992–3997, Orlando, Florida.

915. B. S. MORDUKHOVICH (1995), Discrete approximations and refined Euler-Lagrange conditions for nonconvex differential inclusions, *SIAM J. Control Optim.* **33**, 882–915.
916. B. S. MORDUKHOVICH (1996), Optimization and finite difference approximations of nonconvex differential inclusions with free time, in *Nonsmooth Analysis and Geometric Methods in Deterministic Optimal Control*, edited by B. S. Mordukhovich and H. J. Sussmann, pp. 153–202, Springer, New York.
917. B. S. MORDUKHOVICH (1997), Coderivatives of set-valued mappings: Calculus and applications, *Nonlinear Anal.* **30**, 3059–3070.
918. B. S. MORDUKHOVICH (1999), Minimax design of constrained parabolic systems, in *Control of Distributed Parameter and Stochastic Systems*, edited by S. Chen et al., pp. 111–118, Kluwer, Boston, Massachusetts.
919. B. S. MORDUKHOVICH (1999), On variational analysis in infinite dimensions, in *Systems Modelling and Optimization*, edited by M. P. Polis et al., pp. 189–197, Chapman and Hall/CRC Press, Boca Raton, Florida.
920. B. S. MORDUKHOVICH (2000), Abstract extremal principle with applications to welfare economics, *J. Math. Anal. Appl.* **251**, 187–216.
921. B. S. MORDUKHOVICH (2000), Optimal control of difference, differential, and differential-difference inclusions, *J. Math. Sci.* **100**, 2613–2632.
922. B. S. MORDUKHOVICH (2001), The extremal principle and its applications to optimization and economics, in *Optimization and Related Topics*, edited by A. Rubinov and B. Glover, Applied Optimization **47**, pp. 343–369, Kluwer, Dordrecht, The Netherlands.
923. B. S. MORDUKHOVICH (2002), Calculus of second-order subdifferentials in infinite dimensions, *Control Cybernet.* **31**, 557–573.
924. B. S. MORDUKHOVICH (2004), Coderivative analysis of variational systems, *J. Global Optim.* **28**, 347–362.
925. B. S. MORDUKHOVICH (2004), Necessary conditions in nonsmooth minimization via lower and upper subgradients, *Set-Valued Anal.* **12**, 163–193.
926. B. S. MORDUKHOVICH (2004), Equilibrium problems with equilibrium constraints via multiobjective optimization, *Optim. Meth. Soft.* **19**, 479–492.
927. B. S. MORDUKHOVICH (2004), Lipschitzian stability of parametric constraint systems in infinite dimensions, in *Generalized Convexity, Generalized Monotonicity and Aplications*, edited by A. Eberhard et al., Nonconvex Optimization and Its Applications, pp. 39–59, Springer, New York.
928. B. S. MORDUKHOVICH (2005), Optimization and equilibrium problems with equilibrium constraints, *OMEGA* **33**, 379–384.
929. B. S. MORDUKHOVICH (2005), Sensitivity analysis for variational systems, in *Variational Analysis and Applications*, edited by F. Giannessi and A. Maugeri, pp. 723–743, Springer, Berlin.
930. B. S. MORDUKHOVICH (2005), Nonlinear prices in nonconvex economies with classical Pareto and strong Pareto optimal allocations, *Positivity* **9**.
931. B. S. MORDUKHOVICH (2005), Sensitivity analysis for generalized variational and hemivariational inequalities, in *Advances in Analysis*, edited by H. G. W. Begehr et al., pp. 305–314, World Scientific Publishing, London, UK.
932. B. S. MORDUKHOVICH (2005), Optimal control of evolution inclusions, *Nonlinear Anal.*, to appear.
933. B. S. MORDUKHOVICH (2005), Coderivative calculus and robust Lipschitzian stability of variational systems, *J. Convex Anal.*, to appear.

934. B. S. MORDUKHOVICH AND N. M. NAM (2005), Variational stability and marginal functions via generalized differentiation, *Math. Oper. Res.* **30**, 1–18.
935. B. S. MORDUKHOVICH AND N. M. NAM (2005), Subgradients of distance functions with some applications, *Math. Progr.*, to appear.
936. B. S. MORDUKHOVICH AND N. M. NAM (2005), Subgradients of distance functions at out-of-state points, *Taiwan. J. Math.*, to appear.
937. B. S. MORDUKHOVICH, N. M. NAM AND N. D. YEN (2005), Subgradients of marginal functions in parametric mathematical programming, *Math. Progr.*, to appear.
938. B. S. MORDUKHOVICH, N. M. NAM AND N. D. YEN (2005), Fréchet subdifferential calculus and optimality conditions in mathematical programming, *Optimization*, to appear.
939. B. S. MORDUKHOVICH AND J. V. OUTRATA (2001), Second-order subdifferentials and their applications, *SIAM J. Optim.* **12**, 139–169.
940. B. S. MORDUKHOVICH, J. V. OUTRATA AND M. ČERVINKA (2005), Equilibrium problems with complementarity constraints: Case study with applications to oligopolistic markets, *Optimization*, to appear.
941. B. S. MORDUKHOVICH AND T. PENNANEN (2005), Epi-convergent approximations for generalized Bolza problems, preprint.
942. B. S. MORDUKHOVICH AND V. M. RAKETSKII (1980), Necessary optimality conditions of the maximum principle type in discrete approximations of nonconvex control problems with constraints on trajectories, *Vest. Beloruss. Univ.*, Ser. 1, No. 3, 72–73; Depon. VINITI #791-79, Moscow.
943. B. S. MORDUKHOVICH AND J.-P. RAYMOND (2004), Dirichlet boundary control of hyperbolic equations in the presence of state constraints, *Appl. Math. Optim.* **49**, 145–157.
944. B. S. MORDUKHOVICH AND J.-P. RAYMOND (2004), Neumann boundary control of hyperbolic equations with pointwise state constraints, *SIAM J. Control Optim.* **43**, 1354–1372.
945. B. S. MORDUKHOVICH AND A. M. SASONKIN (1985), Duality and optimality conditions in control problems for functional-differential systems of neutral type, *Diff. Eq.* **21**, 532–340.
946. B. S. MORDUKHOVICH AND Y. SHAO (1995), Differential characterizations of covering, metric regularity, and Lipschitzian properties of multifunctions between Banach spaces, *Nonlinear Anal.* **25**, 1401–1424.
947. B. S. MORDUKHOVICH AND Y. SHAO (1995), On nonconvex subdifferential calculus in Banach spaces, *J. Convex Anal.* **2**, 211–227.
948. B. S. MORDUKHOVICH AND Y. SHAO (1996), Extremal characterizations of Asplund spaces, *Proc. Amer. Math. Soc.* **124**, 197–205.
949. B. S. MORDUKHOVICH AND Y. SHAO (1996), Nonsmooth sequential analysis in Asplund spaces, *Trans. Amer. Math. Soc.* **348**, 1235–1280.
950. B. S. MORDUKHOVICH AND Y. SHAO (1996), Nonconvex coderivative calculus for infinite-dimensional multifunctions, *Set-Valued Anal.* **4**, 205–236.
951. B. S. MORDUKHOVICH AND Y. SHAO (1997), Stability of multifunctions in infinite dimensions: Point criteria and applications, *SIAM J. Control Optim.* **35**, 285–314.
952. B. S. MORDUKHOVICH AND Y. SHAO (1997), Fuzzy calculus for coderivatives of multifunctions, *Nonlinear Anal.* **29**, 605–626.
953. B. S. MORDUKHOVICH AND Y. SHAO (1998), Mixed coderivatives of set-valued mappings in variational analysis, *J. Appl. Anal.* **4**, 269–294.

954. B. S. Mordukhovich, Y. Shao and Q. J. Zhu (2000), Viscosity coderivatives and their limiting behavior in smooth Banach spaces, *Positivity* **4**, 1–39.
955. B. S. Mordukhovich and I. A. Shvartsman (2002), Discrete maximum principle for nonsmooth optimal control problems with delays, *Cybernet. Systems Anal.* **38**, 255–264.
956. B. S. Mordukhovich and I. A. Shvartsman (2004), The approximate maximum principle in constrained optimal control, *SIAM J. Control Optim.* **43**, 1037–1062.
957. B. S. Mordukhovich and I. A. Shvartsman (2004), Optimization and feedback control of constrained parabolic systems under uncertain perturbations, in *Optimal Control, Stabilization and Nonsmooth Analysis*, edited by M. de Queiroz, M. Malisoff and P. Wolenski, Lecture Notes Cont. Inf. Sci. **301**, pp. 121–132, Springer, New York.
958. B. S. Mordukhovich, J. S. Treiman and Q. J. Zhu (2003), An extended extremal principle with applications to multiobjective optimization, *SIAM J. Optim.* **14**, 359–379.
959. B. S. Mordukhovich and R. Trubnik (2001), Stability of discrete approximations and necessary optimality conditions for delay-differential inclusions, *Ann. Oper. Res.* **101**, 149–170.
960. B. S. Mordukhovich and B. Wang (2000), On variational characterizations of Asplund spaces, in *Constructive, Experimental and Nonlinear Analysis*, edited by M. Théra, Canad. Math. Soc. Conf. Proc. **27**, 245–254.
961. B. S. Mordukhovich and B. Wang (2001), Sequential normal compactness in variational analysis, *Nonlinear Anal.* **47**, 717–728.
962. B. S. Mordukhovich and B. Wang (2002), Necessary optimality and suboptimality conditions in nondifferentiable programming via variational principles, *SIAM J. Control Optim.* **41**, 623–640.
963. B. S. Mordukhovich and B. Wang (2002), Extensions of generalized differential calculus in Asplund spaces, *J. Math. Anal. Appl.* **272**, 164–186.
964. B. S. Mordukhovich and B. Wang (2003), Calculus of sequential normal compactness in variational analysis, *J. Math. Anal. Appl.* **282**, 63–84.
965. B. S. Mordukhovich and B. Wang (2003), Differentiability and regularity of Lipschitzian mappings, *Proc. Amer. Math. Soc.* **131**, 389–399.
966. B. S. Mordukhovich and B. Wang (2004), Generalized differentiation for moving objects, in *Optimal Control, Stabilization and Nonsmooth Analysis*, edited by M. de Queiroz, M. Malisoff and P. Wolenski, Lecture Notes Cont. Inf. Sci. **301**, pp. 351-361, Springer, New York.
967. B. S. Mordukhovich and B. Wang (2004), Restrictive metric regularity and generalized differential calculus in Banach spaces, *Int. J. Maths. Math. Sci.* **50**, 2650–2683.
968. B. S. Mordukhovich and B. Wang (2005), Restrictive metric regularity in variational analysis, *Nonlinear Anal.*, to appear.
969. B. S. Mordukhovich and B. Wang (2005), Generalized differential and normal compactness calculi for moving objects, preprint.
970. B. S. Mordukhovich and D. Wang (2005), Optimal control of semilinear unbounded differential inclusions, *Nonlinear Anal.*, to appear.
971. B. S. Mordukhovich and D. Wang (2005), Optimal control of semilinear evolution inclusions via discrete approximations, *Control Cybernet.* **34**, No. 3.

972. B. S. Mordukhovich and L. Wang (2002), Optimal control of hereditary differential inclusions, in *Proc. 41st IEEE Conf. Decis. Control*, pp. 1107–1112, Las Vegas, Nevada.
973. B. S. Mordukhovich and L. Wang (2003), Optimal control of constrained delay-differential inclusions with multivalued initial conditions, *Control Cybernet.* **32**, 585–609.
974. B. S. Mordukhovich and L. Wang (2004), Optimal control of neutral functional-differential inclusions, *SIAM J. Control Optim.* **43**, 111–136.
975. B. S. Mordukhovich and L. Wang (2004), Optimal control of differential-algebraic inclusions, in *Optimal Control, Stabilization and Nonsmooth Analysis*, edited by M. de Queiroz, M. Malisoff and P. Wolenski, Lecture Notes Cont. Inf. Sci. **301**, pp. 73–83, Springer, New York.
976. B. S. Mordukhovich and L. Wang (2005), Optimal control of delay systems with differential and algebraic dynamic constraints, *ESAIM: Control Optim. Calc. Var.* **11**, 285–309.
977. B. S. Mordukhovich and L. Wang (2005), Optimal control of nonautonomous functional-differential inclusions of neutral type, *Nonlinear Anal.*, to appear.
978. B. S. Mordukhovich and K. Zhang (1997), Minimax control of parabolic equations with Dirichlet boundary conditions and state constraints, *Appl. Math. Optim.* **36**, 323–360.
979. B. S. Mordukhovich and K. Zhang (1998), Dirichlet boundary control of parabolic systems with pointwise state constraints, *Int. Ser. Numer. Math.* **126**, 223–236.
980. J.-J. Moreau (1962), Fonctions convexes duales et points proximaux dans un espace hilbertien, *C. R. Acad. Sci. Paris* **255**, 2897–2899.
981. J.-J. Moreau (1963), Fonctionelles sous-différentiables, *C. R. Acad. Sci. Paris* **257**, 4117–4119.
982. J.-J. Moreau (1965), Proximité dualité dans espace hilbertien, *Bull. Soc. Math. France* **93**, 273–299.
983. J.-J. Moreau (1977), Evolution problems associated with moving coinvex set in a Hilbert space, *J. Diff. Eq.* **26**, 347–374.
984. J.-J. Moreau (1988), Bounded variation in time, in *Topics in Nonsmooth Mechanics*, edited by J.-J. Moreau, P. D. Panagiotopoulos and G. Strang, pp. 1–74, Birkhäuser, Boston, Massachusetts.
985. D. Motreanu and N. Pavel (1999), *Tangency, Flow-Invariance for Differential Equations, and Optimization Problems*, Marcel Dekker, New York.
986. M. Moussaoui and A. Seeger (1994), Sensitivity analysis of optimal value functions of convex parametric programs with possibly empty solution sets, *SIAM J. Optim.* **3**, 659–675.
987. M. Moussaoui and A. Seeger (1996), Epsilon-maximum principle of Pontryagin type and perturbation analysis of convex optimal control problems, *SIAM J. Control Optim.* **34**, 407–427.
988. M. Mureşan (2001), *Analiză Nenetedăsi Aplicatii*, Risoprint, Cluj-Napoca, Romania.
989. A. D. Myshkis (1951), *Linear Differential Equations with Retarded Argument*, GITTL, Moscow.
990. N. Nagumo (1942), Ber diw lage der integralkurven gewohlicher differenialgleichungen, *Proc. Phys. Math. Soc. Japan* **24**, 551–559.

991. N. M. Nam and N. D. Yen (2005), Relationships between approximate Jacobians and coderivatives, *J. Nonlinear Convex Anal.*, to appear.
992. I. Namioka and R. R. Phelps (1975), Banach spaces which are Asplund spaces, *Duke Math. J.* **42**, 735–750.
993. Z. Naniewicz (2005), Pseudo-monotone approach to economic equilibrium problem in reflexive Banach space, preprint.
994. Z. Naniewicz and P. D. Panagiotopoulos (1995), *Mathematical Theory of Hemivariational Inequalities and Applications*, Marcel Dekker, New York.
995. J. F. Nash (1951), Non-cooperative games, *Annals Math.* **54**, 286–295.
996. M. H. Nayakkankuppam and M. L. Overton (1999), Conditioning of semilinear programs, *Math. Progr.* **85**, 525–540.
997. P. Neittaaanmäki and D. Tiba (1994), *Optimal Control of Nonlinear Parabolic Systems*, Marcel Dekker, New York.
998. Y. E. Nesterov (2005), Lexicographical differentiation of nonsmooth functions, *Math. Progr.*, to appear.
999. Y. E. Nesterov and A. S. Nemirovsky (1993), *Interior Point Methods in Convex Optimization: Theory and Applications*, SIAM Publications, Philadelphia, Pennsylvania.
1000. J. von Neumann and O. Morgenstern (1944), *Theory of Games and Economic Behaviour*, Princeton University Press, Princeton, New Jersey.
1001. L. W. Neustadt (1966), An abstract variational theory with applications to a broad class of optimization problems, I: General theory, *SIAM J. Control* **4**, 505–527.
1002. L. W. Neustadt (1976), *Optimization: a Theory of Necessary Conditions*, Princeton University Press, Princeton, New Jersey.
1003. K. F. Ng and W. H. Yang (2004), Regularities and their relations to error bounds, *Math. Progr.* **99**, 521–538.
1004. K. F. Ng and X. Y. Zheng (2001), Error bounds for lower semicontinuous functions in normed spaces, SIAM J. Optim. **12**, 1–17.
1005. K. F. Ng and X. Y. Zheng (2003), Global weak sharp minima in Banach spaces, *SIAM J. Control Optim.* **41**, 1858–1885.
1006. H. V. Ngai, D. T. Luc and M. Théra(2000), Approximate convex functions, *J. Nonlinear Convex Anal.* **1**, 155–176.
1007. H. V. Ngai, D. T. Luc and M. Théra (2002), Extensions of Fréchet ε-subdifferential calculus and applications, *J. Math. Anal. Appl.* **268**, 266–290.
1008. N. V. Ngai and M. Théra (2001), Metric regularity, subdifferential calculus and applications, *Set-Valued Anal.* **9**, 187–216.
1009. N. V. Ngai and M. Théra (2002), A fuzzy necessary optimality condition for non-Lipschitz optimization in Asplund spaces, *SIAM J. Optim.* **12**, 656–668.
1010. N. V. Ngai and M. Théra (2004), Error bounds, and implicit multifunction theorem in smooth Banach spaces and applications to optimization, *Set-Valued Anal.* **12**, 195–223.
1011. A. Nijenhuis (1974), Strong derivatives and inverse mappings, *Amer. Math. Monthly* **81**, 969–980.
1012. M. S. Nikolskii (2004), Difference approximation of some optimization problems, *Comput. Maths. Math. Phys.* **44**, 467–475.
1013. K. Nitka-Styczen (1988), Difference approximations of optimal periodic control problems, *Int. J. Control* **47**, 1893–1904.

1014. K. NITKA-STYCZEN (1989), Approximate discrete maximum principle for the discrete approximation of optimal periodic control problems, *Int. J. Control* **50**, 1863–1871.

1015. K. NITKA-STYCZEN (1999), *Optimal Periodic Control of Hereditary Processes. Theory, Algorithms and Applications*, Technical University of Wrocław, Wrocław, Poland.

1016. A. NOWAKOWSKI AND I. NOWAKOWSKA (2005), Dirichlet problem for semilinear hyperbolic equation, *Nonlinear Anal.*, to appear.

1017. E. A. NURMINSKII (1979), *Methods of Solutions to Deterministic and Stochastic Minimax Problems*, Naukova Dumka, Kiev.

1018. M. N. OĞUZTÖRELI (1966), *Time-Lag Control Systems*, Academic Press, New York.

1019. A. OLBROT (1976), Control of retarded systems with function space constraints: Necessary optimality conditions, *Control Cybernet.* **5**, 5–31.

1020. C. OLECH (1976), Existence theory in optimal control problems, in *Control Theory and Topics in Functional Analysis*, Vol. 1, pp. 291–328, International Atomic Energy Agency, Vienna.

1021. N. ORTIZ (2005), Necessary conditions for the neutral problem of Bolza with continuously varying time delay, *J. Math. Anal. Appl.* **305**, 513–527.

1022. N. ORTIZ AND P. R. WOLENSKI (2004), The decoupling technique for continuously varying time delay systems, *Set-Valued Anal.* **12**, 225–239.

1023. Y. S. OSIPOV, L. PANDOLFI AND V. I. MAKSIMOV (2001), Problems of dynamical reconstruction and robust boundary control: The case of the Dirichlet boundary conditions, *J. Inverse Ill-Posed Probl.* **9**, 149–162.

1024. J. V. OUTRATA (1999), Optimality conditions for a class of mathematical programs with equilibrium constraints, *Math. Oper. Res.* **24**, 627–644.

1025. J. V. OUTRATA (1999), Optimality conditions for a class of mathematical programs with equilibrium constraints: Strongly regular case, *Kybernetika* **35**, 177–193.

1026. J. V. OUTRATA (2000), On mathematical programs with complementarity constraints, *Optim. Meth. Soft.* **14**, 117–137.

1027. J. V. OUTRATA (2000), A generalized mathematical program with equilibrium constraints, *SIAM J. Control Optim.* **38**, 1623–1638.

1028. J. V. OUTRATA (2001), On constraint qualifications for mathematical programs with mixed complementarity constraints, in *Complementarity, Applications and Extensions*, edited by M. C. Ferris, O. L. Mangasarian and J.-S. Pang, pp. 253–272, Kluwer, Dordrecht, The Netherlands.

1029. J. V. OUTRATA (2004), A note on a class of equilibrium problems with equilibrium constraints, *Kybernetika* **40**, 585–594.

1030. J. V. OUTRATA (2005), Mathematical programs with equilibrium constraints: Theory and numerical methods, in *Nonsmooth Mechanics of Solids, CISM Lecture Notes*, edited by J. Haslinger and G. E. Stavroulakis, Springer, New York.

1031. J. V. OUTRATA, M. KOČVARA AND J. ZOWE (1998), *Nonsmooth Approach to Optimization Problems with Equilibrium Constraints*, Kluwer, Dordrecht, The Netherlands.

1032. J. V. OUTRATA AND W. RÖMISCH (2005), On optimality conditions for some nonsmooth optimization problems over L^p spaces, *Optim. Theory Appl.* **126**, 1–28.

1033. M. L. Overton (1988), On minimizing the maximum eigenvalue of a symmetric matrix, *SIAM J. Matrix Anal. Appl.* **9**, 256–268.
1034. M. L. Overton and R. S. Womersley (1993), Optimality conditions and duality theory for minimizing sums of the largest eigenvalue of symmetric matrices, *Math. Progr.* **62**, 321–337.
1035. A. E. Ozdaglar and D. P. Bertsekas (2004), The relation between pseudonormality and quasiregularity in constrained optimization, *Optim. Meth. Soft.* **19**, 493–506.
1036. Z. Páles and V. Zeidan (1994), First and second order necessary conditions for control problems with constraints, *Trans. Amer. Math. Soc.* **346**, 421–453.
1037. Z. Páles and V. Zeidan (1996), Generalized Hessians for $C^{1,1}$ functions in infinite dimensional normed spaces, *Math. Progr.* **74**, 59–78.
1038. Z. Páles and V. Zeidan (2004), Critical and critical tangent cones in optimization problems, *Set-Valued Anal.* **12**, 241–258.
1039. D. Pallaschke, P. Recht and R. Urbański (1987), On extensions of the second-order derivative, *Bull. Polish Acad. Sci., Ser. Math.* **35**, 750–763.
1040. D. Pallaschke and S. Rolewicz (1998), *Foundations of Mathematical Optimization: Convex Analysis without Linearity*, Kluwer, Dordrecht, The Netherlands.
1041. D. Pallaschke and R. Urbański (2002), *Pairs of Compact Convex Sets*, Kluwer, Dordrecht, The Netherlands.
1042. P. D. Panagiotopoulos (1993), *Hemivariational Inequalities: Applications to Mechanics and Engineering*, Springer, Berlin.
1043. J.-S. Pang (1990), Solution differentiability and continuation of Newton's method for variational inequality problems over polyhedral sets, *J. Optim. Theory Appl.* **66**, 121–135.
1044. J.-S. Pang (1993), A degree-theoretical approach to parametric nonsmooth equations with multivalued perturbed solution sets, *Math. Progr.* **62**, 359–383.
1045. J.-S. Pang (1997), Error bounds in mathematical programming, *Math. Progr.* **79**, 299–332.
1046. J.-S. Pang and D. Stewart (2005), Differential variational inequalities, *Math. Progr.*, to appear.
1047. J.-S. Pang and L. Qi (1993), Nonsmooth equations: Motivations and algorithms, *SIAM J. Optim.* **3**, 443–465.
1048. C. Pantelides, D. Gritsis, K. P. Morison and R. W. H. Sargent (1988), The mathematical modelling of transient systems using differential-algebraic equations, *Comput. Chem. Eng.* **12**, 449–454.
1049. N. S. Papageorgiou (1995), On parametric evolution inclusions of the subdifferential type with applications to optimal control problems, *Trans. Amer. Math. Soc.* **347**, 203–231.
1050. M. Papi and S. Sbaraglia (2003), Regularity properties of constrained set-valued mappings, *Nonlinear Anal.* **54**, 1337–1353.
1051. M. Papi and S. Sbaraglia (2005), Lipschitzian estimates in discrete-time constrained stochastic optimal control, *Dynamics Cont. Disc. Impuls. Syst.*, to appear.
1052. M. Papi and S. Sbaraglia (2005), Optimal asset-liability management with constraints: A dynamic programming approach, *Appl. Math. Comput.*, to appear.
1053. V. Pareto (1909), *Manuel d'Economie Politique*, Giard, Paris.

1054. G. PEANO (1892), Sur la définition de la dérivée, *Mathesis* **2**, 12–14.
1055. J. PEÑA (2000), Understanding the geometry of infeasible perturbations of a conic linear system, *SIAM J. Optim.* **10**, 534–550.
1056. J. PEÑA (2001), Conditioning of convex programs from a primal-dual perspectives, *Math. Oper. Res.* **26**, 206–220.
1057. J. PEÑA (2003), A characterization of the distance to infeasibility under structured perturbations, *Linear Algebra Appl.* **370**, 193–216.
1058. J. PEÑA (2004), Conic systems and sublinear mappings: Equivalent approaches, *Operations Research Letters* **32** (2004), 463–467.
1059. J. PEÑA (2005), On the block-structured distance to non-surjectivity of sublinear mappings, *Math. Progr.* **103**, 561–573.
1060. T. PENNANEN (1999), Graph-convex mappings and K-convex functions, *J. Convex Anal.* **6**, 235–266.
1061. T. PENNANEN (2005), Epi-convergent discretizations of multistage stochastic programs, *Math. Oper. Res.* **30**, 245–256.
1062. T. PENNANEN (2005), Epi-convergent discretizations of multistage stochastic programs via integration quadratures, *Math. Progr.*, to appear.
1063. J.-P. PENOT (1974), Sous-différentieles de fonctions numériques nonconvexes, *C. R. Acad. Paris* **278**, 1553–1555.
1064. J.-P. PENOT (1978), Calcul sous-différentiel et optimisation, *J. Funct. Anal.* **27**, 248–276.
1065. J.-P. PENOT (1981), A characterization of tangential regularity, *Nonlinear Anal.* **5**, 625–643.
1066. J.-P. PENOT (1989), Metric regularity, openness and Lipschitzian behavior of multifunctions, *Nonlinear Anal.* **13**, 629–643.
1067. J.-P. PENOT (1994), Sub-hessians, super-hessions and conjugation, *Nonlinear Anal.* **23**, 689–702.
1068. J.-P. PENOT (1995), Inverse function theorems for mappings and multimappings, *Southeast. Asian Bull. Math.* **19**, 1–16.
1069. J.-P. PENOT (1996), Favorable classes of mappings and multimappings in nonlinear analysis and optimization, *J. Convex Anal.* **3**, 97–116.
1070. J.-P. PENOT (1997), Metric estimates for the calculus of multimappings, *Set-Valued Anal.* **5**, 291–308.
1071. J.-P. PENOT (1998), Compactness properties, openness criteria and coderivatives, *Set-Valued Anal.* **6**, 363–380.
1072. J.-P. PENOT (1998), Are generalized derivatives useful for generalized convex functions?, *Generalized Convexity, Generalized Monotonicity: Recent Results*, edited by J.-P. Crouzeix, J. E. Martinez-Legaz and M. Volle, pp. 3–59, Kluwer, Dordrecht, The Netherlands.
1073. R. R. PHELPS (1993), *Convex Functions, Monotone Operators and Differentiability*, 2nd edition, Springer, Berlin.
1074. M. D. R. DE PINHO (2003), Mixed constrained control problems, *J. Math. Anal. Appl.* **278**, 293–307.
1075. M. D. R. DE PINHO, M. M. A. FERREIRA AND F. A. C. C. FONTES (2002), An Euler-Lagrange inclusion for optimal control problems with state constraints, *J. Dynam. Control Systems* **8**, 23–45.
1076. M. D. R. DE PINHO, M. M. A. FERREIRA AND F. A. C. C. FONTES (2005), Unmaximized inclusion necessary conditions for nonconvex constrained optimal control problems, *ESAIM: Control Optim. Calc. Var.* **11**.

1077. M. D. R. DE PINHO AND A. ILCHMANN (2002), Weak maximum principle for optimal control problems with mixed constraints, *Nonlinear Anal.* **48**, 1179–1196.
1078. M. D. R. DE PINHO AND R. B. VINTER (1995), An Euler-Lagrange inclusion for optimal control problems, *IEEE Trans. Autom. Control* **40**, 1191–1198.
1079. M. D. R. DE PINHO AND R. B. VINTER (1997), Necessary conditions for optimal control problems involving nonlinear differential algebraic equations, *J. Math. Anal. Appl.* **212**, 493–516.
1080. M. D. R. DE PINHO, R. B. VINTER AND H. ZHENG (2001), A maximum principle for optimal control problems with mixed constraints, *IMA J. Math. Control Inf.* **18**, 189–205.
1081. S. PLASKASZ (1992), On the solution sets for differential inclusions, *Boll. Un. Mat. Ital.* **6**, 387–394.
1082. V. A. PLOTNIKOV, A. V. PLOTNIKOV AND A. N. VITYUK (1999), *Differential Equations with a Multivalued Right-Hand Side: Asymptotic Methods*, AstroPrint, Odessa, Ukraine.
1083. V. I. PLOTNIKOV (1972), Necessary and sufficient conditions for optimality and conditions for uniqueness of the optimizing functions for control systems of general form, *Izv. Akad. Nauk SSSR, Ser. Mat.* **36**, 652–679.
1084. V. I. PLOTNIKOV AND M. I. SUMIN (1982), The construction of minimizing sequences in control problems for systems with distributed paratemers, *USSR Comput. Maths. Math. Phys.* **22**, 292–296.
1085. V. I. PLOTNIKOV AND V. I. SUMIN (1972), Optimization of objects with distributed parameters that can be described by Coursat-Darboux systems, *USSR Comput. Maths. Math. Phys.* **12**, 49–57.
1086. E. POLAK (1999), *Optimization: Algorithms and Consistent Approximations*, Springer, New York.
1087. R. A. POLIQUIN (1990), Subgradient monotonicity and convex functions, *Nonlinear Anal.* **14**, 305–317.
1088. R. A. POLIQUIN (1992), An extension of Attouch's theorem and its application to second-order epi-differentiation of convexly composite functions, *Trans. Amer. Math. Soc.* **332**, 861–874.
1089. R. A. POLIQUIN AND R. T. ROCKAFELLAR (1992), Amenable functions in optimization, in *Nonsmooth Optimization Methods and Applications*, edited by F. Giannessi, pp. 338–353, Gordon and Breach, Philadelphia, Pennsylvania.
1090. R. A. POLIQUIN AND R. T. ROCKAFELLAR (1996), Prox-regular functions in variational analysis, *Trans. Amer. Math. Soc.* **348**, 1805–1838.
1091. R. A. POLIQUIN AND R. T. ROCKAFELLAR (1996), Generalized Hessian properties of regularized nonsmooth functions, *SIAM J. Optim.* **6**, 1121–1137.
1092. R. A. POLIQUIN AND R. T. ROCKAFELLAR (1998), Tilt stability of a local minimum, *SIAM J. Optim.* **8**, 287–299.
1093. R. A. POLIQUIN, R. T. ROCKAFELLAR AND L. THIBAULT (2000), Local differentiability of distance functions, *Trans. Amer. Math. Soc.* **332**, 5231–5249.
1094. E. S. POLOVINKIN AND G. V. SMIRNOV (1986), An approach to differentiation of multifunctions and necessary optimality conditions for differential inclusions, *Diff. Eq.* **22**, 660–668.
1095. E. S. POLOVINKIN AND G. V. SMIRNOV (1986), Time-optimal problem for differential inclusions, *Diff. Eq.* **22**, 940–952.
1096. B. T. POLYAK (1969), Semicontinuity of intergral functionals and existence theorems on extremal problems, *Math. Sbornik* **7**, 59–77.

1097. B. T. Polyak (1987), *Introduction to Optimization*, Optimization Software, New York.

1098. B. T. Polyak (1998), Convexity of quadratic transformations and its use in control and optimization, *J. Optim. Theory Appl.* **99**, 553–583.

1099. B. T. Polyak (2000), History of mathematical programming in the USSR: Analyzing the phenomenon, *Math. Progr.* **91**, 401–416.

1100. B. T. Polyak (2001), Convexity of nonlinear image of a small ball with applicatiobs to optimization, *Set-Valued Anal.* **9**, 159–168.

1101. D. Pompeiu (1905), Fonctions de variables complexes, *Ann. Facul. Sci. Univ. Toulouse* **7**, 265–315.

1102. L. S. Pontryagin, V. G. Boltyanskii, R. V. Gamkrelidze and E. F. Mishchenko (1962), *The Mathematical Theory of Optimal Processes*, Wiley, New York, 1962.

1103. D. Preiss (1984), Gâteaux differentiable functions are somewhere Fréchet differentiable, *Rend. Circ. Mat. Palermo* **33**, 122–133.

1104. D. Preiss (1990), Differentiability of Lipschitz functions on Banach spaces, *J. Funct. Anal.* **91**, 312–345.

1105. A. I. Propoi (1973), *Elements of the Theory of Optimal Discrete Processes*, Nauka, Moscow.

1106. B. N. Pshenichnyi (1971), *Necessary Conditions for an Extremum*, Marcel Dekker, New York.

1107. B. N. Pshenichnyi (1976), Necessary conditions for an extremum for differential inclusions, *Kibernetika* **12**, 60–73.

1108. B. N. Pshenichnyi (1977), On necessary extremality conditions for nonsmooth functions, *Kibernetika* **13**, 92–96.

1109. B. N. Pshenichnyi (1980), *Convex Analysis and Extremal Problems*, Nauka, Moscow.

1110. V. Pták (1974), A quantitative refinement of the closed graph theorem, *Czech. Math. J.* **24**, 503–506.

1111. L. Qi and J. Sun (1993), A nonsmooth version of Newton's method, *Math. Progr.* **58**, 353–368.

1112. Y. P. Qiu and T. L. Magnanti (1992), Sensitivity analysis for variational inequalities, *Math. Oper. Res.* **17**, 61–76.

1113. M. Quinzii (1992), *Increasing Returns and Efficiency*, Oxford University Press, Oxford, UK.

1114. H. Rademacher (1919), Uber partielle und totale differenzierbarkeit von funktionen mehrerer varieabeln und über die transformation der doppelintegrale, *Math. Ann.* **79**, 340–359.

1115. D. Ralph (1994), A chain rule for nonsmooth composite functions via minimization, *Bull. Austral. Math. Soc.* **49**, 129–137.

1116. D. Ralph (2002), A stable homotopy approach to horizontal linear complementarity problems, *Control Cybernet.* **31**, 575–599.

1117. D. Ralph and S. J. Wright (2004), Some properties of regularization and penalization schemes for MPECs, *Optim. Meth. Soft.* **5**, 527–556.

1118. F. Rampazzo and R. B. Vinter (2000), Degenerate optimal control problems with state constraints, *SIAM J. Control Optim.* **39**, 989–1007.

1119. T. Rapcsák (1997), *Smooth Nonlinear Optimization in* $I\!R^n$, Kluwer, Dordrecht, The Netherlands.

1120. J.-P. Raymond (1997), Nonlinear boundary control of semilinear parabolic problems with pointwise state constraints, *Disc. Cont. Dynam. Systems* **3**, 341–370.
1121. J.-P. Raymond and H. Zidani (1998), Pontryagin's principle for state-constrained control problems governed by parabolic equations with unbounded controls, *SIAM J. Control Optim.* **38**, 1853–1879.
1122. J. Renegar (1995), Incorporating condition measures into the complexity theory of linear programming, *SIAM J. Optim.* **5**, 506–524.
1123. J. Renegar (1995), Linear programming, complexity theory and elementary functional analysis, *Math. Progr.* **70**, 279–540.
1124. L. Rifford (2002), Semiconcave control Lyapunov functions and stabilizing feedback, *SIAM J. Control Optim.* **41**, 659–681.
1125. S. M. Robinson (1972), Normed convex processes, *Trans. Amer. Math. Soc.* **174**, 127–140.
1126. S. M. Robinson (1973), Bounds for error in the solution set of a perturbed linear program, *Linear Algebra Appl.* **6**, 69–81.
1127. S. M. Robinson (1975), An application of error bounds for convex programming in a linear space, *SIAM J. Control* **13**, 271–273.
1128. S. M. Robinson (1976), Stability theory for systems of inequalities, II: Differentiable nonlinear systems, *SIAM J. Numer. Anal.* **13**, 130–143.
1129. S. M. Robinson (1976), Regularity and stability for convex multivalued functions, *Math. Oper. Res.* **1**, 130–143.
1130. S. M. Robinson (1979), Generalized equations and their solutions, I: Basic theory, *Math. Progr. Study* **10**, 128–141.
1131. S. M. Robinson (1980), Strongly regular generalized equations, *Math. Oper. Res.* **5**, 43–62.
1132. S. M. Robinson (1981), Some continuity properties of polyhedral multifunctions, *Math. Progr. Study* **14**, 206–214.
1133. S. M. Robinson (1982), Generalized equations and their solutions, II: Applications to nonlinear programming, *Math. Progr. Study* **19**, 200–221.
1134. S. M. Robinson (1983), Generalized equations, in *Mathematical Programming: The State of the Art*, edited by A. Bachem et al., pp. 346–367, Springer, Berlin.
1135. S. M. Robinson (1987), Local epi-continuity and local optimization, *Math. Progr. Study* **37**, 208–222.
1136. S. M. Robinson (1991), An implicit function theorem for a class of nonsmooth functions, *Math. Oper. Res.* **16**, 292–309.
1137. S. M. Robinson (2003), Constraint nondegeneracy in variational analysis, *Math. Oper. Res.* **28**, 201–232.
1138. S. M. Robinson (2003), Variational conditions with smooth constraints: Structure and analysis, *Math. Progr.* **97**, 245–265.
1139. S. M. Robinson (2004), Localized normal maps and the stability of variational conditions, *Set-Valued Anal.* **12**, 259–274.
1140. R. T. Rockafellar (1963), *Convex Functions and Dual Extremum Problems*, Ph.D. dissertation, Department of Mathematics, Harvard University, Cambridge, Massachusetts.
1141. R. T. Rockafellar (1966), Characterization of subdifferentials of convex functions, *Pacific J. Math.* **17**, 497–510.
1142. R. T. Rockafellar (1970), *Convex Analysis*, Princeton University Press, Princeton, New Jersey.

1143. R. T. ROCKAFELLAR (1970), Conjugate convex functions in optimal control and the calculus of variations, *J. Math. Anal. Appl.* **32**, 174–222.
1144. R. T. ROCKAFELLAR (1970), Generalized Hamiltonian equations for convex problems of Lagrange, *Pacific J. Math.* **33**, 411–427.
1145. R. T. ROCKAFELLAR (1971), Existence and duality theorems for convex problems of Bolza, *Trans. Amer. Math. Soc.* **159**, 1–39.
1146. R. T. ROCKAFELLAR (1979), Clarke's tangent cone and the boundaries of closed sets in $I\!R^n$, *Nonlinear Anal.* **3**, 145–154.
1147. R. T. ROCKAFELLAR (1979), Directional Lipschitzian functions and subdifferential calculus, *Proc. London Math. Soc.* **39**, 331–355.
1148. R. T. ROCKAFELLAR (1980), Generalized directional derivatives and subgradients of nonconvex functions, *Canad. J. Math.* **32**, 157–180.
1149. R. T. ROCKAFELLAR (1981), *The Theory of Subgradients and Its Applications to Problems of Optimization: Convex and Nonconvex Functions*, Helderman Verlag, Berlin.
1150. R. T. ROCKAFELLAR (1981), Proximal subgradients, marginal values and augmented Lagrangians in nonconvex optimization, *Math. Oper. Res.* **6**, 424–436.
1151. R. T. ROCKAFELLAR (1982), Favorable classes of Lipschitz continuous functions in subgradient optimization, in *Progress in Nondifferentiable Optimization*, edited by E. A. Nurminskii, pp. 125–143, IIASA, Laxenburg, Austria.
1152. R. T. ROCKAFELLAR (1982), Lagrange multipliers and subderivatives of optimal value functions in nonlinear programming, *Math. Progr. Study* **17**, 28–66.
1153. R. T. ROCKAFELLAR (1985), Maximal monotone relations and the second derivatives of nonsmooth functions, *Ann. Inst. H. Poincaré: Analyse Non Linéaire* **2**, 167–184.
1154. R. T. ROCKAFELLAR (1985), Lipschitzian properties of multifunctions, *Nonlinear Anal.* **9**, 867–885.
1155. R. T. ROCKAFELLAR (1985), Extensions of subgradient calculus with applications to optimization, *Nonlinear Anal.* **9**, 665–698.
1156. R. T. ROCKAFELLAR (1988), First- and second-order epi-differentiability in nonlinear programming, *Trans. Amer. Math. Soc.* **307**, 75–108.
1157. R. T. ROCKAFELLAR (1989), Letter to B. S. Mordukhovich of March 6, 1989, Seattle, Washington.
1158. R. T. ROCKAFELLAR (1989), Derivation of some improved formulas in subdifferential calculus, privately circulated note.
1159. R. T. ROCKAFELLAR (1989), Proto-differentiability of set-valued mappings and its applications in optimization, in *Analyse Non Linéaire*, edited by H. Attouch et al., pp. 449–482, Gauthier-Villars, Paris.
1160. R. T. ROCKAFELLAR (1993), Lagrange multipliers and optimality, *SIAM Rev.* **35**, 183–238.
1161. R. T. ROCKAFELLAR (1993), Dualization of subgradient conditions for optimality, *Nonlinear Anal.* **20**, 627–642.
1162. R. T. ROCKAFELLAR (1996), Equivalent subgradient versions of Hamiltonian and Euler-Lagrange equations in variational analysis, *SIAM J. Control Optim.* **34**, 1300–1315.
1163. R. T. ROCKAFELLAR (2000), Second-order convex analysis, *J. Nonlinear Convex Anal.* **1**, 1–16.
1164. R. T. ROCKAFELLAR (2004), Hamilton–Jacobi theory and parametric analysis in fully convex problems of optimal control, *J. Global Optim.* **28**, 419–431.

1165. R. T. Rockafellar and R. J-B. Wets (1998), *Variational Analysis*, Springer, Berlin.
1166. R. T. Rockafellar and P. R. Wolenski (2000), Convexity in Hamilton-Jacobi theory, I: Dynamics and duality, *SIAM J. Control Optim.* **39**, 1323–1350.
1167. R. T. Rockafellar and P. R. Wolenski (2000), Convexity in Hamilton-Jacobi theory, II: Envelope representations, *SIAM J. Control Optim.* **39**, 1351–1372.
1168. R. T. Rockafelar and D. Zagrodny (1997), A derivative-coderivative inclusion in second-order nonsmooth analysis, *Set-Valued Anal.* **5**, 1–17.
1169. S. Rolewicz (1979), On paraconvex multifunctions, *Oper. Res. Verfahren* **31**, 540–546.
1170. S. Rolewicz (1979), On γ-paraconvex multifunctions, *Math. Japon.* **24**, 293–300.
1171. S. Rolewicz (2005), On differentiability of strongly $\alpha(\cdot)$-paraconvex functions in non-separable Asplund spaces, *Studia Math.*, **167**, 235–244.
1172. S. Rolewicz (2005), Paraconvex analysis on $C^{1,u}_{\mathbf{E}}$ manifolds, *Optimization*, to appear.
1173. J. F. Rosenblueth and R. B. Vinter (1991), Relaxation procedures for time delay systems, *J. Math. Anal. Appl.* **162**, 542–563.
1174. J. F. Rosenblueth, J. Warga and Q. J. Zhu (1997), On the characterization of properly relaxed delayed controls, *J. Math. Anal. Appl.* **209**, 274–290.
1175. I. M. Ross and F. Fahroo (2004), Legendre pseudospectral approximations of optimal control problems, *Lecture Notes Cont. Inf. Sci.* **295**, pp. 327–342, Springer, New York.
1176. R. Rossi and G. Savaré (2005), Gradient flows of nonconvex functionals in Hilbert spaces and applications, *ESAIM: Control Optim. Cal. Var.*, to appear.
1177. T. Roubiček (1997), *Relaxation in Optimization Theory and Variational Calculus*, De Gruyter, Berlin.
1178. J. D. L. Rowland and R. B. Vinter (1992), Pontryagin type conditions for differential inclusions, *J. Math. Anal. Appl.* **165**, 587–597.
1179. J. D. L. Rowland and R. B. Vinter (1993), Dynamic optimization problems with free-time and active state constraints, *SIAM J. Control Optim.* **31**, 677–697.
1180. L. I. Rozonoér (1959), L. S. Pontryagin maximum principle in the theory of optimal systems, I–III, *Autom. Remote Control* **20**, 1288–1302, 1405–1421, 1517–1532.
1181. A. M. Rubinov (2000), *Abstract Convexity and Global Optimization*, Kluwer, Dordrecht, The Netherlands.
1182. A. M. Rubinov (2002), Equilibrium with restriction on exchange, *Cybernetics and Systems Analysis*, No. 2, 55–70.
1183. J.-J. Rückmann (1999), On existence and uniqueness of stationary points in semi-infinite programming, *Math. Progr.* **86**, 387–415.
1184. D. L. Russell (1966), Optimal regulation of linear symmetruc hyperbolic systems with finite dimensional control, *SIAM J. Control* **4**, 276–294.
1185. L. D. Sabbagh (1969), Variational problems with lags, *J. Optim. Theory Appl.* **3**, 34–51.
1186. S. Saks (1937), *Theory of the Integral*, 2nd edition, Hafner Publishing Co., New York.

1187. D. Salamon (1984), *Control and Observation of Neutral Systems*, Pitman, Harlow, Essex, UK.

1188. P. A. Samuelson (1947), *Foundations of Economic Analysis*, Harvard University Press, Cambridge, Massachusetts.

1189. P. A. Samuelson (1954), The pure theory of public expenditures, *Rev. Econ. Stat.* **36**, 387–389.

1190. J. Schauder (1930), Über die umkerhrung linearer, stetiger funktionaloperationen, *Studia Math.* **2**, 1–6.

1191. H. Scheel and S. Scholtes (2000), Mathematical programs with equilibrium constraints: Stationarity, optimality and sensitivity, *Math. Oper. Res.* **25**, 1–22.

1192. S. Scholtes (2001), Convergence properties of a regularization scheme for mathematical programs with complementarity constraints, *SIAM J. Optim.* **11**, 818–936.

1193. S. Scholtes (2002), On the existence and computation of EPEC solutions, talk given at the ICCP Conference in Cambridge, UK.

1194. S. Schotes and M. Stöhr (1999), Exact penalization of mathematical programs with equilibrium constraints, *SIAM J. Control Optim.* **37**, 617–652.

1195. L. Scrimali (2004), Variational inequalities and optimal equilibrium distributions in transportation networks, *Math. Ineq. Appl.* **7**, 439–451.

1196. R. Schultz (2000), Some aspects of stability in stochastic programming, *Ann. Oper. Res.* **100**, 55–84.

1197. L. Schwartz (1966), *Théorie des Distributions*, Hermann, Paris.

1198. A. Seeger (1992), Limiting behavior of the approximate second-order subdifferential of a convex function, *J. Optim. Theory Appl.* **74**, 527–544.

1199. A. Seeger (1994), Approximate Euler-Lagrange inclusion, approximate transversality condition, and sensitivity analysis of convex parametric problems of calculus of variations, *Set-Valued Anal.* **2**, 307–325.

1200. H. S. Sendov (2000), *Variational Spectral Analysis*, Ph.D. dissertation, Department of Combinatorics and Optimization, University of Waterloo, http://etd/uwaterloo.ca/etd/hssendov2000.pdf, Waterloo, Canada.

1201. B. Sendov and V. A. Popov (1988), *The Averaged Moduli of Smoothness*, Wiley, New York.

1202. F. Severi (1930), Su alcune questioni di topologia infinitesimale, *Ann. Soc. Polon. Math.* **9**, 97–108.

1203. A. Shapiro (1988) Sensitivity analysis of nonlinear programs and differentiability property of metric projections, *SIAM J. Control Optim.* **26**, 628–645.

1204. A. Shapiro (1990), On concepts of directional defferentiability, *J. Math. Anal. Appl.* **66**, 477–487.

1205. A. Shapiro (1994), Sensitivity analysis of parameterized programs via generalized equations, *SIAM J. Control Optim.* **32**, 553–571.

1206. A. Shapiro, T. Homem-de-Mello and J. Kim (2002), Conditioning of convex pievewise linear stochastic programs, *Math. Progr.* **94**, 1–19.

1207. N. Z. Shor (1972), On a class of almost-differentiable functions and on a minimization method for functions from this class, *Kibernetika*, No. 4. 65–70.

1208. N. Z. Shor (1985), *Minimization Methods for Non-Differentiable Functions*, Springer, Berlin.

1209. I. A. Shvartsman (2005), New approximation method in the proof of the maximum principle for nonsmooth optimal control problems with state constraints, *J. Math. Anal. Appl.*, to appear.

1210. I. A. SHVARTSMAN AND R. B. VINTER (2005), Regularity properties of optimal controls for state constrained problems with time-varying control constraints, *Nonlinear Anal.*, to appear.
1211. G. H. SILVA AND R. B. VINTER (1998), Necessary conditions for optimal impulsive control problems, *SIAM J. Control Optim.* **35**, 1829–1846.
1212. J. SIMON (1987), Compact sets in the space $L^p(0,T;B)$, *Ann Mat. Pura Appl.* **146**, 65–96.
1213. S. SIMONS (1998), *Minimax and Monotonicity*, Springer, Berlin.
1214. I. SINGER (1997), *Abstract Convex Analysis*, Wiley, New York.
1215. G. V. SMIRNOV (1991), Discrete approximations and optimal solutions to differential inclusions, *Cybernetics* **27**, 101–107.
1216. G. V. SMIRNOV (2001), *Introduction to the Theory of Differential Inclusions*, American Mathematical Society, Providence, Rhode Island.
1217. A. SMITH (1776), *An Inquiry into the Nature and Causes of the Wealth of Nations*, Clarendon Press, Oxford, England.
1218. S. L. SOBOLEV (1963), *Applications of Functional Analysis in Mathematical Physics*, American Mathematical Society, Providence, Rhode Island.
1219. M. V. SOLODOV AND B. F. SVAITER (2000), Error bounds for proximal point subproblems and associated inexasct proximal point algorithms, *Math. Progr.* **88**, 371–389.
1220. E. SONTAG (1999), Stability and stabilization: Discontinuity and the effect of disturbances, in *Nonlinear Analysis, Differential Equations, and Control*, edited by F. H. Clarke and R. J. Stern, pp. 551–598, Kluwer, Dordrecht, The Netherlands.
1221. V. A. SROCHKO (1984), Optimality conditions of the maximum principle type in Goursat-Darboux systems, *Siberian J. Math.* **25**, 126–132.
1222. H. VON STACKELBERG (1934), *Marktform und Gleichgewicht*, Springer, Berlin.
1223. G. STAMPACCHIA (1964), Formes bilinéaires coercitives sur les ensemble convexes, *C. R. Acad. Sci. Paris* **258**, 4413–4416.
1224. C. STEGALL (1978), Optimization of functions on certain subsets of Banach spaces, *Trans. Amer. Math. Soc.* **236**, 171–176.
1225. O. STEIN (2003), *Bilevel Strategies in Semi-Infinite Programming*, Kluwer, Boston, Massachusetts.
1226. A. S. STREKALOVSKY (1987), On the problem of global extremum, *Soviet Math. Dokl.* **35**, 194–198.
1227. A. S. STREKALOVSKY (1998), Global optimality conditions for nonconvex optimization, *J. Global Optim.* **12**, 415–434.
1228. A. S. STREKALOVSKY (2003), *Elements of Nonconvex Optimization*, Nauka, Novosibirsk.
1229. M. STUDNIARSKI AND D. E. WARD (1999), Weak sharp minima: Characterizations and sufficient conditions, *SIAM J. Control Optim.* **38** (1999), 219–236.
1230. A. I. SUBBOTIN (1995), *Generalized Solutions of First-Order PDEs*, Birkhäuser, Boston, Massuchusetts.
1231. N. N. SUBBOTINA (1989), The maximum principle and the superdifferential of the value function, *Prob. Control Inform. Theory* **18**, 151–160.
1232. F. SULLIVAN (1981), A characterization of complete metric spaces, *Proc. Amer. Math. Soc.* **83**, 345–346.
1233. M. I. SUMIN (2000), Optimal control of semilinear elliptic equations with state constraints: Maximum principle for minimizing sequences, regularity, normality, sensitivity, *Control Cybernet.* **29**, 449–472.

1234. M. I. SUMIN (2001), Suboptimal control of a semilinear elliptic equation with a phase constraint and a boundary control, *Diff. Eq.* **37**, 281–300.
1235. H. J. SUSSMANN (1994), A strong version of the Lojasiewicz maximum principle, in *Optimal Control and Differential Equations*, edited by N. N. Pavel, pp. 293–309, Marcel Dekker, New York.
1236. H. J. SUSSMANN (1998), Geometry and optimal control, in *Mathematical Control Theory*, edited by J. Baillieul and J. C. Willems, pp. 140-198, Springer, New York.
1237. H. J. SUSSMANN (2000), New theories of set-valued differentials and new versions of the maximum principle, in *Nonlinear Control in the Year 2000*, edited by A. Isidori, F. Lamnabhi-Lagarrigue and W. Respondek, pp. 487–526, Springer, Berlin.
1238. H. J. SUSSMANN (2002), Needle variations and almost lower semicontinuous differential inclusions, *Set-Valued Anal.* **10**, 233–285.
1239. H. J. SUSSMANN AND J. V. WILLEMS (1997), 300 years of optimal control: From the brachystochrone to the maximum principle, *IEEE Control Syst. Magaz.* **17**, 32–44.
1240. A. ŚWIECH (1994), Unbounded second order partial differential equations in infinite dimensional Hilbert spaces, *Comm. Part. Diff. Eq.* **19**, 1999-2036.
1241. A. ŚWIECH (1996), Sub- and superoptimality principles of dynamic programming revisited, *Nonlinear Anal.* **26**, 1429–1436.
1242. T. TADUMADZE AND L. ALKHAZISHVILI (2003), Formulas of variations of solutions for nonlinear controlled delay differential equations with discontinuous initial conditions, *Mem. Diff. Eq. Math. Phys.* **29**, 125–150.
1243. W. TAKAHASHI (2000), *Nonlinear Functional Analysis*, Yokohama Publishers, Yokohama, Japan.
1244. L. THIBAULT (1976), Probléme de Bolza dans un espace de Banach séparable, *C. R. Acad. Sci. Paris* **282**, 1303–1306.
1245. L. THIBAULT (1978), Sous-différentiel de fonctions vertorielles compactement lipschitziennes, *C. R. Acad. Sci. Paris* **286**, 995–998.
1246. L. THIBAULT (1980), Subdifferentials of compactly Lipschitzian vector functions, *Ann. Mat. Pura Appl.* **125**, 157–192.
1247. L. THIBAULT (1982), On generalized differentials and subdifferentials of Lipschitz vector-valued functions, *Nonlinear Anal.* **6**, 1037–1053.
1248. L. THIBAULT (1983), Tangent cones and quasi-interiorly tangent cones to multifunctions, *Trans. Amer. Math. Soc.* **277**, 601–621.
1249. L. THIBAULT (1991), On subdifferentials of optimal value functions, *SIAM J. Control Optim.* **29**, 1019–1036.
1250. L. THIBAULT (1992), Lagrange-Kuhn-Tucker multipliers for general mathematical programming problems, in *Optimization and Nonlinear Analysis*, edited by A. Ioffe, L. Marcus and S. Reich, Pitman Researh Notes Math. Ser. **244**, pp. 311–315, Longman, Harlow, Essex, UK.
1251. L. THIBAULT (1995), A note on the Zagrodny mean value theorem, *Optimization* **35**,127–130.
1252. L. THIBAULT (1997), On compactly Lipschitzian mappings, in *Recent Advances in Optimization*, edited by P. Gritzmann et al., Lecture Notes Econ. Math. Syst. **456**, pp. 356–364, Springer, Berlin.
1253. L. THIBAULT (2003), Sweeping process with regular and nonregular sets, *J. Diff. Eq.* **193**, 1–26.

1254. L. Thibault and D. Zagrodny (1995), Integration of subdifferentials of lower semicontinuous functions, *J. Math. Anal. Appl.* **189**, 22–58.
1255. D. Tiba (1990), *Optimal Control of Nonsmooth Distributed Parameter Systems*, Springer, Berlin.
1256. V. M. Tikhomirov (1990), *Stories about Maxima and Minima*, American Mathematical Society, Providence, Rhode Island.
1257. V. M. Tikhomirov (1997), *Elements of the Theory of Extrema*, Tinbergen Institute, TI 97-048/4, Amsterdam, The Netherlands.
1258. A. A. Tolstonogov (2000), *Differential Inclusions in a Banach Spaces*, Kluwer, Dordrecht, The Netherlands.
1259. A. A. Tolstonogov (2004), Properties of the attainable sets of evolution inclusions and control systems of subdifferential type, *Siberian Math. J.* **45**, 763–784.
1260. L. Tonelli (1923), *Fondamenti di Calcolo delle Variazioni*, published in two volumes, Zanichelli, Bologna, Italy.
1261. R. Tourky (1998), A new approach to the limit theorem on the core of an economy in vector lattices, *J. Econ. Theory* **78**, 321–328.
1262. J. S. Treiman (1983), Characterization of Clarke's tangent and normal cones in finite and infinite dimensions, *Nonlinear Anal.* **7**, 771–783.
1263. J. S. Treiman (1986), Clarke's gradients and epsilon-subgradients in Banach spaces, *Trans. Amer. Math. Soc.* **294**, 65–78.
1264. J. S. Treiman (1988), Shrinking generalized gradients, *Nonlinear Anal.* **12**, 1429–1450.
1265. J. S. Treiman (1989), Finite dimensional optimality conditions: B-gradients, *J. Optim. Theory Appl.* **62**, 771–783.
1266. J. S. Treiman (1990), Optimal control with small generalized gradients, *SIAM J. Control Optim.* **28**, 720–732.
1267. J. S. Treiman (1995), The linear nonconvex generalized gradients and Lagrange multipliers, *SIAM J. Optim.* **5**, 670–680.
1268. J. S. Treiman (1999), Lagrange multipliers for nonconvex generalized gradients with equality, inequality, and set constraints, *SIAM J. Control Optim.* **37**, 1313–1329.
1269. J. S. Treiman (2002), The linear generalized gradient in infinite dimensions, *Nonlinear Anal.* **48**, 427–443.
1270. A. A. Tret'yakov and J. E. Marsden (2003), Factor-analysis of nonlinear mappings: p-Regularity theory, *Commun. Pure Appl. Anal.* **2**, 425–445.
1271. F. Tröltzsch (1984), *Optimality Conditions for Parabolic Control Problems and Applications*, Teubner Texte, Leipzig.
1272. I. Tsevendorj (2001), Piecewise-convex maximization problems, *J. Global Optim.* **21**, 1–14.
1273. H. D. Tuan (1995), On controllability and extremality in nonconvex differential inclusions, *J. Optim. Theory Appl.* **85**, 435–472.
1274. A. Uderzo (2002), Notes on metric regularity and open covering, Technical Report MS2.7.8, pp. 1–26, Dipartimento di Sistemi ed Istituzioni per l'Economia, Univ. di L'Aquila.
1275. C. Ursescu (1975), Multifunctions with closed convex graphs, *Czech. Math. J.* **25**, 438–441.
1276. C. Ursescu (1982), Tangent sets' calculus and necessary conditions for extremality, *SIAM J. Control Optim.* **20**, 563–574.

1277. M. VALADIER (1990), Young measures, in *Methods of Convex Analysis*, Lecture Notes Math. **1446**, pp. 152–188, Springer, Berlin.
1278. P. P. VARAJAY (1967), Nonlinear programming in Banach spaces, *SIAM J. Appl. Math.* **15**, 284–293.
1279. F. P. VASILIEV (1969), Optimality conditions for certain classes of systems which are not solved with respect to the derivative, *Soviet Math. Dokl.* **10**, 224–227.
1280. F. P. VASILIEV (1981), *Solution Methods for Extremal Problems*, Nauka, Moscow.
1281. O. V. VASILIEV (1996), *Optimization Methods*, World Federation Publishers, Atlanta, Georgia.
1282. V. VELIOV (1994), Differential inclusions with stable subinclusions, *Nonlinear Anal.* **23**, 1027–1038.
1283. V. VELIOV (1997), Lipschitz continuity of the value function in optimal control, *J. Optim. Theory Appl.* **94**, 335–363.
1284. V. VELIOV (1997), On the time-discretization of control systems, *SIAM J. Control Optim.* **35**, 1470–1486.
1285. I. Y. VERCHENKO AND A. N. KOLMOGOROV (1934), Continuation of investigation on discontinuity points for functions of two variables, *Dokl. Acad. Nauk SSSR* **4**, 361–364.
1286. J.-P. VIAL (1983), Strong and weak convexity of sets and functions, *Math. Oper. Res.* **8**, 231–259.
1287. A. VILLAR (2000), *Equilibrium and Efficiency in Production Economies*, Springer, Berlin.
1288. A. VILLAR (2001), On the efficiency of market equilibrium in production economies, *J. Global Optim.* **20**, 375–389.
1289. R. B. VINTER (2000), *Optimal Control*, Birkhäuser, Boston, Massachusetts.
1290. R. B. VINTER AND G. PAPPAS (1982), A maximum principle for non-smooth optimal control problems with state constraints, *J. Math. Anal. Appl.* **89**, 212–232.
1291. R. B. VINTER AND F. L. PEREIRA (1988), A maximum principle for optimal processes with discontinuous trajectories, *SIAM J. Control Optim.* **26**, 205–229.
1292. R. B. VINTER AND P. R. WOLENSKI (1990), Coextremals and the value function for control problems with data measurable in time, *J. Optim. Theory Appl.* **153**, 37–51.
1293. R. B. VINTER AND P. D. WOODFORD (1997), On the occurrence of intermediate local minimizers that are not strong local minimizers, *Systems Cont. Lett.* **31**, 235–342.
1294. R. B. VINTER AND H. ZHENG (1997), The extended Euler-Lagrange condition for nonconvex variational problems, *SIAM J. Control Optim.* **35**, 56–77.
1295. R. B. VINTER AND H. ZHENG (1998), The extended Euler-Lagrange condition for nonconvex variational problems with state constraints, *Trans. Amer. Math. Soc.* **350**, 1181–1204.
1296. R. B. VINTER AND H. ZHENG (2000), Necessary conditions for free end-time, measurable time dependent optimal control problems with state constraints, *Set-Valued Anal.* **8**, 10–29.
1297. R. B. VINTER AND H. ZHENG (2003), Some finance problems solved with nonsmooth optimization techniques, *J. Optim. Theory Appl.* **119**, 1–18.

1298. V. Volterra (1931), *Théories Mathématoque de la Lutte pour la Vie*, Guathoer-Villar, Paris.
1299. D. W. Walkup and R. J-B. Wets (1969), A Lipschitzian characterization of convex polyhedra, *Proc. Amer. Math. Soc.* **20**, 167–173.
1300. L. Walras (1874-7), *Eléments d'Economie Politique Pure*, L. Corbaz and Company, Lausanne.
1301. L. Wang (2005), Discrete approximations to optimization of neutral functional differential inclusions, *J. Math. Anal. Appl.* **309**, 474–488.
1302. P. K. C. Wang (1964), Control of distributed parameter systems, in *Advances in Control Systems, Theory and Applications*, edited by C. T. Leondes, Vol. 1, pp. 75–172, Academic Press, New York.
1303. X. Wang (2005), Subdifferentiability of real functions, *Real Anal. Exchange* **30**, 137–171.
1304. X. Wang (2005), Extremal characterizations of reflexive spaces, *Pacific J. Optim.*, to appear.
1305. X. Wang and V. Jeyakumar (2000), A sharp Lagrange multiplier rule for nonsmooth mathematical programming problems involving equality constraints, *SIAM J. Optim.* **10**, 1136–1148.
1306. D. E. Ward (1993), Calculus for parabolic second-order derivatives, *Set-Valued Anal.* **1**, 213–246.
1307. D. E. Ward (1994), A chain rule for parabolic second-order epiderivatives, *Optimization* **28**, 223–236.
1308. D. E. Ward (1995), A comparison of second-order epiderivatives: Calculus and optimality conditions, *J. Math. Anal. Appl.* **193**, 465–482.
1309. D. E. Ward (1996), Dini derivatives of the marginal functions of a non-Lipschitzian program, *SIAM J. Optim.* **6**, 198–211.
1310. D. E. Ward (1999), Second-order necessary conditions in nonsmooth programming, in *Systems Modelling and Optimization*, edited by M. P. Polis et al., pp. 216–224, Chapman and Hall/CRC Press, Boca Raton, Florida.
1311. D. E. Ward and J. M. Borwein (1987), Nonconvex calculus in finite dimensions, *SIAM J. Control Optim.* **25**, 1312–1340.
1312. D. E. Ward and G. M. Lee (2001), Generalized properly efficient solutions of vector optimization problems, *Math. Oper. Res.* **53**, 215–232.
1313. J. Warga (1962), Relaxed variational problems, *J. Math. Anal. Appl.* **4**, 111–128.
1314. J. Warga (1962), Necessary conditions for minimum in relaxed variational problems, *J. Math. Anal. Appl.* **4**, 129–145.
1315. J. Warga (1972), *Optimal Control of Differential and Functional Equations*, Academic Press, New York.
1316. J. Warga (1975), Necessary optimality conditions without differentiability assumptions in optimal control, *J. Diff. Eq.* **15**, 41–62.
1317. J. Warga (1976), Necessary conditions withiut differentiability assumptions in unilatateral control problems, *J. Diff. Eq.* **21**, 25–38.
1318. J. Warga (1976), Derivate containers, inverse functions, and controllability, in *Calculus of Variations and Control Theory*, edited by D. L. Russel, pp. 13–46, Academic Press, New York.
1319. J. Warga (1978), Controllability and a multiplier rule for nondifferentiable optimization problems, *SIAM J. Control Optim.* **16**, 803–812.
1320. J. Warga (1981), Fat homeomorphisms and unbounded derivate containers, *J. Math. Anal. Appl.* **81**, 545–560.

1321. J. Warga (1983), Controllability, extremality and abnormality in nonsmooth optimal control, *J. Optim. Theory Appl.* **41**, 239–260.

1322. J. Warga (1988), Homeomorphisms and local $\mathcal{C}^1$ approximations, *Nonlinear Anal.* **12**, 593–597.

1323. J. Warga (1995), A proper relaxation of control with variable shifts, *J. Math. Anal. Appl.* **196**, 783–793.

1324. D. Washburn (1979), A bound on the boundary input map for parabolic equations with applications to time optimal control, *SIAM J. Control Optim.* **17**, 652–671.

1325. T. Ważewski (1961), Systémes de commandes et équations au contingent, *Bull. Acad. Pol. Sci., Ser. Math. Ast. Phys.* **9**, 151–155.

1326. K. Weierstrass (1927), *Vorlesungen über Variationsrechnung*, Akademische Verlagsgesellschaft, Leipzig.

1327. R. J-B. Wets (1996), Challenges in stochastic programming, *Math. Progr.* **75**, 115–135.

1328. L. W. White (1985), Distributed control of a hyperbolic problem with control and stress constraints, *J. Math. Anal. Appl.* **106**, 41–53.

1329. P. R. Wolenski (1990), The exponential formula for the reachable set of a Lipschitz differential inclusion, *SIAM J. Control Optim.* **28**, 1148–1161.

1330. P. R. Wolenski and Y. Zhuang (1998), Proximal analysis and the minimal time function, *SIAM J. Control Optim.* **36**, 1048–1072.

1331. P. D. Woodford (1997), *Optimal Control for Nonsmooth Dynamic Systems*, Ph.D. dissertation, Department of Electric. Electron. Eng., Imperial College, London, UK.

1332. M. H. Wright (1999), Ill-conditioning and computational error in interior methods for nonlinear programming, *SIAM J. Optim.* **9**, 84–111.

1333. S. E. Wright (1995), Consistency of primal-dual approximations for convex optimal control, *SIAM J. Control Optim.* **33**, 1489–1509.

1334. Z. Wu and J. J. Ye (2002), On error bounds for lower semicontinuous functions, *Math. Progr.* **92**, 421–435.

1335. Z. Wu and J. J. Ye (2003), First-order and second-order conditions for error bounds, *SIAM J. Optim.* **14**, 621–645.

1336. Z. Wu and J. J. Ye (2003), Equivalence between various derivatives and subdifferentials of the distance function, *J. Math. Anal. Appl.* **282**, 629–647.

1337. X. Q. Yang and V. Jeyakumar (1992), Generalized second-order directional derivatives and optimization with $C^{1,1}$ functions, *Optimization* **26**, 165–185.

1338. J. J. Ye (1999), Optimality conditions for optimization problems with complementarity constraints, *SIAM J. Control Optim.* **9**, 374–387.

1339. J. J. Ye (2000), Constraint qualifications and necessary optimality conditions for optimization problems with variational inequality constraints, *SIAM J. Optim.* **10**, 943–962.

1340. J. J. Ye (2001), Multiplier rules under mixed assumptions on differentiability and Lipschitz continuity, *SIAM J. Control Optim.* **39**, 1441–1460.

1341. J. J. Ye (2004), Nondifferentiable multiplier rules for optimization and bilevel optimization problems, *SIAM J. Optim.* **15**, 252–274.

1342. J. J. Ye (2005), Necessary and sufficient optimality conditions for mathematical programs with equilibrium constraints, *J. Math. Anal. Appl.* **307**, 350–369.

1343. J. J. Ye and X. Y. Ye (1997), Necessary optimality conditions for optimization problems with variational inequality constraints, *Math. Oper. Res.* **22**, 977–997.
1344. J. J. Ye, D. L. Zhu and Q. J. Zhu (1997), Exact penalization and necessary optimality conditions for generalized bilevel programming problems, *SIAM J. Optim.* **7**, 481–507.
1345. J. J. Ye and Q. J. Zhu (2003), Multiobjective optimization problems with variational inequality constraints, *Math. Progr.* **96**, 139–160.
1346. N. D. Yen (1995), Lipschitz continuity of solutions of variational inequalities with a parametric polyhedral constraint, *Math. Oper. Res.* **20**, 695–708.
1347. N. D. Yen (1995), Hölder continuity of solutions to a parametric variational inequality, *Appl. Math. Optim.* **31**, 245–255.
1348. D. Yost (1993), Asplund spaces for beginners, *Acta Univ. Carolinae, Ser. Math. Phys.* **34**, 159–177.
1349. L. C. Young (1933), On approximation by polygons in the calculus of variations, *Proc. Royal Soc.* (A) **141**, 325–341.
1350. L. C. Young (1937), Generalized curves and the existence of an attained absolute minimum in the calculus of variations, *C. R. Soc. Sci. Lett. Varsovie*, Cl. III, **30**, 212–234.
1351. L. C. Young (1969), *Lectures on the Calculus of Variations and Optimal Control Theory*, Saunders, Philadelphia, Pennsylvania.
1352. D. Zagrodny (1988), Approximate mean value theorem for upper subderivatives, *Nonlinear Anal.* **12**, 1413–1428.
1353. C. Zălinescu (2002), *Convex Analysis in General Vector Spaces*, World Scientific, Singapore.
1354. W. I. Zangwill (1967), Nonlinear programming via penalty functions, *Management Sci.* **13**, 344–358.
1355. S. C. Zaremba (1936), Sur les équations au paratingent, *Bull. Sci. Math.* **60**, 139–160.
1356. A. J. Zaslavski (2000), Generic well-posedness of optimal control problems without convexity assumptions, *SIAM J. Control Optim.* **39**, 250–280.
1357. A. J. Zaslavski (2006), *Turnpike Properties in the Calculus of Variations and Optimal Control*, Springer, New York.
1358. V. M. Zeidan (2001), New second-order optimality conditions for variational problems with C^2-Hamiltonians, *SIAM J. Control Optim.* **40**, 577–609.
1359. M. I. Zelikin and N. B. Melnikov (2004), The Bellman function and optimal synthesis in control problems with nonsmooth constraints, *J. Math. Sci.* **121**, 2281–2294.
1360. R. Zhang (1994), Problems of hierarchical optimization in finite dimensions, *SIAM J. Optim.* **4**, 521–536.
1361. R. Zhang (2003), Multistage bilevel programming problems, *Optimization* **52**, 605–616.
1362. R. Zhang (2005), Weakly upper Lipschitzian multifunctions and applications to parametric optimization, *Math. Progr.* **102**, 153–166.
1363. R. Zhang and J. S. Treiman (1995), Upper-Lipschitz multifunctions and inverse subdifferentials, *Nonlinear Anal.* **24**, 273–286.
1364. X. Y. Zheng and K. F. Ng (2005), The Fermat rule for multifunctions in Banach spaces, *Math. Progr.* **104**, 69–90.

1365. X. Y. Zheng and K. F. Ng (2005), Metric regularity and constraint qualifications for generalized equations in Banach spaces, *SIAM J. Optim.*, to appear.

1366. X. Y. Zhou (1990), Maximum principle, dynamic programming and their connections in deterministic control, *J. Optim. Theory Appl.* **65**, 363–373.

1367. X. Y. Zhou (1995), Deterministic near-optimal controls, part I: Necessary and sufficient conditions for near optimality, *J. Optim. Theory Appl.* **85**, 473–488.

1368. X. Y. Zhou (1996), Deterministic near-optimal controls, part II: Dynamic programming and viscosity solution approach, *Math. Oper. Res.* **21**, 655–674.

1369. X. Y. Zhou (1998), Stochastic near-optimal controls: Necessary and sufficient conditions for near-optimality, *SIAM J. Control Optim.* **36**, 929–947.

1370. Q. J. Zhu (1996), Necessary optimality conditions for nonconvex differential inclusions with endpoint constraints, *J. Diff. Eq.* **124**, 186–204.

1371. Q. J. Zhu (1998), The equivalence of several basic theorems for subdifferentials, *Set-Valued Anal.* **6**, 171–185.

1372. Q. J. Zhu (2000), Hamiltonian necessary conditions for a multiobjective optimal control problem, *SIAM J. Control Optim.* **39**, 97–112.

1373. Q. J. Zhu (2002), Necessary conditions for optimization problems in smooth Banach spaces, *SIAM J. Optim.* **12**, 1032–1047.

1374. Q. J. Zhu (2003), Lower semicontinuous Lyapunov functions and stability, *J. Nonlinear Convex Anal.* **4**, 325–332.

1375. Q. J. Zhu (2004), Nonconvex separation theorem for multifunctions, subdifferential calculus and applications, *Set-Valued Anal.* **12**, 275–290.

1376. T. Zolezzi (2002), On the distance theorem in quadratic optimization, *J. Convex Anal.* **9**, 693–700.

1377. T. Zolezzi (2003), Condition number theorems in optimization, *SIAM J. Optim.* **14**, 507–516.

1378. J. Zowe and S. Kurcyusz (1979), Regularity and stability for the mathematical programming problem in Banach spaces, *Appl. Math. Optim.* **5**, 49–62.

1379. E. Zuazua (2005), Propagation, observation, and control of waves approximated by finite difference methods, *SIAM Rev.* **47**, 197–243.

List of Statements

Chapter 1

Chapter 2

2.5.2C Discussion:	Specific structures: Viscosity subdifferentials
2.5.2D Discussion:	Specific structures: Proximal constructions
2.5.2E Discussion:	Specific structures: Derivate Sets
2.46 Theorem:	Derivate sets and Fréchet ε-subgradients
2.47 Corollary:	Relationship between Fréchet subgradients and screens
2.48 Corollary:	Relationship between Fréchet subgradients and derivate containers
2.49 Example:	Computing subgradients of Lipschitzian functions
2.50 Definition:	Abstract sequential normal compactness
2.51 Theorem:	Abstract versions of the extremal principle
2.52 Corollary:	Prenormal and normal structures at boundary points
2.6.1 Comment:	The origin of the extremal principle
2.6.2 Comment:	The extremal principle in Fréchet smooth spaces and separable reduction
2.6.3 Comment:	Asplund spaces
2.6.4 Comment:	The extremal principle in Asplund spaces
2.6.5 Comment:	The Ekeland variational principle
2.6.6 Comment:	Subdifferential variational principles
2.6.7 Comment:	Smooth variational principles
2.6.8 Comment:	Limiting normal and subgradient representations in Asplund spaces
2.6.9 Comment:	Other subdifferential structures and abstract versions of the extremal principle

Chapter 3

3.1 Lemma:	A fuzzy intersection rule from the extremal principle
3.2 Definition:	Basic qualification conditions for sets
3.3 Definition:	PSNC properties in product spaces
3.4 Theorem:	Basic normals to set intersections in product spaces
3.5 Corollary:	Intersection rule under the SNC condition
3.6 Example:	Intersection rule with no CEL assumption
3.7 Theorem:	Sum rules for generalized normals
3.8 Theorem:	Basic normals to inverse images
3.9 Corollary:	Inverse images under metrically regular mappings
3.10 Theorem:	Sum rules for coderivatives
3.11 Corollary:	Coderivative sum rule for Lipschitz-like multifunctions
3.12 Proposition:	Coderivatives of special sums
3.13 Theorem:	Chain rules for coderivatives
3.14 Theorem:	Zero chain rule for mixed coderivatives
3.15 Corollary:	Coderivative chain rules for Lipschitz-like and metrically regular mappings
3.16 Corollary:	Coderivative chain rules with strictly differentiable inner mappings
3.17 Corollary:	Partial coderivatives
3.18 Theorem:	Coderivatives of h-compositions
3.19 Corollary:	Inner product rule for coderivatives
3.20 Proposition:	Coderivative intersection rule
3.21 Remark:	Fuzzy coderivative calculus

Chapter 4

Chapter 5

Chapter 6

Chapter 7

Chapter 8

Glossary of Notation

Operations and Symbols

$:=$ and $=:$	equal by definition
$\equiv$	identically equal
$*$	indication of some dual/adjoint/polar operation
$\langle\cdot,\cdot\rangle$	canonical pairing between space X and its topological dual X^*
$x \to \bar{x}$	x converges to $\bar{x}$ strongly (by norm)
$x \stackrel{w}{\to} \bar{x}$	x converges to $\bar{x}$ weakly (in weak topology)
$x \stackrel{w^*}{\to} \bar{x}$	x converges to $\bar{x}$ weak* (in weak* topology)
$x \stackrel{\Omega}{\to} \bar{x}$	x converges to $\bar{x}$ with $x \in \Omega$
$\liminf$	lower limit for real numbers
$\limsup$	upper limit for real numbers
$\mathop{\rm Lim\,inf}$	lower/inner limit for set-valued mappings
$\mathop{\rm Lim\,sup}$	upper/outer limit for set-valued mappings
$\dim X$ and $\mathrm{codim}\, X$	dimension and codimension of X, respectively
$\prec$	preference relation
$\|\cdot\|$ or $\lvert\cdot\rvert$ or $\lvert\lvert\lvert\cdot\rvert\rvert\rvert$	norms
$\mathrm{haus}(\Omega_1, \Omega_2)$	Pompieu-Hausdorff distance between sets
$\mathrm{lip}\, F(\bar{x},\bar{y})$	exact Lipschitzian bound of F around $(\bar{x},\bar{y})$
$\mathrm{reg}\, F(\bar{x},\bar{y})$	exact metric regularity bound of F around $(\bar{x},\bar{y})$
$\mathrm{cov}\, F(\bar{x},\bar{y})$	exact covering/linear openness bound of F around $(\bar{x},\bar{y})$
$\mathrm{rad}\, F(\bar{x},\bar{y})$	radius of metric regularity of F around $(\bar{x},\bar{y})$
$\triangle$	end of proof

Spaces

$I\!\!R := (-\infty, \infty)$	real line
$\overline{I\!\!R} := [-\infty, \infty]$	extended real line
$I\!\!R^n$	n-dimensional Euclidean space
$I\!\!R^n_+$ and $I\!\!R^n_-$	nonnegative and nonpositive orthant of $I\!\!R^n$, respectively

$\mathcal{C}([a,b];X)$	space of X-valued continuous mappings with the supremum norm on $[a,b]$
$\mathcal{C}(K)$	space of continuous functions on the compact set K
$\mathcal{C}[0,\omega_1]$	continuous functions on $[0,\omega_1]$, where ω_1 is the first uncountable ordinal
$\mathcal{C}_0$	continuous functions with compact supports
$\mathcal{C}^k$, $1 \le k \le \infty$,	k times differentiable functions with all continuous derivatives
$\mathcal{C}^{1,1}$	continuously differentiable functions with Lipschitzian derivatives
$L^p([a,b];X)$, $1 \le p \le \infty$,	standard Lebesgue spaces of X-valued mappings
$W^{1,p}$ and H^p	standard Sobolev spaces
$\mathcal{M}$ and $\mathcal{M}_b$	measure spaces (dual to spaces of continuous functions)
BV	functions of bounded variation
c	space of real number sequences with the supremum norm
c_0	subspace of c with sequences converging to zero
ℓ^p, $1 \le p \le \infty$,	sequences of real numbers with standard p–norms

Sets

$\emptyset$	empty set
$\mathbb{N}$	set of natural numbers
$B_r(x)$	ball centered at x with radius r
$\mathbb{B}_X$	closed unit ball of space X
$\mathbb{B}$ and $\mathbb{B}^*$	closed unit balls of the space and duals space in question
S and S^*	unit spheres of the space and dual space in question
$\operatorname{int}\Omega$ and $\operatorname{ri}\Omega$	interior and relative interior, respectively
$\operatorname{cl}\Omega$ and $\operatorname{cl}^*\Omega$	closure and weak* topological closure, respectively
$\operatorname{bd}\Omega$ or $\partial\Omega$	set boundary
$\operatorname{co}\Omega$ and $\operatorname{clco}\Omega$	convex hull and closed convex hull, respectively
$\operatorname{cone}\Omega$	conic hull
$\operatorname{aff}\Omega$ and $\overline{\operatorname{aff}}\,\Omega$	affine hull and closed affine hull, respectively
$\operatorname{mes}\Omega$ or $\mathcal{L}^n(\Omega)$	Lebesgue (n-dimensional) measure
$\Pi(x;\Omega)$	projection of x to Ω
$T(\bar{x};\Omega)$	contingent cone to Ω at $\bar{x}$
$T_W(\bar{x};\Omega)$	weak contingent cone to Ω at $\bar{x}$
$T_C(\bar{x};\Omega)$	Clarke tangent cone to Ω at $\bar{x}$
$N(\bar{x};\Omega)$	basic/limiting normal cone to Ω at $\bar{x}$
$N_+(\bar{x};\Omega(\bar{y}))$	extended limiting normal cone to $\Omega(\bar{y})$ at $\bar{x}$
$\widehat{N}(\bar{x};\Omega)$	prenormal cone or Fréchet normal cone to Ω at $\bar{x}$
$N_C(\bar{x};\Omega)$	Clarke normal cone to Ω at $\bar{x}$
$N_G(\bar{x};\Omega)$ and $\widetilde{N}_G(\bar{x};\Omega)$	approximate G-normal cone and its nucleus to Ω at $\bar{x}$

$N_P(\bar{x};\Omega)$	proximal normal cone to Ω at $\bar{x}$
$\widehat{N}_\varepsilon(\bar{x};\Omega)$	sets of ε-normals to Ω at $\bar{x}$
$S_\varepsilon(\bar{x};\Omega)$	ε-support to Ω at $\bar{x}$

Functions

$\delta(\cdot;\Omega)$	set indicator function
$\operatorname{dist}(\cdot;\Omega)$ or $d_\Omega(\cdot)$	distance function
$\rho(x,y) := \operatorname{dist}(y;F(x))$	extended distance function
$\operatorname{dom}\varphi$	domain of $\varphi\colon X \to \overline{I\!R}$
$\operatorname{epi}\varphi$, $\operatorname{hypo}\varphi$, and $\operatorname{gph}\varphi$	epigraph, hypergraph, and graph of φ, respectively
$x \xrightarrow{\varphi} \bar{x}$	$x \to \bar{x}$ with $\varphi(x) \to \varphi(\bar{x})$
$\mathcal{H}$	Hamiltonian function in optimal control
H	Hamilton-Pontryagin function in optimal control
L	Lagrangian function in optimization
L_Ω	essential Lagrangian relative to Ω
$\tau(F;h)$	averaged modulus of continuity
$\varphi'(\bar{x})$ or $\nabla\varphi(\bar{x})$	Fréchet derivative/gradient of φ at $\bar{x}$
$\varphi'_\beta(\bar{x})$ or $\nabla_\beta\varphi(\bar{x})$	derivative/gradient of φ at $\bar{x}$ with respect to some bornology
$\|\nabla\varphi\|(\bar{x})$	(strong) slope of φ at $\bar{x}$
$\varphi'(\bar{x};v)$	classical directional derivative of φ at $\bar{x}$ in direction v
$\varphi^\circ(\bar{x};v)$ and $\varphi^\uparrow(\bar{x};v)$	generalized directional derivative and subderivative of φ
$d^-\varphi(\bar{x};v)$ and $d^+\varphi(\bar{x};v)$	Dini-Hadamard lower and upper directional derivative of φ
$\partial\varphi(\bar{x})$	basic/limiting subdifferential of φ at $\bar{x}$
$\partial^+\varphi(\bar{x})$	upper subdifferential of φ at $\bar{x}$
$\partial^0\varphi(\bar{x})$	symmetric subdifferential of φ at $\bar{x}$
$\partial_\geq\varphi(\bar{x})$	right-sided subdifferential of φ at $\bar{x}$
$\partial^\infty\varphi(\bar{x})$	singular subdifferential of φ at $\bar{x}$
$\widehat{\partial}\varphi(\bar{x})$ and $\widehat{\partial}^+\varphi(\bar{x})$	Fréchet subdifferential and upper subdifferential of φ at $\bar{x}$, respectively
$\partial_A\varphi(\bar{x})$ and $\partial_G\varphi(\bar{x})$	approximate A-subdifferential and G-subdifferential of φ at $\bar{x}$
$\partial_C\varphi(\bar{x})$	Clarke subdifferential/generalized gradient of φ at $\bar{x}$
$\partial_\beta\varphi(\bar{x})$	viscosity (bornological) β-subdifferential of φ at $\bar{x}$
$\partial_P\varphi(\bar{x})$	proximal subdifferential of φ at $\bar{x}$
$\widehat{\partial}_\varepsilon\varphi(\bar{x})$, $\widehat{\partial}_{a\varepsilon}\varphi(\bar{x})$, and $\widehat{\partial}_{g\varepsilon}\varphi(\bar{x})$	Fréchet-type ε-subdifferentials of φ at $\bar{x}$
$\partial_\varepsilon^-\varphi(\bar{x})$	Dini ε-subdifferential of φ at $\bar{x}$
$\nabla^2\varphi(\bar{x})$	classical Hessian (matrix of second derivatives if in $I\!R^n$) of φ at $\bar{x}$
$\partial^2\varphi$, $\partial^2_N\varphi$, and $\partial^2_M\varphi$	second-order subdifferentials (generalized Hessians) of φ

Mappings

$f: X \to Y$	single-valued mappings from X to Y
$F: X \rightrightarrows Y$	set-valued mappings from X to Y
$\operatorname{dom} F$	domain of F
$\operatorname{rge} F$	range of F
$\operatorname{gph} F$	graph of F
$\ker F$	kernel of F
$F^{-1}: Y \rightrightarrows X$	inverse mapping to $F: X \rightrightarrows Y$
$F(\Omega)$ and $F^{-1}(\Omega)$	image and inverse image/preimage of Ω under F
$F \circ G$	composition of mappings
$F \overset{h}{\circ} G$	h-composition of mappings
$\Delta(\cdot; \Omega)$	set indicator mapping
Ω_ρ	set enlargement mapping
E_φ	epigraphical mapping
$\mathcal{E}(f, \Theta)$	generalized epigraph of $f: X \to Y$ with respect to $\Theta \subset Y$
$DF(\bar{x}, \bar{y})$	graphical/contingent derivative of F at $(\bar{x}, \bar{y}) \in \operatorname{gph} F$
$D^*F(\bar{x}, \bar{y})$	(basic) coderivative of F at $(\bar{x}, \bar{y}) \in \operatorname{gph} F$
$D^*_N F(\bar{x}, \bar{y})$	normal coderivative of F at $(\bar{x}, \bar{y}) \in \operatorname{gph} F$
$D^*_M F(\bar{x}, \bar{y})$ and $\widetilde{D}^*_M F(\bar{x}, \bar{y})$	mixed and reversed mixed coderivative of F at $(\bar{x}, \bar{y})$, respectively
$\widehat{D}^* F(\bar{x}, \bar{y})$ and $\widehat{D}^*_\varepsilon F(\bar{x}, \bar{y})$	Fréchet coderivative and ε-coderivative of F at $(\bar{x}, \bar{y})$, respectively
$Jf(\bar{x})$	generalized Jacobian of f at $\bar{x}$
$\Lambda f(\bar{x})$	derivate container of f at $\bar{x}$

Subject Index

Grundlehren der mathematischen Wissenschaften

A Series of Comprehensive Studies in Mathematics

A Selection

246. Naimark/Stern: Theory of Group Representations
247. Suzuki: Group Theory I
248. Suzuki: Group Theory II
249. Chung: Lectures from Markov Processes to Brownian Motion
250. Arnold: Geometrical Methods in the Theory of Ordinary Differential Equations
251. Chow/Hale: Methods of Bifurcation Theory
252. Aubin: Nonlinear Analysis on Manifolds. Monge-Ampère Equations
253. Dwork: Lectures on ρ-adic Differential Equations
254. Freitag: Siegelsche Modulfunktionen
255. Lang: Complex Multiplication
256. Hörmander: The Analysis of Linear Partial Differential Operators I
257. Hörmander: The Analysis of Linear Partial Differential Operators II
258. Smoller: Shock Waves and Reaction-Diffusion Equations
259. Duren: Univalent Functions
260. Freidlin/Wentzell: Random Perturbations of Dynamical Systems
261. Bosch/Güntzer/Remmert: Non Archimedian Analysis – A System Approach to Rigid Analytic Geometry
262. Doob: Classical Potential Theory and Its Probabilistic Counterpart
263. Krasnosel'skiĭ/Zabreĭko: Geometrical Methods of Nonlinear Analysis
264. Aubin/Cellina: Differential Inclusions
265. Grauert/Remmert: Coherent Analytic Sheaves
266. de Rham: Differentiable Manifolds
267. Arbarello/Cornalba/Griffiths/Harris: Geometry of Algebraic Curves, Vol. I
268. Arbarello/Cornalba/Griffiths/Harris: Geometry of Algebraic Curves, Vol. II
269. Schapira: Microdifferential Systems in the Complex Domain
270. Scharlau: Quadratic and Hermitian Forms
271. Ellis: Entropy, Large Deviations, and Statistical Mechanics
272. Elliott: Arithmetic Functions and Integer Products
273. Nikol'skiĭ: Treatise on the shift Operator
274. Hörmander: The Analysis of Linear Partial Differential Operators III
275. Hörmander: The Analysis of Linear Partial Differential Operators IV
276. Liggett: Interacting Particle Systems
277. Fulton/Lang: Riemann-Roch Algebra
278. Barr/Wells: Toposes, Triples and Theories
279. Bishop/Bridges: Constructive Analysis
280. Neukirch: Class Field Theory
281. Chandrasekharan: Elliptic Functions
282. Lelong/Gruman: Entire Functions of Several Complex Variables
283. Kodaira: Complex Manifolds and Deformation of Complex Structures
284. Finn: Equilibrium Capillary Surfaces
285. Burago/Zalgaller: Geometric Inequalities
286. Andrianaov: Quadratic Forms and Hecke Operators

287. Maskit: Kleinian Groups
288. Jacod/Shiryaev: Limit Theorems for Stochastic Processes
289. Manin: Gauge Field Theory and Complex Geometry
290. Conway/Sloane: Sphere Packings, Lattices and Groups
291. Hahn/O'Meara: The Classical Groups and K-Theory
292. Kashiwara/Schapira: Sheaves on Manifolds
293. Revuz/Yor: Continuous Martingales and Brownian Motion
294. Knus: Quadratic and Hermitian Forms over Rings
295. Dierkes/Hildebrandt/Küster/Wohlrab: Minimal Surfaces I
296. Dierkes/Hildebrandt/Küster/Wohlrab: Minimal Surfaces II
297. Pastur/Figotin: Spectra of Random and Almost-Periodic Operators
298. Berline/Getzler/Vergne: Heat Kernels and Dirac Operators
299. Pommerenke: Boundary Behaviour of Conformal Maps
300. Orlik/Terao: Arrangements of Hyperplanes
301. Loday: Cyclic Homology
302. Lange/Birkenhake: Complex Abelian Varieties
303. DeVore/Lorentz: Constructive Approximation
304. Lorentz/v. Golitschek/Makovoz: Construcitve Approximation. Advanced Problems
305. Hiriart-Urruty/Lemaréchal: Convex Analysis and Minimization Algorithms I. Fundamentals
306. Hiriart-Urruty/Lemaréchal: Convex Analysis and Minimization Algorithms II. Advanced Theory and Bundle Methods
307. Schwarz: Quantum Field Theory and Topology
308. Schwarz: Topology for Physicists
309. Adem/Milgram: Cohomology of Finite Groups
310. Giaquinta/Hildebrandt: Calculus of Variations I: The Lagrangian Formalism
311. Giaquinta/Hildebrandt: Calculus of Variations II: The Hamiltonian Formalism
312. Chung/Zhao: From Brownian Motion to Schrödinger's Equation
313. Malliavin: Stochastic Analysis
314. Adams/Hedberg: Function spaces and Potential Theory
315. Bürgisser/Clausen/Shokrollahi: Algebraic Complexity Theory
316. Saff/Totik: Logarithmic Potentials with External Fields
317. Rockafellar/Wets: Variational Analysis
318. Kobayashi: Hyperbolic Complex Spaces
319. Bridson/Haefliger: Metric Spaces of Non-Positive Curvature
320. Kipnis/Landim: Scaling Limits of Interacting Particle Systems
321. Grimmett: Percolation
322. Neukirch: Algebraic Number Theory
323. Neukirch/Schmidt/Wingberg: Cohomology of Number Fields
324. Liggett: Stochastic Interacting Systems: Contact, Voter and Exclusion Processes
325. Dafermos: Hyperbolic Conservation Laws in Continuum Physics
326. Waldschmidt: Diophantine Approximation on Linear Algebraic Groups
327. Martinet: Perfect Lattices in Euclidean Spaces
328. Van der Put/Singer: Galois Theory of Linear Differential Equations
329. Korevaar: Tauberian Theory. A Century of Developments
330. Mordukhovich: Variational Analysis and Generalized Differentiation I: Basic Theory